STUDENT'S SOLUTIONS
MANUAL TO ACCOMPANY

PHYSICAL
CHEMISTRY

EIGHTH EDITION

STUDENT'S SOLUTIONS
MANUAL TO ACCOMPANY

PHYSICAL
CHEMISTRY

Eighth Edition

P. W. Atkins

Professor of Chemistry, University of Oxford
and Fellow of Lincoln College

C. A. Trapp

Professor of Chemistry, University of Louisville,
Louisville, Kentucky, USA

M. P. Cady

Professor of Chemistry,
Indiana University Southeast, New Albany Indiana, USA

C. Giunta

Professor of Chemistry,
Le Moyne College, Syracuse, NY, USA

W. H. Freeman and Company
New York

Student's solutions manual to accompany Physical Chemistry, Eighth Edition
© Oxford University Press, 2006

ISBN-13: 978–0–7167–6206–5
ISBN-10: 0–7167–6206–4

Published in Great Britain by Oxford University Press
This edition has been authorized by Oxford University Press for sale in the
United States and Canada only and not for export therefrom.

Library of Congress Cataloging in Publication Data
Data available

Second printing

W. H. Freeman and Company
41 Madison Avenue
New York, NY 10010
www.whfreeman.com

Preface

This manual provides detailed solutions to all the end-of-chapter (**b**) Exercises, and to the odd-numbered Discussion Questions and Problems. Solutions to Exercises and Problems carried over from previous editions have been reworked, modified, or corrected when needed.

The solutions to the Problems in this edition rely more heavily on the mathematical and molecular modelling software that is now generally accessible to physical chemistry students, and this is particularly true for many of the new Problems that request the use of such software for their solutions. But almost all of the Exercises and many of the Problems can still be solved with a modern hand-held scientific calculator. When a quantum chemical calculation or molecular modelling process has been called for, we have usually provided the solution with PC Spartan Pro^{TM} because of its common availability.

In general, we have adhered rigorously to the rules for significant figures in displaying the final answers. However, when intermediate answers are shown, they are often given with one more figure than would be justified by the data. These excess digits are indicated with an overline.

We have carefully cross-checked the solutions for errors and expect that most have been eliminated. We would be grateful to any readers who bring any remaining errors to our attention.

We warmly thank our publishers for their patience in guiding this complex, detailed project to completion.

P. W. A.

C. A. T.

M. P. C.

C. G.

Contents

PART 1 Equilibrium

1 The properties of gases

Answers to discussion questions

D1.1 An equation of state is an equation that relates the variables that define the state of a system to each other. Boyle, Charles, and Avogadro established these relations for gases at low pressures (perfect gases) by appropriate experiments. Boyle determined how volume varies with pressure ($V \propto 1/p$), Charles how volume varies with temperature ($V \propto T$), and Avogadro how volume varies with amount of gas ($V \propto n$). Combining all of these proportionalities into one we find

$$V \propto \frac{nT}{p}.$$

Inserting the constant of proportionality, R, yields the perfect gas equation

$$V = \frac{RnT}{p} \quad \text{or} \quad pV = nRT.$$

D1.3 Consider three temperature regions:

(1) $T < T_B$. At very low pressures, all gases show a compression factor, $Z \approx 1$. At high pressures, all gases have $Z > 1$, signifying that they have a molar volume greater than a perfect gas, which implies that repulsive forces are dominant. At intermediate pressures, most gases show $Z < 1$, indicating that attractive forces reducing the molar volume below the perfect value are dominant.

(2) $T \approx T_B$. $Z \approx 1$ at low pressures, slightly greater than 1 at intermediate pressures, and significantly greater than 1 only at high pressures. There is a balance between the attractive and repulsive forces at low to intermediate pressures, but the repulsive forces predominate at high pressures where the molecules are very close to each other.

(3) $T > T_B$. $Z > 1$ at all pressures because the frequency of collisions between molecules increases with temperature.

D1.5 The van der Waals equation 'corrects' the perfect gas equation for both attractive and repulsive interactions between the molecules in a real gas. See *Justification 1.1* for a fuller explanation.

The Bertholet equation accounts for the volume of the molecules in a manner similar to the van der Waals equation but the term representing molecular attractions is modified to account for the effect of temperature. Experimentally one finds that the van der Waals a decreases with increasing temperature. Theory (see Chapter 18) also suggests that intermolecular attractions can decrease with temperature.

This variation of the attractive interaction with temperature can be accounted for in the equation of state by replacing the van der Waals a with a/T.

Solutions to exercises

E1.1(b)   (a) The perfect gas law is

$$pV = nRT$$

implying that the pressure would be

$$p = \frac{nRT}{V}$$

All quantities on the right are given to us except n, which can be computed from the given mass of Ar.

$$n = \frac{25 \text{ g}}{39.95 \text{ g mol}^{-1}} = 0.62\bar{6} \text{ mol}$$

$$\text{so } p = \frac{(0.62\bar{6} \text{ mol}) \times (8.31 \times 10^{-2} \text{ dm}^3 \text{ bar K}^{-1}\text{mol}^{-1}) \times (30 + 273 \text{ K})}{1.5 \text{ dm}^3} = \boxed{10.\bar{5} \text{ bar}}$$

not 2.0 bar.

(b) The van der Waals equation is

$$p = \frac{RT}{V_m - b} - \frac{a}{V_m^2}$$

$$\text{so } p = \frac{(8.31 \times 10^{-2} \text{ dm}^3 \text{ bar K}^{-1}\text{mol}^{-1}) \times (30 + 273) \text{ K}}{(1.53 \text{ dm}^3/0.62\bar{6} \text{ mol}) - 3.20 \times 10^{-2} \text{ dm}^3 \text{ mol}^{-1}}$$

$$- \frac{(1.337 \text{ dm}^6\text{atm mol}^{-2}) \times (1.013 \text{ bar atm}^{-1})}{(1.5 \text{ dm}^3/0.62\bar{6} \text{ mol})^2} = \boxed{10.\bar{4} \text{ bar}}$$

E1.2(b)   (a) Boyle's law applies:

$$pV = \text{constant} \quad \text{so} \quad p_f V_f = p_i V_i$$

and

$$p_i = \frac{p_f V_f}{V_i} = \frac{(1.97 \text{ bar}) \times (2.14 \text{ dm}^3)}{(2.14 + 1.80) \text{ dm}^3} = \boxed{1.07 \text{ bar}}$$

(b) The original pressure in bar is

$$p_i = (1.07 \text{ bar}) \times \left(\frac{1 \text{ atm}}{1.013 \text{ bar}}\right) \times \left(\frac{760 \text{ Torr}}{1 \text{ atm}}\right) = \boxed{803 \text{ Torr}}$$

E1.3(b)   The relation between pressure and temperature at constant volume can be derived from the perfect gas law

$$pV = nRT \quad \text{so} \quad p \propto T \quad \text{and} \quad \frac{p_i}{T_i} = \frac{p_f}{T_f}$$

The final pressure, then, ought to be

$$p_f = \frac{p_i T_f}{T_i} = \frac{(125 \text{ kPa}) \times (11 + 273) \text{ K}}{(23 + 273) \text{ K}} = \boxed{120 \text{ kPa}}$$

E1.4(b)   According to the perfect gas law, one can compute the amount of gas from pressure, temperature, and volume. Once this is done, the mass of the gas can be computed from the amount and the molar mass using

$$pV = nRT$$

so $n = \dfrac{pV}{RT} = \dfrac{(1.00 \text{ atm}) \times (1.013 \times 10^5 \text{ Pa atm}^{-1}) \times (4.00 \times 10^3 \text{ m}^3)}{(8.3145 \text{ J K}^{-1}\text{mol}^{-1}) \times (20 + 273) \text{ K}} = 1.66 \times 10^5 \text{mol}$

and $m = (1.66 \times 10^5 \text{ mol}) \times (16.04 \text{ g mol}^{-1}) = 2.67 \times 10^6 \text{g} = \boxed{2.67 \times 10^3 \text{ kg}}$

E1.5(b)   Identifying p_{ex} in the equation $p = p_{ex} + \rho gh$ [1.3] as the pressure at the top of the straw and p as the atmospheric pressure on the liquid, the pressure difference is

$$p - p_{ex} = \rho gh = (1.0 \times 10^3 \text{ kg m}^{-3}) \times (9.81 \text{ m s}^{-2}) \times (0.15 \text{ m})$$

$$= \boxed{1.5 \times 10^3 \text{ Pa}} \, (= 1.5 \times 10^{-2} \text{ atm})$$

E1.6(b)   The pressure in the apparatus is given by

$$p = p_{atm} + \rho gh \, [1.3]$$

$$p_{atm} = 760 \text{ Torr} = 1 \text{ atm} = 1.013 \times 10^5 \text{ Pa}$$

$$\rho gh = 13.55 \text{ g cm}^{-3} \times \left(\frac{1 \text{ kg}}{10^3 \text{ g}}\right) \times \left(\frac{10^6 \text{ cm}^3}{\text{m}^3}\right) \times 0.100 \text{ m} \times 9.806 \text{ m s}^{-2} = 1.33 \times 10^4 \text{ Pa}$$

$$p = 1.013 \times 10^5 \text{ Pa} + 1.33 \times 10^4 \text{ Pa} = 1.146 \times 10^5 \text{ Pa} = \boxed{115 \text{ kPa}}$$

E1.7(b)   All gases are perfect in the limit of zero pressure. Therefore the extrapolated value of pV_m/T will give the best value of R.

The molar mass is obtained from $pV = nRT = \dfrac{m}{M}RT$

which upon rearrangement gives $M = \dfrac{m}{V}\dfrac{RT}{p} = \rho\dfrac{RT}{p}$

The best value of M is obtained from an extrapolation of ρ/p versus p to $p = 0$; the intercept is M/RT.

Draw up the following table

$p/$atm	$(pV_m/T)/(\text{dm}^3 \text{ atm K}^{-1}\text{mol}^{-1})$	$(\rho/p)/(\text{dm}^{-3}\text{atm}^{-1})$
0.750 000	0.082 0014	1.428 59
0.500 000	0.082 0227	1.428 22
0.250 000	0.082 0414	1.427 90

From Figure 1.1(a), $\left(\dfrac{pV_m}{T}\right)_{p=0} = \boxed{0.082\,061\,5 \text{ dm}^3 \text{ atm K}^{-1} \text{ mol}^{-1}}$

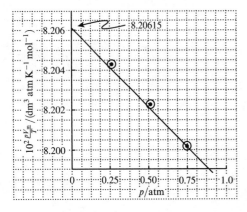

Figure 1.1(a)

From Figure 1.1(b), $\left(\dfrac{\rho}{p}\right)_{p=0} = 1.427\ 55\ \text{g dm}^{-3}\ \text{atm}^{-1}$

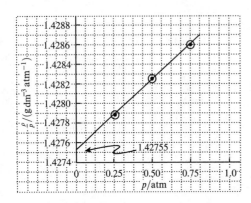

Figure 1.1(b)

$$M = RT \left(\frac{\rho}{p}\right)_{p=0} = (0.082\ 061\ 5\ \text{dm}^3\ \text{atm mol}^{-1}\ \text{K}^{-1}) \times (273.15\ \text{K}) \times (1.42755\ \text{g dm}^{-3}\text{atm}^{-1})$$

$$= \boxed{31.9987\ \text{g mol}^{-1}}$$

The value obtained for R deviates from the accepted value by 0.005 percent. The error results from the fact that only three data points are available and that a linear extrapolation was employed. The molar mass, however, agrees exactly with the accepted value, probably because of compensating plotting errors.

E1.8(b) The mass density ρ is related to the molar volume V_m by

$$V_m = \frac{M}{\rho}$$

where M is the molar mass. Putting this relation into the perfect gas law yields

$$pV_m = RT \quad \text{so} \quad \frac{pM}{\rho} = RT$$

Rearranging this result gives an expression for M; once we know the molar mass, we can divide by the molar mass of phosphorus atoms to determine the number of atoms per gas molecule

$$M = \frac{RT\rho}{p} = \frac{(8.314 \text{ Pa m}^3 \text{ mol}^{-1}) \times [(100 + 273) \text{ K}] \times (0.6388 \text{ kg m}^{-3})}{1.60 \times 10^4 \text{ Pa}}$$

$$= 0.124 \text{ kg mol}^{-1} = 124 \text{ g mol}^{-1}$$

The number of atoms per molecule is

$$\frac{124 \text{ g mol}^{-1}}{31.0 \text{ g mol}^{-1}} = 4.00$$

suggesting a formula of $\boxed{P_4}$

E1.9(b)   Use the perfect gas equation to compute the amount; then convert to mass.

$$pV = nRT \quad \text{so} \quad n = \frac{pV}{RT}$$

We need the partial pressure of water, which is 53 percent of the equilibrium vapor pressure at the given temperature and standard pressure.

$$p = (0.53) \times (2.69 \times 10^3 \text{ Pa}) = 1.4\bar{3} \times 10^3 \text{ Pa}$$

$$\text{so } n = \frac{(1.4\bar{3} \times 10^3 \text{ Pa}) \times (250 \text{ m}^3)}{(8.3145 \text{ J K}^{-1} \text{ mol}^{-1}) \times (23 + 273) \text{ K}} = 1.4\bar{5} \times 10^2 \text{ mol}$$

$$\text{or } m = (1.4\bar{5} \times 10^2 \text{ mol}) \times (18.0 \text{ g mol}^{-1}) = 2.6\bar{1} \times 10^3 \text{ g} = \boxed{2.6\bar{1} \text{ kg}}$$

E1.10(b)   **(a)** The volume occupied by each gas is the same, since each completely fills the container. Thus solving for V we have (assuming a perfect gas)

$$V = \frac{n_J RT}{p_J} \quad n_{Ne} = \frac{0.225 \text{ g}}{20.18 \text{ g mol}^{-1}}$$

$$= 1.11\bar{5} \times 10^{-2} \text{ mol}, \quad p_{Ne} = 8.87 \text{ kPa}, \quad T = 300 \text{ K}$$

$$V = \frac{(1.11\bar{5} \times 10^{-2} \text{ mol}) \times (8.314 \text{ dm}^3 \text{ kPa K}^{-1} \text{ mol}^{-1}) \times 300 \text{ K})}{8.87 \text{ kPa}} = 3.13\bar{7} \text{ dm}^3$$

$$= \boxed{3.14 \text{ dm}^3}$$

(b) The total pressure is determined from the total amount of gas, $n = n_{CH_4} + n_{Ar} + n_{Ne}$.

$$n_{CH_4} = \frac{0.320 \text{ g}}{16.04 \text{ g mol}^{-1}} = 1.99\bar{5} \times 10^{-2} \text{mol} \quad n_{Ar} = \frac{0.175 \text{ g}}{39.95 \text{ g mol}^{-1}} = 4.38 \times 10^{-3} \text{mol}$$

$$n = (1.99\bar{5} + 0.438 + 1.11\bar{5}) \times 10^{-2} \text{mol} = 3.54\bar{8} \times 10^{-2} \text{mol}$$

$$p = \frac{nRT}{V} [1.8] = \frac{(3.54\bar{8} \times 10^{-2} \text{ mol}) \times (8.314 \text{ dm}^3 \text{ kPa K}^{-1} \text{ mol}^{-1}) \times (300 \text{ K})}{3.13\bar{7} \text{ dm}^3}$$

$$= \boxed{28.2 \text{ kPa}}$$

E1.11(b) This is similar to Exercise 1.11(a) with the exception that the density is first calculated.

$$M = \rho \frac{RT}{p} \text{ [Exercise 1.8(a)]}$$

$$\rho = \frac{33.5\,\text{mg}}{250\,\text{cm}^3} = 0.134\overline{0}\,\text{g dm}^{-3}, \quad p = 152\,\text{Torr}, \quad T = 298\,\text{K}$$

$$M = \frac{(0.134\overline{0}\,\text{g dm}^{-3}) \times (62.36\,\text{dm}^3\,\text{Torr K}^{-1}\,\text{mol}^{-1}) \times (298\,\text{K})}{152\,\text{Torr}} = \boxed{16.14\,\text{g mol}^{-1}}$$

E1.12(b) This exercise is similar to Exercise 1.12(a) in that it uses the definition of absolute zero as that temperature at which the volume of a sample of gas would become zero if the substance remained a gas at low temperatures. The solution uses the experimental fact that the volume is a linear function of the Celsius temperature.

Thus $V = V_0 + \alpha V_0 \theta = V_0 + b\theta$, $b = \alpha V_0$

At absolute zero, $V = 0$, or $0 = 20.00\,\text{dm}^3 + 0.0741\,\text{dm}^3\,^\circ\text{C}^{-1} \times \theta(\text{abs. zero})$

$$\theta(\text{abs. zero}) = -\frac{20.00\,\text{dm}^3}{0.0741\,\text{dm}^3\,^\circ\text{C}^{-1}} = \boxed{-270\,^\circ\text{C}}$$

which is close to the accepted value of $-273\,^\circ\text{C}$.

E1.13(b) **(a)** $p = \dfrac{nRT}{V}$

$n = 1.0\,\text{mol}$

$T = $ (i) 273.15 K; (ii) 500 K

$V = $ (i) 22.414 dm^3; (ii) 150 cm^3

(i) $$p = \frac{(1.0\,\text{mol}) \times (8.206 \times 10^{-2}\,\text{dm}^3\,\text{atm K}^{-1}\,\text{mol}^{-1}) \times (273.15\,\text{K})}{22.414\,\text{dm}^3}$$

$$= \boxed{1.0\,\text{atm}}$$

(ii) $$p = \frac{(1.0\,\text{mol}) \times (8.206 \times 10^{-2}\,\text{dm}^3\,\text{atm K}^{-1}\,\text{mol}^{-1}) \times (500\,\text{K})}{0.150\,\text{dm}^3}$$

$$= \boxed{270\,\text{atm}} \text{ (2 significant figures)}$$

(b) From Table (1.6) for H_2S

$$a = 4.484\,\text{dm}^6\,\text{atm mol}^{-1} \qquad b = 4.34 \times 10^{-2}\,\text{dm}^3\,\text{mol}^{-1}$$

$$p = \frac{nRT}{V - nb} - \frac{an^2}{V^2}$$

(i) $$p = \frac{(1.0\,\text{mol}) \times (8.206 \times 10^{-2}\,\text{dm}^3\,\text{atm K}^{-1}\,\text{mol}^{-1}) \times (273.15\,\text{K})}{22.414\,\text{dm}^3 - (1.0\,\text{mol}) \times (4.34 \times 10^{-2}\,\text{dm}^3\,\text{mol}^{-1})}$$

$$- \frac{(4.484\,\text{dm}^6\,\text{atm mol}^{-1}) \times (1.0\,\text{mol})^2}{(22.414\,\text{dm}^3)^2}$$

$$= \boxed{0.99\,\text{atm}}$$

(ii) $$p = \frac{(1.0\,\text{mol}) \times (8.206 \times 10^{-2}\,\text{dm}^3\,\text{atm}\,\text{K}^{-1}\,\text{mol}^{-1}) \times (500\,\text{K})}{0.150\,\text{dm}^3 - (1.0\,\text{mol}) \times (4.34 \times 10^{-2}\,\text{dm}^3\,\text{mol}^{-1})}$$

$$- \frac{(4.484\,\text{dm}^6\text{atm}\,\text{mol}^{-1}) \times (1.0\,\text{mol})^2}{(0.150\,\text{dm}^3)^2}$$

$$= 18\overline{5.6}\,\text{atm} \approx \boxed{190\,\text{atm}}\ (2\ \text{significant figures}).$$

E1.14(b)   The conversions needed are as follows:

$$1\,\text{atm} = 1.013 \times 10^5\,\text{Pa}; \quad 1\,\text{Pa} = 1\,\text{kg}\,\text{m}^{-1}\,\text{s}^{-2}; \quad 1\,\text{dm}^6 = 10^{-6}\,\text{m}^6; \quad 1\,\text{dm}^3 = 10^{-3}\,\text{m}^3$$

Therefore,

$a = 1.32\,\text{atm}\,\text{dm}^6\,\text{mol}^{-2}$ becomes, after substitution of the conversions

$a = \boxed{1.34 \times 10^{-1}\,\text{kg}\,\text{m}^5\text{s}^{-2}\text{mol}^{-2}}$, and

$b = 0.0436\,\text{dm}^3\,\text{mol}^{-1}$ becomes

$b = \boxed{4.36 \times 10^{-5}\,\text{m}^3\text{mol}^{-1}}$

E1.15(b)   The compression factor is

$$Z = \frac{pV_\text{m}}{RT} = \frac{V_\text{m}}{V_\text{m}^\text{o}}$$

(a) Because $V_\text{m} = V_\text{m}^\text{o} + 0.12\,V_\text{m}^\text{o} = (1.12)V_\text{m}^\text{o}$, we have $Z = \boxed{1.12}$ $\boxed{\text{Repulsive}}$ forces dominate.

(b) The molar volume is

$$V = (1.12)V_\text{m}^\text{o} = (1.12) \times \left(\frac{RT}{p}\right)$$

$$V = (1.12) \times \left(\frac{(0.08206\,\text{dm}^3\,\text{atm}\,\text{K}^{-1}\,\text{mol}^{-1}) \times (350\,\text{K})}{12\,\text{atm}}\right) = \boxed{2.7\,\text{dm}^3\,\text{mol}^{-1}}$$

E1.16(b)   **(a)** $$V_\text{m}^\text{o} = \frac{RT}{p} = \frac{(8.314\,\text{J}\,\text{K}^{-1}\,\text{mol}^{-1}) \times (298.15\,\text{K})}{(200\,\text{bar}) \times (10^5\,\text{Pa}\,\text{bar}^{-1})}$$

$$= 1.24 \times 10^{-4}\,\text{m}^3\,\text{mol}^{-1} = \boxed{0.124\,\text{dm}^3\,\text{mol}^{-1}}$$

(b) The van der Waals equation is a cubic equation in V_m. The most direct way of obtaining the molar volume would be to solve the cubic analytically. However, this approach is cumbersome, so we proceed as in Example 1.4. The van der Waals equation is rearranged to the cubic form

$$V_\text{m}^3 - \left(b + \frac{RT}{p}\right)V_\text{m}^2 + \left(\frac{a}{p}\right)V_\text{m} - \frac{ab}{p} = 0 \ \text{ or }\ x^3 - \left(b + \frac{RT}{p}\right)x^2 + \left(\frac{a}{p}\right)x - \frac{ab}{p} = 0$$

with $x = V_\text{m}/(\text{dm}^3\,\text{mol}^{-1})$.

The coefficients in the equation are evaluated as

$$b + \frac{RT}{p} = (3.183 \times 10^{-2}\,\text{dm}^3\,\text{mol}^{-1}) + \frac{(8.206 \times 10^{-2}\,\text{dm}^3\,\text{mol}^{-1}) \times (298.15\,\text{K})}{(200\,\text{bar}) \times (1.013\,\text{atm}\,\text{bar}^{-1})}$$

$$= (3.183 \times 10^{-2} + 0.120\overline{8})\,\text{dm}^3\,\text{mol}^{-1} = 0.152\overline{6}\,\text{dm}^3\text{mol}^{-1}$$

$$\frac{a}{p} = \frac{1.360\,\text{dm}^6\,\text{atm}\,\text{mol}^{-2}}{(200\,\text{bar}) \times (1.013\,\text{atm}\,\text{bar}^{-1})} = 6.71 \times 10^{-3}(\text{dm}^3\,\text{mol}^{-1})^2$$

$$\frac{ab}{p} = \frac{(1.360\,\text{dm}^6\,\text{atm}\,\text{mol}^{-2}) \times (3.183 \times 10^{-2}\text{dm}^3\,\text{mol}^{-1})}{(200\,\text{bar}) \times (1.013\,\text{atm}\,\text{bar}^{-1})} = 2.13\overline{7} \times 10^{-4}(\text{dm}^3\,\text{mol}^{-1})^3$$

Thus, the equation to be solved is $x^3 - 0.152\overline{6}x^2 + (6.71 \times 10^{-3})x - (2.13\overline{7} \times 10^{-4}) = 0$. Calculators and computer software for the solution of polynomials are readily available. In this case we find

$$x = 0.112 \quad \text{or} \quad V_m = \boxed{0.112\,\text{dm}^3\,\text{mol}^{-1}}$$

The difference is about 15 percent.

E1.17(b)   The molar volume is obtained by solving $Z = pV_m/RT$ [1.17], for V_m, which yields

$$V_m = \frac{ZRT}{p} = \frac{(0.86) \times (0.08206\,\text{dm}^3\,\text{atm}\,\text{K}^{-1}\,\text{mol}^{-1}) \times (300\,\text{K})}{20\,\text{atm}} = 1.05\overline{9}\,\text{dm}^3\,\text{mol}^{-1}$$

(a) Then, $V = nV_m = (8.2 \times 10^{-3}\,\text{mol}) \times (1.05\overline{9}\,\text{dm}^3\,\text{mol}^{-1}) = 8.7 \times 10^{-3}\,\text{dm}^3 = \boxed{8.7\,\text{cm}^3}$

(b) An approximate value of B can be obtained from eqn 1.19 by truncation of the series expansion after the second term, B/V_m, in the series. Then,

$$B = V_m \left(\frac{pV_m}{RT} - 1\right) = V_m \times (Z - 1)$$

$$= (1.05\overline{9}\,\text{dm}^3\,\text{mol}^{-1}) \times (0.86 - 1) = \boxed{-0.15\,\text{dm}^3\text{mol}^{-1}}$$

E1.18(b)   **(a)** Mole fractions are

$$x_N = \frac{n_N}{n_{\text{total}}} = \frac{2.5\,\text{mol}}{(2.5 + 1.5)\,\text{mol}} = \boxed{0.63}$$

Similarly, $x_H = \boxed{0.37}$

(c) According to the perfect gas law

$$p_{\text{total}} V = n_{\text{total}} RT$$

$$\text{so } p_{\text{total}} = \frac{n_{\text{total}} RT}{V}$$

$$= \frac{(4.0\,\text{mol}) \times (0.08206\,\text{dm}^3\,\text{atm}\,\text{mol}^{-1}\,\text{K}^{-1}) \times (273.15\,\text{K})}{22.4\,\text{dm}^3} = \boxed{4.0\,\text{atm}}$$

(b) The partial pressures are

$$p_N = x_N p_{tot} = (0.63) \times (4.0\,\text{atm}) = \boxed{2.5\,\text{atm}}$$

and $p_H = (0.37) \times (4.0\,\text{atm}) = \boxed{1.5\,\text{atm}}$

E1.19(b)   The critical volume of a van der Waals gas is

$$V_c = 3b$$

so $b = \frac{1}{3}V_c = \frac{1}{3}(148\,\text{cm}^3\,\text{mol}^{-1}) = 49.3\,\text{cm}^3\,\text{mol}^{-1} = \boxed{0.0493\,\text{dm}^3\,\text{mol}^{-1}}$

By interpreting b as the excluded volume of a mole of spherical molecules, we can obtain an estimate of molecular size. The centers of spherical particles are excluded from a sphere whose radius is the diameter of those spherical particles (i.e. twice their radius); that volume times the Avogadro constant is the molar excluded volume b

$$b = N_A \left(\frac{4\pi(2r)^3}{3} \right) \quad \text{so} \quad r = \frac{1}{2} \left(\frac{3b}{4\pi N_A} \right)^{1/3}$$

$$r = \frac{1}{2} \left(\frac{3(49.3\,\text{cm}^3\,\text{mol}^{-1})}{4\pi(6.022 \times 10^{23}\,\text{mol}^{-1})} \right)^{1/3} = 1.94 \times 10^{-8}\,\text{cm} = \boxed{1.94 \times 10^{-10}\,\text{m}}$$

The critical pressure is

$$p_c = \frac{a}{27b^2}$$

so $a = 27 p_c b^2 = 27(48.20\,\text{atm}) \times (0.0493\,\text{dm}^3\,\text{mol}^{-1})^2 = \boxed{3.16\,\text{dm}^6\,\text{atm}\,\text{mol}^{-2}}$

But this problem is overdetermined. We have another piece of information

$$T_c = \frac{8a}{27Rb}$$

According to the constants we have already determined, T_c should be

$$T_c = \frac{8(3.16\,\text{dm}^6\,\text{atm}\,\text{mol}^{-2})}{27(0.08206\,\text{dm}^3\,\text{atm}\,\text{K}^{-1}\,\text{mol}^{-1}) \times (0.0493\,\text{dm}^3\,\text{mol}^{-1})} = 231\,\text{K}$$

However, the reported T_c is 305.4 K, suggesting our computed a/b is about 25 percent lower than it should be.

E1.20(b)   **(a)** The Boyle temperature is the temperature at which $\lim_{V_m \to \infty} dZ/(d(1/V_m))$ vanishes. According to the van der Waals equation

$$Z = \frac{pV_m}{RT} = \frac{\left(\dfrac{RT}{V_m - b} - \dfrac{a}{V_m^2} \right) V_m}{RT} = \frac{V_m}{V_m - b} - \frac{a}{V_m RT}$$

so $\dfrac{dZ}{d(1/V_m)} = \left(\dfrac{dZ}{dV_m}\right) \times \left(\dfrac{dV_m}{d(1/V_m)}\right)$

$= -V_m^2 \left(\dfrac{dZ}{dV_m}\right) = -V_m^2 \left(\dfrac{-V_m}{(V_m - b)^2} + \dfrac{1}{V_m - b} + \dfrac{a}{V_m^2 RT}\right)$

$= \dfrac{V_m^2 b}{(V_m - b)^2} - \dfrac{a}{RT}$

In the limit of large molar volume, we have

$$\lim_{V_m \to \infty} \dfrac{dZ}{d(1/V_m)} = b - \dfrac{a}{RT} = 0 \quad \text{so} \quad \dfrac{a}{RT} = b$$

and $T = \dfrac{a}{Rb} = \dfrac{\left(4.484\, \text{dm}^6\, \text{atm}\, \text{mol}^{-2}\right)}{(0.08206\, \text{dm}^3\, \text{atm}\, \text{K}^{-1}\, \text{mol}^{-1}) \times (0.0434\, \text{dm}^3\, \text{mol}^{-1})} = \boxed{1259\, \text{K}}$

(b) By interpreting b as the excluded volume of a mole of spherical molecules, we can obtain an estimate of molecular size. The centres of spherical particles are excluded from a sphere whose radius is the diameter of those spherical particles (i.e. twice their radius); the Avogadro constant times the volume is the molar excluded volume b

$$b = N_A \left(\dfrac{4\pi (2r)^3}{3}\right) \quad \text{so} \quad r = \dfrac{1}{2}\left(\dfrac{3b}{4\pi N_A}\right)^{1/3}$$

$$r = \dfrac{1}{2}\left(\dfrac{3(0.0434\, \text{dm}^3\, \text{mol}^{-1})}{4\pi (6.022 \times 10^{23}\, \text{mol}^{-1})}\right)^{1/3} = 1.286 \times 10^{-9}\, \text{dm} = 1.29 \times 10^{-10}\, \text{m} = \boxed{0.129\, \text{nm}}$$

E1.21(b)   States that have the same reduced pressure, temperature, and volume are said to correspond. The reduced pressure and temperature for N_2 at 1.0 atm and 25 °C are

$$p_r = \dfrac{p}{p_c} = \dfrac{1.0\, \text{atm}}{33.54\, \text{atm}} = 0.030 \quad \text{and} \quad T_r = \dfrac{T}{T_c} = \dfrac{(25 + 273)\, \text{K}}{126.3\, \text{K}} = 2.36$$

The corresponding states are

(a) For H_2S

$$p = p_r p_c = (0.030) \times (88.3\, \text{atm}) = \boxed{2.6\, \text{atm}}$$

$$T = T_r T_c = (2.36) \times (373.2\, \text{K}) = \boxed{881\, \text{K}}$$

(Critical constants of H_2S obtained from *Handbook of Chemistry and Physics*.)

(b) For CO_2

$$p = p_r p_c = (0.030) \times (72.85\, \text{atm}) = \boxed{2.2\, \text{atm}}$$

$$T = T_r T_c = (2.36) \times (304.2\, \text{K}) = \boxed{718\, \text{K}}$$

(c) For Ar

$$p = p_r p_c = (0.030) \times (48.00\, \text{atm}) = \boxed{1.4\, \text{atm}}$$

$$T = T_r T_c = (2.36) \times (150.72\, \text{K}) = \boxed{356\, \text{K}}$$

E1.22(b)   The van der Waals equation is

$$p = \frac{RT}{V_m - b} - \frac{a}{V_m^2}$$

which can be solved for b

$$b = V_m - \frac{RT}{p + \dfrac{a}{V_m^2}} = 4.00 \times 10^{-4}\,\text{m}^3\,\text{mol}^{-1} - \frac{(8.3145\,\text{J K}^{-1}\text{mol}^{-1}) \times (288\,\text{K})}{4.0 \times 10^6\,\text{Pa} + \left(\dfrac{0.76\,\text{m}^6\,\text{Pa mol}^{-2}}{(4.00 \times 10^{-4}\,\text{m}^3\,\text{mol}^{-1})^2}\right)}$$

$$= \boxed{1.3 \times 10^{-4}\,\text{m}^3\,\text{mol}^{-1}}$$

The compression factor is

$$z = \frac{pV_m}{RT} = \frac{(4.0 \times 10^6\,\text{Pa}) \times (4.00 \times 10^{-4}\,\text{m}^3\,\text{mol}^{-1})}{(8.3145\,\text{J K}^{-1}\,\text{mol}^{-1}) \times (288\,\text{K})} = \boxed{0.67}$$

Solutions to problems

Solutions to numerical problems

P1.1   Since the Neptunians know about perfect gas behavior, we may assume that they will write $pV = nRT$ at both temperatures. We may also assume that they will establish the size of their absolute unit to be the same as the °N, just as we write $1\,\text{K} = 1°\text{C}$. Thus

$$pV(T_1) = 28.0\,\text{dm}^3\,\text{atm} = nRT_1 = nR \times (T_1 + 0°\text{N}),$$

$$pV(T_2) = 40.0\,\text{dm}^3\,\text{atm} = nRT_2 = nR \times (T_1 + 100°\text{N}),$$

or $T_1 = \dfrac{28.0\,\text{dm}^3\,\text{atm}}{nR}$, $T_1 + 100°\text{N} = \dfrac{40.0\,\text{dm}^3\,\text{atm}}{nR}$.

Dividing, $\dfrac{T_1 + 100°\text{N}}{T_1} = \dfrac{40.0\,\text{dm}^3\,\text{atm}}{28.0\,\text{dm}^3\,\text{atm}} = 1.42\bar{9}$ or $T_1 + 100°\text{N} = 1.42\bar{9}T_1$, $T_1 = 233$ absolute units.
As in the relationship between our Kelvin scale and Celsius scale $T = \theta -$ absolute zero(°N) so absolute zero (°N) $= \boxed{-233°\text{N}}$.

COMMENT. To facilitate communication with Earth students we have converted the Neptunians' units of the pV product to units familiar to humans, which are $\text{dm}^3\,\text{atm}$. However, we see from the solution that only the ratio of pV products is required, and that will be the same in any civilization.

Question. If the Neptunians' unit of volume is the lagoon (L), their unit of pressure is the poseidon (P), their unit of amount is the nereid (n), and their unit of absolute temperature is the titan (T), what is the value of the Neptunians' gas constant (R) in units of L, P, n, and T?

P1.3   The value of absolute zero can be expressed in terms of α by using the requirement that the volume of a perfect gas becomes zero at the absolute zero of temperature. Hence

$$0 = V_0[1 + \alpha\theta(\text{abs. zero})].$$

Then θ (abs. zero) $= -\dfrac{1}{\alpha}$.

All gases become perfect in the limit of zero pressure, so the best value of α and, hence, θ (abs. zero) is obtained by extrapolating α to zero pressure. This is done in Fig. 1.2. Using the extrapolated value, $\alpha = 3.6637 \times 10^{-3\circ}C^{-1}$, or

$$\theta(\text{abs. zero}) = -\dfrac{1}{3.6637 \times 10^{-3\circ}C^{-1}} = \boxed{-272.95°C},$$

which is close to the accepted value of $-273.15°C$.

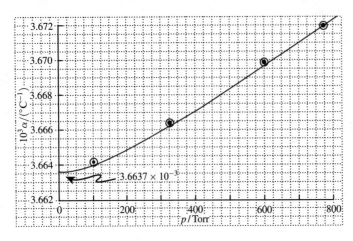

Figure 1.2

P1.5 $\dfrac{p}{T} = \dfrac{nR}{V} = $ constant, if n and V are constant. Hence, $\dfrac{p}{T} = \dfrac{p_3}{T_3}$, where p is the measured pressure at temperature, T, and p_3 and T_3 are the triple point pressure and temperature, respectively. Rearranging,

$$p = \left(\dfrac{p_3}{T_3}\right) T.$$

The ratio $\dfrac{p_3}{T_3}$ is a constant $= \dfrac{6.69\,\text{kPa}}{273.16\,\text{K}} = 0.0245\,\text{kPa K}^{-1}$. Thus the change in p, Δp, is proportional to the change in temperature, $\Delta T : \Delta p = (0.0245\,\text{kPa K}^{-1}) \times (\Delta T)$.

(a) $\Delta p = (0.0245\,\text{kPa K}^{-1}) \times (1.00\,\text{K}) = \boxed{0.0245\,\text{kPa}}$.

(b) Rearranging, $p = \left(\dfrac{T}{T_3}\right) p_3 = \left(\dfrac{373.16\,\text{K}}{273.16\,\text{K}}\right) \times (6.69\,\text{kPa}) = \boxed{9.14\,\text{kPa}}$.

(c) Since $\dfrac{p}{T}$ is a constant at constant n and V, it always has the value $0.0245\,\text{kPa K}^{-1}$; hence

$$\Delta p = p_{374.15\,\text{K}} - p_{373.15\,\text{K}} = (0.0245\,\text{kPa K}^{-1}) \times (1.00\,\text{K}) = \boxed{0.0245\,\text{kPa}}.$$

P1.7 **(a)** $V_m = \dfrac{RT}{p} = \dfrac{(8.206 \times 10^{-2}\,\text{dm}^3\,\text{atm K}^{-1}\,\text{mol}^{-1}) \times (350\,\text{K})}{2.30\,\text{atm}} = \boxed{12.5\,\text{dm}^3\,\text{mol}^{-1}}$.

(b) From $p = \dfrac{RT}{V_m - b} - \dfrac{a}{V_m^2}$ [1.21b], we obtain $V_m = \dfrac{RT}{\left(p + \dfrac{a}{V_m^2}\right)} + b$ [rearrange 1.21b].

Then, with a and b from Table 1.6,

$$V_m \approx \frac{(8.206 \times 10^{-2}\,\mathrm{dm^3\,atm\,K^{-1}\,mol^{-1}}) \times (350\,\mathrm{K})}{(2.30\,\mathrm{atm}) + \left((6.260\,\mathrm{dm^6\,atm\,mol^{-2}})/(12.5\,\mathrm{dm^3\,mol^{-1}})^2\right)} + (5.42 \times 10^{-2}\,\mathrm{dm^3\,mol^{-1}})$$

$$\approx \frac{28.7\bar{2}\,\mathrm{dm^3\,mol^{-1}}}{2.34} + \left(5.42 \times 10^{-2}\,\mathrm{dm^3\,mol^{-1}}\right) \approx \boxed{12.3\,\mathrm{dm^3\,mol^{-1}}}.$$

Substitution of $12.3\,\mathrm{dm^3\,mol^{-1}}$ into the denominator of the first expression again results in $V_m = 12.3\,\mathrm{dm^3\,mol^{-1}}$, so the cycle of approximation may be terminated.

P1.9 As indicated by eqns 1.18 and 1.19 the compression factor of a gas may be expressed as either a virial expansion in p or in $\left(\dfrac{1}{V_m}\right)$. The virial form of the van der Waals equation is derived in Exercise 1.20(a) and is $p = \dfrac{RT}{V_m}\left\{1 + \left(b - \dfrac{a}{RT}\right) \times \left(\dfrac{1}{V_m}\right) + \cdots\right\}$

Rearranging, $Z = \dfrac{pV_m}{RT} = 1 + \left(b - \dfrac{a}{RT}\right) \times \left(\dfrac{1}{V_m}\right) + \cdots$

On the assumption that the perfect gas expression for V_m is adequate for the second term in this expansion, we can readily obtain Z as a function of p.

$$Z = 1 + \left(\frac{1}{RT}\right) \times \left(b - \frac{a}{RT}\right)p + \cdots$$

(a) $T_c = 126.3\,\mathrm{K}.$

$$V_m = \left(\frac{RT}{p}\right) \times Z = \frac{RT}{p} + \left(b - \frac{a}{RT}\right) + \cdots$$

$$= \frac{(0.08206\,\mathrm{dm^3\,atm\,K^{-1}\,mol^{-1}}) \times (126.3\,\mathrm{K})}{10.0\,\mathrm{atm}}$$

$$+ \left\{(0.0387\,\mathrm{dm^3\,mol^{-1}}) - \left(\frac{1.352\,\mathrm{dm^6\,atm\,mol^{-2}}}{(0.08206\,\mathrm{dm^3\,atm\,K^{-1}mol^{-1}}) \times (126.3\,\mathrm{K})}\right)\right\}$$

$$= (1.036 - 0.092)\,\mathrm{dm^3\,mol^{-1}} = \boxed{0.944\,\mathrm{dm^3\,mol^{-1}}}.$$

$$Z = \left(\frac{p}{RT}\right) \times (V_m) = \frac{(10.0\,\mathrm{atm}) \times (0.944\,\mathrm{dm^3\,mol^{-1}})}{(0.08206\,\mathrm{dm^3\,atm\,K^{-1}\,mol^{-1}}) \times (126.3\,\mathrm{K})} = 0.911.$$

(b) The Boyle temperature corresponds to the temperature at which the second virial coefficient is zero, hence correct to the first power in p, $Z = 1$, and the gas is close to perfect. However, if we assume that N_2 is a van der Waals gas, when the second virial coefficient is zero,

$$\left(b - \frac{a}{RT_B}\right) = 0, \quad \text{or} \quad T_B = \frac{a}{bR}.$$

$$T_B = \frac{1.352\,\mathrm{dm^6\,atm\,mol^{-2}}}{(0.0387\,\mathrm{dm^3\,mol^{-1}}) \times (0.08206\,\mathrm{dm^3\,atm\,K^{-1}\,mol^{-1}})} = 426\,\mathrm{K}.$$

The experimental value (Table 1.5) is 327.2 K. The discrepancy may be explained by two considerations.

1. Terms beyond the first power in p should not be dropped in the expansion for Z.
2. Nitrogen is only approximately a van der Waals gas.

When $Z = 1$, $V_m = \dfrac{RT}{p}$, and using $T_B = 327.2$ K

$$= \frac{(0.08206 \, \text{dm}^3 \, \text{atm} \, \text{K}^{-1}\text{mol}^{-1}) \times 327.2 \, \text{K}}{10.0 \, \text{atm}}$$

$$= \boxed{2.69 \, \text{dm}^3 \, \text{mol}^{-1}}$$

and this is the ideal value of V_m. Using the experimental value of T_B and inserting this value into the expansion for V_m above, we have

$$V_m = \frac{0.08206 \, \text{dm}^3 \, \text{atm} \, \text{K}^{-1}\text{mol}^{-1} \times 327.2 \, \text{K}}{10.0 \, \text{atm}}$$

$$+ \left\{ 0.0387 \, \text{dm}^3 \text{mol}^{-1} - \left(\frac{1.352 \, \text{dm}^6 \text{atm} \, \text{mol}^{-2}}{0.08206 \, \text{dm}^3 \, \text{atm} \, \text{K}^{-1}\text{mol}^{-1} \times 327.2 \, \text{K}} \right) \right\}$$

$$= (2.68\overline{5} - 0.012) \, \text{dm}^3 \text{mol}^{-1} = \boxed{2.67 \, \text{dm}^3 \, \text{mol}^{-1}}$$

and $Z = \dfrac{V_m}{V_m^{\circ}} = \dfrac{2.67 \, \text{dm}^3 \, \text{mol}^{-1}}{2.69 \, \text{dm}^3 \, \text{mol}^{-1}} = 0.992 \approx 1.$

(c) $T_I = 621$ K [Table 2.9].

$$V_m = \frac{0.08206 \, \text{dm}^3 \text{atm} \, \text{K}^{-1}\text{mol}^{-1} \times 621 \, \text{K}}{10.0 \, \text{atm}}$$

$$+ \left\{ 0.0387 \, \text{dm}^3 \, \text{mol}^{-1} - \left(\frac{1.352 \, \text{dm}^6 \text{atm} \, \text{mol}^{-2}}{0.08206 \, \text{dm}^3 \, \text{atm} \, \text{K}^{-1}\text{mol}^{-1} \times 621 \, \text{K}} \right) \right\}$$

$$= (5.09\overline{6} + 0.012) \, \text{dm}^3 \, \text{mol}^{-1} = \boxed{5.11 \, \text{dm}^3 \, \text{mol}^{-1}}$$

and $Z = \dfrac{5.11 \, \text{dm}^3 \text{mol}^{-1}}{5.10 \, \text{dm}^3 \, \text{mol}^{-1}} = 1.002 \approx 1.$

Based on the values of T_B and T_I given in Tables 1.4 and 2.9 and assuming that N_2 is a van der Waals gas, the calculated value of Z is closest to 1 at $\boxed{T_I}$, but the difference from the value at T_B is less than the accuracy of the method.

P1.11   **(a)** $V_m = \dfrac{\text{molar mass}}{\text{density}} = \dfrac{M}{\rho} = \dfrac{18.02 \, \text{g mol}^{-1}}{1.332 \times 10^2 \, \text{g dm}^{-3}} = \boxed{0.1353 \, \text{dm}^3 \, \text{mol}^{-1}}.$

(b) $Z = \dfrac{pV_m}{RT}$ [1.17b] $= \dfrac{(327.6 \, \text{atm}) \times (0.1353 \, \text{dm}^3 \, \text{mol}^{-1})}{(0.08206 \, \text{dm}^3 \, \text{atm} \, \text{K}^{-1} \, \text{mol}^{-1}) \times (776.4 \, \text{K})} = \boxed{0.6957}.$

(c) Two expansions for Z based on the van der Waals equation are given in Problem 1.9. They are

$$Z = 1 + \left(b - \frac{a}{RT}\right) \times \left(\frac{1}{V_m}\right) + \cdots$$

$$= 1 + \left\{(0.0305 \text{ dm}^3 \text{ mol}^{-1}) - \left(\frac{5.464 \text{ dm}^6 \text{ atm mol}^{-2}}{(0.08206 \text{ dm}^3 \text{ atm K}^{-1} \text{ mol}^{-1}) \times (776.4 \text{ K})}\right)\right\}$$

$$\times \frac{1}{0.1353 \text{ dm}^3 \text{ mol}^{-1}} = 1 - 0.4084 = 0.5916 \approx 0.59.$$

$$Z = 1 + \left(\frac{1}{RT}\right) \times \left(b - \frac{a}{RT}\right) \times (p) + \cdots$$

$$= 1 + \frac{1}{(0.08206 \text{ dm}^3 \text{ atm K}^{-1} \text{ mol}^{-1}) \times (776.4 \text{ K})}$$

$$\times \left\{(0.0305 \text{ dm}^3 \text{ mol}^{-1}) - \left(\frac{5.464 \text{ dm}^6 \text{ atm mol}^{-2}}{(0.08206 \text{ dm}^3 \text{ atm K}^{-1} \text{ mol}^{-1}) \times (776.4 \text{ K})}\right)\right\} \times 327.6 \text{ atm}$$

$$= 1 - 0.2842 \approx \boxed{0.72}.$$

In this case the expansion in p gives a value close to the experimental value; the expansion in $\frac{1}{V_m}$ is not as good. However, when terms beyond the second are included the results from the two expansions for Z converge.

P1.13 $V_c = 2b, \qquad T_c = \frac{a}{4bR}$ [Table 1.7]

Hence, with V_c and T_c from Table 1.5, $b = \frac{1}{2}V_c = \frac{1}{2} \times (118.8 \text{ cm}^3 \text{ mol}^{-1}) = \boxed{59.4 \text{ cm}^3 \text{ mol}^{-1}}$.

$$a = 4bRT_c = 2RT_c V_c$$

$$= (2) \times (8.206 \times 10^{-2} \text{ dm}^3 \text{ atm K}^{-1} \text{ mol}^{-1}) \times (289.75 \text{ K}) \times (118.8 \times 10^{-3} \text{ dm}^3 \text{ mol}^{-1})$$

$$= \boxed{5.649 \text{ dm}^6 \text{ atm mol}^{-2}}.$$

Hence

$$p = \frac{RT}{V_m - b} e^{-a/RTV_m} = \frac{nRT}{V - nb} e^{-na/RTV}$$

$$= \frac{(1.0 \text{ mol}) \times (8.206 \times 10^{-2} \text{ dm}^3 \text{ atm K}^{-1} \text{ mol}^{-1}) \times (298 \text{ K})}{(1.0 \text{ dm}^3) - (1.0 \text{ mol}) \times (59.4 \times 10^{-3} \text{ dm}^3 \text{ mol}^{-1})}$$

$$\times \exp\left(\frac{-(1.0 \text{ mol}) \times (5.649 \text{ dm}^6 \text{ atm mol}^{-2})}{(8.206 \times 10^{-2} \text{ dm}^3 \text{ atm K}^{-1} \text{ mol}^{-1}) \times (298 \text{ K}) \times (1.0 \text{ dm}^6 \text{ atm mol}^{-1})}\right)$$

$$= 26.\bar{0} \text{ atm} \times e^{-0.23\bar{1}} = \boxed{21 \text{ atm}}.$$

Solutions to theoretical problems

P1.15 This expansion has already been given in the solutions to Exercise 1.20(a) and Problem 1.14; the result is

$$p = \frac{RT}{V_m}\left(1 + \left[b - \frac{a}{RT}\right]\frac{1}{V_m} + \frac{b^2}{V_m^2} + \cdots\right).$$

Compare this expansion with $p = \dfrac{RT}{V_m}\left(1 + \dfrac{B}{V_m} + \dfrac{C}{V_{m^2}} + \cdots\right)$ [1.19]

and hence find $\boxed{B = b - \dfrac{a}{RT}}$ and $\boxed{C = b^2}$.

Since $C = 1200\,\text{cm}^6\,\text{mol}^{-2}$, $b = C^{1/2} = \boxed{34.6\,\text{cm}^3\,\text{mol}^{-1}}$

$$a = RT(b - B) = (8.206 \times 10^{-2}) \times (273\,\text{dm}^3\,\text{atm}\,\text{mol}^{-1}) \times (34.6 + 21.7)\,\text{cm}^3\,\text{mol}^{-1}$$

$$= (22.4\overline{0}\,\text{dm}^3\,\text{atm}\,\text{mol}^{-1}) \times (56.3 \times 10^{-3}\,\text{dm}^3\,\text{mol}^{-1}) = \boxed{1.26\,\text{dm}^6\,\text{atm}\,\text{mol}^{-2}}.$$

P1.17 The critical point corresponds to a point of zero slope that is simultaneously a point of inflection in a plot of pressure versus molar volume. A critical point exists if there are values of p, V, and T that result in a point that satisfies these conditions.

$$p = \frac{RT}{V_m} - \frac{B}{V_m^2} + \frac{C}{V_m^3}.$$

$$\left.\begin{array}{l}\left(\dfrac{\partial p}{\partial V_m}\right)_T = -\dfrac{RT}{V_m^2} + \dfrac{2B}{V_m^3} - \dfrac{3C}{V_m^4} = 0 \\[4mm] \left(\dfrac{\partial^2 p}{\partial V_m^2}\right)_T = \dfrac{2RT}{V_m^3} - \dfrac{6B}{V_m^4} + \dfrac{12C}{V_m^5} = 0\end{array}\right\} \text{ at the critical point.}$$

That is, $\left.\begin{array}{l}-RT_c V_c^2 + 2BV_c - 3C = 0 \\ RT_c V_c^2 - 3BV_c + 6C = 0\end{array}\right\}$

which solve to $V_c = \boxed{\dfrac{3C}{B}}$, $\boxed{T_c = \dfrac{B^2}{3RC}}$.

Now use the equation of state to find p_c

$$p_c = \frac{RT_c}{V_c} - \frac{B}{V_c^2} + \frac{C}{V_c^3} = \left(\frac{RB^2}{3RC}\right) \times \left(\frac{B}{3C}\right) - B\left(\frac{B}{3C}\right)^2 + C\left(\frac{B}{3C}\right)^3 = \boxed{\frac{B^3}{27C^2}}.$$

It follows that $Z_c = \dfrac{p_c V_c}{RT_c} = \left(\dfrac{B^3}{27C^2}\right) \times \left(\dfrac{3C}{B}\right) \times \left(\dfrac{1}{R}\right) \times \left(\dfrac{3RC}{B^2}\right) = \boxed{\dfrac{1}{3}}.$

P1.19 For a real gas we may use the virial expansion in terms of p [1.18]

$$p = \frac{nRT}{V}(1 + B'p + \cdots) = \rho\frac{RT}{M}(1 + B'p + \cdots)$$

which rearranges to $\dfrac{p}{\rho} = \dfrac{RT}{M} + \dfrac{RT\,B'}{M}p + \cdots$.

Therefore, the limiting slope of a plot of $\dfrac{p}{\rho}$ against p is $\dfrac{B'RT}{M}$. From Fig. 1.3 the limiting slope is

$$\frac{B'RT}{M} = \frac{(5.84 - 5.44) \times 10^4 \, \text{m}^2 \, \text{s}^{-2}}{(10.132 - 1.223) \times 10^4 \, \text{Pa}} = 4.4 \times 10^{-2} \, \text{kg}^{-1} \, \text{m}^3.$$

From Fig. 1.3, $\dfrac{RT}{M} = 5.40 \times 10^4 \, \text{m}^2 \, \text{s}^{-2}$; hence

$$B' = \frac{4.4 \times 10^{-2} \, \text{kg}^{-1} \, \text{m}^3}{5.40 \times 10^4 \, \text{m}^2 \, \text{s}^{-2}} = 0.81 \times 10^{-6} \, \text{Pa}^{-1},$$

$$B' = (0.81 \times 10^{-6} \, \text{Pa}^{-1}) \times (1.0133 \times 10^5 \, \text{Pa}\,\text{atm}^{-1}) = \boxed{0.082 \, \text{atm}^{-1}}.$$

$B = RTB'$ [Problem 1.18]

$$= (8.206 \times 10^{-2} \, \text{dm}^3 \, \text{atm} \, \text{K}^{-1} \, \text{mol}^{-1}) \times (298 \, \text{K}) \times (0.082 \, \text{atm}^{-1})$$

$$= \boxed{2.0 \, \text{dm}^3 \, \text{mol}^{-1}}.$$

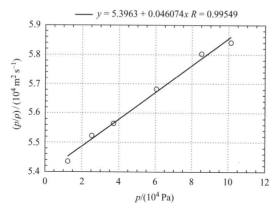

$$y = 5.3963 + 0.046074x \quad R = 0.99549$$

Figure axis: vertical $(p/\rho)/(10^4 \, \text{m}^2 \, \text{s}^{-1})$, horizontal $p/(10^4 \, \text{Pa})$

Figure 1.3

P1.21 The critical temperature is that temperature above which the gas cannot be liquefied by the application of pressure alone. Below the critical temperature two phases, liquid and gas, may coexist at equilibrium, and in the two-phase region there is more than one molar volume corresponding to the same conditions of temperature and pressure. Therefore, any equation of state that can even approximately describe this situation must allow for more than one real root for the molar volume at some values of T and p, but as the temperature is increased above T_c, allows only one real root. Thus, appropriate equations of state must be equations of odd degree in V_m.

The equation of state for gas A may be rewritten $V_m^2 - (RT/p)V_m - (RTb/p) = 0$, which is a quadratic and never has just one real root. Thus, this equation can never model critical behavior. It could possibly model in a very crude manner a two-phase situation, since there are some conditions under which a quadratic has two real positive roots, but not the process of liquefaction.

The equation of state of gas B is a first-degree equation in V_m and therefore can never model critical behavior, the process of liquefaction, or the existence of a two-phase region.

A cubic equation is the equation of lowest degree that can show a cross-over from more than one real root to just one real root as the temperature increases. The van der Waals equation is a cubic equation in V_m.

P1.23   The two masses represent the same volume of gas under identical conditions, and therefore, the same number of molecules (Avogadro's principle) and moles, n. Thus, the masses can be expressed as

$$nM_N = 2.2990\,g$$

for 'chemical nitrogen' and

$$n_{Ar}M_{Ar} + n_N M_N = n[x_{Ar}M_{Ar} + (1 - x_{Ar})M_N] = 2.3102\,g$$

for 'atmospheric nitrogen'. Dividing the latter expression by the former yields

$$\frac{x_{Ar}M_{Ar}}{M_N} + (1 - x_{Ar}) = \frac{2.3102}{2.2990} \quad so \quad x_{Ar}\left(\frac{M_{Ar}}{M_N} - 1\right) = \frac{2.3102}{2.2990} - 1$$

and $x_{Ar} = \dfrac{(2.3102/2.2990) - 1}{(M_{Ar}/M_N) - 1} = \dfrac{(2.3102/2.2990) - 1}{(39.95\,g\,mol^{-1})/(28.013\,g\,mol^{-1} - 1)} = \boxed{0.011}$.

COMMENT.   This value for the mole fraction of argon in air is close to the modern value.

Solutions to applications

P1.25 $1\,t = 10^3$ kg. Assume 300 t per day.

$$n(SO_2) = \frac{300 \times 10^3\,kg}{64 \times 10^{-3}\,kg\,mol^{-1}} = 4.7 \times 10^6\,mol.$$

$$V = \frac{nRT}{p} = \frac{(4.7 \times 10^6\,mol) \times (0.082\,dm^3\,atm\,K^{-1}mol^{-1}) \times 1073\,K}{1.0\,atm} = \boxed{4.1 \times 10^8\,dm^3}.$$

P1.27   The pressure at the base of a column of height H is $p = \rho g H$ (Example 1.1). But the pressure at any altitude h within the atmospheric column of height H depends only on the air above it; therefore

$$p = \rho g(H - h) \text{ and } dp = -\rho g\,dh.$$

Since $\rho = \dfrac{pM}{RT}$ [Problem 1.2], $dp = -\dfrac{pMg\,dh}{RT}$, implying that $\dfrac{dp}{p} = -\dfrac{Mg\,dh}{RT}$

This relation integrates to $p = p_0 e^{-Mgh/RT}$

For air $M \approx 29\,g\,mol^{-1}$ and at 298 K

$$\frac{Mg}{RT} \approx \frac{(29 \times 10^{-3}\,kg\,mol^{-1}) \times (9.81\,m\,s^{-2})}{2.48 \times 10^3\,J\,mol^{-1}} = 1.1\bar{5} \times 10^{-4}m^{-1}\,[1\,J = 1\,kg\,m^2\,s^{-2}].$$

(a) $h = 15$ cm.
$$p = p_0 \times e^{(-0.15\,\text{m}) \times (1.1\overline{5} \times 10^{-4}\,\text{m}^{-1})} = 0.99\overline{998}\,p_0; \quad \frac{p - p_0}{p_0} = \boxed{0.00}.$$

(b) $h = 11$ km $= 1.1 \times 10^4$ m.
$$p = p_0 \times e^{(-1.1 \times 10^{-4}) \times (1.1\overline{5} \times 10^{-4}\,\text{m}^{-1})} = 0.28\,p_0; \quad \frac{p - p_0}{p_0} = \boxed{-0.72}.$$

P1.29   Refer to Fig. 1.4.

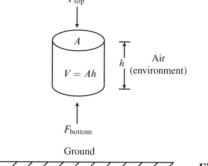

Figure 1.4

The buoyant force on the cylinder is

$$F_{\text{buoy}} = F_{\text{bottom}} - F_{\text{top}}$$
$$= A(p_{\text{bottom}} - p_{\text{top}})$$

according to the barometric formula.

$$p_{\text{top}} = p_{\text{bottom}} e^{-Mgh/RT}$$

where M is the molar mass of the environment (air). Since h is small, the exponential can be expanded in a Taylor series around $h = 0$ $\left(e^{-x} = 1 - x + \frac{1}{2!}x^2 + \cdots \right)$. Keeping the first-order term only yields

$$p_{\text{top}} = p_{\text{bottom}} \left(1 - \frac{Mgh}{RT} \right).$$

The buoyant force becomes

$$F_{\text{buoy}} = A p_{\text{bottom}} \left(1 - 1 + \frac{Mgh}{RT} \right) = Ah \left(\frac{p_{\text{bottom}} M}{RT} \right) g$$
$$= \left(\frac{p_{\text{bottom}} VM}{RT} \right) g = nMg \quad \left[n = \frac{p_{\text{bottom}} V}{RT} \right]$$

n is the number of moles of the environment (air) displaced by the balloon, and $nM = m$, the mass of the displaced environment. Thus $F_{\text{buoy}} = mg$. The net force is the difference between the buoyant force and the weight of the balloon. Thus

$$F_{\text{net}} = mg - m_{\text{balloon}}\, g = (m - m_{\text{balloon}})g$$

This is Archimedes' principle.

2 The First Law

Answers to discussion questions

D2.1   Work is a precisely defined mechanical concept. It is produced from the application of a force through a distance. The technical definition is based on the realization that both force and displacement are vector quantities and it is the component of the force acting in the direction of the displacement that is used in the calculation of the amount of work, that is, work is the scalar product of the two vectors. In vector notation $w = -\boldsymbol{f} \cdot \boldsymbol{d} = -fd \cos\theta$, where θ is the angle between the force and the displacement. The negative sign is inserted to conform to the standard thermodynamic convention.

Heat is associated with a non-adiabatic process and is defined as the difference between the adiabatic work and the non-adiabatic work associated with the same change in state of the system. This is the formal (and best) definition of heat and is based on the definition of work. A less precise definition of heat is the statement that heat is the form of energy that is transferred between bodies in thermal contact with each other by virtue of a difference in temperature.

At the molecular level, work is a transfer of energy that results in orderly motion of the atoms and molecules in a system; heat is a transfer of energy that results in disorderly motion. See *Molecular interpretation* 2.1 for a more detailed discussion.

D2.3   The difference results from the definition $H = U + PV$; hence $\Delta H = \Delta U + \Delta(PV)$. As $\Delta(PV)$ is not usually zero, except for isothermal processes in a perfect gas, the difference between ΔH and ΔU is a non-zero quantity. As shown in Sections 2.4 and 2.5 of the text, ΔH can be interpreted as the heat associated with a process at constant pressure, and ΔU as the heat at constant volume.

D2.5   In the Joule experiment, the change in internal energy of a gas at low pressures (a perfect gas) is zero. Hence in the calculation of energy changes for processes in a perfect gas one can ignore any effect due to a change in volume. This greatly simplifies the calculations involved because one can drop the first term of eqn 2.40 and need work only with $dU = C_V\,dT$. In a more sensitive apparatus, Joule would have observed a small temperature change upon expansion of the 'real' gas. Joule's result holds exactly only in the limit of zero pressure where all gases can be considered perfect.

The solution to Problem 2.33 shows that the Joule–Thomson coefficient can be expressed in terms of the parameters representing the attractive and repulsive interactions in a real gas. If the attractive forces predominate, then expanding the gas will reduce its energy and hence its temperature. This reduction in temperature could continue until the temperature of the gas falls below its condensation point. This is the principle underlying the liquefaction of gases with the Linde refrigerator, which utilizes the Joule–Thomson effect. See Section 2.12 for a more complete discussion.

D2.7   The vertical axis of a thermogram represents C_p, and the baselines represent the heat capacity associated with simple heating in the absence of structural transformations or similar transitions. In the example shown in Fig. 2.16, the sample undergoes a structural change between T_1 and T_2, so there is no reason to expect C_p after the transition to return to its value before the transition. Just as diamond and graphite have different heat capacities because of their different structures, the structural changes that occur during the measurement of a thermogram can also give rise to a change in heat capacity.

Solutions to exercises

E2.1(b)   The physical definition of work is $dw = -F\,dz$ [2.4]

In a gravitational field the force is the weight of the object, which is $F = mg$

If g is constant over the distance the mass moves, dw may be intergrated to give the total work

$$w = -\int_{z_i}^{z_f} F\,dz = -\int_{z_i}^{z_f} mg\,dz = -mg(z_f - z_i) = -mgh \quad \text{where} \quad h = (z_f - z_i)$$

$$w = -(0.120\,\text{kg}) \times (9.81\,\text{m s}^{-2}) \times (50\,\text{m}) = -59\,\text{J} = \boxed{59\,\text{J needed}}$$

E2.2(b)   This is an expansion against a constant external pressure; hence $w = -p_{ex}\Delta V$ [2.8]

The change in volume is the cross-sectional area times the linear displacement:

$$\Delta V = (50.0\,\text{cm}^2) \times (15\,\text{cm}) \times \left(\frac{1\,\text{m}}{100\,\text{cm}}\right)^3 = 7.5 \times 10^{-4}\,\text{m}^3,$$

so $w = -(121 \times 10^3\,\text{Pa}) \times (7.5 \times 10^{-4}\,\text{m}^3) = \boxed{-91\,\text{J}}$ as $1\,\text{Pa m}^3 = 1\,\text{J}$.

E2.3(b)   For all cases $\Delta U = 0$, since the internal energy of a perfect gas depends only on temperature. (See *Molecular interpretation* 2.2 and Section 2.11(b) for a more complete discussion.) From the definition of enthalpy, $H = U + pV$, so $\Delta H = \Delta U + \Delta(pV) = \Delta U + \Delta(nRT)$ (perfect gas). Hence, $\Delta H = 0$ as well, at constant temperature for all processes in a perfect gas.

(a) $\boxed{\Delta U = \Delta H = 0}$

$$w = -nRT \ln\left(\frac{V_f}{V_i}\right) \text{[2.11]}$$

$$= -(2.00\,\text{mol}) \times (8.3145\,\text{J K}^{-1}\,\text{mol}^{-1}) \times (22 + 273)\,\text{K} \times \ln\frac{31.7\,\text{dm}^3}{22.8\,\text{dm}^3} = \boxed{-1.62 \times 10^3\,\text{J}}$$

$$q = -w = \boxed{1.62 \times 10^3\,\text{J}}$$

(b) $\boxed{\Delta U = \Delta H = 0}$

$$w = -p_{ex}\Delta V \text{ [2.8]}$$

where p_{ex} in this case can be computed from the perfect gas law

$$pV = nRT$$

$$\text{so } p = \frac{(2.00\,\text{mol}) \times (8.3145\,\text{J K}^{-1}\text{mol}^{-1}) \times (22 + 273)\,\text{K}}{31.7\,\text{dm}^3} \times (10\,\text{dm m}^{-1})^3 = 1.55 \times 10^5\,\text{Pa}$$

$$\text{and } w = \frac{-(1.55 \times 10^5\,\text{Pa}) \times (31.7 - 22.8)\,\text{dm}^3}{(10\,\text{dm m}^{-1})^3} = \boxed{-1.38 \times 10^3\,\text{J}}$$

$$q = -w = \boxed{1.38 \times 10^3\,\text{J}}$$

(c) $\qquad \boxed{\Delta U = \Delta H = 0}$

$$\boxed{w = 0}\,\text{[free expansion]} \quad q = \Delta U - w = 0 - 0 = \boxed{0}$$

COMMENT. An isothermal free expansion of a perfect gas is also adiabatic.

E2.4(b) The perfect gas law leads to

$$\frac{p_1 V}{p_2 V} = \frac{nRT_1}{nRT_2} \quad \text{or} \quad p_2 = \frac{p_1 T_2}{T_1} = \frac{(111\,\text{kPa}) \times (356\,\text{K})}{277\,\text{K}} = \boxed{143\,\text{kPa}}$$

There is no change in volume, so $\boxed{w = 0}$. The heat flow is

$$q = \int C_V\,dT \approx C_V \Delta T = (2.5) \times (8.3145\,\text{J K}^{-1}\,\text{mol}^{-1}) \times (2.00\,\text{mol}) \times (356 - 277)\,\text{K}$$

$$= \boxed{3.28 \times 10^3\,\text{J}}$$

$$\Delta U = q + w = \boxed{3.28 \times 10^3\,\text{J}}$$

E2.5(b)   **(a)** $\quad w = -p_{\text{ex}}\Delta V = \dfrac{-(7.7 \times 10^3\,\text{Pa}) \times (2.5\,\text{dm}^3)}{(10\,\text{dm m}^{-1})^3} = \boxed{-19\,\text{J}}$

       (b) $\quad w = -nRT \ln\left(\dfrac{V_f}{V_i}\right)$ [2.11]

$$w = -\left(\frac{6.56\,\text{g}}{39.95\,\text{g mol}^{-1}}\right) \times \left(8.3145\,\text{J K}^{-1}\text{mol}^{-1}\right) \times (305\text{K}) \times \ln \frac{(2.5 + 18.5)\,\text{dm}^3}{18.5\,\text{dm}^3}$$

$$= \boxed{-52.8\,\text{J}}$$

E2.6(b) $\qquad \Delta H = \Delta_{\text{cond}}H = -\Delta_{\text{vap}}H = -(2.00\,\text{mol}) \times (35.3\,\text{kJ mol}^{-1}) = \boxed{-70.6\,\text{kJ}}$

Since the condensation is done isothermally and reversibly, the external pressure is constant at 1.00 atm. Hence,

$$q = q_p = \Delta H = \boxed{-70.6\,\text{kJ}}$$

$$w = -p_{\text{ex}}\Delta V \;\text{[2.8]} \quad \text{where} \quad \Delta V = V_{\text{liq}} - V_{\text{vap}} \approx -V_{\text{vap}} \quad \text{because} \quad V_{\text{liq}} \ll V_{\text{vap}}$$

On the assumption that methanol vapor is a perfect gas, $V_{\text{vap}} = nRT/p$ and $p = p_{\text{ex}}$, since the condensation is done reversibly. Hence,

$$w \approx nRT = (2.00\,\text{mol}) \times (8.3145\,\text{J K}^{-1}\,\text{mol}^{-1}) \times (64 + 273)\,\text{K} = \boxed{5.60 \times 10^3\,\text{J}}$$

and $\Delta U = q + w = (-70.6 + 5.60)\,\text{kJ} = \boxed{-65.0\,\text{kJ}}$

E2.7(b) The reaction is

$$Zn + 2H^+ \rightarrow Zn^{2+} + H_2$$

so it liberates 1 mol of $H_2(g)$ for every 1 mol Zn used. Work at constant pressure is

$$w = -p_{ex}\Delta V = -pV_{gas} = -nRT$$

$$= -\left(\frac{5.0\,g}{65.4\,g\,mol^{-1}}\right) \times \left(8.3145\,J\,K^{-1}mol^{-1}\right) \times (23+273)\,K = \boxed{-188\,J}$$

E2.8(b) **(a)** At constant pressure, $q = \Delta H$.

$$q = \int C_p dT = \int_{0+273\,K}^{100+273\,K} [20.17 + (0.4001)T/K]\,dT\,J\,K^{-1}$$

$$= \left[(20.17)\,T + \frac{1}{2}(0.4001) \times \left(\frac{T^2}{K}\right)\right]\Bigg|_{273\,K}^{373\,K}\,J\,K^{-1}$$

$$= \left[(20.17) \times (373 - 273) + \frac{1}{2}(0.4001) \times (373^2 - 273^2)\right]J = \boxed{14.9 \times 10^3\,J} = \Delta H$$

$$w = -p\Delta V = -nR\Delta T = -(1.00\,mol) \times \left(8.3145\,J\,K^{-1}\,mol^{-1}\right) \times (100\,K) = \boxed{-831\,J}$$

$$\Delta U = q + w = (14.9 - 0.831)\,kJ = \boxed{14.1\ kJ}$$

(b) The energy and enthalpy of a perfect gas depend on temperature alone. Thus, $\Delta H = \boxed{14.9\,kJ}$ and $\Delta U = \boxed{14.1\,kJ}$ as above. At constant volume, $w = \boxed{0}$ and $\Delta U = q$, so $q = \boxed{+14.1\ kJ}$.

E2.9(b) For reversible adiabatic expansion

$$T_f = T_i \left(\frac{V_i}{V_f}\right)^{1/c} \quad [2.28a]$$

where

$$c = \frac{C_{V,m}}{R} = \frac{C_{p,m} - R}{R} = \frac{(37.11 - 8.3145)\ J\,K^{-1}mol^{-1}}{8.3145\,J\,K^{-1}mol^{-1}} = 3.463,$$

so the final temperature is

$$T_f = (298.15\,K) \times \left(\frac{500 \times 10^{-3}\,dm^3}{2.00\,dm^3}\right)^{1/3.463} = \boxed{200\,K}$$

E2.10(b) Reversible adiabatic work is

$$w = C_V \Delta T \quad [2.27] = n(C_{p,m} - R) \times (T_f - T_i)$$

where the temperatures are related by [solution to Exercise 2.15(b)]

$$T_f = T_i \left(\frac{V_i}{V_f}\right)^{1/c} \quad [2.28a] \quad \text{where} \quad c = \frac{C_{V,m}}{R} = \frac{C_{p,m} - R}{R} = 2.503$$

So $T_f = [(23.0 + 273.15) \text{ K}] \times \left(\dfrac{400 \times 10^{-3} \text{dm}^3}{2.00 \text{ dm}^3}\right)^{1/2.503} = 156 \text{ K}$

and $w = \left(\dfrac{3.12 \text{ g}}{28.0 \text{ g mol}^{-1}}\right) \times (29.125 - 8.3145) \text{ J K}^{-1} \text{mol}^{-1} \times (156 - 296) \text{ K} = \boxed{-325 \text{ J}}$

E2.11(b) For reversible adiabatic expansion

$$p_f V_f^\gamma = p_i V_i^\gamma \text{ [2.29]} \quad \text{so} \quad p_f = p_i \left(\dfrac{V_i}{V_f}\right)^\gamma = (8.73 \text{ Torr}) \times \left(\dfrac{500 \times 10^{-3} \text{ dm}^3}{3.0 \text{ dm}^3}\right)^{1.3} = \boxed{8.5 \text{ Torr}}$$

E2.12(b) $q_p = n C_{p,m} \Delta T$ [2.24]

$C_{p,m} = \dfrac{q_p}{n \Delta T} = \dfrac{178 \text{ J}}{1.9 \text{ mol} \times 1.78 \text{ K}} = \boxed{53 \text{ J K}^{-1} \text{mol}^{-1}}$

$C_{V,m} = C_{p,m} - R = (53 - 8.3) \text{ J K}^{-1} \text{mol}^{-1} = \boxed{45 \text{ J K}^{-1} \text{mol}^{-1}}$

E2.13(b) $\Delta H = q_p = C_p \Delta T \text{ [2.23b, 2.24]} = n C_{p,m} \Delta T$

$\Delta H = q_p = (2.0 \text{ mol}) \times (37.11 \text{ J K}^{-1} \text{mol}^{-1}) \times (277 - 250) \text{ K} = \boxed{2.0 \times 10^3 \text{ J mol}^{-1}}$

$\Delta H = \Delta U + \Delta(pV) = \Delta U + nR\Delta T \quad \text{so} \quad \Delta U = \Delta H - nR\Delta T$

$\Delta U = 2.0 \times 10^3 \text{ J mol}^{-1} - (2.0 \text{ mol}) \times (8.3145 \text{ J K}^{-1} \text{mol}^{-1}) \times (277 - 250) \text{ K}$

$\qquad = \boxed{1.6 \times 10^3 \text{ J mol}^{-1}}$

E2.14(b) In an adiabatic process, $q = \boxed{0}$. Work against a constant external pressure is

$$w = -p_{ex} \Delta V = \dfrac{-(78.5 \times 10^3 \text{ Pa}) \times (4 \times 15 - 15) \text{ dm}^3}{(10 \text{ dm m}^{-1})^3} = \boxed{-3.5 \times 10^3 \text{ J}}$$

$\Delta U = q + w = \boxed{-3.5 \times 10^3 \text{ J}}$

One can also relate adiabatic work to ΔT (eqn 2.27):

$$w = C_V \Delta T = n(C_{p,m} - R)\Delta T \quad \text{so} \quad \Delta T = \dfrac{w}{n(C_{p,m} - R)},$$

$\Delta T = \dfrac{-3.5 \times 10^3 \text{ J}}{(5.0 \text{ mol}) \times (37.11 - 8.3145) \text{ J K}^{-1} \text{mol}^{-1}} = \boxed{-24 \text{ K}}.$

$\Delta H = \Delta U + \Delta(pV) = \Delta U + nR\Delta T,$

$\qquad = -3.5 \times 10^3 \text{ J} + (5.0 \text{ mol}) \times (8.3145 \text{ J K}^{-1} \text{mol}^{-1}) \times (-24 \text{ K}) = \boxed{-4.5 \times 10^3 \text{ J}}$

E2.15(b) In an adiabatic process, the initial and final pressures are related by (eqn 2.29)

$$p_f V_f^\gamma = p_i V_i^\gamma \quad \text{where} \quad \gamma = \dfrac{C_{p,m}}{C_{V,m}} = \dfrac{C_{p,m}}{C_{p,m} - R} = \dfrac{20.8 \text{ J K}^{-1} \text{mol}^{-1}}{(20.8 - 8.31) \text{ J K}^{-1} \text{mol}^{-1}} = 1.67$$

Find V_i from the perfect gas law:

$$V_i = \frac{nRT_i}{p_i} = \frac{(1.5\,\text{mol})(8.31\,\text{J K}^{-1}\,\text{mol}^{-1})(315\,\text{K})}{230 \times 10^3\,\text{Pa}} = 0.017\bar{1}\,\text{m}^3$$

so $V_f = V_i \left(\dfrac{p_i}{p_f}\right)^{1/\gamma} = (0.017\bar{1}\,\text{m}^3)\left(\dfrac{230\,\text{kPa}}{170\,\text{kPa}}\right)^{1/1.67} = \boxed{0.020\bar{5}\,\text{m}^3}$.

Find the final temperature from the perfect gas law:

$$T_f = \frac{p_f V_f}{nR} = \frac{(170 \times 10^3\,\text{Pa}) \times (0.020\bar{5}\,\text{m}^3)}{(1.5\,\text{mol})(8.31\,\text{J K}^{-1}\,\text{mol}^{-1})} = \boxed{27\bar{9}\,\text{K}}$$

Adiabatic work is (eqn 2.27)

$$w = C_V \Delta T = (20.8 - 8.31)\,\text{J K}^{-1}\,\text{mol}^{-1} \times 1.5\,\text{mol} \times (27\bar{9} - 315)\,\text{K} = \boxed{-6.\bar{7} \times 10^2\,\text{J}}$$

E2.16(b) At constant pressure

$$q = \Delta H = n\Delta_{\text{vap}}H^{\ominus} = (0.75\,\text{mol}) \times (32.0\,\text{kJ mol}^{-1}) = \boxed{24.\bar{0}\,\text{kJ}}$$

and $w = -p\Delta V \approx -pV_{\text{vapor}} = -nRT = -(0.75\,\text{mol}) \times (8.3145\,\text{J K}^{-1}\,\text{mol}^{-1}) \times (260\,\text{K})$

$$w = -1.6 \times 10^3\,\text{J} = \boxed{-1.6\,\text{kJ}}$$

$$\Delta U = w + q = 24.\bar{0} - 1.6\,\text{kJ} = \boxed{22.\bar{4}\,\text{kJ}}$$

COMMENT. Because the vapor is here treated as a perfect gas, the specific value of the external pressure provided in the statement of the exercise does not affect the numerical value of the answer.

E2.17(b) The reaction is

$$C_6H_5OH(l) + 7O_2(g) \rightarrow 6CO_2(g) + 3H_2O(l)$$

$$\Delta_c H^{\ominus} = 6\Delta_f H^{\ominus}(CO_2) + 3\Delta_f H^{\ominus}(H_2O) - \Delta_f H^{\ominus}(C_6H_5OH) - 7\Delta_f H^{\ominus}(O_2)$$

$$= [6(-393.15) + 3(-285.83) - (-165.0) - 7(0)]\,\text{kJ mol}^{-1} = \boxed{-3053.6\,\text{kJ mol}^{-1}}$$

E2.18(b) We need $\Delta_f H^{\ominus}$ for the reaction

(4) $2B(s) + 3H_2(g) \rightarrow B_2H_6(g)$

reaction(4) = reaction(2) + 3 × reaction(3) − reaction(1)

Thus, $\Delta_f H^{\ominus} = \Delta_r H^{\ominus}\{\text{reaction}(2)\} + 3 \times \Delta_r H^{\ominus}\{\text{reaction}(3)\} - \Delta_r H^{\ominus}\{\text{reaction}(1)\}$

$$= [-2368 + 3 \times (-241.8) - (-1941)]\,\text{kJ mol}^{-1} = \boxed{-1152\,\text{kJ mol}^{-1}}$$

E2.19(b)   For anthracene the reaction is

$$C_{14}H_{10}(s) + \tfrac{33}{2}O_2(g) \rightarrow 14CO_2(g) + 5H_2O(l)$$

$$\Delta_c U^{\ominus} = \Delta_c H^{\ominus} - \Delta n_g RT \ [2.21], \quad \Delta n_g = -\tfrac{5}{2} \ \text{mol}$$

$$\Delta_c U^{\ominus} = -7061 \ \text{kJ mol}^{-1} - \left(-\tfrac{5}{2} \times 8.3 \times 10^{-3} \ \text{kJ K}^{-1} \text{mol}^{-1} \times 298 \ \text{K}\right)$$

$$= -7055 \ \text{kJ mol}^{-1}$$

$$|q| = |q_V| = |n\Delta_c U^{\ominus}| = \left(\frac{2.25 \times 10^{-3} \ \text{g}}{172.23 \ \text{g mol}^{-1}}\right) \times \left(7055 \ \text{kJ mol}^{-1}\right) = 0.0922 \ \text{kJ}$$

$$C = \frac{|q|}{\Delta T} = \frac{0.0922 \ \text{kJ}}{1.35 \ \text{K}} = 0.0683 \ \text{kJ K}^{-1} = \boxed{68.3 \ \text{J K}^{-1}}$$

When phenol is used the reaction is

$$C_6H_5OH(s) + \tfrac{15}{2}O_2(g) \rightarrow 6CO_2(g) + 3H_2O(l)$$

$$\Delta_c H^{\ominus} = -3054 \ \text{kJ mol}^{-1} \ \text{[Table 2.5]}$$

$$\Delta_c U = \Delta_c H - \Delta n_g RT, \quad \Delta n_g = -\tfrac{3}{2}$$

$$= (-3054 \ \text{kJ mol}^{-1}) + (\tfrac{3}{2}) \times (8.314 \times 10^{-3} \ \text{kJ K}^{-1} \text{mol}^{-1}) \times (298 \ \text{K})$$

$$= -3050 \ \text{kJ mol}^{-1}$$

$$|q| = \left(\frac{135 \times 10^{-3} \ \text{g}}{94.12 \ \text{g mol}^{-1}}\right) \times \left(3050 \ \text{kJ mol}^{-1}\right) = 4.37\overline{5} \ \text{kJ}$$

$$\Delta T = \frac{|q|}{C} = \frac{4.37\overline{5} \ \text{kJ}}{0.0683 \ \text{kJ K}^{-1}} = \boxed{+64.1 \ \text{K}}$$

COMMENT. In this case $\Delta_c U^{\ominus}$ and $\Delta_c H^{\ominus}$ differed by about 0.1 percent. Thus, to within 3 significant figures, it would not have mattered if we had used $\Delta_c H^{\ominus}$ instead of $\Delta_c U^{\ominus}$, but for very precise work it would.

E2.20(b)   The reaction is $AgBr(s) \rightarrow Ag^+(aq) + Br^-(aq)$

$$\Delta_{sol} H^{\ominus} = \Delta_f H^{\ominus}(Ag^+, aq) + \Delta_f H^{\ominus}(Br^-, aq) - \Delta_f H^{\ominus}(AgBr, s)$$

$$= [105.58 + (-121.55) - (-100.37)] \ \text{kJ mol}^{-1} = \boxed{+84.40 \ \text{kJ mol}^{-1}}$$

E2.21(b)   The combustion products of graphite and diamond are the same, so the transition $C(gr) \rightarrow C(d)$ is equivalent to the combustion of graphite plus the reverse of the combustion of diamond, and

$$\Delta_{trans} H^{\ominus} = [-393.51 - (395.41)] \ \text{kJ mol}^{-1} = \boxed{+1.90 \ \text{kJ mol}^{-1}}$$

E2.22(b)   **(a)** $\text{reaction}(3) = (-2) \times \text{reaction}(1) + \text{reaction}(2)$ and $\Delta n_g = -1$

The enthalpies of reactions are combined in the same manner as the equations (Hess's law).

$$\Delta_r H^{\ominus}(3) = (-2) \times \Delta_r H^{\ominus}(1) + \Delta_r H^{\ominus}(2)$$
$$= [(-2) \times (52.96) + (-483.64)]\,\text{kJ mol}^{-1}$$
$$= \boxed{-589.56\,\text{kJ mol}^{-1}}$$

$$\Delta_r U^{\ominus} = \Delta_r H^{\ominus} - \Delta n_g RT$$
$$= -589.56\,\text{kJ mol}^{-1} - (-3) \times (8.314\,\text{J K}^{-1}\text{mol}^{-1}) \times (298\,\text{K})$$
$$= -589.56\,\text{kJ mol}^{-1} + 7.43\,\text{kJ mol}^{-1} = \boxed{-582.13\,\text{kJ mol}^{-1}}$$

(b) $\Delta_f H^{\ominus}$ refers to the formation of one mole of the compound, so

$$\Delta_f H^{\ominus}(\text{HI}) = \tfrac{1}{2}\left(52.96\,\text{kJ mol}^{-1}\right) = \boxed{26.48\,\text{kJ mol}^{-1}}$$
$$\Delta_f H^{\ominus}(\text{H}_2\text{O}) = \tfrac{1}{2}\left(-483.64\,\text{kJ mol}^{-1}\right) = \boxed{-241.82\,\text{kJ mol}^{-1}}$$

E2.23(b) $\Delta_r H^{\ominus} = \Delta_r U^{\ominus} + RT\Delta n_g$ [2.21]
$$= -772.7\,\text{kJ mol}^{-1} + (5) \times (8.3145 \times 10^{-3}\,\text{kJ K}^{-1}\text{mol}^{-1}) \times (298\,\text{K})$$
$$= \boxed{-760.3\,\text{kJ mol}^{-1}}$$

E2.24(b)   Combine the reactions in such a way that the combination is the desired formation reaction. The enthalpies of the reactions are then combined in the same way as the equations to yield the enthalpy of formation.

	$\Delta_r H^{\ominus}/(\text{kJ mol}^{-1})$
$\tfrac{1}{2}\text{N}_2(g) + \tfrac{1}{2}\text{O}_2(g) \rightarrow \text{NO}(g)$	$+90.25$
$\text{NO}(g) + \tfrac{1}{2}\text{Cl}_2(g) \rightarrow \text{NOCl}(g)$	$-\tfrac{1}{2}(75.5)$
$\tfrac{1}{2}\text{N}_2(g) + \tfrac{1}{2}\text{O}_2(g) + \tfrac{1}{2}\text{Cl}_2(g) \rightarrow \text{NOCl}(g)$	$+52.5$

Hence, $\Delta_f H^{\ominus}(\text{NOCl, g}) = \boxed{+52.5\,\text{kJ mol}^{-1}}$

E2.25(b)   According to Kirchhoff's law [2.36]

$$\Delta_r H^{\ominus}(100°\text{C}) = \Delta_r H^{\ominus}(25\,°\text{C}) + \int_{25°\text{C}}^{100°\text{C}} \Delta_r C_p^{\ominus}\,dT$$

where Δ_r as usual signifies a sum over product and reactant species weighted by stoichiometric coefficients. Because $C_{p,m}$ can frequently be parametrized as

$$C_{p,m} = a + bT + c/T^2$$

the indefinite integral of $C_{p,m}$ has the form

$$\int C_{p,m} dT = aT + \tfrac{1}{2}bT^2 - c/T$$

Combining this expression with our original integral, we have

$$\Delta_r H^{\ominus}(100\,^{\circ}\mathrm{C}) = \Delta_r H^{\ominus}(25\,^{\circ}\mathrm{C}) + (T\Delta_r a + \tfrac{1}{2}T^2\Delta_r b - \Delta_r c/T)\Big|_{298\,\mathrm{K}}^{373\,\mathrm{K}}$$

Now for the pieces

$$\Delta_r H^{\ominus}(25\,^{\circ}\mathrm{C}) = 2(-285.83\,\mathrm{kJ\,mol^{-1}}) - 2(0) - 0 = -571.66\,\mathrm{kJ\,mol^{-1}}$$

$$\Delta_r a = [2(75.29) - 2(27.28) - (29.96)]\,\mathrm{J\,K^{-1}\,mol^{-1}} = 0.06606\,\mathrm{kJ\,K^{-1}\,mol^{-1}}$$

$$\Delta_r b = [2(0) - 2(3.29) - (4.18)] \times 10^{-3}\,\mathrm{J\,K^{-2}\,mol^{-1}} = -10.76 \times 10^{-6}\,\mathrm{kJ\,K^{-2}\,mol^{-1}}$$

$$\Delta_r c = [2(0) - 2(0.50) - (-1.67)] \times 10^5\,\mathrm{J\,K\,mol^{-1}} = 67\,\mathrm{kJ\,K\,mol^{-1}}$$

$$\Delta_r H^{\ominus}(100\,^{\circ}\mathrm{C}) = \Big[-571.66 + (373 - 298) \times (0.06606) + \frac{1}{2}(373^2 - 298^2)$$

$$\times(-10.76 \times 10^{-6}) - (67) \times \left(\frac{1}{373} - \frac{1}{298} \right) \Big]\,\mathrm{kJ\,mol^{-1}}$$

$$= \boxed{-566.93\,\mathrm{kJ\,mol^{-1}}}$$

E2.26(b)   The hydrogenation reaction is

(1) $C_2H_2(g) + H_2(g) \rightarrow C_2H_4(g)$ $\Delta_r H^{\ominus}(T) = ?$

The reactions and accompanying data which are to be combined in order to yield reaction (1) and $\Delta_r H^{\ominus}(T)$ are

(2) $H_2(g) + \tfrac{1}{2}O_2(g) \rightarrow H_2O(l)$ $\Delta_c H^{\ominus}(2) = -285.83\,\mathrm{kJ\,mol^{-1}}$

(3) $C_2H_4(g) + 3O_2(g) \rightarrow 2H_2O(l) + 2CO_2(g)$ $\Delta_c H^{\ominus}(3) = -1411\,\mathrm{kJ\,mol^{-1}}$

(4) $C_2H_2(g) + \tfrac{5}{2}O_2(g) \rightarrow H_2O(l) + 2CO_2(g)$ $\Delta_c H^{\ominus}(4) = -1300\,\mathrm{kJ\,mol^{-1}}$

reaction (1) = reaction (2) − reaction (3) + reaction (4)

(a) Hence, at 298 K:

$$\Delta_r H^{\ominus} = \Delta_c H^{\ominus}(2) - \Delta_c H^{\ominus}(3) + \Delta_c H^{\ominus}(4)$$

$$= [(-285.83) - (-1411) + (-1300)]\,\mathrm{kJ\,mol^{-1}} = \boxed{-175\,\mathrm{kJ\,mol^{-1}}}$$

$$\Delta_r U^{\ominus} = \Delta_r H^{\ominus} - \Delta n_g RT \quad [2.21]; \quad \Delta n_g = -1$$

$$= -175\,\mathrm{kJ\,mol^{-1}} - (-1) \times (2.48\,\mathrm{kJ\,mol^{-1}}) = \boxed{-173\,\mathrm{kJ\,mol^{-1}}}$$

(b) At 348 K:

$$\Delta_r H^\ominus(348\,K) = \Delta_r H^\ominus(298\,K) + \Delta_r C_p^\ominus(348\,K - 298\,K) \quad \text{[Example 2.6]}$$

$$\Delta_r C_p = \sum_J \nu_J C_{p,m}^\ominus(J)\,[2.37] = C_{p,m}^\ominus(C_2H_4, g) - C_{p,m}^\ominus(C_2H_2, g) - C_{p,m}^\ominus(H_2, g)$$

$$= (43.56 - 43.93 - 28.82) \times 10^{-3}\,kJ\,K^{-1}\,mol^{-1} = -29.19 \times 10^{-3}\,kJ\,K^{-1}\,mol^{-1}$$

$$\Delta_r H^\ominus(348\,K) = (-175\,kJ\,mol^{-1}) - (29.19 \times 10^{-3}\,kJ\,K^{-1}\,mol^{-1}) \times (50\,K)$$

$$= \boxed{-176\,kJ\,mol^{-1}}$$

E2.27(b) NaCl, AgNO$_3$, and NaNO$_3$ are strong electrolytes; therefore the net ionic equation is

$$Ag^+(aq) + Cl^-(aq) \rightarrow AgCl(s)$$
$$\Delta_r H^\ominus = \Delta_f H^\ominus(AgCl) - \Delta_f H^\ominus(Ag^+) - \Delta_f H^\ominus(Cl^-)$$
$$= [(-127.07) - (105.58) - (-167.16)]\,kJ\,mol^{-1} = \boxed{-65.49\,kJ\,mol^{-1}}$$

E2.28(b) The cycle is shown in Figure 2.1.

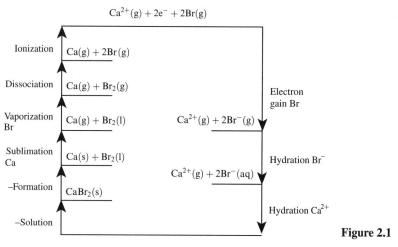

Figure 2.1

$$-\Delta_{hyd}H^\ominus(Ca^{2+}) = -\Delta_{soln}H^\ominus(CaBr_2) - \Delta_f H^\ominus(CaBr_2, s) + \Delta_{sub}H^\ominus(Ca)$$
$$+ \Delta_{vap}H^\ominus(Br_2) + \Delta_{diss}H^\ominus(Br_2) + \Delta_{ion}H^\ominus(Ca)$$
$$+ \Delta_{ion}H^\ominus(Ca^+) + 2\Delta_{eg}H^\ominus(Br) + 2\Delta_{hyd}H^\ominus(Br^-)$$
$$= [-(-103.1) - (-682.8) + 178.2 + 30.91 + 192.9$$
$$+ 589.7 + 1145 + 2(-331.0) + 2(-337)]\,kJ\,mol^{-1}$$
$$= \boxed{1587\,kJ\,mol^{-1}}$$

so $\Delta_{hyd}H^\ominus(Ca^{2+}) = \boxed{-1587\,kJ\,mol^{-1}}$

E2.29(b) The Joule–Thomson coefficient μ is the ratio of temperature change to pressure change under conditions of isenthalpic expansion. So

$$\mu = \left(\frac{\partial T}{\partial p}\right)_H \approx \frac{\Delta T}{\Delta p} = \frac{-10\,\text{K}}{(1.00 - 22)\,\text{atm}} = \boxed{0.48\ \text{K atm}^{-1}}$$

E2.30(b) The internal energy is a function of temperature and volume, $U_m = U_m(T, V_m)$, so

$$dU_m = \left(\frac{\partial U_m}{\partial T}\right)_{V_m} dT + \left(\frac{\partial U_m}{\partial V_m}\right)_T dV_m \qquad \left[\pi_T = \left(\frac{\partial U_m}{\partial V}\right)_T\right]$$

For an isothermal expansion $dT = 0$; hence

$$dU_m = \left(\frac{\partial U_m}{\partial V_m}\right)_T dV_m = \pi_T\, dV_m = \frac{a}{V_m^2}\, dV_m$$

$$\Delta U_m = \int_{V_{m,1}}^{V_{m,2}} dU_m = \int_{V_{m,1}}^{V_{m,2}} \frac{a}{V_m^2}\, dV_m = a\int_{1.00\,\text{dm}^3\,\text{mol}^{-1}}^{22.1\,\text{dm}^3\,\text{mol}^{-1}} \frac{dV_m}{V_m^2} = -\frac{a}{V_m}\bigg|_{1.00\,\text{dm}^3\,\text{mol}^{-1}}^{22.1\,\text{dm}^3\,\text{mol}^{-1}}$$

$$= -\frac{a}{22.1\,\text{dm}^3\,\text{mol}^{-1}} + \frac{a}{1.00\,\text{dm}^3\,\text{mol}^{-1}} = \frac{21.1a}{22.1\,\text{dm}^3\,\text{mol}^{-1}} = 0.954\overline{75}a\,\text{dm}^{-3}\,\text{mol}$$

From Table 1.6, $a = 1.337\,\text{dm}^6\,\text{atm mol}^{-1}$

$$\Delta U_m = (0.95475\,\text{mol dm}^3) \times (1.337\,\text{atm dm}^6\,\text{mol}^{-2})$$

$$= (1.27\overline{65}\,\text{atm dm}^3\,\text{mol}^{-1}) \times (1.01325 \times 10^5\,\text{Pa atm}^{-1}) \times \left(\frac{1\,\text{m}^3}{10^3\,\text{dm}^3}\right)$$

$$= 129\,\text{Pa m}^3\,\text{mol}^{-1} = \boxed{129\,\text{J mol}^{-1}}$$

$$w = -\int p\, dV_m \quad \text{where} \quad p = \frac{RT}{V_m - b} - \frac{a}{V_m^2} \text{ for a van der Waals gas.}$$

Hence,

$$w = -\int \left(\frac{RT}{V_m - b}\right) dV_m + \int \frac{a}{V_m^2}\, dV_m = -q + \Delta U_m$$

Thus

$$q = \int_{1.00\,\text{dm}^3\,\text{mol}^{-1}}^{22.1\,\text{dm}^3\,\text{mol}^{-1}} \left(\frac{RT}{V_m - b}\right) dV_m = RT\ln(V_m - b)\bigg|_{1.00\,\text{dm}^3\,\text{mol}^{-1}}^{22.1\,\text{dm}^3\,\text{mol}^{-1}}$$

$$= (8.314\,\text{J K}^{-1}\,\text{mol}^{-1}) \times (298\,\text{K}) \times \ln\left(\frac{22.1 - 3.20 \times 10^{-2}}{1.00 - 3.20 \times 10^{-2}}\right) = \boxed{+7.74\overline{65}\,\text{kJ mol}^{-1}}$$

and $w = -q + \Delta U_m = -(774\overline{7}\,\text{J mol}^{-1}) + (129\,\text{J mol}^{-1}) = \boxed{-761\overline{8}\,\text{J mol}^{-1}} = \boxed{-7.62\,\text{kJ mol}^{-1}}$

E2.31(b)   The expansion coefficient is

$$\alpha = \frac{1}{V}\left(\frac{\partial V}{\partial T}\right)_p = \frac{V'(3.7 \times 10^{-4}\,\mathrm{K^{-1}} + 2 \times 1.52 \times 10^{-6}\,T\,\mathrm{K^{-2}})}{V}$$

$$= \frac{V'[3.7 \times 10^{-4} + 2 \times 1.52 \times 10^{-6}\,(T/\mathrm{K})]\,\mathrm{K^{-1}}}{V'[0.77 + 3.7 \times 10^{-4}(T/\mathrm{K}) + 1.52 \times 10^{-6}(T/\mathrm{K})^2]}$$

$$= \frac{[3.7 \times 10^{-4} + 2 \times 1.52 \times 10^{-6}(310)]\,\mathrm{K^{-1}}}{0.77 + 3.7 \times 10^{-4}(310) + 1.52 \times 10^{-6}(310)^2} = \boxed{1.27 \times 10^{-3}\,\mathrm{K^{-1}}}$$

E2.32(b)   Isothermal compressibility is

$$\kappa_T = -\frac{1}{V}\left(\frac{\partial V}{\partial p}\right)_T \approx -\frac{\Delta V}{V\Delta p} \qquad \text{so} \qquad \Delta p = -\frac{\Delta V}{V\kappa_T}$$

A density increase of 0.08 percent means $\Delta V/V = -0.0008$. So the additional pressure that must be applied is

$$\Delta p = \frac{0.0008}{2.21 \times 10^{-6}\,\mathrm{atm^{-1}}} = \boxed{3.\overline{6} \times 10^2\,\mathrm{atm}}$$

E2.33(b)   The isothermal Joule–Thomson coefficient is

$$\left(\frac{\partial H}{\partial p}\right)_T = -\mu C_p = -(1.11\,\mathrm{K\,atm^{-1}}) \times (37.11\,\mathrm{J\,K^{-1}\,mol^{-1}}) = \boxed{-41.2\,\mathrm{J\,atm^{-1}\,mol^{-1}}}$$

If this coefficient is constant in an isothermal Joule–Thomson experiment, then the heat which must be supplied to maintain constant temperature is ΔH in the following relationship

$$\frac{\Delta H/n}{\Delta p} = -41.2\,\mathrm{J\,atm^{-1}\,mol^{-1}} \qquad \text{so} \qquad \Delta H = -(41.2\,\mathrm{J\,atm^{-1}\,mol^{-1}})n\Delta p$$

$$\Delta H = -(41.2\,\mathrm{J\,atm^{-1}\,mol^{-1}}) \times (12.0\,\mathrm{mol}) \times (-55\,\mathrm{atm}) = \boxed{27.\overline{2} \times 10^3\,\mathrm{J}}$$

Solutions to problems

Assume all gases are perfect unless stated otherwise. Unless otherwise stated, thermochemical data are for 298 K.

Solutions to numerical problems

P2.1   The temperatures are readily obtained from the perfect gas equation, $T = \dfrac{pV}{nR}$,

$$T_1 = \frac{(1.00\,\mathrm{atm}) \times (22.4\,\mathrm{dm^3})}{(1.00\,\mathrm{mol}) \times (0.0821\,\mathrm{dm^3\,atm\,mol^{-1}\,K^{-1}})} = \boxed{273\,\mathrm{K}} = T_3 \text{ [isotherm]}.$$

Similarly, $T_2 = \boxed{546\,\mathrm{K}}$.

In the solutions that follow all steps in the cycle are considered to be reversible.

Step 1 → 2

$$w = -p_{ex}\Delta V = -p\Delta V = -nR\Delta T \quad [\Delta(pV) = \Delta(nRT)],$$

$$w = -(1.00\,\text{mol}) \times (8.314\,\text{J K}^{-1}\,\text{mol}^{-1}) \times (546 - 273)\,\text{K} = \boxed{-2.27 \times 10^3\,\text{J}}.$$

$$\Delta U = nC_{V,m}\Delta T = \boxed{(1.00\,\text{mol}) \times \frac{3}{2} \times (8.314\,\text{J K}^{-1}\,\text{mol}^{-1}) \times (273\,\text{K}) = +3.40 \times 10^3\,\text{J}.}$$

$$q = \Delta U - w = +3.40 \times 10^3\,\text{J} - (-2.27 \times 10^3\,\text{J}) = \boxed{+5.67 \times 10^3\,\text{J}}.$$

$$\Delta H = q_p = \boxed{+5.67 \times 10^3\,\text{J}}.$$

If this step is not reversible, then w, q, and ΔH would be indeterminate.

Step 2 → 3

$$\boxed{w = 0} \text{ [constant volume]}.$$

$$q_V = \Delta U = nC_{V,m}\Delta T = (1.00\,\text{mol}) \times (\frac{3}{2}) \times (8.314\,\text{J K}^{-1}\,\text{mol}^{-1}) \times (-273\,\text{K})$$

$$= \boxed{-3.40 \times 10^3\,\text{J}}.$$

From $H \equiv U + pV$

$$\Delta H = \Delta U + \Delta(pV) = \Delta U + \Delta(nRT) = \Delta U + nR\Delta T$$

$$= (-3.40 \times 10^3\,\text{J}) + (1.00\,\text{mol}) \times (8.314\,\text{J K}^{-1}\,\text{mol}^{-1}) \times (-273\,\text{K}) = \boxed{-5.67 \times 10^3\,\text{J}}.$$

Step 3 → 1

ΔU and ΔH are $\boxed{\text{zero}}$ for an isothermal process in a perfect gas; hence for the reversible compression

$$-q = w = -nRT\,\ln\frac{V_1}{V_3} = (-1.00\,\text{mol}) \times (8.314\,\text{J K}^{-1}\,\text{mol}^{-1}) \times (273\,\text{K}) \times \ln\left(\frac{22.4\,\text{dm}^3}{44.8\,\text{dm}^3}\right)$$

$$= \boxed{+1.57 \times 10^3\,\text{J}}, \quad q = \boxed{-1.57 \times 10^3\,\text{J}}.$$

If this step is not reversible, then q and w would have different values which would be determined by the details of the process.

Total cycle

State	p/atm	V/dm^3	T/K
1	1.00	22.44	273
2	1.00	44.8	546
3	0.50	44.8	273

Thermodynamic quantities calculated for reversible steps

Step	Process	q/kJ	w/kJ	ΔU/kJ	ΔH/kJ
$1 \to 2$	p constant $= p_{ex}$	+5.67	−2.27	+3.40	+5.67
$2 \to 3$	V constant	−3.40	0	−3.40	−5.67
$3 \to 1$	Isothermal, reversible	−1.57	+1.57	0	0
Cycle		+0.70	−0.70	0	0

COMMENT. All values can be determined unambiguously for the reversible cycle. The net result of the overall process is that 700 J of heat has been converted to work.

P2.3   Since the volume is fixed, $\boxed{w = 0}$.

Since $\Delta U = q$ at constant volume, $\boxed{\Delta U = +2.35 \text{ kJ}}$.

$$\Delta H = \Delta U + \Delta(pV) = \Delta U + V \Delta p \ \ [\Delta V = 0].$$

From the van der Waals equation [Table 1.6]

$$p = \frac{RT}{V_m - b} - \frac{a}{V_m^2} \quad \text{so} \quad \Delta p = \frac{R \Delta T}{V_m - b} \quad [\Delta V_m = 0 \text{ at constant volume}].$$

Therefore, $\Delta H = \Delta U + \dfrac{R V \Delta T}{V_m - b}$.

From the data,

$$V_m = \frac{15.0 \text{ dm}^3}{2.0 \text{ mol}} = 7.5 \text{ dm}^3 \text{ mol}^{-1}, \ \ \Delta T = (341 - 300) \text{ K} = 41 \text{ K}.$$

$$V_m - b = (7.5 - 4.3 \times 10^{-2}) \text{ dm}^3 \text{ mol}^{-1} = 7.4\bar{6} \text{ dm}^3 \text{ mol}^{-1}.$$

$$\frac{R V \Delta T}{V_m - b} = \frac{(8.314 \text{ J K}^{-1} \text{ mol}^{-1}) \times (15.0 \text{ dm}^3) \times (41 \text{ K})}{7.4\bar{6} \text{ dm}^3 \text{ mol}^{-1}} = 0.68 \text{ kJ}.$$

Therefore, $\Delta H = (2.35 \text{ kJ}) + (0.68 \text{ kJ}) = \boxed{+3.03 \text{ kJ}}$.

P2.5   This cycle is represented in Figure 2.2. Assume that the initial temperature is 298 K .

(a) First, note that $\boxed{w = 0}$ (constant volume). Then calculate ΔU since ΔT is known ($\Delta T = 298$ K) and then calculate q from the First Law.

$$\Delta U = nC_{V,m} \Delta T \text{ [2.16b]}; \ \ C_{V,m} = C_{p,m} - R = \frac{7}{2}R - R = \frac{5}{2}R,$$

$$\Delta U = (1.00 \text{ mol}) \times \left(\frac{5}{2}\right) \times (8.314 \text{ J K}^{-1} \text{ mol}^{-1}) \times (298 \text{ K}) = 6.19 \times 10^3 \text{ J} = \boxed{+6.19 \text{ kJ}}.$$

$$q = q_V = \Delta U - w = 6.19 \text{ kJ} - 0 = \boxed{+6.19 \text{ kJ}}.$$

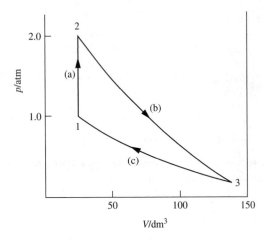

Figure 2.2

$$\Delta H = \Delta U + \Delta(pV) = \Delta U + \Delta(nRT) = \Delta U + nR\Delta T$$

$$= (6.19\,\text{kJ}) + (1.00\,\text{mol}) \times (8.31 \times 10^{-3}\,\text{kJ mol}^{-1}) \times (298\,\text{K}) = \boxed{+8.67\,\text{kJ}}.$$

(b) $\boxed{q = 0}$ (adiabatic).

Because the energy and enthalpy of a perfect gas depend on temperature alone,

$$\Delta U(\text{b}) = -\Delta U(\text{a}) = \boxed{-6.19\,\text{kJ}}, \text{ since } \Delta T(\text{b}) = -\Delta T(\text{a}).$$

Likewise $\Delta H(\text{b}) = -\Delta H(\text{a}) = \boxed{-8.67\,\text{kJ}}$.

$$w = \Delta U = \boxed{-6.19\,\text{kJ}} \text{ [First Law with } q = 0\text{]}.$$

(c) $\Delta U = \Delta H = 0$ [isothermal process in perfect gas].

$$q = -w \text{ [First Law with } \Delta U = 0\text{]}; \quad w = -nRT_1 \ln \frac{V_1}{V_3} \text{ [2.11]}.$$

$$V_2 = V_1 = \frac{nRT_1}{p_1} = \frac{(1.00\,\text{mol}) \times (0.08206\,\text{dm}^3\,\text{atm K}^{-1}\,\text{mol}^{-1}) \times (298\,\text{K})}{1.00\,\text{atm}} = 24.4\bar{5}\,\text{dm}^3.$$

$$V_2 T_2^c = V_3 T_3^c \text{ [2.28b]; hence } V_3 = V_2 \left(\frac{T_2}{T_3}\right)^c \text{ where } c = \frac{C_{V,m}}{R} = \frac{5}{2}$$

$$\text{so } V_3 = (24.4\bar{5}\,\text{dm}^3) \times \left(\frac{(2) \times (298\,\text{K})}{298\,\text{K}}\right)^{5/2} = 138.3\,\text{dm}^3.$$

$$w = (-1.00\,\text{mol}) \times (8.314\,\text{J K}^{-1}\,\text{mol}^{-1}) \times (298\,\text{K}) \times \ln\left(\frac{22.4\bar{5}\,\text{dm}^3}{138.3\,\text{dm}^3}\right)$$

$$= 4.29 \times 10^3\,\text{J} = \boxed{+4.29\,\text{kJ}}.$$

$$q = \boxed{-4.29\,\text{kJ}}.$$

P2.7   The formation reaction is

$$2C(s) + 3H_2(g) \rightarrow 2C_6H_6(g), \qquad \Delta_f H^{\ominus}(298\ K) = -84.68\ kJ\ mol^{-1}.$$

In order to determine $\Delta_f H^{\ominus}(350\ K)$ we employ Kirchhoff's law [2.36] with $T_2 = 350\ K, T_1 = 298\ K$,

$$\Delta_f H^{\ominus}(T_2) = \Delta_f H^{\ominus}(T_1) + \int_{T_1}^{T_2} \Delta_r C_p\ dT$$

where $\Delta_r C_p = \sum_J \nu_J C_{p,m}(J) = C_{p,m}(C_6H_6) - 2C_{p,m}(C) - 3C_{p,m}(H_2)$.

From Table 2.2

$$C_{p,m}(C_6H_6)/(J\ K^{-1}\ mol^{-1}) = 14.73 + \left(\frac{0.1272}{K}\right)T,$$

$$C_{p,m}(C,\ s)/(J\ K^{-1}\ mol^{-1}) = 16.86 + \left(\frac{4.77 \times 10^{-3}}{K}\right)T - \left(\frac{8.54 \times 10^5\ K^2}{T^2}\right),$$

$$C_{p,m}(H_2,\ g)/(J\ K^{-1}\ mol^{-1}) = 27.28 + \left(\frac{3.26 \times 10^{-3}}{K}\right)T - \left(\frac{0.50 \times 10^5\ K^2}{T^2}\right),$$

$$\Delta_r C_p/(J\ K^{-1}\ mol^{-1}) = -100.83 + \left(\frac{0.1079T}{K}\right) - \left(\frac{1.56 \times 10^6\ K^2}{T^2}\right).$$

$$\int_{T_1}^{T_2} \frac{\Delta_r C_p\ dT}{J\ K^{-1}\ mol^{-1}} = -100.83 \times (T_2 - T_1) + \left(\frac{1}{2}\right)\left(0.1079\ K^{-1}\right)(T_2^2 - T_1^2)$$

$$- (1.56 \times 10^6\ K^2)\left(\frac{1}{T_2} - \frac{1}{T_1}\right)$$

$$= -100.83 \times (52\ K) + \left(\frac{1}{2}\right)(0.1079)(350^2 - 298^2)\ K$$

$$- (1.56 \times 10^6)\left(\frac{1}{350} - \frac{1}{298}\right)\ K$$

$$= -2.65 \times 10^3\ K.$$

Multiplying by the units $J\ K^{-1}mol^{-1}$, we obtain

$$\int_{T_1}^{T_2} \Delta_r C_p dT = -(2.65 \times 10^3\ K) \times (J\ K^{-1}mol^{-1}) = -2.65 \times 10^3\ J\ mol^{-1}$$

$$= -2.65\ kJ\ mol^{-1}.$$

Hence $\Delta_f H^{\ominus}(350\ K) = \Delta_f H^{\ominus}(298\ K) - 2.65\ kJ\ mol^{-1}$

$$= -84.68\ kJ\ mol^{-1} - 2.65\ kJ\ mol^{-1} = \boxed{-87.33\ kJ\ mol^{-1}}.$$

P2.9 $Cr(C_6H_6)_2(s) \rightarrow Cr(s) + 2C_6H_6(g), \quad \Delta n_g = +2\ mol.$

$$\Delta_r H^{\ominus} = \Delta_r U^{\ominus} + 2RT,\ \text{from [2.21]}$$

$$= (8.0\ kJ\ mol^{-1}) + (2) \times (8.314\ J\ K^{-1}\ mol^{-1}) \times (583\ K) = \boxed{+17.7\ kJ\ mol^{-1}}.$$

In terms of enthalpies of formation

$$\Delta_r H^\ominus = (2) \times \Delta_f H^\ominus(\text{benzene}, 583 \text{ K}) - \Delta_f H^\ominus(\text{metallocene}, 583 \text{ K})$$

or $\Delta_r H^\ominus(\text{metallocene}, 583 \text{ K}) = 2\Delta_f H^\ominus(\text{benzene, } 583 \text{ K}) - 17.7 \text{ kJ mol}^{-1}$.

The enthalpy of formation of benzene gas at 583 K is related to its value at 298 K by

$$\Delta_f H^\ominus(\text{benzene}, 583 \text{ K}) = \Delta_f H^\ominus(\text{benzene}, 298 \text{ K})$$
$$+ (T_b - 298 \text{ K})C_{p,m}(l) + \Delta_{vap}H^\ominus + (583 \text{ K} - T_b)C_{p,m}(g)$$
$$- 6 \times (583 \text{ K} - 298 \text{ K})C_{p,m}(gr) - 3 \times (583 \text{ K} - 298 \text{ K})C_{p,m}(H_2, \text{ g})$$

where T_b is the boiling temperature of benzene (353 K). We shall assume that the heat capacities of graphite and hydrogen are approximately constant in the range of interest and use their values from Table 2.7.

$$\Delta_f H^\ominus(\text{benzene}, 583 \text{ K}) = (49.0 \text{ kJ mol}^{-1}) + (353 - 298) \text{ K} \times (136.1 \text{ J K}^{-1} \text{ mol}^{-1})$$
$$+ (30.8 \text{ kJ mol}^{-1}) + (583 - 353) \text{ K} \times (81.67 \text{ J K}^{-1} \text{ mol}^{-1})$$
$$- (6) \times (583 - 298) \text{ K} \times (8.53 \text{ J K}^{-1} \text{ mol}^{-1})$$
$$- (3) \times (583 - 298) \text{ K} \times (28.82 \text{ J K}^{-1} \text{ mol}^{-1})$$
$$= \{(49.0) + (7.49) + (18.78) + (30.8) - (14.59) - (24.64)\} \text{ kJ mol}^{-1}$$
$$= +66.8 \text{ kJ mol}^{-1}.$$

Therefore $\Delta_f H^\ominus$ (metallocene, 583 K) = $(2 \times 66.8 - 17.7)$ kJ mol^{-1} = $\boxed{+ 116.0 \text{ kJ mol}^{-1}}$.

P2.11   (a) and (b). The table displays computed enthalpies of formation (semi-empirical, PM3 level, PC Spartan Pro^{TM}), enthalpies of combustion based on them (and on experimental enthalpies of formation of $H_2O(l)$ and $CO_2(g)$, -285.83 and -393.51 kJ mol^{-1} respectively), experimental enthalpies of combustion (Table 2.5), and the relative error in enthalpy of combustion.

Compound	$\Delta_f H^\ominus$/kJ mol^{-1}	$\Delta_c H^\ominus$/kJ mol^{-1}(calc.)	$\Delta_c H^\ominus$/kJ mol^{-1}(expt.)	% error
$CH_4(g)$	-54.45	-910.72	-890	2.33
$C_2H_6(g)$	-75.88	-1568.63	-1560	0.55
$C_3H_8(g)$	-98.84	-2225.01	-2220	0.23
$C_4H_{10}(g)$	-121.60	-2881.59	-2878	0.12
$C_5H_{12}(g)$	-142.11	-3540.42	-3537	0.10

The combustion reactions can be expressed as:

$$C_nH_{2n+2}(g) + \left(\frac{3n+1}{2}\right) O_2(g) \rightarrow n \, CO_2(g) + (n+1) \, H_2O(l).$$

The enthalpy of combustion, in terms of enthalpies of reaction, is

$$\Delta_c H^\ominus = n\Delta_f H^\ominus(CO_2) + (n+1)\Delta_f H^\ominus(H_2O) - \Delta_f H^\ominus(C_nH_{2n+2}),$$

where we have left out $\Delta_f H^{\ominus}(O_2) = 0$. The % error is defined as:

$$\%error = \frac{\Delta_c H^{\ominus}(calc.) - \Delta_c H^{\ominus}(expt.)}{\Delta_c H^{\ominus}(expt.)} \times 100\%.$$

The agreement is quite good.

(c) If the enthalpy of combustion is related to the molar mass by

$$\Delta_c H^{\ominus} = k[M/(g\ mol^{-1})]^n$$

then one can take the natural log of both sides to obtain:

$$\ln \left| \Delta_c H^{\ominus} \right| = \ln |k| + n \ln M/(g\ mol^{-1}).$$

Thus, if one plots $\ln \left| \Delta_c H^{\ominus} \right|$ vs. $\ln [M/(g\ mol^{-1})]$, one ought to obtain a straight line with slope n and y-intercept $\ln |k|$. Draw up the following table.

Compound	$M/(g\ mol^{-1})$	$\Delta_c H/kJ\ mol^{-1}$	$\ln M/(g\ mol^{-1})$	$\ln \left| \Delta_c H^{\ominus}/kJ\ mol^{-1} \right|$
$CH_4(g)$	16.04	−910.72	2.775	6.814
$C_2H_6(g)$	30.07	−1568.63	3.404	7.358
$C_3H_8(g)$	44.10	−2225.01	3.786	7.708
$C_4H_{10}(g)$	58.12	−2881.59	4.063	7.966
$C_5H_{12}(g)$	72.15	−3540.42	4.279	8.172

The plot is shown in Fig 2.3.

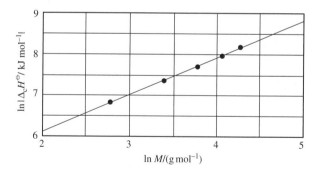

Figure 2.3

The linear least-squares fit equation is:

$$\ln |\Delta_c H^{\ominus}/kJ\ mol^{-1}| = 4.30 + 0.903\ \ln M/(g\ mol^{-1})\quad R^2 = 1.00$$

These compounds support the proposed relationships, with

$$n = \boxed{0.903} \quad \text{and} \quad k = -e^{4.30}\ kJ\ mol^{-1} = \boxed{-73.7\ kJ\ mol^{-1}}.$$

The agreement of these theoretical values of k and n with the experimental values obtained in P2.10 is rather good.

P2.13   The reaction is

$$C_{60}(s) + 60O_2(g) \rightarrow 60CO_2(g).$$

Because the reaction does not change the number of moles of gas, $\Delta_c H = \Delta_c U$ [2.21]. Therefore

$$\Delta_c H^\ominus = (-36.0334\,\text{kJ}\,\text{g}^{-1}) \times (60 \times 12.011\,\text{g}\,\text{mol}^{-1}) = \boxed{25968\,\text{kJ}\,\text{mol}^{-1}}.$$

Now relate the enthalpy of combustion to enthalpies of formation and solve for that of C_{60}.

$$\Delta_c H^\ominus = 60\Delta_f H^\ominus(CO_2) - 60\Delta_f H^\ominus(O_2) - \Delta_f H^\ominus(C_{60}),$$

$$\Delta_f H^\ominus(C_{60}) = 60\Delta_f H^\ominus(CO_2) - 60\Delta_f H^\ominus(O_2) - \Delta_c H^\ominus$$

$$= [60(-393.51) - 60(0) - (-25968)]\,\text{kJ}\,\text{mol}^{-1} = \boxed{2357\,\text{kJ}\,\text{mol}^{-1}}.$$

P2.15   **(a)** $\Delta_r H^\ominus = \Delta_f H^\ominus(SiH_2) + \Delta_f H^\ominus(H_2) - \Delta_f H^\ominus(SiH_4)$

$$= (274 + 0 - 34.3)\,\text{kJ}\,\text{mol}^{-1} = \boxed{240\,\text{kJ}\,\text{mol}^{-1}}.$$

(b) $\Delta_r H^\ominus = \Delta_f H^\ominus(SiH_2) + \Delta_f H^\ominus(SiH_4) - \Delta_f H^\ominus(Si_2H_6)$

$$= (274 + 34.3 - 80.3)\,\text{kJ}\,\text{mol}^{-1} = \boxed{228\,\text{kJ}\,\text{mol}^{-1}}.$$

P2.17   The temperatures and volumes in reversible adiabatic expansion are related by eqn 2.28a:

$$T_f = T_i\left(\frac{V_f}{V_i}\right)^{1/c} \quad \text{where } c = \frac{C_{V,m}}{R}.$$

From eqn 2.29, we can relate the pressures and volumes:

$$p_f = p_i\left(\frac{V_f}{V_i}\right)^{\gamma} \quad \text{where } \gamma = \frac{C_{p,m}}{C_{V,m}}.$$

We are looking for $C_{p,m}$, which can be related to c and γ.

$$c\gamma = \left(\frac{C_{V,m}}{R}\right) \times \left(\frac{C_{p,m}}{C_{V,m}}\right) = \frac{C_{p,m}}{R}.$$

Solving both relationships for the ratio of volumes, we have

$$\left(\frac{p_f}{p_i}\right)^{1/\gamma} = \frac{V_f}{V_i} = \left(\frac{T_f}{T_i}\right)^{c} \quad \text{so} \quad \frac{p_f}{p_i} = \left(\frac{T_f}{T_i}\right)^{c\gamma}.$$

Therefore

$$C_{p,m} = R\frac{\ln\left(\dfrac{p_f}{p_i}\right)}{\ln\left(\dfrac{T_f}{T_i}\right)} = (8.314\,\text{J}\,\text{K}^{-1}\,\text{mol}^{-1}) \times \left(\frac{\ln\left(\dfrac{202.94\,\text{kPa}}{81.840\,\text{kPa}}\right)}{\ln\left(\dfrac{298.15\,\text{K}}{248.44\,\text{K}}\right)}\right) = \boxed{41.40\,\text{J}\,\text{K}^{-1}\,\text{mol}^{-1}}.$$

P2.19 $H_m = H_m(T, p)$.

$$dH_m = \left(\frac{\partial H_m}{\partial T}\right)_p dT + \left(\frac{\partial H_m}{\partial p}\right)_T dp.$$

Since $dT = 0$,

$$dH_m = \left(\frac{\partial H_m}{\partial p}\right)_T dp \quad \text{where} \quad \left(\frac{\partial H_m}{\partial p}\right)_T = -\mu C_{p,m} \ [2.53] = -\left(\frac{2a}{RT} - b\right).$$

$$\Delta H_m = \int_{p_i}^{p_f} dH_m = -\int_{p_i}^{p_f}\left(\frac{2a}{RT} - b\right)dp = -\left(\frac{2a}{RT} - b\right)(p_f - p_i)$$

$$= -\frac{(2) \times (1.352 \text{ dm}^6 \text{ atm mol}^{-2})}{(0.08206 \text{ dm}^3 \text{ atm K}^{-1} \text{ mol}^{-1}) \times (300 \text{ K})} - (0.0387 \text{ dm}^3 \text{ mol}^{-1})$$

$$\times (1.00 \text{ atm} - 500 \text{ atm})$$

$$= (35.5 \text{ atm}) \times \left(\frac{1 \text{ m}}{10 \text{ dm}}\right)^3 \times \left(\frac{1.013 \times 10^5 \text{ Pa}}{1 \text{ atm}}\right) = 3.60 \times 10^3 \text{ J} = \boxed{+3.60 \text{ kJ}}.$$

COMMENT. Note that it is not necessary to know the value of $C_{p,m}$.

Solutions to theoretical problems

P2.21   **(a)** $dz = \left(\frac{\partial z}{\partial x}\right)_y dx + \left(\frac{\partial z}{\partial y}\right)_x dy$ [definition of total differential].

$$\left(\frac{\partial z}{\partial x}\right)_y = (2x - 2y + 2), \quad \left(\frac{\partial z}{\partial y}\right)_x = (4y - 2x - 4),$$

$$dz = \boxed{(2x - 2y + 2)\, dx + (4y - 2x - 4)dy}.$$

(b) $\frac{\partial}{\partial y}\left(\frac{\partial z}{\partial x}\right) = \frac{\partial}{\partial y}(2x - 2y + 2) = -2, \quad \frac{\partial}{\partial x}\left(\frac{\partial z}{\partial y}\right) = \frac{\partial}{\partial x}(4y - 2x - 4) = -2.$

(c) $\left(\frac{\partial z}{\partial x}\right)_y = \left(y + \frac{1}{x}\right), \quad \left(\frac{\partial z}{\partial y}\right)_x = (x - 1),$

$$dz = \boxed{\left(y + \frac{1}{x}\right)dx + (x - 1)\, dy}.$$

A differential is exact if it satisfies the condition

$$\frac{\partial}{\partial x}\left(\frac{\partial z}{\partial y}\right) = \frac{\partial}{\partial y}\left(\frac{\partial z}{\partial x}\right),$$

$$\frac{\partial}{\partial y}\left(\frac{\partial z}{\partial x}\right) = \frac{\partial}{\partial y}\left(y + \frac{1}{x}\right) = 1, \quad \frac{\partial}{\partial x}\left(\frac{\partial z}{\partial y}\right) = \frac{\partial}{\partial x}(x - 1) = 1.$$

COMMENT. The total differential of a function is necessarily exact.

P2.23   **(a)** $U = U(T, V)$ so $dU = \left(\dfrac{\partial U}{\partial T}\right)_V dT + \left(\dfrac{\partial U}{\partial V}\right)_T dV = C_V dT + \left(\dfrac{\partial U}{\partial V}\right)_T dV.$

For U = constant, $dU = 0$, and

$$C_V dT = -\left(\dfrac{\partial U}{\partial V}\right)_T dV \quad \text{or} \quad C_V = -\left(\dfrac{\partial U}{\partial V}\right)_T \left(\dfrac{dV}{dT}\right)_U = -\left(\dfrac{\partial U}{\partial V}\right)_T \left(\dfrac{\partial V}{\partial T}\right)_U.$$

This relationship is essentially Euler's chain relation [*Further information* 2.2].

(b) $H = H(T, p)$ so $dH = \left(\dfrac{\partial H}{\partial T}\right)_p dT + \left(\dfrac{\partial H}{\partial p}\right)_T dp = C_p dT + \left(\dfrac{\partial H}{\partial p}\right)_T dp.$

According to Euler's chain relation

$$\left(\dfrac{\partial H}{\partial p}\right)_T \left(\dfrac{\partial p}{\partial T}\right)_H \left(\dfrac{dT}{dH}\right)_p = -1$$

so, using the reciprocal identity [*Further information* 2.2],

$$\left(\dfrac{\partial H}{\partial p}\right)_T = -\left(\dfrac{\partial T}{\partial p}\right)_H \left(\dfrac{dH}{dT}\right)_p = \boxed{-\mu C_p}.$$

P2.25   **(a)** $H = U + pV$ so $\left(\dfrac{\partial H}{\partial U}\right)_p = 1 + p\left(\dfrac{\partial V}{\partial U}\right)_p = \boxed{1 + \dfrac{p}{(\partial U/\partial V)_p}}.$

(b) $\left(\dfrac{\partial H}{\partial U}\right)_p = \dfrac{(\partial H/\partial V)_p}{(\partial U/\partial V)_p} = \dfrac{\left(\dfrac{\partial(U + pV)}{\partial V}\right)_p}{(\partial U/\partial V)_p} = \dfrac{(\partial U/\partial V)_p + p}{(\partial U/\partial V)_p}$

so $\left(\dfrac{\partial H}{\partial U}\right)_p = 1 + \dfrac{p}{(\partial U/\partial V)_p} = \boxed{1 + p\left(\dfrac{\partial V}{\partial U}\right)_p}.$

P2.27 $w = -\displaystyle\int_{V_1}^{V_2} p\, dV.$

Inserting $\dfrac{V}{n} = V_m$ into the virial equation for p we obtain

$$p = nRT\left(\dfrac{1}{V} + \dfrac{nB}{V^2} + \dfrac{n^2 C}{V^3} + \ldots\right).$$

Therefore, $w = -nRT \displaystyle\int_{V_1}^{V_2} \left(\dfrac{1}{V} + \dfrac{nB}{V^2} + \dfrac{n^2 C}{V^3} + \ldots\right) dV,$

$$w = -nRT \ln\dfrac{V_2}{V_1} + n^2 RTB\left(\dfrac{1}{V_2} - \dfrac{1}{V_1}\right) + \dfrac{1}{2}n^3 RTC\left(\dfrac{1}{V_2^2} - \dfrac{1}{V_2^2}\right) + \ldots.$$

For $n = 1$ mol: $nRT = (1.0 \text{ mol}) \times (8.314 \text{ J K}^{-1} \text{ mol}^{-1}) \times (273 \text{ K}) = 2.2\overline{7} \text{ kJ}.$

From Table 1.4, $B = -21.7 \text{ cm}^3 \text{ mol}^{-1}$ and $C = 1200 \text{ cm}^6 \text{ mol}^{-2}$, so

$$n^2 BRT = (1.0 \text{ mol}) \times (-21.7 \text{ cm}^3 \text{ mol}^{-1}) \times (2.2\overline{7} \text{ kJ}) = -49.\overline{3} \text{ kJ cm}^3,$$

$$\frac{1}{2} n^3 CRT = \frac{1}{2}(1.0 \text{ mol})^2 \times (1200 \text{ cm}^6 \text{ mol}^{-2}) \times (2.2\overline{7} \text{ kJ}) = +1362 \text{ kJ cm}^6.$$

Therefore,

(a) $w = -2.2\overline{7} \text{ kJ} \ln 2 - (49.\overline{3} \text{ kJ}) \times \left(\dfrac{1}{1000} - \dfrac{1}{500} \right) + (136\overline{2} \text{ kJ}) \times \left(\dfrac{1}{1000^2} - \dfrac{1}{500^2} \right)$

$\qquad = (-1.5\overline{7}) + (0.049) - (4.1 \times 10^{-3}) \text{ kJ} = -1.5\overline{2} \text{ kJ} = \boxed{-1.5 \text{ kJ}}.$

(b) A perfect gas corresponds to the first term of the expansion of p, so

$$w = -1.5\overline{7} \text{ kJ} = \boxed{-1.6 \text{ kJ}}.$$

P2.29 $\mu = \left(\dfrac{\partial T}{\partial p} \right)_H = -\dfrac{1}{C_p} \left(\dfrac{\partial H}{\partial p} \right)_T$ [2.51 and 2.53],

$\qquad \mu = \dfrac{1}{C_p} \left\{ T \left(\dfrac{\partial V}{\partial T} \right)_p - V \right\}$ [See problem 2.34 for this result].

But $V = \dfrac{nRT}{p} + nb$ or $\left(\dfrac{\partial V}{\partial T} \right)_p = \dfrac{nR}{p}$.

Therefore,

$$\mu = \frac{1}{C_p} \left\{ \frac{nRT}{p} - V \right\} = \frac{1}{C_p} \left\{ \frac{nRT}{p} - \frac{nRT}{p} - nb \right\} = \frac{-nb}{C_p}.$$

Since $b > 0$ and $C_p > 0$, we conclude that for this gas $\mu < 0$ or $\left(\dfrac{\partial T}{\partial p} \right)_H < 0$. This says that when the pressure drops during a Joule–Thomson expansion the temperature must $\boxed{\text{increase}}$.

P2.31 $p = \dfrac{nRT}{V - nb} - \dfrac{n^2 a}{V^2}$ [Table 1.7].

Hence $\boxed{T = \left(\dfrac{p}{nR} \right) \times (V - nb) + \left(\dfrac{na}{RV^2} \right) \times (V - nb)},$

$\boxed{\left(\dfrac{\partial T}{\partial p} \right)_V = \dfrac{V - nb}{nR}} = \dfrac{V_m - b}{R} = \dfrac{1}{\left(\dfrac{\partial p}{\partial T} \right)_V}.$

For Euler's chain relation, we need to show that $\left(\dfrac{\partial T}{\partial p} \right)_V \left(\dfrac{\partial p}{\partial V} \right)_T \left(\dfrac{\partial V}{\partial T} \right)_p = -1.$

Hence, in addition to $\left(\dfrac{\partial T}{\partial p}\right)_V$ we need $\left(\dfrac{\partial p}{\partial V}\right)_T$ and $\left(\dfrac{\partial V}{\partial T}\right)_p = \dfrac{1}{\left(\dfrac{\partial T}{\partial V}\right)_p}$.

$$\left(\frac{\partial p}{\partial V}\right)_T = \frac{-nRT}{(V-nb)^2} + \frac{2n^2a}{V^3}$$

which can be found from

$$\left(\frac{\partial T}{\partial V}\right)_p = \left(\frac{p}{nR}\right) + \left(\frac{na}{RV^2}\right) - \left(\frac{2na}{RV^3}\right) \times (V-nb),$$

$$\left(\frac{\partial T}{\partial V}\right)_p = \left(\frac{T}{V-nb}\right) - \left(\frac{2na}{RV^3}\right) \times (V-nb).$$

Therefore,

$$\left(\frac{\partial T}{\partial p}\right)_V \left(\frac{\partial p}{\partial V}\right)_T \left(\frac{\partial V}{\partial T}\right)_p = \frac{\left(\dfrac{\partial T}{\partial p}\right)_V \left(\dfrac{\partial p}{\partial V}\right)_T}{\left(\dfrac{\partial T}{\partial V}\right)_p}$$

$$= \frac{\left(\dfrac{V-nb}{nR}\right) \times \left(\dfrac{-nRT}{(V-nb)^2} + \dfrac{2n^2a}{V^3}\right)}{\left(\dfrac{T}{V-nb}\right) - \left(\dfrac{2na}{RV^3}\right) \times (V-nb)}$$

$$= \frac{\left(\dfrac{-T}{V-nb}\right) + \left(\dfrac{2na}{RV^3}\right) \times (V-nb)}{\left(\dfrac{T}{V-nb}\right) - \left(\dfrac{2na}{RV^3}\right) \times (V-nb)}$$

$$= -1.$$

P2.33 $\quad \mu C_p = T\left(\dfrac{\partial V}{\partial T}\right)_p - V = \dfrac{T}{\left(\dfrac{\partial T}{\partial V}\right)_p} - V$ [reciprocal identity, *Further information* 2.2]

$$\left(\frac{\partial T}{\partial V}\right)_p = \frac{T}{V-nb} - \frac{2na}{RV^3}(V-nb) \text{ [Problem 2.31]}$$

Introduction of this expression followed by rearrangement leads to

$$\mu C_p = \frac{(2na) \times (V-nb)^2 - nbRTV^2}{RTV^3 - 2na(V-nb)^2} \times V.$$

Then, introducing $\zeta = \dfrac{RTV^3}{2na(V-nb)^2}$ to simplify the appearance of the expression

$$\boxed{\mu C_p = \left(\frac{1 - \dfrac{nb\zeta}{V}}{\zeta - 1}\right) V = \left(\frac{1 - \dfrac{b\zeta}{V_m}}{\zeta - 1}\right) V.}$$

For xenon, $V_m = 24.6\,\mathrm{dm^3\,mol^{-1}}, T = 298\,\mathrm{K}, a = 4.137\,\mathrm{dm^6\,atm\,mol^{-2}}, b = 5.16 \times 10^{-2}\,\mathrm{dm^3\,mol^{-1}}$,

$$\frac{nb}{V} = \frac{b}{V_m} = \frac{5.16 \times 10^{-1}\,\mathrm{dm^3\,mol^{-1}}}{24.6\,\mathrm{dm^3\,mol^{-1}}} = 2.09 \times 10^{-3},$$

$$\zeta = \frac{(8.206 \times 10^{-2}\,\mathrm{dm^3\,atm\,K^{-1}\,mol^{-1}}) \times (298\,\mathrm{K}) \times (24.6\,\mathrm{dm^3\,mol^{-1}})^3}{(2) \times (4.137\,\mathrm{dm^6\,atm\,mol^{-2}}) \times (24.6\,\mathrm{dm^3\,mol^{-1}} - 5.16 \times 10^{-2}\,\mathrm{dm^3\,mol^{-1}})^2} = 73.0.$$

Therefore, $\mu C_p = \dfrac{1 - (73.0) \times (2.09 \times 10^{-3})}{72.0} \times (24.6\,\mathrm{dm^3\,mol^{-1}}) = 0.290\,\mathrm{dm^3\,mol^{-1}}$.

$C_p = 20.79\,\mathrm{J\,K^{-1}\,mol^{-1}}$ [Table 2.7], so

$$\mu = \frac{0.290\,\mathrm{dm^3\,mol^{-1}}}{20.79\,\mathrm{J\,K^{-1}\,mol^{-1}}} = \frac{0.290 \times 10^{-3}\,\mathrm{m^3\,mol^{-1}}}{20.79\,\mathrm{J\,K^{-1}\,mol^{-1}}}$$

$$= 1.39\overline{3} \times 10^{-5}\,\mathrm{K\,m^3\,J^{-1}} = 1.39\overline{3} \times 10^{-5}\,\mathrm{K\,Pa^{-1}}$$

$$= (1.39\overline{3} \times 10^{-5}) \times (1.013 \times 10^5\,\mathrm{K\,atm^{-1}}) = \boxed{1.41\ \mathrm{K\,atm^{-1}}}.$$

The value of μ changes at $T = T_I$ and when the sign of the numerator $1 - \dfrac{nb\zeta}{V}$ changes sign ($\zeta - 1$ is positive). Hence

$$\frac{b\zeta}{V_m} = 1 \text{ at } T = T_I \quad \text{or} \quad \frac{RTbV^3}{2na(V - nb)^2 V_m} = 1 \quad \text{implying that } T_I = \frac{2a(V_m - b)^2}{RbV_m^2},$$

that is, $T_I = \left(\dfrac{2a}{Rb}\right) \times \left(1 - \dfrac{b}{V_m}\right)^2 = \boxed{\dfrac{27}{4} T_c \left(1 - \dfrac{b}{V_m}\right)^2}$.

For xenon, $\dfrac{2a}{Rb} = \dfrac{(2) \times (4.137\,\mathrm{dm^6\,atm\,mol^{-2}})}{(8.206 \times 10^{-2}\,\mathrm{dm^3\,atm\,K^{-1}\,mol^{-1}}) \times (5.16 \times 10^{-2}\,\mathrm{dm^3\,mol^{-1}})} = 1954\,\mathrm{K}$

and so $T_I = (1954\,\mathrm{K}) \times \left(1 - \dfrac{5.16 \times 10^{-2}}{24.6}\right)^2 = \boxed{1946\ \mathrm{K}}$.

Question. An approximate relationship for μ of a van der Waals gas was obtained in Problem 2.30. Use it to obtain an expression for the inversion temperature, calculate it for xenon, and compare to the result above.

P2.35 $\qquad C_{p,m} - C_{V,m} = \dfrac{\alpha^2 TV}{n\kappa_T}$ [2.49] $= \dfrac{\alpha TV}{n}\left(\dfrac{\partial p}{\partial T}\right)_V$ [2.57].

$$\left(\frac{\partial p}{\partial T}\right)_V = \frac{nR}{V - nb} \text{ [Problem 2.31].}$$

$$\alpha V = \left(\frac{\partial V}{\partial T}\right)_p = \frac{1}{\left(\dfrac{\partial T}{\partial V}\right)_p}.$$

Substituting,

$$C_{p,m} - C_{V,m} = \frac{T \left(\frac{\partial p}{\partial T} \right)_V}{n \left(\frac{\partial T}{\partial V} \right)_p}.$$

Also $\left(\frac{\partial T}{\partial V} \right)_p = \frac{T}{V - nb} - \frac{2na}{RV^3}(V - nb)$ [Problem 2.31].

Substituting,

$$C_{p,m} - C_{V,m} = \frac{\dfrac{RT}{(V - nb)}}{\dfrac{T}{(V - nb)} - \dfrac{2na}{RV^3} \times (V - nb)} = \lambda R$$

with $\lambda = \dfrac{1}{1 - \dfrac{2na}{(RTV^3)} \times (V - nb)^2}$ or $\dfrac{1}{\lambda} = 1 - \dfrac{2a(V_m - b)^3}{RTV_m^3}$.

Now introduce the reduced variables and use $T_c = \dfrac{8a}{27Rb}$, $V_c = 3b$.

After rearrangement,

$$\boxed{\frac{1}{\lambda} = 1 - \frac{(3V_r - 1)^2}{4T_r V_r^3}}.$$

For xenon, $V_c = 118.1$ cm^3 mol^{-1}, $T_c = 289.8$ K. The perfect gas value for V_m may be used as any error introduced by this approximation occurs only in the correction term for $\dfrac{1}{\lambda}$.

Hence, $V_m \approx 2.45$ dm^3, $V_c = 118.8$ cm^3 mol^{-1}, $T_c = 289.8$ K, and $V_r = 20.6$ and $T_r = 1.03$; therefore

$$\frac{1}{\lambda} = 1 - \frac{(61.8 - 1)^2}{(4) \times (1.03) \times (20.6)^3} = 0.90, \text{ giving } \lambda \approx 1.1$$

and

$$C_{p,m} - C_{V,m} \approx 1.1\,R = \boxed{9.2 \, \text{J K}^{-1} \, \text{mol}^{-1}}.$$

P2.37 **(a)** $\mu = -\dfrac{1}{C_p} \left(\dfrac{\partial H}{\partial p} \right)_T = \dfrac{1}{C_p} \left\{ T \left(\dfrac{\partial V_m}{\partial T} \right)_p - V_m \right\}$ [2.53 and Problem 2.34].

$$V_m = \frac{RT}{p} + aT^2 \text{ so } \left(\frac{\partial V_m}{\partial T} \right)_p = \frac{R}{p} + 2aT.$$

$$\mu = \frac{1}{C_p} \left\{ \frac{RT}{p} + 2aT^2 - \frac{RT}{p} - aT^2 \right\} = \boxed{\frac{aT^2}{C_p}}.$$

(b)
$$C_V = C_p - \alpha T V_m \left(\frac{\partial p}{\partial T}\right)_V = C_p - T \left(\frac{\partial V_m}{\partial T}\right)_p \left(\frac{\partial p}{\partial T}\right)_V.$$

But, $p = \dfrac{RT}{V_m - aT^2}$.

$$\left(\frac{\partial p}{\partial T}\right)_V = \frac{R}{V_m - aT^2} - \frac{RT(-2aT)}{(V_m - aT^2)^2}$$

$$= \frac{R}{(RT/p)} + \frac{2aRT^2}{(RT/p)^2} = \frac{p}{T} + \frac{2ap^2}{R}.$$

Therefore

$$C_V = C_p - T \left(\frac{R}{p} + 2aT\right) \times \left(\frac{p}{T} + \frac{2ap^2}{R}\right)$$

$$= C_p - \frac{RT}{p}\left(1 + \frac{2apT}{R}\right) \times \left(1 + \frac{2apT}{R}\right) \times \left(\frac{p}{T}\right),$$

$$\boxed{C_V = C_p - R\left(1 + \frac{2apT}{R}\right)^2.}$$

Solutions to applications

P2.39   Taking the specific enthalpy of digestible carbohydrates to be $17\,\text{kJ g}^{-1}$ (*Impact* I2.2), the serving of pasta yields

$$q = (40\,\text{g}) \times (17\,\text{kJ g}^{-1}) = 680\,\text{kJ}.$$

Converting to Calories (kcal) gives:

$$q = (680\,\text{kJ}) \times \frac{1\,\text{Cal}}{4.184\,\text{kJ}} = 16\overline{2}\,\text{Cal}.$$

As a percentage of a 2200-Calorie diet, this serving is

$$\frac{16\overline{2}\,\text{Cal}}{2200\,\text{Cal}} \times 100\% = \boxed{7.4\%}.$$

P2.41   **(a)** $q = n\Delta_c H^{\ominus} = \dfrac{1.5\,\text{g}}{342.3\,\text{g mol}^{-1}} \times (-5645\,\text{kJ mol}^{-1}) = \boxed{-25\,\text{kJ}}.$

(b) Effective work available is $\approx 25\,\text{kJ} \times 0.25 = 6.2\,\text{kJ}.$
Because $w = mgh$, and $m \approx 65\,\text{kg}$

$$h \approx \frac{6.2 \times 10^3\,\text{J}}{65\,\text{kg} \times 9.81\,\text{m s}^{-2}} = \boxed{9.7\,\text{m}}.$$

(c) The energy released as heat is

$$q = -\Delta_r H = -n\Delta_c H^{\ominus} = -\left(\frac{2.5\,\text{g}}{180\,\text{g mol}^{-1}}\right) \times (-2808\,\text{kJ mol}^{-1}) = \boxed{39\,\text{kJ}}.$$

(d) If one-quarter of this energy were available as work a 65 kg person could climb to a height h given by

$$\frac{1}{4}q = w = mgh \quad \text{so} \quad h = \frac{q}{4mg} = \frac{39 \times 10^3\,\text{J}}{4(65\,\text{kJ}) \times (9.8\,\text{m s}^{-2})} = \boxed{15\,\text{m}}.$$

P2.43   First, with the pure sample, record a thermogram over a temperature range within which P undergoes a structural change, as can be inferred from a peak in the thermogram. The area under the thermogram is the enthalpy change associated with the structural change for the given quantity of P. Then, with an identical mass of the suspected sample, record a thermogram over the same temperature range. Assuming that the impurities in P undergo no structural change over the temperature range—a reasonable assumption if the impurities are monomers or oligomers and if the temperature range is sufficiently narrow—then the peak in the test sample thermogram is attributable only to P. The ratio of areas under the curve in the test sample to the pure sample is a measure of the purity of the test sample.

P2.45   The coefficient of thermal expansion is

$$\alpha = \frac{1}{V}\left(\frac{\partial V}{\partial T}\right)_p \approx \frac{\Delta V}{V \Delta T} \quad \text{so} \quad \Delta V \approx \alpha V \Delta T.$$

This change in volume is equal to the change in height (sea level rise, Δh) times the area of the ocean (assuming that area remains constant). We will use α of pure water, although the oceans are complex solutions. For a 2°C rise in temperature

$$\Delta V = (2.1 \times 10^{-4}\,\text{K}^{-1}) \times (1.37 \times 10^9\,\text{km}^3) \times (2.0\,\text{K}) = 5.8 \times 10^5\,\text{km}^3$$

so $\Delta h = \dfrac{\Delta V}{A} = 1.6 \times 10^{-3}\,\text{km} = \boxed{1.6\,\text{m}}$.

Since the rise in sea level is directly proportional to the rise in temperature, $\Delta T = 1°\text{C}$ would lead to $\Delta h = \boxed{0.80\,\text{m}}$ and $\Delta T = 3.5°\text{C}$ would lead to $\Delta h = \boxed{2.8\,\text{m}}$.

COMMENT. More detailed models of climate change predict somewhat smaller rises, but the same order of magnitude.

P2.47   We compute μ from

$$\mu = -\frac{1}{C_p}\left(\frac{\partial H}{\partial p}\right)_T$$

and we estimate $\left(\dfrac{\partial H}{\partial p}\right)_T$ from the enthalpy and pressure data. We are given both enthalpy and heat capacity data on a mass basis rather than a molar basis; however, the masses will cancel, so we need not convert to a molar basis.

(a) At 300 K.

The regression analysis gives the slope as $-18.0 \, \text{J g}^{-1} \, \text{MPa}^{-1} \approx \left(\dfrac{\partial H}{\partial p}\right)_T$,

so $\mu = -\dfrac{-18.0 \, \text{kJ kg}^{-1} \, \text{MPa}^{-1}}{0.7649 \, \text{kJ kg}^{-1} \, \text{K}^{-1}} = \boxed{23.5 \, \text{K MPa}^{-1}}$.

(b) At 350 K.

The regression analysis gives the slope as $-14.5 \, \text{J g}^{-1} \, \text{MPa}^{-1} \approx \left(\dfrac{\partial H}{\partial p}\right)_T$,

so $\mu = -\dfrac{-14.5 \, \text{kJ kg}^{-1} \, \text{MPa}^{-1}}{1.0392 \, \text{kJ kg}^{-1} \, \text{K}^{-1}} = \boxed{14.0 \, \text{K MPa}^{-1}}$.

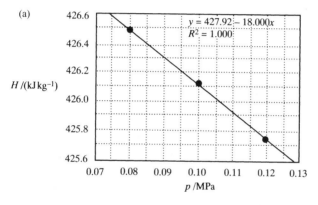

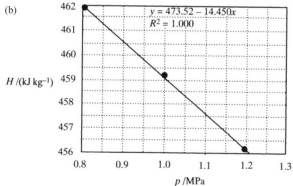

Figure 2.4

3

The Second Law

Answers to discussion questions

D3.1   We must remember that the Second Law of Thermodynamics states only that the total entropy of both the system (here, the molecules organizing themselves into cells) and the surroundings (here, the medium) must increase in a naturally occurring process. It does not state that entropy must increase in a portion of the universe that interacts with its surroundings. In this case, the cells grow by using chemical energy from their surroundings (the medium) and in the process the increase in the entropy of the medium outweighs the decrease in entropy of the system. Hence, the Second Law is not violated.

D3.3   All of these expressions are obtained from a combination of the First Law of Thermodynamics with the Clausius inequality in the form $T dS \geq dq$, as was done at the start of *Justification* 3.2. It may be written as

$$-dU - p_{ex}dV + dw_{add} + TdS \geq 0$$

where we have divided the work into pressure–volume work and additional work. Under conditions of constant energy and volume and no additional work, that is, an isolated system, this relation reduces to

$$dS \geq 0$$

which is equivalent to $\Delta S_{tot} = \Delta S_{universe} \geq 0$. (The universe is an isolated system.)

Under conditions of constant entropy and volume and no additional work, the fundamental relation reduces to

$$dU \leq 0.$$

Under conditions of constant temperature and volume, with no additional work, the relation reduces to

$$dA \leq 0,$$

where A is defined as $U - TS$.

Under conditions of constant temperature and pressure, with no additional work, the relation reduces to

$$dG \leq 0,$$

where G is defined as $U + pV - TS = H - TS$.

In all of the these relations, choosing the inequality provides the criteria for *spontaneous change*. Choosing the equal sign gives us the criteria for *equilibrium* under the conditions specified.

D3.5   The Maxwell relations are relations between partial derivatives all of which are expressed in terms of functions of state (properties of the system). Partial derivatives can be thought of as a kind of shorthand for an experiment. Therefore, the partial derivative $(\partial S/\partial V)_T$ tells us how the entropy of the system changes when we change its volume under constant-temperature conditions. But, as entropy is not a property that can be measured directly (there are no entropy meters), it is important that the derivative (and hence the experiment) be transformed into a form that involves directly measurable properties. That is what the following Maxwell relation does for us.

$$\left(\frac{\partial S}{\partial V}\right)_T = \left(\frac{\partial p}{\partial T}\right)_V.$$

Pressure, temperature, and volume are easily measured properties.

D3.7   The relation $(\partial G/\partial p)_T = V$ shows that the Gibbs function of a system increases with p at constant T in proportion to the magnitude of its volume. This makes good sense when one considers the definition of G, which is $G = U + pV - TS$. Hence, G is expected to increase with p in proportion to V when T is constant.

Solutions to exercises

Assume that all gases are perfect and that data refer to 298.15 K unless otherwise stated.

E3.1(b) $$\Delta S = \int \frac{dq_{rev}}{T} = \frac{q}{T}$$

(a) $$\Delta S = \frac{50 \times 10^3\,\text{J}}{273\,\text{K}} = \boxed{1.8 \times 10^2\,\text{J K}^{-1}}$$

(b) $$\Delta S = \frac{50 \times 10^3\,\text{J}}{(70 + 273)\,\text{K}} = \boxed{1.5 \times 10^2\,\text{J K}^{-1}}$$

E3.2(b)   At 250 K, the entropy is equal to its entropy at 298 K plus ΔS where

$$\Delta S = \int \frac{dq_{rev}}{T} = \int \frac{C_{V,m}\,dT}{T} = C_{V,m} \ln \frac{T_f}{T_i}$$

so $$S = 154.84\,\text{J K}^{-1}\,\text{mol}^{-1} + [(20.786 - 8.3145)\,\text{J K}^{-1}\text{mol}^{-1}] \times \ln \frac{250\,\text{K}}{298\,\text{K}}$$

$$S = \boxed{152.65\,\text{J K}^{-1}\,\text{mol}^{-1}}$$

E3.3(b)   However the change occurred ΔS has the same value as if the change happened by reversible heating at constant pressure (step 1) followed by reversible isothermal compression (step 2)

$$\Delta S = \Delta S_1 + \Delta S_2$$

For the first step

$$\Delta S_1 = \int \frac{dq_{rev}}{T} = \int \frac{C_{p,m}\,dT}{T} = C_{p,m}\ln\frac{T_f}{T_i}$$

$$\Delta S_1 = (2.00\,\text{mol}) \times \left(\frac{7}{2}\right) \times (8.3145\,\text{J K}^{-1}\,\text{mol}^{-1}) \times \ln\frac{(135+273)\,\text{K}}{(25+273)\,\text{K}} = 18.3\,\text{J K}^{-1}$$

and for the second

$$\Delta S_2 = \int \frac{dq_{rev}}{T} = \frac{q_{rev}}{T}$$

where $q_{rev} = -w = \int p\,dV = nRT\ln\dfrac{V_f}{V_i} = nRT\ln\dfrac{p_i}{p_f}$

so $\Delta S_2 = nR\ln\dfrac{p_i}{p_f} = (2.00\,\text{mol}) \times (8.3145\,\text{J K}^{-1}\,\text{mol}^{-1}) \times \ln\dfrac{1.50\,\text{atm}}{7.00\,\text{atm}} = -25.6\,\text{J K}^{-1}$

$$\Delta S = (18.3 - 25.6)\,\text{J K}^{-1} = \boxed{-7.3\,\text{J K}^{-1}}$$

The heat lost in step 2 was more than the heat gained in step 1, resulting in a net loss of entropy. Or the ordering represented by confining the sample to a smaller volume in step 2 overcame the disordering represented by the temperature rise in step 1. A negative entropy change is allowed for a system as long as an increase in entropy elsewhere results in $\Delta S_{total} > 0$.

E3.4(b) $q = q_{rev} = 0$ [adiabatic reversible process]

$$\Delta S = \int_i^f \frac{dq_{rev}}{T} = \boxed{0}$$

$$\Delta U = nC_{V,m}\Delta T = (2.00\,\text{mol}) \times (27.5\,\text{J K}^{-1}\,\text{mol}^{-1}) \times (300 - 250)\,\text{K}$$

$$= 2750\,\text{J} = \boxed{+2.75\,\text{kJ}}$$

$$w = \Delta U - q = 2.75\,\text{kJ} - 0 = \boxed{2.75\,\text{kJ}}$$

$$\Delta H = nC_{p,m}\Delta T$$

$$C_{p,m} = C_{V,m} + R = (27.5\,\text{J K}^{-1}\,\text{mol}^{-1} + 8.314\,\text{J K}^{-1}\,\text{mol}^{-1}) = 35.81\overline{4}\,\text{J K}^{-1}\,\text{mol}^{-1}$$

So $\Delta H = (2.00\,\text{mol}) \times (35.81\overline{4}\,\text{J K}^{-1}\,\text{mol}^{-1}) \times (+50\,\text{K}) = 358\overline{1.4}\,\text{J} = \boxed{3.58\,\text{kJ}}$

E3.5(b)   Since the masses are equal and the heat capacity is assumed constant, the final temperature will be the average of the two initial temperatures,

$$T_f = \tfrac{1}{2}(200\,^\circ\text{C} + 25\,^\circ\text{C}) = 112.\overline{5}\,^\circ\text{C}$$

The heat capacity of each block is

$$C = mC_s \quad \text{where } C_s \text{ is the specific heat capacity}$$

so ΔH (individual) $= mC_s\Delta T = 1.00 \times 10^3\,\text{g} \times 0.449\,\text{J K}^{-1}\,\text{g}^{-1} \times (\pm87.\overline{5}\,\text{K}) = \pm39\,\text{kJ}$

These two enthalpy changes add up to zero: $\boxed{\Delta H_{\text{tot}} = 0}$

$$\Delta S = mC_s \ln\left(\frac{T_f}{T_i}\right); \ 200\,°C = 473.2\,\text{K}; \ 25\,°C = 298.2\,\text{K}; \ 112.\overline{5}\,°C = 385.\overline{7}\,\text{K}$$

$$\Delta S_1 = (1.00 \times 10^3\,\text{g}) \times (0.449\,\text{J K}^{-1}\,\text{g}^{-1}) \times \ln\left(\frac{385.7}{298.2}\right) = 115.\overline{5}\,\text{J K}^{-1}$$

$$\Delta S_2 = (1.00 \times 10^3\,\text{g}) \times (0.449\,\text{J K}^{-1}\,\text{g}^{-1}) \times \ln\left(\frac{385.7}{473.2}\right) = -91.80\overline{2}\,\text{J K}^{-1}$$

$$\Delta S_{\text{total}} = \Delta S_1 + \Delta S_2 = \boxed{24\,\text{J K}^{-1}}$$

E3.6(b) **(a)** $q = 0$ [adiabatic]

(b) $w = -p_{\text{ex}}\Delta V = -(1.5\,\text{atm}) \times \left(\dfrac{1.01 \times 10^5\,\text{Pa}}{\text{atm}}\right) \times (100.0\,\text{cm}^2) \times (15\,\text{cm}) \times \left(\dfrac{1\,\text{m}^3}{10^6\,\text{cm}^3}\right)$

$$= -22\overline{7}.2\,\text{J} = \boxed{-230\,\text{J}}$$

(c) $\Delta U = q + w = 0 - 230\,\text{J} = \boxed{-230\,\text{J}}$

(d) $\Delta U = nC_{V,m}\Delta T$

$$\Delta T = \frac{\Delta U}{nC_{V,m}} = \frac{-22\overline{7}.2\,\text{J}}{(1.5\,\text{mol}) \times (28.8\,\text{J K}^{-1}\,\text{mol}^{-1})}$$

$$= \boxed{-5.3\,\text{K}}$$

(e) Entropy is a state function, so we can compute it by any convenient path. Although the specified transformation is adiabatic, a more convenient path is constant-volume cooling followed by isothermal expansion. The entropy change is the sum of the entropy changes of these two steps:

$$\Delta S = \Delta S_1 + \Delta S_2 = nC_{V,m} \ln\left(\frac{T_f}{T_i}\right) + nR \ln\left(\frac{V_f}{V_i}\right) \quad [3.19 \text{ and } 3.13]$$

$$T_f = 288.\overline{15}\,\text{K} - 5.26\,\text{K} = 282.\overline{9}\,\text{K}$$

$$V_i = \frac{nRT}{p_i} = \frac{(1.5\,\text{mol}) \times (8.206 \times 10^{-2}\,\text{dm}^3\,\text{atm K}^{-1}\,\text{mol}^{-1}) \times (288.\overline{2}\,\text{K})}{9.0\,\text{atm}}$$

$$= 3.9\overline{42}\,\text{dm}^3$$

$$V_f = 3.9\overline{42}\,\text{dm}^3 + (100\,\text{cm}^2) \times (15\,\text{cm}) \times \left(\frac{1\,\text{dm}^3}{1000\,\text{cm}^3}\right)$$

$$= 3.9\overline{42}\,\text{dm}^3 + 1.5\,\text{dm}^3 = 5.4\overline{4}\,\text{dm}^3$$

$$\Delta S = (1.5 \,\text{mol}) \times \left\{ (28.8 \,\text{J K}^{-1}\,\text{mol}^{-1}) \times \ln\left(\frac{282.\overline{9}}{288.\overline{2}}\right) \right.$$

$$\left. + (8.314 \,\text{J K}^{-1}\,\text{mol}^{-1}) \times \ln\left(\frac{5.4\overline{4}}{3.9\overline{42}}\right) \right\}$$

$$= 1.5 \,\text{mol}(-0.534\overline{6} \,\text{J K}^{-1}\,\text{mol}^{-1} + 2.67\overline{8} \,\text{J K}^{-1}\,\text{mol}^{-1}) = \boxed{3.2 \,\text{J K}^{-1}}$$

E3.7(b)   **(a)** $\Delta_{\text{vap}}S = \dfrac{\Delta_{\text{vap}}H}{T_{\text{b}}} = \dfrac{35.27 \times 10^3 \,\text{J mol}^{-1}}{(64.1 + 273.15) \,\text{K}} = +104.5\overline{8} \,\text{J K}^{-1} = \boxed{104.6 \,\text{J K}^{-1}}$

(b) If vaporization occurs reversibly, as is generally assumed

$$\Delta S_{\text{sys}} + \Delta S_{\text{sur}} = 0 \quad \text{so} \quad \Delta S_{\text{sur}} = \boxed{-104.6 \,\text{J K}^{-1}}$$

E3.8(b)   **(a)** $\Delta_{\text{r}}S^{\ominus} = S_{\text{m}}^{\ominus}(\text{Zn}^{2+}, \text{aq}) + S_{\text{m}}^{\ominus}(\text{Cu}, \text{s}) - S_{\text{m}}^{\ominus}(\text{Zn}, \text{s}) - S_{\text{m}}^{\ominus}(\text{Cu}^{2+}, \text{aq})$

$$= [-112.1 + 33.15 - 41.63 + 99.6] \,\text{J K}^{-1}\,\text{mol}^{-1} = \boxed{-21.0 \,\text{J K}^{-1}\text{mol}^{-1}}$$

(b) $\Delta_{\text{r}}S^{\ominus} = 12 S_{\text{m}}^{\ominus}(\text{CO}_2, \text{g}) + 11 S_{\text{m}}^{\ominus}(\text{H}_2\text{O}, \text{l}) - S_{\text{m}}^{\ominus}(\text{C}_{12}\text{H}_{22}\text{O}_{11}, \text{s}) - 12 S_{\text{m}}^{\ominus}(\text{O}_2, \text{g})$

$$= [(12 \times 213.74) + (11 \times 69.91) - 360.2 - (12 \times 205.14)] \,\text{J K}^{-1}\,\text{mol}^{-1}$$

$$= \boxed{+512.0 \,\text{J K}^{-1}\,\text{mol}^{-1}}$$

E3.9(b)   **(a)** $\Delta_{\text{r}}H^{\ominus} = \Delta_{\text{f}}H^{\ominus}(\text{Zn}^{2+}, \text{aq}) - \Delta_{\text{f}}H^{\ominus}(\text{Cu}^{2+}, \text{aq})$

$$= -153.89 - 64.77 \,\text{kJ mol}^{-1} = -218.66 \,\text{kJ mol}^{-1}$$

$$\Delta_{\text{r}}G^{\ominus} = -218.66 \,\text{kJ mol}^{-1} - (298.15 \,\text{K}) \times (-21.0 \,\text{J K}^{-1}\,\text{mol}^{-1}) = \boxed{-212.40 \,\text{kJ mol}^{-1}}$$

(b) $\Delta_{\text{r}}H^{\ominus} = \Delta_{\text{c}}H^{\ominus} = -5645 \,\text{kJ mol}^{-1}$

$$\Delta_{\text{r}}G^{\ominus} = -5645 \,\text{kJ mol}^{-1} - (298.15 \,\text{K}) \times (512.0 \,\text{J K}^{-1}\,\text{mol}^{-1}) = \boxed{-5798 \,\text{kJ mol}^{-1}}$$

E3.10(b)   **(a)** $\Delta_{\text{r}}G^{\ominus} = \Delta_{\text{f}}G^{\ominus}(\text{Zn}^{2+}, \text{aq}) - \Delta_{\text{f}}G^{\ominus}(\text{Cu}^{2+}, \text{aq})$

$$= -147.06 - 65.49 \,\text{kJ mol}^{-1} = \boxed{-212.55 \,\text{kJ mol}^{-1}}$$

(b) $\Delta_{\text{r}}G^{\ominus} = 12 \Delta_{\text{f}}G^{\ominus}(\text{CO}_2, \text{g}) + 11 \Delta_{\text{f}}G^{\ominus}(\text{H}_2\text{O}, \text{l}) - \Delta_{\text{f}}G^{\ominus}(\text{C}_{12}\text{H}_{22}\text{O}_{11}, \text{s}) - 12 \Delta_{\text{f}}G^{\ominus}(\text{O}_2, \text{g})$

$$= [12 \times (-394.36) + 11 \times (-237.13) - (-1543) - 12 \times 0] \,\text{kJ mol}^{-1}$$

$$= \boxed{-5798 \,\text{kJ mol}^{-1}}$$

COMMENT. In each case these values of $\Delta_{\text{r}}G^{\ominus}$ agree closely with the calculated values in Exercise 3.9(b).

E3.11(b) $CO(g) + CH_3OH(l) \rightarrow CH_3COOH(l)$

$$\Delta_r H^\ominus = \sum_{\text{Products}} \nu \Delta_f H^\ominus - \sum_{\text{Reactants}} \nu \Delta_f H^\ominus \text{ [2.32]}$$

$$= -484.5 \text{ kJ mol}^{-1} - (-238.66 \text{ kJ mol}^{-1}) - (-110.53 \text{ kJ mol}^{-1})$$

$$= -135.3\overline{1} \text{ kJ mol}^{-1}$$

$$\Delta_r S^\ominus = \sum_{\text{Products}} \nu S_m^\ominus - \sum_{\text{Reactants}} \nu S_m^\ominus \text{ [3.21]}$$

$$= 159.8 \text{ J K}^{-1} \text{ mol}^{-1} - 126.8 \text{ J K}^{-1} \text{ mol}^{-1} - 197.67 \text{ J K}^{-1} \text{ mol}^{-1}$$

$$= -164.6\overline{7} \text{ J K}^{-1} \text{ mol}^{-1}$$

$$\Delta_r G^\ominus = \Delta_r H^\ominus - T \Delta_r S^\ominus$$

$$= -135.3\overline{1} \text{ kJ mol}^{-1} - (298 \text{ K}) \times (-164.6\overline{7} \text{ J K}^{-1} \text{ mol}^{-1})$$

$$= -135.3\overline{1} \text{ kJ mol}^{-1} + 49.07\overline{2} \text{ kJ mol}^{-1} = \boxed{-86.2 \text{ kJ mol}^{-1}}$$

E3.12(b)   The formation reaction of urea is

$$C(gr) + \tfrac{1}{2}O_2(g) + N_2(g) + 2H_2(g) \rightarrow CO(NH_2)_2(s)$$

The combustion reaction is

$$CO(NH_2)_2(s) + \tfrac{3}{2}O_2(g) \rightarrow CO_2(g) + 2H_2O(l) + N_2(g)$$

$$\Delta_c H = \Delta_f H^\ominus(CO_2, g) + 2\Delta_f H^\ominus(H_2O, l) - \Delta_f H^\ominus(CO(NH_2)_2, s)$$

$$\Delta_f H^\ominus(CO(NH_2)_2, s) = \Delta_f H^\ominus(CO_2, g) + 2\Delta_f H^\ominus(H_2O, l) - \Delta_c H(CO(NH_2)_2, s)$$

$$= -393.51 \text{ kJ mol}^{-1} + (2) \times (-285.83 \text{ kJ mol}^{-1}) - (-632 \text{ kJ mol}^{-1})$$

$$= -333.17 \text{ kJ mol}^{-1}$$

$$\Delta_f S^\ominus = S_m^\ominus(CO(NH_2)_2, s) - S_m^\ominus(C, gr) - \tfrac{1}{2}S_m^\ominus(O_2, g) - S_m^\ominus(N_2, g) - 2S_m^\ominus(H_2, g)$$

$$= 104.60 \text{ J K}^{-1} \text{ mol}^{-1} - 5.740 \text{ J K}^{-1} \text{ mol}^{-1} - \tfrac{1}{2}(205.138 \text{ J K}^{-1} \text{ mol}^{-1})$$

$$- 191.61 \text{ J K}^{-1} \text{ mol}^{-1} - 2(130.684 \text{ J K}^{-1} \text{ mol}^{-1})$$

$$= -456.68\overline{7} \text{ J K}^{-1} \text{ mol}^{-1}$$

$$\Delta_f G^\ominus = \Delta_f H^\ominus - T \Delta_f S^\ominus$$

$$= -333.1\overline{7} \text{ kJ mol}^{-1} - (298 \text{ K}) \times (-456.68\overline{7} \text{ J K}^{-1} \text{ mol}^{-1})$$

$$= -333.1\overline{7} \text{ kJ mol}^{-1} + 136.\overline{093} \text{ kJ mol}^{-1}$$

$$= \boxed{-197 \text{ kJ mol}^{-1}}$$

E3.13(b) **(a)** $\Delta S(\text{gas}) = nR \ln \left(\dfrac{V_f}{V_i} \right)$ [3.13] $= \left(\dfrac{21\,\text{g}}{39.95\,\text{g}\,\text{mol}^{-1}} \right) \times (8.314\,\text{J}\,\text{K}^{-1}\,\text{mol}^{-1}) \ln 2$

$$= 3.0\overline{29}\,\text{J}\,\text{K}^{-1} = \boxed{3.0\,\text{J}\,\text{K}^{-1}}$$

$\Delta S(\text{surroundings}) = -\Delta S(\text{gas}) = \boxed{-3.0\,\text{J}\,\text{K}^{-1}}$ [reversible]

$\Delta S(\text{total}) = \boxed{0}$

(b) $\Delta S(\text{gas}) = \boxed{+3.0\,\text{J}\,\text{K}^{-1}}$ [S is a state function]

$\Delta S(\text{surroundings}) = \boxed{0}$ [no change in surroundings]

$\Delta S(\text{total}) = \boxed{+3.0\,\text{J}\,\text{K}^{-1}}$

(c) $q_{\text{rev}} = 0$ so $\Delta S(\text{gas}) = \boxed{0}$

$\Delta S(\text{surroundings}) = \boxed{0}$ [No heat is transfered to the surroundings]

$\Delta S(\text{total}) = \boxed{0}$

E3.14(b) $C_3H_8(g) + 5O_2(g) \rightarrow 3CO_2(g) + 4H_2O(l)$

$\Delta_r G^{\ominus} = 3\Delta_f G^{\ominus}(CO_2, g) + 4\Delta_f G^{\ominus}(H_2O, l) - \Delta_f G^{\ominus}(C_3H_8, g) - 0$

$$= 3(-394.36\,\text{kJ}\,\text{mol}^{-1}) + 4(-237.13\,\text{kJ}\,\text{mol}^{-1}) - 1(-23.49\,\text{kJ}\,\text{mol}^{-1})$$

$$= -2108.11\,\text{kJ}\,\text{mol}^{-1}$$

The maximum non-expansion work is $\boxed{2108.11\,\text{kJ}\,\text{mol}^{-1}}$ since $|w_{\text{add}}| = |\Delta G|$.

E3.15(b) **(a)** $\varepsilon = 1 - \dfrac{T_c}{T_h}$ [3.10] $= 1 - \dfrac{500\,\text{K}}{1000\,\text{K}} = \boxed{0.500}$

(b) Maximum work $= \varepsilon|q_h| = (0.500) \times (1.0\,\text{kJ}) = \boxed{0.50\,\text{kJ}}$

(c) $\varepsilon_{\text{max}} = \varepsilon_{\text{rev}}$ and $|w_{\text{max}}| = |q_h| - |q_{c,\text{min}}|$

$|q_{c,\text{min}}| = |q_h| - |w_{\text{max}}|$

$$= 1.0\,\text{kJ} - 0.50\,\text{kJ}$$

$$= \boxed{0.5\,\text{kJ}}$$

E3.16(b) $\Delta G = nRT \ln \left(\dfrac{p_f}{p_i} \right)$ [3.56] $= nRT \ln \left(\dfrac{V_i}{V_f} \right)$ [Boyle's law]

$\Delta G = (2.5 \times 10^{-3}\,\text{mol}) \times (8.314\,\text{J}\,\text{K}^{-1}\,\text{mol}^{-1}) \times (298\,\text{K}) \times \ln \left(\dfrac{72}{100} \right) = \boxed{-2.0\,\text{J}}$

E3.17(b) $\left(\dfrac{\partial G}{\partial T} \right)_p = -S$ [3.50]; hence $\left(\dfrac{\partial G_f}{\partial T} \right)_p = -S_f$, and $\left(\dfrac{\partial G_i}{\partial T} \right)_p = -S_i$

$\Delta S = S_f - S_i = -\left(\dfrac{\partial G_f}{\partial T} \right)_p + \left(\dfrac{\partial G_i}{\partial T} \right)_p = -\left(\dfrac{\partial(G_f - G_i)}{\partial T} \right)_p$

$$= -\left(\dfrac{\partial \Delta G}{\partial T} \right)_p = -\dfrac{\partial}{\partial T} \left(-73.1\,\text{J} + 42.8\,\text{J} \times \dfrac{T}{\text{K}} \right)$$

$$= \boxed{-42.8\,\text{J}\,\text{K}^{-1}}$$

E3.18(b) $dG = -S\,dT + V\,dp$ [3.49]; at constant T, $dG = V\,dp$; therefore

$$\Delta G = \int_{p_i}^{p_f} V\,dp$$

The change in volume of a condensed phase under isothermal compression is given by the isothermal compressibility (eqn 2.44).

$$\kappa_T = \frac{1}{V}\left(\frac{\partial V}{\partial p}\right)_T = 1.26 \times 10^{-9}\,\text{Pa}^{-1}$$

This small isothermal compressibility (typical of condensed phases) tells us that we can expect a small change in volume from even a large increase in pressure. So we can make the following approximations to obtain a simple expression for the volume as a function of the pressure

$$\kappa_T \approx \frac{1}{V}\left(\frac{V - V_i}{p - p_i}\right) \approx \frac{1}{V_i}\left(\frac{V - V_i}{p}\right) \quad \text{so} \quad V = V_i(1 - \kappa_T p),$$

where V_i is the volume at 1 atm, namely the sample mass over the density, m/ρ.

$$\Delta G = \int_{100\,\text{kPa}}^{100\,\text{MPa}} \frac{m}{\rho}(1 - \kappa_T p)\,dp$$

$$= \frac{m}{\rho}\left(\int_{100\,\text{kPa}}^{100\,\text{MPa}} dp - \kappa_T \int_{100\,\text{kPa}}^{100\,\text{MPa}} p\,dp\right)$$

$$= \frac{m}{\rho}\left(p\Big|_{100\,\text{kPa}}^{100\,\text{MPa}} - \frac{1}{2}\kappa_T p^2\Big|_{100\,\text{kPa}}^{100\,\text{MPa}}\right)$$

$$= \frac{25\,\text{g}}{0.791\,\text{g cm}^{-3}}\left(9.99 \times 10^7\,\text{Pa} - \frac{1}{2}(1.26 \times 10^{-9}\,\text{Pa}^{-1}) \times (1.00 \times 10^{16}\,\text{Pa}^2)\right)$$

$$= 31.\overline{6}\,\text{cm}^3 \times \left(\frac{1\,\text{m}}{100\,\text{cm}}\right)^3 \times 9.36 \times 10^7\,\text{Pa}$$

$$= 2.9\overline{6} \times 10^3\,\text{J} = \boxed{3.0\,\text{kJ}}$$

E3.19(b) $\Delta G_m = G_{m,f} - G_{m,i} = RT \ln\left(\dfrac{p_f}{p_i}\right)$ [3.56]

$$= (8.314\,\text{J K}^{-1}\,\text{mol}^{-1}) \times (323\,\text{K}) \times \ln\left(\frac{252.0}{92.0}\right) = \boxed{2.71\,\text{kJ mol}^{-1}}$$

E3.20(b)   For an ideal gas, $G_m^O = G_m^{\ominus} + RT \ln\left(\dfrac{p}{p^{\ominus}}\right)$ [3.56 with $G_m = G_m^O$]

But for a real gas, $G_m = G_m^{\ominus} + RT \ln\left(\dfrac{f}{p^{\ominus}}\right)$ [3.58]

So $G_m - G_m^O = RT \ln\dfrac{f}{p}$ [3.58 minus 3.56]; $\dfrac{f}{p} = \phi$

$$= RT \ln \phi = (8.314\,\text{J K}^{-1}\,\text{mol}^{-1}) \times (290\,\text{K}) \times (\ln 0.68) = \boxed{-0.93\,\text{kJ mol}^{-1}}$$

E3.21(b) $\Delta G = nV_m \Delta p \ [3.55] = V\Delta p$

$$\Delta G = (1.0\,\text{dm}^3) \times \left(\frac{1\,\text{m}^3}{10^3\,\text{dm}^3}\right) \times (200 \times 10^3\,\text{Pa}) = 200\,\text{Pa}\,\text{m}^3 = \boxed{200\,\text{J}}$$

E3.22(b) $\Delta G_m = RT \ln\left(\dfrac{p_f}{p_i}\right) = (8.314\,\text{J K}^{-1}\,\text{mol}^{-1}) \times (500\,\text{K}) \times \ln\left(\dfrac{100.0\,\text{kPa}}{50.0\,\text{kPa}}\right) = \boxed{+2.88\,\text{kJ mol}^{-1}}$

Solutions to problems

Solutions to numerical problems

P3.1   (a) Because entropy is a state function $\Delta_{trs}S(l \rightarrow s, -5°C)$ may be determined indirectly from the following cycle

$$H_2O(l, 0°C) \xrightarrow{\ \Delta_{trs}S(l\rightarrow s,0°C)\ } H_2O(s, 0°C)$$

$$\Delta S_1 \uparrow \qquad\qquad\qquad \downarrow \Delta S_s$$

$$H_2O(l, -5°C) \xrightarrow{\ \Delta_{trs}S(l\rightarrow s,-5°C)\ } H_2O(s, -5°C).$$

Thus $\Delta_{trs}S(l \rightarrow s, -5°C) = \Delta S_1 + \Delta_{trs}S(l \rightarrow s, 0°C) + \Delta S_s,$

where $\Delta S_1 = C_{p,m}(l) \ln\dfrac{T_f}{T}$ [3.19; $\theta_f = 0°C$, $\theta = -5°C$]

and $\Delta S_s = C_{p,m}(s) \ln\dfrac{T}{T_f}.$

$$\Delta S_1 + \Delta S_s = -\Delta C_p \ln\frac{T}{T_f} \quad \text{with } \Delta C_p = C_{p,m}(l) - C_{p,m}(s) = +37.3\,\text{J K}^{-1}\text{mol}^{-1}.$$

$$\Delta_{trs}S(l \rightarrow s, T_f) = \frac{-\Delta_{fus}H}{T_f} \ [3.16].$$

Thus, $\Delta_{trs}S(l \rightarrow s, T) = \dfrac{-\Delta_{fus}H}{T_f} - \Delta C_p \ln\dfrac{T}{T_f}.$

$$\Delta_{trs}S(l \rightarrow s, 5°C) = \frac{-6.01 \times 10^3\,\text{J mol}^{-1}}{273\text{K}} - (37.3\,\text{J K}^{-1}\text{mol}^{-1}) \times \ln\frac{268}{273}$$

$$= \boxed{-21.3\,\text{J K}^{-1}\,\text{mol}^{-1}}.$$

$$\Delta S_{sur} = \frac{\Delta_{fus}H(T)}{T}.$$

$$\Delta_{fus}H(T) = -\Delta H_1 + \Delta_{fus}H(T_f) - \Delta H_s.$$

$$\Delta H_1 + \Delta H_s = C_{p,m}(l)(T_f - T) + C_{p,m}(s)(T - T_f) = \Delta C_p(T_f - T).$$

$$\Delta_{fus}H(T) = \Delta_{fus}H(T_f) - \Delta C_p(T_f - T).$$

Thus, $\Delta S_{\text{sur}} = \dfrac{\Delta_{\text{fus}}H(T)}{T} = \dfrac{\Delta_{\text{fus}}H(T_{\text{f}})}{T} + \Delta C_p \dfrac{(T - T_{\text{f}})}{T}$,

$$\Delta S_{\text{sur}} = \frac{6.01 \text{ kJ mol}^{-1}}{268 \text{ K}} + (37.3 \text{ J K}^{-1}\text{mol}^{-1}) \times \left(\frac{268 - 273}{268} \right)$$

$$= \boxed{+21.7 \text{ J K}^{-1} \text{ mol}^{-1}}.$$

$$\Delta S_{\text{total}} = \Delta S_{\text{sur}} + \Delta S = (21.7 - 21.3) \text{ J K}^{-1}\text{mol}^{-1} = \boxed{+0.4 \text{ J K}^{-1} \text{ mol}^{-1}}.$$

Since $\Delta S_{\text{total}} > 0$, the transition $l \rightarrow s$ is spontaneous at $-5°C$.

(b) A similar cycle and analysis can be set up for the transition liquid $\rightarrow$ vapor at 95°C. However, since the transformation here is to the high temperature state (vapor) from the low temperature state (liquid), which is the opposite of part **(a)**, we can expect that the analogous equations will occur with a change of sign.

$$\Delta_{\text{trs}}S(1 \rightarrow g, T) = \Delta_{\text{trs}}S(1 \rightarrow g, T_{\text{b}}) + \Delta C_p \ln \frac{T}{T_{\text{b}}}$$

$$= \frac{\Delta_{\text{vap}}H}{T_{\text{b}}} + \Delta C_p \ln \frac{T}{T_{\text{b}}}, \qquad \Delta C_p = -41.9 \text{ J K}^{-1}\text{mol}^{-1}.$$

$$\Delta_{\text{trs}}S(1 \rightarrow g, T) = \frac{40.7 \text{ kJ mol}^{-1}}{373 \text{ K}} - (41.9 \text{ J K}^{-1} \text{ mol}^{-1}) \times \ln \left(\frac{368}{373} \right)$$

$$= \boxed{+109.7 \text{ J K}^{-1} \text{ mol}^{-1}}.$$

$$\Delta S_{\text{sur}} = \frac{-\Delta_{\text{vap}}H(T)}{T} = -\frac{\Delta_{\text{vap}}H(T_{\text{b}})}{T} - \frac{\Delta C_p(T - T_{\text{b}})}{T}$$

$$= \left(\frac{-40.7 \text{ kJ mol}^{-1}}{368 \text{ K}} \right) - (-41.9 \text{ J K}^{-1} \text{ mol}^{-1}) \times \left(\frac{368 - 373}{368} \right)$$

$$= \boxed{-111.2 \text{ J K}^{-1} \text{ mol}^{-1}}.$$

$$\Delta S_{\text{total}} = (109.7 - 111.2) \text{ J K}^{-1} \text{ mol}^{-1} = \boxed{-1.5 \text{ J K}^{-1} \text{ mol}^{-1}}.$$

Since $\Delta S_{\text{total}} < 0$, the reverse transition, $g \rightarrow l$, is spontaneous at 95°C.

P3.3 **(a)** $q(\text{total}) = q(H_2O) + q(Cu) = 0$; hence $- q(H_2O) = q(Cu)$.

$q(H_2O) = n(-\Delta_{\text{vap}}H) + nC_{p,\text{m}}(H_2O, 1) \times (\theta - 100°C)$

where θ is the final temperature of the water and copper.

$$q(Cu) = mC_s(\theta - 0) = mC_s\theta, \qquad C_s = 0.385 \text{ J K}^{-1}\text{g}^{-1}.$$

Setting $-q(H_2O) = q(Cu)$ allows us to solve for θ.

$$n(\Delta_{\text{vap}}H) - nC_{p,\text{m}}(H_2O, 1) \times (\theta - 100°C) = mC_s\theta$$

Solving for θ yields:

$$\theta = \frac{n\{\Delta_{vap}H + C_{p,m}(H_2O,l) \times 100°C\}}{mC_s + nC_{p,m}(H_2O,l)}$$

$$= \frac{(1.00\,mol) \times (40.656 \times 10^3\,J\,mol^{-1} + 75.3°C^{-1}mol^{-1} \times 100°C)}{2.00 \times 10^3\,g \times 0.385\,J\,°C^{-1}g^{-1} + 1.00\,mol \times 75.3\,J\,°C^{-1}mol^{-1}}$$

$$= 57.0°C = 330.2\,K.$$

$$q(Cu) = (2.00 \times 10^3\,g) \times (0.385\,J\,K^{-1}\,g^{-1}) \times (57.0\,K) = 4.39 \times 10^4\,J = \boxed{43.9\,kJ}.$$

$$q(H_2O) = \boxed{-43.9\,kJ}.$$

$$\Delta S(total) = \Delta S(H_2O) + \Delta S(Cu).$$

$$\Delta S(H_2O) = \frac{-n\Delta_{vap}H}{T_b}\,[3.16] + nC_{p,m}\ln\left(\frac{T_f}{T_i}\right)\,[3.19]$$

$$= -\frac{(1.00\,mol) \times (40.656 \times 10^3\,J\,mol^{-1})}{373.2\,K}$$

$$+ (1.00\,mol) \times (75.3\,J\,K^{-1}\,mol^{-1}) \times \ln\left(\frac{330.2\,K}{373.2\,K}\right)$$

$$= -108.\bar{9}\,J\,K^{-1} - 9.22\,J\,K^{-1} = \boxed{-118.\bar{1}\,J\,K^{-1}}.$$

$$\Delta S(Cu) = mC_s\ln\frac{T_f}{T_i} = (2.00 \times 10^3\,g) \times (0.385\,J\,K^{-1}\,g^{-1}) \times \ln\left(\frac{330.2\,K}{273.2\,K}\right)$$

$$= \boxed{145.\bar{9}\,J\,K^{-1}}.$$

$$\Delta S(total) = -118.\bar{1}\,J\,K^{-1} + 145.\bar{9}\,J\,K^{-1} = \boxed{28\,J\,K^{-1}}.$$

This process is spontaneous since ΔS(surroundings) is zero and, hence,

$$\Delta S(universe) = \Delta S(total) > 0.$$

(b) The volume of the container may be calculated from the perfect gas law.

$$V = \frac{nRT}{p} = \frac{(1.00\,mol) \times (0.08206\,dm^3\,atm\,K^{-1}\,mol^{-1}) \times (373.2\,K)}{1.00\,atm} = 30.6\,dm^3$$

At 57°C the vapor pressure of water is 130 Torr (*Handbook of Chemistry and Physics*). The amount of water vapor present at equilibrium is then

$$n = \frac{pV}{RT} = \frac{(130\,Torr) \times \left(\dfrac{1\,atm}{760\,Torr}\right) \times (30.6\,dm^3)}{(0.08206\,dm^3\,atm\,K^{-1}\,mol^{-1}) \times (330.2\,K)} = 0.193\,mol.$$

This is a substantial fraction of the original amount of water and cannot be ignored. Consequently the calculation needs to be redone taking into account the fact that only a part, n_1, of the vapor condenses into a liquid while the remainder $(1.00\,\text{mol} - n_1)$ remains gaseous. The heat flow involving water, then, becomes

$$q(H_2O) = -n_1 \Delta_{vap}H + n_1 C_{p,m}(H_2O, l)\Delta T(H_2O)$$
$$+ (1.00\,\text{mol} - n_1)C_{p,m}(H_2O, g)\Delta T(H_2O).$$

Because n_1 depends on the equilibrium temperature through $n_1 = 1.00\,\text{mol} - \dfrac{pV}{RT}$, where p is the vapor pressure of water, we will have two unknowns (p and T) in the equation $-q(H_2O) = q(Cu)$. There are two ways out of this dilemma: (1) p may be expressed as a function of T by use of the Clapeyron equation (Chapter 4), or (2) by use of successive approximations. Redoing the calculation yields:

$$\theta = \frac{n_1 \Delta_{vap}H + n_1 C_{p,m}(H_2O, l) \times 100°C + (1.00 - n_1)C_{p,m}(H_2O, g) \times 100°C}{mC_s + nC_{p,m}(H_2O, l) + (1.00 - n_1)C_{p,m}(H_2O, g)}.$$

With

$$n_1 = (1.00\,\text{mol}) - (0.193\,\text{mol}) = 0.80\bar{7}\,\text{mol}$$

(noting that $C_{p,m}(H_2O, g) = 33.6\,\text{J mol}^{-1}\,\text{K}^{-1}$ [Table 2.7]), $\theta = 47.2°C$. At this temperature, the vapor pressure of water is 80.41 Torr, corresponding to

$$n_1 = (1.00\,\text{mol}) - (0.123\,\text{mol}) = 0.87\bar{7}\,\text{mol}.$$

This leads to $\theta = 50.8°C$. The successive approximations eventually converge to yield a value of $\theta = \boxed{49.9°C = 323.1\,\text{K}}$ for the final temperature. (At this temperature, the vapor pressure is 0.123 bar.) Using this value of the final temperature, the heat transferred and the various entropies are calculated as in part (a).

$$q(Cu) = (2.00 \times 10^3\,\text{g}) \times (0.385\,\text{J K}^{-1}\,\text{g}^{-1}) \times (49.9\,\text{K}) = \boxed{38.4\,\text{kJ}} = -q(H_2O).$$

$$\Delta S(H_2O) = \frac{-n\Delta_{vap}H}{T_b} + nC_{p,m} \ln\left(\frac{T_f}{T_i}\right) = \boxed{-119.\bar{8}\,\text{J K}^{-1}}.$$

$$\Delta S(Cu) = mC_s \ln \frac{T_f}{T_i} = \boxed{129.\bar{2}\,\text{J K}^{-1}}.$$

$$\Delta S(\text{total}) = -119.\bar{8}\,\text{J K}^{-1} + 129.\bar{2}\,\text{J K}^{-1} = \boxed{9\,\text{J K}^{-1}}.$$

P3.5

	Step 1	Step 2	Step 3	Step 4	Cycle
q	+11.5 kJ	0	−5.74 kJ	0	−5.8 kJ
w	−11.5 kJ	−3.74 kJ	+5.74 kJ	+3.74 kJ	−5.8 kJ
ΔU	0	−3.74 kJ	0	+3.74 kJ	0
ΔH	0	−6.23 kJ	0	+6.23 kJ	0
ΔS	+19.1 J K^{-1}	0	−19.1 J K^{-1}	0	0
ΔS_{tot}	0	0	0	0	0
ΔG	−11.5 kJ	?	+11.5 kJ	?	0

Step 1

$\Delta U = \Delta H = \boxed{0}$ [isothermal].

$$w = -nRT \ln \left(\frac{V_f}{V_i} \right) = nRT \ln \left(\frac{p_f}{p_i} \right) \text{ [2.11, and Boyle's law]}$$

$$= (1.00 \, \text{mol}) \times (8.314 \, \text{J K}^{-1} \, \text{mol}^{-1}) \times (600 \, \text{K}) \times \ln \left(\frac{1.00 \, \text{atm}}{10.0 \, \text{atm}} \right) = \boxed{-11.5 \, \text{kJ}}.$$

$q = -w = \boxed{11.5 \, \text{kJ}}$.

$$\Delta S = nR \ln \left(\frac{V_f}{V_i} \right) \text{ [3.15]} = -nR \ln \left(\frac{p_f}{p_i} \right) \text{ [Boyle's law]}$$

$$= -(1.00 \, \text{mol}) \times (8.314 \, \text{J K}^{-1} \, \text{mol}^{-1}) \times \ln \left(\frac{1.00 \, \text{atm}}{10.0 \, \text{atm}} \right) = \boxed{+19.1 \, \text{J K}^{-1}}.$$

$\Delta S(\text{sur}) = -\Delta S(\text{system}) \text{ [reversible process]} = -19.1 \, \text{J K}^{-1}.$

$\Delta S_{\text{tot}} = \Delta S(\text{system}) + \Delta S(\text{sur}) = \boxed{0}$.

$\Delta G = \Delta H - T\Delta S = 0 - (600\text{K}) \times (19.1 \, \text{J K}^{-1}) = \boxed{-11.5 \, \text{kJ}}$.

Step 2

$q = \boxed{0}$ [adiabatic] .

$\Delta U = nC_{V,m}\Delta T$ [2.16b]

$$= (1.00 \, \text{mol}) \times \left(\frac{3}{2} \right) \times (8.314 \, \text{J K}^{-1} \, \text{mol}^{-1}) \times (300 \, \text{K} - 600 \, \text{K}) = \boxed{-3.74 \, \text{kJ}}.$$

$w = \Delta U = \boxed{-3.74 \, \text{kJ}}$.

$\Delta H = \Delta U + \Delta(pV) = \Delta U + nR\Delta T$

$$= (-3.74 \, \text{kJ}) + (1.00 \, \text{mol}) \times (8.314 \, \text{J K}^{-1} \, \text{mol}^{-1}) \times (-300 \, \text{K})$$

$$= \boxed{-6.23 \, \text{kJ}}.$$

$\Delta S = \Delta S(\text{sur}) = \boxed{0}$ [reversible adiabatic process].

$\Delta S_{\text{tot}} = \boxed{0}$.

$\Delta G = \Delta(H - TS) = \Delta H - S\Delta T$ [no change in entropy].

Although the change in entropy is known to be zero, the entropy itself is not known, so ΔG is $\boxed{\text{indeterminate}}$.

Step 3

These quantities may be calculated in the same manner as for *Step 1* or more easily as follows.

$\Delta U = \Delta H = \boxed{0}$ [isothermal].

$\varepsilon_{\text{rev}} = 1 - \dfrac{T_c}{T_h}$ [3.10] $= 1 - \dfrac{300\,\text{K}}{600\,\text{K}} = 0.500 = 1 + \dfrac{q_c}{q_h}$ [3.9].

$q_c = -0.500\, q_h = -(0.500) \times (11.5\,\text{kJ}) = -5.74\,\text{kJ}.$

$q_c = \boxed{-5.74\,\text{kJ}}, \qquad w = -q_c = \boxed{5.74\,\text{kJ}}.$

$\Delta S = \dfrac{q_{\text{rev}}}{T}$ [isothermal] $= \dfrac{-5.74 \times 10^3\,\text{J}}{300\,\text{K}} = \boxed{-19.1\,\text{J K}^{-1}}.$

$\Delta S(\text{sur}) = -\Delta S(\text{system}) = +19.1\,\text{J K}^{-1}.$

$\Delta S_{\text{tot}} = \boxed{0}.$

$\Delta G = \Delta H - T\Delta S = 0 - (300\,\text{K}) \times (-19.1\,\text{J K}^{-1}) = \boxed{+11.5\,\text{kJ}}.$

Step 4

ΔU and ΔH are the negative of their values in *Step 2*. (Initial and final temperatures reversed.)

$\Delta U = \boxed{+3.74\,\text{kJ}}, \qquad \Delta H = \boxed{+6.23\,\text{kJ}}, \qquad q = \boxed{0}\text{ [adiabatic].}$

$w = \Delta U = \boxed{+3.74\,\text{kJ}}.$

$\Delta S = \Delta S(\text{sur}) = \boxed{0}\text{ [reversible adiabatic process].}$

$\Delta S_{\text{tot}} = \boxed{0}.$

Again $\Delta G = \Delta(H - TS) = \Delta H - S\Delta T$ [no change in entropy]

but S is not known, so ΔG is $\boxed{\text{indeterminate}}$.

Cycle

$\Delta U = \Delta H = \Delta S = \Delta G = \boxed{0}$ [Δ(state function) $= 0$ for any cycle].

$\Delta S(\text{sur}) = 0$ [all reversible processes].

$\Delta S_{\text{tot}} = \boxed{0}.$

$q(\text{cycle}) = (11.5 - 5.74)\,\text{kJ} = \boxed{5.8\,\text{kJ}}, w(\text{cycle}) = -q(\text{cycle}) = \boxed{-5.8\,\text{kJ}}.$

P3.7 $S_m^{\ominus}(T) = S_m^{\ominus}(298\,\text{K}) + \Delta S.$

$\Delta S = \displaystyle\int_{T_1}^{T_2} C_{p,m}\,\dfrac{\mathrm{d}T}{T} = \int_{T_1}^{T_2} \left(\dfrac{a}{T} + b + \dfrac{c}{T^3}\right) \mathrm{d}T = a \ln \dfrac{T_2}{T_1} + b(T_2 - T_1) - \dfrac{1}{2}c\left(\dfrac{1}{T_2^2} - \dfrac{1}{T_1^2}\right).$

(a)
$$S_m^\ominus(373\,\mathrm{K}) = (192.45\,\mathrm{J\,K^{-1}mol^{-1}}) + (29.75\,\mathrm{J\,K^{-1}\,mol^{-1}}) \times \ln\left(\frac{373}{298}\right)$$

$$+ (25.10 \times 10^{-3}\,\mathrm{J\,K^{-2}\,mol^{-1}}) \times (75.0\,\mathrm{K})$$

$$+ \left(\frac{1}{2}\right) \times (1.55 \times 10^5\,\mathrm{J\,K^{-1}mol^{-1}}) \times \left(\frac{1}{(373.15)^2} - \frac{1}{(298.15)^2}\right)$$

$$= \boxed{200.7\,\mathrm{J\,K^{-1}\,mol^{-1}}}.$$

(b)
$$S_m^\ominus(773\,\mathrm{K}) = (192.45\,\mathrm{J\,K^{-1}mol^{-1}}) + (29.75\,\mathrm{J\,K^{-1}mol^{-1}}) \times \ln\left(\frac{773}{298}\right)$$

$$+ (25.10 \times 10^{-3}\,\mathrm{J\,K^{-2}\,mol^{-1}}) \times (475\mathrm{K})$$

$$+ \left(\frac{1}{2}\right) \times (1.55 \times 10^5\,\mathrm{J\,K^{-1}mol^{-1}}) \times \left(\frac{1}{773^2} - \frac{1}{298^2}\right)$$

$$= \boxed{232.0\,\mathrm{J\,K^{-1}\,mol^{-1}}}.$$

P3.9
$$\Delta S = \boxed{nC_{p,m} \ln\frac{T_f}{T_h} + nC_{p,m} \ln\frac{T_f}{T_c}} \;[3.19]\; \left[T_f \text{ is the final temperature, } T_f = \frac{1}{2}(T_h + T_c)\right].$$

In the present case, $T_f = \dfrac{1}{2}(500\,\mathrm{K} + 250\,\mathrm{K}) = 375\,\mathrm{K}$.

$$\Delta S = nC_{p,m} \ln\frac{T_f^2}{T_h T_c} = nC_{p,m} \ln\frac{(T_h + T_c)^2}{4T_h T_c} = \left(\frac{500\,\mathrm{g}}{63.54\,\mathrm{g\,cm^{-3}}}\right) \times (24.4\,\mathrm{J\,K^{-1}\,mol^{-1}})$$

$$\times \ln\left(\frac{375^2}{500 \times 250}\right) = \boxed{+22.6\,\mathrm{J\,K^{-1}}}.$$

P3.11
$$S_m(T) = S_m(0) + \int_0^T \frac{C_{p,m}\,dT}{T} \;[3.18].$$

From the data, draw up the following table.

T/ K	10	15	20	25	30	50
$\dfrac{C_{p,m}}{T}$ /(J K^{-2} mol^{-1})	0.28	0.47	0.540	0.564	0.550	0.428

T/ K	70	100	150	200	250	298
$\dfrac{C_{p,m}}{T}$ /(J K^{-2} mol^{-1})	0.333	0.245	0.169	0.129	0.105	0.089

Plot $C_{p,m}/T$ against T (Fig. 3.1). This has been done on two scales. The region 0 to 10 K has been constructed using $C_{p,m} = aT^3$, fitted to the point at $T = 10\,\mathrm{K}$, at which $C_{p,m} = 2.8\,\mathrm{J\,K^{-1}mol^{-1}}$, so $a = 2.8 \times 10^{-3}\,\mathrm{J\,K^{-4}\,mol^{-1}}$. The area can be determined (primitively) by counting squares. Area A = 38.28 J K^{-1}mol^{-1}.

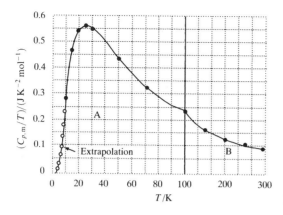

Figure 3.1

Area B up to 0°C = 25.60 J K⁻¹mol⁻¹; area B up to 25°C = 27.80 J K⁻¹ mol⁻¹. Hence

(a) $S_m(273 \text{ K}) = S_m(0) + \boxed{63.88 \text{ J K}^{-1} \text{ mol}^{-1}}$,

(b) $S_m(298 \text{ K}) = S_m(0) + \boxed{66.08 \text{ J K}^{-1} \text{ mol}^{-1}}$.

P3.13 $S_m(T) = S_m(0) + \int_0^T \frac{C_{p,m} \, dT}{T}$ [3.18].

Perform a graphical integration by plotting $C_{p,m}/T$ against T and determining the area under the curve. Draw up the following table. (The last two columns come from determining areas under the curves described below.)

T/K	$\dfrac{C_{p,m}}{\text{J K}^{-1}\,\text{mol}^{-1}}$	$\dfrac{C_{p,m}/T}{\text{J K}^{-2}\,\text{mol}^{-1}}$	$\dfrac{S_m^{\ominus} - S_m^{\ominus}(0)}{\text{J K}^{-1}\,\text{mol}^{-1}}$	$\dfrac{H_m^{\ominus} - H_m^{\ominus}(0)}{\text{kJ mol}^{-1}}$
0.00	0.00	0.00	0.00	0.00
10.00	2.09	0.21	0.80	0.01
20.00	14.43	0.72	5.61	0.09
30.00	36.44	1.21	15.60	0.34
40.00	62.55	1.56	29.83	0.85
50.00	87.03	1.74	46.56	1.61
60.00	111.00	1.85	64.62	2.62
70.00	131.40	1.88	83.29	3.84
80.00	149.40	1.87	102.07	5.26
90.00	165.30	1.84	120.60	6.84
100.00	179.60	1.80	138.72	8.57
110.00	192.80	1.75	156.42	10.44
150.00	237.60	1.58	222.91	19.09
160.00	247.30	1.55	238.54	21.52
170.00	256.50	1.51	253.79	24.05
180.00	265.10	1.47	268.68	26.66
190.00	273.00	1.44	283.21	29.35
200.00	280.30	1.40	297.38	32.13

Plot $C_{p,m}/T$ against T (Fig. 3.2(a)). Extrapolate to $T = 0$ using $C_{p,m} = aT^3$ fitted to the point at $T = 10\,\text{K}$, which gives $a = 2.09\,\text{mJ K}^{-4}\,\text{mol}^{-1}$. Determine the area under the graph up to each T and plot S_m against T (Fig. 3.2(b)).

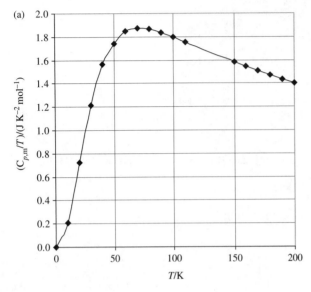

(a)

Figure 3.2(a)

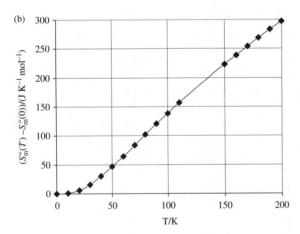

(b)

Figure 3.2(b)

The molar enthalpy is determined in a similar manner from a plot of $C_{p,m}$ against T by determining the area under the curve (Fig. 3.3)

$$H_m^{\ominus}(200\,\text{K}) - H_m^{\ominus}(0) = \int_0^{200\,\text{K}} C_{p,m}\,dT = \boxed{32.1\,\text{kJ mol}^{-1}}.$$

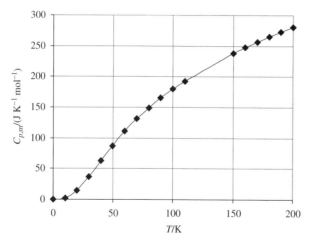

Figure 3.3

P3.15   The entropy at 200 K is calculated from

$$S_m^{\ominus}(200\text{K}) = S_m^{\ominus}(100\text{ K}) + \int_{100\text{ K}}^{200\text{ K}} \frac{C_{p,m}\,dT}{T}.$$

The integrand may be evaluated at each of the data points; the transformed data appear below. The numerical integration can be carried out by a standard procedure such as the trapezoid rule (taking the integral within any interval as the mean value of the integrand times the length of the interval). Programs for performing this integration are readily available for personal computers. Many graphing calculators will also perform this numerical integration.

T/K	100	120	140	150	160	180	200
$C_{p,m}/(\text{J K}^{-1}\text{ mol}^{-1})$	23.00	23.74	24.25	24.44	24.61	24.89	25.11
$\dfrac{C_{p,m}}{T}/(\text{J K}^{-2}\text{ mol}^{-1})$	0.230	0.1978	0.1732	0.1629	0.1538	0.1383	0.1256

Integration by the trapezoid rule yields

$$S_m^{\ominus}(200\text{ K}) = (29.79 + 16.81)\text{ J K}^{-1}\text{mol}^{-1} = \boxed{46.60\text{ J K}^{-1}\text{ mol}^{-1}}.$$

Taking $C_{p,m}$ constant yields

$$S_m^{\ominus}(200\text{ K}) = S_m^{\ominus}(100\text{ K}) + C_{p,m}\ln(200\text{ K}/100\text{ K})$$

$$= [29.79 + 24.44\ln(200\text{ K}/100\text{ K})]\text{ J K}^{-1}\text{mol}^{-1} = \boxed{46.60\text{ J K}^{-1}\text{ mol}^{-1}}.$$

The difference is slight.

P3.17   The Gibbs–Helmholtz equation [3.52] may be recast into an analogous equation involving ΔG and ΔH, since

$$\left(\frac{\partial \Delta G}{\partial T}\right)_p = \left(\frac{\partial G_f}{\partial T}\right)_p - \left(\frac{\partial G_i}{\partial T}\right)_p$$

and $\Delta H = H_f - H_i$.

Thus, $\left(\dfrac{\partial}{\partial T}\dfrac{\Delta_r G^{\ominus}}{T}\right)_p = -\dfrac{\Delta_r H^{\ominus}}{T^2}$,

$$d\left(\dfrac{\Delta_r G^{\ominus}}{T}\right) = \left(\dfrac{\partial}{\partial T}\dfrac{\Delta_r G^{\ominus}}{T}\right)_p dT\,[\text{constant pressure}] = -\dfrac{\Delta_r H^{\ominus}}{T^2}\,dT,$$

$$\Delta\left(\dfrac{\Delta_r G^{\ominus}}{T}\right) = -\int_{T_c}^{T}\dfrac{\Delta_r H^{\ominus}\,dT}{T^2}$$

$$\approx -\Delta_r H^{\ominus}\int_{T_c}^{T}\dfrac{dT}{T^2} = \Delta_r H^{\ominus}\left(\dfrac{1}{T}-\dfrac{1}{T_c}\right)\quad [\Delta_r H^{\ominus}\text{ assumed constant}].$$

Therefore, $\dfrac{\Delta_r G^{\ominus}(T)}{T} - \dfrac{\Delta_r G^{\ominus}(T_c)}{T_c} \approx \Delta_r H^{\ominus}\left(\dfrac{1}{T}-\dfrac{1}{T_c}\right)$

and so $\Delta_r G^{\ominus}(T) = \dfrac{T}{T_c}\Delta_r G^{\ominus}(T_c) + \left(1-\dfrac{T}{T_c}\right)\Delta_r H^{\ominus}(T_c)$

$$= \tau\Delta_r G^{\ominus}(T_c) + (1-\tau)\Delta_r H^{\ominus}(T_c)\quad\text{where}\quad \tau = \dfrac{T}{T_c}.$$

For the reaction

$$N_2(g) + 3H_2(g) \rightarrow 2NH_3(g),\quad \Delta_r G^{\ominus} = 2\Delta_f G^{\ominus}(NH_3, g).$$

(a) At 500 K, $\tau = \dfrac{500}{298} = 1.67\bar{8}$,

so $\Delta_r G^{\ominus}(500\,\text{K}) = \{(1.67\bar{8}) \times 2 \times (-16.45) + (1 - 1.67\bar{8}) \times 2 \times (-46.11)\}\,\text{kJ mol}^{-1}$

$$\boxed{= -7\,\text{kJ mol}^{-1}}.$$

(b) At 1000 K, $\tau = \dfrac{1000}{298} = 3.35\bar{6}$,

so $\Delta_r G^{\ominus}(1000\,\text{K}) = \{(3.35\bar{6}) \times 2 \times (-16.45) + (1 - 3.35\bar{6}) \times 2 \times (-46.11)\}\,\text{kJ mol}^{-1}$

$$\boxed{= +107\,\text{kJ mol}^{-1}}.$$

Solutions to theoretical problems

P3.19 The isotherms correspond to $T = $ constant, and the reversibly traversed adiabats correspond to $S = $ constant. Thus we can represent the cycle as in Fig. 3.4.

In this figure, paths 1, 2, 3, and 4 correspond to the four stages of the Carnot cycle listed in the text following eqn 3.6. The area within the rectangle is

$$\text{Area} = \oint T\,dS = (T_h - T_c) \times (S_2 - S_1) = (T_h - T_c)\Delta S = (T_h - T_c)nR\ln\dfrac{V_B}{V_A}$$

(isothermal expansion from V_A to V_B, stage 1).

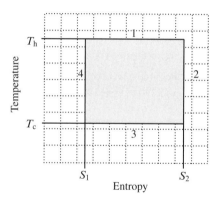

Temperature (y-axis), Entropy (x-axis)

Figure 3.4

But, $w(\text{cycle}) = \varepsilon q_h = \left(\dfrac{T_h - T_c}{T_h}\right) n R T_h \ln \dfrac{V_B}{V_A}$ [text Fig. 3.6] $= n R (T_h - T_c) \ln \dfrac{V_B}{V_A}$.

Therefore, the area is equal to the net work done in the cycle.

P3.21   The thermodynamic temperature scale defines a temperature T^a (where the superscript a is used to distinguish this absolute thermodynamic temperature from the perfect gas temperature) in terms of the reversible heat flows of a heat engine operating between it and an arbitrary fixed temperature T^a_h (eqn 3.11)

$$T^a = (1 - \varepsilon_{rev}) T^a_h$$

where the efficiency of a heat engine is given in terms of work and heat flows:

$$\varepsilon = \frac{|w|}{q_h} \text{ [3.8]} = 1 + \frac{q_c}{q_h} \text{ [3.9]}.$$

The problem asks us to show that the thermodynamic and perfect gas temperatures differ by at most a constant numerical factor. That amounts to showing that

$$\frac{T^a_c}{T^a_h} = \frac{T^g_c}{T^g_h}$$

where the superscipt g indicates the perfect gas temperature defined by the perfect gas law. The subscripts c and h represent two reservoirs between which one might run a heat engine and whose temperatures one might characterize using either temperature scale. *Justification* 3.1 relates the ratio of two perfect-gas temperatures to reversible isothermal heat flows in a Carnot cycle run between the two temperatures.

$$\frac{q_h}{q_c} = -\frac{T^g_h}{T^g_c} \text{ [3.7, \textit{Justification} 3.1]} \quad \text{so} \quad \frac{T^g_c}{T^g_h} = -\frac{q_c}{q_h}.$$

The corresponding ratio of thermodynamic temperatures is:

$$\frac{T^a_c}{T^a_h} = 1 - \varepsilon_{rev} = 1 - \left(1 - \frac{|w|}{q_h}\right)_{rev} = -\left(\frac{q_c}{q_h}\right)_{rev}.$$

As Section 3.2(b) shows, the efficiency of any reversible heat engine (including one that uses a perfect gas as a working fluid) is the same, and therefore the ratio of heat flows to the two reservoirs is the same. That is, the ratio $\dfrac{q_c}{q_h}$ is the same in the expression for the perfect-gas temperature ratio and the thermodynamic ratio; since the two ratios are equal to the same heat ratio, they are equal to each other. The constant numerical factor becomes 1 if T_h and T_h^a are both assigned the same value, say 273.16 at the triple point of water.

P3.23 $\left(\dfrac{\partial S}{\partial V}\right)_T = \left(\dfrac{\partial p}{\partial T}\right)_V$ [Table 3.5].

(a) For a van der Waals gas

$$p = \frac{nRT}{V - nb} - \frac{n^2 a}{V^2} = \frac{RT}{V_m - b} - \frac{a}{V_m^2}.$$

Hence, $\left(\dfrac{\partial S}{\partial V}\right)_T = \left(\dfrac{\partial p}{\partial T}\right)_V = \boxed{\dfrac{R}{V_m - b}}.$

(b) For a Dieterici gas

$$p = \frac{RT e^{-a/RTV_m}}{V_m - b},$$

$$\left(\frac{\partial S}{\partial V}\right)_T = \left(\frac{\partial p}{\partial T}\right)_V = \boxed{\frac{R\left(1 + \dfrac{a}{RV_m T}\right) e^{-a/RV_m T}}{V_m - b}}.$$

For an isothermal expansion, $\Delta S = \displaystyle\int_{V_i}^{V_f} dS = \int_{V_i}^{V_f} \left(\frac{\partial S}{\partial V}\right)_T dV,$

so we can simply compare $\left(\dfrac{\partial S}{\partial V}\right)_T$ expressions for the three gases. For a perfect gas,

$$p = \frac{nRT}{V} = \frac{RT}{V_m} \quad \text{so} \quad \left(\frac{\partial S}{\partial V}\right)_T = \left(\frac{\partial p}{\partial T}\right)_V = \frac{R}{V_m}.$$

$\left(\dfrac{\partial S}{\partial V}\right)_T$ is certainly greater for a van der Waals gas than for a perfect gas, for the denominator is smaller for the van der Waals gas. To compare the van der Waals gas to the Dieterici gas, we assume that both have the same parameter b. (That is reasonable, for b is an excluded volume in both equations of state.) In that case,

$$\left(\frac{\partial S}{\partial V}\right)_{T,\text{Die}} = \frac{R\left(1 + \dfrac{a}{RV_m T}\right) e^{-a/RV_m T}}{V_m - b} = \left(\frac{\partial S}{\partial V}\right)_{T,\text{vdW}} \left(1 + \frac{a}{RV_m T}\right) e^{-a/RV_m T}.$$

Now notice that the additional factor in $\left(\dfrac{\partial S}{\partial V}\right)_{T,\text{Die}}$ has the form $(1 + x)e^{-x}$, where $x > 0$. This factor is always less than 1. Clearly $(1 + x)e^{-x} < 1$ for large x, for then the exponential dominates. But

$(1+x)e^{-x} < 1$ even for small x, as can be seen by using the power series expansion for the exponential: $(1 + x)(1 - x + x^2/2 + \cdots) = 1 - x^2/2 + \cdots$ So $\left(\dfrac{\partial S}{\partial V}\right)_{T,\text{Die}} < \left(\dfrac{\partial S}{\partial V}\right)_{T,\text{vdW}}$. To summarize, for isothermal expansions:

$$\boxed{\Delta S_{\text{vdW}} > \Delta S_{\text{Die}}} \quad \text{and} \quad \boxed{\Delta S_{\text{vdW}} > \Delta S_{\text{perfect}}}.$$

The comparison between a perfect gas and a Dieterici gas depends on particular values of the constants a and b and on the physical conditions.

P3.25 $H \equiv U + pV$.

$$dH = dU + p\,dV + V\,dp = T\,dS - p\,dV\ [3.43] + p\,dV + V\,dp = T\,dS + V\,dp.$$

Since H is a state function, dH is exact, and it follows that

$$\left(\frac{\partial H}{\partial S}\right)_p = T \quad \text{and} \quad \boxed{\left(\frac{\partial V}{\partial S}\right)_p = \left(\frac{\partial T}{\partial p}\right)_S}.$$

Similarly, $A \equiv U - TS$.

$$dA = dU - T\,dS - S\,dT = T\,dS - p\,dV\ [3.43] - T\,dS - S\,dT = -p\,dV - S\,dT.$$

Since dA is exact,

$$\boxed{\left(\frac{\partial S}{\partial V}\right)_T = \left(\frac{\partial p}{\partial T}\right)_V}.$$

P3.27 $\left(\dfrac{\partial S}{\partial V}\right)_T = \left(\dfrac{\partial p}{\partial T}\right)_V$ [Maxwell relation]; $\left(\dfrac{\partial p}{\partial T}\right)_V = \left\{\dfrac{\partial}{\partial T}\left(\dfrac{nRT}{V}\right)\right\}_V = \dfrac{nR}{V}$.

$$dS = \left(\frac{\partial S}{\partial V}\right)_T dV\ [\text{constant temperature}] = nR\,\frac{dV}{V} = nR\,d\ln V.$$

$$S = \int dS = \int nR\,d\ln V.$$

$$S = nR\ln V + \text{constant} \quad \text{or} \quad S \propto R\ln V.$$

P3.29 $\pi_T = T\left(\dfrac{\partial p}{\partial T}\right)_V - p$ [3.48].

$$p = \frac{RT}{V_m} + \frac{BRT}{V_m^2} \text{ [first two terms of the virial expansion, 1.19]}.$$

$$\left(\frac{\partial p}{\partial T}\right)_V = \frac{R}{V_m} + \frac{BR}{V_m^2} + \frac{RT}{V_m^2}\left(\frac{\partial B}{\partial T}\right)_V = \frac{p}{T} + \frac{RT}{V_m^2}\left(\frac{\partial B}{\partial T}\right)_V.$$

Hence, $\pi_T = \frac{RT^2}{V_m^2}\left(\frac{\partial B}{\partial T}\right)_V \approx \frac{RT^2 \Delta B}{V_m^2 \Delta T}.$

Since π_T represents a (usually) small deviation from perfect gas behaviour, we may approximate V_m.

$$V_m \approx \frac{RT}{p} \qquad \boxed{\pi_T \approx \frac{p^2}{R} \times \frac{\Delta B}{\Delta T}}.$$

From the data $\Delta B = \{(-15.6) - (-28.0)\} \text{ cm}^3\text{mol}^{-1} = +12.4 \text{ cm}^3\text{mol}^{-1}$

Hence,

(a) $\qquad \pi_T = \dfrac{(1.0\,\text{atm})^2 \times \left(12.4 \times 10^{-3}\,\text{dm}^3\,\text{mol}^{-1}\right)}{\left(8.206 \times 10^{-2}\,\text{dm}^3\,\text{atm}\,\text{K}^{-1}\,\text{mol}^{-1}\right) \times (50\,\text{K})} = \boxed{3.0 \times 10^{-3}\,\text{atm}}.$

(b) $\qquad \pi_T \propto p^2; \quad$ so at $p = 10.0\,\text{atm}, \ \pi_T = \boxed{0.30\,\text{atm}}.$

COMMENT. In **(a)** π_T is 0.3 per cent of p; in **(b)** it is 3 per cent. Hence at these pressures the approximation for V_m is justified. At 100 atm it would not be.

Question. How would you obtain a reliable estimate of π_T for argon at 100 atm?

P3.31 $\qquad \pi_T = T\left(\dfrac{\partial p}{\partial T}\right)_V - p$ [3.48].

$$p = \frac{nRT}{V - nb} \times e^{-an/RTV} \text{ [Table 1.7]}.$$

$$T\left(\frac{\partial p}{\partial T}\right)_V = \frac{nRT}{V - nb} \times e^{-an/RTV} + \frac{na}{RTV} \times \frac{nRT}{V - nb} \times e^{-an/RTV} = p + \frac{nap}{RTV}.$$

Hence, $\boxed{\pi_T = \dfrac{nap}{RTV}}.$

$\pi_T \to 0$ as $p \to 0$, $V \to \infty$, $a \to 0$, and $T \to \infty$. The fact that $\pi_T > 0$ (because $a > 0$) is consistent with a representing attractive contributions, since it implies that $\left(\dfrac{\partial U}{\partial V}\right)_T > 0$ and the internal energy rises as the gas expands (so decreasing the average attractive interactions).

P3.33   If $S = S(T, p)$

then $dS = \left(\dfrac{\partial S}{\partial T}\right)_p dT + \left(\dfrac{\partial S}{\partial p}\right)_T dp,$

$$T dS = T \left(\dfrac{\partial S}{\partial T}\right)_p dT + T \left(\dfrac{\partial S}{\partial p}\right)_T dp.$$

Use $\left(\dfrac{\partial S}{\partial T}\right)_p = \left(\dfrac{\partial S}{\partial H}\right)_p \left(\dfrac{\partial H}{\partial T}\right)_p = \dfrac{1}{T} \times C_p$ $\left[\left(\dfrac{\partial H}{\partial S}\right)_p = T, \;\; \text{Problem 3.25}\right],$

$\left(\dfrac{\partial S}{\partial p}\right)_T = -\left(\dfrac{\partial V}{\partial T}\right)_p$ [Maxwell relation].

Hence, $T dS = C_p dT - T \left(\dfrac{\partial V}{\partial T}\right)_p dp = \boxed{C_p dT - \alpha T V dp}.$

For reversible, isothermal compression, $T dS = dq_{rev}$ and $dT = 0$; hence

$$dq_{rev} = -\alpha T V dp.$$

$$q_{rev} = \int_{p_i}^{p_f} -\alpha T V \, dp = \boxed{-\alpha T V \, \Delta p} \;\; [\alpha \text{ and } V \text{ assumed constant}].$$

For mercury

$$q_{rev} = \left(-1.82 \times 10^{-4} \, \text{K}^{-1}\right) \times (273 \, \text{K}) \times (1.00 \times 10^{-4} \, \text{m}^{-3}) \times (1.0 \times 10^8 \, \text{Pa}) = \boxed{-0.50 \, \text{kJ}}.$$

P3.35 $\ln \phi = \displaystyle\int_0^p \left(\dfrac{Z-1}{p}\right) dp \;\; [3.60],$

$$Z = 1 + \dfrac{B}{V_m} + \dfrac{C}{V_m^2} = 1 + B'p + C'p^2 + \cdots$$

with $B' = \dfrac{B}{RT},$ $C' = \dfrac{C - B^2}{R^2 T^2}$ [Problem 1.18].

$$\dfrac{Z-1}{p} = B' + C'p + \cdots .$$

Therefore, $\ln \phi = \displaystyle\int_0^p B' \, dp + \int_0^p C'p \, dp + \cdots = B'p + \dfrac{1}{2}C'p^2 + \cdots = \boxed{\dfrac{Bp}{RT} + \dfrac{(C - B^2)p^2}{2R^2 T^2} + \cdots}.$

For argon, $\dfrac{Bp}{RT} = \dfrac{(-21.13 \times 10^{-3} \, \text{dm}^3 \, \text{mol}^{-1}) \times (1.00 \, \text{atm})}{(8.206 \times 10^{-2} \, \text{dm}^3 \, \text{atm} \, \text{K}^{-1} \, \text{mol}^{-1}) \times (273 \, \text{K})} = -9.43 \times 10^{-4},$

$$\dfrac{(C - B^2)p^2}{2R^2 T^2} = \dfrac{\{(1.054 \times 10^{-3} \, \text{dm}^6 \, \text{mol}^{-2}) - (-21.13 \times 10^{-3} \, \text{dm}^3 \, \text{mol}^{-1})^2\} \times (1.00 \, \text{atm})^2}{(2) \times \{(8.206 \times 10^{-2} \, \text{dm}^3 \, \text{atm} \, \text{K}^{-1} \, \text{mol}^{-1}) \times (273 \, \text{K})\}^2}$$

$$= 6.05 \times 10^{-7}.$$

Therefore, $\ln \phi = (-9.43 \times 10^{-4}) + (6.05 \times 10^{-7}) = -9.42 \times 10^{-4}$; $\phi = 0.9991$.

Hence, $f = (1.00\,\text{atm}) \times (0.9991) = \boxed{0.9991\,\text{atm}}$.

Solutions to applications

P3.37 $w_{\text{add, max}} = \Delta_r G$ [3.38].

$$\Delta_r G^{\ominus}(37°\text{C}) = \tau\ \Delta_r G^{\ominus}(T_c) + (1 - \tau)\Delta_r H^{\ominus}(T_c) \quad \left[\text{Problem 3.17, } \tau = \frac{T}{T_c} \right]$$

$$= \left(\frac{310\,\text{K}}{298.15\,\text{K}} \right) \times (-6333\,\text{kJ mol}^{-1}) + \left(1 - \frac{310\,\text{K}}{298.15\,\text{K}} \right) \times (-5797\,\text{kJ mol}^{-1})$$

$$= -6354\,\text{kJ mol}^{-1}.$$

The difference is

$$\Delta_r G^{\ominus}(37°\text{C}) - \Delta_r G^{\ominus}(T_c) = \{-6354 - (-6333)\}\,\text{kJ mol}^{-1} = \boxed{-21\,\text{kJ mol}^{-1}}. \text{ Therefore an}$$
additional 21 kJ mol^{-1} of non-expansion work may be done at the higher temperature.

COMMENT.   As shown by Problem 3.16, increasing the temperature does not necessarily increase the maximum non-expansion work. The relative magnitude of $\Delta_r G^{\ominus}$ and $\Delta_r H^{\ominus}$ is the determining factor.

P3.39   The relative increase in water vapor in the atmosphere at constant relative humidity is the same as the relative increase in the equilibrium vapor pressure of water. Examination of the molar Gibbs function will help us estimate this increase. At equilibrium, the vapor and liquid have the same molar Gibbs function. So at the current temperature

$$G_{m,\text{liq}}(T_0) = G_{m,\text{vap}}(T_0) \quad \text{so} \quad G^{\ominus}_{m,\text{liq}}(T_0) = G^{\ominus}_{m,\text{vap}}(T_0) + RT_0 \ln p_0,$$

where the subscript 0 refers to the current equilibrium and p is the pressure divided by the standard pressure. The Gibbs function changes with temperature as follows

$$(\partial G/\partial T) = -S \quad \text{so} \quad G^{\ominus}_{m,\text{liq}}(T_1) = G^{\ominus}_{m,\text{liq}}(T_0) - (\Delta T)S^{\ominus}_{\text{liq}}$$

and similarly for the vapor. Thus, at the higher temperature

$$G^{\ominus}_{m,\text{liq}}(T_0) - (\Delta T)S^{\ominus}_{\text{liq}} = G^{\ominus}_{m,\text{vap}}(T_0) - (\Delta T)S^{\ominus}_{\text{vap}} + R(T_0 + \Delta T) \ln p.$$

Solving both of these expressions for $G^{\ominus}_{m,\text{liq}}(T_0) - G^{\ominus}_{m,\text{vap}}(T_0)$ and equating them leads to

$$(\Delta T)(S^{\ominus}_{\text{liq}} - S^{\ominus}_{\text{vap}}) + R(T_0 + \Delta T) \ln p = RT_0 \ln p_0.$$

Isolating p leads to

$$\ln\ p = \frac{(\Delta T)(S^{\ominus}_{\text{vap}} - S^{\ominus}_{\text{liq}})}{R(T_0 + \Delta T)} + \frac{T_0 \ln p_0}{T_0 + \Delta T},$$

$$p = \exp \left(\frac{(\Delta T)(S^{\ominus}_{\text{vap}} - S^{\ominus}_{\text{liq}})}{R(T_0 + \Delta T)} \right) p_0^{(T_0/(T_0 + \Delta T))}.$$

So $p = \exp\left(\dfrac{(2.0\ \text{K}) \times (188.83 - 69.91)\ \text{J}\,\text{mol}^{-1}\ \text{K}^{-1}}{(8.3145\ \text{J}\,\text{mol}^{-1}\ \text{K}^{-1}) \times (290 + 2.0)\ \text{K}}\right) \times (0.0189)^{(290\text{K}/(290+2.0)\text{K})},$

$p = 0.0214$ which represents a $\boxed{13\ \text{per cent}}$ increase.

P3.41   The change in the Helmholtz energy equals the maximum work associated with stretching the polymer. Then

$$dw_{\text{max}} = dA = -f\,dl.$$

For stretching at constant T

$$f = -\left(\frac{\partial A}{\partial l}\right)_T = -\left(\frac{\partial U}{\partial l}\right)_T + T\left(\frac{\partial S}{\partial l}\right)_T.$$

Assuming that $(\partial U/\partial l)_T = 0$ (valid for rubbers)

$$f = T\left(\frac{\partial S}{\partial l}\right)_T = T\left(\frac{\partial}{\partial l}\right)_T\left\{-\frac{3k_B l^2}{2Na^2} + C\right\}$$

$$= T\left\{-\frac{3k_B l}{Na^2}\right\} = -\left(\frac{3k_B T}{Na^2}\right) l.$$

This tensile force has the Hooke's law form $f = -k_H l$ with $k_H = 3k_B T/Na^2$.

P3.43   The Otto cycle is represented in Fig. 3.5. Assume one mole of air.

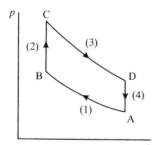

V **Figure 3.5**

$$\varepsilon = \frac{|w|_{\text{cycle}}}{|q_2|}\ [3.8].$$

$w_{\text{cycle}} = w_1 + w_3 = \Delta U_1 + \Delta U_3\ [q_1 = q_3 = 0] = C_V(T_B - T_A) + C_V(T_D - T_C)\ [2.27].$

$q_2 = \Delta U_2 = C_V(T_C - T_B).$

$\varepsilon = \dfrac{|T_B - T_A + T_D - T_C|}{|T_C - T_B|} = 1 - \left(\dfrac{T_D - T_A}{T_C - T_B}\right).$

We know that

$$\frac{T_A}{T_B} = \left(\frac{V_B}{V_A}\right)^{1/c} \quad \text{and} \quad \frac{T_D}{T_C} = \left(\frac{V_C}{V_D}\right)^{1/c} \quad [2.28a].$$

Since $V_B = V_C$ and $V_A = V_D$, $\dfrac{T_A}{T_B} = \dfrac{T_D}{T_C}$, or $T_D = \dfrac{T_A T_C}{T_B}$.

Then $\varepsilon = 1 - \dfrac{\dfrac{T_A T_C}{T_B} - T_A}{T_C - T_B} = 1 - \dfrac{T_A}{T_B}$ or $\boxed{\varepsilon = 1 - \left(\dfrac{V_B}{V_A}\right)^{1/c}}$.

Given that $C_{p,m} = \dfrac{7}{2}R$, we have $C_{V,m} = \dfrac{5}{2}R$ [2.26] and $c = \dfrac{2}{5}$.

For $\dfrac{V_A}{V_B} = 10$, $\varepsilon = 1 - \left(\dfrac{1}{10}\right)^{2/5} = \boxed{0.47}$.

$$\Delta S_1 = \Delta S_3 = \Delta S_{\text{sur},1} = \Delta S_{\text{sur},3} = \boxed{0} \quad \text{[adiabatic reversible steps]}.$$

$$\Delta S_2 = C_{V,m} \ln\left(\frac{T_C}{T_B}\right).$$

At constant volume $\left(\dfrac{T_C}{T_B}\right) = \left(\dfrac{p_C}{p_B}\right) = 5.0.$

$$\Delta S_2 = \left(\frac{5}{2}\right) \times (8.314\,\text{J K}^{-1}\,\text{mol}^{-1}) \times (\ln 5.0) = \boxed{+33\,\text{J K}^{-1}}.$$

$$\Delta S_{\text{sur},2} = -\Delta S_2 = \boxed{-33\,\text{J K}^{-1}}.$$

$$\Delta S_4 = -\Delta S_2 \left[\frac{T_C}{T_D} = \frac{T_B}{T_A}\right] = \boxed{-33\,\text{J K}^{-1}}.$$

$$\Delta S_{\text{sur},4} = -\Delta S_4 = \boxed{+33\,\text{J K}^{-1}}.$$

P3.45   In case (**a**), the electric heater converts 1.00 kJ of electrical energy into heat, providing $\boxed{1.00\,\text{kJ}}$ of energy as heat to the room. (The Second Law places no restriction on the complete conversion of work to heat—only on the reverse process.) In case (**b**), we want to find the heat deposited in the room $|q_h|$:

$$|q_h| = |q_c| + |w| \quad \text{where} \quad \frac{|q_c|}{|w|} = c = \frac{T_c}{T_h - T_c}\,[\textit{Impact} \text{ I3.1}]$$

so $|q_c| = \dfrac{|w|T_c}{T_h - T_c} = \dfrac{1.00\,\text{kJ} \times 260\,\text{K}}{(295 - 260)\text{K}} = 7.4\,\text{kJ}.$

The heat transferred to the room is $|q_h| = |q_c| + |w| = 7.4 \text{ kJ} + 1.00 \text{ kJ} = \boxed{8.4 \text{ kJ}}$. Most of the thermal energy the heat pump deposits into the room comes from outdoors. Difficult as it is to believe on a cold winter day, the intensity of thermal energy (that is, the absolute temperature) outdoors is a substantial fraction of that indoors. The work put into the heat pump is not simply converted to heat, but is 'leveraged' to transfer additional heat from outdoors.

Physical transformations of pure substances

Answers to discussion questions

D4.1 Consider two phases of a system, labeled α and β. The phase with the lower chemical potential under the given set of conditions is the more stable phase. First consider the variation of μ with temperature at a fixed pressure. We have

$$\left(\frac{\partial \mu_\alpha}{\partial T}\right)_p = -S_\alpha \quad \text{and} \quad \left(\frac{\partial \mu_\beta}{\partial T}\right)_p = -S_\beta.$$

Therefore, if S_β is larger in magnitude than S_α, the β phase will be favored over the α phase as temperature increases because its chemical potential decreases more rapidly with temperature than for the α phase. We also have

$$\left(\frac{\partial \mu_\alpha}{\partial P}\right)_T = V_{m,\alpha} \quad \text{and} \quad \left(\frac{\partial \mu_\beta}{\partial P}\right)_T = V_{m,\beta}.$$

Therefore, if $V_{m,\alpha}$ is larger than $V_{m,\beta}$ the β phase will be favored over the α phase as pressure increases because the chemical potential of the β phase will not increase as rapidly with pressure as that of the α phase. See Example 4.1.

D4.3 Refer to Fig. 4.1 below and Fig. 4.5 in the text. Starting at point A and continuing clockwise on path $p(T)$ toward point B, we see a gaseous phase only within the container with water at pressures and temperatures $p(T)$. Upon reaching point B on the vapor pressure curve, liquid appears on the bottom of the container and a phase boundary or meniscus is evident between the liquid and less dense gas above it. The liquid and gaseous phases are at equilibrium at this point. Proceeding clockwise away from the vapor pressure curve the meniscus disappears and the system becomes wholly liquid. Continuing along $p(T)$ to point C at the critical temperature no abrupt changes are observed in the isotropic fluid. Before point C is reached, it is possible to return to the vapor pressure curve and a liquid–gas equilibrium by reducing the pressure isothermally. Continuing clockwise from point C along path $p(T)$ back to point A, no phase boundary is observed even though we now consider the water to have returned to the gaseous state. Additionally, if the pressure is isothermally reduced at any point after point C, it is impossible to return to a liquid–gas equilibrium.

When the path $p(T)$ is chosen to be very close to the critical point, the water appears opaque. At near critical conditions, densities and refractive indices of both the liquid and gas phases are nearly identical. Furthermore, molecular fluctuations cause spatial variations of densities and refractive indices on a scale large enough to strongly scatter visible light. This is called critical opalescence.

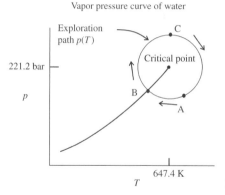

Figure 4.1   Vapor pressure curve of water

D4.5   The supercritical fluid extractor consists of a pump to pressurize the solvent (e.g. CO_2), an oven with extraction vessel, and a trapping vessel. Extractions are performed dynamically or statically. Supercritical fluid flows continuously through the sample within the extraction vessel when operating in dynamic mode. Analytes extracted into the fluid are released through a pressure-maintaining restrictor into a trapping vessel. In static mode the supercritical fluid circulates repetitively through the extraction vessel until being released into the trapping vessel after a period of time. Supercritical carbon dioxide volatilizes when decompression occurs upon release into the trapping vessel.

Advantages	Disadvantages	Current uses
Dissolving power of SCF can be adjusted with selection of T and p	Elevated pressures are required and the necessary apparatus expensive	Extraction of caffeine, fatty acids, spices, aromas, flavors, and biological materials from natural sources
Select SCFs are inexpensive and non-toxic. They reduce pollution	Cost may prohibit large scale applications	Extraction of toxic salts (with a suitable chelation agent) and organics from contaminated water
Thermally unstable analytes may be extracted at low temperature	Modifiers like methanol (1–10%) may be required to increase solvent polarity	Extraction of herbicides from soil
The volatility of scCO$_2$ makes it easy to isolate analyte	scCO$_2$ is toxic to whole cells in biological applications (CO$_2$ is not toxic to the environment)	scH$_2$O oxidation of toxic, intractable organic waste during water treatment
SCFs have high diffusion rates, low viscosity, and low surface tension		Synthetic chemistry, polymer synthesis and crystallization, textile processing
O_2 and H_2 are completely miscible with scCO$_2$. This reduces multi-phase reaction problems		Heterogeneous catalysis for green chemistry processes

D4.7   First-order phase transitions show discontinuities in the first derivative of the Gibbs energy with respect to temperature. They are recognized by finite discontinuities in plots of H, U, S, and V against temperature and by an infinite discontinuity in C_p. Second-order phase transitions show discontinuities in the second derivatives of the Gibbs energy with respect to temperature, but the first derivatives are continuous. The second-order transitions are recognized by kinks in plots of H, U, S, and V against temperature, but most easily by a finite discontinuity in a plot of C_p against temperature. A λ-transition shows characteristics of both first- and second-order transitions and, hence, is difficult to classify by the Ehrenfest scheme. It resembles a first-order transition in a plot of C_p against T, but appears to be a higher-order transition with respect to other properties.

At the molecular level first-order transitions are associated with discontinuous changes in the interaction energies between the atoms or molecules constituting the system and in the volume they occupy. One kind of second-order transition may involve only a continuous change in the arrangement of the atoms from one crystal structure (symmetry) to another while preserving their orderly arrangement. In one kind of 8-transition, called an order–disorder transition, randomness is introduced into the atomic arrangement. See Figs. 4.18 and 4.19 of the text.

Solutions to exercises

E4.1(b)   Assume vapor is a perfect gas and $\Delta_{vap}H$ is independent of temperature

$$\ln\frac{p^*}{p} = +\frac{\Delta_{vap}H}{R}\left(\frac{1}{T} - \frac{1}{T^*}\right)$$

$$\frac{1}{T} = \frac{1}{T^*} + \frac{R}{\Delta_{vap}H}\ln\frac{p^*}{p}$$

$$= \frac{1}{293.2\,\text{K}} + \frac{8.314\,\text{J K}^{-1}\,\text{mol}^{-1}}{32.7\times10^3\,\text{J mol}^{-1}}\times\ln\left(\frac{58.0}{66.0}\right)$$

$$= 3.378\times10^{-3}\,\text{K}^{-1}$$

$$T = \frac{1}{3.3\overline{7}8\times10^{-3}\,\text{K}^{-1}} = 296\,\text{K} = \boxed{23\,°\text{C}}$$

E4.2(b) $$\frac{dp}{dT} = \frac{\Delta_{fus}S}{\Delta_{fus}V}$$

$$\Delta_{fus}S = \Delta_{fus}V\left(\frac{dp}{dT}\right) \approx \Delta_{fus}V\frac{\Delta p}{\Delta T}$$

assuming $\Delta_{fus}S$ and $\Delta_{fus}V$ independent of temperature.

$$\Delta_{fus}S = (152.6\,\text{cm}^3\,\text{mol}^{-1} - 142.0\,\text{cm}^3\,\text{mol}^{-1})\times\frac{(1.2\times10^6\,\text{Pa}) - (1.01\times10^5\,\text{Pa})}{429.26\,\text{K} - 427.15\,\text{K}}$$

$$= (10.6\,\text{cm}^3\,\text{mol}^{-1})\times\left(\frac{1\,\text{m}^3}{10^6\,\text{cm}^3}\right)\times(5.21\times10^5\,\text{Pa K}^{-1})$$

$$= 5.52\,\text{Pa m}^3\,\text{K}^{-1}\,\text{mol}^{-1} = \boxed{5.5\,\text{J K}^{-1}\text{mol}^{-1}}$$

$$\Delta_{fus}H = T_f \Delta S = (427.15 \, \text{K}) \times (5.5\overline{2} \, \text{J K}^{-1} \, \text{mol}^{-1})$$

$$= \boxed{2.4 \, \text{kJ mol}^{-1}}$$

E4.3(b) Use $\displaystyle \int d \ln p = \int \frac{\Delta_{vap}H}{RT^2} \, dT$

$$\ln p = \text{constant} - \frac{\Delta_{vap}H}{RT}$$

Terms with $1/T$ dependence must be equal, so

$$-\frac{3036.8 \, \text{K}}{T/K} = -\frac{\Delta_{vap}H}{RT}$$

$$\Delta_{vap}H = (3036.8 \, \text{K})R = (8.314 \, \text{J K}^{-1} \, \text{mol}^{-1}) \times (3036.8 \, \text{K})$$

$$= \boxed{25.25 \, \text{kJ mol}^{-1}}$$

E4.4(b) **(a)** $\log p = \text{constant} - \Delta_{vap}H/(RT(2.303))$

Thus

$$\Delta_{vap}H = (1625 \, \text{K}) \times (8.314 \, \text{J K}^{-1} \, \text{mol}^{-1}) \times (2.303)$$

$$= \boxed{31.11 \, \text{kJ mol}^{-1}}$$

(b) Normal boiling point corresponds to $p = 1.000 \, \text{atm} = 760 \, \text{Torr}$

$$\log(760) = 8.750 - \frac{1625}{T/K}$$

$$\frac{1625}{T/K} = 8.750 - \log(760)$$

$$T/K = \frac{1625}{8.750 - \log(760)} = 276.8\overline{7}$$

$$T_b = \boxed{276.9 \, \text{K}}$$

E4.5(b) $\displaystyle \Delta T = \frac{\Delta_{fus}V}{\Delta_{fus}S} \times \Delta p = \frac{T_f \Delta_{fus}V}{\Delta_{fus}H} \times \Delta p = \frac{T_f \Delta p M}{\Delta_{fus}H} \times \Delta \left(\frac{1}{\rho} \right)$

$[T_f = -3.65 + 273.15 = 269.50 \, \text{K}]$

$$\Delta T = \frac{(269.50 \, \text{K}) \times (99.9 \, \text{MPa})M}{8.68 \, \text{kJ mol}^{-1}} \times \left(\frac{1}{0.789 \, \text{g cm}^{-3}} - \frac{1}{0.801 \, \text{g cm}^{-3}} \right)$$

$$= (3.10\overline{17} \times 10^6 \, \text{K Pa J}^{-1} \, \text{mol}) \times (M) \times (+0.01\overline{899} \, \text{cm}^3/\text{g}) \times \left(\frac{\text{m}^3}{10^6 \, \text{cm}^3} \right)$$

$$= (+5.\overline{889} \times 10^{-2} \, \text{K Pa m}^3 \, \text{J}^{-1} \, \text{g}^{-1} \, \text{mol})M = (+5.\overline{889} \times 10^{-2} \, \text{K g}^{-1} \, \text{mol})M$$

$$\Delta T = (46.07 \, \text{g mol}^{-1}) \times (+5.\overline{889} \times 10^{-2} \, \text{K g}^{-1} \, \text{mol})$$

$$= +2.7\overline{1} \, \text{K}$$

$$T_f = 269.50 \, \text{K} + 2.7\overline{1} \, \text{K} = \boxed{272 \, \text{K}}$$

E4.6(b)

$$\frac{dm}{dt} = \frac{dn}{dt} \times M_{H_2O} \text{ where } n = \frac{q}{\Delta_{vap}H}$$

$$\frac{dn}{dt} = \frac{dq/dt}{\Delta_{vap}H} = \frac{(0.87 \times 10^3 \text{ W m}^{-2}) \times (10^4 \text{ m}^2)}{44.0 \times 10^3 \text{ J mol}^{-1}}$$

$$= 197.\overline{7} \text{ J s}^{-1} \text{ J}^{-1} \text{ mol}$$

$$= 200 \text{ mol s}^{-1}$$

$$\frac{dm}{dt} = (197.\overline{7} \text{ mol s}^{-1}) \times (18.02 \text{ g mol}^{-1})$$

$$= \boxed{3.6 \text{ kg s}^{-1}}$$

E4.7(b)   The vapor pressure of ice at $-5\,°C$ is 0.40 kPa. Therefore, the frost will sublime. A partial pressure of 0.40 kPa or more will ensure that the frost remains.

E4.8(b)   **(a)** According to Trouton's rule (Section 3.3(b), eqn 3.16)

$$\Delta_{vap}H = (85 \text{ J K}^{-1} \text{ mol}^{-1}) \times T_b = (85 \text{ J K}^{-1} \text{ mol}^{-1}) \times (342.2 \text{ K}) = \boxed{29.\overline{1} \text{ kJ mol}^{-1}}$$

(b) Use the Clausius–Clapeyron equation [Exercise 4.8(a)]

$$\ln\left(\frac{p_2}{p_1}\right) = \frac{\Delta_{vap}H}{R} \times \left(\frac{1}{T_1} - \frac{1}{T_2}\right)$$

At $T_2 = 342.2$ K, $p_2 = 1.000$ atm; thus at $25\,°C$

$$\ln p_1 = -\left(\frac{29.\overline{1} \times 10^3 \text{ J mol}^{-1}}{8.314 \text{ J K}^{-1} \text{ mol}^{-1}}\right) \times \left(\frac{1}{298.2 \text{ K}} - \frac{1}{342.2 \text{ K}}\right) = -1.50\overline{9}$$

$$p_1 = \boxed{0.22 \text{ atm}} = 16\overline{8} \text{ Torr}$$

At $60\,°C$,

$$\ln p_1 = -\left(\frac{29.\overline{1} \times 10^3 \text{ J mol}^{-1}}{8.314 \text{ J K}^{-1} \text{ mol}^{-1}}\right) \times \left(\frac{1}{333.2 \text{ K}} - \frac{1}{342.2 \text{ K}}\right) = -0.27\overline{6}$$

$$p_1 = \boxed{0.76 \text{ atm}} = 57\overline{6} \text{ Torr}$$

E4.9(b) $$\Delta T = T_{fus}(10 \text{ MPa}) - T_{fus}(0.1 \text{ MPa}) = \frac{T_{fus}\Delta pM}{\Delta_{fus}H}\Delta\left(\frac{1}{\rho}\right) \text{ [See Exercise 4.5(b)]}$$

$$\Delta_{fus}H = 6.01 \text{ kJ mol}^{-1}$$

$$\Delta T = \left\{\frac{(273.15 \text{ K}) \times (9.9 \times 10^6 \text{ Pa}) \times (18 \times 10^{-3} \text{ kg mol}^{-1})}{6.01 \times 10^3 \text{ J mol}^{-1}}\right\}$$

$$\times \left\{\frac{1}{9.98 \times 10^2 \text{ kg m}^{-3}} - \frac{1}{9.15 \times 10^2 \text{ kg m}^{-3}}\right\}$$

$$= -0.74 \text{ K}$$

$$T_{fus}(10 \text{ MPa}) = 273.15 \text{ K} - 0.74 \text{ K} = \boxed{272.41 \text{ K}}$$

E4.10(b) $\Delta_{vap}H = \Delta_{vap}U + \Delta_{vap}(pV)$

$\Delta_{vap}H = 43.5 \text{ kJ mol}^{-1}$

$\Delta_{vap}(pV) = p\Delta_{vap}V = p(V_{gas} - V_{liq}) = pV_{gas} = RT$ [per mole, perfect gas]

$\Delta_{vap}(pV) = (8.314 \text{ J K}^{-1} \text{ mol}^{-1}) \times (352 \text{ K}) = 292\bar{7} \text{ J mol}^{-1}$

$$\text{Fraction} = \frac{\Delta_{vap}(pV)}{\Delta_{vap}H} = \frac{2.92\bar{7} \text{ kJ mol}^{-1}}{43.5 \text{ kJ mol}^{-1}}$$

$$= \boxed{6.73 \times 10^{-2}} = 6.73 \text{ percent}$$

Solutions to problems

Solutions to numerical problems

P4.1   At the triple point, T_3, the vapor pressures of liquid and solid are equal; hence

$$10.5916 - \frac{1871.2 \text{ K}}{T_3} = 8.3186 - \frac{1425.7 \text{ K}}{T_3}; \quad T_3 = \boxed{196.0 \text{ K}}.$$

$$\log(p_3/\text{Torr}) = \frac{-1871.2 \text{ K}}{196.0 \text{ K}} + 10.5916 = 1.044\bar{7}; \quad p_3 = \boxed{11.1 \text{ Torr}}.$$

P4.3   **(a)** $\dfrac{dp}{dT} = \dfrac{\Delta_{vap}S}{\Delta_{vap}V} = \dfrac{\Delta_{vap}H}{T_b \Delta_{vap}V}$ [4.6, Clapeyron equation]

$$= \frac{14.4 \times 10^3 \text{ J mol}^{-1}}{(180 \text{ K}) \times (14.5 \times 10^{-3} - 1.15 \times 10^{-4}) \text{ m}^3 \text{ mol}^{-1}} = \boxed{+5.56 \text{ kPa K}^{-1}}.$$

(b) $\dfrac{dp}{dT} = \dfrac{\Delta_{vap}H}{RT^2} \times p \left[4.11, \text{ with d } \ln p = \dfrac{dp}{p} \right]$

$$= \frac{(14.4 \times 10^3 \text{ J mol}^{-1}) \times (1.013 \times 10^5 \text{ Pa})}{(8.314 \text{ J K}^{-1} \text{ mol}^{-1}) \times (180 \text{ K})^2} = +5.42 \text{ kPa K}^{-1}.$$

The percentage error is $\boxed{2.5 \text{ per cent}}$.

P4.5   **(a)** $\left(\dfrac{\partial \mu(1)}{\partial p} \right)_T - \left(\dfrac{\partial \mu(s)}{\partial p} \right)_T = V_m(1) - V_m(s)$ [4.13] $= M\Delta \left(\dfrac{1}{\rho} \right)$

$$= (18.02 \text{ g mol}^{-1}) \times \left(\frac{1}{1.000 \text{ g cm}^{-3}} - \frac{1}{0.917 \text{ g cm}^{-3}} \right)$$

$$= \boxed{-1.63 \text{ cm}^3 \text{ mol}^{-1}}.$$

(b) $\left(\dfrac{\partial \mu(g)}{\partial p} \right)_T - \left(\dfrac{\partial \mu(1)}{\partial p} \right)_T = V_m(g) - V_m(1)$

$$= (18.02 \text{ g mol}^{-1}) \times \left(\frac{1}{0.598 \text{ g dm}^{-3}} - \frac{1}{0.958 \times 10^3 \text{ g dm}^{-3}} \right)$$

$$= \boxed{+30.1 \text{ dm}^3 \text{ mol}^{-1}}.$$

At $1.\overline{0}$ atm and $100°C$, $\mu(l) = \mu(g)$; therefore, at 1.2 atm and $100°C$, $\mu(g) - \mu(l) \approx \Delta V_{vap}\Delta p =$ (as in Problem 4.4)

$$(30.1 \times 10^{-3} \text{ m}^3 \text{ mol}^{-1}) \times (0.2) \times (1.013 \times 10^5 \text{ Pa}) \approx \boxed{+ 0.6 \text{ kJ mol}^{-1}}.$$

Since $\mu(g) > \mu(l)$, the gas tends to condense into a liquid.

P4.7   The amount (moles) of water evaporated is $n_g = \dfrac{p_{H_2O}V}{RT}$.

The heat leaving the water is $q = n\Delta_{vap}H$.

The temperature change of the water is $\Delta T = \dfrac{-q}{nC_{p,m}}$, $n =$ amount of liquid water.

Therefore, $\Delta T = \dfrac{-p_{H_2O} V \Delta_{vap}H}{RTnC_{p,m}}$

$$= \frac{- (3.17 \text{ kPa}) \times (50.0 \text{ dm}^3) \times (44.0 \times 10^3 \text{ J mol}^{-1})}{(8.314 \text{ kPa dm}^3 \text{ K}^{-1} \text{ mol}^{-1}) \times (298.15 \text{ K}) \times (75.5 \text{ J K}^{-1} \text{ mol}^{-1}) \times \left(\frac{250 \text{ g}}{18.02 \text{ g mol}^{-1}}\right)}$$

$$= -2.7\text{K}.$$

The final temperature will be about $\boxed{22°C}$.

P4.9   (a) Follow the procedure in Problem 4.8, but note that $T_b = \boxed{227.5°C}$ is obvious from the data.

(b) Draw up the following table.

$\theta/°C$	57.4	100.4	133.0	157.3	203.5	227.5
T/K	330.6	373.6	406.2	430.5	476.7	500.7
$1000 \text{ K}/T$	3.02	2.68	2.46	2.32	2.10	2.00
$\ln p/\text{Torr}$	0.00	2.30	3.69	4.61	5.99	6.63

The points are plotted in Fig. 4.2. The slope is

$$-6.4 \times 10^3 \text{ K, so } \frac{-\Delta_{vap}H}{R} = -6.4 \times 10^3 \text{ K},$$

implying that $\Delta_{vap}H = \boxed{+53 \text{ kJ mol}^{-1}}$.

P4.11   (a) The phase diagram is shown in Fig. 4.3.

(b) The standard melting point is the temperature at which solid and liquid are in equilibrium at 1 bar. That temperature can be found by solving the equation of the solid–liquid coexistence curve for the temperature

$$1 = p_3/\text{bar} + 1000(5.60 + 11.727x)x.$$

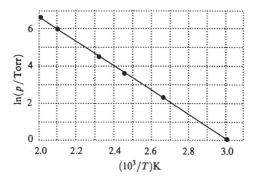

Figure 4.2

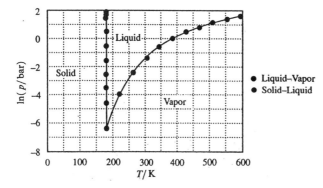

Figure 4.3

So $11727x^2 + 5600x + (4.362 \times 10^{-7} - 1) = 0$.

The quadratic formula yields

$$x = \frac{-5600 \pm \left\{(5600)^2 - 4\,(11\,727) \times (-1)\right\}^{1/2}}{2\,(11\,727)} = \frac{-1 \pm \left\{1 + (4(11\,727)/5600^2)\right\}^{1/2}}{2\,((11\,727)/5600)}.$$

The square root is rewritten to make it clear that the square root is of the form $\{1 + a\}^{1/2}$, with $a \ll 1$; thus the numerator is approximately $-1 + \left(1 + \frac{1}{2}a\right) = \frac{1}{2}a$, and the whole expression reduces to

$$x \approx 1/5600 = 1.79 \times 10^{-4}.$$

Thus, the melting point is

$$T = (1 + x)\,T_3 = (1.000179) \times (178.15\ \text{K}) = \boxed{178.18\ \text{K}}.$$

(c) The standard boiling point is the temperature at which the liquid and vapor are in equilibrium at 1 bar. That temperature can be found by solving the equation of the liquid–vapor coexistence curve for the temperature. This equation is too complicated to solve analytically, but not difficult to solve numerically with a spreadsheet. The calculated answer is $\boxed{T = 383.6\ \text{K}}$.

(d) The slope of the liquid–vapor coexistence curve is given by

$$\frac{dp}{dT} = \frac{\Delta_{vap}H^{\ominus}}{T\Delta_{vap}V^{\ominus}}$$

so $\Delta_{vap}H^{\ominus} = T\Delta_{vap}V^{\ominus}\dfrac{dp}{dT}$.

The slope can be obtained by differentiating the equation for the coexistence curve.

$$\frac{dp}{dT} = p\frac{d\ln p}{dT} = p\frac{d\ln p}{dy}\frac{dy}{dT},$$

$$\frac{dp}{dT} = \left(\frac{10.413}{y^2} - 15.996 + 2(14.015)y - 3(5.0120)y^2 - (1.70) \times (4.7224) \times (1-y)^{0.70}\right)$$

$$\times \left(\frac{p}{T_c}\right).$$

At the boiling point, $y = 0.6458$, so

$$\frac{dp}{dT} = 2.851 \times 10^{-2} \text{ bar K}^{-1} = 2.851 \text{ kPa K}^{-1}$$

and $\Delta_{vap}H^{\ominus} = (383.6 \text{ K}) \times \left(\dfrac{(30.3 - 0.12) \text{ dm}^3 \text{ mol}^{-1}}{1000 \text{ dm}^3 \text{ m}^{-3}}\right) \times \left(2.851 \text{ kPa K}^{-1}\right)$

$$= \boxed{33.0 \text{ kJ mol}^{-1}}.$$

Solutions to theoretical problems

P4.13

$$\left(\frac{\partial \Delta G}{\partial p}\right)_T = \left(\frac{\partial G_\beta}{\partial p}\right)_T - \left(\frac{\partial G_\alpha}{\partial p}\right)_T = V_\beta - V_\alpha.$$

Therefore, if $V_\beta = V_\alpha$, ΔG is independent of pressure. In general, $V_\beta \neq V_\alpha$, so that ΔG is nonzero, though small, since $V_\beta - V_\alpha$ is small.

P4.15 Amount of gas bubbled through liquid $= \dfrac{pV}{RT}$

(p = initial pressure of gas and emerging gaseous mixture).

Amount of vapor carried away $= \dfrac{m}{M}$.

Mole fraction of vapor in gaseous mixture $= \dfrac{m/M}{(m/M) + (pV/RT)}$.

Partial pressure of vapor $= p = \dfrac{m/M}{(m/M) + (pV/RT)} \times p = \dfrac{p(mRT/PVM)}{(mRT/PVM) + 1} = \dfrac{mPA}{mA + 1}$, $A = \dfrac{RT}{PVM}$.

For geraniol, $M = 154.2 \text{ g mol}^{-1}$, $T = 383 \text{ K}$, $V = 5.00 \text{ dm}^3$, $p = 1.00 \text{ atm}$, and $m = 0.32 \text{ g}$, so

$$A = \frac{(8.206 \times 10^{-2} \text{ dm}^3 \text{ atm K}^{-1} \text{ mol}^{-1}) \times (383 \text{ K})}{(1.00 \text{ atm}) \times (5.00 \text{ dm}^3) \times (154.2 \times 10^{-3} \text{ kg mol}^{-1})} = 40.7\overline{6} \text{ kg}^{-1}.$$

Therefore

$$p = \frac{(0.32 \times 10^{-3} \text{ kg}) \times (760 \text{ Torr}) \times (40.76 \text{ kg}^{-1})}{(0.32 \times 10^{-3} \text{ kg}) \times (40.76 \text{ kg}^{-1}) + 1} = \boxed{9.8 \text{ Torr}}.$$

P4.17   In each phase the slopes are given by

$$\left(\frac{\partial \mu}{\partial T} \right)_p = -S_m \text{ [4.1]}.$$

The curvatures of the graphs of μ against T are given by

$$\left(\frac{\partial^2 \mu}{\partial T^2} \right)_p = -\left(\frac{\partial S_m}{\partial T} \right)_p = \boxed{-\frac{1}{T} \times C_{p,m}} \text{ [Problem 3.26]}.$$

Since $C_{p,m}$ is necessarily positive, the curvatures in all states of matter are necessarily negative. $C_{p,m}$ is often largest for the liquid state, though not always, but it is the ratio $C_{p,m}/T$ that determines the magnitude of the curvature, so no precise answer can be given for the state with the greatest curvature. It depends upon the substance.

P4.19 $S_m = S_m(T,p)$.

$$dS_m = \left(\frac{\partial S_m}{\partial T} \right)_p dT + \left(\frac{\partial S_m}{\partial p} \right)_T dp.$$

$$\left(\frac{\partial S_m}{\partial T} \right)_p = \frac{C_p}{T} \text{ [Problem 3.26]}, \quad \left(\frac{\partial S_m}{\partial p} \right)_T = -\left(\frac{\partial V_m}{\partial T} \right)_p \text{ [Maxwell relation]}.$$

$$dq_{rev} = T \, dS_m = C_p \, dT - T \left(\frac{\partial V_m}{\partial T} \right)_p dp.$$

$$C_S = \left(\frac{\partial q}{\partial T} \right)_S = C_p - TV_m \alpha \left(\frac{\partial p}{\partial T} \right)_S = C_p - \alpha V_m \times \frac{\Delta_{trs} H}{\Delta_{trs} V} \text{ [4.7]}.$$

Solutions to applications

P4.21   (a) The Dieterici equation of state is purported to have good accuracy near the critical point. It does fail badly at high densities where V_m begins to approach the value of the Dieterici coefficient b. We will use it to derive a practical equation for the computations.

$$p_r = \frac{e^2 T_r e^{-2/T_r V_r}}{2V_r - 1} \text{ [Table 1.7]}.$$

Substitution of the Dieterici equation of state derivative $(\partial p_r / \partial T_r)_{V_r} = (2 + T_r V_r) p_r / T_r^2 V_r$ into the reduced form of eqn 3.48 gives

$$\left(\frac{\partial U_r}{\partial V_r} \right)_{T_r} = T_r \left(\frac{\partial p_r}{\partial T_r} \right)_{V_r} - p_r = \frac{2p_r}{T_r V_r} \quad (U_r = U/p_c \, V_c).$$

Integration along the isotherm T_r from an infinite volume to V_r yields the practical computational equation.

$$\Delta U_r(T_r, V_r) = -\int_{T_r\text{constant}}^{\infty} \frac{2p_r(T_r, V_r)}{T_r V_r} dV_r.$$

The integration is performed with mathematical software.

(b) See Fig. 4.4(a).

(c) $\delta(T_r, V_r) = \sqrt{-p_c \Delta U_r / V_r}$ where $p_c = 72.9$ atm.
Carbon dioxide should have solvent properties similar to liquid carbon tetrachloride ($8 \leq \delta \leq 9$) when the reduced pressure is in the approximate range $\boxed{0.85 \text{ to } 0.90 \text{ when } T_r = 1}$. See Fig. 4.4(b).

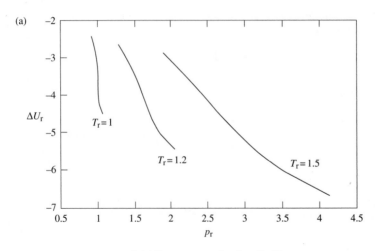

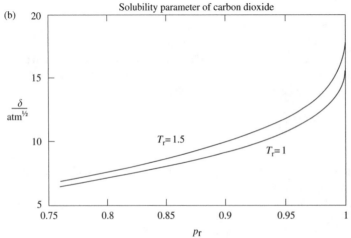

Figure 4.4

P4.23 C (graphite) $\rightleftharpoons$ C (diamond), $\Delta_r G^\ominus = 2.8678 \, \text{kJ mol}^{-1}$ at $\mathbf{T}$. $\mathbf{T} = 25\,°C$.

We want the pressure at which $\Delta_r G = 0$; above that pressure the reaction will be spontaneous. Equation 3.50 determines the rate of change of $\Delta_r G$ with p at constant T.

(1) $\qquad \left(\dfrac{\partial \Delta_r G}{\partial p} \right)_T = \Delta_r V = (V_D - V_G)M$

where M is the molar mass of carbon; V_D and V_G are the specific volumes of diamond and graphite, respectively.

$\Delta_r G(\mathbf{T}, p)$ may be expanded in a Taylor series around the pressure $p^\ominus = 100 \, \text{kPa}$ at $\mathbf{T}$.

(2) $\qquad \Delta_r G(\mathbf{T}, p) = \Delta_r G^\ominus(\mathbf{T}, p^\ominus) + \left(\dfrac{\partial \Delta_r G(\mathbf{T}, p^\ominus)}{\partial p} \right)_T (p - p^\ominus)$

$\qquad\qquad + \dfrac{1}{2} \left(\dfrac{\partial^2 \Delta_r G^\ominus(T, p^\ominus)}{\partial p^2} \right)_T (p - p^\ominus)^2 + \theta \, (p - p^\ominus)^3.$

We will neglect the third- and higher-order terms; the derivative of the first-order term can be calculated with eqn 1. An expression for the derivative of the second-order term can be derived with eqn 1.

(3) $\qquad \left(\dfrac{\partial^2 \Delta_r G}{\partial p^2} \right)_T = \left\{ \left(\dfrac{\partial V_D}{\partial p} \right)_T - \left(\dfrac{\partial V_G}{\partial p} \right)_T \right\} M = \{V_G \kappa_T\,(G) - V_D \kappa_T\,(D)\}\, M$ [2.52 with 2.22].

Calculating the derivatives of eqns 1 and 2 at $\mathbf{T}$ and $p^\ominus$,

(4) $\qquad \left(\dfrac{\partial \Delta_r G(\mathbf{T}, p^\ominus)}{\partial p} \right)_T = (0.284 - 0.444) \times \left(\dfrac{\text{cm}^3}{\text{g}} \right) \times \left(\dfrac{12.01 \, \text{g}}{\text{mol}} \right) = -1.92 \, \text{cm}^3 \, \text{mol}^{-1},$

(5) $\qquad \left(\dfrac{\partial^2 \Delta_r G(\mathbf{T}, p^\ominus)}{\partial p^2} \right)_T = \{0.444(3.04 \times 10^{-8}) - 0.284(0.187 \times 10^{-8})\}$

$\qquad\qquad\qquad \times \left(\dfrac{\text{cm}^3 \, \text{kPa}^{-1}}{\text{g}} \right) \times \left(\dfrac{12.01 \, \text{g}}{\text{mol}} \right)$

$\qquad\qquad = 1.56 \times 10^{-7} \, \text{cm}^3 \, (\text{kPa})^{-1} \, \text{mol}^{-1}.$

It is convenient to convert the value of $\Delta_r G^\ominus$ to the units $\text{cm}^3 \, \text{kPa mol}^{-1}$,

$$\Delta_r G^\ominus = 2.8678 \, \text{kJ mol}^{-1} \left\{ \dfrac{8.315 \times 10^{-2} \, \text{dm}^3 \, \text{bar K}^{-1}\text{mol}^{-1}}{8.315 \text{J K}^{-1}\text{mol}^{-1}} \times \left(\dfrac{10^3 \, \text{cm}^3}{\text{dm}^3} \right) \times \left(\dfrac{10^5 \, \text{Pa}}{\text{bar}} \right) \right\}$$

(6) $\Delta_r G^\ominus = 2.8678 \times 10^6 \, \text{cm}^3 \, \text{kPa mol}^{-1}.$

Setting $\chi = p - p^\ominus$, eqns 2 and 3–6 give

$2.8678 \times 10^6 \, \text{cm}^3 \, \text{kPa mol}^{-1} - (1.92 \, \text{cm}^3 \text{mol}^{-1})\chi + (7.80 \times 10^{-8} \, \text{cm}^3 \, \text{kPa}^{-1} \, \text{mol}^{-1})\chi^2 = 0$

when $\Delta_r G(\mathcal{T}, p) = 0$. One real root of this equation is

$$\chi = 1.60 \times 10^6 \, \text{kPa} = p - p^{\ominus} \text{ or}$$

$$p = 1.60 \times 10^6 \, \text{kPa} - 10^2 \, \text{kPa}$$

$$= 1.60 \times 10^6 \, \text{kPa} = \boxed{1.60 \times 10^4 \, \text{bar}}.$$

Above this pressure the reaction is spontaneous. The other real root is much higher: $2.3 \times 10^7 \, \text{kPa}$.

Question. What interpretation might you give to the other real root?

Answers to discussion questions

D5.1   At equilibrium, the chemical potentials of any component in both the liquid and vapor phases must be equal. This is justified by the requirement that, for systems at equilibrium under constant temperature and pressure conditions, with no additional work, $\Delta G = 0$ [see Section 3.5(e) and the answer to Discussion question 3.3]. Here $\Delta G = \mu_i(v) - \mu_i(l)$, for all components, i, of the solution; hence their chemical potentials must be equal in the liquid and vapor phases.

D5.3   All of the colligative properties are a function of the concentration of the solute, which implies that the concentration can be determined by a measurement of these properties. See eqns 5.33, 5.34, 5.36, 5.37, and 5.40. Knowing the mass of the solute in solution then allows for a calculation of its molar mass. For example, the mole fraction of the solute is related to its mass as follows:

$$x_B = \frac{m_B/M_B}{m_B/M_B + m_A/M_A}.$$

The only unknown in this expression is M_B which is easily solved for. See Example 5.4 for the details of how molar mass is determined from osmotic pressure.

D5.5   A regular solution has excess entropy of zero, but an excess enthalpy that is non-zero and dependent on composition, perhaps in the manner of eqn 5.30. We can think of a regular solution as one in which the different molecules of the solution are distributed randomly, as in an ideal solution, but have different energies of interaction with each other.

Solutions to exercises

E5.1(b)   Total volume $V = n_A V_A + n_B V_B = n(x_A V_A + x_B V_B)$

Total mass $m = n_A M_A + n_B M_B$

$$= n(x_A M_A + (1 - x_A)M_B) \quad \text{where } n = n_A + n_B$$

$$\frac{m}{x_A M_A + (1 - x_A)M_B} = n$$

$$n = \frac{1.000\,\text{kg}(10^3\,\text{g/kg})}{(0.3713) \times (241.1\,\text{g/mol}) + (1 - 0.3713) \times (198.2\,\text{g/mol})} = 4.670\overline{1}\,\text{mol}$$

$$V = n(x_A V_A + x_B V_B)$$

$$= (4.670\overline{1} \text{ mol}) \times [(0.3713) \times (188.2 \text{ cm}^3 \text{ mol}^{-1}) + (1 - 0.3713) \times (176.14 \text{ cm}^3 \text{ mol}^{-1})]$$

$$= \boxed{843.5 \text{ cm}^3}$$

E5.2(b)   Let A denote water and B ethanol. The total volume of the solution is $V = n_A V_A + n_B V_B$

We know V_B; we need to determine n_A and n_B in order to solve for V_A.

Assume we have 100 cm^3 of solution; then the mass is

$$m = \rho V = (0.9687 \text{ g cm}^{-3}) \times (100 \text{ cm}^3) = 96.87 \text{ g}$$

of which $(0.20) \times (96.87 \text{ g}) = 19.\overline{374}$ g is ethanol and $(0.80) \times (96.87 \text{ g}) = 77.\overline{496}$ g is water.

$$n_A = \frac{77.\overline{496} \text{ g}}{18.02 \text{ g mol}^{-1}} = 4.3\overline{0} \text{ mol H}_2\text{O}$$

$$n_B = \frac{19.374 \text{ g}}{46.07 \text{ g mol}^{-1}} = 0.42\overline{05} \text{ mol ethanol}$$

$$\frac{V - n_B V_B}{n_A} = V_A = \frac{100 \text{ cm}^3 - (0.42\overline{05} \text{ mol}) \times (52.2 \text{ cm}^3 \text{ mol}^{-1})}{4.3\overline{0} \text{ mol}}$$

$$= 18.\overline{15} \text{ cm}^3$$

$$= \boxed{18 \text{ cm}^3}$$

E5.3(b)   Check that $p_B/x_B = $ a constant (K_B)

x_B	0.010	0.015	0.020
$(p_B/x_B)/\text{kPa}$	8.2×10^3	8.1×10^3	8.3×10^3

$K_B = p/x$, average value is $\boxed{8.2 \times 10^3 \text{ kPa}}$

E5.4(b)   In Exercise 5.3(b), the Henry's law constant was determined for concentrations expressed in mole fractions. Thus the concentration in molality must be converted to mole fraction.

$$m(A) = 1000 \text{ g}, \quad \text{corresponding to } n(A) = \frac{1000 \text{ g}}{74.1 \text{ g mol}^{-1}} = 13.5\overline{0} \text{ mol} \quad n(B) = 0.25 \text{ mol}$$

Therefore,

$$x_B = \frac{0.25 \text{ mol}}{0.25 \text{ mol} + 13.5\overline{0} \text{ mol}} = 0.018\overline{2}$$

using $K_B = 8.2 \times 10^3$ kPa [Exercise 5.3(b)]

$$p = 0.018\overline{2} \times 8.2 \times 10^3 \text{ kPa} = \boxed{1.5 \times 10^2 \text{ kPa}}$$

E5.5(b)   We assume that the solvent, 2-propanol, is ideal and obeys Raoult's law.

$$x_A(\text{solvent}) = p/p^* = \frac{49.62}{50.00} = 0.9924$$

$$M_A(C_3H_8O) = 60.096 \text{ g mol}^{-1}$$

$$n_A = \frac{250 \text{ g}}{60.096 \text{ g mol}^{-1}} = 4.16\overline{00} \text{ mol}$$

$$x_A = \frac{n_A}{n_A + n_B} \qquad n_A + n_B = \frac{n_A}{x_A}$$

$$n_B = n_A \left(\frac{1}{x_A} - 1 \right)$$

$$= 4.16\overline{00} \text{ mol} \left(\frac{1}{0.9924} - 1 \right) = 3.1\overline{86} \times 10^{-2} \text{ mol}$$

$$M_B = \frac{8.69 \text{ g}}{3.186 \times 10^{-2} \text{ mol}} = 27\overline{3} \text{ g mol}^{-1} = \boxed{270 \text{ g mol}^{-1}}$$

E5.6(b) $K_f = 6.94$ for naphthalene

$$M_B = \frac{\text{mass of B}}{n_B}$$

$$n_B = \text{mass of naphthalene} \cdot b_B$$

$$b_B = \frac{\Delta T}{K_f} \quad \text{so} \quad M_B = \frac{(\text{mass of B}) \times K_f}{(\text{mass of naphthalene}) \times \Delta T}$$

$$M_B = \frac{(5.00 \text{ g}) \times (6.94 \text{ K kg mol}^{-1})}{(0.250 \text{ kg}) \times (0.780 \text{ K})} = \boxed{178 \text{ g mol}^{-1}}$$

E5.7(b) $\Delta T = K_f b_B \quad \text{and} \quad b_B = \dfrac{n_B}{\text{mass of water}} = \dfrac{n_B}{V\rho}$

$$\rho = 10^3 \text{ kg m}^{-3} \text{ (density of solution} \approx \text{density of water)}$$

$$n_B = \frac{\Pi V}{RT} \quad \Delta T = K_f \frac{\Pi}{RT\rho} \quad K_f = 1.86 \text{ K mol}^{-1} \text{ kg}$$

$$\Delta T = \frac{(1.86 \text{ K kg mol}^{-1}) \times (99 \times 10^3 \text{ Pa})}{(8.314 \text{ J K}^{-1} \text{ mol}^{-1}) \times (288 \text{ K}) \times (10^3 \text{ kg m}^{-3})} = 7.7 \times 10^{-2} \text{ K}$$

$$T_f = \boxed{-0.077\,°C}$$

E5.8(b) $\Delta_{\text{mix}} G = nRT(x_A \ln x_A + x_B \ln x_B)$

$$n_{Ar} = n_{Ne}, \quad x_{Ar} = x_{Ne} = 0.5, \quad n = n_{Ar} + n_{Ne} = \frac{pV}{RT}$$

$$\Delta_{\text{mix}} G = pV(\tfrac{1}{2} \ln \tfrac{1}{2} + \tfrac{1}{2} \ln \tfrac{1}{2}) = -pV \ln 2$$

$$= -(100 \times 10^3 \text{ Pa}) \times (250 \text{ cm}^3) \left(\frac{1 \text{ m}^3}{10^6 \text{ cm}^3} \right) \ln 2$$

$$= -17.3 \text{ Pa m}^3 = \boxed{-17.3 \text{ J}}$$

$$\Delta_{\text{mix}} S = \frac{-\Delta_{\text{mix}} G}{T} = \frac{17.3 \text{ J}}{273 \text{ K}} = \boxed{6.34 \times 10^{-2} \text{ J K}^{-1}}$$

E5.9(b)

$$\Delta_{mix}G = nRT \sum_J x_J \ln x_J \text{ [5.18]}$$

$$\Delta_{mix}S = -nR \sum_J x_J \ln x_J \text{ [5.19]} = \frac{-\Delta_{mix}G}{T}$$

$$n = 1.00 \text{ mol} + 1.00 \text{ mol} = 2.00 \text{ mol}$$

$$x(\text{Hex}) = x(\text{Hep}) = 0.500$$

Therefore,

$$\Delta_{mix}G = (2.00 \text{ mol}) \times (8.314 \text{ J K}^{-1} \text{ mol}^{-1}) \times (298 \text{ K}) \times (0.500 \ln 0.500 + 0.500 \ln 0.500)$$

$$= \boxed{-3.43 \text{ kJ}}$$

$$\Delta_{mix}S = \frac{+3.43 \text{ kJ}}{298 \text{ K}} = \boxed{+11.5 \text{ J K}^{-1}}$$

$\Delta_{mix}H$ for an ideal solution is zero as it is for a solution of perfect gases [7.20]. It can be demonstrated from

$$\Delta_{mix}H = \Delta_{mix}G + T\Delta_{mix}S = (-3.43 \times 10^3 \text{ J}) + (298 \text{ K}) \times (11.5 \text{ J K}^{-1}) = \boxed{0}$$

E5.10(b)   Benzene and ethylbenzene form nearly ideal solutions, so

$$\Delta_{mix}S = -nR(x_A \ln x_A + x_B \ln x_B)$$

To find maximum $\Delta_{mix}S$, differentiate with respect to x_A and find value of x_A at which the derivative is zero.

Note that $x_B = 1 - x_A$ so

$$\Delta_{mix}S = -nR(x_A \ln x_A + (1 - x_A) \ln(1 - x_A))$$

use $\dfrac{d \ln x}{dx} = \dfrac{1}{x}$:

$$\frac{d}{dx}(\Delta_{mix}S) = -nR(\ln x_A + 1 - \ln(1 - x_A) - 1) = -nR \ln \frac{x_A}{1 - x_A}$$

$$= 0 \quad \text{when } x_A = \frac{1}{2}$$

Thus the maximum entropy of mixing is attained by mixing equal molar amounts of two components.

$$\frac{n_B}{n_E} = 1 = \frac{m_B/M_B}{m_E/M_E} \times \frac{m_E}{m_B} = \frac{M_E}{M_B} = \frac{106.169}{78.115} = 1.3591$$

$$\frac{m_B}{m_E} = \boxed{0.7358}$$

E5.11(b)   With concentrations expressed in molalities, Henry's law [5.26] becomes $p_B = b_B K$.

Solving for b, the molality, we have $b_B = p_B/K = x p_{total}/K$ and $p_{total} = p_{atm}$

For N_2, $K = 1.56 \times 10^5$ kPa kg mol^{-1} [Table 5.1]

$$b = \frac{0.78 \times 101.3 \text{ kPa}}{1.56 \times 10^5 \text{ kPa kg mol}^{-1}} = \boxed{0.51 \text{ mmol kg}^{-1}}$$

For O_2, $K = 7.92 \times 10^4$ kPa kg mol^{-1} [Table 5.1]

$$b = \frac{0.21 \times 101.3 \text{ kPa}}{7.92 \times 10^4 \text{ kPa kg mol}^{-1}} = \boxed{0.27 \text{ mmol kg}^{-1}}$$

E5.12(b) $$b_B = \frac{p_B}{K} = \frac{2.0 \times 101.3 \text{ kPa}}{3.01 \times 10^3 \text{ kPa kg mol}^{-1}} = 0.067 \text{ mol kg}^{-1}$$

The molality will be about 0.067 mol kg^{-1} and, since molalities and molar concentrations for dilute aqueous solutions are approximately equal, the molar concentration is about $\boxed{0.067 \text{ mol dm}^{-3}}$

E5.13(b)   The procedure here is identical to Exercise 5.13(a).

$$\ln x_B = \frac{\Delta_{\text{fus}} H}{R} \times \left(\frac{1}{T^*} - \frac{1}{T} \right) \text{ [5.39; B, the solute, is lead]}$$

$$= \left(\frac{5.2 \times 10^3 \text{ J mol}^{-1}}{8.314 \text{ J K}^{-1} \text{ mol}^{-1}} \right) \times \left(\frac{1}{600 \text{ K}} - \frac{1}{553 \text{ K}} \right)$$

$$= -0.088\overline{6}, \text{ implying that } x_B = 0.92$$

$$x_B = \frac{n(\text{Pb})}{n(\text{Pb}) + n(\text{Bi})}, \text{ implying that } n(\text{Pb}) = \frac{x_B n(\text{Bi})}{1 - x_B}$$

For 1 kg of bismuth, $n(\text{Bi}) = \dfrac{1000 \text{ g}}{208.98 \text{ g mol}^{-1}} = 4.785 \text{ mol}$

Hence, the amount of lead that dissolves in 1 kg of bismuth is

$$n(\text{Pb}) = \frac{(0.92) \times (4.785 \text{ mol})}{1 - 0.92} = 55 \text{ mol}, \quad \text{or} \quad \boxed{11 \text{ kg}}$$

COMMENT. It is highly unlikely that a solution of 11 kg of lead and 1 kg of bismuth could in any sense be considered ideal. The assumptions upon which eqn 5.39 is based are not likely to apply. The answer above must then be considered an order of magnitude result only.

E5.14(b)   Proceed as in Exercise 5.14(a). The data are plotted in Figure 5.1, and the slope of the line is 1.78 cm/(mg cm^{-3}) = 1.78 cm/(g dm^{-3}) = 1.78×10^{-2} m^4 kg^{-1}.

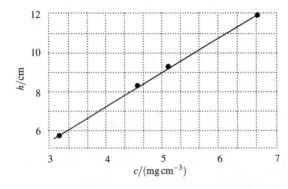

Figure 5.1

Therefore,

$$M = \frac{(8.314 \text{ J K}^{-1} \text{ mol}^{-1}) \times (293.15 \text{ K})}{(1.000 \times 10^3 \text{ kg m}^{-3}) \times (9.81 \text{ m s}^{-2}) \times (1.78 \times 10^{-2} \text{ m}^4 \text{ kg}^{-1})} = \boxed{14.0 \text{ kg mol}^{-1}}$$

E5.15(b)   Let A = water and B = solute.

$$a_A = \frac{p_A}{p_A^*} \text{ [5.43]} = \frac{0.02239 \text{ atm}}{0.02308 \text{ atm}} = \boxed{0.9701}$$

$$\gamma_A = \frac{a_A}{x_A} \quad \text{and} \quad x_A = \frac{n_A}{n_A + n_B}$$

$$n_A = \frac{0.920 \text{ kg}}{0.01802 \text{ kg mol}^{-1}} = 51.0\overline{5} \text{ mol} \quad \text{and} \quad n_B = \frac{0.122 \text{ kg}}{0.241 \text{ kg mol}^{-1}} = 0.506 \text{ mol}$$

$$x_A = \frac{51.0\overline{5}}{51.05 + 0.506} = 0.990 \quad \text{and} \quad \gamma_A = \frac{0.9701}{0.990} = \boxed{0.980}$$

E5.16(b)   B = Benzene $\mu_B(l) = \mu_B^*(l) + RT \ln x_B$ [5.25, ideal solution]

$$RT \ln x_B = (8.314 \text{ J K}^{-1} \text{ mol}^{-1}) \times (353.3 \text{ K}) \times (\ln 0.30) = \boxed{-353\overline{6} \text{ J mol}^{-1}}$$

Thus, its chemical potential is lowered by this amount.

$$p_B = a_B p_B^* \text{ [5.43]} = \gamma_B x_B p_B^* = (0.93) \times (0.30) \times (760 \text{ Torr}) = \boxed{212 \text{ Torr}}$$

Question. What is the lowering of the chemical potential in the nonideal solution with $\gamma = 0.93$?

E5.17(b) $$y_A = \frac{p_A}{p_A + p_B} = \frac{p_A}{101.3 \text{ kPa}} = 0.314$$

$$p_A = (101.3 \text{ kPa}) \times (0.314) = 31.8 \text{ kPa}$$

$$p_B = 101.3 \text{ kPa} - 31.8 \text{ kPa} = 69.5 \text{ kPa}$$

$$a_A = \frac{p_A}{p_A^*} = \frac{31.8 \text{ kPa}}{73.0 \text{ kPa}} = \boxed{0.436}$$

$$a_B = \frac{p_B}{p_B^*} = \frac{69.5 \text{ kPa}}{92.1 \text{ kPa}} = \boxed{0.755}$$

$$\gamma_A = \frac{a_A}{x_A} = \frac{0.436}{0.220} = \boxed{1.98}$$

$$\gamma_B = \frac{a_B}{x_B} = \frac{0.755}{0.780} = \boxed{0.968}$$

E5.18(b) $I = \frac{1}{2} \sum_i (b_i/b^\ominus) z_i^2$ [5.71]

and for an $M_p X_q$ salt, $b_+/b^\ominus = pb/b^\ominus$, $b_-/b^\ominus = qb/b^\ominus$, so

$$I = \frac{1}{2}(pz_+^2 + qz_-^2)b/b^\ominus$$

$$I = I(K_3[Fe(CN)_6]) + I(KCl) + I(NaBr) = \frac{1}{2}(3 + 3^2)\frac{b(K_3[Fe(CN)_6])}{b^\ominus} + \frac{b(KCl)}{b^\ominus} + \frac{b(NaBr)}{b^\ominus}$$

$$= (6) \times (0.040) + (0.030) + (0.050) = \boxed{0.320}$$

Question. Can you establish that the statement in the comment following the solution to Exercise 5.18(a) holds for the solution of this exercise?

E5.19(b) $I = I(KNO_3) = \dfrac{b}{b^\ominus}(KNO_3) = 0.110$

Therefore, the ionic strengths of the added salts must be 0.890.

(a) $I(KNO_3) = \dfrac{b}{b^\ominus}$, so $b(KNO_3) = 0.890 \text{ mol kg}^{-1}$

and $(0.890 \text{ mol kg}^{-1}) \times (0.500 \text{ kg}) = 0.445 \text{ mol KNO}_3$

So $(0.445 \text{ mol}) \times (101.11 \text{ g mol}^{-1}) = \boxed{45.0 \text{ g KNO}_3}$ must be added.

(b) $I(Ba(NO_3)_2) = \dfrac{1}{2}(2^2 + 2 \times 1^2)\dfrac{b}{b^\ominus} = 3\dfrac{b}{b^\ominus} = 0.890$

$$b = \frac{0.890}{3}b^\ominus = 0.296\overline{7} \text{ mol kg}^{-1}$$

and $(0.296\overline{7} \text{ mol kg}^{-1}) \times (0.500 \text{ kg}) = 0.148\overline{4} \text{ mol Ba(NO}_3)_2$

So $(0.148\overline{4} \text{ mol}) \times (261.32 \text{ g mol}^{-1}) = \boxed{38.8 \text{ g Ba(NO}_3)_2}$

E5.20(b)   Since the solutions are dilute, use the Debye–Hückel limiting law

$$\log \gamma_\pm = -|z_+ z_-|A I^{1/2}$$

$$I = \frac{1}{2}\sum_i z_i^2 (b_i/b^\ominus) = \frac{1}{2}\{1 \times (0.020) + 1 \times (0.020) + 4 \times (0.035) + 2 \times (0.035)\}$$

$$= 0.125$$

$$\log \gamma_\pm = -1 \times 1 \times 0.509 \times (0.125)^{1/2} = -0.179\overline{96}$$

(For NaCl) $\gamma_\pm = 10^{-0.179\overline{96}} = \boxed{0.661}$

E5.21(b)   The extended Debye–Hückel law is $\log \gamma_\pm = -\dfrac{A|z_+ z_-| I^{1/2}}{1 + BI^{1/2}}$

Solving for B

$$B = -\left(\frac{1}{I^{1/2}} + \frac{A|z_+ z_-|}{\log \gamma_\pm}\right) = -\left(\frac{1}{(b/b^\ominus)^{1/2}} + \frac{0.509}{\log \gamma_\pm}\right)$$

Draw up the following table

$b/(\mathrm{mol\ kg^{-1}})$	5.0×10^{-3}	10.0×10^{-3}	50.0×10^{-3}
$\gamma_\pm$	0.927	0.902	0.816
B	1.3$\overline{2}$	1.3$\overline{6}$	1.2$\overline{9}$

$$B = \boxed{1.3}$$

Solutions to problems

Solutions to numerical problems

P5.1 $p_A = y_A p$ and $p_B = y_B p$ (Dalton's law). Hence, draw up the following table.

p_A/kPa	0	1.399	3.566	5.044	6.996	7.940	9.211	10.105	11.287	12.295
x_A	0	0.0898	0.2476	0.3577	0.5194	0.6036	0.7188	0.8019	0.9105	1
y_A	0	0.0410	0.1154	0.1762	0.2772	0.3393	0.4450	0.5435	0.7284	1

p_B/kPa	0	4.209	8.487	11.487	15.462	18.243	23.582	27.334	32.722	36.066
x_B	0	0.0895	0.1981	0.2812	0.3964	0.4806	0.6423	0.7524	0.9102	1
y_B	0	0.2716	0.4565	0.5550	0.6607	0.7228	0.8238	0.8846	0.9590	1

The data are plotted in Fig. 5.2.

We can assume, at the lowest concentrations of both A and B, that Henry's law will hold. The Henry's law constants are then given by

$$K_A = \frac{p_A}{x_A} = \boxed{15.58 \text{ kPa}} \text{ from the point at } x_A = 0.0898.$$

$$K_B = \frac{p_B}{x_B} = \boxed{47.03 \text{ kPa}} \text{ from the point at } x_B = 0.0895.$$

P5.3 $V_{\text{salt}} = \left(\dfrac{\partial V}{\partial b}\right)_{H_2O} \mathrm{mol^{-1}}$ [Problem 5.2]

$$= 69.38(b - 0.070)\mathrm{cm^3\ mol^{-1}} \quad \text{with } b \equiv b/(\mathrm{mol\ kg^{-1}}).$$

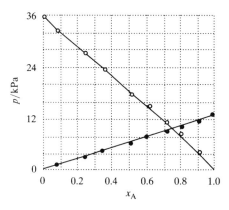

x_A

Figure 5.2

Therefore, at $b = 0.050 \text{ mol kg}^{-1}$, $V_{\text{salt}} = \boxed{-1.4 \text{ cm}^3 \text{ mol}^{-1}}$.

The total volume at this molality is

$$V = (1001.21) + (34.69) \times (0.02)^2 \text{ cm}^3 = 1001.22 \text{ cm}^3.$$

Hence, as in Problem 5.2,

$$V(\text{H}_2\text{O}) = \frac{(1001.22 \text{ cm}^3) - (0.050 \text{ mol}) \times (-1.4 \text{ cm}^3 \text{mol}^{-1})}{55.49 \text{ mol}} = \boxed{18.04 \text{ cm}^2 \text{ mol}^{-1}}.$$

Question. What meaning can be ascribed to a negative partial molar volume?

P5.5   Let E denote ethanol and W denote water; then

$$V = n_E V_E + n_W V_W \text{ [5.3]}.$$

For a 50 per cent mixture by mass, $m_E = m_W$, implying that

$$n_E M_E = n_W M_W, \text{ or } n_W = \frac{n_E M_E}{M_W}.$$

Hence, $V = n_E V_E + \dfrac{n_E M_E V_W}{M_W}$

which solves to $n_E = \dfrac{V}{V_E + \dfrac{M_E V_W}{M_W}}, \quad n_W = \dfrac{M_E V}{V_E M_W + M_E V_W}.$

Furthermore, $x_E = \dfrac{n_E}{n_E + n_W} = \dfrac{1}{1 + \dfrac{M_E}{M_W}}.$

Since $M_E = 46.07 \text{ g mol}^{-1}$ and $M_W = 18.02 \text{ g mol}^{-1}$, $\dfrac{M_E}{M_W} = 2.557.$ Therefore

$$x_E = 0.2811, \quad x_W = 1 - x_E = 0.7189.$$

At this composition

$$V_E = 56.0 \, \text{cm}^3 \text{mol}^{-1}, \qquad V_W = 17.5 \, \text{cm}^3 \, \text{mol}^{-1} \, [\text{Fig.5.1 of the text}].$$

Therefore, $n_E = \dfrac{100 \, \text{cm}^3}{(56.0 \, \text{cm}^3 \, \text{mol}^{-1}) + (2.557) \times (17.5 \, \text{cm}^3 \, \text{mol}^{-1})} = 0.993 \, \text{mol},$

$n_W = (2.557) \times (0.993 \, \text{mol}) = 2.54 \, \text{mol}.$

The fact that these amounts correspond to a mixture containing 50 per cent by mass of both components is easily checked as follows:

$$m_E = n_E M_E = (0.993 \, \text{mol}) \times (46.07 \, \text{g mol}^{-1}) = 45.7 \, \text{g ethanol},$$

$$m_W = n_W M_W = (2.54 \, \text{mol}) \times (18.02 \, \text{g mol}^{-1}) = 45.7 \, \text{g water}.$$

At 20°C the densities of ethanol and water are,

$$\rho_E = 0.789 \, \text{g cm}^{-3}, \quad \rho_W = 0.997 \, \text{g cm}^{-3}. \text{ Hence,}$$

$$V_E = \frac{m_E}{\rho_E} = \frac{45.7 \, \text{g}}{0.789 \, \text{g cm}^{-3}} = \boxed{57.9 \, \text{cm}^3} \text{ of ethanol},$$

$$V_W = \frac{m_W}{\rho_W} = \frac{45.7 \, \text{g}}{0.997 \, \text{g cm}^{-3}} = \boxed{45.8 \, \text{cm}^3} \text{ of water}.$$

The change in volume upon adding a small amount of ethanol can be approximated by

$$\Delta V = \int dV \approx \int V_E dn_E \approx V_E \Delta n_E$$

where we have assumed that both V_E and V_W are constant over this small range of n_E. Hence

$$\Delta V \approx (56.0 \, \text{cm}^3 \text{mol}^{-1}) \times \left(\frac{(1.00 \, \text{cm}^3) \times (0.789 \, \text{g cm}^{-3})}{(46.07 \, \text{g mol}^{-1})} \right) = \boxed{+0.96 \, \text{cm}^3}.$$

P5.7 $\quad b_B = \dfrac{\Delta T}{K_f} = \dfrac{0.0703 \, \text{K}}{1.86 \, \text{K}/(\text{mol kg}^{-1})} = 0.0378 \, \text{mol kg}^{-1}.$

Since the solution molality is nominally $0.0096 \, \text{mol kg}^{-1}$ in $\text{Th(NO}_3)_4$, each formula unit supplies $\dfrac{0.0378}{0.0096} \approx \boxed{4 \text{ ions}}$. (More careful data, as described in the original reference gives $\nu \approx 5$ to 6.)

P5.9 The data are plotted in Figure 5.3. The regions where the vapor pressure curves show approximate straight lines are denoted R for Raoult and H for Henry. A and B denote acetic acid and benzene respectively.

As in Problem 5.8, we need to form $\gamma_A = \dfrac{p_A}{x_A p_A^*}$ and $\gamma_B = \dfrac{p_B}{x_B p_B^*}$ for the Raoult's law activity coefficients

and $\gamma_B = \dfrac{p_B}{x_B K}$ for the activity coefficient of benzene on a Henry's law basis, with K determined by extrapolation. We use $p^*_A = 7.3 \, \text{kPa}$, $p_B^* = 35.2 \, \text{kPa}$, and $K^*_B = 80.0 \, \text{kPa}$ to draw up the following table.

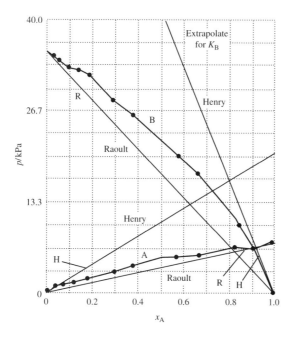

Figure 5.3

x_A	0	0.2	0.4	0.6	0.8	1.0
p_A/kPa	0	2.7	4.0	5.1	6.7	7.3
p_B/kPa	35.2	30.4	25.3	20.0	12.4	0
$a_A(R)$	0	0.36	0.55	0.69	0.91	$1.00[p_A/p_A^*]$
$a_B(R)$	1.00	0.86	0.72	0.57	0.35	$0[p_B/p_B^*]$
$\gamma_A(R)$	—	1.82	1.36	1.15	1.14	$1.00[p_A/x_A p_A^*]$
$\gamma_B(R)$	1.00	1.08	1.20	1.42	1.76	$-[p_B/x_B p_B^*]$
$a_B(H)$	0.44	0.38	0.32	0.25	0.16	$0[p_B/K_B]$
$\gamma_B(H)$	0.44	0.48	0.53	0.63	0.78	$1.00[p_B/x_B K_B]$

G^E is defined as [Section 5.4]

$$G^E = \Delta_{mix}G(\text{actual}) - \Delta_{mix}G(\text{ideal}) = nRT(x_A \ln a_A + x_B \ln a_B) - nRT(x_A \ln x_A + x_B \ln x_B)$$

and, with $a = \gamma x$,

$$G^E = nRT(x_A \ln \gamma_A + x_A \ln \gamma_B).$$

For $n = 1$, we can draw up the following table from the information above and $RT = 2.69\,\text{kJ mol}^{-1}$.

x_A	0	0.2	0.4	0.6	0.8	1.0
$x_A \ln \gamma_A$	0	0.12	0.12	0.08	0.10	0
$x_B \ln \gamma_B$	0	0.06	0.11	0.14	0.11	0
$G^E/(\text{kJ mol}^{-1})$	0	0.48	0.62	0.59	0.56	0

P5.11   (a) The volume of an ideal mixture is

$$V_{ideal} = n_1 V_{m,1} + n_2 V_{m,2}$$

so the volume of a real mixture is

$$V = V_{ideal} + V^E.$$

We have an expression for excess molar volume in terms of mole fractions. To compute partial molar volumes, we need an expression for the actual excess volume as a function of moles.

$$V^E = (n_1 + n_2)V_m^E = \frac{n_1 n_2}{n_1 + n_2}\left(a_0 + \frac{a_1(n_1 - n_2)}{n_1 + n_2}\right)$$

so $V = n_1 V_{m,1} + n_2 V_{m,2} + \dfrac{n_1 n_2}{n_1 + n_2}\left(a_0 + \dfrac{a_1(n_1 - n_2)}{n_1 + n_2}\right).$

The partial molar volume of propionic acid is

$$V_1 = \left(\frac{\partial V}{\partial n_1}\right)_{p,T,n_2} = V_{m,1} + \frac{a_0 n_2^2}{(n_1 + n_2)^2} + \frac{a_1(3n_1 - n_2)n_2^2}{(n_1 + n_2)^3},$$

$$\boxed{V_1 = V_{m,1} + a_0 x_2^2 + a_1(3x_1 - x_2)x_2^2}.$$

That of oxane is

$$\boxed{V_2 = V_{m,2} + a_0 x_1^2 + a_1(x_1 - 3x_2)x_1^2}.$$

(b) We need the molar volumes of the pure liquids,

$$V_{m,1} = \frac{M_1}{\rho_1} = \frac{74.08 \text{ g mol}^{-1}}{0.97174 \text{ g cm}^{-3}} = 76.23 \text{ cm}^3\text{mol}^{-1}$$

and $V_{m,2} = \dfrac{86.13 \text{ g mol}^{-1}}{0.86398 \text{ g cm}^{-3}} = 99.69 \text{ cm}^3\text{mol}^{-1}.$

In an equimolar mixture, the partial molar volume of propionic acid is

$$V_1 = 76.23 + (-2.4697) \times (0.500)^2 + (0.0608) \times [3(0.5) - 0.5] \times (0.5)^2 \text{cm}^3\text{mol}^{-1}$$

$$= \boxed{75.63 \text{ cm}^3 \text{ mol}^{-1}}$$

and that of oxane is

$$V_2 = 99.69 + (-2.4697) \times (0.500)^2 + (0.0608) \times [0.5 - 3(0.5)] \times (0.5)^2 \text{ cm}^3 \text{ mol}^{-1}$$

$$= \boxed{99.06 \text{ cm}^3 \text{ mol}^{-1}}.$$

P5.13   Henry's law constant is the slope of a plot of p_B versus x_B in the limit of zero x_B (Fig. 5.4). The partial pressures of CO_2 are almost but not quite equal to the total pressures reported.

$$p_{CO_2} = p y_{CO_2} = p(1 - y_{cyc}).$$

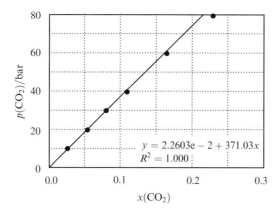

Figure 5.4

Linear regression of the low-pressure points gives $K_H = \boxed{371 \text{ bar}}$.

The activity of a solute is

$$a_B = \frac{p_B}{K_H} = x_B \gamma_B$$

so the activity coefficient is

$$\gamma_B = \frac{p_B}{x_B K_H} = \frac{y_B p}{x_B K_H}$$

where the last equality applies Dalton's law of partial pressures to the vapor phase. A spreadsheet applied this equation to the above data to yield

p/bar	y_{cyc}	x_{cyc}	γ_{CO_2}
10.0	0.0267	0.9741	1.01
20.0	0.0149	0.9464	0.99
30.0	0.0112	0.9204	1.00
40.0	0.00947	0.892	0.99
60.0	0.00835	0.836	0.98
80.0	0.00921	0.773	0.94

P5.15 $$G^E = RTx(1-x)\{0.4857 - 0.1077(2x - 1) + 0.0191(2x - 1)^2\}$$

with $x = 0.25$ gives $G^E = 0.1021RT$. Therefore, since

$$\Delta_{\text{mix}}G(\text{actual}) = \Delta_{\text{mix}}G(\text{ideal}) + nG^E,$$

$$\Delta_{\text{mix}}G = nRT(x_A \ln x_A + x_B \ln x_B) + nG^E = nRT(0.25 \ln 0.25 + 0.75 \ln 0.75) + nG^E$$

$$= -0.562nRT + 0.1021nRT = -0.460nRT.$$

Since $n = 4$ mol and $RT = (8.314 \text{ J K}^{-1} \text{ mol}^{-1}) \times (303.15 \text{ K}) = 2.52 \text{ kJ mol}^{-1}$,

$$\Delta_{\text{mix}}G = (-0.460) \times (4 \text{ mol}) \times (2.52 \text{ kJ mol}^{-1}) = \boxed{-4.6 \text{ kJ}}.$$

Solutions to theoretical problems

P5.17
$$\mu_A = \left(\frac{\partial G}{\partial n_A}\right)_{n_B} [5.4] = \mu_A^\circ + \left(\frac{\partial}{\partial n_A}(nG^E)\right)_{n_B} \quad [\mu_A^\circ \text{ is ideal value} = \mu_A^* + RT \ln x_A],$$

$$\left(\frac{\partial nG^E}{\partial n_A}\right)_{n_B} = G^E + n\left(\frac{\partial G^E}{\partial n_A}\right)_{n_B} = G^E + n\left(\frac{\partial x_A}{\partial n_A}\right)_B\left(\frac{\partial G^E}{\partial x_A}\right)_B$$

$$= G^E + n \times \frac{x_B}{n} \times \left(\frac{\partial G^E}{\partial x_A}\right)_B \quad [\partial x_A/\partial n_A = x_B/n]$$

$$= gRTx_A(1 - x_A) + (1 - x_A)gRT(1 - 2x_A)$$

$$= gRT(1 - x_A)^2 = gRTx_B^2.$$

Therefore, $\boxed{\mu_A = \mu_A^* + RT \ln x_A + gRTx_B^2}$.

P5.19
$$n_A dV_A + n_B dV_B = 0 \text{ [Example 5.1]}.$$

Hence $\dfrac{n_A}{n_B} dV_A = -dV_B.$

Therefore, by integration,

$$V_B(x_A) - V_B(0) = -\int_{V_A(0)}^{V_A(x_A)} \frac{n_A}{n_B} dV_A = -\int_{V_A(0)}^{V_A(x_A)} \frac{x_A\, dV_A}{1 - x_A} \quad [n_A = x_A n, \ n_B = x_B n].$$

Therefore, $V_B(x_A, x_B) = V_B(0, 1) - \displaystyle\int_{V_A(0)}^{V_A(x_A)} \frac{x_A\, dV_A}{1 - x_A}.$

We should now plot $x_A/(1 - x_A)$ against V_A and estimate the integral. For the present purpose we integrate up to $V_A(0.5, 0.5) = 74.06$ cm^3 mol^{-1} [Fig. 5.5], and use the data to construct the following table.

V_A (cm^3 mol^{-1})	74.11	73.96	73.50	72.74
x_A	0.60	0.40	0.20	0
$x_A/(1 - x_A)$	1.50	0.67	0.25	0

The points are plotted in Fig. 5.5, and the area required is 0.30. Hence,

$$V(\text{CHCl}_3; 0.5, 0.5) = 80.66 \text{ cm}^3 \text{ mol}^{-1} - 0.30 \text{ cm}^3 \text{ mol}^{-1}$$

$$= \boxed{80.36 \text{ cm}^3 \text{ mol}^{-1}}.$$

P5.21
$$\phi = -\frac{\ln a_A}{r}. \tag{a}$$

Therefore, $d\phi = -\dfrac{1}{r} d\ln a_A + \dfrac{1}{r^2} \ln a_A dr,$

$$d\ln a_A = \frac{1}{r} \ln a_A dr - rd\phi. \tag{b}$$

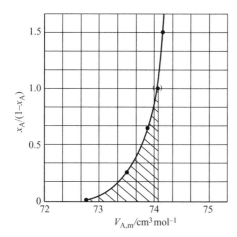

Figure 5.5

From the Gibbs–Duhem equation, $x_A \, d\mu_A = x_B \, d\mu_B = 0$, which implies that (since $\mu = \mu^{\ominus} + RT \ln a$, $d\mu_A = RT d \ln a_A$, $d\mu_B = RT d \ln a_B$)

$$d \ln a_B = -\frac{x_A}{x_B} d \ln a_A = -\frac{d \ln a_A}{r}$$

$$= -\frac{1}{r^2} \ln a_A \, dr = d\phi \; [\text{from(b)}] = \frac{1}{r} \phi dr = dr = d\phi \; [\text{from(a)}]$$

$$= \phi d \ln r + d\phi.$$

Subtract $d \ln r$ from both sides, to obtain

$$d \ln \frac{a_B}{r} = (\phi - 1) \, d \ln r + d\phi = \frac{(\phi - 1)}{r} dr + d\phi.$$

Then, by integration and noting that $\ln \left(\dfrac{a_B}{r} \right)_{r=0} = \ln \left(\dfrac{\gamma_B x_B}{r} \right)_{r=0} = \ln (\gamma_B)_{r=0} = \ln 1 = 0$,

$$\ln \frac{a_B}{r} = \boxed{\phi - \phi(0) = \int_0^r \left(\frac{\phi - 1}{r} \right) dr}.$$

P5.23 $A(s) \rightleftharpoons A(l).$

$$\mu_A^*(s) = \mu_A^*(l) + RT \ln a_A$$

and $\Delta_{fus} G = \mu_A^*(l) - \mu_A^*(s) = -RT \ln a_A.$

Hence, $\ln a_A = \dfrac{-\Delta_{fus} G}{RT}.$

$$\frac{d \ln a_A}{dT} = -\frac{1}{R} \frac{d}{dT} \left(\frac{\Delta_{fus} G}{T} \right) = \frac{\Delta_{fus} H}{RT^2} \; [\text{Gibbs–Helmholtz eqn}].$$

For $\Delta T = T_f^* - T, d\Delta T = -dT$ and

$$\frac{d \ln a_A}{d\Delta T} = \frac{-\Delta_{fus} H}{RT^2} \approx \frac{-\Delta_{fus} H}{RT_f^2}.$$

But $K_f = \dfrac{RT_f^2 M_A}{\Delta_{fus}H}$.

Therefore,

$$\frac{d \ln a_A}{d\Delta T} = \frac{-M_A}{K_f} \quad \text{and} \quad d \ln a_A = \frac{-M_A d\Delta T}{K_f}.$$

According to the Gibbs–Duhem equation

$$n_A d\mu_A + n_B d\mu_B = 0$$

which implies that

$$n_A d \ln a_A + n_B d \ln a_B = 0 \; [\mu = \mu^\ominus + RT \ln a]$$

and hence that $d \ln a_A = -\dfrac{n_B}{n_A} d \ln a_B$.

Hence, $\dfrac{d \ln a_B}{d\Delta T} = \dfrac{n_A M_A}{n_B K_f} = \dfrac{1}{b_B K_f}$ [for $n_A M_A = 1$ kg]

We know from the Gibbs–Duhem equation that

$$x_A \, d \ln a_A + x_B \, d \ln a_B = 0$$

and hence that $\displaystyle\int d \ln a_A = -\int \frac{x_B}{x_A} \, d \ln a_B$.

Therefore $\ln a_A = -\displaystyle\int \frac{x_B}{x_A} d \ln a_B$.

The osmotic coefficient was defined in Problem 5.21 as

$$\phi = -\frac{1}{r} \ln a_A = -\frac{x_A}{x_B} \ln a_A.$$

Therefore,

$$\phi = \frac{x_A}{x_B} \int \frac{x_B}{x_A} d \ln a_B = \frac{1}{b} \int_0^b b \, d \ln a_B = \frac{1}{b} \int_0^b b \, d \ln \gamma b = \frac{1}{b} \int_0^b b \, d \ln b + \frac{1}{b} \int_0^b b \, d \ln \gamma$$

$$= 1 + \frac{1}{b} \int_0^b b \, d \ln \gamma.$$

From the Debye–Hückel limiting law,

$$\ln \gamma = -A' b^{1/2} \qquad [A' = 2.303A].$$

Hence, $d \ln \gamma = -\dfrac{1}{2} A' b^{-1/2} db$ and so

$$\phi = 1 + \frac{1}{b} \left(-\frac{1}{2} A' \right) \int_0^b b^{1/2} \, db = 1 - \frac{1}{2} \left(\frac{A'}{b} \right) \times \frac{2}{3} b^{3/2} = \boxed{1 - \frac{1}{3} A' b^{1/2}}.$$

COMMENT.   For the depression of the freezing point in a 1,1-electrolyte

$$\ln a_A = \frac{-\Delta_{fus}G}{RT} + \frac{\Delta_{fus}G}{RT^*}$$

and hence $-r\phi = \dfrac{-\Delta_{fus}H}{R}\left(\dfrac{1}{T} - \dfrac{1}{T*}\right)$.

Therefore, $\phi = \dfrac{\Delta_{fus}H x_A}{R x_B}\left(\dfrac{1}{T} - \dfrac{1}{T*}\right) = \dfrac{\Delta_{fus}H x_A}{R x_B}\left(\dfrac{T* - T}{TT*}\right) \approx \dfrac{\Delta_{fus}H x_A \, \Delta T}{R x_B T*^2}$

$\approx \dfrac{\Delta_{fus}H \Delta T}{\nu R b_B T*^2 M_A}$

where $\nu = 2$. Therefore, since $K_f = \dfrac{MRT*^2}{\Delta_{fus}H}$,

$$\boxed{\phi = \dfrac{\Delta T}{2 b_B K_f}}.$$

Solutions to applications

P5.25   In this case it is convenient to rewrite the Henry's law expression as

$$\text{mass of } N_2 = p_{N_2} \times \text{mass of } H_2O \times K_{N_2}.$$

(1) At $p_{N_2} = 0.78 \times 4.0 \text{ atm} = 3.1 \text{ atm}$,

$$\text{mass of } N_2 = 3.1 \text{ atm} \times 100 \text{ g } H_2O \times 0.18 \,\mu\text{g } N_2/(\text{g } H_2O \text{ atm}) = \boxed{56 \,\mu\text{g } N_2}.$$

(2) At $p_{N_2} = 0.78 \text{ atm}$, mass of $N_2 = \boxed{14 \,\mu\text{g } N_2}$.

(3) In fatty tissue the increase in N_2 concentration from 1 atm to 4 atm is

$$4 \times (56 - 14) \,\mu\text{g } N_2 = \boxed{1.7 \times 10^2 \,\mu\text{g } N_2}.$$

P5.27   **(a)** i = 1 only, $N_1 = 4, K_1 = 1.0 \times 10^7 \text{ dm}^3 \text{ mol}^{-1}$,

$$\dfrac{\nu}{[A]} = \dfrac{4 \times 10 \text{ dm}^3 \,\mu\text{mol}^{-1}}{1 + 10 \text{ dm}^3 \mu\text{mol}^{-1} \times [A]}.$$

The plot is shown in Fig. 5.6(a).

(b) $i = 1; N_1 = 4, N_2 = 2; K_1 = 1.0 \times 10^5 \text{ dm}^3 \text{ mol}^{-1} = 0.10 \text{ dm}^3 \,\mu\text{mol}^{-1}$,

$K_2 = 2.0 \times 10^6 \text{ dm}^3 \text{ mol}^{-1} = 2.0 \text{ dm}^3 \,\mu\text{mol}^{-1}$.

$$\dfrac{\nu}{[A]} = \dfrac{4 \times 0.10 \text{ dm}^3 \mu\text{mol}^{-1}}{1 + 0.10 \text{ dm}^3 \,\mu\text{mol}^{-1} \times [A]} + \dfrac{2 \times 2.0 \text{ dm}^3 \,\mu\text{mol}^{-1}}{1 + 2.0 \text{ dm}^3 \,\mu\text{mol}^{-1} \times [A]}.$$

The plot is shown in Fig. 5.6(b).

P5.29   By the van't Hoff equation [5.40],

$$\Pi = [B]RT = \dfrac{cRT}{M}.$$

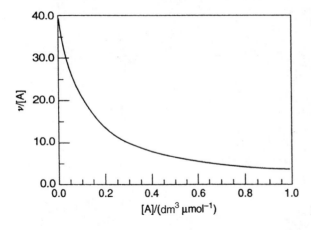

Figure 5.6(a)

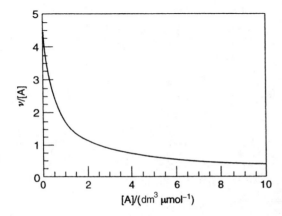

Figure 5.6(b)

Division by the standard acceleration of free fall, g, gives

$$\frac{\Pi}{g} = \frac{c(R/g)T}{M}.$$

(a) This expression may be written in the form

$$\Pi' = \frac{cR'T}{M},$$

which has the same form as the van't Hoff equation, but the unit of osmotic pressure (Π') is now

$$\frac{\text{force/area}}{\text{length/time}^2} = \frac{(\text{mass length})/(\text{area time}^2)}{\text{length/time}^2} = \frac{\text{mass}}{\text{area}}.$$

This ratio can be specified in g cm^{-2}. Likewise, the constant of proportionality (R') would have the units of R/g,

$$\frac{\text{energy K}^{-1}\text{ mol}^{-1}}{\text{length/time}^2} = \frac{(\text{mass length}^2/\text{time}^2)\text{ K}^{-1}\text{ mol}^{-1}}{\text{length/time}^2} = \text{mass length K}^{-1}\text{mol}^{-1}.$$

This result may be specified in $\boxed{\text{g cm K}^{-1}\text{ mol}^{-1}}$.

$$R' = \frac{R}{g} = \frac{8.314\,47\,\text{J K}^{-1}\text{ mol}^{-1}}{9.806\,65\,\text{m s}^{-2}}$$

$$= 0.847\,840\,\text{kg m K}^{-1}\text{ mol}^{-1}\left(\frac{10^3\,\text{g}}{\text{kg}}\right)\times\left(\frac{10^2\,\text{cm}}{\text{m}}\right)$$

$$\boxed{R' = 84784.0\ \text{g cm K}^{-1}\text{ mol}^{-1}}.$$

In the following we will drop the primes giving

$$\Pi = \frac{cRT}{M}$$

and use the Π units of g cm^{-2} and the R units $\text{g cm K}^{-1}\text{mol}^{-1}$.

(b) By extrapolating the low concentration plot of Π/c versus c (Fig. 5.7(a)) to $c = 0$ we find the intercept $230\,\text{g cm}^{-2}/\text{g cm}^{-3}$. In this limit the van't Hoff equation is valid so

$$\frac{RT}{M} = \text{intercept}\ \text{ or }\ M = \frac{RT}{\text{intercept}},$$

$$M = \frac{(84\,784.0\,\text{g cm K}^{-1}\text{ mol}^{-1})\times(298.15\,\text{K})}{(230\,\text{g cm}^{-2})/(\text{g cm}^{-3})},$$

$$\boxed{M = 1.1\times10^5\ \text{g mol}^{-1}}.$$

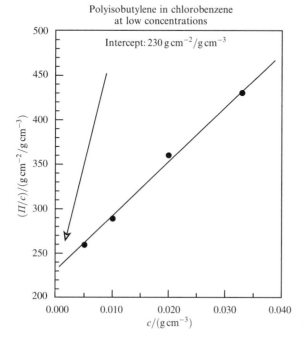

Figure 5.7(a)

(c) The plot of Π/c versus c for the full concentration range (Fig. 5.7(b)) is very nonlinear. We may conclude that the solvent is good. This may be due to the nonpolar nature of both solvent and solute.

(d) $\Pi/c = (RT/M)(1 + B'c + C'c^2)$.

Since RT/M has been determined in part **(b)** by extrapolation to $c = 0$, it is best to determine the second and third virial coefficients with the linear regression fit

$$\frac{(\Pi/c)/(RT/M) - 1}{c} = B' + C'c,$$

$R = 0.9791$.

$$\boxed{\begin{array}{ll} B' = 21.4\,\text{cm}^3\,\text{g}^{-1}, & \text{standard deviation} = 2.4\,\text{cm}^3\,\text{g}^{-1}. \\ C' = 211\text{cm}^6\,\text{g}^{-2}, & \text{standard deviation} = 15\,\text{cm}^6\,\text{g}^{-2}. \end{array}}$$

(e) Using 1/4 for g and neglecting terms beyond the second power, we may write

$$\left(\frac{\Pi}{c}\right)^{1/2} = \left(\frac{RT}{M}\right)^{1/2} \left(1 + \frac{1}{2}B'c\right).$$

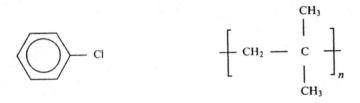

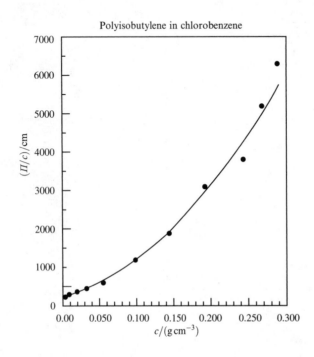

Polyisobutylene in chlorobenzene

Figure 5.7(b)

We can solve for B'; then $g(B')^2 = C'$,

$$\left[\frac{\left(\dfrac{\Pi}{c}\right)^{1/2}}{\left(\dfrac{RT}{M}\right)^{1/2}} \right] - 1 = \frac{1}{2}B'c.$$

RT/M has been determined above as 230 g cm^{-2}/g cm^{-3}. We may analytically solve for B' from one of the data points, say, $\Pi/c = 430$ g cm^{-2}/g cm^{-3} at $c = 0.033$ g cm^{-3}.

$$\left(\frac{430 \text{ g cm}^{-2}/\text{g cm}^{-3}}{230 \text{ g cm}^{-2}/\text{g cm}^{-3}} \right)^{1/2} - 1 = \frac{1}{2}B' \times (0.033 \text{ g cm}^{-3}).$$

$$B' = \frac{2 \times (1.367 - 1)}{0.033 \text{ g cm}^{-3}} = 22.\overline{2} \text{ cm}^3 \text{ g}^{-1}.$$

$$C' = g(B')^2 = 0.25 \times (22.\overline{2} \text{ cm}^3 \text{ g}^{-1})^2 = 12\overline{3} \text{ cm}^6 \text{ g}^{-2}.$$

Better values of B' and C' can be obtained by plotting $\left(\dfrac{\Pi}{c}\right)^{1/2} \Big/ \left(\dfrac{RT}{M}\right)^{1/2}$ against c. This plot is shown in Fig. 5.7(c). The slope is $14.\overline{03}$ cm^3 g^{-1}. $B' = 2 \times$ slope $= \boxed{28.\overline{0} \text{ cm}^3 \text{ g}^{-1}}$. C' is then $\boxed{19\overline{6} \text{ cm}^6 \text{ g}^{-2}}$. The intercept of this plot should theoretically be 1.00, but it is in fact 0.916 with a standard deviation of 0.066. The overall consistency of the values of the parameters confirms that g is roughly 1/4 as assumed.

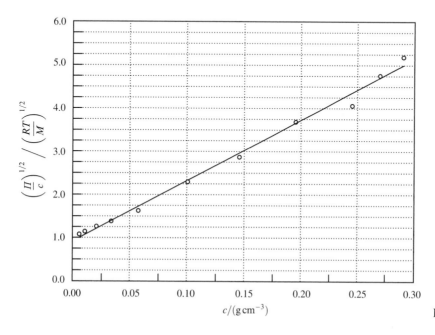

Figure 5.7(c)

6 Phase diagrams

Answers to discussion questions

D6.1   Phase: a state of matter that is uniform throughout, not only in chemical composition but also in physical state.

Constituent: any chemical species present in the system.

Component: a chemically independent constituent of the system. It is best understood in relation to the phrase 'number of components' which is the minimum number of independent species necessary to define the composition of all the phases present in the system.

Degree of freedom (or variance): the number of intensive variables that can be changed without disturbing the number of phases in equilibrium.

D6.3   See Figs. 6.1(a) and (b).

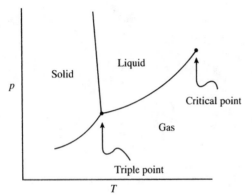

Figure 6.1(a)

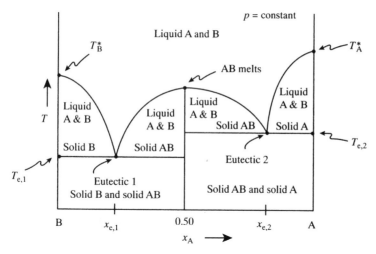

p = constant

Liquid A and B

T_B^*

AB melts

T_A^*

Liquid A & B

Liquid A & B

Liquid A & B

Liquid A & B

T

Liquid A & B

Solid AB   Solid A

$T_{e,2}$

Solid B

Solid AB

Eutectic 2

$T_{e,1}$

Eutectic 1
Solid B and solid AB

Solid AB and solid A

B $x_{e,1}$ 0.50 $x_{e,2}$ A

x_A ⟶

Figure 6.1(b)

D6.5   See Fig. 6.2.

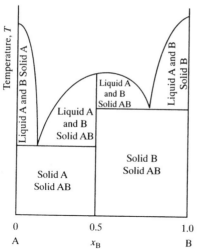

Temperature, T

Liquid A and B Solid A

Liquid A and B Solid B

Liquid A and B Solid AB

Liquid A and B Solid AB

Liquid A and B Solid AB

Solid B Solid AB

Solid A Solid AB

0   0.5   1.0
A x_B B   **Figure 6.2**

Solutions to exercises

E6.1(b) $p = p_A + p_B = x_A p_A^* + (1 - x_A) p_B^*$

$$x_A = \frac{p - p_B^*}{p_A^* - p_B^*}$$

$$x_A = \frac{19\,\text{kPa} - 18\,\text{kPa}}{20\,\text{kPa} - 18\,\text{kPa}} = \boxed{(0.5)} \quad \text{A is 1, 2-dimethylbenzene}$$

$$y_A = \frac{x_A p_A^*}{p_B^* + (p_A^* - p_B^*)x_A} = \frac{(0.5) \times (20\,\text{kPa})}{18\,\text{kPa} + (20\,\text{kPa} - 18\,\text{kPa})0.5} = 0.5\overline{26} \approx \boxed{0.5}$$

$$y_B = 1 - 0.5\overline{26} = 0.4\overline{74} \approx 0.5$$

E6.2(b) $\quad p_A = y_A p = 0.612p = x_A p_A^* = x_A (68.8\,\text{kPa})$

$$p_B = y_B p = (1 - y_A)p = 0.388p = x_B p_B^* = (1 - x_A) \times 82.1\,\text{kPa}$$

$$\frac{y_A p}{y_B p} = \frac{x_A p_A^*}{x_B p_B^*} \quad \text{and} \quad \frac{0.612}{0.388} = \frac{68.8 x_A}{82.1(1 - x_A)}$$

$$(0.388) \times (68.8)x_A = (0.612) \times (82.1) - (0.612)(82.1)x_A$$

$$26.6\overline{94}x_A = 50.2\overline{45} - 50.2\overline{45}x_A$$

$$x_A = \frac{50.2\overline{45}}{26.6\overline{94} + 50.2\overline{45}} = \boxed{0.653} \quad x_B = 1 - 0.653 = \boxed{0.347}$$

$$p = x_A p_A^* + x_B p_B^* = (0.653) \times (68.8\,\text{kPa}) + (0.347) \times (82.1\,\text{kPa}) = \boxed{73.4\,\text{kPa}}$$

E6.3(b)   (a) If Raoult's law holds, the solution is ideal.

$$p_A = x_A p_A^* = (0.4217) \times (110.1\,\text{kPa}) = 46.43\,\text{kPa}$$

$$p_B = x_B p_B^* = (1 - 0.4217) \times (94.93\,\text{kPa}) = 54.90\,\text{kPa}$$

$$p = p_A + p_B = (46.43 + 54.90)\,\text{kPa} = 101.33\,\text{kPa} = 1.000\,\text{atm}$$

Therefore, Raoult's law correctly predicts the pressure of the boiling liquid and the solution is ideal .

(b) $\quad y_A = \dfrac{p_A}{p} = \dfrac{46.43\,\text{kPa}}{101.33\,\text{kPa}} = \boxed{0.4582}$

$$y_B = 1 - y_A = 1.000 - 0.4582 = \boxed{0.5418}$$

E6.4(b)   Let B = benzene and T = toluene. Since the solution is equimolar $z_B = z_T = 0.500$

(a) Initially $x_B = z_B$ and $x_T = z_T$; thus

$$p = x_B p_B^* + x_T p_T^* \quad [6.3] = (0.500) \times (9.9\,\text{kPa}) + (0.500) \times (2.9\,\text{kPa})$$

$$= 4.9\overline{5}\,\text{kPa} + 1.4\overline{5}\,\text{kPa} = \boxed{6.4 \ \text{kPa}}$$

(b) $\quad y_B = \dfrac{p_B}{p} \quad [6.4] = \dfrac{4.9\overline{5}\,\text{kPa}}{6.4\,\text{kPa}} = \boxed{0.77} \quad y_T = 1 - 0.77 = \boxed{0.23}$

(c) Near the end of the distillation

$$y_B = z_B = 0.500 \quad \text{and} \quad y_T = z_T = 0.500$$

Equation 6.5 may be solved for x_A [A = benzene = B here]

$$x_B = \frac{y_B p_T^*}{p_B^* + \left(p_T^* - p_B^*\right) y_B} = \frac{(0.500) \times (2.9 \, \text{kPa})}{(9.9 \, \text{kPa}) + (2.9 - 9.9) \, \text{kPa} \times (0.500)} = 0.23$$

$$x_T = 1 - 0.23 = 0.77$$

This result for the special case of $z_B = z_T = 0.500$ could have been obtained directly by realizing that

$$y_B \, (\text{initial}) = x_T \, (\text{final}); \, y_T \, (\text{initial}) = x_B \, (\text{final})$$

$$p(\text{final}) = x_B p_B^* + x_T p_T^* = (0.23) \times (9.9 \, \text{kPa}) + (0.77) \times (2.9 \, \text{kPa}) = \boxed{4.5 \, \text{kPa}}$$

Thus in the course of the distillation the vapor pressure fell from 6.4 kPa to 4.5 kPa

E6.5(b)   See the phase diagram in Figure 6.3.

(a) $y_A = \boxed{0.81}$

(b) $x_A = \boxed{0.67}$ $y_A = \boxed{0.925}$

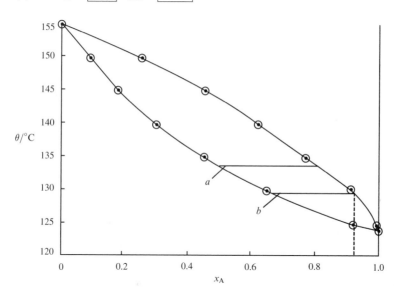

Figure 6.3

E6.6(b) Al^{3+}, H^+, $AlCl_3$, $Al(OH)_3$, OH^-, Cl^-, H_2O giving seven species. There are also three equilibria

$$AlCl_3 + 3H_2O \rightleftharpoons Al(OH)_3 + 3HCl$$

$$AlCl_3 \rightleftharpoons Al^{3+} + 3Cl^-$$

$$H_2O \rightleftharpoons H^+ + OH^-$$

and one condition of electrical neutrality

$$[H^+] + 3[Al^{3+}] = [OH^-] + [Cl^-]$$

Hence, the number of independent components is

$$C = 7 - (3 + 1) = \boxed{3}$$

E6.7(b) $NH_4Cl(s) \rightleftharpoons NH_3(g) + HCl(g)$

(a) For this system $\boxed{C = 1}$ [Example 6.1] and $\boxed{P = 2}$ (s and g).

(b) If ammonia is added before heating, $\boxed{C = 2}$ (because NH_4Cl, NH_3 are now independent) and $\boxed{P = 2}$ (s and g).

E6.8(b) (a) Still $\boxed{C = 2}$ (Na_2SO_4, H_2O), but now there is no solid phase present, so $\boxed{P = 2}$ (liquid solution, vapor).

(b) The variance is $F = 2 - 2 + 2 = \boxed{2}$. We are free to change any two of the three variables, amount of dissolved salt, pressure, or temperature, but not the third. If we change the amount of dissolved salt and the pressure, the temperature is fixed by the equilibrium condition between the two phases.

E6.9(b) See Figure 6.4.

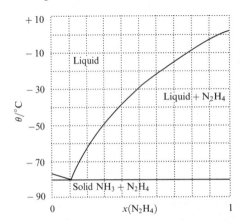

Figure 6.4

E6.10(b) See Figure 6.5. The phase diagram should be labeled as in figure 6.5. (a) Solid Ag with dissolved Sn begins to precipitate at a_1, and the sample solidifies completely at a_2. (b) Solid Ag with dissolved Sn begins to precipitate at b_1, and the liquid becomes richer in Sn. The peritectic reaction occurs at b_2, and

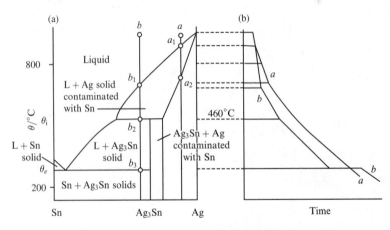

Figure 6.5

as cooling continues Ag_3Sn is precipitated and the liquid becomes richer in Sn. At b_3 the system has its eutectic composition (e) and freezes without further change.

E6.11(b)   See Figure 6.6. The feature denoting incongruent melting is circled. Arrows on the tie line indicate the decomposition products. There are two eutectics: one at $x_B = \boxed{0.53}$, $T = \boxed{T_2}$; another at $x_B = \boxed{0.82}$, $T = \boxed{T_3}$.

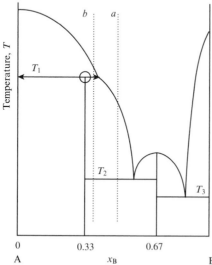

Figure 6.6

E6.12(b)   The cooling curves corresponding to the phase diagram in Figure 6.7(a) are shown in Figure 6.7(b). Note the breaks (abrupt change in slope) at temperatures corresponding to points $a_1, b_1,$ and b_2. Also note the eutectic halts at a_2 and b_3.

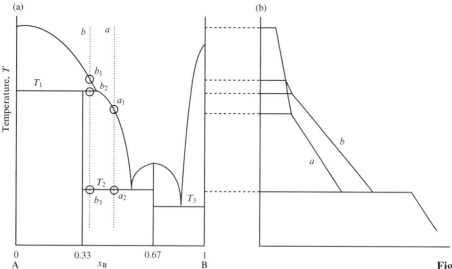

Figure 6.7

E6.13(b)   Rough estimates based on Figure 6.41 of the text are

(a) $x_B \approx \boxed{0.75}$ (b) $x_{AB_2} \approx \boxed{0.8}$ (c) $x_{AB_2} \approx \boxed{0.6}$

E6.14(b)   The phase diagram is shown in Figure 6.8. The given data points are circled. The lines are schematic at best.

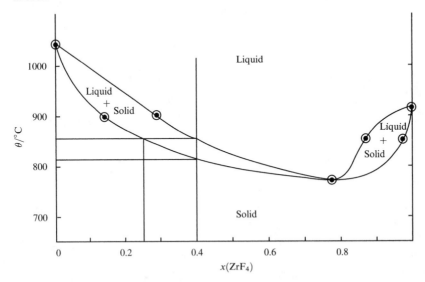

Figure 6.8

A solid solution with $x(ZrF_4) = 0.24$ appears at 855 °C. The solid solution continues to form, and its ZrF_4 content increases until it reaches $x(ZrF_4) = 0.40$ and 820 °C. At that temperature, the entire sample is solid.

E6.15(b)   The phase diagram for this system (Figure 6.9) is very similar to that for the system methyl ethyl ether and diborane of Exercise 6.9(a). The regions of the diagram contain analogous substances. The solid compound begins to crystallize at 120 K. The liquid becomes progressively richer in diborane until the liquid composition reaches 0.90 at 104 K. At that point the liquid disappears as heat is removed. Below 104 K the system is a mixture of solid compound and solid diborane.

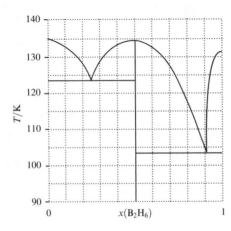

Figure 6.9

E6.16(b)   Refer to the phase diagram in the solution to Exercise 6.14(a). The cooling curves are sketched in Figure 6.10.

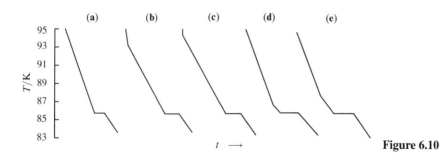

Figure 6.10

E6.17(b)   **(a)** When x_A falls to 0.47, a second liquid phase appears. The amount of new phase increases as x_A falls and the amount of original phase decreases until, at $x_A = 0.314$, only one liquid remains.

(b) The mixture has a single liquid phase at all compositions.
The phase diagram is sketched in Figure 6.11.

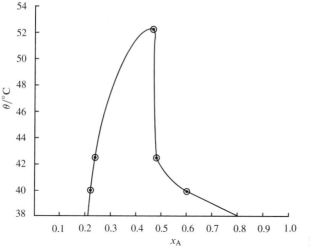

Figure 6.11

Solutions to problems

Solutions to numerical problems

P6.1   **(a)** The data, including that for pure chlorobenzene, are plotted in Fig. 6.12.

(b) The smooth curve through the x, T data crosses $x = 0.300$ at $\boxed{391.0 \text{ K}}$, the boiling point of the mixture.

(c) We need not interpolate data, for 393.94 K is a temperature for which we have experimental data. The mole fraction of 1-butanol in the liquid phase is 0.1700 and in the vapor phase 0.3691. According

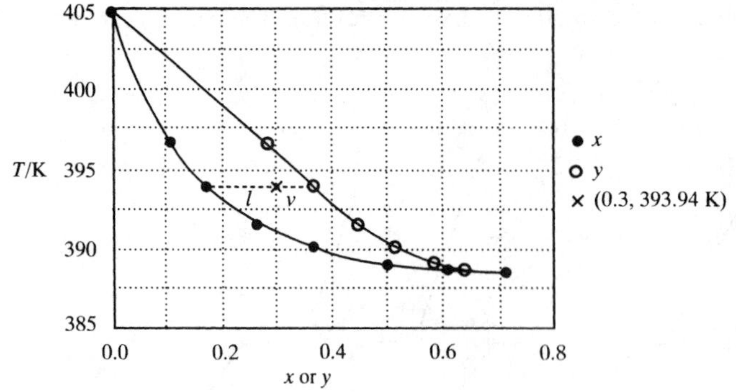

Figure 6.12

to the lever rule, the proportions of the two phases are in an inverse ratio of the distances their mole fractions are from the composition point in question. That is,

$$\frac{n_{\text{liq}}}{n_{\text{vap}}} = \frac{v}{l} = \frac{0.3691 - 0.300}{0.300 - 0.1700} = \boxed{0.532}.$$

P6.3

$$p_A = a_A p_A^* = \gamma_A x_A p_A^* \; [5.45].$$

$$\gamma_A = \frac{p_A}{x_A p_A^*} = \frac{y_A p}{x_A p_A^*}.$$

Sample calculation at 80 K:

$$\gamma_{O_2}(80 \text{ K}) = \frac{0.11(100 \text{ kPa})}{0.34(225 \text{ Torr})} \left(\frac{760 \text{ Torr}}{101.325 \text{ kPa}} \right),$$

$$\gamma_{O_2}(80 \text{ K}) = 1.079.$$

Summary:

T/K	77.3	78	80	82	84	86	88	90.2
γ_{O_2}	—	0.877	1.079	1.039	0.995	0.993	0.990	0.987

To within the experimental uncertainties the solution appears to be ideal ($\gamma = 1$). The low value at 78 K may be caused by nonideality; however, the larger relative uncertainty in $y(O_2)$ is probably the origin of the low value.

A temperature–composition diagram is shown in Fig. 6.13(a). The near ideality of this solution is, however, best shown in the pressure–composition diagram of Fig. 6.13(b). The liquid line is essentially a straight line as predicted for an ideal solution.

P6.5

A compound with $\boxed{\text{probable formula A}_3\text{B exists}}$. It melts incongruently at 700°C, undergoing the peritectic reaction

$$\text{A}_3\text{B}(s) \rightarrow \text{A}(s) + (\text{A} + \text{B}, \text{l}).$$

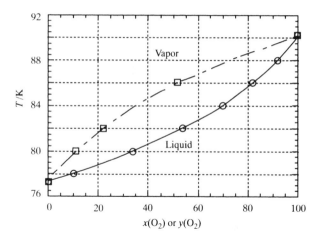

Figure 6.13(a)

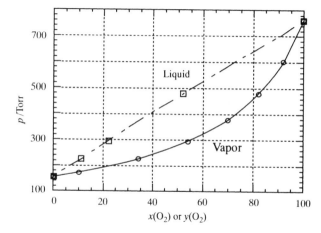

Figure 6.13(b)

The proportions of A and B in the product are dependent upon the overall composition and the temperature. A eutectic exists at 400°C and $x_B \approx 0.83$. See Fig. 6.14.

P6.7 The information has been used to construct the phase diagram in Fig. 6.15(a). In $MgCu_2$ the mass percentage of Mg is $(100) \times \dfrac{24.3}{24.3 + 127} = \boxed{16}$, and in Mg_2Cu it is $(100) \times \dfrac{48.6}{48.6 + 63.5} = \boxed{43}$. The initial point is a_1, corresponding to a liquid single-phase system. At a_2 (at 720°C) $MgCu_2$ begins to come out of solution and the liquid becomes richer in Mg, moving toward e_2. At a_3 there is solid $MgCu_2$ + liquid of composition e_2 (33 per cent by mass of Mg). This solution freezes without further change. The cooling curve will resemble that shown in Fig. 6.15(b).

P6.9 **(a)** $\boxed{\textbf{Eutectic}: 40.2 \text{ at\% Si at } 1268°C}$, $\boxed{\textbf{Eutectic}: 69.4 \text{ at\% Si at } 1030°C}$.

$\boxed{\begin{array}{l}\textbf{Congruent melting compounds}: Ca_2Si \text{ mp} = 1314°C \\ \qquad\qquad\qquad\qquad\qquad\quad CaSi \text{ mp} = 1324°C\end{array}}$

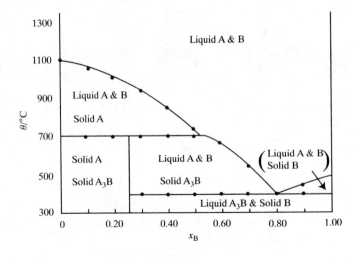

Figure 6.14

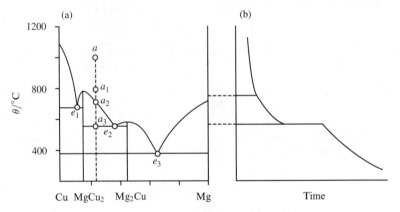

Figure 6.15

Incongruent melting compound: $CaSi_2$ mp $= 1040°C$ melts into $CaSi(s)$ and liquid (68 at% Si).

(b) At $1000°C$ the phases at equilibrium will be $Ca(s)$ and liquid (13 at% Si). The **lever rule** gives the relative amounts:

$$\frac{n_{Ca}}{n_{liq}} = \frac{l_{liq}}{l_{Ca}} = \frac{0.2 - 0}{0.2 - 0.13} = \boxed{2.86}.$$

(c) When an 80 at% Si melt it cooled in a manner that maintains equilibrium, Si(s) begins to appear at about $1250°C$. Further cooling causes more Si(s) to freeze out of the melt so that the melt becomes more concentrated in Ca. There is a 69.4 at% Si eutectic at $1030°C$. Just before the eutectic is reached, the lever rule says that the relative amounts of the Si(s) and liquid (69.4% Si) phases are:

$$\frac{n_{Si}}{n_{liq}} = \frac{l_{liq}}{l_{Si}} = \frac{0.80 - 0.694}{1.0 - 0.80} = \boxed{0.53 = \text{relative amounts at T slightly higher than } 1030°C}.$$

Just before $1030°C$, the Si(s) is 34.6 mol% of the total heterogeneous mixture, the eutectic liquid is 65.4 mol%.

At the eutectic temperature a third phase appears—$CaSi_2(s)$. As the melt cools at this temperature, both $Si(s)$ and $CaSi_2(s)$ freeze out of the melt while the concentration of the melt remains constant. At a temperature slightly below 1030°C, all the melt will have frozen to $Si(s)$ and $CaSi_2(s)$ with the relative amounts:

$$\frac{n_{Si}}{n_{CaSi_2}} = \frac{l_{CaSi_2}}{l_{Si}} = \frac{0.80 - 0.667}{1.0 - 0.80}$$

$$= \boxed{0.665 = \text{relative amounts of } T \text{ slightly higher than } 1030°C}.$$

Just under 1030°C, the $Si(s)$ is 39.9 mol% of the total heterogeneous mixture; the $CaSi_2(s)$ is 60.1 mol%.

A graph of mol% $Si(s)$ and mol% $CaSi_2$ (s) vs. mol% eutectic liquid is a convenient way to show relative amounts of the three phases as the eutectic liquid freezes. See Fig. 6.16. Equations for the graph are derived with the law of conservation of mass. For the silicon mass,

$$nz_{Si} = n_{liq}w_{Si} + n_{Si}x_{Si} + n_{CaSi_2}y_{Si}$$

where n = total number of moles.

w_{Si} = Si fraction in eutectic liquid = 0.694
x_{Si} = Si fraction in Si(s) = 1.000
y_{Si} = Si fraction in $CaSi_2(s)$ = 0.667
z_{Si} = Si fraction in melt = 0.800

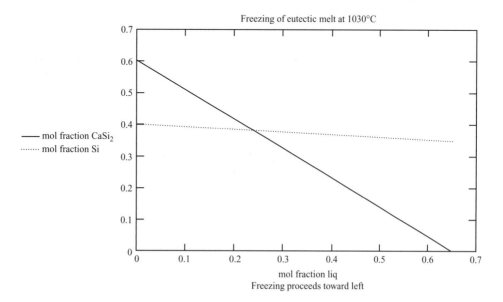

Freezing of eutectic melt at 1030°C

— mol fraction $CaSi_2$
······· mol fraction Si

mol fraction liq
Freezing proceeds toward left

Figure 6.16

This equation may be rewritten in mole fractions of each phase by dividing by n:

$$z_{Si} = (\text{mol fraction liq})w_{Si} + (\text{mol fraction Si})x_{Si} + (\text{mol fraction CaSi}_2)y_{Si}.$$

Since, (mol fraction liq) + (mol fraction Si) + (mol fraction $CaSi_2$) = 1
or (mol fraction $CaSi_2$) = 1 − (mol fraction liq + mol fraction Si), we may write:

$$z_{Si} = (\text{mol fraction liq})w_{Si} + (\text{mol fraction Si})x_{Si}$$
$$+ [1 - (\text{mol fraction liq} + \text{mol fraction Si})]y_{Si}.$$

Solving for mol fraction Si:

$$\text{mol fraction Si} := \frac{(z_{Si} - y_{Si}) - (w_{Si} - y_{Si})(\text{mol fraction liq})}{x_{Si} - y_{Si}},$$

$$\text{mol fraction CaSi}_2 := 1 - (\text{mol fraction liq} + \text{mol fraction Si}).$$

These two eqns are used to prepare plots of the mol fraction of Si and mol fraction of $CaSi_2$ against the mol fraction of the melt in the range 0–0.65.

Solutions to theoretical problems

P6.11   The general condition of equilibrium in an isolated system is $dS = 0$. Hence, if α and β constitute an isolated system, which are in thermal contact with each other

$$dS = dS_\alpha + dS_\beta = 0. \tag{a}$$

Entropy is an additive property and may be expressed in terms of U and V.

$$S = S(U, V).$$

The implication of this problem is that energy in the form of heat may be transferred from one phase to another, but that the phases are mechanically rigid, and hence their volumes are constant. Thus, $dV = 0$, and

$$dS = \left(\frac{\partial S_\alpha}{\partial U_\alpha}\right)_V dU_\alpha + \left(\frac{\partial S_\beta}{\partial U_\beta}\right)_V dU_\beta = \frac{1}{T_\alpha}dU_\alpha + \frac{1}{T_\beta}dU_\beta \text{ [3.45]}.$$

But, $dU_\alpha = -dU_\beta$; therefore $\dfrac{1}{T_\alpha} = \dfrac{1}{T_\beta}$ or $\boxed{T_\alpha = T_\beta}$.

Solutions to applications

P6.13   (i) Below a denaturant concentration of 0.1 only the native and unfolded forms are stable.

(ii) At denaturant concentration of 0.15 only the native form is stable below a temperature of about 0.70. At temperature 0.70 the native and molten-globule forms are at equilibrium. Heating above 0.70 causes all native forms to become molten-globules. At temperature 0.90, equilibrium between molten-globule and unfolded protein is observed and above this temperature only the unfolded form is stable.

P6.15 $C = 1;$ hence, $F = C - P + 2 = 3 - P.$

Since the tube is sealed there will always be some gaseous compound in equilibrium with the condensed phases. Thus when liquid begins to form upon melting, $P = 3$ (s, l, and g) and $F = 0$, corresponding to a definite melting temperature. At the transition to a normal liquid, $P = 3$ $(l, l',$ and g) as well, so again $F = 0$.

P6.17   To examine the process of zone levelling with the phase diagram below, Fig. 6.17, consider a solid on the isopleth through a_1 and heat the sample without coming to overall equilibrium. If the temperature rises to a_2, a liquid of composition b_2 forms and the remaining solid is at a_2'. Heating that solid down an isopleth passing through a_2' forms a liquid of composition b_3 and leaves the solid at a_3'. This sequence of heater passes shows that in a pass the impurities at the end of a sample are reduced while being transferred to the liquid phase which moves with the heater down the length of the sample. With enough passes the dopant, which is initially at the end of the sample, is distributed evenly throughout.

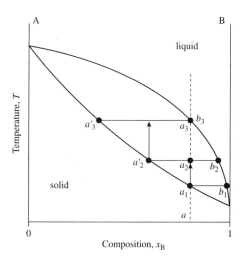

Figure 6.17

P6.19   The data are plotted in Fig. 6.18.

(a) As the solid composition $x(MgO) = 0.3$ is heated, liquid begins to form when the solid (lower) line is reached $\boxed{\text{at } 2150°C}$.

(b) From the tie line at 2200°C, the liquid composition is $y(MgO) = \boxed{0.18}$ and the solid $x(MgO) = \boxed{0.35}$.
   The proportions of the two phases are given by the lever rule,

$$\frac{l_1}{l_2} = \frac{n(\text{liq})}{n(\text{sol})} = \frac{0.05}{0.12} = \boxed{0.4}.$$

(c) Solidification begins at point c, corresponding to $\boxed{2640°C}$.

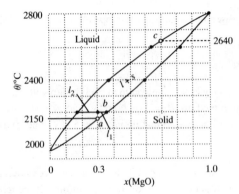

Figure 6.18

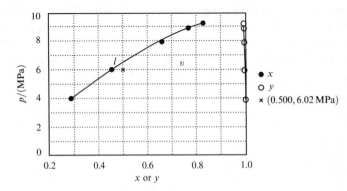

Figure 6.19

P6.21   (a) The data are plotted in Fig. 6.19.

   (b) We need not interpolate data, for 6.02 MPa is a pressure for which we have experimental data. The mole fraction of CO_2 in the liquid phase is 0.4541 and in the vapor phase 0.9980. The proportions of the two phases are in an inverse ratio of the distance their mole fractions are from the composition point in question, according to the lever rule

$$\frac{n_{\text{liq}}}{n_{\text{vap}}} = \frac{v}{l} = \frac{0.9980 - 0.5000}{0.5000 - 0.4541} = \boxed{10.85}.$$

7 Chemical equilibrium

Answers to discussion questions

D7.1   The position of equilibrium is always determined by the condition that the reaction quotient, Q must equal the equilibrium constant, K. If the mixing in of an additional amount of reactant or product destroys that equality, then the reacting system will shift in such a way as to restore the equality. That implies that some of the added reactant or product must be removed by the reacting system and the amounts of other components will also be affected. These adjustments restore the concentrations to their (new) equilibrium values.

D7.3   **(1)** Response to change in pressure. The equilibrium constant is independent of pressure, but the individual partial pressures can change as the total pressure changes. This will happen when there is a difference, Δn_g, between the sums of the number of moles of gases on the product and reactant sides of the chemical equation. The requirement of an unchanged equilibrium constant implies that the side with the smaller number of moles of gas be favored as pressure increases.

(2) Response to change in temperature. Equation 7.23a shows that K decreases with increasing temperature when the reaction is exothermic; thus the reaction shifts to the left, the opposite occurs in endothermic reactions. See Section 7.4 (a) for a more detailed discussion.

D7.5   **(a)** Consider the metals M and Z, which, for the sake of simplifying discussion, form 1:1 oxides having the formulas MO and ZO. Z will spontaneously reduce MO provided that the ZO line upon the Ellingham diagram lies above the MO line (this statement assumes that the vertical $\Delta_r G$ axis decreases upward). In this case the standard Gibbs energy for the reaction $MO(s) + Z(s) \rightarrow M(s) + ZO(s)$ will be negative. Figure 7.10 of the text indicates that Fe will reduce PbO, CuO, and Ag_2O.

(b) Using $\Delta_f G^\ominus(ZnO) = -318 \text{ kJ mol}^{-1}$ at 25°C (Table 2.7) and a slope that is common for all the oxides, we may add the approximate line for ZnO in the Ellingham diagram as shown in Fig.7.1. The ZnO curve passes under the reaction (iii) curve at about 1300°C so that is the estimate of the lowest temperature at which zinc oxide can be reduced to the metal by carbon. See Fig. 7.1.

D7.7   Electrode combinations that produce identical cell compartments with differing concentrations only (electrolyte concentration cells) have a cell potential dependence upon the liquid junction potential and the concentration difference. If the cell has identical compartments with either gaseous or amalgam electrodes (electrode concentration cell), the cell potential will depend upon the gas pressure differences or the amalgam concentration differences but will not have a liquid junction potential. Other electrode combinations produce cells for which the cell potential depends upon the half-reaction reduction potentials.

D7.9   The pH of an aqueous solution can in principle be measured with any electrode having an emf that is sensitive to $H^+(aq)$ concentration (activity). In principle, the hydrogen gas electrode is the simplest

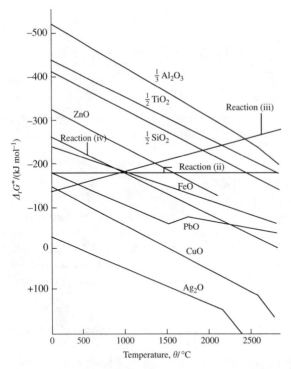

Figure 7.1

and most fundamental. A cell is constructed with the hydrogen electrode being the right-hand electrode and any reference electrode with known potential as the left-hand electrode. A common choice is the saturated calomel electrode. The pH can then be obtained by measuring the emf (zero-current potential difference), E, of the cell. The hydrogen gas electrode is not convenient to use, so in practice glass electrodes are used because of ease of handling.

Solutions to exercises

E7.1(b) $N_2O_4(g) \rightleftharpoons 2NO_2(g)$

Amount at equilibrium $(1 - \alpha)n$ $2\alpha n$

Mole fraction $\dfrac{1 - \alpha}{1 + \alpha}$ $\dfrac{2\alpha}{1 + \alpha}$

Partial pressure $\dfrac{(1 - \alpha)P}{1 + \alpha}$ $\dfrac{2\alpha P}{1 + \alpha}$

Assuming that the gases are perfect, $a_J = \dfrac{p_J}{p^{\ominus}}$

$$K = \frac{(p_{NO_2}/p^{\ominus})^2}{(p_{N_2O_4}/p^{\ominus})} = \frac{4\alpha^2 p}{(1 - \alpha^2)p^{\ominus}}$$

For $p = p^{\ominus}, K = \dfrac{4\alpha^2}{1 - \alpha^2}$

(a) $\boxed{\Delta_r G = 0}$ at equilibrium

(b) $\alpha = 0.201$ $K = \dfrac{4(0.201)^2}{1 - 0.201^2} = \boxed{0.16841}$

(c) $\Delta_r G^\ominus = -RT \ln K = -(8.314 \, \text{J K}^{-1} \, \text{mol}^{-1}) \times (298 \, \text{K}) \times \ln(0.168\overline{41})$

$$= \boxed{4.41 \, \text{kJ mol}^{-1}}$$

E7.2(b)   (a) $Br_2(g) \rightleftharpoons 2Br(g)$ $\alpha = 0.24$

Amount at equilibrium $(1 - \alpha)n$ $2\alpha n$

Mole fraction $\dfrac{1 - \alpha}{1 + \alpha}$ $\dfrac{2\alpha}{1 + \alpha}$

Partial pressure $\dfrac{(1 - \alpha)P}{1 + \alpha}$ $\dfrac{2\alpha P}{1 + \alpha}$

Assuming both gases are perfect $a_J = \dfrac{p_J}{p^\ominus}$

$$K = \frac{(p_{Br}/p^\ominus)^2}{p_{Br_2}/p^\ominus} = \frac{4\alpha^2 p}{(1 - \alpha^2)p^\ominus} = \frac{4\alpha}{1 - \alpha} \quad [p = p^\ominus]$$

$$= \frac{4(0.24)^2}{1 - (0.24)^2} = 0.24\overline{45} = \boxed{0.24}$$

(b) $\Delta_r G^\ominus = -RT \ln K = -(8.314 \, \text{J K}^{-1} \, \text{mol}^{-1}) \times (1600 \, \text{K}) \times \ln(0.24\overline{45})$

$$\boxed{= 19 \, \text{kJ mol}^{-1}}$$

(c) $\ln K(2273 \, \text{K}) = \ln K(1600 \, \text{K}) - \dfrac{\Delta_r H^\ominus}{R}\left(\dfrac{1}{2273 \, \text{K}} - \dfrac{1}{1600 \, \text{K}}\right)$

$$= \ln(0.24\overline{45}) - \left(\frac{112 \times 10^3 \, \text{mol}^{-1}}{8.314 \, \text{J K}^{-1} \, \text{mol}^{-1}}\right) \times (-1.851 \times 10^{-4})$$

$$= 1.08\overline{4}$$

$$K(2273 \, \text{K}) = e^{1.08\overline{4}} = \boxed{2.96}$$

E7.3(b) $\nu(CHCl_3) = 1, \quad \nu(HCl) = 3, \quad \nu(CH_4) = -1, \quad \nu(Cl_2) = -3$

(a) $\Delta_r G^\ominus = \Delta_f G^\ominus(CHCl_3, l) + 3\Delta_f G^\ominus(HCl, g) - \Delta_f G^\ominus(CH_4, g)$

$$= (-73.66 \, \text{kJ mol}^{-1}) + (3) \times (-95.30 \, \text{kJ mol}^{-1}) - (-50.72 \, \text{kJ mol}^{-1})$$

$$= \boxed{-308.84 \, \text{kJ mol}^{-1}}$$

$$\ln K = -\frac{\Delta_r G^\ominus}{RT}[7.8] = \frac{-(-308.84 \times 10^3 \, \text{J mol}^{-1})}{(8.3145 \, \text{J K}^{-1} \, \text{mol}^{-1}) \times (298.15 \, \text{K})} = 124.58\overline{4}$$

$$K = \boxed{1.3 \times 10^{54}}$$

(b) $\Delta_r H^\ominus = \Delta_f H^\ominus(CHCl_3, l) + 3\Delta_f H^\ominus(HCl, g) - \Delta_f H^\ominus(CH_4, g)$

$$= (-134.47\,kJ\,mol^{-1}) + (3) \times (-92.31\,kJ\,mol^{-1}) - (-74.81\,kJ\,mol^{-1})$$

$$= -336.59\,kJ\,mol^{-1}$$

$$\ln K(50\,^\circ C) = \ln K(25\,^\circ C) - \frac{\Delta_r H^\ominus}{R}\left(\frac{1}{323.2\,K} - \frac{1}{298.2\,K}\right)\;[7.25]$$

$$= 124.58\bar{4} - \left(\frac{-336.59 \times 10^3\,J\,mol^{-1}}{8.3145\,J\,K^{-1}\,mol^{-1}}\right) \times (-2.594 \times 10^{-4}\,K^{-1}) = 114.08\bar{3}$$

$$K(50\,^\circ C) = \boxed{3.5 \times 10^{49}}$$

$$\Delta_r G^\ominus(50\,^\circ C) = -RT\ln K(50\,^\circ C)\;[7.17] = -(8.3145\,J\,K^{-1}\,mol^{-1}) \times (323.15\,K) \times (114.08\bar{3})$$

$$= \boxed{-306.52\,kJ\,mol^{-1}}$$

E7.4(b)   Draw up the following table.

	A	+	B	$\rightleftharpoons$	C	+	2D	Total
Initial amounts/mol	2.00		1.00		0		3.00	6.00
Stated change/mol					+0.79			
Implied change/mol	−7.09		−7.09		+7.09		+1.58	
Equilibrium amounts/mol	1.21		0.21		0.79		4.58	6.79
Mole fractions	0.178$\bar{2}$		0.030$\bar{2}$		0.116$\bar{2}$		0.674$\bar{2}$	0.9999

(a) Mole fractions are given in the table.

(b) $K_x = \prod_J x_J^{\nu_J},$

$$K_x = \frac{(0.116\bar{3}) \times (0.674\bar{5})^2}{(0.178\bar{2}) \times (0.030\bar{9})} = \boxed{9.6}$$

(c) $p_J = x_J p$. Assuming the gases are perfect, $a_J = p_J/p^\ominus$, so

$$K = \frac{(p_C/p^\ominus) \times (p_D/p^\ominus)^2}{(p_A/p^\ominus) \times (p_B/p^\ominus)} = K_x\left(\frac{p}{p^\ominus}\right) = K_x \quad \text{when } p = 1.00\,bar$$

$$K = K_x = \boxed{9.6}$$

(d) $\Delta_r G^\ominus = -RT\ln K = -(8.314\,J\,K^{-1}\,mol^{-1}) \times (298\,K) \times \ln(9.60\bar{9})$

$$= \boxed{-5.6\,kJ\,mol^{-1}}$$

E7.5(b)   At 1120 K, $\Delta_r G^{\ominus} = +22 \times 10^3 \, \text{J mol}^{-1}$

$$\ln K(1120\,\text{K}) = \frac{\Delta_r G^{\ominus}}{RT} = -\frac{(22 \times 10^3 \, \text{J mol}^{-1})}{(8.314 \, \text{J K}^{-1} \, \text{mol}^{-1}) \times (1120\,\text{K})} = -2.3\overline{63}$$

$$K = e^{-2.3\overline{63}} = 9.\overline{41} \times 10^{-2}$$

$$\ln K_2 = \ln K_1 - \frac{\Delta_r H^{\ominus}}{R}\left(\frac{1}{T_2} - \frac{1}{T_1}\right)$$

Solve for T_2 at $\ln K_2 = 0$ ($K_2 = 1$)

$$\frac{1}{T_2} = \frac{R \ln K_1}{\Delta_r H^{\ominus}} + \frac{1}{T_1} = \frac{(8.314 \, \text{J K}^{-1} \, \text{mol}^{-1}) \times (-2.3\overline{63})}{(125 \times 10^3 \text{J mol}^{-1})} + \frac{1}{1120\,\text{K}} = 7.3\overline{6} \times 10^{-4}$$

$$T_2 = \boxed{1.4 \times 10^3 \, \text{K}}$$

E7.6(b)   Use $\dfrac{d(\ln K)}{d(1/T)} = \dfrac{-\Delta_r H^{\ominus}}{R}$

We have $\ln K = -2.04 - 1176\,\text{K}\left(\dfrac{1}{T}\right) + 2.1 \times 10^7 \, \text{K}^3 \left(\dfrac{1}{T}\right)^3$

$$-\frac{\Delta_r H^{\ominus}}{R} = -1176\,\text{K} + (2.1 \times 10^7 \, \text{K}^3) \times 3\left(\frac{1}{T}\right)^2$$

$T = 450\,\text{K}$ so

$$-\frac{\Delta_r H^{\ominus}}{R} = -1176\,\text{K} + (2.1 \times 10^7 \, \text{K}^3) \times 3\left(\frac{1}{450\,\text{K}}\right)^2 = -86\overline{5}\,\text{K}$$

$$\Delta_r H^{\ominus} = +(86\overline{5}\,\text{K}) \times (8.314 \, \text{J mol}^{-1} \, \text{K}^{-1}) = \boxed{7.1\overline{91}\,\text{kJ mol}^{-1}}$$

Find $\Delta_r S^{\ominus}$ from $\Delta_r G^{\ominus}$

$$\Delta_r G^{\ominus} = -RT \ln K$$

$$= -(8.314 \, \text{J K}^{-1} \, \text{mol}^{-1}) \times (450\,\text{K}) \times \left\{-2.04 - \frac{1176\,\text{K}}{450\,\text{K}} + \frac{2.1 \times 10^7 \, \text{K}^3}{(450\,\text{K})^3}\right\}$$

$$= 16.\overline{55}\,\text{kJ mol}^{-1}$$

$$\Delta_r G^{\ominus} = \Delta_r H^{\ominus} - T\Delta_r S^{\ominus}$$

$$\Delta_r S^{\ominus} = \frac{\Delta_r H^{\ominus} - \Delta_r G^{\ominus}}{T} = \frac{7.1\overline{91}\,\text{kJ mol}^{-1} - 16.5\overline{5}\,\text{kJ mol}^{-1}}{450\,\text{K}} = -20.\overline{79}\,\text{J K}^{-1} \, \text{mol}^{-1}$$

$$= \boxed{-21\,\text{J K}^{-1} \, \text{mol}^{-1}}$$

E7.7(b) $U(s) + \frac{3}{2}H_2(g) \rightleftharpoons UH_3(s), \quad \Delta_f G^{\ominus} = -RT \ln K$

At this low pressure, hydrogen is nearly a perfect gas, $a(H_2) = (p/p^{\ominus})$. The activities of the solids are 1.

Hence, $\ln K = \ln \left(\dfrac{p}{p^{\ominus}}\right)^{-3/2} = -\dfrac{3}{2}\ln \dfrac{p}{p^{\ominus}}$

$$\Delta_f G^{\ominus} = \frac{3}{2}RT \ln \frac{p}{p^{\ominus}}$$

$$= \left(\frac{3}{2}\right) \times (8.314\,\mathrm{J\,K^{-1}\,mol^{-1}}) \times (500\,\mathrm{K}) \times \ln\left(\frac{139\,\mathrm{Pa}}{1.00 \times 10^5\,\mathrm{Pa}}\right)$$

$$= \boxed{-41.0\,\mathrm{kJ\,mol^{-1}}}$$

E7.8(b) $\qquad K_x = \displaystyle\prod_J x_J{}^{\nu_J}$ [analogous to 7.16]

The relation of K_x to K is established in *Illustration 7.5*

$$K_x = \prod_J \left(\frac{p_J}{p^{\ominus}}\right)^{\nu_J} \left[7.16 \text{ with } a_J = \frac{p_J}{p^{\ominus}}\right]$$

$$= \prod_J x_J^{\nu_J} \left(\frac{p}{p^{\ominus}}\right)^{\sum_J \nu_J} [p_J = x_J p] = K_x \times \left(\frac{p}{p^{\ominus}}\right)^{\nu} \left[\nu \equiv \sum_J \nu_J\right]$$

Therefore, $K_x = K\left(p/p^{\ominus}\right)^{-\nu}$, $K_x \propto p^{-\nu}$ [K and $p^{\ominus}$ are constants]

$\nu = 1 + 1 - 1 - 1 = 0$, thus $\boxed{K_x(2\text{ bar}) = K_x(1\text{ bar})}$

E7.9(b) $\qquad N_2(g) + O_2(g) \rightleftharpoons 2NO(g) \quad K = 1.69 \times 10^{-3}$ at 2300 K

Initial moles $N_2 = \dfrac{5.0\,\mathrm{g}}{28.01\,\mathrm{g\,mol^{-1}}} = 0.23\overline{80}\,\mathrm{mol}\,N_2$

Initial moles $O_2 = \dfrac{2.0\,\mathrm{g}}{32.00\,\mathrm{g\,mol^{-1}}} = 6.2\overline{50} \times 10^{-2}\,\mathrm{mol}\,O_2$

	N_2	O_2	NO	Total
Initial amount/mol	$0.23\overline{80}$	$0.062\overline{5}$	0	0.300
Change/mol	$-z$	$-z$	$+2z$	0
Equilibrium amount/mol	$0.23\overline{80} - z$	$0.062\overline{5} - z$	$2z$	0.300
Mole fractions	$\dfrac{0.23\overline{80} - z}{0.300}$	$\dfrac{0.062\overline{5} - z}{0.300}$	$\dfrac{2z}{0.300}$	(1)

$$K = K_x \left(\frac{p}{p^{\ominus}}\right)^{\nu}\left[\nu = \sum_J \nu_J = 0\right], \text{ then}$$

$$K = K_x = \frac{(2z/0.300)^2}{\left(\dfrac{0.2380 - z}{0.300}\right) \times \left(\dfrac{0.0625 - z}{0.300}\right)}$$

$$= \frac{4z^2}{(0.2380 - z)(0.0625 - z)} = 1.69 \times 10^{-3}$$

$$4z^2 = 1.69 \times 10^{-3}\{0.014\overline{88} - 0.30\overline{05}z + z^2\}$$

$$= 2.5\overline{14} \times 10^{-5} - (5.0\overline{78} \times 10^{-4})z + (1.69 \times 10^{-3})z^2$$

$$4.00 - 1.69 \times 10^{-3} = 4.00 \quad \text{so}$$

$$4z^2 + (5.0\overline{78} \times 10^{-4})z - 2.5\overline{14} \times 10^{-5} = 0$$

$$z = \frac{-5.0\overline{78} \times 10^{-4} \pm \{(5.078 \times 10^{-4})^2 - 4 \times (4) \times (-2.514 \times 10^{-5})\}^{1/2}}{8}$$

$$= \frac{1}{8}(-5.078 \times 10^{-4} \pm 2.0\overline{06} \times 10^{-2})$$

$$z > 0 \quad [z < 0 \text{ is physically impossible}] \text{ so}$$

$$z = 2.4\overline{44} \times 10^{-3}$$

$$x_{NO} = \frac{2z}{0.300} = \frac{2(2.4\overline{44} \times 10^{-3})}{0.300} = \boxed{1.6 \times 10^{-2}}$$

E7.10(b) $\qquad \ln\dfrac{K'}{K} = \dfrac{\Delta_f H^{\ominus}}{R}\left(\dfrac{1}{T} - \dfrac{1}{T'}\right) \quad \text{so} \quad \Delta_f H^{\ominus} = \dfrac{R\ln\left(\dfrac{K'}{K}\right)}{\left(\dfrac{1}{T} - \dfrac{1}{T'}\right)}$

$$T = 310\,K, \quad T' = 325\,K; \quad \text{let} \frac{K'}{K} = \kappa$$

Now $\Delta_f H^{\ominus} = \dfrac{(8.314\,\text{J K}^{-1}\,\text{mol}^{-1})}{((1/310\,\text{K}) - (1/325\,\text{K}))} \times \ln\kappa = 55.\overline{84}\,\text{kJ mol}^{-1}\ln\kappa$

(a) $\kappa = 2 \qquad \Delta_f H^{\ominus} = (55.\overline{84}\,\text{kJ mol}^{-1}) \times (\ln 2) = \boxed{39\,\text{kJ mol}^{-1}}$

(b) $\kappa = \dfrac{1}{2} \qquad \Delta_r H^{\ominus} = (55.\overline{84}\,\text{kJ mol}^{-1}) \times (\ln\tfrac{1}{2}) = \boxed{-39\,\text{kJ mol}^{-1}}$

E7.11(b) $\qquad \text{NH}_4\text{Cl}(s) \rightleftharpoons \text{NH}_2(g) + \text{HCl}(g)$

$$p = p(\text{NH}_3) + p(\text{HCl}) = 2p(\text{NH}_3) \quad [p(\text{NH}_3) = p(\text{HCl})]$$

(a) $\qquad K = \displaystyle\prod_J a_J^{v_j}\ [7.16]; \qquad a(\text{gases}) = \dfrac{p_J}{p^{\ominus}}; \qquad a(\text{NH}_4\text{Cl, s}) = 1$

$$K = \left(\frac{p(\text{NH}_3)}{p^{\ominus}}\right) \times \left(\frac{p(\text{HCl})}{p^{\ominus}}\right) = \frac{p(\text{NH}_3)^2}{p^{\ominus 2}} = \frac{1}{4} \times \left(\frac{p}{p^{\ominus}}\right)^2$$

At $427\,^{\circ}\text{C}$ (700 K), $\quad K = \dfrac{1}{4} \times \left(\dfrac{608\,\text{kPa}}{100\,\text{kPa}}\right)^2 = \boxed{9.24}$

At $459\,^{\circ}\text{C}$ (732 K), $\quad K = \dfrac{1}{4} \times \left(\dfrac{1115\,\text{kPa}}{100\,\text{kPa}}\right)^2 = \boxed{31.08}$

(b) $\Delta_r G^\circ = -RT \ln K$ [7.8] $= (-8.314 \, \text{J K}^{-1}\text{mol}^{-1}) \times (700 \, \text{K}) \times (\ln 9.24)$

$$= \boxed{-12.9 \, \text{kJ mol}^{-1}} \quad (\text{at } 427^\circ \, \text{C})$$

(c) $\Delta_r H^\circ \approx \dfrac{R \ln(K'/K)}{(1/T - 1/T')}$ [7.25]

$$\approx \frac{(8.314 \, \text{J K}^{-1}\text{mol}^{-1}) \times \ln (31.08/9.24)}{(1/700 \, \text{K}) - (1/732 \, \text{K})} = \boxed{+161 \, \text{kJ mol}^{-1}}$$

(d) $\Delta_r S^\circ = \dfrac{\Delta_r H^\circ - \Delta_r G^\circ}{T} = \dfrac{(161 \, \text{kJ mol}^{-1}) - (-12.9 \, \text{kJ mol}^{-1})}{700 \, \text{K}} = \boxed{+248 \, \text{J K}^{-1}\, \text{mol}^{-1}}$

E7.12(b) The reaction is

$$\text{CuSO}_4 \cdot 5\text{H}_2\text{O}(s) \rightleftharpoons \text{CuSO}_4(s) + 5\text{H}_2\text{O}(g)$$

For the purposes of this exercise we may assume that the required temperature is that temperature at which $K=1$, which corresponds to a pressure of 1 bar for the gaseous products. For $K = 1, \ln K = 0$, and $\Delta_r G^\circ = 0$.

$$\Delta_r G^\circ = \Delta_r H^\circ - T \Delta_r S^\circ = 0 \quad \text{when } \Delta_r H^\circ = T \Delta_r S^\circ$$

Therefore, the decomposition temperature (when $K = 1$) is

$$T = \frac{\Delta_r H^\circ}{\Delta_r S^\circ}$$

$$\text{CuSO}_4 \cdot 5\text{H}_2\text{O}\,(s) \rightleftharpoons \text{CuSO}_4\,(s) + 5\text{H}_2\text{O}\,(g)$$

$$\Delta_r H^\circ = [(-771.36) + (5) \times (-241.82) - (-2279.7)] \, \text{kJ mol}^{-1} = +299.2 \, \text{kJ mol}^{-1}$$

$$\Delta_r S^\circ = [(109) + (5) \times (188.83) - (300.4)] \, \text{J K}^{-1} \, \text{mol}^{-1} = 752.\overline{8} \, \text{J K}^{-1} \, \text{mol}^{-1}$$

Therefore, $T = \dfrac{299.2 \times 10^3 \, \text{J mol}^{-1}}{752.8 \, \text{J K}^{-1} \, \text{mol}^{-1}} = \boxed{397 \, \text{K}}$

Question. What would the decomposition temperature be for decomposition defined as the state at which $K = 1/2$?

E7.13(b) $\text{PbI}_2(s) \rightleftharpoons \text{PbI}_2(\text{aq}) \qquad K_S = 1.4 \times 10^{-8}$

$$\Delta_r G^\circ = -RT \ln K_S = -(8.314 \, \text{J K}^{-1} \, \text{mol}^{-1}) \times (298.15 \, \text{K}) \times \ln\left(1.4 \times 10^{-8}\right)$$

$$= 44.83 \, \text{kJ mol}^{-1}$$

$$\Delta_r G^\circ = \Delta_f G^\circ \, (\text{PbI}_2, \text{aq}) - \Delta_f G^\circ \, (\text{PbI}_2, s)$$

$$\Delta_f G^\circ \, (\text{PbI}_2, \text{aq}) = \Delta_r G^\circ \Delta + \Delta_f G^\circ \, (\text{PbI}_2, s)$$

$$= 44.8\overline{3} \, \text{kJ mol}^{-1} - 173.64 \, \text{kJ mol}^{-1}$$

$$= \boxed{-128.8 \, \text{kJ mol}^{-1}}$$

E7.14(b)   The cell notation specifies the right and left electrodes. Note that for proper cancellation we must equalize the number of electrons in half-reactions being combined.

For the calculation of the standard emfs of the cells we have used $E^\ominus = E_R^\ominus - E_L^\ominus$, with standard electrode potentials from Table 7.2.

(a)   R : $Ag_2CrO_4(s) + 2e^- \rightarrow 2Ag(s) + CrO_4^{2-}(aq)$ +0.45 V

   L : $Cl_2(g) + 2e^- \rightarrow 2Cl^-(aq)$ +1.36 V

   Overall (R − L) : $Ag_2CrO_4(s) + 2Cl^-(aq) \rightarrow 2Ag(s) + CrO_4^{2-}(aq) + (Cl_2 g)$ −0.91 V

(b)   R : $Sn^{4+}(aq) + 2e^- \rightarrow Sn^{2+}(aq)$ +0.15 V

   L : $2Fe^{3+}(aq) + 2e^- \rightarrow 2Fe^{2+}(aq)$ +0.77 V

   Overall (R − L) : $Sn^{4+}(aq) + 2Fe^{2+}(aq) \rightarrow Sn^{2+}(aq) + 2Fe^{3+}(aq)$ −0.62 V

(c)   R : $MnO_2(s) + 4H^+(aq) + 2e^- \rightarrow Mn^{2+}(aq) + 2Fe^{3+}(aq)$ +1.23 V

   L : $Cu^{2+}(aq) + 2e^- \rightarrow Cu(s)$ +0.34 V

   Overall (R − L) : $Cu(s) + MnO_2(s) + 4H^+(aq) \rightarrow Cu^{2+}(aq) + Mn^{2+}(aq)$
                                                   $+2H_2O(l)$ +0.89 V

COMMENT.   Those cells for which $E^\ominus > 0$ may operate as spontaneous galvanic cells under standard conditions. Those for which $E^\ominus < 0$ may operate as nonspontaneous electrolytic cells. Recall that $E^\ominus$ informs us of the spontaneity of a cell under standard conditions only. For other conditions we require E.

E7.15(b)   The conditions (concentrations, etc.) under which these reactions occur are not given. For the purposes of this exercise we assume standard conditions. The specification of the right and left electrodes is determined by the direction of the reaction as written. As always, in combining half-reactions to form an overall cell reaction we must write half-reactions with equal number of electrons to ensure proper cancellation. We first identify the half-reactions, and then set up the corresponding cell.

(a) R : $2H_2O(l) + 2e^- \rightarrow 2OH^-(aq) + H_2(g)$ − 0.83 V

   L : $2Na^+(aq) + 2e^- \rightarrow 2Na(s)$ − 2.71 V

   and the cell is

   $Na(s)|Na^+(aq)|, OH^-(aq) |H_2(g) \, Pt$ $\boxed{+1.88 \text{ V}}$

   or more simply

   $\boxed{Na(s)|NaOH(aq)|H_2(g)|Pt}$

(b) R : $I_2(s) + 2e^- \rightarrow 2I^-(aq)$ + 0.54 V

   L : $2H^+(aq) + 2e^- \rightarrow H_2(g)$ 0

   and the cell is

   $Pt \,|H_2(g)| \, H^+(aq), I^-(aq) \,|I_2(s)| \, Pt$ $\boxed{+0.54 \text{ V}}$

   or more simply

   $\boxed{Pt|H_2(g)|HI(aq)| \, I_2(s)| \, Pt}$

(c) R : $\quad 2H^+(aq) + 2e^- \rightarrow H_2(g)$ $\qquad\qquad$ 0.00 V

$\quad$ L : $\quad 2H_2O(1) + 2e^- \rightarrow H_2(g) + 2OH^-(aq)$ $\qquad$ 0.083 V

and the cell is

$$Pt\ |H_2(g)|\ H^+(aq), OH^-(aq)|H_2\ (g)|Pt \qquad\qquad \boxed{0.083\ V}$$

or more simply

$$\boxed{Pt|H_2(g)|H_2O(l)|H_2(g)|Pt}$$

COMMENT. All of these cells have $E^\ominus > 0$, corresponding to a spontaneous cell reaction under standard conditions. If $E^\ominus$ had turned out to be negative, the spontaneous reaction would have been the reverse of the one given, with the right and left electrodes of the cell also reversed.

E7.16(b) **(a)** $\qquad E = E^\ominus - \dfrac{RT}{\nu F} \ln Q \quad \nu = 2$

$$Q = \prod_J a_J^{\nu_J} = a_{H^+}^2 a_{Cl^-}^2 \quad [\text{all other activities } = 1]$$

$$= a_+^2 a_-^2 = (\gamma_+ b_+)^2 \times (\gamma_- b_-)^2 \quad \left[b \equiv \frac{b}{b^\ominus} \text{here and below} \right]$$

$$= (\gamma_+ \gamma_-)^2 \times (b_+ b_-)^2 = \gamma_\pm^4 b^4 \quad [5.66,\ b_+ = b,\ b_- = b]$$

Hence, $E = E^\ominus - \dfrac{RT}{2F} \ln \left(\gamma_\pm^4 b^4 \right) = \boxed{E^\ominus - \dfrac{2RT}{F} \ln \left(\gamma_\pm b \right)}$

(b) $\quad \Delta_r G = -\nu F E\ [7.27] = -(2) \times \left(9.6485 \times 10^4\ C\,mol^{-1} \right) \times (0.4658\ V) = \boxed{-89.89\ kJ\,mol^{-1}}$

(c) $\quad \log \gamma_\pm = -|z_+ z_-| A I^{1/2}[5.69] = -(0.509) \times (0.010)^{1/2} \left[I = b \text{ for HCl(aq)} \right] = -0.0509$

$$\gamma_\pm = 0.889$$

$$E^\ominus = E + \frac{2RT}{F} \ln \left(\gamma_\pm b \right) = (0.4658\ V) + (2) \times \left(25.693 \times 10^{-3}\ V \right) \times \ln (0.889 \times 0.010)$$

$$= \boxed{+0.223\ V}$$

The value compares favorably to that given in Table 7.2.

E7.17(b) In each case $\ln K = \dfrac{\nu F E^\ominus}{RT}$ [7.30]

(a) $\quad Sn(s) + CuSO_4(aq) \rightleftharpoons Cu(s) + SnSO_4(aq)$

$\qquad$ R : $\quad Cu^{2+}(aq) + 2e^- \rightarrow Cu(s) \quad +0.34\ V$ $\qquad \left. \right\} +0.48\ V$
$\qquad$ L : $\quad Sn^{2+}(aq) + 2e^- \rightarrow Sn(s) \quad -0.14\ V$

$$\ln K = \frac{(2) \times (0.48\ V)}{25.693\ mV} = +37.\overline{4}, \qquad K = \boxed{1.7 \times 10^{16}}$$

(b) $Cu^{2+}(aq) + Cu(s) \rightleftharpoons 2Cu^{+}(aq)$

$$\left.\begin{array}{llll} R: & Cu^{2+}(aq) + e^{-} \rightarrow Cu(aq) & +0.16\,V \\ L: & Cu^{+}(aq) + e^{-} \rightarrow Cu(s) & +0.52\,V \end{array}\right\} -0.36\,V$$

$$\ln K = \frac{-0.36\,V}{25.693\,mV} = -14.\overline{0}, \qquad K = \boxed{8.2 \times 10^{-7}}$$

E7.18(b)   R: $2Bi^{3+}(aq) + 6e^{-} \rightarrow 2Bi(s)$

L: $Bi_2S_3(s) + 6e^{-} \rightarrow 2Bi(s) + 3S^{2-}(aq)$

Overall $(R - L)$: $2Bi^{3+}(aq) + 3S^{2-}(aq) \rightarrow Bi_2S_3(s) \quad \nu = 6$

$$\ln K = \frac{\nu FE^{\ominus}}{RT} = \frac{6(0.96\,V)}{(25.693 \times 10^{-3}\,V)} = 22\overline{4}$$

$$K = e^{22\overline{4}}$$

(a) $K = \dfrac{a_{Bi_2S_3(s)}}{a_{Bi^{3+}(aq)}^2 \, a_{S^{2-}(aq)}^3} = \dfrac{M^5}{[Bi^{3+}]^2 [S^{2-}]^3} = e^{22\overline{4}}$

In the above equation the activity of the solid equals 1 and, since the solution is extremely dilute, the activity coefficients of dissolved ions also equals 1. Substituting $[S^{2-}] = 1.5[Bi^{3+}]$ and solving for $[Bi^{3+}]$ gives $[Bi^{3+}] = 2.7 \times 10^{-20}$ M. Bi_2S_3 has a solubility equal to $\boxed{1.4 \times 10^{-20}\,M.}$

(b) The solubility equilibrium is written as the reverse of the cell reaction. Therefore,

$$K_S = K^{-1} = 1/e^{22\overline{4}} = \boxed{5.2 \times 10^{-98}}.$$

Solutions to problems

Solutions to numerical problems

P7.1   **(a)** $\Delta_r G^{\ominus} = -RT \ln K = -(8.314\,J\,K^{-1}\,mol^{-1}) \times (298\,K) \times (\ln 0.164) = 4.48 \times 10^{3}\,J\,mol^{-1}$

$$= \boxed{+4.48\,kJ\,mol^{-1}}.$$

(b) Draw up the following equilibrium table.

	I_2	Br_2	IBr
Amounts	—	$(1-\alpha)n$	$2\alpha n$
Mole fractions	—	$\dfrac{(1-\alpha)}{(1+\alpha)}$	$\dfrac{2\alpha}{(1+\alpha)}$
Partial pressure	—	$\dfrac{(1-\alpha)p}{(1+\alpha)}$	$\dfrac{2\alpha p}{(1+\alpha)}$

$$K = \prod_J a_J^{\nu_J} \, [7.16] = \frac{(p_{IBr}/p^{\ominus})^2}{p_{Br_2}/p^{\ominus}} \, [\text{perfect gases}] = \frac{\{(2\alpha)^2(p/p^{\ominus})\}}{(1-\alpha) \times (1+\alpha)} = \frac{(4\alpha^2(p/p^{\ominus}))}{1 - \alpha^2} = 0.164.$$

With $p = 0.164$ atm,

$$4\alpha^2 = 1 - \alpha^2, \qquad \alpha^2 = \frac{1}{5}, \qquad \alpha = 0.447.$$

$$p_{IBr} = \frac{2\alpha}{1+\alpha} \times p = \frac{(2) \times (0.447)}{1 + 0.447} \times (0.164 \text{ atm}) = \boxed{0.101 \text{ atm}}.$$

(c) The equilibrium table needs to be modified as follows.

$$p = p_{I_2} + p_{Br_2} + p_{IBr},$$

$$p_{Br_2} = x_{Br_2}p, \qquad p_{IBr} = x_{IBr}p, \qquad p_{I_2} = x_{I_2}p$$

with $x_{Br_2} = \dfrac{(1-\alpha)n}{(1+\alpha)n + n_{I_2}}$ [n = amount of Br_2 introduced into container]

and $x_{IBr} = \dfrac{2\alpha n}{(1+\alpha)n + n_{I_2}}$.

K is constructed as above [7.16], but with these modified partial pressures. In order to complete the calculation additional data are required, namely, the amount of Br_2 introduced, n, and the equilibrium vapor pressure of $I_2(s)$. n_{I_2} can be calculated from a knowledge of the volume of the container at equilibrium which is most easily determined by successive approximations since p_{I_2} is small.

Question. What is the partial pressure of IBr(g) if 0.0100 mol of Br_2(g) is introduced into the container? The partial pressure of $I_2(s)$ at 25°C is 0.305 Torr.

P7.3 $U(s) + \dfrac{3}{2}H_2(g) \rightleftharpoons UH_3(s)$, $K = (p/p^{\ominus})^{-3/2}$ [Exercise 7.7(b)].

$$\Delta_f H^{\ominus} = RT^2 \frac{d \ln K}{dT} \text{ [7.23a]} = RT^2 \frac{d}{dT}\left(-\frac{3}{2}\ln p/p^{\ominus}\right)$$

$$= -\frac{3}{2}RT^2 \frac{d \ln p}{dT}$$

$$= -\frac{3}{2}RT^2 \left(\frac{1.464 \times 10^4 \text{K}}{T^2} - \frac{5.65}{T}\right)$$

$$= -\frac{3}{2}R(1.464 \times 10^4 \text{K} - 5.65T)$$

$$= \boxed{-(2.196 \times 10^4 \text{ K} - 8.48T)R}.$$

$$d\left(\Delta_f H^{\ominus}\right) = \Delta_r C_p^{\ominus} dT \text{ [from 2.36]}$$

or $\Delta_r C_p^{\ominus} = \left(\dfrac{\partial \Delta_f H^{\ominus}}{\partial T}\right)_p = \boxed{8.48\,R}$.

P7.5 $CaCl_2 \cdot NH_3(s) \rightleftharpoons CaCl_2(s) + NH_3(g)$, $K = \dfrac{p}{p^{\ominus}}$.

Since $\Delta_r G^{\ominus}$ and $\ln K$ are related as above, the dependence of $\Delta_r G^{\ominus}$ on temperature can be determined from the dependence of $\ln K$ on temperature

$$\Delta_r G^{\ominus} = -RT \ln K = -RT \ln \frac{p}{p^{\ominus}}$$

$$= -(8.314 \, \text{J K}^{-1} \text{mol}^{-1}) \times (400 \, \text{K}) \times \ln \left(\frac{17.1 \, \text{kPa}}{100.0 \, \text{kPa}} \right) \, [p^{\ominus} = 1 \text{bar}]$$

$$= +13.5 \, \text{kJ mol}^{-1} \, \text{at 400 K.}$$

$$\frac{\Delta_r G^{\ominus}(T)}{T} - \frac{\Delta_r G^{\ominus}(T')}{T'} = \Delta_r H^{\ominus} \left(\frac{1}{T} - \frac{1}{T'} \right) \, [7.25].$$

Therefore, taking $T' = 400$ K,

$$\Delta_r G^{\ominus}(T) = \left(\frac{T}{400 \, \text{K}} \right) \times (13.5 \, \text{kJ mol}^{-1}) + (78 \, \text{kJ mol}^{-1}) \times \left(1 - \frac{T}{400 \, \text{K}} \right)$$

$$= (78 \, \text{kJ mol}^{-1}) + \left(\frac{(13.5 - 78) \, \text{kJ mol}^{-1}}{400} \right) \times \left(\frac{T}{\text{K}} \right).$$

That is, $\Delta_r G^{\ominus}(T)/(\text{kJ mol}^{-1}) = \boxed{78 - 0.161 \times (T/\text{K})}$.

P7.7   The equilibrium we need to consider is $A_2(g) \rightleftharpoons 2A(g)$. A = acetic acid. It is convenient to express the equilibrium constant in terms of α, the degree of dissociation of the dimer, which is the predominant species at low temperatures.

	A	A_2	Total
At equilibrium	$2\alpha n$	$(1 + \alpha)n$	$(1 + \alpha)n$
Mole fraction	$\dfrac{2\alpha}{1 + \alpha}$	$\dfrac{1 - \alpha}{1 + \alpha}$	1
Partial pressure	$\dfrac{2\alpha p}{1 + \alpha}$	$\left(\dfrac{1 - \alpha}{1 + \alpha} \right) p$	p

The equilibrium constant for the dissociation is

$$K = \frac{\left(\dfrac{p_A}{p^{\ominus}} \right)^2}{\dfrac{p_{A_2}}{p^{\ominus}}} = \frac{p_A^2}{p_{A_2} p^{\ominus}} = \frac{4\alpha^2 \left(\dfrac{p}{p^{\ominus}} \right)}{1 - \alpha^2}.$$

We also know that

$$pV = n_{\text{total}} RT = (1 + \alpha)nRT, \text{ implying that } \alpha = \frac{pV}{nRT} - 1 \text{ and } n = \frac{m}{M}.$$

In the first experiment,

$$\alpha = \frac{pVM}{mRT} - 1 = \frac{(101.9 \, \text{kPa}) \times (21.45 \times 10^{-3} \text{dm}^3) \times (120.1 \, \text{g mol}^{-1})}{(0.0519 \, \text{g}) \times (8.314 \, \text{kPa dm}^3 \, \text{K}^{-1} \, \text{mol}^{-1}) \times (437 \, \text{K})} - 1 = 0.392.$$

Hence, $K = \dfrac{(4) \times (0.392)^2 \times (764.3/750.1)}{1 - (0.392)^2} = \boxed{0.740}$.

In the second experiment,

$$\alpha = \frac{pVM}{mRT} - 1 = \frac{(101.9 \text{ kPa}) \times (21.45 \times 10^{-3} \text{dm}^3) \times (120.1 \text{ g mol}^{-1})}{(0.038 \text{ g}) \times (8.314 \text{ kPa dm}^3 \text{ K}^{-1} \text{ mol}^{-1}) \times (471 \text{ K})} - 1 = 0.764.$$

Hence, $K = \dfrac{(4) \times (0.764)^2 \times \left(\dfrac{764.3}{750.1}\right)}{1 - (0.764)^2} = \boxed{5.71}$.

The enthalpy of dissociation is

$$\Delta_r H^{\ominus} = \frac{R \ln \dfrac{K'}{K}}{\left(\dfrac{1}{T} - \dfrac{1}{T'}\right)} [7.25, \text{Exercise 7.10(a)}] = \frac{R \ln \left(\dfrac{5.71}{0.740}\right)}{\left(\dfrac{1}{437\text{K}} - \dfrac{1}{471\text{K}}\right)} = +103 \text{ kJ mol}^{-1}.$$

The enthalpy of dimerization is the negative of this value, or $\boxed{-103 \text{ kJ mol}^{-1}}$ (i.e. per mole of dimer).

P7.9 The equilibrium $I_2(g) \rightleftharpoons 2I(g)$ is described by the equilibrium constant

$$K = \frac{x(I)^2}{x(I_2)^2} \times \frac{p}{p^{\ominus}} = \frac{4\alpha^2 \left(\dfrac{p}{p^{\ominus}}\right)}{1 - \alpha^2} \text{ [Problem 7.7]}.$$

If $p^{\circ} = \dfrac{nRT}{V}$, then $p = (1 + \alpha)p^{\circ}$, implying that

$$\alpha = \frac{p - p^{\circ}}{p^{\circ}}.$$

We therefore draw up the following table.

	937 K	1073 K	1173 K	
p/atm	0.06244	0.07500	0.09181	
$10^4 n_1$	2.4709	2.4555	2.4366	
p°/atm	0.05757	0.06309	0.06844	$\left[p^{\circ} = \dfrac{nRT}{V}\right]$
α	0.08459	0.1888	0.3415	
K	$\boxed{1.800 \times 10^{-3}}$	$\boxed{1.109 \times 10^{-2}}$	$\boxed{4.848 \times 10^{-2}}$	

$$\Delta H^{\ominus} = RT^2 \times \left(\frac{d \ln K}{dT}\right) = (8.314 \text{ J K}^{-1}\text{mol}^{-1}) \times (1073 \text{ K}^2) \times \left(\frac{-3.027 - (-6.320)}{200 \text{ K}}\right)$$

$$= \boxed{+158 \text{ kJ mol}^{-1}}.$$

P7.11   The reaction is

$$Si(s) + H_2(g) \rightleftharpoons SiH_2(g).$$

The equilibrium constant is

$$K = \exp\left(-\frac{\Delta_r G^{\ominus}}{RT}\right) = \exp\left(\frac{-\Delta_r H^{\ominus}}{RT}\right)\exp\left(\frac{-\Delta_r S^{\ominus}}{R}\right).$$

Let h be the uncertainty in $\Delta_f H^{\ominus}$ so that the high value is $h +$ the low value. The K based on the low value is

$$K_{\text{low}H} = \exp\left(\frac{-\Delta H_{\text{low}}^{\ominus}}{RT}\right)\exp\left(\frac{\Delta_r S^{\ominus}}{R}\right) = \exp\left(\frac{-\Delta_r H_{\text{high}}^{\ominus}}{RT}\right)\exp\left(\frac{h}{RT}\right)\exp\left(\frac{\Delta_r S^{\ominus}}{R}\right)$$

$$= \exp\left(\frac{h}{RT}\right)K_{\text{high}H}.$$

So $\dfrac{K_{\text{low}H}}{K_{\text{high}H}} == \exp\left(\dfrac{h}{RT}\right).$

(a) At 298 K, $\dfrac{K_{\text{low}H}}{K_{\text{high}H}} = \exp\left(\dfrac{(289 - 243)\ \text{kJ mol}^{-1}}{(8.3145 \times 10^{-3}\ \text{kJ K}^{-1}\ \text{mol}^{-1}) \times (298\ \text{K})}\right) = \boxed{1.2 \times 10^8}.$

(b) At 700 K, $\dfrac{K_{\text{low}H}}{K_{\text{high}H}} = \exp\left(\dfrac{(289 - 243)\ \text{kJ mol}^{-1}}{(8.3145 \times 10^{-3}\ \text{kJ K}^{-1}\ \text{mol}^{-1}) \times (700\ \text{K})}\right) = \boxed{2.7 \times 10^3}.$

P7.13   **(a)** $I = \dfrac{1}{2}\left\{\left(\dfrac{b}{b^{\ominus}}\right)_+ z_+^2 + \left(\dfrac{b}{b^{\ominus}}\right)_- z_-^2\right\}$ [5.71] $= 4\left(\dfrac{b}{b^{\ominus}}\right).$

For CuSO₄, $I = (4) \times (1.0 \times 10^{-3}) = \boxed{4.0 \times 10^{-3}}.$

For ZnSO₄, $I = (4) \times (3.0 \times 10^{-3}) = \boxed{1.2 \times 10^{-2}}.$

(b) $\log \gamma_{\pm} = -|z_+ z_-|AI^{1/2}.$

$\log \gamma_{\pm}(\text{CuSO}_4) = -(4) \times (0.509) \times (4.0 \times 10^{-3})^{1/2} = -0.12\overline{88},$

$\gamma_{\pm}(\text{CuSO}_4) = \boxed{0.74}.$

$\log \gamma_{\pm}(\text{ZnSO}_4) = -(4) \times (0.509) \times (1.2 \times 10^{-2})^{1/2} = -0.22\overline{30},$

$\gamma_{\pm}(\text{ZnSO}_4) = \boxed{0.60}.$

(c) The reaction in the Daniell cell is

$$Cu^{2+}(aq) + SO_4^{2-}(aq) + Zn(s) \rightarrow Cu(s) + Zn^{2+}(aq) + SO_4^{2-}(aq).$$

Hence, $Q = \dfrac{a(Zn^{2+})a(SO_4^{2-},R)}{a(Cu^{2+})a(SO_4^{2-},L)}$

$$= \frac{\gamma_+ b_+(Zn^{2+})\gamma_- b_-(SO_4^{2-},R)}{\gamma_+ b_+(Cu^{2+})\gamma_- b_-(SO_4^{2-},L)} \quad \left[b \equiv \frac{b}{b^\ominus} \text{ here and below}\right]$$

where the designations R and L refer to the right and left sides of the equation for the cell reaction and all b are assumed to be unitless, that is, $b/b^\ominus$.

$$b_+(Zn^{2+}) = b_-(SO_4^{2-},R) = b(ZnSO_4).$$

$$b_+(Cu^{2+}) = b_-(SO_4^{2-},L) = b(CuSO_4).$$

Therefore,

$$Q = \frac{\gamma_\pm^2(ZnSO_4)b^2(ZnSO_4)}{\gamma_\pm^2(CuSO_4)b^2(CuSO_4)} = \frac{(0.60)^2 \times (3.0 \times 10^{-3})^2}{(0.74)^2 \times (1.0 \times 10^{-3})^2} = 5.9\bar{2} = \boxed{5.9}.$$

(d) $E^\ominus = -\dfrac{\Delta_r G^\ominus}{\nu F}[7.28] = \dfrac{-(-212.7 \times 10^3 \text{ J mol}^{-1})}{(2) \times (9.6485 \times 10^4 \text{ C mol}^{-1})} = \boxed{+1.102 \text{ V}}.$

(e) $E = E^\ominus = -\dfrac{25.693 \times 10^{-3}}{\nu} \text{ V } \ln Q = (1.102 \text{ V}) - \left(\dfrac{25.693 \times 10^{-3}}{2} \text{ V}\right)\ln(5.9\bar{2})$

$$= (1.102\text{V}) - (0.023\text{V}) = \boxed{+1.079 \text{ V}}.$$

P7.15   The electrode half-reactions and their potentials are

	$E^\ominus$
R: $Q(aq) + 2H^+(aq) + 2e^- \rightarrow QH_2(aq)$	0.6994V
L: $Hg_2Cl_2(s) + 2e^- \rightarrow 2Hg(l) + 2Cl^-(aq)$	0.2676V
Overall (R − L): $Q(aq) + 2H^+(aq) \rightarrow QH_2(aq) + Hg_2Cl_2(s)$,	0.4318 V

$$Q(\text{reaction quotient}) = \frac{a(QH_2)}{a(Q)a^2(H^+)a^2(Cl^-)}.$$

Since quinhydrone is an equimolecular complex of Q and QH_2, $m(Q) = m(QH_2)$ and, since their activity coefficients are assumed to be 1 or to be equal, we have $a(QH_2) \approx a(Q)$. Thus

$$Q = \frac{1}{a^2(H^+)a^2(Cl^-)}, \qquad E = E^\ominus - \frac{25.7 \text{ mV}}{\nu} \ln Q \text{ [Illustration 7.10]}.$$

$$\ln Q = \frac{\nu(E^\ominus - E)}{25.7 \text{ mV}} = \frac{(2) \times (0.4318 - 0.190) \text{ V}}{25.7 \times 10^{-3} \text{ V}} = 18.8\bar{2}, \qquad Q = 1.\overline{49} \times 10^8.$$

$$a^2(H^+) = (\gamma_+ b_+)^2; \quad a^2(Cl^-) = (\gamma_- b_-)^2 \quad \left[b \equiv \frac{b}{b^\ominus}\right].$$

For HCl(aq), $b_+ = b_- = b$ and, if the activity coefficients are assumed equal, $a^2(H^+) = a^2(Cl^-)$; hence

$$Q = \frac{1}{a^2(H^+)a^2(Cl^-)} = \frac{1}{a^4(H^+)}.$$

Thus, $a(H^+) = \left(\frac{1}{Q}\right)^{1/4} = \left(\frac{1}{1.49 \times 10^8}\right)^{1/4} = 9 \times 10^{-3}$,

$$pH = -\log a(H^+) = \boxed{2.0}.$$

P7.17 $H_2(g)|HCl(aq)|Hg_2Cl_2(s)|Hg(l)$.

$$E = E^\ominus - \frac{RT}{F} \ln a(H^+)a(Cl^-) \text{ [Section 7.8]}.$$

$$a(H^+) = \gamma_+ b_+ = \gamma_+ b; \quad a(Cl^-) = \gamma_- b_- = \gamma_- b \left[b = \frac{b}{b^\ominus} \text{here and below}\right].$$

$$a(H^+)a(Cl^-) = \gamma_+\gamma_- b^2 = \gamma_\pm^2 b^2.$$

$$E = E^\ominus - \frac{2RT}{F} \ln b - \frac{2RT}{F} \ln \gamma_\pm. \tag{a}$$

Converting from natural logarithms to common logarithms (base 10) in order to introduce the Debye–Hückel expression, we obtain

$$E = E^\ominus - \frac{(2.303) \times 2RT}{F} \log b - \frac{(2.303) \times 2RT}{F} \log \gamma_\pm$$

$$= E^\ominus - (0.1183 \text{ V}) \log b - (0.1183 \text{ V}) \log \gamma_\pm$$

$$= E^\ominus - (0.1183 \text{ V}) \log b - (0.1183 \text{ V}) \left[-|z_+z_-|AI^{1/2}\right]$$

$$= E^\ominus - (0.1183 \text{ V}) \log b + (0.1183 \text{ V}) \times A \times b^{1/2} [I = b].$$

Rearranging,

$$E + (0.1183 \text{ V}) \log b = E^\ominus + \text{ constant } \times b^{1/2}.$$

Therefore, plot $E + (0.1183 \text{ V}) \log b$ against $b^{1/2}$, and the intercept at $b = 0$ is $E^\ominus/V$. Draw up the following table.

$b/(\text{mmol kg}^{-1})$	1.6077	3.0769	5.0403	7.6938	10.9474
$\left(\dfrac{b}{b^\ominus}\right)^{1/2}$	0.04010	0.05547	0.07100	0.08771	0.1046
$E/V + (0.1183) \log b$	0.27029	0.27109	0.27186	0.27260	0.27337

The points are plotted in Fig. 7.2. The intercept is at 0.26840, so $E^\ominus = +0.26840$ V. A least–squares best fit gives $E^\ominus = \boxed{+0.26843 \text{ V}}$ and a coefficient of determination equal to 0.99895.

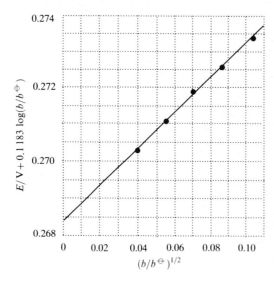

Figure 7.2

For the activity coefficients we obtain from equation (a)

$$\ln \gamma_\pm = \frac{E^\ominus - E}{2RT/F} - \ln \frac{b}{b^\ominus} = \frac{0.26843 - E/V}{0.05139} - \ln \frac{b}{b^\ominus}$$

and we draw up the following table.

$b/(\text{mmol kg}^{-1})$	1.6077	3.0769	5.0403	7.6938	10.9474
$\ln \gamma_\pm$	−0.3465	−0.05038	−0.6542	−0.07993	−0.09500
$\gamma_\pm$	0.9659	0.9509	0.9367	0.9232	0.9094

P7.19   The cells described in the problem are back-to-back pairs of cells each of the type

Ag (s) |AgX (s) |MX (b_1) |M_xHg (s) .

R: $M^+ (b_1) + e^- \xrightarrow{\text{Hg}} M_x\text{Hg (s)}$ (Reduction of M^+ and formation of amalgam)

L: $\text{AgX (s)} + e^- \rightarrow \text{Ag (s)} + X^- (b_1)$

R − L : $\text{Ag (s)} + M^+ (b_1) + X^- (b_1) \xrightarrow{\text{Hg}} M_x\text{Hg (s)} + \text{AgX (s)}, \quad \nu = 1.$

$$Q = \frac{a\left(M_x\text{Hg}\right)}{a\left(M^+\right) a\left(X^-\right)}.$$

$$E = E^\ominus - \frac{RT}{F} \ln Q.$$

For a pair of such cells back to back,

$$\text{Ag (s)} \,|\text{AgX (s)} \,|\text{MX} (b_1) \,|\text{M}_x\text{Hg (s)} \,|\text{MX} (b_2) \,|\text{AgX (s)} \,|\text{Ag (s)},$$

$$E_R = E^\ominus - \frac{Rt}{F} \ln Q_R, \quad E_L = E^\ominus - \frac{RT}{F} \ln Q_L,$$

$$E = \frac{-RT}{F} \ln \frac{Q_L}{Q_R} = \frac{RT}{F} \ln \frac{(a(\text{M}^+) a(\text{X}^-))_L}{(a(\text{M}^+) a(\text{X}^-))_R}.$$

(Note that the unknown quantity $a(\text{M}_x\text{Hg})$ drops out of the expression for E.)

$$a(\text{M}^+) a(\text{X}^-) = \left(\frac{\gamma_+ b_+}{b^\ominus}\right)\left(\frac{\gamma_- b_-}{b^\ominus}\right) = \gamma_\pm^2 \left(\frac{b}{b^\ominus}\right)^2 (b_+ = b_-).$$

With $L = (1)$ and $R = (2)$ we have

$$E = \frac{2RT}{F} \ln \frac{b_1}{b_2} + \frac{2RT}{F} \ln \frac{\gamma_\pm(1)}{\gamma_\pm(2)}.$$

Take $b_2 = 0.09141 \text{ mol kg}^{-1}$ (the reference value), and write $b = \dfrac{b_1}{b^\ominus}$.

$$E = \frac{2RT}{F} \left(\ln \frac{b}{0.09141} + \ln \frac{\gamma_\pm}{\gamma_\pm(\text{ref})}\right).$$

For $b = 0.09141$, the extended Debye–Hückel law gives

$$\log \gamma_\pm(\text{ref}) = \frac{(-1.461) \times (0.09141)^{1/2}}{(1) + (1.70) \times (0.09141)^{1/2}} + (0.20) \times (0.09141) = -0.273\overline{5},$$

$$\gamma_\pm(\text{ref}) = 0.532\overline{8}.$$

Then $E = (0.05139 \text{ V}) \times \left(\ln \dfrac{b}{0.09141} + \ln \dfrac{\gamma_\pm}{0.5328}\right),$

$$\ln \gamma_\pm = \frac{E}{0.05139 \text{ V}} - \ln \frac{b}{(0.09141) \times (0.05328)}.$$

We then draw up the following table.

$b/ (\text{mol/kg}^{-1})$	0.0555	0.09141	0.1652	0.2171	1.040	1.350
E/V	−0.0220	0.0000	0.0263	0.0379	0.1156	0.1336
γ	0.572	0.533	0.492	0.469	0.444	0.486

A more precise procedure is described in the original references for the temperature dependence of $E^\ominus (\text{Ag, AgCl, Cl}^-)$; see Problem 7.20.

P7.21   **(a)** From $\left(\dfrac{\partial G}{\partial p}\right)_T = V$ [3.50],

we obtain $\left(\dfrac{\partial \Delta_r G}{\partial p}\right)_T = \Delta_r V.$

Substituting $\Delta_r G = -\nu F E$ [7.27] yields

$$\boxed{\left(\dfrac{\partial E}{\partial p}\right)_{T,n} = -\dfrac{\Delta_r V}{\nu F}}.$$

(b) The plot (Fig. 7.3) of E against p appears to fit a straight line very closely. A linear regression analysis yields

Slope $\boxed{= 2.480 \times 10^{-3}\ \text{mV atm}^{-1}}$,   standard deviation $= 3 \times 10^{-6}\ \text{mV atm}^{-1}$.

Intercept$= 8.5583$ mV,   standard deviation $= 2.8 \times 10^{-3}$ mV.

$R = 0.99999701$ (an extremely good fit).

From $\Delta_r V$

$$\left(\dfrac{\partial E}{\partial p}\right)_{T,n} = -\dfrac{\left(-2.66\overline{6} \times 10^{-6}\ \text{m}^3\,\text{mol}^{-1}\right)}{1 \times 9.6485 \times 10^4\ \text{C mol}^{-1}}.$$

Since $J = VC = Pa\,m^3,\quad C = \dfrac{Pa\,m^3}{V}\quad$ or $\quad \dfrac{m^3}{C} = \dfrac{V}{Pa}.$

Therefore

$$\left(\dfrac{\partial E}{\partial P}\right)_{T,n} = \left(\dfrac{2.66\overline{6} \times 10^{-6}}{9.648\overline{5} \times 10^4}\right)\dfrac{V}{Pa} \times \dfrac{1.01325 \times 10^5\ \text{Pa}}{\text{atm}} = 2.80 \times 10^{-6}\ \text{V atm}^{-1}$$

$$= \boxed{2.80 \times 10^{-3}\ \text{mV atm}^{-1}}.$$

This compares closely to the result from the potential measurements.

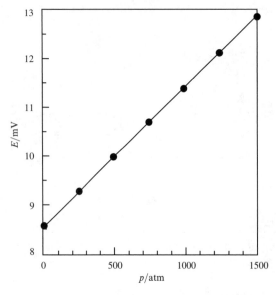

Figure 7.3

(c) A fit to a second-order polynomial of the form

$$E = a + bp + cp^2$$

yields

$a = 8.5592$ mV,	standard deviation $= 0.0039$ mV
$b = 2.835 \times 10^{-3}$ mV atm^{-1},	standard deviation $= 0.012 \times 10^{-3}$ mV atm^{-1}
$c = 3.02 \times 10^{-9}$ mV atm^{-2},	standard deviation $= 7.89 \times 10^{-9}$ mV atm^{-1}
$R = 0.999\ 997\ 11.$	

This regression coefficient is only marginally better than that for the linear fit, but the uncertainty in the quadratic term is > 200 per cent.

$$\left(\frac{\partial E}{\partial p}\right)_T = b + 2cp.$$

The slope changes from $\left(\dfrac{\partial E}{\partial p}\right)_{\min} = b = 2.835 \times 10^{-3}$ mV atm^{-1}

to $\left(\dfrac{\partial E}{\partial p}\right)_{\max} = b + 2c(1500\,\text{atm}) = 2.836 \times 10^{-3}$ mV atm^{-1}.

We conclude that the linear fit and constancy of $\left(\dfrac{\partial E}{\partial p}\right)$ are very good.

(d) We can obtain an order of magnitude value for the isothermal compressibility from the value of c.

$$\frac{\partial^2 E}{\partial p^2} = -\frac{1}{vF}\left(\frac{\partial \Delta_r V}{\partial p}\right)_T = 2c.$$

$$(\kappa_T)_{\text{cell}} = -\frac{1}{V}\left(\frac{\partial \Delta_r V}{\partial p}\right)_T = \frac{2vcF}{V}.$$

$$(\kappa_T)_{\text{cell}} = \frac{2(1) \times \left(3.02 \times 10^{-12}\ \text{V atm}^{-2}\right) \times \left(9.6485 \times 10^4\ \text{C mol}^{-1}\right) \times \left(\dfrac{82.058\ \text{cm}^3\ \text{atm}}{8.3145\ \text{J}}\right)}{\left(1\ \text{cm}^3/0.996\ \text{g}\right) \times \left(\dfrac{18.016\ \text{g}}{1\ \text{mol}}\right)}$$

$$= \boxed{3.2 \times 10^{-7}\ \text{atm}^{-1}} \quad \text{standard deviation} \approx 200 \text{ per cent}$$

where we have assumed the density of the cell to be approximately that of water at 30°C.

COMMENT. It is evident from these calculations that the effect of pressure on the potentials of cells involving only liquids and solids is not important; for this reaction the change is only $\approx 3 \times 10^{-6}$ V atm^{-1}. The effective isothermal compressibility of the cell is of the order of magnitude typical of solids rather than liquids; other than that, little significance can be attached to the calculated numerical value.

P7.23 We need to obtain $\Delta_r H^{\ominus}$ for the reaction

$$\tfrac{1}{2}H_2\,(\text{g}) + Uup^+\,(\text{aq}) \rightarrow Uup\,(\text{s}) + H^+\,(\text{aq}).$$

We draw up the thermodynamic cycle shown in Fig. 7.4.

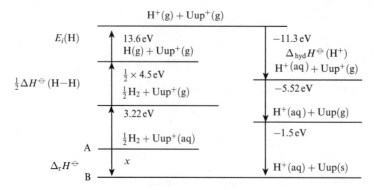

Figure 7.4

Data are obtained from Tables 10.3, 10.4, 11.4, 2.7, and 2.7b. The conversion factor between eV and kJ mol^{-1} is

$$1\,\text{eV} = 96.485\,\text{kJ mol}^{-1}$$

The distance from A to B in the cycle is given by

$$\Delta_r H^\ominus = x = (3.22\,\text{eV}) + \left(\frac{1}{2}\right) \times (4.5\,\text{eV}) + (13.6\,\text{eV}) - (11.3\,\text{eV}) - (5.52\,\text{eV}) - (1.5\,\text{eV})$$

$$= 0.75\,\text{eV}.$$

$$\Delta_r S^\ominus = S^\ominus\,(\text{Uup, s}) + S^\ominus\,(\text{H}^+, \text{aq}) - \tfrac{1}{2}S^\ominus\,(\text{H}_2, \text{g}) - S^\ominus\,(\text{Uup}^+, \text{aq})$$

$$= (0.69) + (0) - \left(\frac{1}{2}\right) \times (1.354) - (1.34)\,\text{meV K}^{-1} = -1.33\,\text{meV K}^{-1}.$$

$$\Delta_r G^\ominus = \Delta_r H^\ominus - T\Delta_r S^\ominus = (0.75\,\text{eV}) + (298.15\,\text{K}) \times \left(1.33\,\text{meV K}^{-1}\right) = +1.1\overline{5}\,\text{eV}$$

which corresponds to $\boxed{+111\,\text{kJ mol}^{-1}}$.

The electrode potential is therefore $\dfrac{-\Delta_r G^\ominus}{\nu F}$, with $\nu = 1$, or $\boxed{-1.1\overline{5}\,\text{V}}$.

Solutions to theoretical problems

P7.25   We draw up the following table using the stoichiometry $A + 3B \rightarrow 2C$ and $\Delta n_J = \nu_J \xi$.

	A	B	C	Total
Initial amount /mol	1	3	0	4
Change, Δn_J/mol	$-\xi$	-3ξ	$+2\xi$	
Equilibrium amount /mol	$1 - \xi$	$3(1 - \xi)$	2ξ	$2(2 - \xi)$
Mole fraction	$\dfrac{1 - \xi}{2(2 - \xi)}$	$\dfrac{3(1 - \xi)}{2(2 - \xi)}$	$\dfrac{\xi}{2 - \xi}$	1

$$K = \frac{(p_C/p^{\ominus})^2}{(p_A/p^{\ominus})(p_B/p^{\ominus})^3} = \frac{x_C^2}{x_A x_B^3} \times \left(\frac{p^{\ominus}}{p}\right)^2 = \frac{\xi^2}{(2-\xi)^2} \times \frac{2(2-\xi)}{1-\xi} \times \frac{2^3(2-\xi)^3}{3^3(1-\xi)^3} \times \left(\frac{p^{\ominus}}{p}\right)^2$$

$$= \frac{16(2-\xi)^2\xi^2}{27(1-\xi)^4} \times \left(\frac{p^{\ominus}}{p}\right)^2.$$

Since K is independent of the pressure

$$\frac{(2-\xi)^2\xi^2}{(1-\xi)^4} = a^2 \left(\frac{p}{p^{\ominus}}\right)^2, \quad a^2 = \frac{27}{16}K, \text{a constant.}$$

Therefore $(2-\xi)\xi = a\left(\frac{p}{p^{\ominus}}\right) \times (1-\xi)^2,$

$$\left(1 + \frac{ap}{p^{\ominus}}\right)\xi^2 - 2\left(1 + \frac{ap}{p^{\ominus}}\right)\xi + \frac{ap}{p^{\ominus}} = 0,$$

which solves to $\boxed{\xi = 1 - \left(\frac{1}{1 + ap/p^{\ominus}}\right)^{1/2}}.$

We choose the root with the negative sign because ξ lies between 0 and 1. The variation of ξ with p is shown in Fig. 7.5.

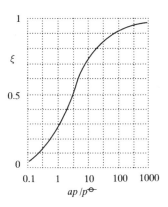

Figure 7.5

$K_s = a(M^+)a(X^-) = b(M^+)b(X^-)\gamma_{\pm}^2; \quad b(M^+) = S', \quad b(X^-) = S' + C$

$\log \gamma_{\pm} = -AI^{1/2} = -AC^{1/2} \quad \ln \gamma_{\pm} = -2.303 AC^{1/2}$

$\gamma_{\pm} = e^{-2.303 AC^{1/2}} \quad \gamma_{\pm}^2 = e^{-4.606 AC^{1/2}}$

$K_s = S'(S' + C) \times e^{-4.606 AC^{1/2}}$

We solve $S'^2 + S'C - \dfrac{K_s}{\gamma_{\pm}^2} = 0$

to get $S' = \dfrac{1}{2}\left(C^2 + \dfrac{4K_s}{\gamma_\pm^2}\right)^{1/2} - \dfrac{1}{2}C \approx \dfrac{K_s}{C\gamma_\pm^2}$

Therefore, since $\gamma_\pm^2 = e^{-4.606\,AC^{1/2}}$ $\boxed{S' \approx \dfrac{K_s e^{-4.606\,AC^{1/2}}}{C}}$

Solutions to applications

P7.29 $\Delta G = \Delta G^\ominus + RT\ln Q$ [7.11].

In equation 7.11 molar solution concentrations are used with 1 M standard states. The standard state ($\ominus$) pH equals zero in contrast to the biological standard state ($\oplus$) of pH 7. For the ATP hydrolysis

$$ATP(aq) + H_2O(l) \rightarrow ADP(aq) + P_i^-(aq) + H_3O^+(aq)$$

we can calculate the standard state free energy given the biological standard free energy of -31 kJ mol^{-1} (*Impact* I7.2).

$\Delta G^\oplus = \Delta G^\ominus + RT\ln(10^{-7}M/1M)$

$\Delta G^\ominus = \Delta G^\oplus - RT\ln(10^{-7}M/1M) = -31$ kJ mol^{-1} $- (8.314$ J mol^{-1} K$^{-1})(310$ K$)\ln(10^{-7})$

$\qquad = -31$ kJmol^{-1} $+ 41.5$ kJ mol^{-1} $= +11$ kJ mol^{-1}.

This calculation shows that under standard conditions the hydrolysis of ATP is not spontaneous! It is endergonic.

The calculation of the ATP hydrolysis free energy with the cell conditions pH = 7, [ATP] = [ADP] = $[P_i^-]$ = 1.0 × 10^{-6} M, is interesting.

$\Delta G = \Delta G^\ominus + RT\ln Q = \Delta G^\ominus + RT\ln\{[ADP][P_i^-][H^+]/[ATP](1M)^2\}$

$\qquad = +11$ kJ mol^{-1} $+ (8.314$ J mol^{-1}K$^{-1})(310$ K$)\ln(10^{-6} \times 10^{-7}) = +11$ kJ mol^{-1} $- 77$ kJ mol^{-1}

$\qquad = -66$ kJ mol^{-1}.

The concentration conditions in biological cells make the hydrolysis of ATP spontaneous and very exergonic. A maximum of 66 kJ of work is available to drive coupled chemical reactions when a mole of ATP is hydrolyzed.

P7.31   Yes, a bacterium can evolve to utilize the ethanol/nitrate pair to exergonically release the free energy needed for ATP synthesis. The ethanol reductant may yield any of the following products.

$$\underset{\text{ethanol}}{CH_3CH_2OH} \rightarrow \underset{\text{ethanal}}{CH_3CHO} \rightarrow \underset{\text{ethanoic acid}}{CH_3COOH} \rightarrow CO_2 + H_2O.$$

The nitrate oxidant may receive electrons to yield any of the following products.

$$\underset{\text{nitrate}}{NO_3^-} \rightarrow \underset{\text{nitrite}}{NO_2^-} \rightarrow \underset{\text{dinitrogen}}{N_2} \rightarrow \underset{\text{ammonia}}{NH_3}.$$

Oxidation of two ethanol molecules to carbon dioxide and water can transfer 8 electrons to nitrate during the formation of ammonia. The half-reactions and net reaction are:

$$2[CH_3CH_2OH(l) \rightarrow 2CO_2(g) + H_2O(l) + 4H^+(aq) + 4e^-]$$
$$NO_3^-(aq) + 9H^+(aq) + 8e^- \rightarrow NH_3(aq) + 3H_2O(l)$$

$$\overline{2CH_3CH_2OH(l) + H^+(aq) + NO_3^-(aq) \rightarrow 4CO_2(g) + 5H_2O(l) + NH_3(aq)}$$

$\Delta_r G^\ominus = -2331.29$ kJ for the reaction as written (a Table 2.5 and 2.7 calculation). Of course, enzymes must evolve that couple this exergonic redox reaction to the production of ATP, which would then be available for carbohydrate, protein, lipid, and nucleic acid synthesis.

P7.33   The half-reactions involved are:

R: $cyt_{ox} + e^- \rightarrow cyt_{red} \quad E^\ominus_{cyt}$

L: $D_{ox} + e^- \rightarrow D_{red} \quad E^\ominus_D$

The overall cell reaction is:

$$R - L = cyt_{ox} + D_{red} \rightleftharpoons cyt_{red} + D_{ox} \quad E^\ominus = E^\ominus_{cyt} - E^\ominus_D$$

(a) The Nernst equation for the cell reaction is

$$E = E^\ominus - \frac{RT}{F} \ln \frac{[cyt_{red}][D_{ox}]}{[cyt_{ox}][D_{red}]}.$$

At equilibrium, $E = 0$; therefore

$$\ln \frac{[cyt_{red}]_{eq}[D_{ox}]_{eq}}{[cyt_{ox}]_{eq}[D_{red}]_{eq}} = \frac{F}{RT}\left(E^\ominus_{cyt} - E^\ominus_D\right),$$

$$\ln\left(\frac{[D_{ox}]_{eq}}{[D_{red}]_{eq}}\right) = \ln\left(\frac{[cyt]_{ox}}{[cyt]_{red}}\right) + \frac{F}{RT}\left(E^\ominus_{cyt} - E^\ominus_D\right).$$

Therefore a plot of $\ln\left(\dfrac{[D_{ox}]_{eq}}{[D_{red}]_{eq}}\right)$ against $\ln\left(\dfrac{[cyt]_{ox}}{[cyt]_{red}}\right)$ is linear with a slope of one and an intercept of $\dfrac{F}{RT}\left(E^\ominus_{cyt} - E^\ominus_D\right)$.

(b) Draw up the following table.

$\ln\left(\dfrac{[D_{ox}]_{eq}}{[D_{red}]_{eq}}\right)$	-5.882	-4.776	-3.661	-3.002	-2.593	-1.436	-0.6274
$\ln\left(\dfrac{[cyt_{ox}]_{eq}}{[cyt_{red}]_{eq}}\right)$	-4.547	-3.772	-2.415	-1.625	-1.094	-0.2120	-0.3293

The plot of $\ln\left(\dfrac{[D_{ox}]_{eq}}{[D_{red}]_{eq}}\right)$ against $\ln\left(\dfrac{[cyt_{ox}]_{eq}}{[cyt_{red}]_{eq}}\right)$ is shown in Fig. 7.6. The intercept is -1.2124. Hence

$$E_{cyt}^{\ominus} = \frac{RT}{F} \times (-1.2124) + 0.237\,\text{V}$$

$$= 0.0257\text{V} \times (-1.2124) + 0.237\,\text{V}$$

$$= \boxed{+0.206\,\text{V}}.$$

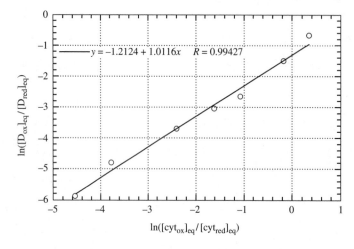

Figure 7.6

P7.35   A reaction proceeds spontaneously if its reaction Gibbs function is negative.

$$\Delta_r G = \Delta_r G^{\ominus} + RT \ln Q$$

Note that under the given conditions, $RT = 1.58\,\text{kJ mol}^{-1}$.

(i) $\Delta_r G/(\text{kJ mol}^{-1}) = \Delta_r G^{\ominus}(1) - RT \ln p_{H_2O} = -23.6 - 1.58 \ln 1.3 \times 10^{-7} = +1.5.$

(ii) $\Delta_r G(\text{kJ mol}^{-1}) = \Delta_r G^{\ominus}(2) - RT \ln(p_{H_2O}p_{HNO_3})$

$$= -57.2 - 1.58 \ln\left[(1.3 \times 10^{-7}) \times (4.1 \times 10^{-10})\right] = +2.0.$$

(iii) $\Delta_r G/(\text{kJ mol}^{-1}) = \Delta_r G^{\ominus}(3) - RT \ln(p_{H_2O}^2 p_{HNO_3})$

$$= -85.6 - 1.58 \ln[(1.3 \times 10^{-7})^2 \times (4.1 \times 10^{-10})] = -1.3.$$

(iv) $\Delta_r G/(\text{kJ mol}^{-1}) = \Delta_r G^{\ominus}(4) - RT \ln(p_{H_2O}^3 p_{HNO_3})$

$$= -85.6 - 1.58 \ln[(1.3 \times 10^{-7})^3 \times (4.1 \times 10^{-10})] = -3.5.$$

So both the dihydrate and trihydrate form spontaneously from the vapor. Does one convert spontaneously into the other? Consider the reaction

$$HNO_3 \cdot 2H_2O(s) + H_2O(g) \rightleftharpoons HNO_3 \cdot 3H_2O(s)$$

which may be considered as reaction (iv) – reaction (iii). Therefore $\Delta_r G$ for this reaction is

$$\Delta_r G = \Delta_r G(4) - \Delta_r G(3) = -2.2 \text{ kJ mol}^{-1}.$$

We conclude that the dihydrate converts spontaneously to the $\boxed{\text{trihydrate}}$, the most stable solid (at least of the four we considered).

PART 2 Structure

8 Quantum theory: introduction and principles

Answers to discussion questions

D8.1   At the end of the nineteenth century and the beginning of the twentieth, there were many experimental results on the properties of matter and radiation that could not be explained on the basis of established physical principles and theories. Here we list only some of the most significant.

(1) The energy density distribution of blackbody radiation as a function of wavelength.
(2) The heat capacities of monatomic solids such as copper metal.
(3) The absorption and emission spectra of atoms and molecules, especially the line spectra of atoms.
(4) The frequency dependence of the kinetic energy of emitted electrons in the photoelectric effect.
(5) The diffraction of electrons by crystals in a manner similar to that observed for X-rays.

D8.3   The heat capacities of monatomic solids are primarily a result of the energy acquired by vibrations of the atoms about their equilibrium positions. If this energy can be acquired continuously, we expect that the equipartition of energy principle should apply. This principle states that, for each direction of motion and for each kind of energy (potential and kinetic), the associated energy should be $\frac{1}{2} kT$. Hence, for three directions and both kinds of motion, a total of $3 kT$, which gives a heat capacity of $3 k$ per atom, or $3 R$ per mole, independent of temperature. But the experiments show a temperature dependence. The heat capacity falls steeply below $3 R$ at low temperatures. Einstein showed that, by allowing the energy of the atomic oscillators to be quantized according to Planck's formula, rather than continuous, this temperature dependence could be explained. The physical reason is that at low temperatures only a few atomic oscillators have enough energy to populate the higher quantized levels; at higher temperatures more of them can acquire the energy to become active.

D8.5   If the wavefunction describing the linear momentum of a particle is precisely known, the particle has a definite state of linear momentum, but then, according to the uncertainty principle, the position of the particle is completely unknown as demonstrated in the derivation leading to eqn 8.21. Conversely, if the position of a particle is precisely known, its linear momentum cannot be described by a single wavefunction, but rather by a superposition of many wavefunctions, each corresponding to a different value for the linear momentum. Thus all knowledge of the linear momentum of the particle is lost. In the limit of an infinite number of superposed wavefunctions, the wavepacket illustrated in Fig. 8.31 turns into the sharply spiked packet shown in Fig. 8.30. But the requirement of the superposition of an infinite number of momentum wavefunctions in order to locate the particle means a complete lack of knowledge of the momentum.

Solutions to exercises

E8.1(b) The de Broglie relation is

$$\lambda = \frac{h}{p} = \frac{h}{mv} \quad \text{so} \quad v = \frac{h}{m\lambda} = \frac{6.626 \times 10^{-34}\,\mathrm{J\,s}}{(1.675 \times 10^{-27}\,\mathrm{kg}) \times (3.0 \times 10^{-2}\,\mathrm{m})}$$

$$v = \boxed{1.3 \times 10^{-5}\ \mathrm{m\,s^{-1}}} \quad \text{extremely slow!}$$

E8.2(b) The moment of a photon is

$$p = \frac{h}{\lambda} = \frac{6.626 \times 10^{-34}\mathrm{J\,s}}{350 \times 10^{-9}\,\mathrm{m}} = \boxed{1.89 \times 10^{-27}\,\mathrm{kg\,m\,s^{-1}}}$$

The momentum of a particle is

$$p = mv \quad \text{so} \quad v = \frac{p}{m} = \frac{1.89 \times 10^{-27}\,\mathrm{kg\,m\,s^{-1}}}{2(1.0078 \times 10^{-3}\,\mathrm{kg\,mol^{-1}}/6.022 \times 10^{23}\,\mathrm{mol^{-1}})}$$

$$v = \boxed{0.565\ \mathrm{m\ s^{-1}}}$$

E8.3(b) The uncertainty principle is

$$\Delta p \Delta x \geq \tfrac{1}{2}\hbar$$

so the minimum uncertainty in position is

$$\Delta x = \frac{\hbar}{2\Delta p} = \frac{\hbar}{2m\Delta v} = \frac{1.0546 \times 10^{-34}\,\mathrm{J\,s}}{2(9.11 \times 10^{-31}\,\mathrm{kg}) \times (0.000\,010) \times (995 \times 10^{3}\,\mathrm{m\,s^{-1}})}$$

$$= \boxed{5.8 \times 10^{-6}\,\mathrm{m}}$$

E8.4(b)
$$E = h\nu = \frac{hc}{\lambda}; \quad E(\text{per mole}) = N_A E = \frac{N_A hc}{\lambda}$$

$$hc = (6.626\,08 \times 10^{-34}\,\mathrm{J\,s}) \times (2.997\,92 \times 10^{8}\,\mathrm{m\,s^{-1}}) = 1.986 \times 10^{-25}\,\mathrm{J\,m}$$

$$N_A hc = (6.022\,14 \times 10^{23}\,\mathrm{mol^{-1}}) \times (1.986 \times 10^{-25}\,\mathrm{J\,m}) = 0.1196\,\mathrm{J\,m\,mol^{-1}}$$

Thus, $E = \dfrac{1.986 \times 10^{-25}\,\mathrm{J\,m}}{\lambda}$; $E(\text{per mole}) = \dfrac{0.1196\,\mathrm{J\,m\,mol^{-1}}}{\lambda}$

We can therefore draw up the following table

λ	E/J	$E/(\mathrm{kJ\ mol^{-1}})$
(a) 200 nm	0.93×10^{-19}	598
(b) 150 pm	1.32×10^{-15}	7.98×10^{5}
(c) 1.00 cm	1.99×10^{-23}	0.012

E8.5(b) Assuming that the ^{4}He atom is free and stationary, if a photon is absorbed, the atom acquires its momentum p achieving a speed v such that $p = mv$.

$$v = \frac{p}{m} \quad m = 4.00 \times 1.6605 \times 10^{-27}\,\text{kg} = 6.64\bar{2} \times 10^{-27}\,\text{kg}$$

$$p = \frac{h}{\lambda}$$

(a) $p = \dfrac{6.626 \times 10^{-34}\,\text{J s}}{200 \times 10^{-9}\,\text{m}} = 3.31\bar{3} \times 10^{-27}\,\text{kg m s}^{-1}$

$v = \dfrac{p}{m} = \dfrac{3.31\bar{3} \times 10^{-27}\,\text{kg m s}^{-1}}{6.642 \times 10^{-27}\,\text{kg}} = \boxed{0.499\,\text{m s}^{-1}}$

(b) $p = \dfrac{6.626 \times 10^{-34}\,\text{J s}}{150 \times 10^{-12}\,\text{m}} = 4.41\bar{7} \times 10^{-24}\,\text{kg m s}^{-1}$

$v = \dfrac{p}{m} = \dfrac{4.41\bar{7} \times 10^{-27}\,\text{kg m s}^{-1}}{6.642 \times 10^{-27}\,\text{kg}} = \boxed{665\ \text{m s}^{-1}}$

(c) $p = \dfrac{6.626 \times 10^{-34}\,\text{J s}}{1.00 \times 10^{-2}\,\text{m}} = 6.626 \times 10^{-32}\,\text{kg m s}^{-1}$

$v = \dfrac{p}{m} = \dfrac{6.626 \times 10^{-32}\,\text{kg m s}^{-1}}{6.642 \times 10^{-27}\,\text{kg}} = \boxed{9.98 \times 10^{-6}\,\text{m s}^{-1}}$

E8.6(b) Each emitted photon increases the momentum of the rocket by h/λ. The final momentum of the rocket will be Nh/λ, where N is the number of photons emitted, so the final speed will be $Nh/\lambda m_{\text{rocket}}$. The rate of photon emission is the power (rate of energy emission) divided by the energy per photon (hc/λ), so

$$N = \frac{tP\lambda}{hc} \quad \text{and} \quad v = \left(\frac{tP\lambda}{hc}\right) \times \left(\frac{h}{\lambda m_{\text{rocket}}}\right) = \frac{tP}{cm_{\text{rocket}}}$$

$$v = \frac{(10.0\,\text{yr}) \times (365\,\text{day yr}^{-1}) \times (24\,\text{h day}^{-1}) \times (3600\,\text{s h}^{-1}) \times (1.50 \times 10^{-3}\,\text{W})}{(2.998 \times 10^8\,\text{m s}^{-1}) \times (10.0\,\text{kg})}$$

$$= \boxed{158\ \text{m s}^{-1}}$$

E8.7(b) Rate of photon emission is rate of energy emission (power) divided by energy per photon (hc/λ)

(a) $\text{rate} = \dfrac{P\lambda}{hc} = \dfrac{(0.10\,\text{W}) \times (700 \times 10^{-9}\,\text{m})}{(6.626 \times 10^{-34}\,\text{J s}) \times (2.998 \times 10^8\,\text{m s}^{-1})} = \boxed{3.52 \times 10^{17}\,\text{s}^{-1}}$

(b) $\text{rate} = \dfrac{(1.0\,\text{W}) \times (700 \times 10^{-9}\,\text{J s})}{(6.626 \times 10^{-34}\,\text{J s}) \times (2.998 \times 10^8\,\text{m s}^{-1})} = \boxed{3.52 \times 10^{18}\,\text{s}^{-1}}$

E8.8(b) Conservation of energy requires

$$E_{\text{photon}} = \Phi + E_K = h\nu = hc/\lambda \quad \text{so} \quad E_K = hc/\lambda - \Phi$$

and $E_K = \frac{1}{2}m_e v^2 \quad \text{so} \quad v = \left(\dfrac{2E_K}{m_e}\right)^{1/2}$

(a) $E_K = \dfrac{(6.626 \times 10^{-34} \, J \, s) \times (2.998 \times 10^8 \, m \, s^{-1})}{650 \times 10^{-9} \, m} - (2.09 \, eV) \times (1.60 \times 10^{-19} \, J \, eV^{-1})$

But this expression is negative, which is unphysical. There is $\boxed{\text{no kinetic energy or velocity}}$ because the photon does not have enough energy to dislodge the electron.

(b) $E_K = \dfrac{(6.626 \times 10^{-34} \, J \, s) \times (2.998 \times 10^8 \, m \, s^{-1})}{195 \times 10^{-9} \, m} - (2.09 \, eV) \times (1.60 \times 10^{-19} \, J \, eV^{-1})$

$= \boxed{6.84 \times 10^{-19} \, J}$

and $v = \left(\dfrac{2(6.84 \times 10^{-19} \, J)}{9.11 \times 10^{-31} \, kg} \right)^{1/2} = \boxed{1.23 \times 10^6 \, m \, s^{-1}}$

E8.9(b) $E = h\nu = h/\tau$, so

(a) $E = 6.626 \times 10^{-34} \, J \, s / 2.50 \times 10^{-15} \, s = \boxed{2.65 \times 10^{-19} \, J = 160 \, kJ \, mol^{-1}}$

(b) $E = 6.626 \times 10^{-34} \, J \, s / 2.21 \times 10^{-15} \, s = \boxed{3.00 \times 10^{-19} \, J = 181 \, kJ \, mol^{-1}}$

(c) $E = 6.626 \times 10^{-34} \, J \, s / 1.0 \times 10^{-3} \, s = \boxed{6.62 \times 10^{-31} \, J = 4.0 \times 10^{-10} \, kJ \, mol^{-1}}$

E8.10(b)   The de Broglie wavelength is

$$\lambda = \frac{h}{p}$$

The momentum is related to the kinetic energy by

$$E_K = \frac{p^2}{2m} \quad \text{so} \quad p = (2mE_K)^{1/2}$$

The kinetic energy of an electron accelerated through 1 V is 1 eV $= 1.60 \times 10^{-19}$ J, so

$$\lambda = \frac{h}{(2mE_K)^{1/2}}$$

(a) $\lambda = \dfrac{6.626 \times 10^{-34} \, J \, s}{(2(9.11 \times 10^{-31} \, kg) \times (100 \, eV) \times (1.60 \times 10^{-19} \, J \, eV^{-1}))^{1/2}}$

$= \boxed{1.23 \times 10^{-10} \, m}$

(b) $\lambda = \dfrac{6.626 \times 10^{-34} \, J \, s}{(2(9.11 \times 10^{-31} \, kg) \times (1.0 \times 10^3 \, eV) \times (1.60 \times 10^{-19} \, J \, eV^{-1}))^{1/2}}$

$= \boxed{3.9 \times 10^{-11} \, m}$

(c) $\lambda = \dfrac{6.626 \times 10^{-34} \, J \, s}{(2(9.11 \times 10^{-31} \, kg) \times (100 \times 10^3 \, eV) \times (1.60 \times 10^{-19} \, J \, eV^{-1}))^{1/2}}$

$= \boxed{3.88 \times 10^{-12} \, m}$

E8.11(b) The upper sign in the following equations represents the math using the $\hat{A} + i\hat{B}$ operator. The lower sign is for the $\hat{A} - i\hat{B}$ operator. τ is a generalized coordinate.

$$\int \psi_i^* |\hat{A} \pm i\hat{B}| \psi_j d\tau = \int \psi_i^* |\hat{A}| \psi_j d\tau \pm i \int \psi_i^* |\hat{B}| \psi_j d\tau$$

$$= \left\{ \int \psi_j^* |\hat{A}| \psi_i d\tau \right\}^* \pm i \left\{ \int \psi_j^* |\hat{B}| \psi_i d\tau \right\}^* \quad \hat{A} \text{ and } \hat{B} \text{ are hermitian [8.30]}$$

$$= \left\{ \int \psi_j^* |\hat{A}| \psi_i d\tau \mp i \int \psi_j^* |\hat{B}| \psi_i d\tau \right\}^*$$

$$= \left\{ \int \psi_j^* |\hat{A} \mp i\hat{B}| \psi_i d\tau \right\}^*$$

This shows that the $\hat{A} \pm i\hat{B}$ operators are not hermitian. If they were hermitian, the result would be $\left\{ \int \psi_j^* |\hat{A} \pm i\hat{B}| \psi_i d\tau \right\}^*$.

E8.12(b) The minimum uncertainty in position is $\boxed{100 \text{ pm}}$. Therefore, since $\Delta x \Delta p \geq \frac{1}{2}\hbar$

$$\Delta p \geq \frac{\hbar}{2\Delta x} = \frac{1.0546 \times 10^{-34} \text{ J s}}{2(100 \times 10^{-12} \text{ m})} = 5.3 \times 10^{-25} \text{ kg m s}^{-1}$$

$$\Delta v = \frac{\Delta p}{m} = \frac{5.3 \times 10^{-25} \text{ kg m s}^{-1}}{9.11 \times 10^{-31} \text{ kg}} = \boxed{5.8 \times 10^5 \text{ m s}^{-1}}$$

E8.13(b) Conservation of energy requires

$$E_{\text{photon}} = E_{\text{binding}} + \frac{1}{2}m_e v^2 = h\nu = hc/\lambda \quad \text{so} \quad E_{\text{binding}} = hc/\lambda - \frac{1}{2}m_e v^2$$

$$\text{and } E_{\text{binding}} = \frac{(6.626 \times 10^{-34} \text{ J s}) \times (2.998 \times 10^8 \text{ m s}^{-1})}{121 \times 10^{-12} \text{ m}}$$

$$- \frac{1}{2}(9.11 \times 10^{-31} \text{ kg}) \times (5.69 \times 10^7 \text{ m s}^{-1})^2$$

$$= \boxed{1.67 \times 10^{-16} \text{ J}}$$

COMMENT. This calculation uses the non-relativistic kinetic energy, which is only about 3 percent less than the accurate (relativistic) value of 1.52×10^{-15} J. In this exercise, however, E_{binding} is a small difference of two larger numbers, so a small error in the kinetic energy results in a larger error in E_{binding}: the accurate value is $E_{\text{binding}} = 1.26 \times 10^{-16}$ J.

E8.14(b) The quality $\hat{\Omega}_1\hat{\Omega}_2 - \hat{\Omega}_2\hat{\Omega}_1$ [*Illustration* 8.3] is referred to as the commutator of the operators $\hat{\Omega}_1$ and $\hat{\Omega}_2$. In obtaining the commutator it is necessary to realize that the operators operate on functions; thus, we form

$$\hat{\Omega}_1\hat{\Omega}_2 f(x) - \hat{\Omega}_2\hat{\Omega}_1 f(x)$$

$$p_x = \frac{\hbar}{i} \frac{d}{dx}$$

Therefore $a = \left(\hat{x} + \hbar\frac{d}{dx} \right)$ and $a^\dagger = \left(\hat{x} - \hbar\frac{d}{dx} \right)$

Then $aa^\dagger f(x) = \dfrac{1}{2}\left(\hat{x} + \hbar\dfrac{d}{dx}\right) \times \left(\hat{x} - \hbar\dfrac{d}{dx}\right)f(x)$

and $a^\dagger a f(x) = \dfrac{1}{2}\left(\hat{x} - \hbar\dfrac{d}{dx}\right) \times \left(\hat{x} + \hbar\dfrac{d}{dx}\right)f(x)$

The terms in $\hat{x}^2$ and $(d/dx)^2$ obviously drop out when the difference is taken and are ignored in what follows; thus

$$aa^\dagger f(x) = \dfrac{1}{2}\left(-\hat{x}\hbar\dfrac{d}{dx} + \hbar\dfrac{d}{dx}x\right)f(x)$$

$$a^\dagger a f(x) = \dfrac{1}{2}\left(x\hbar\dfrac{d}{d}x - \hbar\dfrac{d}{dx}x\right)f(x)$$

These expressions are the negative of each other, therefore

$$(aa^\dagger - a^\dagger a)f(x) = \hbar\dfrac{d}{dx}\hat{x}f(x) - \hbar\hat{x}\dfrac{d}{dx}f(x)$$

$$= \hbar\left(\dfrac{d}{dx}\hat{x} - \hat{x}\dfrac{d}{dx}\right)f(x) = \hbar f(x)$$

Therefore, $(aa^\dagger - a^\dagger a) = \boxed{\hbar}$

Solutions to problems

Solutions to numerical problems

P8.1   A cavity approximates an ideal black body; hence the Planck distribution applies,

$$\rho = \frac{8\pi hc}{\lambda^5}\left(\frac{1}{e^{hc/\lambda kT} - 1}\right) \text{ [8.5]}.$$

Since the wavelength range is small (5 nm) we may write as a good approximation

$$\Delta E = \rho\Delta\lambda, \quad \lambda \approx 652.5\,\text{nm}.$$

$$\frac{hc}{\lambda k} = \frac{(6.626 \times 10^{-34}\,\text{J s}) \times (2.998 \times 10^8\,\text{m s}^{-1})}{(6.525 \times 10^{-7}\,\text{m}) \times (1.381 \times 10^{-23}\,\text{J K}^{-1})} = 2.205 \times 10^4\,\text{K}.$$

$$\frac{8\pi hc}{\lambda^5} = \frac{(8\pi) \times (6.626 \times 10^{-34}\,\text{J s}) \times (2.998 \times 10^8\,\text{m s}^{-1})}{(652.5 \times 10^{-9}\,\text{m})^5} = 4.221 \times 10^7\,\text{J m}^{-4}.$$

$$\Delta E = (4.221 \times 10^7\,\text{J m}^{-4}) \times \left(\frac{1}{e^{(2.205\times 10^4\,\text{K})/T} - 1}\right) \times (5 \times 10^{-9}\,\text{m}).$$

(a) $T = 298\,\text{K}, \Delta E = \dfrac{0.211\,\text{J m}^{-3}}{e^{(2.205\times 10^4)/298} - 1} = \boxed{1.6 \times 10^{-33}\,\text{J m}^{-3}}.$

(b) $T = 3273\,\text{K}, \Delta E = \dfrac{0.211\,\text{J m}^{-3}}{e^{(2.205\times 10^4)/3273} - 1} = \boxed{2.5 \times 10^{-4}\,\text{J m}^{-3}}.$

COMMENT. The energy density in the cavity does not depend on the volume of the cavity, but the total energy in any given wavelength range does, as well as the total energy over all wavelength ranges.

Question. What is the total energy in this cavity within the range 650–655 nm at the stated temperatures?

P8.3
$$\theta_E = \frac{h\nu}{k}, \qquad [\theta_E] = \frac{J\,s \times s^{-1}}{J\,K^{-1}} = K.$$

In terms of θ_E the Einstein equation [8.7] for the heat capacity of solids is

$$C_V = 3R \left(\frac{\theta_E}{T}\right)^2 \times \left(\frac{e^{\theta_E/2T}}{e^{\theta_E/T} - 1}\right)^2, \quad \text{classical value} = 3R.$$

It reverts to the classical value when $T \gg \theta_E$ or when $\dfrac{h\nu}{kT} \ll 1$ as demonstrated in the text (Section 8.1). The criterion for classical behavior is therefore that $\boxed{T \gg \theta_E}$.

$$\theta_E = \frac{h\nu}{k} = \frac{(6.626 \times 10^{-34}\,J\,Hz^{-1}) \times \nu}{1.381 \times 10^{-23}\,J\,K^{-1}} = 4.798 \times 10^{-11}\,(\nu/Hz)K.$$

(a) For $\nu = 4.65 \times 10^{13}$ Hz, $\theta_E = (4.798 \times 10^{-11}) \times (4.65 \times 10^{13}\,K) = \boxed{2231\ K}$.

(b) For $\nu = 7.15 \times 10^{12}$ Hz, $\theta_E = (4.798 \times 10^{-11}) \times (7.15 \times 10^{12}\,K) = \boxed{343\ K}$.

Hence

(a) $\dfrac{C_V}{3R} = \left(\dfrac{2231K}{298K}\right)^2 \times \left(\dfrac{e^{2231/(2\times298)}}{e^{2231/298} - 1}\right)^2 = \boxed{0.031}$.

(b) $\dfrac{C_V}{3R} = \left(\dfrac{343\ K}{298K}\right)^2 \times \left(\dfrac{e^{2231/(2\times298}}{e^{343/298} - 1}\right)^2 = \boxed{0.897}$.

COMMENT. For many metals the classical value is approached at room temperature; consequently, the failure of classical theory became apparent only after methods for achieving temperatures well below 25°C were developed in the latter part of the nineteenth century.

P8.5
The hydrogen atom wavefunctions are obtained from the solution of the Schrödinger equation in Chapter 10. Here we need only the wavefunction that is provided. It is the square of the wavefunction that is related to the probability (Section 8.4).

$$\psi^2 = \frac{1}{\pi a_0^3}e^{-2r/a_0}, \qquad \delta\tau = \frac{4}{3}\pi r_0^3, \quad r_0 = 1.0\,\text{pm}.$$

If we assume that the volume $\delta\tau$ is so small that ψ does not vary within it, the probability is given by

$$\psi^2\delta\tau = \frac{4r_0^3}{3a_0^3}e^{-2r/a_0} = \frac{4}{3} \times \left(\frac{1.0}{53}\right)^3 e^{-2r/a_0}.$$

(a) $r = 0$: $\psi^2 \delta\tau = \dfrac{4}{3}\left(\dfrac{1.0}{53}\right)^3 = \boxed{9.0 \times 10^{-6}}$.

(b) $r = a_0$: $\psi^2 \delta\tau = \dfrac{4}{3}\left(\dfrac{1.0}{53}\right)^3 e^{-2} = \boxed{1.2 \times 10^{-6}}$.

Question. If there is a nonzero probability that the electron can be found at $r = 0$ how does it avoid destruction at the nucleus? (*Hint.* See Chapter 10 for part of the solution to this difficult question.)

P8.7   According to the uncertainty principle,

$$\Delta p \Delta q \geq \frac{1}{2}\hbar,$$

where Δq and Δp are root-mean-square deviations:

$$\Delta q = \left(\langle x^2\rangle - \langle x\rangle^2\right)^{1/2} \quad\text{and}\quad \Delta p = \left(\langle p^2\rangle - \langle p\rangle^2\right)^{1/2}.$$

To verify whether the relationship holds for the particle in a state whose wavefunction is

$$\psi = (2a/\pi)^{1/4} e^{-ax^2},$$

we need the quantum-mechanical averages $\langle x\rangle$, $\langle x^2\rangle$, $\langle p\rangle$, and $\langle p^2\rangle$.

$$\langle x\rangle = \int \psi^* x^2 \psi\, d\tau = \int_{\infty-}^{\infty} \left(\frac{2a}{\pi}\right)^{1/4} e^{-ax^2} x \left(\frac{2a}{\pi}\right)^{1/4} e^{-ax^2}\, dx,$$

$$\langle x\rangle = \left(\frac{2a}{\pi}\right)^{1/2} \int_{-\infty}^{\infty} x e^{-2ax^2}\, dx = 0;$$

$$\langle x^2\rangle = \int_{-\infty}^{\infty} \left(\frac{2a}{\pi}\right)^{1/4} e^{-ax^2} x^2 \left(\frac{2a}{\pi}\right)^{1/4} e^{-ax^2}\, dx = \left(\frac{2a}{\pi}\right)^{1/2} \int_{-\infty}^{\infty} x^2 e^{-2ax^2}\, dx,$$

$$\langle x^2\rangle = \left(\frac{2a}{\pi}\right)^{1/2} \frac{\pi^{1/2}}{2(2a)^{3/2}} = \frac{1}{4a};$$

so $\Delta q = \dfrac{1}{2a^{1/2}}$.

$$\langle p\rangle = \int_{-\infty}^{\infty} \psi^* \left(\frac{\hbar}{i}\frac{d\psi}{dx}\right) dx \quad\text{and}\quad \langle p^2\rangle = \int_{-\infty}^{\infty} \psi^* \left(-\hbar^2 \frac{d^2\psi}{dx^2}\right) dx.$$

We need to evaluate the derivatives:

$$\frac{d\psi}{dx} = \left(\frac{2a}{\pi}\right)^{1/4} (-2ax) e^{-ax^2}$$

and $\quad \dfrac{d^2\psi}{dx^2} = \left(\dfrac{2a}{\pi}\right)^{1/4} [(-2ax)^2 e^{-ax^2} + (-2a)e^{-ax^2}] = \left(\dfrac{2a}{\pi}\right)^{1/4} (4a^2x^2 - 2a)e^{-ax^2}.$

So $\quad \langle p \rangle = \displaystyle\int_{-\infty}^{\infty} \left(\dfrac{2a}{\pi}\right)^{1/4} e^{-ax^2} \left(\dfrac{\hbar}{i}\right) \left(\dfrac{2a}{\pi}\right)^{1/4} (-2ax)e^{-ax^2} dx$

$\qquad = -\dfrac{2\hbar}{i} \left(\dfrac{2a}{\pi}\right)^{1/2} \displaystyle\int_{-\infty}^{\infty} xe^{-2ax^2} dx = 0;$

$\left\langle p^2 \right\rangle = \displaystyle\int_{-\infty}^{\infty} \left(\dfrac{2a}{\pi}\right)^{1/4} e^{-ax^2} (-\hbar^2) \left(\dfrac{2a}{\pi}\right)^{1/4} (4a^2x^2 - 2a)e^{-ax^2} dx,$

$\left\langle p^2 \right\rangle = (-2a\hbar^2)\dfrac{2a}{\pi}^{1/2} \displaystyle\int_{-\infty}^{\infty} (2ax^2 - 1)e^{-2ax^2} dx,$

$\left\langle p^2 \right\rangle = (-2a\hbar^2)\left(\dfrac{2a}{\pi}\right)^{1/2} \left(2a\dfrac{\pi^{1/2}}{2(2a)^{3/2}} - \dfrac{\pi^{1/2}}{(2a)^{1/2}}\right) = a\hbar^2;$

and $\quad \Delta p = a^{1/2}\hbar.$

Finally, $\quad \Delta q \Delta p = \dfrac{1}{2a^{1/2}} \times a^{1/2}\hbar = \dfrac{1}{2}\hbar,$

which is the minimum product consistent with the uncertainty principle.

Solutions to theoretical problems

P8.9 $\qquad \rho = \dfrac{8\pi hc}{\lambda^5} \left(\dfrac{1}{e^{hc/\lambda kT} - 1}\right)$ [8.5]

As λ increases, $hc/\lambda kT$ decreases, and at very long wavelength $hc/\lambda kT \ll 1$. Hence we can expand the exponential in a power series. Let $x = hc/\lambda kT$, then

$e^x = 1 + x + \dfrac{1}{2!}x^2 + \dfrac{1}{3!}x^3 + \cdots,$

$\rho = \dfrac{8\pi hc}{\lambda^5} \left[\dfrac{1}{1 + x + \dfrac{1}{2!}x^2 + \dfrac{1}{3!}x^3 + \cdots - 1}\right],$

$\displaystyle\lim_{\lambda \to \infty} \rho = \dfrac{8\pi hc}{\lambda^5} \left[\dfrac{1}{1 + x - 1}\right] = \dfrac{8\pi hc}{\lambda^5} \left(\dfrac{1}{hc/\lambda kT}\right).$

$\qquad = \dfrac{8\pi kT}{\lambda^4}$

This is the Rayleigh–Jeans law [8.3].

P8.11
$$\mathcal{E} = \int_0^\infty \rho(\lambda)\, d\lambda = 8\pi hc \int_0^\infty \frac{d\lambda}{\lambda^5 \left(e^{hc/\lambda kT} - 1\right)} \cdot [8.5].$$

Let $x = \dfrac{hc}{\lambda kT}$. Then, $dx = -\dfrac{hc}{\lambda^2 kT}\, d\lambda$ or $d\lambda = -\dfrac{\lambda^2 kT}{hc}\, dx$.

$$\mathcal{E} = 8\pi kT \int_0^\infty \frac{\lambda^2 dx}{\lambda^5 (e^x - 1)} = 8\pi kT \int_0^\infty \frac{dx}{\lambda^3 (e^x - 1)}$$

$$= 8\pi kT \left(\frac{kT}{hc}\right)^3 \int_0^\infty \frac{x^3 dx}{(e^x - 1)} = 8\pi kT \left(\frac{kT}{hc}\right)^3 \left(\frac{\pi^4}{15}\right).$$

$$\mathcal{E} = \left(\frac{8\pi^5 k^4}{15 h^3 c^3}\right) T^4 = \left(\frac{4}{c}\right)\sigma T^4$$

where $\sigma = \boxed{(2\pi^5 k^4 / 15 h^3 c^2)}$ is the Stefan–Boltzmann constant.

P8.13
We require $\int \psi^* \psi\, d\tau = 1$, and so write $\psi = Nf$ and find N for the given f.

(a) $N^2 \displaystyle\int_0^L \sin^2 \frac{n\pi x}{L}\, dx = \frac{1}{2} N^2 \int_0^L \left(1 - \cos\frac{2n\pi x}{L}\right) dx$ [trigonometric identity]

$$= \frac{1}{2} N^2 \left(x - \frac{L}{2n\pi}\sin\frac{2n\pi x}{L}\right)\Big|_0^L$$

$$= \frac{L}{2} N^2 = 1 \quad \text{if} \boxed{N = \left(\frac{2}{L}\right)^{1/2}}.$$

(b) $N^2 \displaystyle\int_{-L}^L c^2\, dx = 2N^2 c^2 L = 1 \text{ if } \boxed{N = \frac{1}{c(2L)^{1/2}}}.$

(c) $N^2 \displaystyle\int_0^\infty e^{-2r/a} r^2\, dr \int_0^\pi \sin\theta\, d\theta \int_0^{2\pi} d\phi \quad [d\tau = r^2 \sin\theta\, dr\, d\theta\, d\phi]$

$$= N^2 \left(\frac{a^3}{4}\right) \times (2) \times (2\pi) = 1 \text{ if } \boxed{N = \frac{1}{(\pi a^3)^{1/2}}}.$$

(d) $N^2 \displaystyle\int_0^\infty r^2 \times r^2 e^{-r/a}\, dr \int_0^\pi \sin^3\theta\, d\theta \int_0^{2\pi} \cos^2\phi\, d\phi \quad [x = r\cos\phi\sin\theta]$

$$= N^2 4! a^5 \times \frac{4}{3} \times \pi = 32\pi a^5 N^2 = 1 \text{ if } \boxed{N = \frac{1}{(32\pi a^5)^{1/2}}}.$$

We have used $\int \sin^3\theta\, d\theta = -\frac{1}{3}(\cos\theta)(\sin^2\theta + 2)$, as found in tables of integrals and

$$\int_0^{2\pi} \cos^2\phi\, d\phi = \int_0^{2\pi} \sin^2\phi\, d\phi$$

by symmetry with $\displaystyle\int_0^{2\pi} (\cos^2\phi + \sin^2\phi)\, d\phi = \int_0^{2\pi} d\phi = 2\pi.$

P8.15 In each case form $\hat{\Omega}f$. If the result is ωf where ω is a constant, then f is an eigenfunction of the operator $\hat{\Omega}$ and ω is the eigenvalue [8.25b].

(a) $\dfrac{d}{dx}e^{ikx} = ik\,e^{ikx};$ $\boxed{\text{yes; eigenvalue } =ik}$.

(b) $\dfrac{d}{dx}\cos kx = -k\sin kx;$ no.

(c) $\dfrac{d}{dx}k = 0;$ $\boxed{\text{yes; eigenvalue } = 0}$.

(d) $\dfrac{d}{dx}kx = k = \dfrac{1}{x}kx;$ no $[1/x$ is not a constant$]$.

(e) $\dfrac{d}{dx}e^{-\alpha x^2} = -2\alpha x\,e^{-\alpha x^2};$ no $[-2\alpha x$ is not a constant$]$.

P8.17 Follow the procedure of Problem 8.15.

(a) $\dfrac{d^2}{dx^2}e^{ikx} = -k^2 e^{ikx};$ yes; eigenvalue $= \boxed{-k^2}$.

(b) $\dfrac{d^2}{dx^2}\cos kx = -k^2\cos kx;$ yes; eigenvalue $= \boxed{-k^2}$.

(c) $\dfrac{d^2}{dx^2}k = 0;$ yes; eigenvalue $= \boxed{0}$.

(d) $\dfrac{d^2}{dx^2}kx = 0;$ yes; eigenvalue $= \boxed{0}$.

(e) $\dfrac{d^2}{dx^2}e^{-\alpha x^2} = (-2\alpha + 4\alpha^2 x^2)e^{-\alpha x^2};$ no.

Hence, $\boxed{\text{(a, b, c, d) are eigenfunctions of } \dfrac{d^2}{dx^2}; \text{ (b, d) are eigenfunctions of } \dfrac{d^2}{dx^2}, \text{ but not of } \dfrac{d}{dx}}$.

P8.19 The kinetic energy operator, $\hat{T}$, is obtained from the operator analog of the classical equation

$$E_K = \frac{p^2}{2m},$$

that is,

$$\hat{T} = \frac{(\hat{p})^2}{2m},$$

$$\hat{p}_x = \frac{\hbar}{i}\frac{d}{dx}\ [8.26]; \quad \text{hence} \quad \hat{p}_x^2 = -\hbar^2\frac{d^2}{dx^2} \quad \text{and} \quad \hat{T} = -\frac{\hbar^2}{2m}\frac{d^2}{dx^2}.$$

Then

$$\langle T\rangle = N^2\int\psi^*\left(\frac{\hat{p}_x^2}{2m}\right)\psi\,d\tau = \frac{\int\psi^*\,(\hat{p}^2/2m)\,\psi\,d\tau}{\int\psi^*\psi\,d\tau}\quad\left[N^2 = \frac{1}{\int\psi^*\psi\,d\tau}\right]$$

$$= \frac{\dfrac{-\hbar^2}{2m}\int\psi^*\dfrac{d^2}{dx^2}(e^{ikx}\cos\chi + e^{-ikx}\sin\chi)\,d\tau}{\int\psi^*\psi\,d\tau}$$

$$= \frac{\dfrac{-\hbar^2}{2m}\int\psi^*(-k^2)\times(e^{ikx}\cos\chi + e^{-ikx}\sin\chi)\,d\tau}{\int\psi^*\psi\,d\tau} = \frac{\hbar^2 k^2\int\psi^*\psi\,d\tau}{2m\int\psi^*\psi\,d\tau} = \boxed{\dfrac{\hbar^2 k^2}{2m}}.$$

P8.21 $\langle r \rangle = N^2 \int \psi^* r \psi \, d\tau, \quad \langle r^2 \rangle = N^2 \int \psi^* r^2 \psi \, d\tau.$

(a) $\psi = \left(2 - \dfrac{r}{a_0}\right) e^{-r/2a_0}, \quad N = \left(\dfrac{1}{32\pi a_0^3}\right)^{1/2}$ [Problem 8.14].

$$\langle r \rangle = \frac{1}{32\pi a_0^3} \int_0^\infty r \left(2 - \frac{r}{a_0}\right)^2 r^2 e^{-r/a_0} dr \times 4\pi \left[\int_0^\pi \sin\theta d\theta \int_0^{2\pi} d\phi = 4\pi\right]$$

$$= \frac{1}{8a_0^3} \int_0^\infty \left(4r^3 - \frac{4r^4}{a_0} + \frac{r^5}{a_0^2}\right) e^{-r/a_0} dr$$

$$= \frac{1}{8a_0^3} (4 \times 3! a_0^4 - 4 \times 4! a_0^4 + 5! a_0^4) = \boxed{6a_0} \left[\int_0^\infty x^n e^{-ax} dx = \frac{n!}{a^{n+1}}\right].$$

$$\langle r^2 \rangle = \frac{1}{8a_0^3} \int_0^\infty \left(4r^4 - \frac{4r^5}{a_0} + \frac{r^6}{a_0^2}\right) e^{-r/a_0} dr = \frac{1}{8a_0^3}(4 \times 4! - 4 \times 5! + 6!)a_0^5$$

$$= \boxed{42\, a_o^2}.$$

(b) $\psi = Nr \sin\theta \cos\phi \, e^{-r/2a_0}, \quad N = \left(\dfrac{1}{32\pi a_0^5}\right)^{1/2}$ [Problem 8.14].

$$\langle r \rangle = \frac{1}{32\pi a_0^5} \int_0^\infty r^5 e^{-r/a_0} dr \times \frac{4\pi}{3} = \frac{1}{24a_0^5} \times 5! a_0^6 = \boxed{5a_0}.$$

$$\langle r^2 \rangle = \frac{1}{24a_0^5} \int_0^\infty r^6 e^{-r/a_0} dr = \frac{1}{24a_0^5} \times 6! a_0^7 = \boxed{30a_0^2}.$$

P8.23 The superpositions of cosine functions of the form $\cos(nx)$ can be chosen with n equal to any integer between 1 and m. For convenience, x can be examined in the range between $-\pi/2$ and $\pi/2$. The normalization constant for each function is determined by integrating the function squared over the range of x [8.15]. Using MathCad to perform the integration, we find:

$$\int_{-\pi/2}^{\pi/2} \cos(n \cdot x)^2 dx \rightarrow \frac{1}{2} \cdot \frac{\left(2 \cdot \cos\left(\frac{1}{2} \cdot \pi \cdot n\right) \sin\left(\frac{1}{2} \cdot \pi \cdot n\right) + \pi \cdot n\right)}{n}$$

When n is an even integer, $\sin(\pi n/2) = 0$ and, when n is an odd integer, $\cos(\pi n/2) = 0$. Consequently, when n is an integer, the above integral equals $\pi/2$ and we select $(2/\pi)^{1/2}$ as the normalization constant for the function $\cos(nx)$. The normalized function is $\phi(n,x)$. The superposition, $\psi(m,x)$, is the sum of these cosine functions from $n = 1$ to $n = m$. Since the cosine functions are orthogonal, $\psi(m,x)$ has a normalization constant equal to $(1/m)^{1/2}$.

$$\phi(n,x) := \sqrt{\frac{2}{\pi}} \cdot \cos(n \cdot x) \qquad \psi(m,x) := \sqrt{\frac{1}{m}} \cdot \sum_{n=1}^{m} \phi(n,x)$$

Constants and variables needed for MathCad plots and computations:

$$N := 1000 \quad x_{\min} := \frac{-\pi}{2} \quad x_{\max} := \frac{\pi}{2} \quad i := 0..N \quad x_i := x_{\min} + \frac{i}{N} \cdot (x_{\max} - x_{\min})$$

Examination of Fig. 8.1(a) reveals that, when the superposition has few terms, the particle position is ill-defined. There is great uncertainty in knowledge of position. However, when many terms are added to the superposition, the uncertainty narrows to a small region around $x = 0$. A plot with m greater than 10 will further confirm this conclusion.

Each function in the superposition has been assigned a weight equal to the normalization constant $(1/m)^{1/2}$ [8.33]. This means that each cosine function in the superposition has an identical probability contribution to expectation value for momentum (see *Justification* 8.4). Each cosine function contributes with a probability equal to $1/m$. Furthermore, each cosine function represents a particle momentum that is proportional to the argument n because $(d^2\phi(n,x)/dx^2)^{1/2}$ (the differential component of the squared momentum operator) is proportional to n. The following plot [Fig. 8.1(b)] of momentum probability against momentum, as represented by n, is an interesting contrast to the plot of probability density against position.

Variables needed for the MathCad plot: $n := 0..15 \quad \text{Prob}(n,m) := \text{if}\left(n < m, \frac{1}{m}, 0\right)$

This plot shows that, when there are many terms in the superposition, the range of possible momentum is very broad even though the range of observed positions becomes narrow. Position and momentum are Probability density plots [Fig. 8.1(a)] for superpositions that have 1, 3, and 10 terms:

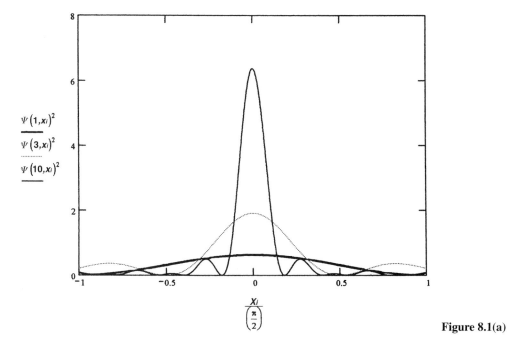

Figure **8.1(a)**

complementary variables. As location becomes more precise with the superposition of many functions, precise knowledge of momentum decreases. This illustrates the Heisenberg uncertainty principle [8.36a].

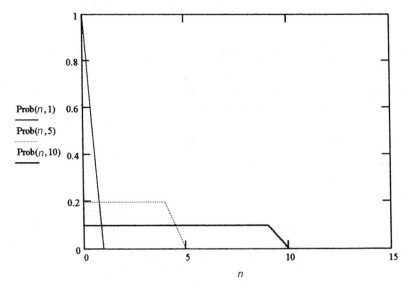

Prob$(n,1)$

Prob$(n,5)$

Prob$(n,10)$

n

Figure 8.1(b)

The plot of probability density against position clearly indicates that the superposition is symmetrical around the point $x = 0$. Consequently, the expectation position for all superpositions is $x = 0$. The expectation value for position is independent of the number of terms in the superposition.

The square root of the expectation value of x^2 is called the root-mean-square value of x, x_{rms}. A plot of x_{rms} against m [Fig. 8.1(c)] indicates that this expectation value depends upon the number of terms in the superposition. However, it does appear to very slowly converge to a very small (zero?) value when the superposition contains many functions.

$$x_{\text{rms}}(m) := \left(\int_{-\pi/2}^{\pi/2} x^2 \cdot \psi(m,x)^2 \mathrm{d}x \right)^{1/2} \qquad m := 1 \dots 50$$

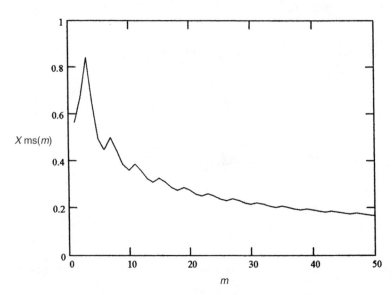

$X\,ms(m)$

m

Figure 8.1(c)

P8.25   **(a)** In the momentum representation $\hat{p}_x = p_x \times$, consequently

$$\left[\hat{x}, \hat{p}_x\right]\phi = \left[\hat{x}, p_x\times\right]\phi = \hat{x}p_x \times \phi - p_x \times \hat{x}\phi = i\hbar\phi \;\; [8.39]$$

Suppose that the position operator has the form $\hat{x} = a(d/dp_x)$ where a is a complex number. Then,

$$a\frac{d}{dp_x}(p_x \times \phi) - p_x \times \left(a\frac{d}{dp_x}\phi\right) = i\hbar\phi,$$

$$\frac{d}{dp_x}(p_x \times \phi) - p_x \times \left(\frac{d}{dp_x}\phi\right) = \frac{i\hbar}{a}\phi,$$

$$\frac{dp_x}{dp_x}\phi + p_x\frac{d\phi}{dp_x} - p_x\frac{d\phi}{dp_x} = \frac{i\hbar}{a}\phi \;\; [\text{Rule for differentiation of } f(x)g(x)].$$

$$\phi = \frac{i\hbar}{a}\phi$$

This is true when $a = i\hbar$. We conclude that $\boxed{\hat{x} = i\hbar(d/dp_x)}$ in the momentum representation.

(b) The fact that integration is the inverse of differentiation suggests the guess that in the momentum representation

$$\hat{x}^{-1}\phi = \left(i\hbar\frac{d}{dp_x}\right)^{-1}\phi = \left(\frac{1}{i\hbar}\int_{-\infty}^{p_x} dp_x\right)\phi = \frac{1}{i\hbar}\int_{-\infty}^{p_x}\phi\,dp_x,$$

where the symbol $\int_{-\infty}^{p_x} dp_x$ is understood to be an integration operator which uses any function on its right side as an integrand. To validate the guess that $\boxed{\hat{x}^{-1} = (1/i\hbar)\int_{-\infty}^{p_x} dp_x}$ we need to confirm the operator relationship $\hat{x}^{-1}\hat{x} = \hat{x}\hat{x}^{-1} = \hat{1}$. Using Leibnitz's rule for differentiation of integrals:

$$\hat{x}\hat{x}^{-1}\phi = \left(i\hbar\frac{d}{dp_x}\right)\left(\frac{1}{i\hbar}\int_{-\infty}^{p_x} dp_x\right)\phi = \left(\frac{d}{dp_x}\right)\left(\int_{-\infty}^{p_x}\phi\,dp_x\right)$$

$$= \left(\int_{-\infty}^{p_x}\frac{d\phi}{dp_x}\,dp_x\right) + \phi(p_x)\lim_{c\to-\infty}\frac{dc}{dp_x} - \phi(-\infty)\frac{dp_x}{dp_x} = \left(\int_{-\infty}^{p_x}\frac{d\phi}{dp_x}\,dp_x\right) - \phi(-\infty).$$

Since $\phi(-\infty)$ must equal zero, we find that

$$\hat{x}\hat{x}^{-1}\phi = \int_{-\infty}^{p_x}\frac{d\phi}{dp_x}\,dp_x = \phi$$

from which we conclude that $\hat{x}\hat{x}^{-1} = \hat{1}$.

$$\hat{x}^{-1}\hat{x}\phi = \left(\frac{1}{i\hbar}\int_{-\infty}^{p_x} dp_x\right)\left(i\hbar\frac{d}{dp_x}\right)\phi = \int_{-\infty}^{p_x} d\phi = \phi(p_x) - \phi(-\infty) = \phi(p_x) = \phi.$$

Solutions to applications

P8.27   **(a)** $\lambda_{\text{non-relativistic}} = \dfrac{h}{(2m_e eV)^{1/2}}$

$$= \frac{6.626 \times 10^{-34}\,\text{J s}}{\left\{2\left(9.109 \times 10^{-31}\text{kg}\right) \times \left(1.602 \times 10^{-19}\,\text{C}\right) \times \left(50.0 \times 10^3\,\text{V}\right)\right\}^{1/2}}.$$

$\lambda_{\text{non-relativistic}} = 5.48$ pm [8.12 and Example 8.2].

$$\lambda_{\text{relativistic}} = \frac{h}{\left\{2m_e eV\left(1 + \dfrac{eV}{2m_e c^2}\right)\right\}^{1/2}}$$

$$= \frac{\lambda_{\text{non-relativistic}}}{\left(1 + \dfrac{eV}{2m_e c^2}\right)^{1/2}} = \frac{5.48 \text{ pm}}{\left\{1 + \dfrac{\left(1.602 \times 10^{-19}\,\text{C}\right)\left(50.0 \times 10^3\,\text{V}\right)}{2\left(9.109 \times 10^{-31}\text{ kg}\right)\left(3.00 \times 10^8\text{ m s}^{-1}\right)^2}\right\}^{1/2}}$$

$$= \frac{5.48 \text{ pm}}{\{1 + 0.0489\}^{1/2}} = \boxed{5.35 \text{ pm}}.$$

(b) For an electron accelerated through 50 kV the non-relativistic de Broglie wavelength is calculated to be high by 2.4%. This error may be insignificant for many applications. However, should an accuracy of 1% or better be required, use the relativistic equation at accelerations through a potential above 20.4 V as demonstrated in the following calculation.

$$\frac{\lambda_{\text{non-relativistic}} - \lambda_{\text{relativistic}}}{\lambda_{\text{relativistic}}} = \frac{\lambda_{\text{non-relativistic}}}{\lambda_{\text{relativistic}}} - 1 = \left(1 + \frac{eV}{2m_e c^2}\right)^{1/2} - 1$$

$$= \cancel{1} + \frac{1}{2}\left(\frac{eV}{2m_e c^2}\right) - \frac{1}{2 \cdot 4}\left(\frac{eV}{2m_e c^2}\right)^2$$

$$+ \frac{1 \cdot 3}{2 \cdot 4 \cdot 6}\left(\frac{eV}{2m_e c^2}\right)^3 - \cdots \cancel{1}$$

$$= \frac{1}{2}\left(\frac{eV}{2m_e c^2}\right) \quad \text{because 2nd and 3rd order terms are very small.}$$

The largest value of V for which the non-relativistic equation yields a value that has less than 1% error:

$$V \simeq 2\left(\frac{2m_e c^2}{e}\right) \times \left(\frac{\lambda_{\text{non-relativistic}} - \lambda_{\text{relativistic}}}{\lambda_{\text{relativistic}}}\right) = 2\left(\frac{2m_e c^2}{e}\right)(0.01) = 20.4\,\text{kV}.$$

P8.29   **(a)** $CH_4(g) \rightarrow C(\text{graphite}) + 2H_2(g).$

$\Delta_r G^{\ominus} = -\Delta_f G^{\ominus}(CH_4) = -(-50.72\,\text{kJ mol}^{-1}) = 50.72\,\text{kJ mol}^{-1}$ at $T = 25°C.$

$\Delta_r H^{\ominus} = -\Delta_f H^{\ominus}(CH_4) = -(-74.81\,\text{kJ mol}^{-1}) = 74.81\,\text{kJ mol}^{-1}$ at $T.$

We want to find the temperature at which $\Delta_r G^{\ominus}(T) = 0$. Below this temperature methane is stable with respect to decomposition into the elements. Above this temperature it is unstable. Assuming that the heat capacities are basically independent of temperature,

$$\Delta_r C_p^{\ominus}(T) \approx \Delta_r C_p^{\ominus}(T^{\ominus}) = [8.527 + 2(28.824) - 35.31]\,\mathrm{J\,K^{-1}\,mol^{-1}}$$
$$\approx 30.865\,\mathrm{J\,K^{-1}\,mol^{-1}}.$$

$$\Delta_r H^{\ominus}(T) = \Delta_r H^{\ominus}(T^{\ominus}) + \int_{T^{\ominus}}^{T} \Delta_r C_p^{\ominus}(T)\,dT \quad [2.36]$$
$$= \Delta_r H^{\ominus}(T^{\ominus}) + \Delta_r C_p^{\ominus} \times (T - T^{\ominus}).$$

$$\left(\frac{\partial}{\partial T}\left(\frac{\Delta_r G^{\ominus}}{T}\right)\right)_p = -\frac{\Delta_r H^{\ominus}}{T^2} \quad [3.52].$$

At constant pressure (the standard pressure)

$$\int_{T^{\ominus}}^{T} d(\Delta_r G^{\ominus}/T) = -\int_{T^{\ominus}}^{T} \frac{\Delta_r H^{\ominus}}{T^2}\,dT,$$

$$\frac{\Delta_r G^{\ominus}(T)}{T} = \frac{\Delta_r G^{\ominus}(T^{\ominus})}{T^{\ominus}} - \int_{T^{\ominus}}^{T} \frac{\Delta_r H^{\ominus}(T^{\ominus}) + \Delta_r C_p^{\ominus} \times (T - T^{\ominus})}{T^2}\,dT$$

$$= \frac{\Delta_r G^{\ominus}(T^{\ominus})}{T^{\ominus}} - [\Delta_r H^{\ominus}(T^{\ominus}) - \Delta_r C_p^{\ominus} \times T^{\ominus}] \int_{T^{\ominus}}^{T} \frac{1}{T^2}\,dT - \Delta_r C_p^{\ominus}\int_{T^{\ominus}}^{T} \frac{1}{T}\,dT$$

$$= \frac{\Delta_r G^{\ominus}(T^{\ominus})}{T^{\ominus}} + [\Delta_r H^{\ominus}(T^{\ominus}) - \Delta_r C_p^{\ominus} \times T^{\ominus}] \times \left[\frac{1}{T} - \frac{1}{T^{\ominus}}\right] - \Delta_r C_p^{\ominus}\ln\left(\frac{T}{T^{\ominus}}\right).$$

The value of T for which $\Delta_r G^{\ominus}(T) = 0$ can be determined by examination of a plot (Fig. 8.2) of $\dfrac{\Delta_r G^{\ominus}(T)}{T}$ against T.

$$\frac{\Delta_r G^{\ominus}(T^{\ominus})}{T^{\ominus}} = 50.72\,\mathrm{kJ\,mol^{-1}}/298.15\,\mathrm{K} = 0.1701\,\mathrm{kJ\,K^{-1}\,mol^{-1}}.$$

$$\Delta_r H^{\ominus}(T^{\ominus}) - \Delta_r C_p^{\ominus} \times T^{\ominus} = 74.81\,\mathrm{kJ\,mol^{-1}} - (30.865\,\mathrm{J\,K^{-1}\,mol^{-1}})$$

$$\times (298\mathrm{K}) \times \left(\frac{10^{-3}\,\mathrm{kJ}}{\mathrm{J}}\right)$$

$$= 65.16\,\mathrm{kJ\,mol^{-1}}.$$

$$\Delta_r C_p^{\ominus} = (30.865\,\mathrm{J\,K^{-1}\,mol^{-1}}) \times \left(\frac{10^{-3}\,\mathrm{kJ}}{\mathrm{J}}\right) = 0.030\,865\,\mathrm{kJ\,K^{-1}\,mol^{-1}}.$$

With the estimate of constant $\Delta_r C_p^{\ominus}$, $\boxed{\text{methane is unstable above 825 K}}$.

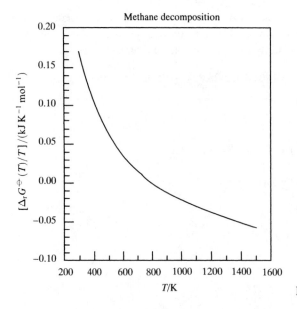

Figure 8.2

(b) $\quad \lambda_{\max} = \dfrac{\frac{1}{5}(1.44\,\text{cm K})}{T}$ [See the solution to Problem 8.10],

$$\lambda_{\max} = \dfrac{\frac{1}{5}(1.44\,\text{cm K})}{1000\,\text{K}} = 2.88 \times 10^{-4}\,\text{cm}\left(\dfrac{10^9\,\text{nm}}{10^2\,\text{cm}}\right),$$

$$\boxed{\lambda_{\max}\,(1000\,\text{K}) = 2880\,\text{nm}}.$$

(c) Excitance ratio $= \dfrac{M(\text{brown dwarf})}{M(\text{Sun})} = \dfrac{\sigma T^4_{\text{brown dwarf}}}{\sigma T^4_{\text{Sun}}}$ [See the solution to Problem 8.11]

$$= \dfrac{(1000\,\text{K})^4}{(6000\,\text{K})^4} = \boxed{7.7 \times 10^{-4}}.$$

Energy density ratio $= \dfrac{\rho(\text{brown dwarf})}{\rho(\text{Sun})}$

$$= \dfrac{8\pi hc/\lambda^5\left(1/(e^{(hc/\lambda kT_{\text{brown dwarf}})} - 1)\right)}{8\pi hc/\lambda^5\left(1/(e^{(hc/\lambda kT_{\text{Sun}})} - 1)\right)}\quad [8.5]$$

$$= \dfrac{e^{(hc/\lambda kT_{\text{Sun}})} - 1}{\left(e^{(hc/\lambda kT_{\text{brown dwarf}})} - 1\right)}.$$

The energy density ratio is a function of λ so we will calculate the ratio at $\lambda_{\max}$ of the brown dwarf.

$$\dfrac{hc}{\lambda_{\text{brown dwarf}}k} = \dfrac{(6.62 \times 10^{-34}\,\text{J s}) \times (3.00 \times 10^8\,\text{m s}^{-1})}{(2880 \times 10^{-9}\,\text{m}) \times (1.381 \times 10^{-23}\,\text{J K}^{-1})}$$

$$= 4998\,\text{K}.$$

$$\text{Energy density ratio} = \frac{e^{4998\,\text{K}/T_{\text{Sun}}} - 1}{e^{4998/T_{\text{brown dwarf}}} - 1}$$

$$= \frac{e^{(4998/6000)} - 1}{e^{(4998/1000)} - 1} = \frac{1.300}{147}$$

$$= \boxed{8.8 \times 10^{-3}}.$$

(d) The wavelength of visible radiation is between about 700 nm (red) and 420 nm (violet). (See text Fig. 8.2.)

$$\text{Fraction of visible energy density} = \frac{1}{aT^4} \left| \int_{700\,\text{nm}}^{420\,\text{nm}} \rho(\lambda)\,\mathrm{d}\lambda \right|$$

[8.3, 8.5, and solution to problem 8.11]

$$= \frac{c}{4\sigma T^4} \left| \int_{700\,\text{nm}}^{420\,\text{nm}} \rho(\lambda)\,\mathrm{d}\lambda \right|.$$

As an estimate, let us suppose that $\rho(\lambda)$ doesn't vary too drastically in the visible at 1000 K. Then,

$$\left| \int_{700\,\text{nm}}^{420\,\text{nm}} \rho(\lambda)\,\mathrm{d}\lambda \right| \sim \rho(560\,\text{nm}) \times (700\,\text{nm} - 420\,\text{nm})$$

$$\sim \left(\frac{8\pi hc}{(560 \times 10^{-9}\,\text{m})^5} \right) \times \left(\frac{1}{e^{((4998\,\text{K}/1000\,\text{K}) \times (2880\,\text{nm}/560\,\text{nm}))} - 1} \right) \times \left(280 \times 10^{-9}\,\text{m} \right)$$

$$= \frac{8\pi(6.626 \times 10^{-34}\,\text{J s}) \times (3.00 \times 10^8\,\text{m s}^{-1})}{1.97 \times 10^{-25}\,\text{m}^4} \left(\frac{1}{e^{25.70} - 1} \right)$$

$$= 1.75 \times 10^{-10}\,\text{J m}^{-3}.$$

$$\text{Fraction of visible energy density} \sim \frac{\left(3.00 \times 10^8\,\text{m s}^{-1} \right) \times \left(1.75 \times 10^{-10}\,\text{J m}^{-3} \right)}{4 \left(5.67 \times 10^{-8}\,\text{W m}^{-2}\text{K}^{-4} \right) \times (1000\,\text{K})^4}$$

$$\sim \boxed{2.31 \times 10^{-7}}.$$

Very little of the brown dwarf's radiation is in the visible. It doesn't shine brightly.

9 Quantum theory: techniques and applications

Answers to discussion questions

D9.1   In quantum mechanics, particles are said to have wave characteristics. The fact of the existence of the particle then requires that the wavelengths of the waves representing it be such that the wave does not experience destructive interference upon reflection by a barrier or in its motion around a closed loop. This requirement restricts the wavelength to values $\lambda = 2/n \times L$, where L is the length of the path and n is a positive integer. Then using the relations $\lambda = h/p$ and $E = p^2/2m$, the energy is quantized at $E = n^2h^2/8mL^2$. This derivation applies specifically to the particle in a box, the derivation is similar for the particle on a ring; the same principles apply (see Section 9.6).

D9.3   The lowest energy level possible for a confined quantum mechanical system is the zero-point energy, and zero-point energy is not zero energy. The system must have at least that minimum amount of energy even at absolute zero. The physical reason is that, if the particle is confined, its position is not completely uncertain, and therefore its momentum, and hence its kinetic energy, cannot be exactly zero. The particle in a box, the harmonic oscillator, the particle on a ring or on a sphere, the hydrogen atom, and many other systems we will encounter, all have zero-point energy.

D9.5   Fermions are particles with half-integral spin, 1/2, 3/2, 5/2, ..., whereas bosons have integral spin, 0, 1, 2, ... All fundamental particles that make up matter have spin 1/2 and are fermions, but composite particles can be either fermions or bosons.

Fermions: electrons, protons, neutrons, ^{3}He, ...

Bosons: photons, deuterons.

Solutions to exercises

E9.1(b) $$E = \frac{n^2h^2}{8m_eL^2} \quad [9.4a]$$

$$\frac{h^2}{8m_eL^2} = \frac{(6.626 \times 10^{-34}\,\text{J s})^2}{8(9.109 \times 10^{-31}\,\text{kg}) \times (1.50 \times 10^{-9}\,\text{m})^2} = 2.67\overline{8} \times 10^{-20}\,\text{J}$$

The conversion factors required are

$$1\,\text{eV} = 1.602 \times 10^{-19}\,\text{J}; \quad 1\,\text{cm}^{-1} = 1.986 \times 10^{-23}\,\text{J}; \quad 1\,\text{eV} = 96.485\,\text{kJ mol}^{-1}$$

(a) $E_3 - E_1 = (9 - 1)\dfrac{h^2}{8m_eL^2} = 8(2.678 \times 10^{-20}\,\text{J})$

$\qquad\qquad = \boxed{2.14 \times 10^{-19}\,\text{J}} = \boxed{1.34\,\text{eV}} = \boxed{1.08 \times 10^4\,\text{cm}^{-1}} = \boxed{129\,\text{kJ}\,\text{mol}^{-1}}$

(b) $E_7 - E_6 = (49 - 36)\dfrac{h^2}{8m_eL^2} = 13(2.678 \times 10^{-20}\,\text{J})$

$\qquad\qquad = \boxed{3.48 \times 10^{-19}\,\text{J}} = \boxed{2.17\,\text{eV}} = \boxed{1.75 \times 10^4\,\text{cm}^{-1}} = \boxed{210\,\text{kJ}\,\text{mol}^{-1}}$

E9.2(b)   The probability is

$$P = \int \psi^*\psi\,dx = \frac{2}{L}\int \sin^2\left(\frac{n\pi x}{L}\right)dx \approx \frac{2\Delta x}{L}\sin^2\left(\frac{n\pi x}{L}\right)$$

where $\Delta x = 0.02L$ and the function is evaluated at $x = 0.66\,L$.

(a)   For $n = 1$ $P = \dfrac{2(0.02L)}{L}\sin^2(0.66\pi) = \boxed{0.03\overline{1}}$

(b)   For $n = 2$ $P = \dfrac{2(0.02L)}{L}\sin^2[2(0.66\pi)] = \boxed{0.02\overline{9}}$

E9.3(b)   The expectation value is

$$\langle\hat{p}\rangle = \int \psi^*\hat{p}\psi\,dx$$

but first we need $\hat{p}\psi$

$$\hat{p}\psi = -i\hbar\frac{d}{dx}\left(\frac{2}{L}\right)^{1/2}\sin\left(\frac{n\pi x}{L}\right) = -i\hbar\left(\frac{2}{L}\right)^{1/2}\frac{n\pi}{L}\cos\left(\frac{n\pi x}{L}\right)$$

so $\langle\hat{p}\rangle = \dfrac{-2i\hbar n\pi}{L^2}\displaystyle\int_0^L \sin\left(\frac{n\pi x}{L}\right)\cos\left(\frac{n\pi x}{L}\right)dx = \boxed{0}$

and $\langle\hat{p}^2\rangle = 2m\langle\hat{H}\rangle = 2mE_n = \dfrac{h^2n^2}{4L^2}$

for all n. So for $n = 2$

$$\langle\hat{p}^2\rangle = \boxed{\dfrac{h^2}{L^2}}$$

E9.4(b)   The zero-point energy is the ground-state energy, that is, with $n_x = n_y = n_z = 1$:

$$E = \frac{(n_x^2 + n_y^2 + n_z^2)h^2}{8mL^2}\quad\text{[9.12b with equal lengths]} = \frac{3h^2}{8mL^2}$$

Set this equal to the rest energy mc^2 and solve for L:

$$mc^2 = \frac{3h^2}{8mL^2}\quad\text{so } L = \boxed{\left(\frac{3}{8}\right)^{1/2}\frac{h}{mc} = \left(\frac{3}{8}\right)^{1/2}\lambda_C}$$

where λ_C is the Compton wavelength of a particle of mass m.

E9.5(b)

$$\psi_5 = \left(\frac{2}{L}\right)^{1/2} \sin\left(\frac{5\pi x}{L}\right)$$

$$P(x) \propto \psi_5^2 \propto \sin^2\left(\frac{5\pi x}{L}\right)$$

Maxima and minima in $P(x)$ correspond to $\dfrac{dP(x)}{dx} = 0$

$$\frac{d}{dx}P(x) \propto \frac{d\psi^2}{dx} \propto \sin\left(\frac{5\pi x}{L}\right)\cos\left(\frac{5\pi x}{L}\right) \propto \sin\left(\frac{10\pi x}{L}\right) \quad [2\sin\alpha\cos\alpha = \sin 2\alpha]$$

$\sin\theta = 0$ when $\theta = 0, \pi, 2\pi, \ldots, n'\pi$ $(n' = 0, 1, 2, \ldots)$

$$\frac{10\pi x}{L} = n'\pi \quad \text{for } n' \le 10 \quad \text{so} \quad x = \frac{n'L}{10}$$

$x = 0, x = L$ are minima. Maxima and minima alternate, so maxima correspond to

$$n' = 1, 3, 5, 7, 9 \quad x = \boxed{\frac{L}{10}}, \boxed{\frac{3L}{10}}, \boxed{\frac{L}{2}}, \boxed{\frac{7L}{10}}, \boxed{\frac{9L}{10}}$$

E9.6(b) The energy levels are

$$E_{n_1, n_2, n_3} = \frac{(n_1^2 + n_2^2 + n_3^2)h^2}{8mL^2} = E_1(n_1^2 + n_2^2 + n_3^2)$$

where E_1 combines all constants besides quantum numbers. The minimum value for all the quantum numbers is 1, so the lowest energy is

$$E_{1,1,1} = 3E_1$$

The question asks about an energy 14/3 times this amount, namely $14E_1$. This energy level can be obtained by any combination of allowed quantum numbers such that

$$n_1^2 + n_2^2 + n_3^2 = 14 = 3^2 + 2^2 + 1^2$$

The degeneracy, then, is $\boxed{6}$, corresponding to $(n_1, n_2, n_3) = (1, 2, 3), (1, 3, 2), (2, 1, 3), (2, 3, 1), (3, 1, 2),$ or $(3, 2, 1)$.

E9.7(b) $E = \frac{3}{2}kT$ is the average translational energy of a gaseous molecule (see Chapter 17).

$$E = \frac{3}{2}kT = \frac{(n_1^2 + n_2^2 + n_3^2)h^2}{8mL^2} = \frac{n^2h^2}{8mL^2}$$

$$E = \left(\frac{3}{2}\right) \times (1.381 \times 10^{-23}\,\text{J K}^{-1}) \times (300\,\text{K}) = 6.21\overline{4} \times 10^{-21}\,\text{J}$$

$$n^2 = \frac{8mL^2}{h^2}E$$

If $L^3 = 1.00\,\text{m}^3$, then $L^2 = 1.00\,\text{m}^2$.

$$\frac{h^2}{8mL^2} = \frac{(6.626 \times 10^{-34}\,\text{J s})^2}{(8) \times \left(\dfrac{0.02802\,\text{kg mol}^{-1}}{6.022 \times 10^{23}\,\text{mol}^{-1}}\right) \times 100\,\text{m}^2} = 1.18\overline{0} \times 10^{-42}\,\text{J}$$

$$n^2 = \frac{6.21\overline{4} \times 10^{-21}\,\text{J}}{1.180 \times 10^{-42}\,\text{J}} = 5.26\overline{5} \times 10^{21}; \quad n = \boxed{7.26 \times 10^{10}}$$

$$\Delta E = E_{n+1} - E_n = E_{7.26 \times 10^{10}+1} - E_{7.26 \times 10^{10}}$$

$$\Delta E = (2n + 1) \times \left(\frac{h^2}{8mL^2}\right) = [(2) \times (7.26 \times 10^{10}) + 1] \times \left(\frac{h^2}{8mL^2}\right) = \frac{14.5\overline{2} \times 10^{10}h^2}{8mL^2}$$

$$= (14.5\overline{2} \times 10^{10}) \times (1.18\overline{0} \times 10^{-42}\,\text{J}) = \boxed{1.71 \times 10^{-31}\,\text{J}}$$

The de Broglie wavelength is obtained from

$$\lambda = \frac{h}{p} = \frac{h}{mv} \quad [8.12]$$

The velocity is obtained from

$$E_K = \tfrac{1}{2}mv^2 = \tfrac{3}{2}kT = 6.21\overline{4} \times 10^{-21}\,\text{J}$$

$$v^2 = \frac{6.21\overline{4} \times 10^{-21}\,\text{J}}{\left(\dfrac{1}{2}\right) \times \left(\dfrac{0.02802\,\text{kg mol}^{-1}}{6.022 \times 10^{23}\,\text{mol}^{-1}}\right)} = 2.67\overline{1} \times 10^5\,\text{m}^2\,\text{s}^{-2}; \quad v = 517\,\text{m s}^{-1}$$

$$\lambda = \frac{6.626 \times 10^{-34}\,\text{J s}}{(4.65 \times 10^{-26}\,\text{kg}) \times (517\,\text{m s}^{-1})} = 2.75 \times 10^{-11}\,\text{m} = \boxed{27.5\,\text{pm}}$$

The conclusion to be drawn from all of these calculations is that the translational motion of the nitrogen molecule can be described classically. The energy of the molecule is essentially continuous,

$$\frac{\Delta E}{E} \ll 1.$$

E9.8(b)   The zero-point energy is

$$E_0 = \frac{1}{2}\hbar\omega = \frac{1}{2}\hbar\left(\frac{k}{m}\right)^{1/2} = \frac{1}{2}(1.0546 \times 10^{-34}\,\text{J s}) \times \left(\frac{285\,\text{N m}^{-1}}{5.16 \times 10^{-26}\,\text{kg}}\right)^{1/2}$$

$$= \boxed{3.92 \times 10^{-21}\,\text{J}}$$

E9.9(b)   The difference in adjacent energy levels is

$$\Delta E = E_{v+1} - E_v = \hbar\omega \; [9.26] = \hbar\left(\frac{k}{m}\right)^{1/2} \quad [9.25]$$

so $k = \dfrac{m(\Delta E)^2}{\hbar^2} = \dfrac{(2.88 \times 10^{-25}\,\text{kg}) \times (3.17 \times 10^{-21}\,\text{J})^2}{(1.0546 \times 10^{-34}\,\text{J s})^2} = \boxed{260\,\text{N m}^{-1}}$

E9.10(b)   The difference in adjacent energy levels, which is equal to the energy of the photon, is

$$\Delta E = \hbar\omega = h\nu \ \text{ so } \ \hbar\left(\frac{k}{m}\right)^{1/2} = \frac{hc}{\lambda}$$

and

$$\lambda = \frac{hc}{\hbar}\left(\frac{k}{m}\right)^{1/2} = 2\pi c\left(\frac{m}{k}\right)^{1/2}$$

$$= 2\pi(2.998 \times 10^8 \, \text{m s}^{-1}) \times \left(\frac{(15.9949 \, \text{u}) \times (1.66 \times 10^{-27} \, \text{kg u}^{-1})}{544 \, \text{N m}^{-1}}\right)^{1/2}$$

$$\lambda = 1.32 \times 10^{-5} \, \text{m} = \boxed{13.2 \, \mu\text{m}}$$

E9.11(b)   The difference in adjacent energy levels, which is equal to the energy of the photon, is

$$\Delta E = \hbar\omega = h\nu \ \text{ so } \ \hbar\left(\frac{k}{m}\right)^{1/2} = \frac{hc}{\lambda}$$

and $\lambda = \dfrac{hc}{\hbar}\left(\dfrac{k}{m}\right)^{1/2} = 2\pi c\left(\dfrac{m}{k}\right)^{1/2}$

Doubling the mass, then, increases the wavelength by a factor of $2^{1/2}$. So taking the result from Exercise 9.10(b), the new wavelength is

$$\lambda = 2^{1/2}(13.2 \, \mu\text{m}) = \boxed{18.7 \, \mu\text{m}}$$

E9.12(b) $\Delta E = \hbar\omega = h\nu$

(a) $\Delta E = h\nu = (6.626 \times 10^{-34} \, \text{J Hz}^{-1}) \times (33 \times 10^3 \, \text{Hz}) = \boxed{2.2 \times 10^{-29} \, \text{J}}$

(b) $\Delta E = \hbar\omega = \hbar\left(\dfrac{k}{m_{\text{eff}}}\right)^{1/2} \ \left[\dfrac{1}{m_{\text{eff}}} = \dfrac{1}{m_1} + \dfrac{1}{m_2} \text{ with } m_1 = m_2\right]$

For a two-particle oscillator m_{eff}, replaces m in the expression for ω. (See Chapter 13 for a more complete discussion of the vibration of a diatomic molecule.)

$$\Delta E = \hbar\left(\frac{2k}{m}\right)^{1/2} = (1.055 \times 10^{-34} \, \text{J s}) \times \left(\frac{(2) \times (1177 \, \text{N m}^{-1})}{(16.00) \times (1.6605 \times 10^{-27} \, \text{kg})}\right)^{1/2}$$

$$= \boxed{3.14 \times 10^{-20} \, \text{J}}$$

E9.13(b)   The first excited-state wavefunction has the form

$$\psi = 2N_1 y \exp\left(-\tfrac{1}{2}y^2\right)$$

where N_1 is a collection of constants and $y \equiv x(m\omega/\hbar)^{1/2}$. To see if it satisfies Schrödinger's equation, we see what happens when we apply the energy operator to this function

$$\hat{H}\psi = -\frac{\hbar^2}{2m}\frac{d^2\psi}{dx^2} + \frac{1}{2}m\omega^2 x^2\psi$$

We need derivatives of ψ

$$\frac{d\psi}{dx} = \frac{d\psi}{dy}\frac{dy}{dx} = \left(\frac{m\omega}{\hbar}\right)^{1/2}(2N_1) \times (1 - y^2) \times \exp\left(-\frac{1}{2}y^2\right)$$

and $\dfrac{d^2\psi}{dx^2} = \dfrac{d^2\psi}{dy^2}\left(\dfrac{dy}{dx}\right)^2 = \left(\dfrac{m\omega}{\hbar}\right) \times (2N_1) \times (-3y + y^3) \times \exp\left(-\dfrac{1}{2}y^2\right) = \left(\dfrac{m\omega}{\hbar}\right) \times (y^2 - 3)\psi$

So $\hat{H}\psi = -\dfrac{\hbar^2}{2m} \times \left(\dfrac{m\omega}{\hbar}\right) \times (y^2 - 3)\psi + \dfrac{1}{2}m\omega^2 x^2 \psi$

$$= -\frac{1}{2}\hbar\omega \times (y^2 - 3) \times \psi + \frac{1}{2}\hbar\omega y^2 \psi = \frac{3}{2}\hbar\omega\psi$$

Thus, ψ is a solution of the Schrödinger equation with energy eigenvalue

$$E = \boxed{\tfrac{3}{2}\hbar\omega}$$

E9.14(b) The harmonic oscillator wavefunctions have the form

$$\psi_v(x) = N_v H_v(y) \exp\left(-\frac{1}{2}y^2\right) \quad \text{with} \quad y = \frac{x}{\alpha} \quad \text{and} \quad \alpha = \left(\frac{\hbar^2}{mk}\right)^{1/4} \quad [9.28]$$

The exponential function approaches zero only as x approaches $\pm\infty$, so the nodes of the wavefunction are the nodes of the Hermite polynomials.

$$H_5(y) = 32y^5 - 160y^3 + 120y = 0 \text{ [Table 9.1]} = 8y(4y^4 - 20y^2 + 15)$$

So one solution is $y = 0$, which leads to $x = 0$. The other factor can be made into a quadratic equation by letting $z = y^2$

$$4z^2 - 20z + 15 = 0$$

so $\quad z = \dfrac{-b \pm \sqrt{b^2 - 4ac}}{2a} = \dfrac{20 \pm \sqrt{20^2 - 4 \times 4 \times 15}}{2 \times 4} = \dfrac{5 \pm \sqrt{10}}{2}$

Evaluating the result numerically yields $z = 0.92$ or 4.08, so $y = \pm 0.96$ or ± 2.02. Therefore $x = \boxed{0, \pm 0.96\alpha, \text{ or } \pm 2.02\alpha}$.

COMMENT. Numerical values could also be obtained graphically by plotting $H_5(y)$.

E9.15(b) The zero-point energy is

$$E_0 = \frac{1}{2}\hbar\omega = \frac{1}{2}\hbar\left(\frac{k}{m_{\text{eff}}}\right)^{1/2}$$

For a homonuclear diatomic molecule, the effective mass is half the mass of an atom, so

$$E_0 = \frac{1}{2}(1.0546 \times 10^{-34}\,\text{J s}) \times \left(\frac{2293.8\,\text{N m}^{-1}}{\frac{1}{2}(14.0031\,\text{u}) \times (1.66054 \times 10^{-27}\,\text{kg u}^{-1})}\right)^{1/2}$$

$$E_0 = \boxed{2.3421 \times 10^{-20}\,\text{J}}$$

E9.16(b)   Orthogonality requires that

$$\int \psi_m^* \psi_n \, d\tau = 0$$

if $m \neq n$.

Performing the integration

$$\int \psi_m^* \psi_n \, d\tau = \int_0^{2\pi} N e^{-im\phi} N e^{in\phi} \, d\phi = N^2 \int_0^{2\pi} e^{i(n-m)\phi} \, d\phi$$

If $m \neq n$, then

$$\int \psi_m^* \psi_n \, d\tau = \frac{N^2}{i(n-m)} e^{i(n-m)\phi} \Big|_0^{2\pi} = \frac{N^2}{i(n-m)}(1-1) = 0$$

Therefore, they are orthogonal.

E9.17(b)   The magnitude of angular momentum is

$$\left\langle \hat{L}^2 \right\rangle^{1/2} = \{l(l+1)\}^{1/2} \hbar \ [9.54a] = \{2(3)\}^{1/2}(1.0546 \times 10^{-34} \, \text{J s}) = \boxed{2.58 \times 10^{-34} \, \text{J s}}$$

Possible projections onto an arbitrary axis are

$$\left\langle \hat{L}_z \right\rangle = m_l \hbar \ [9.54b]$$

where $m_l = 0$ or ± 1 or ± 2. So possible projections include

$$\boxed{0, \ \pm 1.0546 \times 10^{-34} \, \text{J s and} \pm 2.1109 \times 10^{-34} \, \text{J s}}$$

E9.18(b)   The cones are constructed as described in Section 9.7(d) and Figure 9.40(b) of the text; their edges are of length $\{6(6+1)\}^{1/2} = 6.48$ and their projections are $m_j = +6, +5, \ldots, -6$. See Figure 9.1(a).

The vectors follow, in units of $\hbar$. From the highest-pointing to the lowest-pointing vectors (Figure 9.1(b)), the values of m_l are $6, 5, 4, 3, 2, 1, 0, -1, -2, -3, -4, -5,$ and -6.

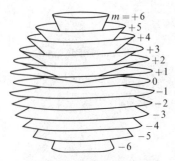

$m = +6$
$+5$
$+4$
$+3$
$+2$
$+1$
0
-1
-2
-3
-4
-5
-6

Figure 9.1(a)

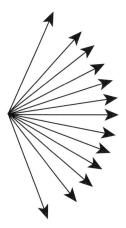

Figure 9.1(b)

Solutions to problems

Solutions to numerical problems

P9.1
$$E = \frac{n^2 h^2}{8mL^2}, \quad E_2 - E_1 = \frac{3h^2}{8mL^2}.$$

We take $m(O_2) = (32.000) \times (1.6605 \times 10^{-27}$ kg), and find

$$E_2 - E_1 = \frac{(3) \times (6.626 \times 10^{-34}\,\text{J s})^2}{(8) \times (32.00) \times (1.6605 \times 10^{-27}\,\text{kg}) \times (5.0 \times 10^{-2}\,\text{m})^2} = \boxed{1.24 \times 10^{-39}\,\text{J}}.$$

We set $E = \dfrac{n^2 h^2}{8mL^2} = \dfrac{1}{2}kT$ and solve for n.

From above, $h^2/8mL^2 = (E_2 - E_1)/3 = 4.13 \times 10^{-40}$ J; then

$$n^2 \times (4.13 \times 10^{-40}\,\text{J}) = \left(\frac{1}{2}\right) \times (1.381 \times 10^{-23}\,\text{J K}^{-1}) \times (300\,\text{K}) = 2.07 \times 10^{-21}\,\text{J}.$$

We find $n = \left(\dfrac{2.07 \times 10^{-21}\,\text{J}}{4.13 \times 10^{-40}\,\text{J}}\right)^{1/2} = \boxed{2.2 \times 10^9}$.

At this level,

$$E_n - E_{n-1} = \{n^2 - (n-1)^2\} \times \frac{h^2}{8mL^2} = (2n-1) \times \frac{h^2}{8mL^2} \approx (2n) \times \frac{h^2}{8mL^2}$$

$$= (4.4 \times 10^9) \times (4.13 \times 10^{-40}\,\text{J}) \approx \boxed{1.8 \times 10^{-30}\,\text{J}} \quad \text{[or } 1.1\,\mu\text{J mol}^{-1}\text{]}.$$

P9.3
$$E = \frac{m_l^2 \hbar^2}{2I}\,[9.38a] = \frac{m_l^2 \hbar^2}{2mr^2} \quad [I = mr^2].$$

$$E_0 = 0 \quad [m_l = 0].$$

$$E_1 = \frac{\hbar^2}{2mr^2} = \frac{(1.055 \times 10^{-34}\,\text{J s})^2}{(2) \times (1.008) \times (1.6605 \times 10^{-27}\,\text{kg}) \times (160 \times 10^{-12}\,\text{m})^2} = \boxed{1.30 \times 10^{-22}\,\text{J}}.$$

The minimum angular momentum is $\boxed{\pm \hbar}$.

P9.5 **(a)** Treat the small step in the potential energy function as a perturbation in the energy operator:

$$H^{(1)} = \begin{cases} 0 & \text{for } 0 \leq x \leq (1/2)(L-a) \text{ and } (1/2)(L+a) \leq x \leq L \\ \varepsilon & \text{for } (1/2)(L-a) \leq x \leq (1/2)(L+a). \end{cases}$$

The first-order correction to the ground-state energy, E_1, is

$$E_1^{(1)} = \int_0^L \psi_1^{(0)*} H^{(1)} \psi_1^{(0)} dx = \int_{(1/2)(L-a)}^{(1/2)(L+a)} \left(\frac{2}{L}\right)^{1/2} \sin\left(\frac{\pi x}{L}\right) \varepsilon \left(\frac{2}{L}\right)^{1/2} \sin\left(\frac{\pi x}{L}\right) dx,$$

$$E_1^{(1)} = \frac{2\varepsilon}{L} \int_{(1/2)(L-a)}^{(1/2)(L+a)} \sin^2\left(\frac{\pi x}{L}\right) dx = \frac{\varepsilon}{L\pi} \left(\pi x - L\cos\left(\frac{\pi x}{L}\right)\sin\left(\frac{\pi x}{L}\right)\right)\Big|_{1/2(L-a)}^{1/2(L+a)},$$

$$E_1^{(1)} = \frac{\varepsilon a}{L} - \frac{\varepsilon}{\pi}\cos\left(\frac{\pi(L+a)}{2L}\right)\sin\left(\frac{\pi(L+a)}{2L}\right) + \frac{\varepsilon}{\pi}\cos\left(\frac{\pi(L-a)}{2L}\right)\sin\left(\frac{\pi(L-a)}{2L}\right).$$

This expression can be simplified considerably with a few trigonometric identities. The product of sine and cosine is related to the sine of twice the angle:

$$\cos\left(\frac{\pi(L\pm a)}{2L}\right)\sin\left(\frac{\pi(L\pm a)}{2L}\right) = \frac{1}{2}\sin\left(\frac{\pi(L\pm a)}{L}\right) = \frac{1}{2}\sin\left(\pi \pm \frac{\pi a}{L}\right),$$

and the sine of a sum can be written in a particularly simple form since one of the terms in the sum is π:

$$\sin\left(\pi \pm \frac{\pi a}{L}\right) = \sin\pi\cos\left(\frac{\pi a}{L}\right) \pm \cos\pi\sin\left(\frac{\pi a}{L}\right) = \mp\sin\left(\frac{\pi a}{L}\right).$$

Thus $E_1^{(1)} = \boxed{\dfrac{\varepsilon a}{L} + \dfrac{\varepsilon}{\pi}\sin\left(\dfrac{\pi a}{L}\right)}.$

(b) If $a = L/10$, the first-order correction to the ground-state energy is

$$E_1^{(1)} = \boxed{\frac{\varepsilon}{10} + \frac{\varepsilon}{\pi}\sin\left(\frac{\pi}{10}\right)} = \boxed{0.1984\,\varepsilon}.$$

P9.7 The second-order correction to the ground-state energy, E_1, is

$$E_1^{(2)} = \sum_{n=2}^{\infty} \frac{\left|\int_L^0 \psi_n^{(0)*} H^{(1)} \psi_1^{(0)} dx\right|^2}{E_1^{(0)} - E_n^{(0)}},$$

where $H^{(1)} = mgx$, $\psi_n^{(0)} = \sin\dfrac{n\pi x}{L}$, and $E_n = \dfrac{n^2 h^2}{8mL^2}$.

The denominator in the sum is

$$E_1^{(0)} - E_n^{(0)} = \frac{h^2}{8mL^2} - \frac{n^2 h^2}{8mL^2} = \frac{(1-n^2)h^2}{8mL^2}.$$

The integral in the sum is

$$\int_0^L \psi_n^{(0)*} H^{(1)} \psi_1^{(0)} \, dx = \int_0^L \left(\frac{2}{L}\right)^{1/2} \sin\left(\frac{n\pi x}{L}\right) mgx \left(\frac{2}{L}\right)^{1/2} \sin\left(\frac{\pi x}{L}\right) dx,$$

$$\int_0^L \psi_n^{(0)*} H^{(1)} \psi_1^{(0)} \, dx = \frac{2mg}{L} \int_0^L x \sin ax \sin bx \, dx,$$

where $a = n\pi/L$ and $b = \pi/L$.

The integral formulas given with the problem allow this integral to be expressed as

$$-\frac{2mg}{L} \frac{d}{da} \int_0^L \cos ax \sin bx \, dx = -mg \frac{d}{da} \left(\frac{\cos(a-b)x}{2(a-b)} - \frac{\cos(a+b)x}{2(a+b)}\right)\Big|_0^L$$

$$= -\frac{2mg}{L} \left(\frac{-x\sin(a-b)x}{2(a-b)} - \frac{\cos(a-b)x}{2(a-b)^2} + \frac{x\sin(a+b)x}{2(a+b)} + \frac{\cos(a+b)x}{2(a+b)^2}\right)\Big|_0^L.$$

The arguments of the trigonometric functions at the upper limit are:

$$(a-b)L = (n-1)\pi \text{ and } (a+b)L = (n+1)\pi.$$

Therefore, the sine terms vanish. Similarly, the cosines are ± 1 depending on whether the argument is an even or odd multiple of π; they simplify to $(-1)^{n+1}$. At the lower limit, the sines are still zero, and the cosines are all 1. The integral, evaluated at its limits with π/L factors pulled out from the as and bs in its denominator, becomes

$$\frac{mgL}{\pi^2} \left(\frac{(-1)^{n+1}-1}{(n-1)^2} - \frac{(-1)^{n+1}-1}{(n+1)^2}\right) = \frac{mgL[(-1)^{n+1}-1]}{\pi^2} \left(\frac{(n+1)^2 - (n-1)^2}{(n-1)^2(n+1)^2}\right),$$

$$= \frac{4mgL[(-1)^{n+1}-1]n}{\pi^2(n^2-1)^2}.$$

The second-order correction, then, is

$$E_1^{(2)} = \sum_{n=2}^{\infty} \frac{\left(\dfrac{4mgL[(-1)^{n+1}-1]n}{\pi^2(n^2-l)^2}\right)^2}{\dfrac{(1-n^2)h^2}{8mL^2}} = -\frac{128m^3g^2L^4}{\pi^4h^2} \sum_{n=2}^{\infty} \frac{[(-1)^n+1]^2 n^2}{(n^2-1)^5}.$$

Note that the terms with odd n vanish. Therefore, the sum can be rewritten, changing n to $2k$, as

$$E_1^{(2)} = -\frac{2048m^3g^2L^4}{\pi^4h^2} \sum_{k=1}^{\infty} \frac{k^2}{(4k^2-1)^5}.$$

The sum converges rapidly to 4.121×10^{-3}, as can easily be verified numerically; in fact, to three significant figures, terms after the first do not affect the sum. So the second-order correction is

$$E_1^{(2)} = \boxed{-\frac{0.08664m^3g^2L^4}{h^2}}.$$

The first-order correction to the ground-state wavefunction is also a sum:

$$\psi_0^{(1)} = \sum_n c_n \psi_n^{(0)},$$

where $c_n = -\dfrac{\int_0^L \psi_n^{(0)*} H^1 \psi_1^{(0)} \, dx}{E_n^{(0)} - E_1^{(0)}} = -\dfrac{\dfrac{4mgL[(-1)^{n+1} - 1]n}{\pi^2(n^2 - 1)^2}}{((n^2 - 1)h^2/8mL^2)} = \boxed{\dfrac{32m^2gL^3[(-1)^n + 1]n}{\pi^2h^2(n^2 - 1)^3}}.$

Once again, the odd n terms vanish.

How does the first-order correction alter the wavefunction? Recall that the perturbation raises the potential energy near the top of the box (near L) much more than near the bottom (near $x = 0$); therefore, we expect the probability of finding the particle near the bottom to be enhanced compared with that of finding it near the top. Because the zero-order ground-state wavefunction is positive throughout the interior of the box, we thus expect the wavefunction itself to be raised near the bottom of the box and lowered near the top. In fact, the correction terms do just this. First, note that the basis wavefunctions with odd n are symmetric with respect to the center of the box; therefore, they would have the same effect near the top of the box as near the bottom. The coefficients of these terms are zero: they do not contribute to the correction. The even-n basis functions all start positive near $x = 0$ and end negative near $x = L$; therefore, such terms must be multiplied by positive coefficients (as the result provides) to enhance the wavefunction near the bottom and diminish it near the top.

Solutions to theoretical problems

P9.9   The text defines the transmission probability and expresses it as the ratio of $|A'|^2/|A|^2$, where the coefficients A and A' are introduced in eqns 9.14 and 9.17. Eqns 9.18 and 9.19 list four equations for the six unknown coefficients of the full wavefunction. Once we realize that we can set B' to zero, these equations in five unknowns are:

(a) $A + B = C + D$,
(b) $Ce^{\kappa L} + De^{-\kappa L} = A'e^{ikL}$,
(c) $ikA - ikB = \kappa C - \kappa D$,
(d) $\kappa Ce^{\kappa L} - \kappa De^{-\kappa L} = ikA'e^{ikL}$.

We need A' in terms of A alone, which means we must eliminate B, C, and D. Notice that B appears only in eqns (a) and (c). Solving these equations for B and setting the results equal to each other yields

$$B = C + D - A = A - \frac{\kappa C}{ik} + \frac{\kappa D}{ik}.$$

Solve this equation for C:

$$C = \frac{2A + D\left(\dfrac{\kappa}{ik} - 1\right)}{\dfrac{\kappa}{ik} + 1} = \frac{2Aik + D(\kappa - ik)}{\kappa + ik}.$$

Now note that the desired A' appears only in (b) and (d). Solve these for A' and set them equal:

$$A' = e^{-ikL}(Ce^{\kappa L} + De^{\kappa L}) = \frac{\kappa e^{-ikL}}{ik}(Ce^{\kappa L} - De^{-\kappa L}).$$

Solve the resulting equation for C, and set it equal to the previously obtained expression for C:

$$C = \frac{\left(\frac{\kappa}{ik} + 1\right)De^{-2\kappa L}}{\frac{\kappa}{ik} - 1} = \frac{(\kappa + ik)De^{-2\kappa L}}{\kappa - ik} = \frac{2Aik + D(\kappa - ik)}{\kappa + ik}.$$

Solve this resulting equation for D in terms of A:

$$\frac{(\kappa + ik)^2 e^{-2\kappa L} - (\kappa - ik)^2}{(\kappa - ik)(\kappa + ik)}D = \frac{2Aik}{\kappa + ik},$$

so $$D = \frac{2Aik(\kappa - ik)}{(\kappa + ik)^2 e^{-2\kappa L} - (\kappa - ik)^2}.$$

Substituting this expression back into an expression for C yields

$$C = \frac{2Aik(\kappa + ik)e^{-2\kappa L}}{(\kappa + ik)^2 e^{-2\kappa L} - (\kappa - ik)^2}.$$

Substituting for C and D in the expression for A' yields

$$A' = e^{-ikL}(Ce^{\kappa L} + De^{-\kappa L}) = \frac{2Aike^{-ikL}}{(\kappa + ik)^2 e^{-2\kappa L} - (\kappa - ik)^2}[(\kappa + ik)e^{-\kappa L} + (\kappa - ik)e^{-\kappa L}],$$

$$A' = \frac{4Aik\kappa e^{-\kappa L}e^{-ikL}}{(\kappa + ik)^2 e^{-2\kappa L} - (\kappa - ik)^2} = \frac{4Aik\kappa e^{-ikL}}{(\kappa + ik)^2 e^{-\kappa L} - (\kappa - ik)^2 e^{\kappa L}}.$$

The transmission coefficient is

$$T = \frac{|A'|^2}{|A|^2} = \left(\frac{4Aik\kappa e^{-ikL}}{(\kappa + ik)^2 e^{-\kappa L} - (\kappa - ik)^2 e^{\kappa L}}\right)\left(\frac{-4Aik\kappa e^{ikL}}{(\kappa - ik)^2 e^{-\kappa L} - (\kappa + ik)^2 e^{\kappa L}}\right).$$

The denominator is worth expanding separately in several steps. It is

$$(\kappa + ik)^2(\kappa - ik)^2 e^{-2\kappa L} - (\kappa - ik)^4 - (\kappa + ik)^4 + (\kappa - ik)^2(\kappa + ik)^2 e^{2\kappa L}$$

$$= (\kappa^2 + k^2)^2(e^{2\kappa L} + e^{-2\kappa L}) - (\kappa^2 - 2i\kappa k - k^2)^2 - (\kappa^2 + 2i\kappa k - k^2)^2$$

$$= (\kappa^4 + 2\kappa^2 k^2 + k^4)(e^{2\kappa L} + e^{-2\kappa L}) - (2\kappa^4 - 12\kappa^2 k^2 + 2k^2).$$

If the $12\kappa^2 k^2$ term were $-4\kappa^2 k^2$ instead, we could collect terms still further (completing the square), but of course we must also account for the difference between those quantities, making the denominator

$$(\kappa^4 + 2\kappa^2 k^2 + k^4)(e^{2\kappa L} - 2 + e^{-2\kappa L}) + 16\kappa^2 k^2 = (\kappa^2 + k^2)^2(e^{2\kappa L} - e^{-2\kappa L})^2 + 16\kappa^2 k^2.$$

So the coefficient is

$$T = \frac{16k^2\kappa^2}{(\kappa^2 + k^2)^2(e^{2\kappa L} - e^{-2\kappa L})^2 + 16\kappa^2 k^2}.$$

We are almost there. To get to eqn 9.20a, we invert the expression

$$T = \left(\frac{(\kappa^2 + k^2)^2(e^{2\kappa L} - e^{-2\kappa L})^2 + 16\kappa^2 k^2}{16k^2\kappa^2}\right)^{-1} = \left(\frac{(\kappa^2 + k^2)^2(e^{2\kappa L} - e^{-2\kappa L})^2}{16k^2\kappa^2} + 1\right)^{-1}$$

Finally, we try to express $(\kappa^2 + k^2)/k^2\kappa^2$ in terms of a ratio of energies, $\varepsilon = E/V$. Eqns 9.14 and 9.16 define k and κ. The factors involving, 2, $\hbar$, and the mass cancel leaving $\kappa \propto (V - E)^{1/2}$ and $k \propto E^{1/2}$, so

$$\frac{(\kappa^2 + k^2)^2}{k^2\kappa^2} = \frac{[E + (V - E)]^2}{E(V - E)} = \frac{V^2}{E(V - E)} = \frac{1}{\varepsilon(1 - \varepsilon)},$$

which makes the transmission coefficient

$$T = \left(\frac{(e^{2\kappa L} - e^{-2\kappa L})^2}{16\varepsilon(1 - \varepsilon)} + 1\right)^{-1}.$$

P9.11   We assume that the barrier begins at $x = 0$ and that the barrier extends in the positive x direction.

(a) $P = \displaystyle\int_{\text{Barrier}} \psi^2 d\tau = \int_0^\infty N^2 e^{-2\kappa x} dx = \boxed{\dfrac{N^2}{2\kappa}}.$

(b) $\langle x \rangle = \displaystyle\int_0^\infty x\psi^2 dx = N^2 \int_0^\infty xe^{-2\kappa x} dx = \dfrac{N^2}{(2\kappa)^2} = \boxed{\dfrac{N^2}{4\kappa^2}}.$

Question. Is N a normalization constant?

P9.13 $\langle E_K \rangle \equiv \langle T \rangle = \displaystyle\int_{-\infty}^{+\infty} \psi^* \hat{T} \psi dx$ with $\hat{T} \equiv \dfrac{\hat{p}^2}{2m}$ and $\hat{p} = \dfrac{\hbar}{i}\dfrac{d}{dx}.$

$$\hat{T} = -\frac{\hbar^2}{2m}\frac{d^2}{dx^2} = -\frac{\hbar^2}{2m\alpha^2}\frac{d^2}{dy^2} = -\frac{1}{2}\hbar\omega\frac{d^2}{dy^2}, \quad \left[x = \alpha y, \; \alpha^2 = \frac{\hbar}{m\omega}\right]$$

which implies that

$$\hat{T}\psi = -\frac{1}{2}\hbar\omega\left(\frac{d^2\psi}{dy^2}\right).$$

We then use $\psi = NHe^{-y^2/2}$, and obtain

$$\frac{d^2\psi}{dy^2} = N\frac{d^2}{dy^2}(He^{-y^2/2}) = N\{H'' - 2yH' - H + y^2H\}e^{-y^2/2}.$$

From Table 9.1

$$H_v'' - 2yH_v' = -2vH_v$$

$$y^2H_v = y\left(\frac{1}{2}H_{v+1} + vH_{v-1}\right) = \frac{1}{2}\left(\frac{1}{2}H_{v+2} + (v+1)H_v\right) + v\left(\frac{1}{2}H_v + (v-1)H_{v-2}\right)$$

$$= \frac{1}{4}H_{v+2} + v(v-1)H_{v-2} + \left(v + \frac{1}{2}\right)H_v.$$

Hence, $\dfrac{d^2\psi}{dy^2} = N\left[\dfrac{1}{4}H_{v+2} + v(v-1)H_{v-2} - \left(v + \dfrac{1}{2}\right)H_v\right]e^{-y^2/2}.$

Therefore,

$$\langle T \rangle = N^2 \left(-\frac{1}{2}\hbar\omega \right) \int_{-\infty}^{+\infty} H_v \left[\frac{1}{4} H_{v+2} + v(v-1)H_{v-2} - \left(v + \frac{1}{2} \right) H_v \right] e^{-y^2} dx$$

$$[dx = \alpha dy]$$

$$= \alpha N^2 \left(-\tfrac{1}{2}\hbar\omega \right) \left[0 + 0 - (v + \tfrac{1}{2})\pi^{1/2}2^v v! \right]$$

$$\left[\left[\int_{-\infty}^{+\infty} H_v H_{v'} e^{-y^2} dy = 0 \quad \text{if } v' \neq v, \text{ Comment 9.2} \right] \right]$$

$$= \boxed{\frac{1}{2} \left(v + \frac{1}{2} \right) \hbar\omega} \left[N_v^2 = \frac{1}{\alpha\pi^{1/2}2^v v!}, \text{ Example 9.3} \right].$$

P9.15   **(a)** $\langle x \rangle = \displaystyle\int_0^L \left(\frac{2}{L} \right)^{1/2} \sin\left(\frac{n\pi x}{L} \right) x \left(\frac{2}{L} \right)^{1/2} \sin\left(\frac{n\pi x}{L} \right) dx$

$$= \left(\frac{2}{L} \right) \int_0^L x \sin^2 ax\, dx \left[a = \frac{n\pi}{L} \right]$$

$$= \left(\frac{2}{L} \right) \times \left(\frac{x^2}{4} - \frac{x\sin 2ax}{4a} - \frac{\cos 2ax}{8a^2} \right)\Big|_0^L = \left(\frac{2}{L} \right) \times \left(\frac{L^2}{4} \right)$$

$$= \frac{L}{2} \text{ [by symmetry also]}.$$

$$\langle x^2 \rangle = \frac{2}{L}\int_0^L x^2 \sin^2 ax\, dx = \left(\frac{2}{L} \right) \times \left[\frac{x^3}{6} - \left(\frac{x^2}{4a} - \frac{1}{8a^3} \right)\sin 2ax - \frac{x\cos 2ax}{4a^2} \right]\Big|_0^L$$

$$= \left(\frac{2}{L} \right) \times \left(\frac{L^3}{6} - \frac{L^3}{4n^2\pi} \right) = \frac{L^2}{3}\left(1 - \frac{1}{6n^2\pi^2} \right).$$

$$\delta x = \left[\frac{L^2}{3}\left(1 - \frac{1}{6n^2\pi^2} \right) - \frac{L^2}{4} \right]^{1/2} = \boxed{L\left(\frac{1}{12} - \frac{1}{2\pi^2 n^2} \right)^{1/2}}.$$

$\langle p \rangle = 0$ [by symmetry, also see Exercise 9.2(a)],

$\langle p^2 \rangle = n^2 h^2/4L^2$ [from $E = p^2/2m$, also Exercise 9.2(a)].

$$\delta p = \left(\frac{n^2 h^2}{4L^2} \right)^{1/2} = \boxed{\frac{nh}{2L}}.$$

$$\delta p \delta x = \frac{nh}{2L} \times L\left(\frac{1}{12} - \frac{1}{2\pi^2 n^2} \right)^{1/2} = \frac{nh}{2\sqrt{3}}\left(1 - \frac{1}{24\pi^2 n^2} \right)^{1/2} > \frac{\hbar}{2}.$$

(b) $\langle x \rangle = \alpha^2 \displaystyle\int_{-\infty}^{+\infty} \psi^2 y\, dy \ [x = \alpha y] = 0$ [by symmetry, y is an odd function].

$$\langle x^2 \rangle = \frac{2}{k}\left\langle \frac{1}{2}kx^2 \right\rangle = \frac{2}{k}\langle V \rangle$$

since $2 \langle T \rangle = b \langle V \rangle$ [9.35, $\langle T \rangle \equiv E_K$] $= 2 \langle V \rangle$ $\left[V = ax^b = \frac{1}{2}kx^2, b = 2 \right]$

or $\langle V \rangle = \langle T \rangle = \frac{1}{2} \left(v + \frac{1}{2} \right) \hbar \omega$ [Problem 9.13].

$$\langle x^2 \rangle = \left(v + \frac{1}{2} \right) \times \left(\frac{\hbar \omega}{k} \right) = \left(v + \frac{1}{2} \right) \times \left(\frac{\hbar}{\omega m} \right) = \left(v + \frac{1}{2} \right) \times \left(\frac{\hbar^2}{mk} \right)^{1/2} \quad [9.32].$$

$$\delta x = \boxed{\left[\left(v + \frac{1}{2} \right) \frac{\hbar}{\omega m} \right]^{1/2}}.$$

$\langle p \rangle = 0$ [by symmetry, or by noting that the integrand is an odd function of x].

$$\langle p^2 \rangle = 2m \langle T \rangle = (2m) \times \left(\frac{1}{2} \right) \times \left(v + \frac{1}{2} \right) \times \hbar \omega \quad \text{[Problem 9.13]}.$$

$$\delta p = \boxed{\left[\left(v + \frac{1}{2} \right) \hbar \omega m \right]^{1/2}}.$$

$$\delta p \delta x = \left(v + \frac{1}{2} \right) \hbar \geq \frac{\hbar}{2}.$$

COMMENT. Both results show a consistency with the uncertainty principle in the form $\Delta p \Delta q \geq \frac{\hbar}{2}$ as given in Section 8.6, eqn 8.36a.

P9.17   Use the first two terms of the Taylor series expansion of cosine:

$$V = V_0 (1 - \cos 3\phi) \approx V_0 \left(1 - 1 + \frac{(3\phi)^2}{2} \right) = \frac{9V_0}{2} \phi^2.$$

The Schrödinger equation becomes

$$-\frac{\hbar^2}{2I} \frac{\partial^2 \psi}{\partial \phi^2} + \frac{9V_0}{2} \phi^2 \psi = E\psi \quad \text{[9.40 with a non-zero potential energy].}$$

This has the form of the harmonic-oscillator wavefunction (eqn 9.24). The difference in adjacent energy levels is:

$$E_1 - E_0 = \hbar \omega \ \text{[9.26]} \quad \text{where} \ \ \omega = \left(\frac{9V_0}{I} \right)^{1/2} \quad \text{[adapting 9.25].}$$

If the displacements are sufficiently large, the potential energy does not rise as rapidly with the angle as would a harmonic potential. Each successive energy level would become lower than that of a harmonic oscillator, so the energy levels will become progressively closer together.

Question. The next term in the Taylor series for the potential energy is $-\frac{(27V_0)}{8} \phi^4$. Treat this as a perturbation to the harmonic oscillator wavefunction and compute the first-order correction to the energy.

P9.19
$$V = -\frac{e^2}{4\pi\varepsilon_0} \cdot \frac{1}{r} \quad [10.4 \text{ with } Z = 1] = \alpha x^b \quad \text{with } b = -1 \quad [x \to r]$$

Since $2\langle T\rangle = b\langle V\rangle$ [9.35, $\langle T\rangle \equiv E_K$],

$$2\langle T\rangle = -\langle V\rangle.$$

Therefore, $\boxed{\langle T\rangle = -\frac{1}{2}\langle V\rangle}$.

P9.21
The elliptical ring to which the particle is confined is defined by the set of all points that obey a certain equation. In Cartesian coordinates, that equation is

$$\frac{x^2}{a^2} + \frac{y^2}{b^2} = 1$$

as you may remember from analytical geometry. An ellipse is similar to a circle, and an appropriate change of variable can transform the ellipse of this problem into a circle. That change of variable is most conveniently described in terms of new Cartesian coordinates (X, Y) where

$$X = x \quad \text{and} \quad Y = ay/b.$$

In this new coordinate system, the equation for the ellipse becomes

$$\frac{x^2}{a^2} + \frac{y^2}{b^2} = 1 \quad \Rightarrow \quad \frac{X^2}{a^2} + \frac{Y^2}{a^2} = 1 \quad \Rightarrow \quad X^2 + Y^2 = a^2,$$

which we recognize as the equation of a circle of radius a centered at the origin of our (X, Y) system. The text found the eigenfunctions and eigenvalues for a particle on a circular ring by transforming from Cartesian coordinates to plane polar coordinates. Consider plane polar coordinates (R, Φ) related in the usual way to (X, Y):

$$X = R\cos\Phi \quad \text{and} \quad Y = R\sin\Phi.$$

In this coordinate system, we can simply quote the results obtained in the text. The energy levels are

$$E = \frac{m_l^2 \hbar}{2I} \quad [9.38a]$$

where the moment of inertia is the mass of the particle times the radius of the circular ring

$$I = ma^2.$$

The eigenfunctions are

$$\psi = \frac{e^{im_l\Phi}}{(2\pi)^{1/2}} \quad [9.38b].$$

It is customary to express results in terms of the original coordinate system, so express Φ in terms first of X and Y, and then substitute the original coordinates:

$$\frac{Y}{X} = \tan\Phi \quad \text{so} \quad \Phi = \tan^{-1}\frac{Y}{X} = \tan^{-1}\frac{ay}{bx}.$$

P9.23   The Schrödinger equation is

$$-\frac{\hbar^2}{2m}\nabla^2\psi = E\psi \ [9.48 \text{ with } V = 0].$$

$$\nabla^2\psi = \frac{1}{r}\frac{\partial^2(r\psi)}{\partial r^2} + \frac{1}{r^2}\Lambda^2\psi \ [\text{Table 8.1}].$$

Since r = constant, the first term is eliminated and the Schrödinger equation may be rewritten

$$-\frac{\hbar^2}{2mr^2}\Lambda^2\psi = E\psi \ \text{ or } \ -\frac{\hbar^2}{2I}\Lambda^2\psi = E\psi \ [I = mr^2] \ \text{ or } \ \Lambda^2\psi = -\frac{2IE\psi}{\hbar^2}$$

where $\Lambda^2 = \dfrac{1}{\sin^2\theta}\dfrac{\partial^2}{\partial\phi^2} + \dfrac{1}{\sin\theta}\dfrac{\partial}{\partial\theta}\sin\theta\dfrac{\partial}{\partial\theta}.$

Now use the specified $\psi = Y_{l,m_l}$ from Table 9.3, and see if they satisfy this equation.

(a) Because $Y_{0,0}$ is a constant, all derivatives with respect to angles are zero, so $\Lambda^2 Y_{0,0} = \boxed{0}$ implying that $E = \boxed{0}$ and angular momentum $= \boxed{0}$ [from $\{l(l+1)\}^{1/2}\hbar$].

(b) $\Lambda^2 Y_{2,-1} = \dfrac{1}{\sin^2\theta}\dfrac{\partial^2 Y_{2,-1}}{\partial\phi^2} + \dfrac{1}{\sin\theta}\dfrac{\partial}{\partial\theta}\sin\theta\dfrac{\partial Y_{2,-1}}{\partial\theta}$ where $Y_{2,-1} = N\cos\theta\sin\theta e^{-i\phi}.$

$$\frac{\partial Y_{2,-1}}{\partial\theta} = Ne^{-i\phi}(\cos^2\theta - \sin^2\theta).$$

$$\frac{1}{\sin\theta}\frac{\partial}{\partial\theta}\sin\theta\frac{\partial Y_{2,-1}}{\partial\theta} = \frac{1}{\sin\theta}\frac{\partial}{\partial\theta}\sin\theta Ne^{-i\phi}(\cos^2\theta - \sin^2\theta)$$

$$= \frac{Ne^{-i\phi}}{\sin\theta}\left(\sin\theta(-4\cos\theta\sin\theta) + \cos\theta(\cos^2\theta - \sin^2\theta)\right)$$

$$= Ne^{-i\phi}\left(-6\cos\theta\sin\theta + \frac{\cos\theta}{\sin\theta}\right) \quad [\cos^3\theta = \cos\theta(1 - \sin^2\theta)].$$

$$\frac{1}{\sin^2\theta}\frac{\partial^2 Y_{2,-1}}{\partial\phi^2} = \frac{-N\cos\theta\sin\theta e^{-i\phi}}{\sin^2\theta} = \frac{-N\cos\theta e^{-i\phi}}{\sin\theta}$$

so $\Lambda^2 Y_{2,-1} = Ne^{-i\phi}(-6\cos\theta\sin\theta) = -6Y_{2,-1} = -2(2+1)Y_{2,-1}$ [i.e. $l = 2$]
and hence

$$-6Y_{2,-1} = -\frac{2IE}{\hbar^2}Y_{2,-1}, \text{ implying that } \boxed{E = \frac{3\hbar^2}{I}}$$

and the angular momentum is $\{2(2+1)\}^{1/2}\hbar = \boxed{6^{1/2}\hbar}.$

(c) $\Lambda^2 Y_{3,3} = \dfrac{1}{\sin^2\theta}\dfrac{\partial^2 Y_{3,3}}{\partial\phi^2} + \dfrac{1}{\sin\theta}\dfrac{\partial}{\partial\theta}\sin\theta\dfrac{\partial Y_{3,3}}{\partial\theta}$ where $Y_{3,3} = N\sin^3\theta e^{3i\phi}.$

$$\frac{\partial Y_{3,3}}{\partial\theta} = 3N\sin^2\theta\cos\theta e^{3i\phi}.$$

$$\frac{1}{\sin\theta}\frac{\partial}{\partial\theta}\sin\theta\frac{\partial Y_{3,3}}{\partial\theta} = \frac{1}{\sin\theta}\frac{\partial}{\partial\theta}3N\sin^3\theta\cos\theta e^{3i\phi}$$

$$= \frac{3Ne^{3i\phi}}{\sin\theta}(3\sin^2\theta\cos^2\theta - \sin^4\theta) = 3Ne^{3i\phi}\sin\theta(3\cos^2\theta - \sin^2\theta)$$

$$= 3Ne^{3i\phi}\sin\theta(3 - 4\sin^2\theta) \quad [\cos^3\theta = \cos\theta(1 - \sin^2\theta)].$$

$$\frac{1}{\sin^2\theta}\frac{\partial^2 Y_{3,3}}{\partial\phi^2} = \frac{-9N\sin^3\theta e^{3i\phi}}{\sin^2\theta} = -9N\sin\theta e^{3i\phi}$$

so $\Lambda^2 Y_{3,3} = -12N\sin^3\theta\ e^{3i\phi} = -12Y_{3,3} = -3(3+1)Y_{3,3}$ [i.e. $l = 3$]
and hence

$$-12Y_{3,3} = -\frac{2IE}{\hbar^2}Y_{3,3}, \quad \text{implying that} \quad \boxed{E = \frac{6\hbar^2}{I}}$$

and the angular momentum is $\{3(3+1)\}^{1/2}\hbar = \boxed{2\sqrt{3}\hbar}$.

P9.25   From the diagram in Fig. 9.2, $\cos\theta = m_l/\{l(l+1)\}^{1/2}$ and hence $\boxed{\theta = \arccos\dfrac{m_l}{\{l(l+1)\}^{1/2}}}$.

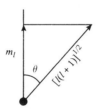

Figure 9.2

For an α electron, $m_s = +\dfrac{1}{2}$, $s = \dfrac{1}{2}$ and (with $m_l \to m_s$, $l \to s$)

$$\theta = \arccos\frac{\dfrac{1}{2}}{\left(\dfrac{3}{4}\right)^{1/2}} = \arccos\frac{1}{\sqrt{3}} = \boxed{54°44'}.$$

The minimum angle occurs for $m_l = l$:

$$\lim_{l\to\infty}\theta_{\min} = \lim_{l\to\infty}\arccos\left(\frac{l}{\{l(l+1)\}^{1/2}}\right) = \lim_{l\to\infty}\arccos\frac{l}{l} = \arccos 1 = \boxed{0}.$$

P9.27
$$\hat{l} = \hat{r}\times\hat{p} = \begin{vmatrix} \mathbf{i} & \mathbf{j} & \mathbf{k} \\ \hat{x} & \hat{y} & \hat{z} \\ \hat{p}_x & \hat{p}_y & \hat{p}_z \end{vmatrix} \quad \text{[see any book treating the vector product of vectors]}$$

$$= \mathbf{i}(\hat{y}\hat{p}_z - \hat{z}\hat{p}_y) + \mathbf{j}(\hat{z}\hat{p}_x - \hat{x}\hat{p}_z) + \mathbf{k}(\hat{x}\hat{p}_y - \hat{y}\hat{p}_x)$$

Therefore,

$$\hat{l}_x = (\hat{y}\hat{p}_z - \hat{z}\hat{p}_y) = \boxed{\frac{\hbar}{i}\left(y\frac{\partial}{\partial z} - z\frac{\partial}{\partial y}\right)},$$

$$\hat{l}_y = (\hat{z}\hat{p}_x - \hat{x}\hat{p}_z) = \boxed{\frac{\hbar}{i}\left(z\frac{\partial}{\partial x} - x\frac{\partial}{\partial z}\right)},$$

$$\hat{l}_z = (\hat{x}\hat{p}_y - \hat{y}\hat{p}_x) = \boxed{\frac{\hbar}{i}\left(x\frac{\partial}{\partial y} - y\frac{\partial}{\partial x}\right)}.$$

We have used $\hat{p}_x = \frac{\hbar}{i}\frac{\partial}{\partial x}$, etc. The commutator of $\hat{l}_x$ and $\hat{l}_y$ is $(\hat{l}_x\hat{l}_y - \hat{l}_y\hat{l}_x)$. We note that the operations always imply operation on a function. We form

$$\hat{l}_x\hat{l}_y f = -\hbar^2\left(y\frac{\partial}{\partial z} - z\frac{\partial}{\partial y}\right)\left(z\frac{\partial}{\partial x} - x\frac{\partial}{\partial z}\right)f$$

$$= -\hbar^2\left(yz\frac{\partial^2 f}{\partial z\partial x} + y\frac{\partial f}{\partial x} - yx\frac{\partial^2 f}{\partial z^2} - z^2\frac{\partial^2 f}{\partial y\partial x} + zx\frac{\partial^2 f}{\partial z\partial y}\right),$$

and $\hat{l}_y\hat{l}_x f = -\hbar^2\left(z\frac{\partial}{\partial x} - x\frac{\partial}{\partial z}\right)\left(y\frac{\partial}{\partial z} - z\frac{\partial}{\partial y}\right)f$

$$= -\hbar^2\left(zy\frac{\partial^2 f}{\partial x\partial z} - z^2\frac{\partial^2 f}{\partial x\partial y} - xy\frac{\partial^2 f}{\partial z^2} + xz\frac{\partial^2 f}{\partial z\partial y} + x\frac{\partial f}{\partial y}\right).$$

Since multiplication and differentiation are each commutative, the results of the operation $\hat{l}_x\hat{l}_y$ and $\hat{l}_y\hat{l}_x$ differ only in one term. For $\hat{l}_y\hat{l}_x f$, $x(\partial f/\partial y)$ replaces $y(\partial f/\partial x)$. Hence, the commutator of the operations, $(\hat{l}_x\hat{l}_y - \hat{l}_y\hat{l}_x)$ is $-\hbar^2\left(y\frac{\partial}{\partial x} - x\frac{\partial}{\partial y}\right)$ or $\boxed{-\frac{\hbar}{i}\hat{l}_z}$.

COMMENT.   We also would find

$$(\hat{l}_y\hat{l}_z - \hat{l}_z\hat{l}_y) = -\frac{\hbar}{i}\hat{l}_x \text{ and } (\hat{l}_z\hat{l}_x - \hat{l}_x\hat{l}_z) = -\frac{\hbar}{i}\hat{l}_y.$$

P9.29   We are to show that $[\hat{l}^2, \hat{l}_z] = [\hat{l}_x^2 + \hat{l}_y^2 + \hat{l}_z^2, \hat{l}_z] = [\hat{l}_x^2, \hat{l}_z] + [\hat{l}_y^2, \hat{l}_z] + [\hat{l}_z^2, \hat{l}_z] = 0$

The three commutators are:

$$[\hat{l}_z^2, \hat{l}_z] = \hat{l}_z^2\hat{l}_z - \hat{l}_z\hat{l}_z^2 = \hat{l}_z^3 - \hat{l}_z^3 = 0,$$

$$[\hat{l}_x^2, \hat{l}_z] = \hat{l}_x^2\hat{l}_z - \hat{l}_z\hat{l}_x^2 = \hat{l}_x^2\hat{l}_z - \hat{l}_x\hat{l}_z\hat{l}_x + \hat{l}_x\hat{l}_z\hat{l}_x - \hat{l}_z\hat{l}_x^2$$

$$= \hat{l}_x(\hat{l}_x\hat{l}_z - \hat{l}_z\hat{l}_x) + (\hat{l}_x\hat{l}_z - \hat{l}_z\hat{l}_x)\hat{l}_x = \hat{l}_x[\hat{l}_x, \hat{l}_z] + [\hat{l}_x, \hat{l}_z]\hat{l}_x$$

$$= \hat{l}_x(-i\hbar\hat{l}_y) + (-i\hbar\hat{l}_y)\hat{l}_x = -i\hbar(\hat{l}_x\hat{l}_y + \hat{l}_y\hat{l}_x) \text{ [9.56a]},$$

$$[\hat{l}_y^2, \hat{l}_z] = \hat{l}_y^2\hat{l}_z - \hat{l}_z\hat{l}_y^2 = \hat{l}_y^2\hat{l}_z - \hat{l}_y\hat{l}_z\hat{l}_y + \hat{l}_y\hat{l}_z\hat{l}_y - \hat{l}_z\hat{l}_y^2$$

$$= \hat{l}_y(\hat{l}_y\hat{l}_z - \hat{l}_z\hat{l}_y) + (\hat{l}_y\hat{l}_z - \hat{l}_z\hat{l}_y)\hat{l}_y = \hat{l}_y[\hat{l}_y, \hat{l}_z] + [\hat{l}_y, \hat{l}_z]\hat{l}_y$$

$$= \hat{l}_y(i\hbar\hat{l}_x) + (i\hbar\hat{l}_x)\hat{l}_y = i\hbar(\hat{l}_y\hat{l}_x + \hat{l}_x\hat{l}_y) \text{ [9.56a]}.$$

Therefore, $[\hat{l}^2, \hat{l}_z] = -i\hbar(\hat{l}_x\hat{l}_y + \hat{l}_y\hat{l}_x) + i\hbar(\hat{l}_x\hat{l}_y + \hat{l}_y\hat{l}_x) + 0 = 0.$

We may also conclude that $[\hat{l}^2, \hat{l}_x] = 0$ and $[\hat{l}^2, \hat{l}_y] = 0$ because $\hat{l}_x, \hat{l}_y,$ and $\hat{l}_z$ occur symmetrically in $\hat{l}^2$.

Solutions to applications

P9.31 (a) The energy levels are given by

$$E_n = \frac{h^2 n^2}{8mL^2},$$

and we are looking for the energy difference between $n = 6$ and $n = 7$:

$$\Delta E = \frac{h^2(7^2 - 6^2)}{8mL^2}.$$

Since there are 12 atoms on the conjugated backbone, the length of the box is 11 times the bond length,

$$L = 11(140 \times 10^{-12}\,\text{m}) = 1.54 \times 10^{-9}\,\text{m},$$

so $\Delta E = \dfrac{(6.626 \times 10^{-34}\,\text{J s})^2(49 - 36)}{8(9.11 \times 10^{-31}\,\text{kg})(1.54 \times 10^{-9}\,\text{m})^2} = \boxed{3.30 \times 10^{-19}\,\text{J}}.$

(b) The relationship between energy and frequency is

$$\Delta E = h\nu \quad \text{so} \quad \nu = \frac{\Delta E}{h} = \frac{3.30 \times 10^{-19}\,\text{J}}{6.626 \times 10^{-34}\,\text{J s}} = \boxed{4.95 \times 10^{-14}\,\text{s}^{-1}}.$$

(c) Look at the terms in the energy expression that change with the number of conjugated atoms, N. The energy (and frequency) are inversely proportional to L^2 and directly proportional to $(n+1)^2 - n^2 = 2n + 1$, where n is the quantum number of the highest occupied state. Since n is proportional to N (equal to $N/2$) and L is approximately proportional to N (strictly to $N-1$), the energy and frequency are approximately proportional to N^{-1}. So *the absorption spectrum of a linear polyene shifts to* $\boxed{lower}$ *frequency as the number of conjugated atoms* $\boxed{increases}$.

P9.33 In effect, we are looking for the vibrational frequency of an O atom bound, with a force constant equal to that of free CO, to an infinitely massive and immobile protein complex. The angular frequency is

$$\omega = \left(\frac{k}{m}\right)^{1/2},$$

where m is the mass of the O atom,

$$m = (16.0\,\text{u})(1.66 \times 10^{-27}\,\text{kg u}^{-1}) = 2.66 \times 10^{-26}\,\text{kg},$$

and k is the same force constant as in Problem 9.2, namely, 1902 N m^{-1}:

$$\omega = \left(\frac{1902 \text{ N m}^{-1}}{2.66 \times 10^{-26} \text{ kg}} \right)^{1/2} = \boxed{2.68 \times 10^{14} \text{ s}^{-1}}.$$

P9.35 The angular momentum states are defined by the quantum number $m_l = 0, \pm 1, \pm 2$, etc. By rearranging eqn 9.42, we see that the energy of state m_l is

$$E_{m_l} = \frac{m_l^2 \hbar^2}{2I}$$

and the angular momentum is

$$l_z = m_l \hbar$$

(a) If there are 22 electrons, two in each of the lowest 11 states, then the highest occupied state is $m_l = \pm 5$, so

$$l_z = \pm 5\hbar = \pm 5 \times (1.055 \times 10^{-34} \text{ J s}) = \boxed{5.275 \times 10^{-34} \text{ J s}}$$

and $E_{\pm 5} = \dfrac{25\hbar^2}{2I}$.

The moment of inertia of an electron on a ring of radius 440 pm is

$$I = mr^2 = 9.11 \times 10^{-31} \text{ kg} \times (440 \times 10^{-12} \text{ m})^2 = 1.76 \times 10^{-49} \text{ kg m}^2.$$

Hence $E_{\pm 5} = \dfrac{25 \times (1.055 \times 10^{-34} \text{ J s})^2}{2 \times (1.76 \times 10^{-49} \text{ kg m}^2)} = \boxed{7.89 \times 10^{-19} \text{ J}}.$

(b) The lowest unoccupied energy level is $m_l = \pm 6$, which has energy

$$E_{\pm 6} = \frac{36 \times (1.055 \times 10^{-34} \text{ J s})^2}{2 \times (1.76 \times 10^{-49} \text{ kg m}^2)} = 1.14 \times 10^{-18} \text{ J}.$$

Radiation that would induce a transition between these levels must have a frequency such that

$$h\nu = \Delta E \quad \text{so} \quad \nu = \frac{\Delta E}{h} = \frac{(11.4 - 7.89) \times 10^{-19} \text{ J}}{6.626 \times 10^{-34} \text{ J s}} = \boxed{5.2 \times 10^{14} \text{ Hz}}.$$

This corresponds to a wavelength of about 570 nm, a wave of visible light.

P9.37 The Coulombic force is

$$F = -\frac{d}{dr} \frac{q_1 q_2}{4\pi\varepsilon_0 r} = \frac{q_1 q_2}{4\pi\varepsilon_0 r^2}.$$

For two electrons 2.0 nm apart, the force is

$$F = \frac{(1.60 \times 10^{-19} \text{ C})^2}{4\pi \times (8.854 \times 10^{-12} \text{ C}^2 \text{ J}^{-1} \text{ m}^{-1}) \times (2.0 \times 10^{-9} \text{ m})^2} = \boxed{5.8 \times 10^{-11} \text{ N}}.$$

P9.39   **(a)** In the sphere, the Schrödinger equation is

$$-\frac{\hbar^2}{2m}\left(\frac{\partial^2}{\partial r^2} + \frac{2}{r}\frac{\partial}{\partial r} + \frac{1}{r^2}\Lambda^2\right)\psi = E\psi \quad [9.51\text{a}]$$

where Λ^2 is an operator that contains derivatives with respect to θ and ϕ only.

Let $\psi(r,\theta,\phi) = X(r)Y(\theta,\phi)$.

Substituting into the Schrödinger equation gives

$$-\frac{\hbar^2}{2m}\left(Y\frac{\partial^2 X}{\partial r^2} + \frac{2Y}{r}\frac{\partial X}{\partial r} + \frac{X}{r^2}\Lambda^2 Y\right) = EXY.$$

Divide both sides by XY:

$$-\frac{\hbar^2}{2m}\left(\frac{1}{X}\frac{\partial^2 X}{\partial r^2} + \frac{2}{Xr}\frac{\partial X}{\partial r} + \frac{1}{Yr^2}\Lambda^2 Y\right) = E.$$

The first two terms in parentheses depend only on r, but the last one depends on both r and angles; however, multiplying both sides of the equation by r^2 will effect the desired separation:

$$-\frac{\hbar^2}{2m}\left(\frac{r^2}{X}\frac{\partial^2 X}{\partial r^2} + \frac{2r}{X}\frac{\partial X}{\partial r} + \frac{1}{Y}\Lambda^2 Y\right) = Er^2$$

Put all of the terms involving angles on the right-hand side and the terms involving distance on the left:

$$-\frac{\hbar^2}{2m}\left(\frac{r^2}{X}\frac{\partial^2 X}{\partial r^2} + \frac{2r}{X}\frac{\partial X}{\partial r}\right) - Er^2 = \frac{\hbar^2}{2mY}\Lambda^2 Y.$$

Note that the right side depends only on θ and ϕ, while the left side depends on r. The only way that the two sides can be equal to each other for all r, θ, and ϕ is if they are both equal to a constant. Call that constant $-(\hbar^2 l(l+1))/2m$ (with l as yet undefined) and we have, from the right side of the equation,

$$\frac{\hbar^2}{2mY}\Lambda^2 Y = -\frac{\hbar^2 l(l+1)}{2m} \qquad \text{so} \qquad \Lambda^2 Y = -l(l+1)Y.$$

From the left side of the equation, we have

$$-\frac{\hbar^2}{2m}\left(\frac{r^2}{X}\frac{\partial^2 X}{\partial r^2} + \frac{2r}{X}\frac{\partial X}{\partial r}\right) - Er^2 = -\frac{\hbar^2 l(l+1)}{2m}.$$

After multiplying both sides by X/r^2 and rearranging, we get the desired radial equation

$$-\frac{\hbar^2}{2m}\left(\frac{\partial^2 X}{\partial r^2} + \frac{2}{r}\frac{\partial X}{\partial r}\right) + \frac{\hbar^2 l(l+1)}{2mr^2}X = EX.$$

Thus the assumption that the wavefunction can be written as a product of functions is a valid one, for we can find separate differential equations for the assumed factors. That is what it means for a partial differential equation to be separable.

(b) The radial equation with $l = 0$ can be rearranged to read

$$\frac{\partial^2 X}{\partial r^2} + \frac{2}{r}\frac{\partial X}{\partial r} = -\frac{2mEX}{\hbar^2}.$$

Form the following derivatives of the proposed solution:

$$\frac{\partial X}{\partial r} = (2\pi R)^{-1/2}\left[\frac{\cos(n\pi r/R)}{r}\left(\frac{n\pi}{R}\right) - \frac{\sin(n\pi r/R)}{r^2}\right]$$

and $\dfrac{\partial^2 X}{\partial r^2} = (2\pi R)^{-1/2}\left[-\dfrac{\sin(n\pi r/R)}{r}\left(\dfrac{n\pi}{R}\right)^2 - \dfrac{2\cos(n\pi r/R)}{r^2}\left(\dfrac{n\pi}{R}\right) + \dfrac{2\sin(n\pi r/R)}{r^3}\right].$

Substituting into the left side of the rearranged radial equation yields

$$(2\pi R)^{-1/2}\left[-\frac{\sin(n\pi r/R)}{r}\left(\frac{n\pi}{R}\right)^2 - \frac{2\cos(n\pi r/R)}{r^2}\left(\frac{n\pi}{R}\right) + \frac{2\sin(n\pi r/R)}{r^3}\right]$$

$$+ (2\pi R)^{-1/2}\left[\frac{2\cos(n\pi r/R)}{r^2}\left(\frac{n\pi}{R}\right) - \frac{2\sin(n\pi r/R)}{r^3}\right]$$

$$= -(2\pi R)^{-1/2}\frac{\sin(n\pi r/R)}{r}\left(\frac{n\pi}{R}\right)^2 = -\left(\frac{n\pi}{R}\right)^2 X.$$

Acting on the proposed solution by taking the prescribed derivatives yields the function back multiplied by a constant, so the proposed solution is in fact a solution.

(c) Comparing this result to the right side of the rearranged radial equation gives an equation for the energy

$$\left(\frac{n\pi}{R}\right)^2 = \frac{2mE}{\hbar^2} \quad \text{so} \quad E = \left(\frac{n\pi}{R}\right)^2\frac{\hbar^2}{2m} = \frac{n^2\pi^2}{2mR^2}\left(\frac{h}{2\pi}\right)^2 = \frac{n^2h^2}{8mR^2}.$$

10 Atomic structure and atomic spectra

Answers to discussion questions

D10.1
The Schrödinger equation for the hydrogen atom is a six-dimensional partial differential equation, three dimensions for each particle in the atom. One cannot directly solve a multidimensional differential equation; it must be broken down into one-dimensional equations. This is the separation of variables procedure. The choice of coordinates is critical in this process. The separation of the Schrödinger equation can be accomplished in a set of coordinates that are natural to the system, but not in others. These natural coordinates are those directly related to the description of the motion of the atom. The atom as a whole (center of mass) can move from point to point in three-dimensional space. The natural coordinates for this kind of motion are the Cartesian coordinates of a point in space. The internal motion of the electron with respect to the proton is most naturally described with spherical polar coordinates. So the six-dimensional Schrödinger equation is first separated into two three-dimensional equations, one for the motion of the center of mass, the other for the internal motion. The separation of the center of mass equation and its solution is fully discussed in Section 9.2. The equation for the internal motion is separable into three one-dimensional equations, one in the angle ϕ, another in the angle θ, and a third in the distance r. The solutions of these three one-dimensional equations can be obtained by standard techniques and were already well known long before the advent of quantum mechanics. Another choice of coordinates would not have resulted in the separation of the Schrödinger equation just described. For the details of the separation procedure, see Sections 10.1 and 9.7.

D10.3
The selection rules are

$$\Delta n = \pm 1, \pm 2, \ldots, \quad \Delta l = \pm 1, \quad \Delta m_l = 0, \pm 1.$$

In a spectroscopic transition the atom emits or absorbs a photon. Photons have a spin angular momentum of 1. Therefore, as a result of the transition, the angular momentum of the electromagnetic field has changed by $\pm 1\hbar$. The principle of the conservation of angular momentum then requires that the angular momentum of the atom has undergone an equal and opposite change in angular momentum. Hence, the selection rule on $\Delta l = \pm 1$. The principal quantum number n can change by any amount since n does not directly relate to angular momentum. The selection rule on Δm_l is harder to account for on basis of these simple considerations alone. One has to evaluate the transition dipole moment between the wavefunctions representing the initial and final states involved in the transition. See *Justification* 10.4 for an example of this procedure.

D10.5   See Section 10.4(d) of the text and any general chemistry book, for example, Sections 1.10–1.13 of P. Atkins and L. Jones, *Chemical principles*, 2nd edn, W. H. Freeman, and Co., New York (2002).

D10.7   In the crudest form of the orbital approximation, the many-electron wavefunctions for atoms are represented as a simple product of one-electron wavefunctions. At a somewhat more sophisticated level, the many-electron wavefunctions are written as linear combinations of such simple product functions that explicitly satisfy the Pauli exclusion principle. Relatively good one-electron functions are generated by the Hartree–Fock self-consistent field method described in Section 10.5. If we place no restrictions on the form of the one-electron functions, we reach the Hartree–Fock limit which gives us the best value of the calculated energy within the orbital approximation. The orbital approximation is based on the disregard of significant portions of the electron–electron interaction terms in the many-electron Hamiltonian, so we cannot expect that it will be quantitatively accurate. By abandoning the orbital approximation, we could in principle obtain essentially exact energies; however, there are significant conceptual advantages to retaining the orbital approximation. Increased accuracy can be obtained by reintroducing the neglected electron–electron interaction terms and including their effects on the energies of the atom by a form of perturbation theory similar to that described in *Further information* 9.2 and Section 10.9. For a more complete discussion consult the references listed under *Further reading*.

Solutions to exercises

E10.1(b)   The energy of the photon that struck the Xe atom goes into liberating the bound electron and giving it any kinetic energy it now possesses

$$E_{photon} = I + E_{kinetic} \qquad I = \text{ionization energy}$$

The energy of a photon is related to its frequency and wavelength

$$E_{photon} = h\nu = \frac{hc}{\lambda}$$

and the kinetic energy of an electron is related to its mass and speed, s

$$E_{kinetic} = \tfrac{1}{2}m_e s^2$$

So $\dfrac{hc}{\lambda} = I + \dfrac{1}{2}m_e s^2 \Rightarrow I = \dfrac{hc}{\lambda} - \dfrac{1}{2}m_e s^2$

$$I = \frac{(6.626 \times 10^{-34}\,\text{J s}) \times (2.998 \times 10^{8}\,\text{m s}^{-1})}{58.4 \times 10^{-9}\,\text{m}} - \frac{1}{2}\left(9.11 \times 10^{-31}\,\text{kg}\right)$$

$$\times \left(1.79 \times 10^{6}\,\text{m s}^{-1}\right)^2$$

$$= \boxed{1.94 \times 10^{-18}\,\text{J}} = 12.1\,\text{eV}$$

E10.2(b)   The radial wavefunction is [Table 10.1]

$$R_{3,0} = A\left(6 - 2\rho + \frac{1}{9}\rho^2\right) e^{-\rho/6} \text{ where } \rho \equiv \frac{2Zr}{a_0}, \text{ and } A \text{ is a collection of constants.}$$

[Note: ρ defined here is $3 \times \rho$ as defined in Table 10.1]

Differentiating with respect to ρ yields

$$\frac{dR_{3,0}}{d\rho} = 0 = A\left(6 - 2\rho + \frac{1}{9}\rho^2\right) \times \left(-\frac{1}{6}\right)e^{-\rho/6} + \left(-2 + \frac{2}{9}\rho\right)Ae^{-\rho/6}$$

$$= Ae^{-\rho/6}\left(-\frac{\rho^2}{54} + \frac{5}{9}\rho - 3\right)$$

This is a quadratic equation

$$0 = a\rho^2 + b\rho + c \qquad \text{where } a = -\frac{1}{54}, \quad b = \frac{5}{9}, \quad \text{and } c = -3.$$

The solution is

$$\rho = \frac{-b \pm (b^2 - 4ac)^{1/2}}{2a} = 15 \pm 3\sqrt{7}$$

$$\text{so } r = \boxed{\left(\frac{15}{2} \pm \frac{3\left(7^{1/2}\right)}{2}\right)\frac{a_0}{Z}}.$$

Numerically, this works out to $\rho = 7.65$ and 2.35, so $r = \boxed{11.5a_0/Z}$ and $\boxed{3.53a_0/Z}$. Substituting $Z = 1$ and $a_0 = 5.292 \times 10^{-11}$ m, $r = \boxed{607 \text{ pm}}$ and $\boxed{187 \text{ pm}}$.

The other maximum in the wavefunction is at $\boxed{r = 0}$. It is a physical maximum, but not a calculus maximum: the first derivative of the wavefunction does not vanish there, so it cannot be found by differentiation.

E10.3(b) The complete radial wavefunction, $R_{4,1}$ is not given in Table 10.1; however in the statement of the exercise we are told that it is proportional to

$$(20 - 10\rho + \rho^2)\rho \quad \text{where} \quad \rho = \frac{2Zr}{a_0} \quad \text{[Note: } \rho \text{ defined here is } n \times \rho \text{ as defined in Table 10.1]}$$

The radial nodes occur where the radial wavefunction vanishes, namely where

$$(20 - 10\rho + \rho^2)\rho = 0.$$

The zeros of this function occur at

$$\rho = 0, \qquad \boxed{r = 0}$$

and when

$$(20 - 10\rho + \rho^2) = 0, \quad \text{with roots } \rho = 2.764, \quad \text{and } \rho = 7.236$$

$$\text{then } r = \frac{\rho a_0}{2Z} = \frac{\rho a_0}{2} = \frac{2.764 a_0}{2} = \boxed{1.382 a_0} \quad \text{and} \quad \frac{7.236 a_0}{2} = \boxed{3.618 a_0}$$

$$\text{or } r = \boxed{7.31 \times 10^{-11} \text{ m}} \text{ and } \boxed{1.917 \times 10^{-10} \text{ m}}$$

E10.4(b)   Normalization requires

$$\int |\psi|^2 \, d\tau = 1 = \int_0^\infty \int_0^\pi \int_0^{2\pi} [N(2 - r/a_0)\, e^{-r/2a_0}]^2 \, d\phi \, \sin\theta \, d\theta \, r^2 \, dr$$

$$1 = N^2 \int_0^\infty e^{-r/a_0} (2 - r/a_0)^2 r^2 \, dr \int_0^\pi \sin\theta \, d\theta \int_0^{2\pi} d\phi$$

Integrating over angles yields

$$1 = 4\pi N^2 \int_0^\infty e^{-r/a_0} (2 - r/a_0)^2 r^2 \, dr$$

$$= 4\pi N^2 \int_0^\infty e^{-r/a_0} (4 - 4r/a_0 + r^2/a_0^2) r^2 \, dr = 4\pi N^2 (8 a_0^3)$$

In the last step, we used

$$\int_0^\infty e^{-r/k} r^2 \, dr = 2k^3, \quad \int_0^\infty e^{-r/k} r^3 \, dr = 6k^4, \quad \text{and} \quad \int_0^\infty e^{-r/k} r^4 \, dr = 24k^5,$$

So $\boxed{N = \dfrac{1}{4\sqrt{2\pi a_0^3}}}$

E10.5(b)   The average kinetic energy is

$$\langle \hat{E}_K \rangle = \int \psi^* \hat{E}_K \psi \, d\tau$$

where $\psi = N(2 - \rho)e^{-\rho/2}$ with $N = \dfrac{1}{4}\left(\dfrac{Z^3}{2\pi a_0^3}\right)^{1/2}$ and $\rho \equiv \dfrac{Zr}{a_0}$ here

$$\hat{E}_K = -\frac{\hbar^2}{2m} \nabla^2 \qquad d\tau = r^2 \sin\theta \, dr \, d\theta \, d\phi = \frac{a_0^3 \rho^2 \sin\theta \, d\rho \, d\theta \, d\phi}{Z^3}$$

In spherical polar coordinates, three of the derivatives in ∇^2 are derivatives with respect to angles, so those parts of $\nabla^2 \psi$ vanish. Thus

$$\nabla^2 \psi = \frac{\partial^2 \psi}{\partial r^2} + \frac{2}{r} \frac{\partial \psi}{\partial r} = \frac{\partial^2 \psi}{\partial \rho^2} \left(\frac{\partial \rho}{\partial r}\right)^2 + \frac{2Z}{\rho a_0} \left(\frac{\partial \psi}{\partial \rho}\right) \frac{\partial \rho}{\partial r} = \left(\frac{Z}{a_0}\right)^2 \times \left(\frac{\partial^2 \psi}{\partial \rho^2} + \frac{2}{\rho} \frac{\partial \psi}{\partial \rho}\right)$$

$$\frac{\partial \rho}{\partial r} = N(2 - \rho) \times \left(-\frac{1}{2}\right) e^{-\rho/2} - N e^{-\rho/2} = N\left(\tfrac{1}{2}\rho - 2\right) e^{-\rho/2}$$

$$\frac{\partial^2 \psi}{\partial \rho^2} = N\left(\tfrac{1}{2}\rho - 2\right) \times \left(-\tfrac{1}{2}\right) e^{-\rho/2} + \tfrac{1}{2} N e^{-\rho/2} = N\left(\tfrac{3}{2} - \tfrac{1}{4}\rho\right) e^{-\rho/2}$$

$$\nabla^2 \psi = \left(\frac{Z}{a_0}\right)^2 N e^{-\rho/2} (-4/\rho + 5/2 - \rho/4)$$

and

$$\langle \hat{E}_K \rangle = \int_0^\infty \int_0^\pi \int_0^{2\pi} N(2-\rho) e^{-\rho/2} \left(\frac{Z}{a_0}\right)^2 \times \left(\frac{-\hbar^2}{2m}\right)$$

$$\times N e^{-\rho/2} (-4/\rho + 5/2 - \rho/4) \frac{a_0^3 \, d\phi \, \sin\theta \, d\theta \, \rho^2 \, d\rho}{Z^3}$$

The integrals over angles give a factor of 4π, so

$$\langle \hat{E}_K \rangle = 4\pi N^2 \left(\frac{a_0}{Z}\right) \times \left(-\frac{\hbar^2}{2m}\right) \int_0^\infty (2-\rho) \times \left(-4 + \frac{5}{2}\rho - \frac{1}{4}\rho^2\right) \rho e^{-\rho} \, d\rho$$

The integral in this last expression works out to -2, using $\int_0^\infty e^{-\rho}\rho^n \, d\rho = n!$ for $n = 1, 2,$ and 3. So

$$\langle \hat{E}_K \rangle = 4\pi \left(\frac{Z^3}{32\pi a_0^3}\right) \times \left(\frac{a_0}{Z}\right) \times \left(\frac{\hbar^2}{m}\right) = \boxed{\frac{\hbar^2 Z^2}{8m a_0^2}}$$

The average potential energy is

$$\langle V \rangle = \int \psi^* V \psi \, d\tau \quad \text{where} \quad V = -\frac{Ze^2}{4\pi\varepsilon_0 r} = -\frac{Z^2 e^2}{4\pi\varepsilon_0 a_0 \rho}$$

and $$\langle V \rangle = \int_0^\infty \int_0^\pi \int_0^{2\pi} N(2-\rho)e^{-\rho/2}\left(-\frac{Z^2 e^2}{4\pi\varepsilon_0 a_0 \rho}\right) N(2-\rho)e^{-\rho/2} \frac{a_0^3 \rho^2 \sin\theta \, d\rho \, d\theta \, d\phi}{Z^3}$$

The integrals over angles give a factor of 4π, so

$$\langle V \rangle = 4\pi N^2 \left(\frac{Z^2 e^2}{4\pi\varepsilon_0 a_0}\right) \times \left(\frac{a_0^3}{Z^3}\right) \int_0^\infty (2-\rho)^2 \rho e^{-\rho} \, d\rho$$

The integral in this last expression works out to 2, using $\int_0^\infty e^{-\rho}\rho^n \, d\rho = n!$ for $n = 1, 2, 3,$ and 4. So

$$\langle V \rangle = 4\pi \left(\frac{Z^3}{32\pi a_0^3}\right) \times \left(-\frac{Z^2 e^2}{4\pi\varepsilon_0 a_0}\right) \times \left(\frac{a_0^3}{Z^3}\right) \times (2) = \boxed{-\frac{Z^2 e^2}{16\pi\varepsilon_0 a_0}}$$

E10.6(b)   The radial distribution function is defined as

$$P = 4\pi r^2 \psi^2 \quad \text{so} \quad P_{3s} = 4\pi r^2 (Y_{0,0} R_{3,0})^2,$$

$$\boxed{P_{3s} = 4\pi r^2 \left(\frac{1}{4\pi}\right) \times \left(\frac{1}{243}\right) \times \left(\frac{Z}{a_0}\right) \times (6 - 6\rho + \rho^2)^2 e^{-\rho}}$$

where $\rho \equiv \dfrac{2Zr}{na_0} = \dfrac{2Zr}{3a_0}$ here.

But we want to find the most likely radius, so it would help to simplify the function by expressing it in terms either of r or ρ, but not both. To find the most likely radius, we could set the derivative of P_{3s} equal to zero; therefore, we can collect all multiplicative constants together (including the factors of a_0/Z needed to turn the initial r^2 into ρ^2) since they will eventually be divided into zero

$$P_{3s} = C^2 \rho^2 (6 - 6\rho + \rho^2)^2 e^{-\rho}$$

Note that not all the extrema of P are maxima; some are minima. But all the extrema of $(P_{3s})^{1/2}$ correspond to maxima of P_{3s}. So let us find the extrema of $(P_{3s})^{1/2}$

$$\frac{d\,(P_{3s})^{1/2}}{d\rho} = 0 = \frac{d}{d\rho}C\rho(6 - 6\rho + \rho^2)e^{-\rho/2}$$

$$= C[\rho(6 - 6\rho + \rho^2) \times \left(-\tfrac{1}{2}\right) + (6 - 12\rho + 3\rho^2)]e^{-\rho/2}$$

$$0 = C\left(6 - 15\rho + 6\rho^2 - \tfrac{1}{2}\rho^3\right)e^{-\rho/2} \quad \text{so} \quad 12 - 30\rho + 12\rho^2 - \rho^3 = 0$$

Numerical solution of this cubic equation yields

$$\rho = 0.49, \;\; 2.79, \;\; \text{and} \;\; 8.72$$

corresponding to

$$r = \boxed{0.74a_0/Z, \;\; 4.19a_0/Z, \;\; \text{and} \;\; 13.08a_0/Z}$$

COMMENT. If numerical methods are to be used to locate the roots of the equation which locates the extrema, then graphical/numerical methods might as well be used to locate the maxima directly. That is, the student may simply have a spreadsheet compute P_{3s} and examine or manipulate the spreadsheet to locate the maxima.

E10.7(b) The most probable radius occurs when the radial wavefunction is a maximum. At this point the derivative of the function wrt either r or ρ equals zero.

$$\left(\frac{dR_{31}}{d\rho}\right)_{\text{max}} = 0 = \left(\frac{d\left((4 - \rho)\,\rho e^{-\rho/2}\right)}{d\rho}\right)_{\text{max}} \quad \text{[Table 10.1]} = \left(4 - 4\rho + \frac{\rho^2}{2}\right)e^{-\rho/2}$$

The function is a maximum when the polynomial equals zero. The quadratic equation gives the roots $\rho = 4 + 2\sqrt{2} = 6.89$ and $\rho = 4 - 2\sqrt{2} = 1.17$. Since $\rho = (2Z/na_0)r$ and $n = 3$, these correspond to $r = 10.3 \times a_0/Z$ and $r = 1.76 \times a_0/Z$. However, $\left|\dfrac{R_{31}\,(\rho_1)}{R_{31}\,(\rho_2)}\right| = \left|\dfrac{R_{31}\,(1.17)}{R_{31}\,(10.3)}\right| = 4.90$. So, we conclude that the function is a maximum at $\rho = 1.17$ which corresponds to $\boxed{r = 1.76a_0/Z.}$

E10.8(b) Orbital angular momentum is

$$\langle\hat{L}^2\rangle^{1/2} = \hbar(l(l + 1))^{1/2}$$

There are l angular nodes and $n - l - 1$ radial nodes

(a) $n = 4, l = 2$, so $\langle\hat{L}^2\rangle^{1/2} = 6^{1/2}\hbar = \boxed{2.45 \times 10^{-34}\,\text{J s}}$ $\boxed{2}$ angular nodes $\boxed{1}$ radial node

(b) $n = 2, l = 1$, so $\langle\hat{L}^2\rangle^{1/2} = 2^{1/2}\hbar = \boxed{1.49 \times 10^{-34}\,\text{J s}}$ $\boxed{1}$ angular nodes $\boxed{0}$ radial nodes

(c) $n = 3, l = 1$, so $\langle\hat{L}^2\rangle^{1/2} = 2^{1/2}\hbar = \boxed{1.49 \times 10^{-34}\,\text{J s}}$ $\boxed{1}$ angular node $\boxed{1}$ radial node

E10.9(b)   For $l > 0$, $j = l \pm 1/2$, so

(a) $l = 1$, so $j = \boxed{1/2 \text{ or } 3/2}$

(b) $l = 5$, so $j = \boxed{9/2 \text{ or } 11/2}$

E10.10(b)   Use the Clebsch–Gordan series in the form

$$J = j_1 + j_2, \ j_1 + j_2 - 1, \ldots, |j_1 - j_2|$$

Then, with $j_1 = 5$ and $j_2 = 3$

$$J = \boxed{8, 7, 6, 5, 4, 3, 2}$$

E10.11(b)   The degeneracy g of a hydrogenic atom with principal quantum number n is $g = n^2$. The energy E of hydrogenic atoms is

$$E = -\frac{hcZ^2 R_H}{n^2} = -\frac{hcZ^2 R_H}{g}$$

so the degeneracy is

$$g = -\frac{hcZ^2 R_H}{E}$$

(a) $g = -\dfrac{hc\,(2)^2\,R_H}{-4hcR_H} = \boxed{1}$

(b) $g = -\dfrac{hc\,(4)^2\,R_H}{-\frac{1}{4}hcR_H} = \boxed{64}$

(c) $g = -\dfrac{hc(5)^2 R_H}{-hcR_H} = \boxed{25}$

E10.12(b)   The letter F indicates that the total orbital angular momentum quantum number L is 3; the superscript 3 is the multiplicity of the term, $2S + 1$, related to the spin quantum number $S = 1$; and the subscript 4 indicates the total angular momentum quantum number J.

E10.13(b)   The radial distribution function varies as

$$P = 4\pi r^2 \psi^2 = \frac{4}{a_0^3} r^2 e^{-2r/a_0}$$

The maximum value of P occurs at $r = a_0$ since

$$\frac{dP}{dr} \propto \left(2r - \frac{2r^2}{a_0}\right) e^{-2r/a_0} = 0 \ \text{at} \ r = a_0 \ \text{and} \ P_{\max} = \frac{4}{a_0} e^{-2}$$

P falls to a fraction f of its maximum given by

$$f = \frac{(4r^2/a_0^3) e^{-2r/a_0}}{(4/a_0) e^{-2}} = \frac{r^2}{a_0^2} e^2 e^{-2r/a_0}$$

and hence we must solve for r in

$$\frac{f^{1/2}}{e} = \frac{r}{a_0} e^{-r/a_0}$$

(a) $\quad f = 0.50$

$$0.260 = \frac{r}{a_0} e^{-r/a_0} \text{ solves to } r = 2.08a_0 = \boxed{110\,\text{pm}} \text{ and to } r = 0.380a_0 = \boxed{20.1\,\text{pm}}$$

(b) $\quad f = 0.75$

$$0.319 = \frac{r}{a_0} e^{-r/a_0} \text{ solves to } r = 1.63a_0 = \boxed{86\,\text{pm}} \text{ and to } r = 0.555a_0 = \boxed{29.4\,\text{pm}}$$

In each case the equation is solved numerically (or graphically) with readily available personal computer software. The solutions above are easily checked by substitution into the equation for f. The radial distribution function is readily plotted and is shown in Figure 10.1.

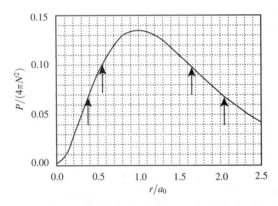

Figure 10.1

E10.14(b) **(a)** $5d \rightarrow 2s$ is $\boxed{\text{not}}$ an allowed transition, for $\Delta l = -2$ (Δl must equal ±1).

(b) $5p \rightarrow 3s$ is $\boxed{\text{allowed}}$, since $\Delta l = -1$.

(c) $5p \rightarrow 3f$ is $\boxed{\text{not}}$ allowed, for $\Delta l = +2$ (Δl must equal ±1).

(d) $6h : l = 5$; maximum occupancy $= \boxed{22}$

E10.15(b) $V^{2+} : 1s^2 2s^2 2p^6 3s^2 3p^6 3d^3 = [Ar]3d^3$

The only unpaired electrons are those in the $3d$ subshell. There are three.

$S = \boxed{\frac{3}{2}}$ and $\frac{3}{2} - 1 = \boxed{\frac{1}{2}}$.

For $S = \frac{3}{2}$, $M_S = \boxed{\pm\frac{1}{2} \text{ and } \pm\frac{3}{2}}$

for $S = \frac{1}{2}$, $M_S = \boxed{\pm\frac{1}{2}}$

E10.16(b) **(a)** Possible values of S for four electrons in different orbitals are $\boxed{2, 1, \text{ and } 0}$; the multiplicity is $2S+1$, so multiplicities are $\boxed{5, 3, \text{ and } 1}$ respectively.

(b) Possible values of S for five electrons in different orbitals are $\boxed{5/2, 3/2 \text{ and } 1/2}$; the multiplicity is $2S + 1$, so multiplicities are $\boxed{6, 4, \text{ and } 2}$ respectively.

E10.17(b) The coupling of a p electron $(l = 1)$ and a d electron $(l = 2)$ gives rise to $L = 3$ (F), 2 (D), and 1 (P) terms. Possible values of S include 0 and 1. Possible values of J (using Russell–Saunders coupling) are 3, 2, and 1 $(S = 0)$ and 4, 3, 2, 1, and 0 $(S = 1)$. The term symbols are

$$\boxed{{}^{1}F_{3}; {}^{3}F_{4}, {}^{3}F_{3}, {}^{3}F_{2}; {}^{1}D_{2}; {}^{3}D_{3}, {}^{3}D_{2}, {}^{3}D_{1}; {}^{1}P_{1}, {}^{3}P_{2}, {}^{3}P_{1}, {}^{3}P_{0}}$$

Hund's rules state that the lowest energy level has maximum multiplicity. Consideration of spin–orbit coupling says the lowest energy level has the lowest value of $J(J + 1) - L(L + 1) - S(S + 1)$. So the lowest energy level is $\boxed{{}^{3}F_{2}}$.

E10.18(b) **(a)** ${}^{3}D$ has $S = 1$ and $L = 2$, so $J = \boxed{3, 2, \text{ and } 1}$ are present. $J = 3$ has $\boxed{7}$ states, with $M_{J} = 0, \pm 1, \pm 2$, or ± 3; $J = 2$ has $\boxed{5}$ states, with $M_{J} = 0, \pm 1$, or ± 2; $J = 1$ has $\boxed{3}$ states, with $M_{J} = 0$, or ± 1.

(b) ${}^{4}D$ has $S = 3/2$ and $L = 2$, so $J = \boxed{7/2, 5/2, 3/2, \text{ and } 1/2}$ are present. $J = 7/2$ has $\boxed{8}$ possible states, with $M_{J} = \pm 7/2, \pm 5/2, \pm 3/2$ or $\pm 1/2$; $J = 5/2$ has $\boxed{6}$ possible states, with $M_{J} = \pm 5/2, \pm 3/2$ or $\pm 1/2$; $J = 3/2$ has $\boxed{4}$ possible states, with $M_{J} = \pm 3/2$ or $\pm 1/2$; $J = 1/2$ has $\boxed{2}$ possible states, with $M_{J} = \pm 1/2$.

(c) ${}^{2}G$ has $S = 1/2$ and $L = 4$, so $J = 9/2$ and $7/2$ are present. $J = 9/2$ has $\boxed{10}$ possible states, with $M_{J} = \pm 9/2, \pm 7/2, \pm 5/2, \pm 3/2$, or $\pm 1/2$; $J = 7/2$ has $\boxed{8}$ possible states, with $M_{J} = \pm 7/2, \pm 5/2, \pm 3/2$, or $\pm 1/2$.

E10.19(b) Closed shells and subshells do not contribute to either L or S and thus are ignored in what follows.

(a) $Sc[Ar]3d^{1}4s^{2}$: $S = \frac{1}{2}, L = 2; J = \frac{5}{2}, \frac{3}{2}$, so the terms are $\boxed{{}^{2}D_{5/2} \text{ and } {}^{2}D_{3/2}}$.

(b) $Br[Ar]3d^{10}4s^{2}4p^{5}$. We treat the missing electron in the $4p$ subshell as equivalent to a single "electron" with $l = 1$, $s = \frac{1}{2}$. Hence $L = 1$, $S = \frac{1}{2}$, and $J = \frac{3}{2}, \frac{1}{2}$, so the terms are $\boxed{{}^{2}P_{3/2} \text{ and } {}^{2}P_{1/2}}$.

Solutions to problems

Solutions to numerical problems

P10.1 All lines in the hydrogen spectrum fit the Rydberg formula

$$\frac{1}{\lambda} = R_{H}\left(\frac{1}{n_{1}^{2}} - \frac{1}{n_{2}^{2}}\right) \quad \left[10.1, \text{ with } \tilde{\nu} = \frac{1}{\lambda}\right] \quad R_{H} = 109677\,\text{cm}^{-1}.$$

Find n_1 from the value of λ_{max}, which arises from the transition $n_1 + 1 \rightarrow n_1$.

$$\frac{1}{\lambda_{max}R_H} = \frac{1}{n_1^2} - \frac{1}{(n_1 + 1)^2} = \frac{2n_1 + 1}{n_1^2(n_1 + 1)^2},$$

$$\lambda_{max}R_H = \frac{n_1^2(n_1 + 1)^2}{2n_1 + 1} = (12\,368 \times 10^{-9}\,\text{m}) \times (109\,677 \times 10^2\,\text{m}^{-1}) = 135.65.$$

Since $n_1 = 1, 2, 3$, and 4 have already been accounted for, try $n_1 = 5, 6, \ldots$. With $n_1 = 6$ we get $\frac{n_1^2(n_1 + 1)^2}{2n_1 + 1} = 136$. Hence, the Humphreys series is $\boxed{n_2 \rightarrow 6}$ and the transitions are given by

$$\frac{1}{\lambda} = (109\,677\,\text{cm}^{-1}) \times \left(\frac{1}{36} - \frac{1}{n_2^2}\right), \quad n_2 = 7, 8, \cdots$$

and occur at 12 372 nm, 7503 nm, 5908 nm, 5129 nm, ..., 3908 nm (at $n_2 = 15$), converging to 3282 nm as $n_2 \rightarrow \infty$, in agreement with the quoted experimental result.

P10.3   A Lyman series corresponds to $n_1 = 1$; hence

$$\tilde{\nu} = R_{Li^{2+}}\left(1 - \frac{1}{n^2}\right), \quad n = 2, 3, \ldots \quad \boxed{\tilde{\nu} = \frac{1}{\lambda}}.$$

Therefore, if the formula is appropriate, we expect to find that $\tilde{\nu}\left(1 - \frac{1}{n^2}\right)^{-1}$ is a constant $(R_{Li^{2+}})$.

We therefore draw up the following table.

n	2	3	4
$\tilde{\nu}/\text{cm}^{-1}$	740747	877924	925933
$\tilde{\nu}\left(1 - \frac{1}{n^2}\right)^{-1}/\text{cm}^{-1}$	987663	987665	987662

Hence, the formula does describe the transitions, and $\boxed{R_{Li^{2+}} = 987\,663\,\text{cm}^{-1}}$. The Balmer transitions lie at

$$\tilde{\nu} = R_{Li^{2+}}\left(\frac{1}{4} - \frac{1}{n^2}\right) \quad n = 3, 4, \ldots$$

$$= (987\,663\,\text{cm}^{-1}) \times \left(\frac{1}{4} - \frac{1}{n^2}\right) = \boxed{137\,175\,\text{cm}^{-1}}, \boxed{185\,187\,\text{cm}^{-1}}, \ldots.$$

The ionization energy of the ground-state ion is given by

$$\tilde{\nu} = R_{Li^{2+}}\left(1 - \frac{1}{n^2}\right), \quad n \rightarrow \infty$$

and hence corresponds to

$$\tilde{\nu} = 987\,663\mathrm{cm}^{-1}, \quad \text{or} \quad \boxed{122.5 \text{ eV}}.$$

P10.5   The $7p$ configuration has just one electron outside a closed subshell. That electron has $l = 1$, $s = 1/2$, and $j = 1/2$ or $3/2$, so the atom has $L = 1$, $S = 1/2$, and $J = 1/2$ or $3/2$. The term symbols are $\boxed{{}^2P_{1/2} \text{ and } {}^2P_{3/2}}$, of which the former has the lower energy. The $6d$ configuration also has just one electron outside a closed subshell; that electron has $l = 2$, $s = 1/2$, and $j = 3/2$ or $5/2$, so the atom has $L = 2$, $S = 1/2$, and $J = 3/2$ or $5/2$. The term symbols are $\boxed{{}^2D_{3/2} \text{ and } {}^2D_{5/2}}$, of which the former has the lower energy. According to the simple treatment of spin–orbit coupling, the energy is given by

$$E_{l,s,j} = \tfrac{1}{2}hcA[j(j+1) - l(l+1) - s(s+1)]$$

where A is the spin–orbit coupling constant. So

$$E({}^2P_{1/2}) = \tfrac{1}{2}hcA[\tfrac{1}{2}(1/2+1) - 1(1+1) - \tfrac{1}{2}(1/2+1)] = -hcA$$

and $E({}^2D_{3/2}) = \tfrac{1}{2}hcA[\tfrac{3}{2}(3/2+1) - 2(2+1) - \tfrac{1}{2}(1/2+1)] = -\tfrac{3}{2}hcA$.

This approach would predict the ground state to be $\boxed{{}^2D_{3/2}}$.

COMMENT.   The computational study cited finds the ${}^2P_{1/2}$ level to be lowest, but the authors caution that the error of similar calculations on Y and Lu is comparable to the computed difference between levels.

P10.7 $$R_{\mathrm{H}} = k\mu_{\mathrm{H}}, \quad R_{\mathrm{D}} = k\mu_{\mathrm{D}}, \quad R = k\mu \quad [10.16]$$

where R corresponds to an infinitely heavy nucleus, with $\mu = m_{\mathrm{e}}$.

Since $\mu = \dfrac{m_{\mathrm{e}}m_{\mathrm{N}}}{m_{\mathrm{e}} + m_{\mathrm{N}}}$ $[N = p$ or $d]$,

$$R_{\mathrm{H}} = k\mu_{\mathrm{H}} = \frac{km_{\mathrm{e}}}{1 + (m_{\mathrm{e}}/m_{\mathrm{p}})} = \frac{R}{1 + (m_{\mathrm{e}}/m_{\mathrm{p}})}.$$

Likewise, $R_{\mathrm{D}} = \dfrac{R}{(1 + (m_{\mathrm{e}}/m_{\mathrm{d}}))}$ where m_{p} is the mass of the proton and m_{d} the mass of the deuteron.

The two lines in question lie at

$$\frac{1}{\lambda_{\mathrm{H}}} = R_{\mathrm{H}}\left(1 - \frac{1}{4}\right) = \frac{3}{4}R_{\mathrm{H}} \qquad \frac{1}{\lambda_{\mathrm{D}}} = R_{\mathrm{D}}\left(1 - \frac{1}{4}\right) = \frac{3}{4}R_{\mathrm{D}}$$

and hence

$$\frac{R_{\mathrm{H}}}{R_{\mathrm{D}}} = \frac{\lambda_{\mathrm{D}}}{\lambda_{\mathrm{H}}} = \frac{\tilde{\nu}_{\mathrm{H}}}{\tilde{\nu}_{\mathrm{D}}}.$$

Then, since

$$\frac{R_H}{R_D} = \frac{1 + (m_e/m_d)}{1 + (m_e/m_p)}, \quad m_d = \frac{m_e}{(1 + (m_e/mp))(R_H/R_D) - 1}$$

and we can calculate m_d from

$$m_d = \frac{m_e}{\left(1 + (m_e/m_p)\right)(\lambda_D/\lambda_H) - 1} = \frac{m_e}{\left(1 + (m_e/m_p)\right)(\tilde{\nu}_H/\tilde{\nu}_D) - 1}$$

$$= \frac{9.10939 \times 10^{-31}\,\text{kg}}{\left(1 + \dfrac{9.1039 \times 10^{-31}\,\text{kg}}{1.67262 \times 10^{-27}\,\text{kg}}\right) \times \left(\dfrac{82259.098\,\text{cm}^{-1}}{82281.476\,\text{cm}^{-1}}\right)^{-1}}$$

$$= \boxed{3.3429 \times 10^{-27}\,\text{kg}}.$$

Since $I = Rhc$,

$$\frac{I_D}{I_H} = \frac{R_D}{R_H} = \frac{\tilde{\nu}_D}{\tilde{\nu}_H} = \frac{82281.476\,\text{cm}^{-1}}{82259.098\,\text{cm}^{-1}} = \boxed{1.000272}.$$

P10.9   (a) The splitting of adjacent energy levels is related to the difference in wavenumber of the spectral lines as follows:

$$hc\Delta\tilde{\nu} = \Delta E = \mu_B B, \quad \text{so } \Delta\tilde{\nu} = \frac{\mu_B B}{hc} = \frac{(9.274 \times 10^{-24}\,\text{J T}^{-1})(2\,\text{T})}{(6.626 \times 10^{-34}\,\text{J s})(2.998 \times 10^{10}\,\text{cm s}^{-1})}$$

$$\Delta\tilde{\nu} = \boxed{0.9\,\text{cm}^{-1}}.$$

(b) Transitions induced by absorbing visible light have wavenumbers in the tens of thousands of reciprocal centimeters, so normal Zeeman splitting is $\boxed{\text{small}}$ compared to the difference in energy of the states involved in the transition. Take a wavenumber from the middle of the visible spectrum as typical:

$$\tilde{\nu} = \frac{1}{\lambda} = \frac{1}{600\,\text{nm}} \left(\frac{10^9\,\text{nm m}^{-1}}{10^2\,\text{cm m}^{-1}}\right) = 1.7 \times 10^4\,\text{cm}^{-1}.$$

Or take the Balmer series as an example, as suggested in the problem; the Balmer wavenumbers are (eqn 10.1):

$$\tilde{\nu} = R_H \left(\frac{1}{2^2} - \frac{1}{n^3}\right).$$

The smallest Balmer wavenumber is

$$\tilde{\nu} = (109\,677\,\text{cm}^{-1}) \times (1/4 - 1/9) = 15\,233\,\text{cm}^{-1}$$

and the upper limit is

$$\tilde{\nu} = (109\,677\,\text{cm}^{-1}) \times (1/4 - 0) = 27\,419\,\text{cm}^{-1}.$$

Solutions to theoretical problems

P10.11 Consider $\psi_{2p_z} = \psi_{2,1,0}$ which extends along the z-axis. The most probable point along the z-axis is where the radial function has its maximum value (for ψ^2 is also a maximum at that point). From Table 10.1 we know that

$$R_{21} \propto \rho e^{-\rho/4}$$

and so $\dfrac{dR}{d\rho} = \left(1 - \dfrac{1}{4}\rho\right) e^{-\rho/4} = 0$ when $\rho = 4$.

Therefore, $r^* = \dfrac{2a_0}{Z}$, and the point of maximum probability lies at $z = \pm\dfrac{2a_0}{Z} = \boxed{\pm 106 \text{ pm}}$.

COMMENT. Since the radial portion of a 2p function is the same, the same result would have been obtained for all of them. The direction of the most probable point would, however, be different.

P10.13 **(a)** We must show that $\int |\psi_{3p_x}|^2 \, d\tau = 1$. The integrations are most easily performed in spherical coordinates (Fig. 8.22).

$$\int |\psi_{3p_x}|^2 \, d\tau = \int_0^{2\pi} \int_0^{\pi} \int_0^{\infty} |\psi_{3p_x}|^2 \, r^2 \sin(\theta) \, dr \, d\theta \, d\phi$$

$$= \int_0^{2\pi} \int_0^{\pi} \int_0^{\infty} \left| R_{31}(\rho) \left\{ \frac{Y_{1-1} - Y_{11}}{\sqrt{2}} \right\} \right|^2 r^2 \sin(\theta) \, dr \, d\theta \, d\phi \quad \text{[Table 10.1, eqn 10.24]}$$

where $\rho = 2r/a_0$, $r = \rho a_0/2$, $dr = (a_0/2) \, d\rho$.

$$= \frac{1}{2} \int_0^{2\pi} \int_0^{\pi} \int_0^{\infty} \left(\frac{a_0}{2}\right)^3 \left| \left[\left(\frac{1}{27(6)^{1/2}}\right) \left(\frac{1}{a_0}\right)^{3/2} \left(4 - \frac{1}{3}\rho\right) \rho e^{-\rho/6} \right] \right.$$

$$\left. \times \left[\left(\frac{3}{8\pi}\right)^{1/2} 2 \sin(\theta) \cos(\phi) \right] \right|^2 \rho^2 \sin(\theta) \, d\rho \, d\theta \, d\phi$$

$$= \frac{1}{46\,656\pi} \int_0^{2\pi} \int_0^{\pi} \int_0^{\infty} \left| \left(4 - \frac{1}{3}\rho\right) \rho e^{-\rho/6} \sin(\theta) \cos(\phi) \right|^2 \rho^2 \sin(\theta) \, d\rho \, d\theta \, d\phi$$

$$= \frac{1}{46\,656\pi} \underbrace{\int_0^{2\pi} \cos^2(\phi) \, d\phi}_{\pi} \underbrace{\int_0^{\pi} \sin^3(\theta) \, d\theta}_{4/3} \underbrace{\int_0^{\infty} \left(4 - \frac{1}{3}\rho\right)^2 \rho^4 e^{-\rho/3} \, d\rho}_{34\,992}$$

$$= 1 \quad \text{Thus, } \psi_{3p_x} \text{ is normalized to 1.}$$

We must also show that $\int \psi_{3p_x} \psi_{3d_{xy}} \, d\tau = 0$.

Using Tables 9.3 and 10.1, we find that

$$\psi_{3p_x} = \frac{1}{54(2\pi)^{1/2}} \left(\frac{1}{a_0}\right)^{3/2} \left(4 - \frac{1}{3}\rho\right) \rho e^{-\rho/6} \sin(\theta) \cos(\phi)$$

$$\psi_{3d_{xy}} = R_{32}\left\{\frac{Y_{22} - Y_{2-2}}{\sqrt{2}i}\right\}$$

$$= \frac{1}{32(2\pi)^{1/2}} \left(\frac{1}{a_0}\right)^{3/2} \rho^2 e^{-\rho/6} \sin^2(\theta) \sin(2\phi)$$

where $\rho = 2r/a_0, r = \rho a_0/2, dr = (a_0/2)\,d\rho$.

$$\int \psi_{3p_x}\psi_{3d_{xy}}\,d\tau = \text{constant} \times \int_0^\infty \rho^5 e^{-\rho/3}\,d\rho \underbrace{\int_0^{2\pi} \cos(\phi)\sin(2\phi)\,d\phi}_{0} \int_0^\pi \sin^4(\theta)\,d\theta$$

Since the integral equals zero, ψ_{3p_x} and $\psi_{3d_{xy}}$ are orthogonal.

(b) Radial nodes are determined by finding the ρ values ($\rho = 2r/a_0$) for which the radial wavefunction equals zero. These values are the roots of the polynomial portion of the wavefunction. For the $3s$ orbital, $6 - 6\rho + \rho^2 = 0$, when $\boxed{\rho_{node} = 3 + \sqrt{3} \text{ and } \rho_{node} = 3 - \sqrt{3}}$.

The $3s$ orbital has these two spherically symmetrical modes. There is no node at $\rho = 0$ so we conclude that there is a finite probability of finding a $3s$ electron at the nucleus.

For the $3p_x$ orbital, $(4 - \rho)(\rho) = 0$, when $\boxed{\rho_{node} = 0 \text{ and } \rho_{node} = 4}$.

There is a zero probability of finding a $3p_x$ electron at the nucleus.

For the $3d_{xy}$ orbital $\boxed{\rho_{node} = 0}$ is the only radial node.

(c)
$$\langle r \rangle_{3s} = \int |R_{10}Y_{00}|^2\, r\,d\tau = \int |R_{10}Y_{00}|^2\, r^3 \sin(\theta)\,dr\,d\theta\,d\phi$$

$$= \int_0^\infty R_{10}^2 r^3\,dr \underbrace{\int_0^{2\pi} \int_0^\pi |Y_{00}|^2 \sin(\theta)\,d\theta\,d\phi}_{1}$$

$$= \frac{a_0}{3888} \underbrace{\int_0^\infty \left(6 - 2\rho + \rho^2/9\right)^2 \rho^3 e^{-\rho/3}\,d\rho}_{52\,488}.$$

$$\boxed{\langle r \rangle_{3s} = \frac{27a_0}{2}}.$$

(d) The plot Fig. 10.2 (a) shows that the $3s$ orbital has larger values of the radial distribution function for $r < a_0$. This penetration of inner core electrons of multielectron atoms means that a $3s$ electron experiences a larger effective nuclear charge and, consequently, has a lower energy than either a $3p$ or $3d_{xy}$ electron. This reasoning also leads us to conclude that a $3p_x$ electron has less energy than a $3d_{xy}$ electron.

$$E_{3s} < E_{3p_x} < E_{3d_{xy}}.$$

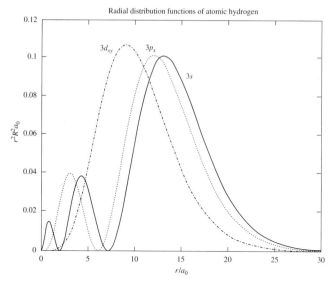

Figure 10.2(a)

(e) Polar plots with $\theta = 90°$.

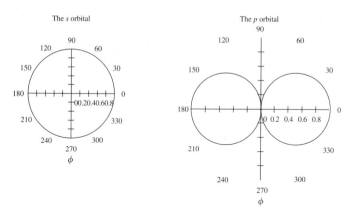

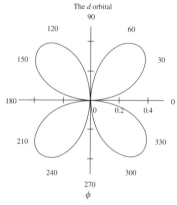

Figure 10.2(b)

Boundary surface plots.

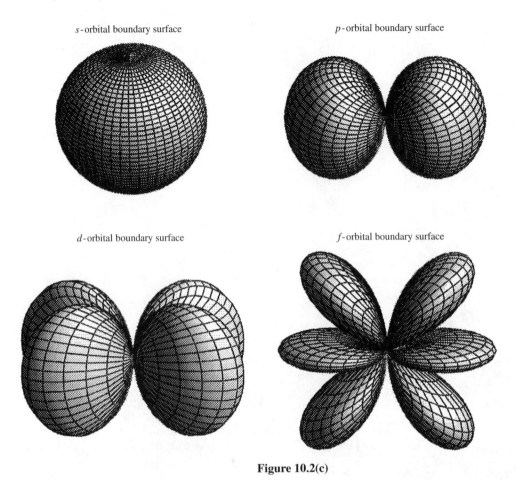

s-orbital boundary surface

p-orbital boundary surface

d-orbital boundary surface

f-orbital boundary surface

Figure 10.2(c)

P10.15 The general rule to use in deciding commutation properties is that operators having no variable in common will commute with each other. We first consider the commutation of $\hat{l}_z$ with the Hamiltonian. This is most easily solved in spherical polar coordinates.

$$\hat{l}_z = \frac{\hbar}{i} \frac{\partial}{\partial \phi} \quad \text{[Problem 9.28 and Section 9.6 and eqn 9.46].}$$

$$H = -\frac{\hbar^2}{2\mu} \nabla^2 + V \text{ [Further information 10.1]} \quad V = -\frac{Ze^2}{4\pi\varepsilon_0 r}$$

Since V has no variable in common with $\hat{l}_z$, this part of the Hamiltonian and $\hat{l}_z$ commute.

$$\nabla^2 = \text{terms in } r \text{ only} + \text{terms in } \theta \text{ only} + \frac{1}{r^2 \sin^2 \theta} \frac{\partial^2}{\partial \phi^2} \quad \text{[Justification 9.7]}$$

The terms in r only and θ only necessarily commute with $\hat{l}_z (\phi$ only). The final term in ∇^2 contains $\frac{\partial^2}{\partial \phi^2}$ which commutes with $\frac{\partial}{\partial \phi}$, since an operator necessarily commutes with itself. By symmetry we can deduce that, if H commutes with $\hat{l}_z$ it must also commute with $\hat{l}_x$ and $\hat{l}_y$ since they are related to each other by a simple transformation of coordinates. This proves useful in establishing the commutation of l^2 and H. We form

$$\hat{l}^2 = \hat{l} \cdot \hat{l} = (\hat{i}\hat{l}_x + \hat{j}\hat{l}_y + \hat{k}\hat{l}_z) \cdot (\hat{i}\hat{l}_x + \hat{j}\hat{l}_y + \hat{k}\hat{l}_z) = \hat{l}_x^2 + \hat{l}_y^2 + \hat{l}_z^2.$$

If H commutes with each of $\hat{l}_x$, $\hat{l}_y$, and $\hat{l}_z$ it must commute with $\hat{l}_x^2$, $\hat{l}_y^2$, and $\hat{l}_z^2$. Therefore it also commutes with $\hat{l}^2$. Thus H commutes with both $\hat{l}^2$ and $\hat{l}_z$.

COMMENT. As described at the end of Section 8.6, the physical properties associated with non-commuting operators cannot be simultaneously known with precision. However, since H, $\hat{l}^2$, and $\hat{l}_z$ commute we may simultaneously have exact knowledge of the energy, the total orbital angular momentum, and the projection of the orbital angular momentum along an arbitrary axis.

P10.17

$$\langle r^m \rangle_{nl} = \int r^m |\psi_{nl}|^2 \, d\tau = \int_0^\infty \int_0^{2\pi} \int_0^\pi r^{m+2} |R_{nl} Y_{l0}|^2 \sin(\theta) \, d\theta \, d\phi \, dr$$

$$= \int_0^\infty r^{m+2} |R_{nl}|^2 \, dr \int_0^{2\pi} \int_0^\pi |Y_{l0}|^2 \sin(\theta) \, d\theta \, d\phi = \int_0^\infty r^{m+2} |R_{nl}|^2 \, dr.$$

With $r = (na_0/2Z)\rho$ and $m = -1$, the expectation value is

$$\langle r^{-1} \rangle_{nl} = \left(\frac{na_0}{2Z} \right)^2 \int_0^\infty \rho \, |R_{nl}|^2 \, d\rho.$$

(a) $\langle r^{-1} \rangle_{1s} = \left(\frac{a_0}{2Z} \right)^2 \left\{ 2 \left(\frac{Z}{a_0} \right)^{3/2} \right\}^2 \int_0^\infty \rho \, e^{-\rho} \, d\rho$ [Table 10.1]

$$= \boxed{\frac{Z}{a_0}} \quad \text{because} \quad \int_0^\infty \rho \, e^{-\rho} \, d\rho = 1.$$

(b) $\langle r^{-1} \rangle_{2s} = \left(\frac{a_0}{Z} \right)^2 \left\{ \frac{1}{8^{1/2}} \left(\frac{Z}{a_0} \right)^{3/2} \right\}^2 \int_0^\infty \rho \, (2 - \rho)^2 \, e^{-\rho} d\rho$ [Table 10.1]

$$= \frac{Z}{8a_0} \, (2) \quad \text{because} \quad \int_0^\infty \rho \, (2 - \rho)^2 \, e^{-\rho} \, d\rho = 2.$$

$$\langle r^{-1} \rangle_{2s} = \boxed{\frac{Z}{4a_0}}.$$

(c) $\langle r^{-1} \rangle_{2p} = \left(\frac{a_0}{Z} \right)^2 \left\{ \frac{1}{24^{1/2}} \left(\frac{Z}{a_0} \right)^{3/2} \right\}^2 \int_0^\infty \rho^3 \, e^{-\rho} d\rho$ [Table 10.1]

$$= \frac{Z}{24a_0} \, (6) \quad \text{because} \quad \int_0^\infty \rho^3 \, e^{-\rho} \, d\rho = 6.$$

$$\langle r^{-1} \rangle_{2p} = \boxed{\frac{Z}{4a_0}}.$$

The general formula for a hydrogenic orbital is $\langle r^{-1} \rangle_{nl} = \dfrac{Z}{n^2 a_0}$.

P10.19   The trajectory is defined, which is not allowed according to quantum mechanics.

The angular momentum of a three-dimensional system is given by $\{l(l+1)\}^{1/2}\hbar$, not by $n\hbar$. In the Bohr model, the ground state possesses orbital angular momentum ($n\hbar$, with $n = 1$), but the actual ground state has no angular momentum ($l = 0$). Moreover, the distribution of the electron is quite different in the two cases. The two models can be distinguished experimentally by (a) showing that there is zero orbital angular momentum in the ground state (by examining its magnetic properties) and (b) examining the electron distribution (such as by showing that the electron and the nucleus do come into contact, Chapter 15).

P10.21   *Justification* 10.4 noted that the transition dipole moment, μ_{fi}, had to be non-zero for a transition to be allowed. The *Justification* examined conditions that allowed the z component of this quantity to be non-zero; now examine the x and y components.

$$\mu_{x,fi} = -e \int \psi_f^* x \psi_i \, d\tau \quad \text{and} \quad \mu_{y,fi} = -e \int \psi_f^* y \psi_i \, d\tau.$$

As in the *Justification*, express the relevant Cartesian variables in terms of the spherical harmonics, $Y_{l,m}$. Start by expressing them in spherical polar coordinates:

$$x = r \sin \theta \cos \phi \quad \text{and} \quad y = r \sin \theta \sin \phi.$$

Note that $Y_{1,1}$ and $Y_{1,-1}$ have factors of $\sin \theta$. They also contain complex exponentials that can be related to the sine and cosine of ϕ through the identities

$$\cos \phi = 1/2(e^{i\phi} + e^{-i\phi}) \quad \text{and} \quad \sin \phi = 1/2i(e^{i\phi} - e^{-i\phi}).$$

These relations motivate us to try linear combinations $Y_{1,1} + Y_{1,-1}$ and $Y_{1,1} - Y_{1,-1}$ (from Table 9.3; note c here corresponds to the normalization constant in the table):

$$Y_{1,1} + Y_{1,-1} = -c \sin \theta (e^{i\phi} + e^{-i\phi}) = -2c \sin \theta \cos \phi = -2cx/r,$$

so $x = -(Y_{1,1} + Y_{1,-1})r/2c;$

$$Y_{1,1} - Y_{1,-1} = c \sin \theta (e^{i\phi} - e^{-i\phi}) = 2ic \sin \theta \sin \phi = 2icy/r,$$

so $y = (Y_{1,1} - Y_{1,-1})r/2ic.$

Now we can express the integrals in terms of radial wavefunctions $R_{n,l}$ and spherical harmonics Y_{l,m_l}

$$\mu_{x,fi} = \frac{e}{2c} \int_0^\infty R_{n_f,l_f} r R_{n_i,l_i} r^2 \, dr \int_0^\pi \int_0^{2\pi} Y_{l_f,m_{l_f}}^* (Y_{1,1} + Y_{1,-1}) Y_{l_i,m_{l_i}} \sin \theta \, d\theta \, d\phi.$$

The angular integral can be broken into two, one of which contains $Y_{1,1}$ and the other $Y_{1,-1}$. According to the 'triple integral' relation in Comment 9.6, the integral

$$\int_0^\pi \int_0^{2\pi} Y_{l_f,m_{l_f}}^* Y_{1,1} Y_{l_i,m_{l_i}} \sin \theta \, d\theta \, d\phi$$

vanishes unless $l_f = l_i \pm 1$ and $m_f = m_i \pm 1$. The integral that contains $Y_{1,-1}$ introduces no further constraints; it vanishes unless $l_f = l_i \pm 1$ and $m_{l_f} = m_{l_i} \pm 1$. Similarly, the y component introduces no further constraints, for it involves the same spherical harmonics as the x component. The whole set of selection rules, then, is that transitions are allowed only if

$$\boxed{\Delta l = \pm 1 \text{ and } \Delta m_l = 0 \text{ or } \pm 1}.$$

P10.23 (a) The Slater wavefunction [10.32] is

$$\psi(1,2,3,\ldots,N) = \frac{1}{(N!)^{1/2}} \begin{vmatrix} \psi_a(1)\alpha(1) & \psi_a(2)\alpha(2) & \psi_a(3)\alpha(3) & \cdots & \psi_a(N)\alpha(N) \\ \psi_a(1)\beta(1) & \psi_a(2)\beta(2) & \psi_a(3)\beta(3) & \cdots & \psi_a(N)\beta(N) \\ \psi_b(1)\alpha(1) & \psi_b(2)\alpha(2) & \psi_b(3)\alpha(3) & \cdots & \psi_b(N)\alpha(N) \\ \vdots & \vdots & \vdots & \vdots & \vdots \\ \psi_z(1)\beta(1) & \psi_z(2)\beta(2) & \psi_z(3)\beta(3) & \cdots & \psi_z(N)\beta(N) \end{vmatrix}.$$

Interchanging any two columns or rows leaves the function unchanged except for a change in sign. For example, interchanging the first and second columns of the above determinant gives:

$$\psi(1,2,3,\ldots,N) = \frac{-1}{(N!)^{1/2}} \begin{vmatrix} \psi_a(2)\alpha(2) & \psi_a(1)\alpha(1) & \psi_a(3)\alpha(3) & \cdots & \psi_a(N)\alpha(N) \\ \psi_a(2)\beta(2) & \psi_a(1)\beta(1) & \psi_a(3)\beta(3) & \cdots & \psi_a(N)\beta(N) \\ \psi_b(2)\alpha(2) & \psi_b(1)\alpha(1) & \psi_b(3)\alpha(3) & \cdots & \psi_b(N)\alpha(N) \\ \vdots & \vdots & \vdots & \vdots & \vdots \\ \psi_z(2)\beta(2) & \psi_z(1)\beta(1) & \psi_z(3)\beta(3) & \cdots & \psi_z(N)\beta(N) \end{vmatrix}$$

$$= -\psi(2,1,3,\ldots,N).$$

This demonstrates that a Slater determinant is antisymmetric under particle exchange.

(b) The possibility that 2 electrons occupy the same orbital with the same spin can be explored by making any two rows of the Slater determinant identical, thereby, providing identical orbital and spin functions to two rows. Rows 1 and 2 are identical in the Slater wavefunction below. Interchanging these two rows causes the sign to change without in any way changing the determinant.

$$\psi(1,2,3,\ldots,N) = \frac{1}{N!^{1/2}} \begin{vmatrix} \psi_a(1)\alpha(1) & \psi_a(2)\alpha(2) & \psi_a(3)\alpha(3) & \cdots & \psi_a(N)\alpha(N) \\ \psi_a(1)\alpha(1) & \psi_a(2)\alpha(2) & \psi_a(3)\alpha(3) & \cdots & \psi_a(N)\alpha(N) \\ \psi_b(1)\alpha(1) & \psi_b(2)\alpha(2) & \psi_b(3)\alpha(3) & \cdots & \psi_b(N)\alpha(N) \\ \vdots & \vdots & \vdots & \vdots & \vdots \\ \psi_z(1)\beta(1) & \psi_z(2)\beta(2) & \psi_z(3)\beta(3) & \cdots & \psi_z(N)\beta(N) \end{vmatrix}$$

$$= -\psi(2,1,3,\ldots,N) = -\psi(1,2,3,\ldots,N).$$

Only the null function satisfies a relationship in which it is the negative of itself so we conclude that, since the null function is inconsistent with existence, the Slater determinant satisfies the Pauli exclusion principle [Section 10.4 b]. No two electrons can occupy the same orbital with the same spin.

Solutions to applications

P10.25   The wavenumber of a spectroscopic transition is related to the difference in the relevant energy levels. For a one-electron atom or ion, the relationship is

$$hc\tilde{v} = \Delta E = \frac{Z^2 \mu_{He} e^4}{32\pi^2 \varepsilon_0^2 \hbar^2 n_1^2} - \frac{Z^2 \mu_{He} e^4}{32\pi^2 \varepsilon_0^2 \hbar^2 n_2^2} = \frac{Z^2 \mu_{He} e^4}{32\pi^2 \varepsilon_0^2 \hbar^2} \left(\frac{1}{n_2^2} - \frac{1}{n_1^2} \right).$$

Solving for $\tilde{v}$, using the definition $\hbar = h/2\pi$ and the fact that $Z = 2$ for He, yields

$$\tilde{v} = \frac{\mu_{He} e^4}{2\varepsilon_0^2 h^3 c} \left(\frac{1}{n_2^2} - \frac{1}{n_1^2} \right).$$

Note that the wavenumbers are proportional to the reduced mass, which is very close to the mass of the electron for both isotopes. In order to distinguish between them, we need to carry lots of significant figures in the calculation.

$$\tilde{v} = \frac{\mu_{He}(1.60218 \times 10^{-19}\text{C})^4}{2(8.85419 \times 10^{-12}\,\text{J}^{-1}\,\text{C}^2\,\text{m}^{-1})^2 \times (6.62607 \times 10^{-34}\,\text{J s})^3 \times (2.99792 \times 10^{10}\,\text{cm s}^{-1})}$$

$$\times \left(\frac{1}{n_2^2} - \frac{1}{n^2} \right)$$

$$\tilde{v}/\text{cm}^{-1} = 4.81870 \times 10^{35} (\mu_{He}/\text{kg}) \left(\frac{1}{n_2^2} - \frac{1}{n_1^2} \right).$$

The reduced masses for the ^{4}He and ^{3}He nuclei are

$$\mu = \frac{m_e m_{nuc}}{m_e + m_{nuc}}$$

where $m_{nuc} = 4.00260\,\text{u}$ for ^{4}He and $3.01603\,\text{u}$ for ^{3}He, or, in kg

$$^4\text{He}\ m_{nuc} = (4.00260\,\text{u}) \times (1.66054 \times 10^{-27}\,\text{kg u}^{-1}) = 6.64648 \times 10^{-27}\,\text{kg},$$

$$^3\text{He}\ m_{nuc} = (3.01603\,\text{u}) \times (1.66054 \times 10^{-27}\,\text{kg u}^{-1}) = 5.00824 \times 10^{-27}\text{kg}.$$

The reduced masses are

$$^4\text{He}\ \mu = \frac{(9.10939 \times 10^{-31}\,\text{kg}) \times (6.64648 \times 10^{-27}\,\text{kg})}{(9.10939 \times 10^{-31} + 6.64648 \times 10^{-27})\,\text{kg}} = 9.10814 \times 10^{-31}\,\text{kg},$$

$$^3\text{He}\ \mu = \frac{(9.10939 \times 10^{-31}\,\text{kg}) \times (5.00824 \times 10^{-27}\,\text{kg})}{(9.10939 \times 10^{-31} + 5.00824 \times 10^{-27})\,\text{kg}} = 9.10773 \times 10^{-31}\,\text{kg}.$$

Finally, the wavenumbers for $n = 3 \rightarrow n = 2$ are

$$^4\text{He}\ \tilde{v} = (4.81870 \times 10^{35}) \times (9.10814 \times 10^{-31}) \times (1/4 - 1/9)\,\text{cm}^{-1} = \boxed{60957.4 \ \text{cm}^{-1}},$$

$$^3\text{He}\ \tilde{v} = (4.81870 \times 10^{35}) \times (9.10773 \times 10^{-31}) \times (1/4 - 1/9)\,\text{cm}^{-1} = \boxed{60954.7\,\text{cm}^{-1}}.$$

The wavenumbers for $n = 2 \rightarrow n = 1$ are

$$^4\mathrm{He}\ \tilde{\nu} = (4.81870 \times 10^{35}) \times (9.10814 \times 10^{-31}) \times (1/1 - 1/4)\,\mathrm{cm}^{-1} = \boxed{329\,170\,\mathrm{cm}^{-1}},$$

$$^3\mathrm{He}\ \tilde{\nu} = (4.81870 \times 10^{35}) \times (9.10773 \times 10^{-31}) \times (1/1 - 1/4)\,\mathrm{cm}^{-1} = \boxed{329\,155\,\mathrm{cm}^{-1}}.$$

P10.27   **(a)** Compute the ratios v_{star}/v for all three lines. We are given wavelength data, so we can use:

$$\frac{v_{\mathrm{star}}}{v} = \frac{\lambda}{\lambda_{\mathrm{star}}}.$$

The ratios are:

$$\frac{438.392\,\mathrm{nm}}{438.882\,\mathrm{nm}} = 0.998884, \qquad \frac{440.510\,\mathrm{nm}}{441.000\,\mathrm{nm}} = 0.998889, \quad \text{and} \quad \frac{441.510\,\mathrm{nm}}{442.020\,\mathrm{nm}} = 0.998846.$$

The frequencies of the stellar lines are all less than those of the stationary lines, so we infer that the star is $\boxed{\text{receding}}$ from earth. The Doppler effect follows:

$$v_{\mathrm{receding}} = vf \quad \text{where } f = \left(\frac{1 - s/c}{1 + s/c}\right)^{1/2}, \quad \text{so}$$

$$f^2(1 + s/c) = (1 - s/c), \quad (f^2 + 1)s/c = 1 - f^2, \quad s = \frac{1 - f^2}{1 + f^2}c.$$

Our average value of f is 0.998873. (Note: the uncertainty is actually greater than the significant figures here imply, and a more careful analysis would treat uncertainty explicitly.) So the speed of recession with respect to the earth is:

$$s = \left(\frac{1 - 0.997747}{1 + 0.997747}\right)c = \boxed{1.128 \times 10^{-3}\,c} = \boxed{3.381 \times 10^5\,\mathrm{m\,s}^{-1}}.$$

(b) One could compute the star's radial velocity with respect to the sun if one knew the earth's speed with respect to the sun along the sun–star vector at the time of the spectral observation. This could be estimated from quantities available through astronomical observation: the earth's orbital velocity times the cosine of the angle between that velocity vector and the earth–star vector at the time of the spectral observation. (The earth–star direction, which is observable by earth-based astronomers, is practically identical to the sun–star direction, which is technically the direction needed.) Alternatively, repeat the experiment half a year later. At that time, the earth's motion with respect to the sun is approximately equal in magnitude and opposite in direction compared to the original experiment. Averaging f values over the two experiments would yield f values in which the earth's motion is effectively averaged out.

P10.29   See Figure 10.3.

Trends:

(i) $I_1 < I_2 < I_3$ because of decreased nuclear shielding as each successive electron is removed.

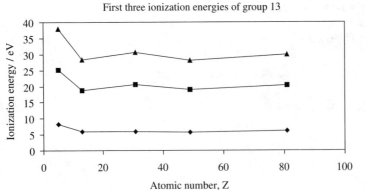

Figure 10.3

(ii) The ionization energies of boron are much larger than those of the remaining group elements because the valence shell of boron is very small and compact with little nuclear shielding. The boron atom is much smaller than the aluminum atom.

(iii) The ionization energies of Al, Ga, In, and Tl are comparable even though successive valence shells are further from the nucleus because the ionization energy decrease expected from large atomic radii is balanced by an increase in effective nuclear charge.

11 Molecular structure

Answers to discussion questions

D11.1   Our comparison of the two theories will focus on the manner of construction of the trial wavefunctions for the hydrogen molecule in the simplest versions of both theories. In the valence bond method, the trial function is a linear combination of two simple product wavefunctions, in which one electron resides totally in an atomic orbital on atom A, and the other totally in an orbital on atom B. See eqns 11.1 and 11.2, as well as Fig. 11.2. There is no contribution to the wavefunction from products in which both electrons reside on either atom A or B. So the valence bond approach undervalues, by totally neglecting, any ionic contribution to the trial function. It is a totally covalent function. The molecular orbital function for the hydrogen molecule is a product of two functions of the form of eqn 11.8, one for each electron, that is,

$$\psi = [A(1) \pm B(1)][A(2) \pm B(2)] = A(1)A(2) + B(1)B(2) + A(1)B(2) + B(1)A(2).$$

This function gives as much weight to the ionic forms as to the covalent forms. So the molecular orbital approach greatly overvalues the ionic contributions. At these crude levels of approximation, the valence bond method gives dissociation energies closer to the experimental values. However, more sophisticated versions of the molecular orbital approach are the methods of choice for obtaining quantitative results on both diatomic and polyatomic molecules. See Sections 11.6–11.8.

D11.3   Both the Pauling and Mulliken methods for measuring the attracting power of atoms for electrons seem to make good chemical sense. If we look at eqn 11.23 (the Pauling scale), we see that if $D(A$—$B)$ were equal to $1/2[D(A$—$A) + D(B$—$B)]$ the calculated electronegativity difference would be zero, as expected for completely non-polar bonds. Hence, any increased strength of the A—B bond over the average of the A—A and B—B bonds, can reasonably be thought of as being due to the polarity of the A—B bond, which in turn is due to the difference in electronegativity of the atoms involved. Therefore, this difference in bond strengths can be used as a measure of electronegativity difference. To obtain numerical values for individual atoms, a reference state (atom) for electronegativity must be established. The value for fluorine is arbitrarily set at 4.0.

The Mulliken scale may be more intuitive than the Pauling scale because we are used to thinking of ionization energies and electron affinities as measures of the electron attracting powers of atoms. The choice of factor 1/2, however, is arbitrary, though reasonable, and no more arbitrary than the specific form of eqn 11.23 that defines the Pauling scale.

D11.5   The Hückel method parameterizes, rather than calculates, the energy integrals, α and β, that arise in molecular orbital theory. They are considered to be adjustable parameters; their numerical values emerge only at the end of the calculation by comparison to experimental energies. The overlap integral is neglected, set equal to zero. Three other rather drastic approximations, listed in Section 11.6(a) of the text, eliminate many terms from the secular determinant and make it easier to solve: all diagonal terms of the determinant are set equal $\alpha - E$; nearest-neighbor terms (that is, between bonded atoms) all have the same value, β; and all other terms are zero. (Ease of solution was important in the early days of quantum chemistry before the advent of computers, and without the use of these approximations, calculations on polyatomic molecules would have been difficult to accomplish.)

The simple Hückel method is usually applied only to the calculation of π-electron energies in conjugated organic systems. The simple method is based on the assumption of the separability of the σ- and π-electron systems in the molecule. This is a very crude approximation and works best when the energy level pattern is determined largely by the symmetry of the molecule. (See Chapter 12.)

D11.7   The ground electronic configurations of the valence electrons are found in Figures 11.31–11.33 and 11.37.

$$N_2 \quad 1\sigma_g^2 1\sigma_u^2 1\pi_u^4 2\sigma_g^2 \qquad b = 3 \qquad 2S + 1 = 0$$
$$O_2 \quad 1\sigma_g^2 1\sigma_u^2 2\sigma_g^2 1\pi_u^4 1\pi_g^2 \quad b = 2 \qquad 2S + 1 = 3$$
$$NO \quad 1\sigma^2 2\sigma^2 3\sigma^2 1\pi^4 2\pi^1 \quad b = 2\frac{1}{2} \quad 2S + 1 = 2$$

The following figures show HOMOs of each. Shaded vs. unshaded atomic orbital lobes represent opposite signs of the wavefunctions. A relatively large atomic orbital represents the major contribution to the molecular orbital.

N_2 2σ molecular orbital

O_2 $1\pi_g$ molecular orbital, doubly degenerate

NO 2π

Dinitrogen with a bond order of three and paired electrons in relatively low energy molecular orbitals is very unreactive. Special biological, or industrial, processes are needed to channel energy for promotion of 2σ electrons into high energy, reactive states. The high energy $1\pi_g$ LUMO is not expected to form stable complexes with electron donors.

Molecular nitrogen is very stable in most biological organisms, and as a result the task of converting plentiful atmospheric N_2 to the fixed forms of nitrogen that can be incorporated into proteins is a difficult one. The fact that N_2 possesses no unpaired electrons is itself an obstacle to facile reactivity, and the great strength (large dissociation energy) of the N_2 bond is another obstacle. Molecular orbital theory explains both of these obstacles by assigning N_2 a configuration that gives rise to a high bond order (triple bond) with all electrons paired. (See Fig. 11.33 of the text.)

Dioxygen is kinetically stable because of a bond order equal to two and a high effective nuclear charge that causes the molecular orbitals to have relatively low energy. But two electrons are in the high energy $1\pi_g$ HOMO level, which is doubly degenerate. These two electrons are unpaired and can contribute to bonding of dioxygen with other species such as the atomic radicals Fe(II) of hemoglobin and Cu(II) of the electron transport chain. When sufficient, though not excessively large, energy is available, biological processes can channel an electron into this HOMO to produce the reactive superoxide anion of bond order $1\frac{1}{2}$. As a result, O_2 is very reactive in biological systems in ways that promote function (such as respiration) and in ways that disrupt it (damaging cells).

Although the bond order of nitric oxide is $2\frac{1}{2}$, the nitrogen nucleus has a smaller effective nuclear change than an oxygen atom would have. Thus, the one electron of the 2π HOMO is a high energy, reactive radical compared to the HOMO of dioxygen. Additionally, the HOMO, being anti-bonding and predominantly centered on the nitrogen atom, is expected to bond through the nitrogen. Oxidation can result from the loss of the radical electron to form the nitrosyl ion, NO^+, which has a bond order equal to 2. Even though it has a rather high bond order, NO is readily converted to the damagingly reactive peroxynitrite ion (ONOO–) by reaction with O_2^-—without breaking the NO linkage.

Solutions to exercises

E11.1(b)   Use Figure 11.23 for H_2^-, 11.33 for N_2, and 11.31 for O_2.

    (a) H_2^- (3 electrons) : $\boxed{1\sigma^2 2\sigma^{*1}}$ $b = 0.5$

    (b) N_2 (10 electrons) : $\boxed{1\sigma^2 2\sigma^{*2} 1\pi^4 3\sigma^2}$ $b = 3$

    (c) O_2 (12 electrons) : $\boxed{1\sigma^2 2\sigma^{*2} 3\sigma^2 1\pi^4 2\pi^{*2}}$ $b = 2$

E11.2(b)   ClF is isoelectronic with F_2, CS with N_2.

    (a)   ClF (14 electrons) : $\boxed{1\sigma^2 2\sigma^{*2} 3\sigma^2 1\pi^4 2\pi^{*4}}$ $b = 1$

    (b)   CS (10 electrons) : $\boxed{1\sigma^2 2\sigma^{*2} 1\pi^4 3\sigma^2}$ $b = 3$

    (c) O_2^- (13 electrons) : $\boxed{1\sigma^2 2\sigma^{*2} 3\sigma^2 1\pi^4 2\pi^{*3}}$ $b = 1.5$

E11.3(b)   Decide whether the electron added or removed increases or decreases the bond order. The simplest procedure is to decide whether the electron occupies or is removed from a bonding or antibonding orbital. We can draw up the following table, which denotes the orbital involved

	N_2	NO	O_2	C_2	F_2	CN
(a) AB^-	$2\pi^*$	$2\pi^*$	$2\pi^*$	3σ	$4\sigma^*$	3σ
Change in bond order	$-1/2$	$-1/2$	$-1/2$	$+1/2$	$-1/2$	$+1/2$
(b) AB^+	3σ	$2\pi^*$	$2\pi^*$	1π	$2\pi^*$	3σ
Change in bond order	$-1/2$	$+1/2$	$+1/2$	$-1/2$	$+1/2$	$-1/2$

(a) Therefore, $\boxed{C_2 \text{ and } CN}$ are stabilized (have lower energy) by anion formation.

(b) $\boxed{NO, O_2 \text{ and } F_2}$ are stabilized by cation formation; in each of these cases the bond order increases.

E11.4(b) Figure 11.1 is based on Figure 11.31 of the text but with Cl orbitals lower than Br orbitals. BrCl is likely to have a shorter bond length than $BrCl^-$; it has a bond order of 1, while $BrCl^-$ has a bond order of 1/2.

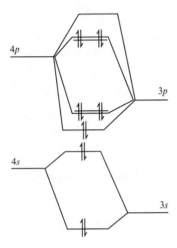

4p

3p

4s

3s

Figure 11.1

E11.5(b)

O_2^+ (11 electrons) : $1\sigma^2 2\sigma^{*2} 3\sigma^2 1\pi^4 2\pi^{*1}$ $b = 5/2$

O_2 (12 electrons) : $1\sigma^2 2\sigma^{*2} 3\sigma^2 1\pi^4 2\pi^{*2}$ $b = 2$

O_2^- (13 electrons) : $1\sigma^2 2\sigma^{*2} 3\sigma^2 1\pi^4 2\pi^{*3}$ $b = 3/2$

O_2^{2-} (14 electrons) : $1\sigma^2 2\sigma^{*2} 3\sigma^2 1\pi^4 2\pi^{*4}$ $b = 1$

Each electron added to O_2^+ is added to an antibonding orbital, thus increasing the length. So the sequence $\boxed{O_2^+, O_2, O_2^-, O_2^{2-}}$ has progressively longer bonds.

E11.6(b) $\int \psi^2 \, d\tau = N^2 \int (\psi_A + \lambda\psi_B)^2 \, d\tau = 1 = N^2 \int (\psi_A^2 + \lambda^2 \psi_B^2 + 2\lambda\psi_A\psi_B) \, d\tau = 1$

$$= N^2(1 + \lambda^2 + 2\lambda S) \qquad \left[\int \psi_A\psi_B d\tau = S\right]$$

Hence $\boxed{N = \left(\dfrac{1}{1 + 2\lambda S + \lambda^2}\right)^{1/2}}$

E11.7(b) We seek an orbital of the form $aA + bB$, where a and b are constants, which is orthogonal to the orbital $N(0.145A + 0.844B)$. Orthogonality implies

$$\int (aA + bB)N(0.145A + 0.844B) \, d\tau = 0$$

$$N \int [0.145aA^2 + (0.145b + 0.844a)AB + 0.844bB^2] \, d\tau = 0$$

The integrals of squares of orbitals are 1 and the integral $\int AB \, d\tau$ is the overlap integral S, so

$$0 = (0.145 + 0.844S)a + (0.145S + 0.844)b \quad \text{so} \quad \boxed{a = -\frac{0.145S + 0.844}{0.145 + 0.844S}b}$$

This would make the orbitals orthogonal, but not necessarily normalized. If $S = 0$, the expression simplifies to

$$a = -\frac{0.844}{0.145}b$$

and the new orbital would be normalized if $a = 0.844N$ and $b = -0.145N$. That is

$$\boxed{N(0.844A - 0.145B)}$$

E11.8(b) The trial function $\psi = x^2(L - 2x)$ does not obey the boundary conditions of a particle in a box, so it is $\boxed{\text{not appropriate}}$. In particular, the function does not vanish at $x = L$.

E11.9(b) The variational principle says that the minimum energy is obtained by taking the derivative of the trial energy with respect to adjustable parameters, setting it equal to zero, and solving for the parameters:

$$E_{\text{trial}} = \frac{3a\hbar^2}{2\mu} - \frac{e^2}{\varepsilon_0}\left(\frac{a}{2\pi^3}\right)^{1/2} \quad \text{so} \quad \frac{dE_{\text{trial}}}{da} = \frac{3\hbar^2}{2\mu} - \frac{e^2}{2\varepsilon_0}\left(\frac{1}{2\pi^3 a}\right)^{1/2} = 0.$$

Solving for a yields:

$$\frac{3\hbar^2}{2\mu} = \frac{e^2}{2\varepsilon_0}\left(\frac{1}{2\pi^3 a}\right)^{1/2} \quad \text{so} \quad a = \left(\frac{\mu e^2}{3\hbar^2 \varepsilon_0}\right)^2\left(\frac{1}{2\pi^3}\right) = \frac{\mu^2 e^4}{18\pi^3 \hbar^4 \varepsilon_0^2}.$$

Substituting this back into the trial energy yields the minimum energy:

$$E_{\text{trial}} = \frac{3\hbar^2}{2\mu}\left(\frac{\mu^2 e^4}{18\pi^3 \hbar^4 \varepsilon_0^2}\right) - \frac{e^2}{\varepsilon_0}\left(\frac{\mu^2 e^4}{18\pi^3 \hbar^4 \varepsilon_0^2 \cdot 2\pi^3}\right)^{1/2} = \boxed{\frac{-\mu e^4}{12\pi^3 \varepsilon_0^2 \hbar^2}}.$$

E11.10(b) Energy is conserved, so when the photon is absorbed, its energy is transferred to the electron. Part of it overcomes the binding energy (ionization energy) and the remainder is manifest as the now freed electron's kinetic energy.

$$E_{\text{photon}} = I + E_{\text{kinetic}}$$

so $\quad E_{\text{kinetic}} = E_{\text{photon}} - I = \dfrac{hc}{\lambda} - I = \dfrac{(6.626 \times 10^{-34}\,\text{J s}) \times (2.998 \times 10^8\,\text{m s}^{-1})}{(584 \times 10^{-12}\,\text{m}) \times (1.602 \times 10^{-19}\,\text{J eV}^{-1})} - 4.69\,\text{eV}$

$$= \boxed{2119\,\text{eV}} = \boxed{3.39 \times 10^{-16}\,\text{J}}$$

E11.11(b) The molecular orbitals of the fragments and the molecular orbitals that they form are shown in Figure 11.2.

E11.12(b) We use the molecular orbital energy level diagram in Figure 11.41. As usual, we fill the orbitals starting with the lowest energy orbital, obeying the Pauli principle and Hund's rule. We then write

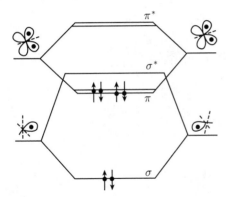

Figure 11.2

(a) $C_6H_6^-$ (7 electrons) : $\boxed{a_{2u}^2 e_{1g}^4 e_{2u}^1}$

$$E = 2(\alpha + 2\beta) + 4(\alpha + \beta) + (\alpha - \beta) = \boxed{7\alpha + 7\beta}$$

(b) $C_6H_6^+$ (5 electrons) : $\boxed{a_{2u}^2 e_{1g}^3}$

$$E = 2(\alpha + 2\beta) + 3(\alpha + \beta) = \boxed{5\alpha + 7\beta}$$

E11.13(b) The secular determinants from E11.13(a) can be diagonalized with the assistance of general-purpose mathematical software. Alternatively, programs specifically designed for Hückel calculations (such as the one at Australia's Northern Territory University, http://www.smps.ntu.edu.au/modules/mod3/interface.html) can be used. In both molecules, 14 π-electrons fill seven orbitals.

(a) In anthracene, the energies of the filled orbitals are $\alpha + 2.414\,21\beta$, $\alpha + 2.000\,00\beta$, $\alpha + 1.414\,21\beta$ (doubly degenerate), $\alpha + 1.000\,00\beta$ (doubly degenerate), and $\alpha + 0.414\,21\beta$, so the total energy is $14\alpha + 19.313\,68\beta$ and the π energy is $\boxed{19.313\,68\beta}$.

(b) For phenanthrene, the energies of the filled orbitals are $\alpha + 2.434\,76\beta$, $\alpha + 1.950\,63\beta$, $\alpha + 1.516\,27\beta$, $\alpha + 1.305\,80\beta$, $\alpha + 1.142\,38\beta$, $\alpha + 0.769\,05\beta$, $\alpha + 0.605\,23\beta$, so the total energy is $14\alpha + 19.448\,24\beta$ and the π energy is $\boxed{19.448\,24\beta}$.

Solutions to problems

Solutions to numerical problems

P11.1 $\psi_A = \cos kx$ measured from A, $\psi_B = \cos k'(x - R)$ measuring x from A.

Then, with $\psi = \psi_A + \psi_B$,

$$\psi = \cos kx + \cos k'(x - R) = \cos kx + \cos k'R \cos k'x + \sin k'R \sin k'x$$

$$[\cos(a - b) = \cos a \cos b + \sin a \sin b].$$

(a) $k = k' = \dfrac{\pi}{2R}$; $\cos k'R = \cos\dfrac{\pi}{2} = 0$; $\sin k'R = \sin\dfrac{\pi}{2} = 1$.

$$\psi = \cos\frac{\pi x}{2R} + \sin\frac{\pi x}{2R}.$$

For the midpoint, $x = \frac{1}{2}R$, so $\psi\left(\frac{1}{2}R\right) = \cos\frac{1}{4}\pi + \sin\frac{1}{4}\pi = 2^{1/2}$ and there is constructive interference ($\psi > \psi_A, \psi_B$).

(b) $k = \frac{\pi}{2R}$, $k' = \frac{3\pi}{2R}$; $\cos k'R = \cos\frac{3\pi}{2} = 0$, $\sin k'R = -1$.

$\psi = \cos\frac{\pi x}{2R} - \sin\frac{3\pi x}{2R}$.

For the midpoint, $x = \frac{1}{2}R$, so $\psi\left(\frac{1}{2}R\right) = \cos\frac{1}{4}\pi - \sin\frac{3}{4}\pi = 0$ and there is destructive interference ($\psi < \psi_A, \psi_B$).

P11.3   The s orbital begins to spread into the region of negative amplitude of the p orbital. When their centers coincide, the region of positive overlap cancels the negative region. Draw up the following table.

R/a_0	0	1	2	3	4	5	6	7	8	9	10
S	0	0.429	0.588	0.523	0.379	0.241	0.141	0.078	0.041	0.021	0.01

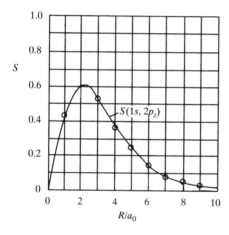

Figure 11.3

The points are plotted in Fig.11.3. The maximum overlap occurs at $\boxed{R = 2.1a_0}$.

P11.5   We obtain the electron densities from $\rho_+ = \psi_+^2$ and $\rho_- = \psi_-^2$ with ψ_+ and ψ_- as given in Problem 11.4.

$$\psi_\pm = N_\pm \left(\frac{1}{\pi a_0^3}\right)^{1/2} \{e^{-|z|/a_0} \pm e^{-|z-R|/a_0}\}$$

where $N_+ = \left(\frac{1}{2(1+S)}\right)^{1/2} = \left(\frac{1}{2(1+0.586)}\right)^{1/2} = 0.561$

and $N_- = \left(\frac{1}{2(1-S)}\right)^{1/2} = \left(\frac{1}{2(1-0.586)}\right)^{1/2} = 1.09\overline{9}$.

Hence $\rho_\pm = N_\pm^2 \left(\frac{1}{\pi a_0^3}\right) \{e^{-|z|/a_0} \pm e^{-|z-R|/a_0}\}^2$.

We evaluate the factors preceding the exponentials in ψ_+ and ψ_-.

$$N_+ \left(\frac{1}{\pi a_0^3}\right)^{1/2} = 0.561 \times \left(\frac{1}{\pi \times (52.9 \, \text{pm})^3}\right)^{1/2} = \frac{1}{1216 \, \text{pm}^{3/2}}.$$

Likewise, $N_- \left(\frac{1}{\pi a_0^3}\right)^{1/2} = \frac{1}{621 \, \text{pm}^{3/2}}.$

Then $\rho_+ = \frac{1}{(1216)^2 \, \text{pm}^3} \{e^{-|z|/a_0} + e^{-|z-R|/a_0}\}^2$

and $\rho_- = \frac{1}{(621)^2 \, \text{pm}^3} \{e^{-|z|/a_0} - e^{-|z-R|/a_0}\}^2.$

The 'atomic' density is

$$\rho = \frac{1}{2}\{\psi_{1s}(A)^2 + \psi_{1s}(B)^2\} = \frac{1}{2} \times \left(\frac{1}{\pi a_0^3}\right) \{e^{-2r_A/a_0} + e^{-2r_B/a_0}\}$$

$$= \frac{e^{-2r_A/a_0} + e^{-2r_B/a_0}}{9.30 \times 10^5 \, \text{pm}^3} = \frac{e^{-2|z|/a_0} + e^{-2|z-R|/a_0}}{9.30 \times 10^5 \, \text{pm}^3}.$$

The difference density is $\delta\rho_\pm = \rho_\pm - \rho$.

Draw up the following table using the information in Problem 14.4.

z/pm	-100	-80	-60	-40	-20	0	20	40
$\rho_+ \times 10^7/\text{pm}^{-3}$	0.20	0.42	0.90	1.92	4.09	8.72	5.27	3.88
$\rho_- \times 10^7/\text{pm}^{-3}$	0.44	0.94	2.01	4.27	9.11	19.40	6.17	0.85
$\rho \times 10^7/\text{pm}^{-3}$	0.25	0.53	1.13	2.41	5.15	10.93	5.47	3.26
$\delta\rho_+ \times 10^7/\text{pm}^{-3}$	-0.05	-0.11	-0.23	-0.49	-1.05	-2.20	-0.20	0.62
$\delta\rho_- \times 10^7/\text{pm}^{-3}$	0.19	0.41	0.87	1.86	3.96	8.47	0.70	-2.40

z/pm	60	80	100	120	140	160	180	200
$\rho_+ \times 10^7/\text{pm}^{-3}$	3.73	4.71	7.42	5.10	2.39	1.12	0.53	0.25
$\rho_- \times 10^7/\text{pm}^{-3}$	0.25	4.02	14.41	11.34	5.32	2.50	1.17	0.55
$\rho \times 10^7/\text{pm}^{-3}$	3.01	4.58	8.88	6.40	3.00	1.41	0.66	0.31
$\delta\rho_+ \times 10^7/\text{pm}^{-3}$	0.70	0.13	-1.46	-1.29	-0.61	-0.29	-0.14	-0.06
$\delta\rho_- \times 10^7/\text{pm}^{-3}$	-2.76	-0.56	5.54	4.95	2.33	1.09	0.51	0.24

The densities are plotted in Fig. 11.4(a) and the difference densities are plotted in Fig. 11.4(b).

P11.7 $P = |\psi|^2 d\tau \approx |\psi|^2 \delta\tau, \quad \delta\tau = 1.00 \, \text{pm}^3$

(a) From Problem 11.5

$$\psi_+^2(z = 0) = \rho_+(z = 0) = 8.7 \times 10^{-7} \, \text{pm}^{-3}.$$

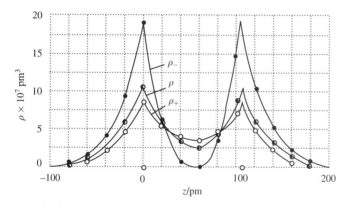

Figure 11.4(a)

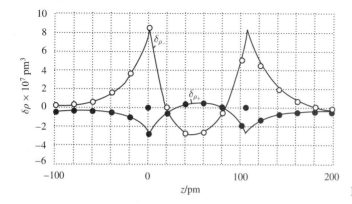

Figure 11.4(b)

Therefore, the probability of finding the electron in the volume $\delta\tau$ at nucleus A is

$$P = 8.6 \times 10^{-7}\,\mathrm{pm}^{-3} \times 1.00\,\mathrm{pm}^3 = \boxed{8.6 \times 10^{-7}}.$$

(b) By symmetry (or by taking $z = 106\,\mathrm{pm}$) $P = \boxed{8.6 \times 10^{-7}}$

(c) From Fig. 11.4(a), $\psi_+^2\left(\frac{1}{2}R\right) = 3.7 \times 10^{-7}\,\mathrm{pm}^{-3}$, so $P = \boxed{3.7 \times 10^{-7}}$

(d) From Fig. 11.5, the point referred to lies at 22.4 pm from A and 86.6 pm from B.

A ● 20.0 pm | 86.0 pm ● B   22.4 pm 10.0 pm 86.6 pm   **Figure 11.5**

Therefore, $\psi = \dfrac{e^{-22.4/52.9} + e^{-86.6/52.9}}{1216\,\mathrm{pm}^{3/2}} = \dfrac{0.65 + 0.19}{1216\,\mathrm{pm}^{3/2}} = 6.98 \times 10^{-4}\,\mathrm{pm}^{-3/2}.$

$$\psi^2 = 4.9 \times 10^{-7}\,\mathrm{pm}^{-3}, \quad \text{so} \quad P = \boxed{4.9 \times 10^{-7}}.$$

For the antibonding orbital, we proceed similarly.

(a) $\psi_-^2(z=0) = 19.6 \times 10^{-7} \text{ pm}^{-3}$ [Problem 11.5], so $\boxed{P = 2.0 \times 10^{-6}}$.

(b) By symmetry, $P = \boxed{2.0 \times 10^{-6}}$.

(c) $\psi_-^2\left(\frac{1}{2}R\right) = 0$, so $P = \boxed{0}$.

(d) We evaluate ψ_- at the point specified in Fig. 11.5.

$$\psi_- = \frac{0.65 - 0.19}{621 \text{ pm}^{3/2}} = 7.41 \times 10^{-4} \text{ pm}^{-3/2}.$$

$$\psi_-^2 = 5.49 \times 10^{-7} \text{ pm}^{-3}, \quad \text{so} \quad \boxed{P = 5.5 \times 10^{-7}}.$$

P11.9 $E_H = E_1 = -hcR_H$ [Section 10.2(b)].

Draw up the following table using the data in question and using

$$\frac{e^2}{4\pi\varepsilon_0 R} = \frac{e^2}{4\pi\varepsilon_0 a_0} \times \frac{a_0}{R} = \frac{e^2}{4\pi\varepsilon_0 \times (4\pi\varepsilon_0\hbar^2/m_e e^2)} \times \frac{a_0}{R}$$

$$= \frac{m_e e^4}{16\pi^2\varepsilon_0^2\hbar^2} \times \frac{a_0}{R} = E_h \times \frac{a_0}{R} \quad \left[E_h \equiv \frac{m_e e^4}{16\pi^2\varepsilon_0^2\hbar^2} = 2hcR_H\right]$$

so that $\dfrac{(e^2/4\pi\varepsilon_0 R)}{E_h} = \dfrac{a_0}{R}$.

Draw up the following table.

R/a_0	0	1	2	3	4	∞
$(e^2/4\pi\varepsilon_0 R)/E_h$	∞	1	0.500	0.333	0.250	0
$(V_1 - V_2)/E_h$	0	−0.007	0.031	0.131	0.158	0
$(E - E_H)/E_h$	∞	1.049	0.425	0.132	0.055	0

The points are plotted in Fig. 11.6. The contribution V_2 decreases rapidly because it depends on the overlap of the two orbitals.

P11.11   The internuclear distance $\langle r\rangle_n \approx n^2 a_0$, would be about twice the average distance ($\approx 1.06 \times 10^6$ pm) of a hydrogenic electron from the nucleus when in the state $n = 100$. This distance is so large that each of the following estimates is applicable.

Resonance integral, $\beta \approx -\delta$ (where $\delta \approx 0$).

Overlap integral, $S \approx \varepsilon$ (where $\varepsilon \approx 0$).

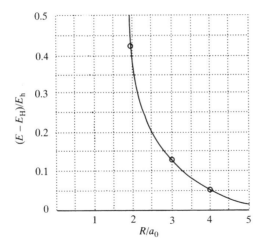

Figure 11.6

Coulomb integral, $\alpha \approx E_{n=100}$ for atomic hydrogen.

Binding energy $= 2\{E_+ - E_{n=100}\}$

$$= 2\left\{\frac{\alpha + \beta}{1 - S} - E_{n=100}\right\}$$

$$= 2\{\alpha - E_{n=100}\}$$

$$\approx 0.$$

Vibrational force constant, $k \approx 0$, because of the weak binding energy.

Rotational constant, $B = \hbar^2/2hcI = \hbar^2/2hc\mu r_{AB}^2 \approx 0$ because r_{AB}^2 is so large.

The binding energy is so small that thermal energies would easily cause the Rydberg molecule to break apart. It is not likely to exist for much longer than a vibrational period.

P11.13   In the simple Hückel approximation

$$\begin{bmatrix} :\!\ddot{O}\!: \\ \| \\ N \\ \diagup \quad \diagdown \\ :\!\ddot{O}. \quad .\ddot{O}\!: \end{bmatrix}^{-} \longleftrightarrow \begin{bmatrix} \overset{1}{O} \\ | \quad 4 \\ N \\ \diagup \quad \diagdown\!\!\diagdown \\ \underset{2}{O} \qquad \underset{3}{O} \end{bmatrix}^{-} \longleftrightarrow \begin{bmatrix} O \\ | \\ N \\ \diagup \quad \diagdown\!\!\diagdown \\ O \qquad O \end{bmatrix}^{-}$$

$$\begin{vmatrix} \alpha_O - E & 0 & 0 & \beta \\ 0 & \alpha_O - E & 0 & \beta \\ 0 & 0 & \alpha_O - E & \beta \\ \beta & \beta & \beta & \alpha_N - E \end{vmatrix} = 0.$$

$$(E - \alpha_O)^2 \times \{(E - \alpha_O) \times (E - \alpha_N) - 3\beta^2\} = 0.$$

Therefore, the roots are

$$E - \alpha_O = 0 \text{ (twice)} \quad \text{and} \quad (E - \alpha_O) \times (E - \alpha_N) - 3\beta^2 = 0.$$

Each equation is easily solved (Fig. 11.7(a)) for the permitted values of E in terms of α_O, α_N, and β. The quadratic equation is applicable in the second case.

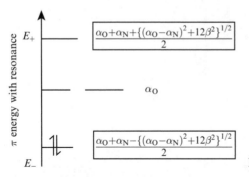

Figure 11.7(a)

In contrast, the π energies in the absence of resonance are derived for just one of the three structures, i.e. for a structure containing a single localized π bond.

$$\begin{vmatrix} \alpha_O - E & \beta \\ \beta & \alpha_N - E \end{vmatrix} = 0.$$

Expanding the determinant and solving for E gives the result in Fig.11.7(b).

Delocalization energy $= 2\{E_-(\text{with resonance}) - E_-(\text{without resonance})\}$

$$= \boxed{\{(\alpha_O - \alpha_N)^2 + 12\beta^2\}^{1/2} - \{(\alpha_O - \alpha_N)^2 + 4\beta^2\}^{1/2}}.$$

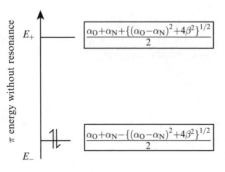

Figure 11.7(b)

If $\beta^2 \ll (\alpha_O - \alpha_N)^2$, then

$$\text{Delocalization energy} \approx \frac{4\beta^2}{(\alpha_O - \alpha_N)}.$$

P11.15   (a) The transitions occur for photons whose energies are equal to the difference in energy between the highest occupied and lowest unoccupied orbital energies:

$$E_{\text{photon}} = E_{\text{LUMO}} - E_{\text{HOMO}}.$$

If N is the number of carbon atoms in these species, then the number of π electrons is also N. These N electrons occupy the first $N/2$ orbitals, so orbital number $N/2$ is the HOMO and orbital number

$1 + N/2$ is the LUMO. Writing the photon energy in terms of the wavenumber, substituting the given energy expressions with this identification of the HOMO and LUMO, gives

$$hc\tilde{v} = \left(\alpha + 2\beta \cos \frac{(\frac{1}{2}N + 1)\pi}{N + 1} \right) - \left(\alpha + 2\beta \cos \frac{\frac{1}{2}N\pi}{N + 1} \right)$$

$$= 2\beta \left(\cos \frac{(\frac{1}{2}N + 1)\pi}{N + 1} - \cos \frac{\frac{1}{2}N\pi}{N + 1} \right).$$

Solving for β yields

$$\beta = \frac{hc\tilde{v}}{2 \left(\cos \frac{(\frac{1}{2}N + 1)\pi}{N + 1} - \cos \frac{\frac{1}{2}N\pi}{N + 1} \right)}.$$

Draw up the following table

Species	N	$\tilde{v}/\text{cm}^{-1}$	estimated β/eV
C_2H_4	2	61500	−3.813
C_4H_6	4	46080	−4.623
C_6H_8	6	39750	−5.538
C_8H_{10}	8	32900	−5.873

(b) The total energy of the π electron system is the sum of the energies of occupied orbitals weighted by the number of electrons that occupy them. In C_8H_{10}, each of the first four orbitals is doubly occupied, so

$$E_\pi = 2 \sum_{k=1}^{4} E_k = 2 \sum_{k=1}^{4} \left(\alpha + 2\beta \cos \frac{k\pi}{9} \right) = 8\alpha + 4\beta \sum_{k=1}^{4} \cos \frac{k\pi}{9} = 8\alpha + 9.518\beta.$$

The delocalization energy is the difference between this quantity and that of four isolated double bonds,

$$E_{\text{deloc}} = E_\pi - 8(\alpha + \beta) = 8\alpha + 9.518\beta - 8(\alpha + \beta) = \boxed{1.158\,\beta}.$$

Using the estimate of β from part **(a)** yields $E_{\text{deloc}} = \boxed{8.913\text{ eV}}$.

(c) Draw up the following table, in which the orbital energy decreases as we go down. For the purpose of comparison, we express orbital energies as $(E_k - \alpha)/\beta$. Recall that β is negative (as is α for that matter), so the orbital with the greatest value of $(E_k - \alpha)/\beta$ has the lowest energy.

	Energy	Coefficients					
Orbital	$(E_k - \alpha)/\beta$	1	2	3	4	5	6
6	−1.8019	0.2319	−0.4179	0.5211	−0.5211	0.4179	−0.2319
5	−1.2470	0.4179	−0.5211	0.2319	0.2319	−0.5211	0.4179
4	−0.4450	0.5211	−0.2319	−0.4179	0.4179	0.2319	−0.5211
3	0.4450	0.5211	0.2319	−0.4179	−0.4179	0.2319	0.5211
2	1.2470	0.4179	0.5211	0.2319	−0.2319	−0.5211	−0.4179
1	1.8019	0.2319	0.4179	0.5211	0.5211	0.4179	0.2319

The orbitals are shown schematically in Fig. 11.8, with each vertical pair of lobes representing a p orbital on one of the carbons in hexatriene. Shaded lobes represent one sign of the wavefunction (say positive) and unshaded lobes the other sign. Where adjacent atoms have atomic orbitals of the same sign, the resulting molecular orbital is bonding with respect to those atoms; where adjacent atoms have different sign, there is a node between the atoms and the resulting molecular orbital is antibonding with respect to them. The lowest energy orbital is totally bonding (no nodes between atoms) and the highest-energy orbital totally antibonding (nodes between each adjacent pair). Note that the orbitals have increasing antibonding character as their energy increases. The size of each atomic

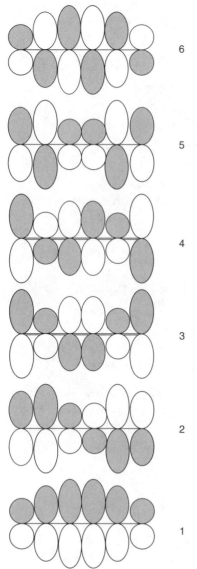

6

5

4

3

2

1

Figure 11.8

p orbital is proportional to the magnitude of the coefficient of that orbital in the molecular orbital. So, for example, in orbitals 1 and 6, the largest lobes are in the center of the molecule, so electrons that occupy those orbitals are more likely to be found near the center of the molecule than on the ends. In the ground state of the molecule, there are two electrons in each of orbitals 1, 2, and 3, with the result that the probability of finding a π electron in hexatriene is uniform over the entire molecule.

P11.17   Mathcad may be used to demonstrate the power of matrix diagonalization techniques. The function eigenvals() performs diagonalization to find the eigenvalues of a matrix while the function eigenvecs() finds the eigenvectors. To show use of the technique, the π conjugated systems of both benzene and hexatriene are analyzed below with the Hückel approximation. The energy levels are reported as x values where $x = (E - \alpha)/\beta$. The molecular orbital for a level is identified by counting nodal planes. Energy increases with the number of nodal planes in the wavefunction. Modern mathematic software makes it very easy to check both normalization of a wavefunction and the orthogonality of two wavefunctions so this is demonstrated for benzene.

Writing the Hückel secular matrix as $M(x)$, the functions eigenvals() and eigenvecs() use $M(0)$ as their arguments. The transpose operator (T) is used to place the coefficients of each eigenfunction along a row. Eigenvalues are sorted with the sort() function. The sum of the squares of the p orbital coefficients equals 1 when the eigenfunction is normalized. When two eigenfunctions are orthogonal, the sum of their multiplied coefficients equals zero.

For benzene:

$$M(x) := \begin{pmatrix} x & 1 & 0 & 0 & 0 & 1 \\ 1 & x & 1 & 0 & 0 & 0 \\ 0 & 1 & x & 1 & 0 & 0 \\ 0 & 0 & 1 & x & 1 & 0 \\ 0 & 0 & 0 & 1 & x & 1 \\ 1 & 0 & 0 & 0 & 1 & x \end{pmatrix} \qquad \text{sort(eigenvals(M(0)))} = \begin{pmatrix} -2 \\ -1 \\ -1 \\ 1 \\ 1 \\ 2 \end{pmatrix} \begin{matrix} x_4 \\ x_3 \\ x_3 \\ x_2 \\ x_2 \\ x_1 \end{matrix}$$

$$\text{eigenvecs(M(0))}^T = \begin{pmatrix} 0.569 & 0.201 & -0.368 & -0.569 & -0.201 & 0.368 \\ -0.096 & -0.541 & -0.445 & 0.096 & 0.541 & 0.445 \\ 0.508 & -0.491 & -0.017 & 0.508 & -0.491 & -0.017 \\ -0.274 & -0.303 & 0.577 & -0.274 & -0.303 & 0.577 \\ 0.408 & 0.408 & 0.408 & 0.408 & 0.408 & 0.408 \\ -0.408 & 0.408 & -0.408 & 0.408 & -0.408 & 0.408 \end{pmatrix} \begin{matrix} \psi_{2a} \\ \psi_{2b} \\ \psi_{3a} \\ \psi_{3b} \\ \psi_1 \\ \psi_4 \end{matrix}$$

Checking the normalization of ψ_{2a} : $\displaystyle\sum_{i=0}^{5} \left[\left(\text{eigenvecs(M(0))}^T \right)_{0,i} \right]^2 = 1$

Checking the orthogonality of ψ_{2a} and ψ_{3a} : $\displaystyle\sum_{i=0}^{5} \left[\left(\text{eigenvecs(M(0))}^T \right)_{0,i} \cdot \left(\text{eigenvecs(M(0))}^T \right)_{2,i} \right] = 0$

Placing a total of 6 π electrons of benzene into the three lowest energy levels gives a total x term energy:

$$2 \cdot \sum_{i=3}^{5} \text{sort(eigenvals (M(0)))}_i = 8$$

$$E_\pi = 2\{E_1 + E_{2a} + E_{2b}\} = 2\{(\alpha + \beta x_1) + (\alpha + \beta x_{2a}) + (\alpha + \beta x_{2b})\}$$

$$= 6\alpha + 2(x_1 + x_{2a} + x_{2b})\beta = 6\alpha + 8\beta.$$

$$E_{\text{deloc}} = E_\pi - 6(\alpha + \beta) = 2\beta.$$

For hexatriene

$$M(x) = \begin{pmatrix} x & 1 & 0 & 0 & 0 & 0 \\ 1 & x & 1 & 0 & 0 & 0 \\ 0 & 1 & x & 1 & 0 & 0 \\ 0 & 0 & 1 & x & 1 & 0 \\ 0 & 0 & 0 & 1 & x & 1 \\ 0 & 0 & 0 & 0 & 1 & x \end{pmatrix} \qquad \text{sort(eigenvals(M(0)))} = \begin{pmatrix} -1.802 \\ -1.247 \\ -0.445 \\ 0.445 \\ 1.247 \\ 1.802 \end{pmatrix} \begin{matrix} E_6 \\ E_5 \\ E_4 \\ E_3 \\ E_2 \\ E_1 \end{matrix}$$

$$\text{eigenvecs(M(0))}^{\mathsf{T}} = \begin{pmatrix} 0.418 & 0.521 & 0.232 & -0.232 & -0.521 & -0.418 \\ -0.521 & -0.232 & 0.418 & 0.418 & -0.232 & -0.521 \\ -0.232 & -0.418 & -0.521 & -0.521 & -0.418 & -0.232 \\ -0.521 & 0.232 & 0.418 & -0.418 & -0.232 & 0.521 \\ -0.418 & 0.521 & -0.232 & -0.232 & 0.521 & -0.418 \\ -0.232 & 0.418 & -0.521 & 0.521 & -0.418 & 0.232 \end{pmatrix} \begin{matrix} \psi_2 \\ \psi_3 \\ \psi_1 \\ \psi_4 \\ \psi_5 \\ \psi_6 \end{matrix}$$

$$2 \cdot \sum_{i=3}^{5} \text{sort(eigenvals(M(0)))}_i = 6.988$$

$$E_\pi = 2\{E_1 + E_2 + E_3\} = 2\{(\alpha + \beta x_1) + (\alpha + \beta x_2) + (\alpha + \beta x_3)\}$$

$$= 6\alpha + 2(x_1 + x_2 + x_3)\beta = \boxed{6\alpha + 6.988\beta}.$$

$$E_{\text{deloc}} = E_\pi - 6(\alpha + \beta) = \boxed{0.988\beta}.$$

Remembering that the resonance integral is negative, we see that the delocalization energy stabilizes the π orbitals of the closed ring conjugated system (benzene) to a greater extent than what is observed in the open chain conjugated system (hexatriene). The unusually large stabilization energy of benzene also demonstrates the validity of the Hückel $4n + 2$ rule for planar, cyclic conjugated π systems.

P11.19   In all of the molecules considered, the HOMO is bonding with respect to the carbon atoms connected by double bonds, but antibonding with respect to the carbon atoms connected by single bonds. (The bond lengths returned by the modeling software suggest that it makes sense to talk about double bonds and single bonds. Despite the electron delocalization, the nominal double bonds are consistently shorter than the nominal single bonds.) The LUMO had just the opposite character, tending to weaken the

C=C bonds but strengthen the C−C bonds. To arrive at this conclusion, examine the nodal surfaces of the orbitals. An orbital has an antibonding effect on atoms between which nodes occur, and it has a binding effect on atoms that lie within regions in which the orbital does not change sign. The $\pi^* \leftarrow \pi$ transition, then, would lengthen and weaken the double bonds and shorten and strengthen the single bonds, bringing the different kinds of polyene bonds closer to each other in length and strength. Since each molecule has more double bonds than single bonds, there is an overall weakening of bonds. (See Fig. 11.9.)

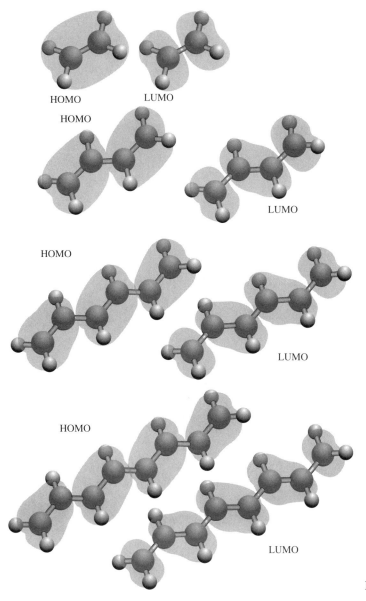

Figure 11.9

Solutions to theoretical problems

P11.21   Since

$$\psi_{2s} = R_{20} Y_{00} = \frac{1}{2\sqrt{2}} \left(\frac{Z}{a_0} \right)^{3/2} \times \left(2 - \frac{\rho}{2} \right) e^{-\rho/4} \times \left(\frac{1}{4\pi} \right)^{1/2} \text{ [Tables 10.1, 9.3]}$$

$$= \frac{1}{4} \left(\frac{1}{2\pi} \right)^{1/2} \times \left(\frac{Z}{a_0} \right)^{3/2} \times \left(2 - \frac{\rho}{2} \right) e^{-\rho/4},$$

$$\psi_{2p_x} = \frac{1}{\sqrt{2}} R_{21}(Y_{1,1} - Y_{1,-1}) \text{ [Section 10.2]}$$

$$= \frac{1}{\sqrt{2}} \left(\frac{Z}{a_0} \right)^{3/2} \frac{\rho}{2} e^{-\rho/4} \left(\frac{3}{8\pi} \right)^{1/2} \sin\theta (e^{i\phi} + e^{-i\phi}) \text{ [Tables 10.1, 9.3]}$$

$$= \frac{1}{\sqrt{2}} \left(\frac{Z}{a_0} \right)^{3/2} \frac{\rho}{2} e^{-\rho/4} \left(\frac{3}{8\pi} \right)^{1/2} \sin\theta \cos\phi$$

$$= \frac{1}{4} \left(\frac{1}{2\pi} \right)^{1/2} \times \left(\frac{Z}{a_0} \right)^{3/2} \frac{\rho}{2} e^{-\rho/4} \sin\theta \cos\phi,$$

$$\psi_{2p_y} = \frac{1}{2i} R_{21}(Y_{1,1} + Y_{1,-1}) \text{ [Section 10.2]}$$

$$= \frac{1}{4} \left(\frac{1}{2\pi} \right)^{1/2} \times \left(\frac{Z}{a_0} \right)^{3/2} \frac{\rho}{2} e^{-\rho/4} \sin\theta \sin\phi \text{ [Tables 10.1, 9.3]}.$$

Therefore,

$$\psi = \frac{1}{\sqrt{3}} \times \frac{1}{4} \times \left(\frac{1}{\pi} \right)^{1/2} \times \left(\frac{Z}{a_0} \right)^{3/2}$$

$$\times \left(\frac{1}{\sqrt{2}} \left(2 - \frac{\rho}{2} \right) - \frac{1}{2} \frac{\rho}{2} \sin\theta \cos\phi + \frac{\sqrt{3}}{2} \frac{\rho}{2} \sin\theta \sin\phi \right) e^{-\rho/4}$$

$$= \frac{1}{4} \left(\frac{1}{6\pi} \right)^{1/2} \times \left(\frac{Z}{a_0} \right)^{3/2} \times \left\{ 2 - \frac{\rho}{2} - \frac{1}{\sqrt{2}} \frac{\rho}{2} \sin\theta \cos\phi + \sqrt{\frac{3}{2}} \frac{\rho}{2} \sin\theta \sin\phi \right\} e^{-\rho/4}$$

$$= \frac{1}{4} \left(\frac{1}{6\pi} \right)^{1/2} \times \left(\frac{Z}{a_0} \right)^{3/2} \times \left\{ 2 - \frac{\rho}{2} \left(1 + \frac{1}{\sqrt{2}} \sin\theta \cos\phi - \sqrt{\frac{3}{2}} \sin\theta \sin\phi \right) \right\} e^{-\rho/4}$$

$$= \frac{1}{4} \left(\frac{1}{6\pi} \right)^{1/2} \times \left(\frac{Z}{a_0} \right)^{3/2} \times \left\{ 2 - \frac{\rho}{2} \left(1 + \frac{[\cos\phi - \sqrt{3}\sin\phi]}{\sqrt{2}} \sin\theta \right) \right\} e^{-\rho/4}.$$

The maximum value of ψ occurs when $\sin\theta$ has its maximum value ($+1$, when $\theta = 90°$; that is, in the xy plane), and the term multiplying $\rho/2$ has its maximum negative value (-1, when $\phi = 120°$).

P11.23   The normalization constants are obtained from

$$\int \psi^2 d\tau = 1, \quad \psi = N(\psi_A \pm \psi_B),$$

$$N^2 \int (\psi_A \pm \psi_B)^2 d\tau = N^2 \int (\psi_A^2 \pm \psi_B^2 \pm 2\psi_A\psi_B) d\tau = N^2(1 + 1 \pm 2S) = 1.$$

Therefore, $N^2 = \dfrac{1}{2(1 \pm S)}$,

$$H = -\frac{\hbar^2}{2m}\nabla^2 - \frac{e^2}{4\pi\varepsilon_0} \cdot \frac{1}{r_A} - \frac{e^2}{4\pi\varepsilon_0} \cdot \frac{1}{r_B} + \frac{e^2}{4\pi\varepsilon_0} \cdot \frac{1}{R}.$$

$H\psi = E\psi$ implies that

$$-\frac{\hbar^2}{2m}\nabla^2\psi - \frac{e^2}{4\pi\varepsilon_0} \cdot \frac{1}{r_A}\psi - \frac{e^2}{4\pi\varepsilon_0} \cdot \frac{1}{r_B}\psi + \frac{e^2}{4\pi\varepsilon_0}\frac{1}{R}\psi = E\psi.$$

Multiply through by $\psi^*(= \psi)$ and integrate using

$$-\frac{\hbar^2}{2m}\nabla^2\psi_A - \frac{e^2}{4\pi\varepsilon_0} \cdot \frac{1}{r_A}\psi_A = E_H\psi_A,$$

$$-\frac{\hbar^2}{2m}\nabla^2\psi_B - \frac{e^2}{4\pi\varepsilon_0} \cdot \frac{1}{r_B}\psi_B = E_H\psi_B.$$

Then for $\psi = N(\psi_A + \psi_B)$

$$N \int \psi \left(E_H\psi_A + E_H\psi_B - \frac{e^2}{4\pi\varepsilon_0} \cdot \frac{1}{r_A}\psi_B - \frac{e^2}{4\pi\varepsilon_0} \cdot \frac{1}{r_B}\psi_A + \frac{e^2}{4\pi\varepsilon_0} \cdot \frac{1}{R}(\psi_A + \psi_B) \right) d\tau = E;$$

hence $E_H \int \psi^2 d\tau + \dfrac{e^2}{4\pi\varepsilon_0} \cdot \dfrac{1}{R} \int \psi^2 d\tau - \dfrac{e^2}{4\pi\varepsilon_0}N \int \psi \left(\dfrac{\psi_B}{r_A} + \dfrac{\psi_A}{r_B} \right) d\tau = E$

and so $E_H + \dfrac{e^2}{4\pi\varepsilon_0} \cdot \dfrac{1}{R} - \dfrac{e^2}{4\pi\varepsilon_0}N^2 \int \left(\psi_A\dfrac{1}{r_A}\psi_B + \psi_B\dfrac{1}{r_A}\psi_B + \psi_A\dfrac{1}{r_B}\psi_A + \psi_B\dfrac{1}{r_A}\psi_A \right) d\tau = E.$

Then use $\displaystyle\int \psi_A\frac{1}{r_A}\psi_B d\tau = \int \psi_B\frac{1}{r_B}\psi_A d\tau$ [by symmetry] $= V_2/(e^2/4\pi\varepsilon_0)$

and $\displaystyle\int \psi_A\frac{1}{r_B}\psi_A d\tau = \int \psi_B\frac{1}{r_A}\psi_B d\tau$ [by symmetry] $= V_1/(e^2/4\pi\varepsilon_0)$

which gives $E_H + \dfrac{e^2}{4\pi\varepsilon_0} \cdot \dfrac{1}{R} - \left(\dfrac{1}{1+S} \right) \times (V_1 + V_2) = E$

or $\qquad E = \boxed{E_H - \dfrac{V_1 + V_2}{1 + S} + \dfrac{e^2}{4\pi\varepsilon_0} \cdot \dfrac{1}{R}} = E_+$

as in Problem 11.8.

The analogous expression for E_- is obtained by starting from

$$\psi = N(\psi_A - \psi_B)$$

with $N^2 = \dfrac{1}{2(1 - S)}$

and following through the step-wise procedure above. The result is

$$E = E_H - \frac{V_1 - V_2}{1 - S} + \frac{e^2}{4\pi\varepsilon_0 R} = E_-$$

as in Problem 11.9.

P11.25 (a) Expanding the determinant yields

$$\begin{vmatrix} \alpha_A - E & \beta \\ \beta & \alpha_B - E \end{vmatrix} = 0 = (\alpha_A - E)(\alpha_B - E) - \beta^2 = E^2 - (\alpha_A + \alpha_B)E + \alpha_A\alpha_B - \beta^2.$$

This is a quadratic equation in E where $a = 1, b = -(\alpha_A + \alpha_B)$, and $c = \alpha_A\alpha_B - \beta^2$. The solution is

$$E_\pm = \frac{-b \pm \sqrt{b^2 - 4ac}}{2a} = \frac{-(\alpha_A + \alpha_B)}{2} \pm \frac{(\alpha_A^2 + 2\alpha_A\alpha_B + \alpha_B^2 - 4\alpha_A\alpha_B + 4\beta^2)^{1/2}}{2}$$

$$= \frac{-(\alpha_A + \alpha_B)}{2} \pm \frac{[(\alpha_A - \alpha_B)^2 + 4\beta^2]^{1/2}}{2}$$

$$= \boxed{\frac{\alpha_A + \alpha_B}{2} \pm \frac{\alpha_A - \alpha_B}{2}\left(1 + \frac{4\beta^2}{(\alpha_A - \alpha_B)^2}\right)^{1/2}}.$$

(b) By hypothesis, $(\alpha_A - \alpha_B) \gg \beta^2$, so the term in parentheses can be expanded:

$$E_\pm = \frac{\alpha_A + \alpha_B}{2} \pm \frac{\alpha_A - \alpha_B}{2}\left(1 + \frac{4\beta^2}{2(\alpha_A - \alpha_B)^2} - \cdots\right)$$

so $\quad E_+ \approx \dfrac{\alpha_A + \alpha_B}{2} + \dfrac{\alpha_A - \alpha_B}{2} + \dfrac{\alpha_A - \alpha_B}{2}\left(\dfrac{4\beta^2}{2(\alpha_A - \alpha_B)^2}\right) = \boxed{\alpha_A + \dfrac{\beta^2}{\alpha_A - \alpha_B}}$

and $\quad E_- \approx \dfrac{\alpha_A + \alpha_B}{2} - \dfrac{\alpha_A - \alpha_B}{2} - \dfrac{\alpha_A - \alpha_B}{2}\left(\dfrac{4\beta^2}{2(\alpha_A - \alpha_B)^2}\right) = \boxed{\alpha_B - \dfrac{\beta^2}{\alpha_A - \alpha_B}}.$

Solutions to applications

P11.27   The secular determinant for a cyclic species H_N^m has the form

$$
\begin{array}{ccccccccc}
1 & 2 & 3 & \cdots & \cdots & \cdots & N-1 & N \\
\end{array}
$$

$$
\begin{vmatrix}
x & 1 & 0 & \cdots & \cdots & \cdots & 0 & 1 \\
1 & x & 1 & \cdots & \cdots & \cdots & 0 & 0 \\
0 & 1 & x & 1 & \cdots & \cdots & 0 & 0 \\
0 & 0 & 1 & x & 1 & \cdots & 0 & 0 \\
\vdots & \vdots & \vdots & \vdots & \vdots & \vdots & \vdots & \vdots \\
\vdots & \vdots & \vdots & \vdots & \vdots & \vdots & \vdots & 1 \\
1 & 0 & 0 & 0 & 0 & \cdots & 1 & x \\
\end{vmatrix}
$$

where $x = (\alpha - E)/\beta$ or $E = \alpha - \beta x$.

Expanding the determinant, finding the roots of the polynomial, and solving for the total binding energy yields the following table. Note that $\alpha < 0$ and $\beta < 0$.

Species	Number of e^-	Permitted x (roots)	Total binding energy
H_4	4	$-2, 0, 0, 2$	$4\alpha + 4\beta$
H_5^+	4	$-2, \frac{1}{2}\left(1-\sqrt{5}\right), \frac{1}{2}\left(1-\sqrt{5}\right), \frac{1}{2}\left(1+\sqrt{5}\right), \frac{1}{2}\left(1+\sqrt{5}\right)$	$4\alpha + \left(3+\sqrt{5}\right)\beta$
H_5	5	$-2, \frac{1}{2}\left(1-\sqrt{5}\right), \frac{1}{2}\left(1-\sqrt{5}\right), \frac{1}{2}\left(1+\sqrt{5}\right), \frac{1}{2}\left(1+\sqrt{5}\right)$	$5\alpha + \frac{1}{2}\left(5+3\sqrt{5}\right)\beta$
H_5^-	6	$-2, \frac{1}{2}\left(1-\sqrt{5}\right), \frac{1}{2}\left(1-\sqrt{5}\right), \frac{1}{2}\left(1+\sqrt{5}\right), \frac{1}{2}\left(1+\sqrt{5}\right)$	$6\alpha + \left(2+2\sqrt{5}\right)\beta$
H_6	6	$-2, -1, -1, 1, 1, 2$	$6\alpha + 8\beta$
H_7^+	6	$-2, -1.248, -1.248, 0.445, 0.445, 1.802, 1.802$	$6\alpha + 8.992\beta$

$$H_4 \rightarrow 2H_2 \qquad \Delta_r U = 4(\alpha + \beta) - (4\alpha + 4\beta) = 0.$$

$$H_5^+ \rightarrow H_2 + H_3^+ \quad \Delta_r U = 2(\alpha + \beta) + (2\alpha + 2\beta) - (4\alpha + 5.236\beta)$$
$$= 0.764\beta < 0.$$

The above $\Delta_r U$ values indicate that H_4 and H_5^+ can fall apart without an energy penalty.

$$H_5^- \rightarrow H_2 + H_3^- \quad \Delta_r U = 2(\alpha + \beta) - (4\alpha + 2\beta) - (6\alpha + 6.472\beta)$$
$$= -2.472\beta > 0.$$

$$H_6 \rightarrow 3H_2 \quad \Delta_r U = 6(\alpha + \beta) - (6\alpha + 8\beta)$$
$$= -2\beta > 0.$$

$$H_7^+ \rightarrow 2H_2 + H_3^+ \quad \Delta_r U = 4(\alpha + \beta) + (2\alpha + 4\beta) - (6\alpha + 8.992\beta)$$
$$= -0.992\beta > 0.$$

The $\Delta_r U$ values for H_5^-, H_6, and H_7^+ suggest that they are stable.

	Satisfies Hückel $4n + 2$ low energy rule	
Species	Correct number e^-	Stable
H_4, $4e^-$	No	No
H_5^+, $4e^-$	No	No
H_5^-, $6e^-$	Yes	Yes
H_6, $6e^-$	Yes	Yes
H_7^+, $6e^-$	Yes	Yes

Hückel's $4n + 2$ rule successfully predicts the stability of hydrogen rings.

P11.29   This question refers to six 1,4-benzoquinones: the unsubstituted, four methyl-substituted, and a dimethyl-dimethoxy species. The table below defines the molecules and displays reduction potentials and computed LUMO energies.

Species	R2	R3	R5	R6	$E^\ominus/V$	E_{LUMO}/eV^*
1	H	H	H	H	−0.078	−1.706
2	CH_3	H	H	H	−0.023	−1.651
3	CH_3	H	CH_3	H	−0.067	−1.583
4	CH_3	CH_3	CH_3	H	−0.165	−1.371
5	CH_3	CH_3	CH_3	CH_3	−0.260	−1.233
6	CH_3	CH_3	CH_3O	CH_3O		−1.446

* Semi-empirical, PM3 level, PC Spartan Pro^{TM}.

(a) The calculations for the species 1–5 are plotted in Fig. 11.10. The figure shows that a linear relationship between the reduction potential and the LUMO energy is consistent with these calculations.

(b) The linear least-squares fit from the plot of E_{LUMO} vs. $E^\ominus$ is

$$E_{LUMO}/eV = -1.621 - 1.435E^\ominus/V \qquad (r^2 = 0.927).$$

Solving for $E^\ominus$ yields

$$E^\ominus/V = -(E_{LUMO}/eV + 1.621)/1.435.$$

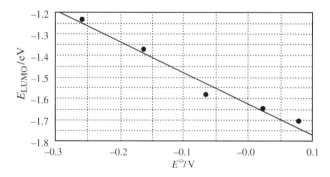

Figure 11.10

Substituting the computed LUMO energy for compound 6 (a model of ubiquinone) yields

$$E^\ominus = [-(-1.446 + 1.621)/1.435] = \boxed{-0.122 \text{ V}}.$$

(c) The model of plastoquinone defined in the problem is compound 4 in the table above. Its experimental reducing potential is known; however, a comparison to the ubiquinone analog based on E_{LUMO} ought to use a computed reducing potential:

$$E^\ominus = [-(-1.371 + 1.621)/1.435] = \boxed{-0.174 \text{ V}}.$$

The better oxidizing agent is the one that is more easily reduced, the one with the less negative reduction potential. Thus we would expect compound 6 to be a better oxidizing agent than compound 4, and $\boxed{\text{ubiquinone a better oxidizing agent than plastoquinone}}$.

(d) *Impact* I7.2 states that coenzyme Q (which is another name for ubiquinone) acts as an oxidizing agent (oxidizing NADH and FADH$_2$) in respiration (i.e. in the overall oxidation of glucose by oxygen). Plastoquinone, on the other hand, acts as a reducing agent (reducing oxidized plastocyanin) in photosynthesis (*Impact* I23.2). Respiration involves the oxidation of glucose by oxygen, while photosynthesis involves a reduction *to* glucose and oxygen. It stands to reason that the better oxidizing agent, ubiquinone, is employed in oxidizing glucose (i.e. in respiration), while the better reducing agent (that is, the poorer oxidizing agent) is used in reduction, i.e. in photosynthesis. (Note, however, that both species are recycled to their original forms: reduced ubiquinone is oxidized by iron (III) and oxidized plastoquinone is reduced by water.)

12 Molecular symmetry

Answers to discussion questions

D12.1 The point group to which a molecule belongs is determined by the symmetry elements it possesses. Therefore the first step is to examine a model (which can be a mental picture) of the molecule for all its symmetry elements. All possible symmetry elements are described in Section 12.1. We list all that apply to the molecule of interest and then follow the assignment procedure summarized by the flow diagram in Figure 12.7 of the text.

D12.3 The dipole moment is a fixed property of a molecule and as a result it must remain unchanged through any symmetry operation of the molecule. Recall that the dipole moment is a vector quantity; therefore both its magnitude and direction must be unaffected by the operation. That can only be the case if the dipole moment is coincident with *all* of the symmetry elements of the molecule. Hence molecules belonging to point groups containing symmetry elements that do not satisfy this criterion can be eliminated. Molecules with a center of symmetry cannot possess a dipole moment because any vector is changed through inversion. Molecules with more than one C_n axis cannot be polar since a vector cannot be coincident with more than one axis simultaneously. If the molecule has a plane of symmetry, the dipole moment must lie in the plane; if it has more than one plane of symmetry, the dipole moment must lie in the axis of intersection of these planes. A molecule can also be polar if it has one plane of symmetry and no C_n. Examination of the character tables at the end of the *Data section* shows that the only point groups that satisfy these restrictions are C_s, C_n, and C_{nv}.

D12.5 A representative is a mathematical operator (usually a matrix) that represents the physical symmetry operation. The set of all these mathematical operators corresponding to all the operations of the group is called a representation. See Section 12.4(a) for examples.

D12.7 Selection rules tell us which transition probabilities between energy levels are non-zero, namely, which spectroscopic transitions will have a non-zero intensity. The intensities are given by the transition moment integral, eqn 12.9, which has the form of the integral of the product of three functions as described by eqn 12.8. Without actually having to perform the integrations involved, group theory can tell us which of these integrals will be non-zero, and hence tell us which are the allowed transitions. Such integrals will be non-zero only if the representation of the triple product in the point group of the molecule spans A_1 or contains a component that spans A_1. In practice, it is usually sufficient to work with the product of the characters of the representations, rather than the matrix representatives themselves. See Examples 12.6 and 12.7.

Solutions to exercises

E12.1(b) CCl_4 has $\boxed{4\,C_3 \text{ axes}}$ (each C—Cl axis), $\boxed{3\,C_2 \text{ axes}}$ (bisecting Cl—C—Cl angles), $\boxed{3S_4 \text{ axes}}$ (the same as the C_2 axes), and $\boxed{6 \text{ dihedral mirror planes}}$ (each Cl—C—Cl plane).

E12.2(b)   Only molecules belonging to $C_s, C_n,$ and C_{nv} groups may be polar, so ...

(a) $CH_3Cl(C_{3v})$ $\boxed{\text{may be polar}}$ along the C—Cl bond;

(b) $HW_2(CO)_{10}(D_{4h})$ $\boxed{\text{may not be polar}}$

(c) $SnCl_4(T_d)$ $\boxed{\text{may not be polar}}$

E12.3(b)   The factors of the integrand have the following characters under the operations of D_{6h}

	E	$2C_6$	$2C_3$	C_2	$3C_2'$	$3C_2''$	i	$2S_3$	$2S_6$	σ_h	$3\sigma_d$	$3\sigma_v$
p_x	2	1	-1	-2	0	0	-2	-1	1	2	0	0
z	1	1	1	1	-1	-1	-1	-1	-1	-1	1	1
p_z	1	1	1	1	-1	-1	-1	-1	-1	-1	1	1
Integrand	2	1	-1	-2	0	0	-2	-1	1	2	0	0

The integrand has the same set of characters as species E_{1u}, so it does not include A_{1g}; therefore the integral $\boxed{\text{vanishes}}$.

E12.4(b)   We need to evaluate the character sets for the product $A_{1g}E_{2u}q$, where $q = x, y,$ or z

	E	$2C_6$	$2C_3$	C_2	$3C_2'$	$3C_2''$	i	$2S_3$	$2S_6$	σ_h	$3\sigma_d$	$3\sigma_v$
A_{1g}	1	1	1	1	1	1	1	1	1	1	1	1
E_{2u}	2	-1	-1	2	0	0	-2	1	1	-2	0	0
(x, y)	2	1	-1	-2	0	0	-2	-1	1	2	0	0
Integrand	4	-1	1	-4	0	0	4	-1	1	-4	0	0

To see whether the totally symmetric species A_{1g} is present, we form the sum over classes of the number of operations times the character of the integrand

$$c(A_{1g}) = (4) + 2(-1) + 2(1) + (-4) + 3(0) + 3(0) + (4)$$
$$+ 2(-1) + 2(1) + (-4) + 3(0) + 3(0) = 0$$

Since the species A_{1g} is absent, the transition is $\boxed{\text{forbidden}}$ for x- or y-polarized light. A similar analysis leads to the conclusion that A_{1g} is absent from the product $A_{1g}E_{2u}z$; therefore the transition is forbidden.

E12.5(b)   The classes of operations for D_2 are: $E, C_2(x), C_2(y),$ and $C_2(z)$. How does the function xyz behave under each kind of operation? E leaves it unchanged. $C_2(x)$ leaves x unchanged and takes y to $-y$ and z to $-z$, leaving the product xyz unchanged. $C_2(y)$ and $C_2(z)$ have similar effects, leaving one axis unchanged and taking the other two into their negatives. These observations are summarized as follows

	E	$C_2(x)$	$C_2(y)$	$C_2(z)$
xyz	1	1	1	1

A look at the character table shows that this set of characters belong to symmetry species $\boxed{A_1}$.

E12.6(b) A molecule cannot be chiral if it has an axis of improper rotation. The point group T_d has $\boxed{S_4 \text{ axes}}$ and $\boxed{\text{mirror planes} (= S_1)}$, which preclude chirality. The T_h group has, in addition, a $\boxed{\text{center of inversion} (= S_2)}$.

E12.7(b) The group multiplication table of group C_{4v} is

	E	C_4^+	C_4^-	C_2	$\sigma_v(x)$	$\sigma_v(y)$	$\sigma_d(xy)$	$\sigma_d(-xy)$
E	E	C_4^+	C_4^-	C_2	$\sigma_v(x)$	$\sigma_v(y)$	$\sigma_d(xy)$	$\sigma_d(-xy)$
C_4^+	C_4^+	C_2	E	C_4^-	$\sigma_d(xy)$	$\sigma_d(-xy)$	$\sigma_v(y)$	$\sigma_v(x)$
C_4^-	C_4^-	E	C_2	C_4^+	$\sigma_d(-xy)$	$\sigma_d(xy)$	$\sigma_v(x)$	$\sigma_v(y)$
C_2	C_2	C_4^-	C_4^+	E	$\sigma_v(y)$	$\sigma_v(x)$	$\sigma_d(-xy)$	$\sigma_d(xy)$
$\sigma_v(x)$	$\sigma_v(x)$	$\sigma_d(-xy)$	$\sigma_d(xy)$	$\sigma_v(y)$	E	C_2	C_4^-	C_4^+
$\sigma_v(y)$	$\sigma_v(y)$	$\sigma_d(xy)$	$\sigma_d(-xy)$	$\sigma_v(x)$	C_2	E	C_4^+	C_4^-
$\sigma_d(xy)$	$\sigma_d(xy)$	$\sigma_v(x)$	$\sigma_v(y)$	$\sigma_d(-xy)$	C_4^+	C_4^-	E	C_2
$\sigma_d(-xy)$	$\sigma_d(-xy)$	$\sigma_v(y)$	$\sigma_v(x)$	$\sigma_d(xy)$	C_4^-	C_4^+	C_2	E

E12.8(b) See Figure 12.1.

(a) Sharpened pencil: E, C_∞, σ_v; therefore $\boxed{C_{\infty v}}$

(b) Propellor: $E, C_3, 3C_2$; therefore $\boxed{D_3}$

(c) Square table: $E, C_4, 4\sigma_v$; therefore $\boxed{C_{4v}}$; Rectangular table: $E, C_2, 2\sigma_v$; therefore $\boxed{C_{2v}}$

(d) Person: E, σ_v (approximately); therefore $\boxed{C_s}$.

E12.9(b) We follow the flow chart in the text (Figure 12.7). The symmetry elements found in order as we proceed down the chart and the point groups are

(a) Naphthalene: $E, C_2, C_2', C_2'', 3\sigma_h, i;$ $\boxed{D_{2h}}$

(b) Anthracene: $E, C_2, C_2', C_2'', 3\sigma_h, i;$ $\boxed{D_{2h}}$

(c) Dichlorobenzenes:

  (i) 1,2-dichlorobenzene: $E, C_2, \sigma_v, \sigma_v';$ $\boxed{C_{2v}}$

  (ii) 1,3-dichlorobenzene: $E, C_2, \sigma_v, \sigma_v';$ $\boxed{C_{2v}}$

  (iii) 1,4-dichlorobenzene: $E, C_2, C_2', C_2'', 3\sigma_h, i;$ $\boxed{D_{2h}}$.

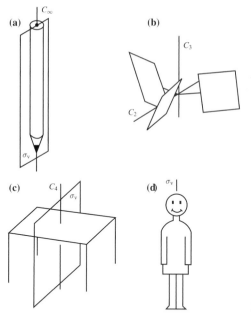

Figure 12.1

E12.10(b) (a) H–F $C_{\infty v}$

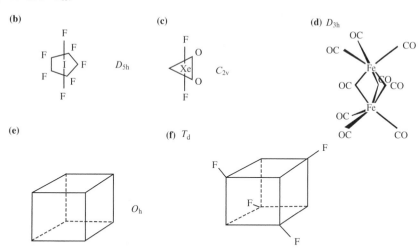

The following responses refer to the text flow chart (Figure 12.7) for assigning point groups.

(a) HF: linear, no i, so $\boxed{C_{\infty v}}$

(b) IF_7: nonlinear, fewer than $2C_n$ with $n > 2, C_5, 5C_2'$ perpendicular to C_5, σ_h, so $\boxed{D_{5h}}$

(c) XeO_2F_2: nonlinear, fewer than $2C_n$ with $n > 2, C_2$, no C_2' perpendicular to C_2, no σ_h, $2\sigma_v$, so $\boxed{C_{2v}}$

(d) $Fe_2(CO)_9$: nonlinear, fewer than $2C_n$ with $n > 2, C_3, 3C_2$ perpendicular to C_3, σ_h, so $\boxed{D_{3h}}$

(e) cubane (C_8H_8): nonlinear, more than $2C_n$ with $n > 2, i$, no C_5, so $\boxed{O_h}$

(f) tetrafluorocubane (**23**): nonlinear, more than $2C_n$ with $n > 2$, no i, so $\boxed{T_d}$.

E12.11(b) **(a)** Only molecules belonging to C_s, C_n, and C_{nv} groups may be polar. In Exercise 12.9(b) $\boxed{ortho\text{-dichlorobenzene}}$ and $\boxed{meta\text{-dichlorobenzene}}$ belong to C_{2v} and so may be polar; in Exercise 12.6(b), $\boxed{\text{HF and XeO}_2\text{F}_2}$ belong to C_{nv} groups, so they may be polar.

(b) A molecule cannot be chiral if it has an axis of improper rotation – including disguised or degenerate axes such as an inversion centre (S_2) or a mirror plane (S_1). In Exercises 12.5(b) and 12.6(b), all the molecules have mirror planes, so $\boxed{\text{none}}$ can be chiral.

E12.12(b) In order to have nonzero overlap with a combination of orbitals that spans E, an orbital on the central atom must itself have some E character, for only E can multiply E to give an overlap integral with a totally symmetric part. A glance at the character table shows that $\boxed{p_x \text{ and } p_y}$ orbitals available to a bonding N atom have the proper symmetry. If d orbitals are available (as in SO_3), $\boxed{\text{all } d \text{ orbitals except } d_{z^2}}$ could have nonzero overlap.

E12.13(b) The product $\Gamma_f \times \Gamma(\mu) \times \Gamma_i$ must contain A_1 (Example 12.7). Then, since $\Gamma_i = B_1, \Gamma(\mu) = \Gamma(y) = B_2$ (C_{2v} character table), we can draw up the following table of characters

	E	C_2	σ_v	σ_v'	
B_2	1	-1	-1	1	
B_1	1	-1	1	-1	
$B_1 B_2$	1	1	-1	-1	$= A_2$

Hence, the upper state is $\boxed{A_2}$, because $A_2 \times A_2 = A_1$.

E12.14(b) **(a)**

Anthracene

D_{2h}

The components of μ span $B_{3u}(x)$, $B_{2u}(y)$, and $B_{1u}(z)$. The totally symmetric ground state is A_g. Since $A_g \times \Gamma = \Gamma$ in this group, the accessible upper terms are $\boxed{B_{3u}}$ (x-polarized), $\boxed{B_{2u}}$ (y-polarized), and $\boxed{B_{1u}}$ (z-polarized).

(b) Coronene, like benzene, belongs to the D_{6h} group. The integrand of the transition dipole moment must be or contain the A_{1g} symmetry species. That integrand for transitions from the ground state is $A_{1g}qf$, where q is x, y, or z and f is the symmetry species of the upper state. Since the ground state is already totally symmetric, the product qf must also have A_{1g} symmetry for the entire integrand to have A_{1g} symmetry. Since the different symmetry species are orthogonal, the only way qf can have A_{1g} symmetry is if q and f have the same symmetry. Such combinations include zA_{2u}, xE_{1u}, and yE_{1u}. Therefore, we conclude that transitions are allowed to states with $\boxed{A_{2u} \text{ or } E_{1u}}$ symmetry.

E12.15(b)

	E	$2C_3$	$3\sigma_v$
A_1	1	1	1
A_2	1	1	−1
E	2	−1	0
$\sin\theta$	1	Linear combinations of	1
$\cos\theta$	1	$\sin\theta$ and $\cos\theta$	−1
Product	1	1	−1

The product does not contain A_1, so $\boxed{\text{yes}}$ the integral vanishes.

Solutions to problems

P12.1   (a) Staggered CH_3CH_3 : E, C_3, C_2, $3\sigma_d$; $\boxed{D_{3d}}$ [see Fig. 12.6(b) of the text].

(b) Chair C_6H_{12}: E, C_3, C_2, $3\sigma_d$; $\boxed{D_{3d}}$.
    Boat C_6H_{12}: E, C_2, σ_v, σ_v'; $\boxed{C_{2v}}$.

(c) B_2H_6: E, C_2, $2C_2'$, σ_h; $\boxed{D_{2h}}$.

(d) $[Co(en)_3]^{3+}$: E, $2C_3$, $3C_2$; $\boxed{D_3}$.

(e) Crown S_8: E, C_4, C_2, $4C_2'$, $4\sigma_d$, $2S_8$; $\boxed{D_{4d}}$.

Only boat C_6H_{12} may be polar, since all the others are D point groups. Only $[Co(en)_3]^{3+}$ belongs to a group without an improper rotation axis ($S_1 = \sigma$), and hence is chiral.

P12.3   Consider Fig. 12.2. The effect of σ_h on a point P is to generate $\sigma_h P$, and the effect of C_2 on $\sigma_h P$ is to generate the point $C_2\sigma_h P$. The same point is generated from P by the inversion i, so $C_2\sigma_h P = iP$ for all points P. Hence, $\boxed{C_2\sigma_h = i}$, and i must be a member of the group.

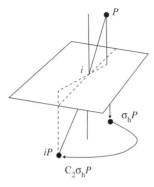

Figure 12.2

P12.5   We examine how the operations of the C_{3v} group affect $l_z = xp_y - yp_x$ when applied to it. The transformations of x, y, and z, and by analogy p_x, p_y, and p_z, are as follows (see Fig. 12.3)

$E(x, y, z) \rightarrow (x, y, z),$

$\sigma_v(x, y, z) \rightarrow (-x, y, z),$

$\sigma_v'(x, y, z) \rightarrow (x, -y, z),$

$\sigma_v''(x, y, z) \rightarrow (x, y, -z),$

$C_3^+(x, y, z) \rightarrow \left(-\frac{1}{2}x + \frac{1}{2}\sqrt{3}y, -\frac{1}{2}\sqrt{3}x - \frac{1}{2}y, z\right),$

$C_3^-(x, y, z) \rightarrow \left(-\frac{1}{2}x - \frac{1}{2}\sqrt{3}y, \frac{1}{2}\sqrt{3}x - \frac{1}{2}y, z\right).$

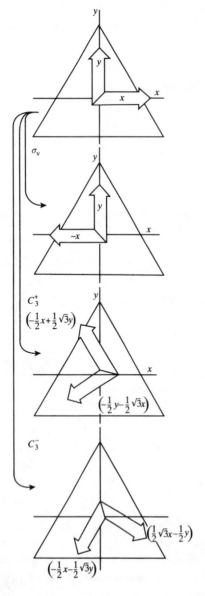

Figure 12.3

The characters of all σ operations are the same, as are those of both C_3 operations (see the C_{3v} character table); hence we need consider only one operation in each class.

$$El_z = xp_y - yp_x = l_z,$$

$$\sigma_v l_z = -xp_y + yp_x = -l_z \quad [(x,y,z) \to (-x,y,z)],$$

$$C_3^+ l_z = (-\tfrac{1}{2}x + \tfrac{1}{2}\sqrt{3}y) \times (-\tfrac{1}{2}\sqrt{3}p_x - \tfrac{1}{2}p_y) - (-\tfrac{1}{2}\sqrt{3}x - \tfrac{1}{2}y) \times (-\tfrac{1}{2}p_x + \tfrac{1}{2}\sqrt{3}p_y)$$

$$[(x,y,z) \to (-\tfrac{1}{2}x + \tfrac{1}{2}\sqrt{3}y, -\tfrac{1}{2}\sqrt{3}x - \tfrac{1}{2}y, z)]$$

$$= \tfrac{1}{4}(\sqrt{3}xp_x + xp_y - 3yp_x - \sqrt{3}yp_y - \sqrt{3}xp_x + 3xp_y - yp_x + \sqrt{3}yp_y)$$

$$= xp_y - yp_x = l_z.$$

The representatives of E, σ_v, and C_3^+ are therefore all one-dimensional matrices with characters 1, -1, 1 respectively. It follows that l_z is a basis for A_2 (see the C_{3v} character table).

P12.7 The multiplication table is

	1	σ_x	σ_y	σ_z
1	1	σ_x	σ_y	σ_z
σ_x	σ_x	1	$i\sigma_x$	$-i\sigma_y$
σ_y	σ_y	$-i\sigma_z$	1	$i\sigma_x$
σ_z	σ_z	$i\sigma_y$	$-i\sigma_z$	1

The matrices $\boxed{\text{do not form a group}}$ since the products $i\sigma_z, i\sigma_y, i\sigma_x$ and their negatives are not among the four given matrices.

P12.9 **(a)** In C_{3v} symmetry the H1s orbitals span the same irreducible representations as in NH_3, which is $A_1 + A_1 + E$. There is an additional A_1 orbital because a fourth H atom lies on the C_3 axis. In C_{3v}, the d orbitals span $A_1 + E + E$ [see the final column of the C_{3v} character table]. Therefore, $\boxed{\text{all five } d \text{ orbitals}}$ may contribute to the bonding.

(b) In C_{2v} symmetry the H1s orbitals span the same irreducible representations as in H_2O, but one 'H_2O' fragment is rotated by $90°$ with respect to the other. Therefore, whereas in H_2O the H1s orbitals span $A_1 + B_2$ [$H_1 + H_2$, $H_1 - H_2$], in the distorted CH_4 molecule they span $A_1 + B_2 + A_1 + B_1$ [$H_1 + H_2$, $H_1 - H_2$, $H_3 + H_4$, $H_3 - H_4$]. In C_{2v} the d orbitals span $2A_1 + B_1 + B_2 + A_2$ [C_{2v} character table]; therefore, $\boxed{\text{all except } A_2(d_{xy})}$ may participate in bonding.

P12.11 **(a)** We work through the flow diagram in the text (Fig. 12.4). We note that this complex with freely rotating CF_3 groups is not linear, it has no C_n axes with $n > 2$, but it does have C_2 axes; in fact it has two C_2 axes perpendicular to whichever C_2 we call principal, and it has a σ_h. Therefore, the point group is $\boxed{D_{2h}}$.

(b) The plane shown in Fig. 12.4 below is a mirror plane so long as each of the CF_3 groups has a CF bond in the plane. (i) If the CF_3 groups are staggered, then the Ag–CN axis is still a C_2 axis; however, there are no other C_2 axes. The Ag–CF_3 axis is an S_2 axis, though, which means that the Ag atom is at an inversion center. Continuing with the flow diagram, there is a σ_h (the plane shown in the figure). So the point group is $\boxed{C_{2h}}$. (ii) If the CF_3 groups are eclipsed, then the axis through the Ag and perpendicular to the plane of the Ag bonds is still a C_2 axis; however, neither of the Ag bond axes is a C_2 axis. There is no σ_h but there are two σ_v planes (the plane shown and the plane perpendicular to it and the Ag bond plane). So the point group is $\boxed{C_{2v}}$.

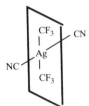

Figure 12.4

P12.13   **(a)** C_{2v}. The functions x^2, y^2, and z^2 are invariant under all operations of the group, and so $z(5z^2 - 3r^2)$ transforms as $z(A_1)$, $y(5y^2 - 3r^2)$ as $y(B_2)$, $x(5x^2 - 3r^2)$ as $x(B_1)$, and likewise for $z(x^2 - y^2)$, $y(x^2 - z^2)$, and $x(z^2 - y^2)$. The function xyz transforms as $B_1 \times B_2 \times A_1 = A_2$. Therefore, in group C_{2v}, $f \rightarrow \boxed{2A_1 + A_2 + 2B_1 + 2B_2}$.

(b) C_{3v}. In C_{3v}, z transforms as A_1, and hence so does z^3. From the C_{3v} character table, $(x^2 - y^2, xy)$ is a basis for E, and so $(xyz, z(x^2 - y^2))$ is a basis for $A_1 \times E = E$. The linear combinations $y(5y^2 - 3r^2) + 5y(x^2 - z^2) \propto y$ and $x(5x^2 - 3r^2) + 5x(z^2 - y^2) \propto x$ are a basis for E. Likewise, the two linear combinations orthogonal to these are another basis for E. Hence, in the group C_{3v}, $f \rightarrow$ $\boxed{A_1 + 3E}$.

(c) T_d. Make the inspired guess that the f orbitals are a basis of dimension $3 + 3 + 1$, suggesting the decomposition $T + T + A$. Is the A representation A_1 or A_2? We see from the character table that the effect of S_4 discriminates between A_1 and A_2. Under S_4, $x \rightarrow y$, $y \rightarrow -x$, $z \rightarrow -z$, and so $xyz \rightarrow xyz$. The character is $\chi = 1$, and so xyz spans A_1. Likewise, $(x^3, y^3, z^3) \rightarrow (y^3, -x^3, -z^3)$ and $\chi = 0 + 0 - 1 = -1$. Hence, this trio spans T_2. Finally,

$$\{x(z^2 - y^2), \ y(z^2 - x^2), \ z(x^2 - y^2)\} \rightarrow \{y(z^2 - x^2), \ -x(z^2 - y^2), \ -z(y^2 - x^2)\}$$

resulting in $\chi = 1$, indicating T_1. Therefore, in T_d, $f \rightarrow \boxed{A_1 + T_1 + T_2}$.

(d) O_h. Anticipate an $A + T + T$ decomposition as in the other cubic group. Since x, y, and z all have odd parity, all the irreducible representatives will be u. Under S_4, $xyz \rightarrow xyz$ (as in **(c)**), and so the representation is $\chi = -1$ (see the character table). Under S_4, $(x^3, y^3, z^3) \rightarrow (y^3, -x^3, -z^3)$, as before, and $\chi = -1$, indicating T_{1u}. In the same way, the remaining three functions span T_{2u}. Hence, in O_h, $f \rightarrow \boxed{A_{2u} + T_{1u} + T_{2u}}$.

(The shapes of the orbitals are shown in *Inorganic Chemistry*, 3rd edn, D. F. Shriver, and P. W. Atkins, Oxford University Press and W. H. Freeman & Co (1999).)

The f orbitals will cluster into sets according to their irreducible representations. Thus **(a)** $f \rightarrow$ $A_1 + T_1 + T_2$ in T_d symmetry, and there is one nondegenerate orbital and two sets of triply degenerate orbitals. **(b)** $f \rightarrow A_{2u} + T_{1u} + T_{2u}$, and the pattern of splitting (but not the order of energies) is the same.

P12.15   We begin by drawing up the following table.

	N2s	N2p$_x$	N2p$_y$	N2p$_z$	O2p$_x$	O2p$_y$	O2p$_z$	O'2p$_x$	O'2p$_y$	O'2p$_z$	χ
E	N2s	N2p$_x$	N2p$_y$	N2p$_z$	O2p$_x$	O2p$_y$	O2p$_z$	O'2p$_x$	O'2p$_y$	O'2p$_z$	10
C_2	N2s	−N2p$_x$	−N2p$_y$	N2p$_z$	−O'2p$_x$	−O'2p$_y$	O'2p$_z$	−O2p$_x$	−O2p$_y$	O2p$_z$	0
σ_v	N2s	N2p$_x$	−N2p$_y$	N2p$_z$	O'2p$_x$	−O'2p$_y$	O'2p$_z$	O2p$_x$	−O2p$_y$	O2p$_z$	2
σ_v'	N2s	−N2p$_x$	N2p$_y$	N2p$_z$	−O2p$_x$	O2p$_y$	O2p$_z$	−O'2p$_x$	O'2p$_y$	O'2p$_z$	4

The character set (10, 0, 2, 4) decomposes into $\boxed{4A_1 + 2B_1 + 3B_2 + A_2}$. We then form symmetry-adapted linear combinations as described in Section 12.5.

$\psi(A_1) = N2s$	(column 1)	$\psi(B_1) = O2p_x + O'2p_x$	(column 5)
$\psi(A_1) = N2p_z$	(column 4)	$\psi(B_2) = N2p_y$	(column 3)
$\psi(A_1) = O2p_z + O'2p_z$	(column 7)	$\psi(B_2) = O2p_y + O'2p_y$	(column 6)
$\psi(A_1) = O2p_y + O'2p_y$	(column 9)	$\psi(A_2) = O2p_z + O'2p_z$	(column 7)
$\psi(B_1) = N2p_x$	(column 2)	$\psi(A_2) = O2p_x + O'2p_x$	(column 5)

(The other columns yield the same combinations.)

P12.17   Consider phenanthrene with carbon atoms as labeled in the stucture below.

(a) The 2p orbitals involved in the π system are the basis we are interested in. To find the irreproducible representations spanned by this basis, consider how each basis is transformed under the symmetry operations of the C_{2v} group. To find the character of an operation in this basis, sum the coefficients of the basis terms that are unchanged by the operation.

	a	a′	b	b′	c	c′	d	d′	e	e′	f	f′	g	g′	χ
E	a	a′	b	b′	c	c′	d	d′	e	e′	f	f′	g	g′	14
C_2	−a′	−a	−b′	−b	−c′	−c	−d′	−d	−e′	−e	−f′	−f	−g′	−g	0
σ_v	a′	a	b′	b	c′	c	d′	d	e′	e	f′	f	g′	g	0
σ_v'	−a	−a′	−b	−b′	−c	−c′	−d	−d′	−e	−e′	−f	−f′	−g	−g′	−14

To find the irreproducible representations that these orbitals span, multiply the characters in the representation of the orbitals by the characters of the irreproducible representations, sum those products, and divide the sum by the order h of the group (as in Section 12.5(a)). The table below illustrates the procedure, beginning at left with the C_{2v} character table.

	E	C_2	σ_v	σ_v'	product	E	C_2	σ_v	σ_v'	sum/h
A_1	1	1	1	1		14	0	0	−14	0
A_2	1	1	−1	−1		14	0	0	14	7
B_1	1	−1	1	−1		14	0	0	14	7
B_2	1	−1	−1	1		14	0	0	−14	0

The orbitals span $\boxed{7A_2 + B_2}$.

To find symmetry-adapted linear combinations (SALCs), follow the procedure described in Section 12.5(c). Refer to the table above that displays the transformations of the original basis orbitals. To find SALCs of a given symmetry species, take a column of the table, multiply each entry by the character of the species' irreproducible representation, sum the terms in the column, and divide by the order of the group. For example, the characters of species A_1 are 1, 1, 1, 1, so the columns to be summed are identical to the columns in the table above. Each column sums to zero, so we conclude that there are no SALCs of A_1 symmetry. (No surprise here: the orbitals span only A_2 and B_1.) An A_2 SALC is obtained by multiplying the characters 1, 1, -1, -1 by the first column:

$$\tfrac{1}{4}(a - a' - a' + a) = \tfrac{1}{2}(a - a').$$

The A_2 combination from the second column is the same. There are seven distinct A_2 combinations in all: $\boxed{\tfrac{1}{2}(a - a'), \tfrac{1}{2}(b - b'), \dots, \tfrac{1}{2}(g - g')}$. The B_1 combination from the first column is

$$\tfrac{1}{4}(a + a' + a' + a) = \tfrac{1}{2}(a + a').$$

The B_1 combination from the second column is the same. There are seven distinct B_1 combinations in all: $\boxed{\tfrac{1}{2}(a + a'), \tfrac{1}{2}(b + b'), \dots, \tfrac{1}{2}(g + g')}$. There are no B_2 combinations, as the columns sum to zero.

(b) The structure is labeled to match the row and column numbers shown in the determinant. The Hückel secular determinant of phenanthrene is:

	a	b	c	d	e	f	g	g'	f'	e'	d'	c'	b'	a'
a	$\alpha - E$	β	0	0	0	0	0	0	0	0	0	0	0	β
b	β	$\alpha - E$	β	0	0	0	β	0	0	0	0	0	0	0
c	0	β	$\alpha - E$	β	0	0	0	0	0	0	0	0	0	0
d	0	0	β	$\alpha - E$	β	0	0	0	0	0	0	0	0	0
e	0	0	0	β	$\alpha - E$	β	0	0	0	0	0	0	0	0
f	0	0	0	0	β	$\alpha - E$	β	0	0	0	0	0	0	0
g	0	β	0	0	0	β	$\alpha - E$	β	0	0	0	0	0	0
g'	0	0	0	0	0	0	β	$\alpha - E$	β	0	0	0	0	0
f'	0	0	0	0	0	0	0	β	$\alpha - E$	β	0	0	0	0
e'	0	0	0	0	0	0	0	0	β	$\alpha - E$	β	0	0	0
d'	0	0	0	0	0	0	0	0	0	β	$\alpha - E$	β	0	0
c'	0	0	0	0	0	0	0	0	0	0	β	$\alpha - E$	β	0
b'	0	0	0	0	0	0	0	β	0	0	0	β	$\alpha - E$	β
a'	β	0	0	0	0	0	0	0	0	0	0	0	β	$\alpha - E$

This determinant has the same eigenvalues as in exercise 11.13b(b).

(c) The ground state of the molecule has $A1$ symmetry by virtue of the fact that its wavefunction is the product of doubly occupied orbitals, and the product of any two orbitals of the same symmetry has A_1 character. If a transition is to be allowed, the transition dipole must be non-zero, which in turn can only happen if the representation of the product $\Psi_f^* \mu \Psi_i$ includes the totally symmetric species A_1. Consider first transitions to another A_1 wavefunction, in which case we need the product $A_1 \mu A_1$.

Now $A_1 A_1 = A_1$, and the only character that returns A_1 when multiplied by A_1 is A_1 itself. The z component of the dipole operator belongs to species A_1, so z-polarized $A_1 \leftarrow A_1$ transitions are allowed. (Note: transitions from the A_1 ground state to an A_1 excited state are transitions from an orbital occupied in the ground state to an excited-state orbital of the same symmetry.) The other possibility is a transition from an orbital of one symmetry (A_2 or B_1) to the other; in that case, the excited-state wavefunction will have symmetry of $A_1 B_1 = B_2$ from the two singly occupied orbitals in the excited state. The symmetry of the transition dipole, then, is $A_1 \mu B_2 = \mu B_2$, and the only species that yields A_1 when multiplied by B_2 is B_2 itself. Now the y component of the dipole operator belongs to species B_2, so these transitions are also allowed (y-polarized).

Solutions to applications

P12.19   The shape of this molecule is shown in Fig. 12.5.

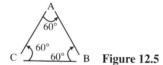

Figure 12.5

(a) Symmetry elements $\boxed{E, 2C_3, 3C_2, \sigma_h, 2S_3, 3\sigma_v}$.
  Point group $\boxed{D_{3h}}$.

(b) $$D(E) = \begin{pmatrix} 1 & 0 & 0 \\ 0 & 1 & 0 \\ 0 & 0 & 1 \end{pmatrix} = D(\sigma_h).$$

$$D(C_3) = \begin{pmatrix} 0 & 0 & 1 \\ 1 & 0 & 0 \\ 0 & 1 & 0 \end{pmatrix}, \quad D(C_3') = D^2(C_3) = \begin{pmatrix} 0 & 1 & 0 \\ 0 & 0 & 1 \\ 1 & 0 & 0 \end{pmatrix}.$$

$$D(S_3) = D(C_3), \quad D(S_3') = D^2(S_3) = D(C_3').$$

C_3' and S_3' are counter clockwise rotations.
σ_v is through A and perpendicular to B–C.
σ_v' is through B and perpendicular to A–C.
σ_v'' is through C and perpendicular to A–B.

$$D(\sigma_v) = \begin{pmatrix} 1 & 0 & 0 \\ 0 & 0 & 1 \\ 0 & 1 & 0 \end{pmatrix}, \quad D(\sigma_v') = \begin{pmatrix} 0 & 0 & 1 \\ 0 & 1 & 0 \\ 1 & 0 & 0 \end{pmatrix},$$

$$D(\sigma_v'') = \begin{pmatrix} 0 & 1 & 0 \\ 1 & 0 & 0 \\ 0 & 0 & 1 \end{pmatrix}.$$

$$D(C_2) = D(\sigma_v), \quad D(C_2') = D(\sigma_v'), \quad D(C_2'') = D(\sigma_v'').$$

(c) Example of elements of group multiplication table

$$D(C_3)D(C_2) = \begin{pmatrix} 0 & 0 & 1 \\ 1 & 0 & 0 \\ 0 & 1 & 0 \end{pmatrix} \begin{pmatrix} 1 & 0 & 0 \\ 0 & 0 & 1 \\ 0 & 1 & 0 \end{pmatrix}$$

$$= \begin{pmatrix} 0 & 1 & 0 \\ 1 & 0 & 0 \\ 0 & 0 & 1 \end{pmatrix} = D(\sigma_v'').$$

$$D(\sigma_v')D(\sigma_v) = \begin{pmatrix} 0 & 0 & 1 \\ 0 & 1 & 0 \\ 1 & 0 & 0 \end{pmatrix} \begin{pmatrix} 1 & 0 & 0 \\ 0 & 0 & 1 \\ 0 & 1 & 0 \end{pmatrix}$$

$$= \begin{pmatrix} 0 & 1 & 0 \\ 0 & 0 & 1 \\ 1 & 0 & 0 \end{pmatrix} = D(C_3').$$

D_{3h}	E	C_3	C_2	σ_v	σ_v'	σ_h	$\dots$
E	E	C_3	C_2	σ_v	σ_v'	σ_h	$\dots$
C_3	C_3	C_3'	σ_v''	σ_v''	σ_v	C_3	$\dots$
C_2	C_2	σ_v'	E	E	C_3	C_2	$\dots$
σ_v	σ_v	σ_v'	E	E	C_3	σ_v	$\dots$
σ_v'	σ_v'	σ_v''	C_3	C_3	E	σ_v'	$\dots$
σ_h	σ_h	C_3	C_2	σ_v	σ_v'	E	$\dots$
$\vdots$	$\vdots$	$\vdots$	$\vdots$	$\vdots$	$\vdots$	$\vdots$	$\ddots$

(d) First, determine the number of s orbitals (the basis has three s orbitals) that have unchanged positions after application of each symmetry species of the D_{3h} point group.

D_{3h}	E	$2C_3$	$3C_2$	σ_h	$2S_3$	$3\sigma_v$
Unchanged basis members	3	0	1	3	0	1

This is not one of the irreducible representations reported in the D_{3h} character table but inspection shows that it is identical to $A_1' + E'$. This allows us to conclude that the three s orbitals span $\boxed{A_1' + E'}$.

COMMENT. The multiplication table in part **(c)** is not strictly speaking *the* group multiplication; it is instead the multiplication table for the matrix representations of the group in the basis under consideration.

P12.21   **(a)** Following the flow chart in Fig. 12.4 of the text, note that the molecule is not linear (at least not in the mathematical sense); there is only one C_n axis (a C_2), and there is a σ_h. The point group, then, is $\boxed{C_{2h}}$.

(b) The $2p_z$ orbitals are transformed under the symmetry operations of the C_{2h} group as follows.

	a	a′	b	b′	c	c′	...	j	j′	k	k′	χ
E	a	a′	b	b′	c	c′	...	j	j′	k	k′	22
C_2	a′	a	b	b	c′	c	...	j′	j	k′	k	0
i	−a′	−a	−b′	−b	−c′	−c	...	−j′	−j′	−k′	−k	0
σ_h	−a	−a′	−b	−b′	−c	−c′	...	−j	−j′	−k	−k′	−22 .

To find the irreproducible representations that these orbitals span, we multiply the characters of orbitals by the characters of the irreproducible representations, sum those products, and divide the sum by the order h of the group (as in Section 12.5(a)). The table below illustrates the procedure, beginning at left with the C_{2h} character table.

	E	C_2	i	σ_h	product	E	C_2	i	σ_h	sum/h
A_g	1	1	1	1		22	0	0	−22	0
A_u	1	1	−1	−1		22	0	0	22	11
B_g	1	−1	1	−1		22	0	0	22	11
B_u	1	−1	−1	1		22	0	0	−22	0

The orbitals span $\boxed{11A_u + 11B_g}$.

To find symmetry-adapted linear combinations (SALCs), follow the procedure described in Section 12.5(c). Refer to the table above that displays the transformations of the original basis orbitals. To find SALCs of a given symmetry species, take a column of the table, multiply each entry by the character of the species' irreproducible representation, sum the terms in the column, and divide by the order of the group. For example, the characters of species A_u are 1, 1, 1, 1, so the columns to be summed are identical to the columns in the table above. Each column sums to zero, so we conclude that there are no SALCs of A_g symmetry. (No surprise: the orbitals span only A_u and B_g.) An A_u SALC is obtained by multiplying the characters 1, 1, −1, −1 by the first column:

$$\tfrac{1}{4}(a + a′ + a′ + a) = \tfrac{1}{2}(a + a′).$$

The A_u combination from the second column is the same. There are 11 distinct A_u combinations in all: $\boxed{\tfrac{1}{2}(a + a′), \tfrac{1}{2}(b + b′), \ldots \tfrac{1}{2}(k + k′)}$. The B_g combination from the first column is

$$\tfrac{1}{4}(a − a′ − a′ + a) = \tfrac{1}{2}(a − a′).$$

The B_g combination from the second column is the same. There are 11 distinct B_g combinations in all: $\boxed{\tfrac{1}{2}(a − a′), \tfrac{1}{2}(b − b′), \ldots \tfrac{1}{2}(k − k′)}$. There are no B_u combinations, as the columns sum to zero.

(c) The structure is labeled to match the row and column numbers shown in the determinant. The Hückel secular determinant is:

	a	b	c	...	i	j	k	k'	j'	i'	...	c'	b'	a'
a	$\alpha - E$	β	0	...	0	0	0	0	0	0	...	0	0	0
b	β	$\alpha - E$	β	...	0	0	0	0	0	0	...	0	0	0
c	0	β	$\alpha - E$	...	0	0	0	0	0	0	...	0	0	0
...	...	...	...	...	...	...	...	...	...	...	...	...	...	...
i	0	0	0	...	$\alpha - E$	β	0	0	0	0	...	0	0	0
j	0	0	0	...	β	$\alpha - E$	β	0	0	0	...	0	0	0
k	0	0	0	...	0	β	$\alpha - E$	β	0	0	...	0	0	0
k'	0	0	0	...	0	0	β	$\alpha - E$	β	0	...	0	0	0
j'	0	0	0	...	0	0	0	β	$\alpha - E$	β	...	0	0	0
i'	0	0	0	...	0	0	0	0	β	$\alpha - l$	...	0	0	0
...	...	...	...	...	...	...	...	...	...	...	...	...	...	...
c'	0	0	0	...	0	0	0	0	0	0	...	$\alpha - E$	β	0
b'	0	0	0	...	0	0	0	0	0	0	...	β	$\alpha - E$	β
a'	0	0	0	...	0	0	0	0	0	0	...	0	β	$\alpha - E$

The energies of the filled orbitals are $\alpha + 1.98137\beta$, $\alpha + 1.92583\beta$, $\alpha + 1.83442\beta$, $\alpha + 1.70884\beta$, $\alpha + 1.55142\beta$, $\alpha + 1.36511\beta$, $\alpha + 1.15336\beta$, $\alpha + 0.92013\beta$, $\alpha + 0.66976\beta$, $\alpha + 0.40691\beta$, and $\alpha + 0.13648\beta$. The π energy is 27.30729β.

(d) The ground state of the molecule has A_g symmetry by virtue of the fact that its wavefunction is the product of doubly occupied orbitals, and the product of any two orbitals of the same symmetry has A_g character. If a transition is to be allowed, the transition dipole must be non-zero, which in turn can only happen if the representation of the product $\Psi_f^* \mu \Psi_i$ includes the totally symmetric species A_g. Consider first transitions to another A_g wavefunction, in which case we need the product $A_g \mu A_g$. Now $A_g A_g = A_g$, and the only character that returns A_g when multiplied by A_g is A_g itself. No component of the dipole operator belongs to species A_g, so no $A_g \leftarrow A_g$ transitions are allowed. (Note: such transitions are transitions from an orbital occupied in the ground state to an excited-state orbital of the same symmetry.) The other possibility is a transition from an orbital of one symmetry (A_u or B_g) to the other; in that case, the excited-state wavefunction will have symmetry of $A_u B_g = B_u$ from the two singly occupied orbitals in the excited state. The symmetry of the transition dipole, then, is $A_g \mu B_u = \mu B_u$, and the only species that yields A_g when multiplied by B_u is B_u itself. The x and y components of the dipole operator belongs to species B_u, so these transitions are allowed.

13 Molecular spectroscopy 1: rotational and vibrational spectra

Answers to discussion questions

D13.1 **(1)** *Doppler broadening.* This contribution to the linewidth is due to the Doppler effect, which shifts the frequency of the radiation emitted or absorbed when the atoms or molecules involved are moving towards or away from the detecting device. Molecules have a wide range of speeds in all directions in a gas and the detected spectral line is the absorption or emission profile arising from all the resulting Doppler shifts. As shown in *Justification* 13.3, the profile reflects the distribution of molecular velocities parallel to the line of sight which is a bell-shaped Gaussian curve.

(2) *Lifetime broadening.* The Doppler broadening is significant in gas phase samples, but lifetime broadening occurs in all states of matter. This kind of broadening is a quantum mechanical effect related to the uncertainty principle in the form of eqn 13.18 and is due to the finite lifetimes of the states involved in the transition. When τ is finite, the energy of the states is smeared out and hence the transition frequency is broadened as shown in eqn 13.19.

(3) *Pressure broadening or collisional broadening.* The actual mechanism affecting the lifetime of energy states depends on various processes, one of which is collisional deactivation and another of which is spontaneous emission. Lowering the pressure can reduce the first of these contributions; the second cannot be changed and results in a natural linewidth.

D13.3 **(1)** *Rotational Raman spectroscopy.* The gross selection rule is that the molecule must be anisotropically polarizable, which is to say that its polarizability, α, depends upon the direction of the electric field relative to the molecule. Non-spherical rotors satisfy this condition. Therefore, linear and symmetric rotors are rotationally Raman active.

(2) *Vibrational Raman spectroscopy.* The gross selection rule is that the polarizability of the molecule must change as the molecule vibrates. All diatomic molecules satisfy this condition as the molecules swell and contract during a vibration, the control of the nuclei over the electrons varies, and the molecular polarizability changes. Hence both homonuclear and heteronuclear diatomics are vibrationally Raman active. In polyatomic molecules it is usually quite difficult to judge by inspection whether or not the molecule is anisotropically polarizable; hence group theoretical methods are relied on for judging the Raman activity of the various normal modes of vibration. The procedure is discussed in Section 13.17(b) and demonstrated in *Illustration* 13.6.

D13.5 The exclusion rule applies to the benzene molecule because it has a center of symmetry. Consequently, none of the normal modes of vibration of benzene can be both infrared and Raman active. If we wish to characterize all the normal modes we must obtain both kinds of spectra. See the solutions to Exercises 13.25(a) and 13.25(b) for specific illustrations of which modes are IR active and which are Raman active.

Solutions to exercises

E13.1(b)   The ratio of coefficients A/B is

(a) $\dfrac{A}{B} = \dfrac{8\pi h v^3}{c^3} = \dfrac{8\pi (6.626 \times 10^{-34} \text{J s}) \times (500 \times 10^6 \text{ s}^{-1})^3}{(2.998 \times 10^8 \text{ m s}^{-1})^3} = \boxed{7.73 \times 10^{-32} \text{ J m}^{-3} \text{ s}}$

(b)   The frequency is

$$v = \frac{c}{\lambda} \text{ so } \frac{A}{B} = \frac{8\pi h}{\lambda^3} = \frac{8\pi (6.626 \times 10^{-34} \text{ J s})}{(3.0 \times 10^{-2} \text{ m })^3} = \boxed{6.2 \times 10^{-28} \text{ J m}^{-3} \text{ s}}$$

E13.2(b)   A source approaching an observer appears to be emitting light of frequency

$$v_{\text{approaching}} = \frac{v}{1 - \dfrac{s}{c}} \quad [13.15, \text{ Section 13.3}]$$

Since $v \propto \dfrac{1}{\lambda}$, $\lambda_{\text{obs}} = (1 - s/c)\,\lambda$

For the light to appear green the speed would have to be

$$s = \left(1 - \frac{\lambda_{\text{obs}}}{\lambda}\right) c = (2.998 \times 10^8 \text{ m s}^{-1}) \times \left(1 - \frac{520 \text{ nm}}{660 \text{ nm}}\right) = \boxed{6.36 \times 10^7 \text{ m s}^{-1}}$$

or about 1.4×10^8 m.p.h.

(Since $s \approx c$, the relativistic expression

$$v_{\text{obs}} = \left(\frac{1 + (s/c)}{1 - (s/c)}\right)^{1/2} v$$

should really be used. It gives $s = 7.02 \times 10^7$ m s^{-1}.)

E13.3(b)   The linewidth is related to the lifetime τ by

$$\delta \tilde{v} = \frac{5.31 \text{ cm}^{-1}}{\tau/\text{ps}} \quad [13.19] \text{ so } \tau = \frac{5.31 \text{ cm}^{-1}}{\delta \tilde{v}} \text{ ps}$$

(a)   We are given a frequency rather than a wavenumber

$$\tilde{v} = v/c \text{ so } \tau = \frac{(5.31 \text{ cm}^{-1}) \times (2.998 \times 10^{10} \text{ cm s}^{-1})}{100 \times 10^6 \text{ s}^{-1}} \text{ps} = 1.59 \times 10^3 \text{ ps}$$

or $\boxed{1.59 \text{ ns}}$

(b) $\tau = \dfrac{5.31 \text{ cm}^{-1}}{2.14 \text{ cm}^{-1}} \text{ps} = \boxed{2.48 \text{ ps}}$

E13.4(b)   The linewidth is related to the lifetime τ by

$$\delta \tilde{v} = \frac{5.31 \text{ cm}^{-1}}{\tau/\text{ps}} \text{ so } \delta v = \frac{(5.31 \text{ cm}^{-1})c}{\tau/\text{ps}}$$

(a) If every collision is effective, then the lifetime is $1/(1.0 \times 10^9 \text{ s}^{-1}) = 1.0 \times 10^{-9} \text{ s} = 1.0 \times 10^3$ ps

$$\delta\tilde{\nu} = \frac{(5.31 \text{ cm}^{-1}) \times (2.998 \times 10^{10} \text{ cm s}^{-1})}{1.0 \times 10^3} = 1.6 \times 10^8 \text{ s}^{-1} = \boxed{160 \text{ MHz}}$$

(b) If only one collision in 10 is effective, then the lifetime is a factor of 10 greater, 1.0×10^4 ps

$$\delta\tilde{\nu} = \frac{(5.31 \text{ cm}^{-1}) \times (2.998 \times 10^{10} \text{ cm s}^{-1})}{1.0 \times 10^4} = 1.6 \times 10^7 \text{ s}^{-1} = \boxed{16 \text{ MHz}}$$

E13.5(b)   The frequency of the transition is related to the rotational constant by

$$h\nu = \Delta E = hc\Delta F = hcB[J(J+1) - (J-1)J] = 2hcBJ$$

where J refers to the upper state ($J = 3$). The rotational constant is related to molecular structure by

$$B = \frac{\hbar}{4\pi cI} = \frac{\hbar}{4\pi c m_{\text{eff}} R^2}$$

where I is moment of inertia, m_{eff} is effective mass, and R is the bond length. Putting these expressions together yields

$$\nu = 2cBJ = \frac{\hbar J}{2\pi m_{\text{eff}} R^2}$$

The reciprocal of the effective mass is

$$m_{\text{eff}}^{-1} = m_{\text{C}}^{-1} + m_{\text{O}}^{-1} = \frac{(12 \text{ u})^{-1} + (15.9949 \text{ u})^{-1}}{1.660\,54 \times 10^{-27} \text{ kg u}^{-1}} = 8.783\,48 \times 10^{25} \text{ kg}^{-1}$$

So $\nu = \dfrac{(8.783\,48 \times 10^{25} \text{ kg}^{-1}) \times (1.0546 \times 10^{-34} \text{ J s}) \times (3)}{2\pi (112.81 \times 10^{-12} \text{m})^2} = \boxed{3.4754 \times 10^{11} \text{ s}^{-1}}$

E13.6(b)   (a) The wavenumber of the transition is related to the rotational constant by

$$hc\tilde{\nu} = \Delta E = hc\Delta F = hcB[J(J+1) - (J-1)J] = 2hcBJ$$

where J refers to the upper state ($J = 1$). The rotational constant is related to molecular structure by

$$B = \frac{\hbar}{4\pi cI}$$

where I is moment of inertia. Putting these expressions together yields

$$\tilde{\nu} = 2BJ = \frac{\hbar J}{2\pi cI} \quad \text{so} \quad I = \frac{\hbar J}{c\tilde{\nu}} = \frac{(1.0546 \times 10^{-34} \text{ J s}) \times (1)}{2\pi (2.998 \times 10^{10} \text{ cm s}^{-1}) \times (16.93 \text{ cm}^{-1})}$$

$$I = \boxed{3.307 \times 10^{-47} \text{ kg m}^2}$$

(b) The moment of inertia is related to the bond length by

$$I = m_{\text{eff}} R^2 \quad \text{so} \quad R = \left(\frac{I}{m_{\text{eff}}}\right)^{1/2}$$

$$m_{\text{eff}}^{-1} = m_{\text{H}}^{-1} + m_{\text{Br}}^{-1} = \frac{(1.0078 \text{ u})^{-1} + (80.9163 \text{ u})^{-1}}{1.660\,54 \times 10^{-27} \text{ kg u}^{-1}} = 6.0494 \times 10^{26} \text{ kg}^{-1}$$

and $R = \{(6.0494 \times 10^{26} \text{ kg}^{-1}) \times (3.307 \times 10^{-47} \text{ kg m}^2)\}^{1/2}$

$= 1.414 \times 10^{-10} \text{m} = \boxed{141.4 \text{ pm}}$

E13.7(b)   The wavenumber of the transition is related to the rotational constant by

$$hc\tilde{v} = \Delta E = hc\Delta F = hcB[J(J+1) - (J-1)J] = 2hcBJ$$

where J refers to the upper state. So wavenumbers of adjacent transitions (transitions whose upper states differ by 1) differ by

$$\Delta\tilde{v} = 2B = \frac{\hbar}{2\pi cI} \quad \text{so } I = \frac{\hbar}{2\pi c\Delta\tilde{v}}$$

where I is moment of inertia, m_{eff} is effective mass, and R is the bond length.

So $I = \dfrac{(1.0546 \times 10^{-34} \text{ J s})}{2\pi(2.9979 \times 10^{10} \text{ cm s}^{-1}) \times (1.033 \text{ cm}^{-1})} = \boxed{5.420 \times 10^{-46} \text{ kg m}^2}$

The moment of inertia is related to the bond length by

$$I = m_{\text{eff}}R^2 \quad \text{so } R = \left(\frac{I}{m_{\text{eff}}}\right)^{1/2}$$

$$m_{\text{eff}}^{-1} = m_{\text{F}}^{-1} + m_{\text{Cl}}^{-1} = \frac{(18.9984 \text{ u})^{-1} + (34.9688 \text{ u})^{-1}}{1.660\,54 \times 10^{-27} \text{ kg u}^{-1}} = 4.891\,96 \times 10^{25} \text{ kg}^{-1}$$

and $R = \{(4.891\,96 \times 10^{25} \text{ kg}^{-1}) \times (5.420 \times 10^{-46} \text{ kg m}^2)\}^{1/2}$

$= 1.628 \times 10^{-10} \text{ m} = \boxed{162.8 \text{ pm}}$

E13.8(b)   The rotational constant is

$$B = \frac{\hbar}{4\pi cI} = \frac{\hbar}{4\pi c(2m_0R^2)} \quad \text{so } R = \left(\frac{\hbar}{8\pi cm_0B}\right)^{1/2}$$

where I is moment of inertia, m_{eff} is effective mass, and R is the bond length.

$$R = \left(\frac{(1.0546 \times 10^{-34} \text{ J s})}{8\pi(2.9979 \times 10^{10} \text{ cm s}^{-1}) \times (15.9949 \text{ u}) \times (1.660\,54 \times 10^{-27} \text{ kg u}^{-1})(0.390\,21)}\right)^{1/2}$$

$= 1.1621 \times 10^{-10} \text{ m} = \boxed{116.21 \text{ pm}}$

E13.9(b) This exercise is analogous to Exercise 13.9(a), but here our solution will employ a slightly different algebraic technique. Let $R = R_{OC}$, $R' = R_{CS}$, $O = {}^{16}O$, $C = {}^{12}C$.

$$I = \frac{\hbar}{4\pi B} \quad [\textit{Comment } 13.4]$$

$$I(OC^{32}S) = \frac{1.054\,57 \times 10^{-34}\,\text{J s}}{(4\pi) \times (6.0815 \times 10^9\,\text{s}^{-1})} = 1.3799 \times 10^{-45}\,\text{kg m}^2 = 8.3101 \times 10^{-19}\,\text{u m}^2$$

$$I(OC^{34}S) = \frac{1.054\,57 \times 10^{-34}\,\text{J s}}{(4\pi) \times (5.9328 \times 10^9\,\text{s}^{-1})} = 1.4145 \times 10^{-45}\,\text{kg m}^2 = 8.5184 \times 10^{-19}\,\text{u m}^2$$

The expression for the moment of inertia given in Table 13.1 may be rearranged as follows.

$$Im = m_A m R^2 + m_C m R'^2 - (m_A R - m_C R')^2$$

$$= m_A m R^2 + m_C m R'^2 - m_A^2 R^2 + 2m_A m_C R R' - m_C^2 R'^2$$

$$= m_A (m_B + m_C) R^2 + m_C (m_A + m_B) R'^2 + 2m_A m_C R R'$$

Let $m_C = m_{32_S}$ and $m_C' = m_{34_S}$

$$\frac{Im}{m_C} = \frac{m_A}{m_C}(m_B + m_C)R^2 + (m_A + m_B)R'^2 + 2m_A R R' \tag{a}$$

$$\frac{I'm'}{m_C'} = \frac{m_A}{m_C'}(m_B + m_C')R^2 + (m_A + m_B)R'^2 + 2m_A R R' \tag{b}$$

Subtracting

$$\frac{Im}{m_C} - \frac{I'm'}{m_C'} = \left[\left(\frac{m_A}{m_C}\right)(m_B + m_C) - \left(\frac{m_A}{m_C'}\right)(m_B + m_C')\right]R^2$$

Solving for R^2

$$R^2 = \frac{\left(\frac{Im}{m_C} - \frac{I'm'}{m_C'}\right)}{\left[\left(\frac{m_A}{m_C}\right)(m_B + m_C) - \left(\frac{m_A}{m_C'}\right)(m_B + m_C')\right]} = \frac{m_C' Im - m_C I'm'}{m_B m_A (m_C' - m_C)}$$

Substituting the masses, with $m_A = m_O$, $m_B = m_C$, $m_C = m_{32_S}$, and $m_C' = m_{34_S}$

$$m = (15.9949 + 12.0000 + 31.9721)\,\text{u} = 59.9670\,\text{u}$$

$$m' = (15.9949 + 12.0000 + 33.9679)\,\text{u} = 61.9628\,\text{u}$$

$$R^2 = \frac{(33.9679\,\text{u}) \times (8.3101 \times 10^{-19}\,\text{u m}^2) \times (59.9670\,\text{u})}{(12.0000\,\text{u}) \times (15.9949\,\text{u}) \times (33.9679\,\text{u} - 31.9721\,\text{u})}$$

$$\quad - \frac{(33.9721\,\text{u}) \times (8.5184 \times 10^{-19}\,\text{u m}^2) \times (61.9628\,\text{u})}{(12.0000\,\text{u}) \times (15.9949\,\text{u}) \times (33.9679\,\text{u} - 31.9721\,\text{u})}$$

$$= \frac{51.6446 \times 10^{-19}\,\text{m}^2)}{383.071} = 1.3482 \times 10^{-20}\,\text{m}^2$$

$$R = 1.161\bar{1} \times 10^{-10}\,\text{m} = \boxed{116.1\,\text{pm}} = R_{OC}$$

Because the numerator of the expression for R^2 involves the difference between two rather large numbers of nearly the same magnitude, the number of significant figures in the answer for R is certainly no greater than 4. Having solved for R, either equation (a) or (b) above can be solved for R'. The result is

$$R' = 1.559 \times 10^{-10} \text{ m} = \boxed{155.9 \text{ pm}} = R_{CS}$$

E13.10(b) The wavenumber of a Stokes line in rotational Raman is

$$\tilde{v}_{\text{Stokes}} = \tilde{v}_i - 2B(2J + 3) \text{ [13.42a]}$$

where J is the initial (lower) rotational state. So

$$\tilde{v}_{\text{Stokes}} = 20\,623 \text{ cm}^{-1} - 2(1.4457 \text{ cm}^{-1}) \times [2(2) + 3] = \boxed{20\,603 \text{ cm}^{-1}}$$

E13.11(b) The separation of lines is $4B$, so $B = \frac{1}{4} \times (3.5312 \text{ cm}^{-1}) = 0.882\,80 \text{ cm}^{-1}$

Then we use $R = \left(\dfrac{\hbar}{4\pi m_{\text{eff}} cB} \right)^{1/2}$ [Exercise 13.8(a)]

with $m_{\text{eff}} = \frac{1}{2} m(^{19}\text{F}) = \frac{1}{2} \times (18.9984 \text{ u}) \times (1.6605 \times 10^{-27} \text{ kg u}^{-1}) = 1.577\,34\bar{2} \times 10^{-26} \text{ kg}$

$$R = \left(\frac{1.0546 \times 10^{-34} \text{ J s}}{4\pi(1.577\,342 \times 10^{-26} \text{ kg}) \times (2.998 \times 10^{10} \text{ cm s}^{-1}) \times (0.882\,80 \text{ cm}^{-1})} \right)^{1/2}$$

$$= 1.417\,8\bar{5} \times 10^{-10} \text{ m} = \boxed{141.78 \text{ pm}}$$

E13.12(b) Polar molecules show a pure rotational absorption spectrum. Therefore, select the polar molecules based on their well-known structures. Alternatively, determine the point groups of the molecules and use the rule that only molecules belonging to C_n, C_{nv}, and C_s may be polar, and in the case of C_n and C_{nv}, that dipole must lie along the rotation axis. Hence all are polar molecules.

Their point group symmetries are

(a) H_2O, C_{2v}, **(b)** H_2O_2, C_2, **(c)** NH_3, C_{3v}, **(d)** N_2O, $C_{\infty v}$

$\boxed{\text{All}}$ show a pure rotational spectrum.

E13.13(b) A molecule must be anisotropically polarizable to show a rotational Raman spectrum; all molecules except spherical rotors have this property. So $\boxed{CH_2Cl_2}$, $\boxed{CH_3CH_3}$, and $\boxed{N_2O}$ can display rotational Raman spectra; SF_6 cannot.

E13.14(b) The angular frequency is

$$\omega = \left(\frac{k}{m} \right)^{1/2} = 2\pi v \quad \text{so} \quad k = (2\pi v)^2 m = (2\pi)^2 \times (3.0 \text{ s}^{-1})^2 \times (2.0 \times 10^{-3} \text{ kg})$$

$$k = \boxed{0.71 \text{ N m}^{-1}}$$

E13.15(b) $\qquad \omega = \left(\dfrac{k}{m_{\text{eff}}} \right)^{1/2} \qquad \omega' = \left(\dfrac{k}{m'_{\text{eff}}} \right)^{1/2}$ [prime $= {}^2\text{H}^{37}\text{Cl}$]

The force constant, k, is assumed to be the same for both molecules. The fractional difference is

$$\frac{\omega' - \omega}{\omega} = \frac{\left(\dfrac{k}{m'_{\text{eff}}}\right)^{1/2} - \left(\dfrac{k}{m_{\text{eff}}}\right)^{1/2}}{\left(\dfrac{k}{m_{\text{eff}}}\right)^{1/2}} = \frac{\left(\dfrac{1}{m'_{\text{eff}}}\right)^{1/2} - \left(\dfrac{1}{m_{\text{eff}}}\right)^{1/2}}{\left(\dfrac{1}{m_{\text{eff}}}\right)^{1/2}} = \left(\frac{m_{\text{eff}}}{m'_{\text{eff}}}\right)^{1/2} - 1$$

$$\frac{\omega' - \omega}{\omega} = \left(\frac{m_{\text{eff}}}{m'_{\text{eff}}}\right)^{1/2} - 1 = \left\{ \frac{m_{\text{H}}m_{\text{Cl}}}{m_{\text{H}} + m_{\text{Cl}}} \times \frac{(m_{2\text{H}} + m_{37\text{Cl}})}{(m_{2\text{H}} \times m_{37\text{Cl}})} \right\}^{1/2} - 1$$

$$= \left\{ \frac{(1.0078 \text{ u}) \times (34.9688 \text{ u})}{(1.0078 \text{ u}) + (34.9688 \text{ u})} \times \frac{(2.0140 \text{ u}) + (36.9651 \text{ u})}{(2.0140 \text{ u}) \times (36.9651 \text{ u})} \right\}^{1/2} - 1$$

$$= -0.284$$

Thus the difference is $\boxed{28.4 \text{ percent}}$

E13.16(b) The fundamental vibrational frequency is

$$\omega = \left(\frac{k}{m_{\text{eff}}}\right)^{1/2} = 2\pi\nu = 2\pi c\tilde{\nu} \quad \text{so} \quad k = (2\pi c\tilde{\nu})^2 m_{\text{eff}}$$

We need the effective mass

$$m_{\text{eff}}^{-1} = m_1^{-1} + m_2^{-1} = (78.9183 \text{ u})^{-1} + (80.9163 \text{ u})^{-1} = 0.025\,029\,8 \text{ u}^{-1}$$

$$k = \frac{[2\pi(2.998 \times 10^{10} \text{ cm s}^{-1}) \times (323.2 \text{ cm}^{-1})]^2 \times (1.66054 \times 10^{-27} \text{ kg u}^{-1})}{0.025\,029\,8 \text{ u}^{-1}}$$

$$= \boxed{245.9 \text{ N m}^{-1}}$$

E13.17(b) The ratio of the population of the ground state (N_0) to the first excited state (N_1) is

$$\frac{N_0}{N_1} = \exp\left(\frac{-h\nu}{kT}\right) = \exp\left(\frac{-hc\tilde{\nu}}{kT}\right)$$

(a) $\dfrac{N_0}{N_1} = \exp\left(\dfrac{-(6.626 \times 10^{-34} \text{ J s}) \times (2.998 \times 10^{10} \text{ cm s}^{-1}) \times (321 \text{ cm}^{-1})}{(1.381 \times 10^{-23} \text{ J K}^{-1}) \times (298 \text{ K})}\right) = \boxed{0.212}$

(b) $\dfrac{N_0}{N_1} = \exp\left(\dfrac{-(6.626 \times 10^{-34} \text{ J s}) \times (2.998 \times 10^{10} \text{ cm s}^{-1}) \times (321 \text{ cm}^{-1})}{(1.381 \times 10^{-23} \text{ J K}^{-1}) \times (800 \text{ K})}\right) = \boxed{0.561}$

E13.18(b) The relation between vibrational frequency and wavenumber is

$$\omega = \left(\frac{k}{m_{\text{eff}}}\right)^{1/2} = 2\pi\nu = 2\pi c\tilde{\nu} \quad \text{so} \quad \tilde{\nu} = \frac{1}{2\pi c}\left(\frac{k}{m_{\text{eff}}}\right)^{1/2} = \frac{(km_{\text{eff}}^{-1})^{1/2}}{2\pi c}$$

The reduced masses of the hydrogen halides are very similar, but not identical

$$m_{\text{eff}}^{-1} = m_{\text{D}}^{-1} + m_{\text{X}}^{-1}$$

We assume that the force constants as calculated in Exercise 13.18(a) are identical for the deuterium halide and the hydrogen halide.

For DF

$$m_{\text{eff}}^{-1} = \frac{(2.0140 \text{ u})^{-1} + (18.9984 \text{ u})^{-1}}{1.66054 \times 10^{-27} \text{ kg u}^{-1}} = 3.3071 \times 10^{26} \text{ kg}^{-1}$$

$$\tilde{\nu} = \frac{\{(3.3071 \times 10^{26} \text{ kg}^{-1}) \times (967.04 \text{ kg s}^{-2})\}^{1/2}}{2\pi(2.9979 \times 10^{10} \text{ cm s}^{-1})} = \boxed{3002.3 \text{ cm}^{-1}}$$

For DCl

$$m_{\text{eff}}^{-1} = \frac{(2.0140 \text{ u})^{-1} + (34.9688 \text{ u})^{-1}}{1.66054 \times 10^{-27} \text{ kg u}^{-1}} = 3.1624 \times 10^{26} \text{ kg}^{-1}$$

$$\tilde{\nu} = \frac{\{(3.1624 \times 10^{26} \text{ kg}^{-1}) \times (515.59 \text{ kg s}^{-2})\}^{1/2}}{2\pi(2.9979 \times 10^{10} \text{ cm s}^{-1})} = \boxed{2143.7 \text{ cm}^{-1}}$$

For DBr

$$m_{\text{eff}}^{-1} = \frac{(2.0140 \text{ u})^{-1} + (80.9163 \text{ u})^{-1}}{1.66054 \times 10^{-27} \text{ kg u}^{-1}} = 3.0646 \times 10^{26} \text{ kg}^{-1}$$

$$\tilde{\nu} = \frac{\{(3.0646 \times 10^{26} \text{ kg}^{-1}) \times (411.75 \text{ kg s}^{-2})\}^{1/2}}{2\pi(2.9979 \times 10^{10} \text{ cm s}^{-1})} = \boxed{1885.8 \text{ cm}^{-1}}$$

For DI

$$m_{\text{eff}}^{-1} = \frac{(2.0140 \text{ u})^{-1} + (126.9045 \text{ u})^{-1}}{1.66054 \times 10^{-27} \text{ kg u}^{-1}} = 3.0376 \times 10^{26} \text{ kg}^{-1}$$

$$\tilde{\nu} = \frac{\{(3.0376 \times 10^{26} \text{ kg}^{-1}) \times (314.21 \text{ kg s}^{-2})\}^{1/2}}{2\pi(2.9979 \times 10^{10} \text{ cm s}^{-1})} = \boxed{1640.1 \text{ cm}^{-1}}$$

E13.19(b) Data on three transitions are provided. Only two are necessary to obtain the value of $\tilde{\nu}$ and x_e. The third datum can then be used to check the accuracy of the calculated values.

$$\Delta G(\nu = 1 \leftarrow 0) = \tilde{\nu} - 2\tilde{\nu}x_e = 2345.15 \text{ cm}^{-1} \text{ [13.57]}$$

$$\Delta G(\nu = 2 \leftarrow 0) = 2\tilde{\nu} - 6\tilde{\nu}x_e = 4661.40 \text{ cm}^{-1} \text{ [13.58]}$$

Multiply the first equation by 3, then subtract the second.

$$\tilde{\nu} = (3) \times (2345.15 \text{ cm}^{-1}) - (4661.40 \text{ cm}^{-1}) = \boxed{2374.05 \text{ cm}^{-1}}$$

Then from the first equation

$$x_e = \frac{\tilde{\nu} - 2345.15 \text{ cm}^{-1}}{2\tilde{\nu}} = \frac{(2374.05 - 2345.15)\text{cm}^{-1}}{(2) \times (2374.05 \text{ cm}^{-1})} = \boxed{6.087 \times 10^{-3}}$$

x_e data are usually reported as $x_e\tilde{\nu}$ which is

$$x_e\tilde{\nu} = 14.45 \text{ cm}^{-1}$$

$$\Delta G(\nu = 3 \leftarrow 0) = 3\tilde{\nu} - 12\nu x_e = (3) \times (2374.05 \text{ cm}^{-1}) - (12) \times (14.45 \text{ cm}^{-1})$$
$$= 6948.74 \text{ cm}^{-1}$$

which is close to the experimental value.

E13.20(b) $\Delta G_{\nu+1/2} = \tilde{\nu} - 2(\nu + 1)x_e\tilde{\nu}$ [13.57]   where $\Delta G_{\nu+1/2} = G(\nu + 1) - G(\nu)$

Therefore, since

$$\Delta G_{\nu+1/2} = (1 - 2x_e)\tilde{\nu} - 2\nu x_e\tilde{\nu}$$

a plot of $\Delta G_{\nu+1/2}$ against ν should give a straight line which gives $(1 - 2x_e)\tilde{\nu}$ from the intercept at $\nu = 0$ and $-2x_e\tilde{\nu}$ from the slope. We draw up the following table

ν	0	1	2	3	4
$G(\nu)/\text{cm}^{-1}$	1144.83	3374.90	5525.51	7596.66	9588.35
$\Delta G_{\nu+1/2}/\text{cm}^{-1}$	2230.07	2150.61	2071.15	1991.69	

The points are plotted in Figure 13.1.

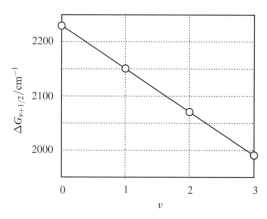

Figure 13.1

The intercept lies at 2230.51 and the slope $= -76.65 \text{ cm}^{-1}$; hence $x_e\tilde{\nu} = 39.83 \text{ cm}^{-1}$.

Since $\tilde{\nu} - 2x_e\tilde{\nu} = 2230.51 \text{ cm}^{-1}$ it follows that $\tilde{\nu} = 2310.16 \text{ cm}^{-1}$

The dissociation energy may be obtained by assuming that a Morse potential describes the molecule and that the constant D_e in the expression for the potential is an adequate first approximation for it. Then

$$D_e = \frac{\tilde{\nu}}{4x_e} \text{ [13.55]} = \frac{\tilde{\nu}^2}{4x_e\tilde{\nu}} = \frac{(2310.16 \text{ cm}^{-1})^2}{(4) \times (39.83 \text{ cm}^{-1})} = 33.50 \times 10^3 \text{ cm}^{-1} = 4.15 \text{ eV}$$

However, the depth of the potential well D_e differs from D_0, the dissociation energy of the bond, by the zero-point energy; hence

$$D_0 = D_e - \frac{1}{2}\tilde{\nu} = (33.50 \times 10^3 \text{ cm}^{-1}) - \left(\frac{1}{2}\right) \times (2310.16 \text{ cm}^{-1})$$

$$= \boxed{3.235 \times 10^4 \text{ cm}^{-1}} = \boxed{4.01 \text{ eV}}$$

E13.21(b) The wavenumber of an R-branch IR transition is

$$\tilde{\nu}_R = \tilde{\nu} + 2B(J + 1) \text{ [13.62c]}$$

where J is the initial (lower) rotational state. So

$$\tilde{\nu}_R = 2308.09 \text{ cm}^{-1} + 2(6.511 \text{ cm}^{-1}) \times (2 + 1) = \boxed{2347.16 \text{ cm}^{-1}}$$

E13.22(b) See Section 13.10. Select those molecules in which a vibration gives rise to a change in dipole moment. It is helpful to write down the structural formulas of the compounds. The infrared active compounds are

(a) CH_3CH_3; **(b)** CH_4; **(c)** $CH_3 Cl$

COMMENT. A more powerful method for determining infrared activity based on symmetry considerations is described in Section 13.15.

E13.23(b) A nonlinear molecule has $3N - 6$ normal modes of vibration, where N is the number of atoms in the molecule; a linear molecule has $3N - 5$.

(a) C_6H_6 has $3(12) - 6 = \boxed{30}$ normal modes.
(b) $C_6H_6CH_3$ has $3(16) - 6 = \boxed{42}$ normal modes.
(c) $HC\equiv C-C\equiv CH$ is linear; it has $3(6) - 5 = \boxed{13}$ normal modes.

E13.24(b) **(a)** A planar AB_3 molecule belongs to the D_{3h} group. Its four atoms have a total of 12 displacements, of which 6 are vibrations. We determine the symmetry species of the vibrations by first determining the characters of the reducible representation of the molecule formed from all 12 displacements and then subtracting from these characters the characters corresponding to translation and rotation. This latter information is directly available in the character table for the group D_{3h}. The resulting set of characters are the characters of the reducible representation of the vibrations. This representation can be reduced to the symmetry species of the vibrations by inspection or by use of the little orthogonality theorem.

D_{3h}	E	σ_h	$2C_3$	$2S_3$	$3C_2'$	$3\sigma_v$
χ (translation)	3	1	0	−2	−1	1
Unmoved atoms	4	4	1	1	2	2
χ (total, product)	12	4	0	−2	−2	2
χ (rotation)	3	−1	0	2	−1	−1
χ (vibration)	6	4	0	−2	0	2

χ (vibration) corresponds to $A'_1 + A''_2 + 2E'$.

Again referring to the character table of D_{3h}, we see that E' corresponds to x and y, A''_2 to z; hence $\boxed{A''_2 \text{ and } E' \text{ are IR active}}$. We also see from the character table that E' and A'_1 correspond to the quadratic terms; hence $\boxed{A'_1 \text{ and } E' \text{ are Raman active}}$.

(b) A trigonal pyramidal AB_3 molecule belongs to the group C_{3v}. In a manner similar to the analysis in part (a) we obtain

C_{3v}	E	$2C_3$	$3\sigma_v$
χ (total)	12	0	2
χ (vibration)	6	−2	2

χ (vibration) corresponds to $2A_1 + 2E$. We see from the character table that $\boxed{A_1 \text{ and } E}$ are IR active and that $\boxed{A_1 + E}$ are also Raman active. Thus all modes are observable in both the IR and the Raman spectra.

E13.25(b) **(b)** The boat-like bending of a benzene ring clearly changes the dipole moment of the ring, for the moving of the C—H bonds out of the plane will give rise to a non-cancelling component of their dipole moments. So the vibration is $\boxed{\text{IR active}}$.

(a) Since benzene has a centre of inversion, the exclusion rule applies: a mode which is IR active (such as this one) must be $\boxed{\text{Raman inactive}}$.

E13.26(b) The displacements span $A_{1g} + A_{1u} + A_{2g} + 2E_{1u} + E_{1g}$. The rotations R_x and R_y span E_{1g}, and the translations span $E_{1u} + A_{1u}$. So the vibrations span $\boxed{A_{1g} + A_{2g} + E_{1u}}$

Solutions to problems

Solutions to numerical problems

P13.1 Use the energy density expression in terms of wavelengths (eqn 8.5)

$$\mathcal{E} = \rho d\lambda \quad \text{where } \rho = \frac{8\pi hc}{\lambda^5 (e^{hc/\lambda kT} - 1)}.$$

Evaluate

$$\mathcal{E} = \int_{400 \times 10^{-9}\,\text{m}}^{700 \times 10^{-9}\,\text{m}} \frac{8\pi hc}{\lambda^5 (e^{hc/\lambda kT} - 1)} d\lambda$$

at three different temperatures. Compare those results to the classical, Rayleigh–Jeans expression (eqn 8.3).

$$\mathcal{E}_{\text{class}} = \rho_{\text{class}} \, d\lambda \quad \text{where } \rho_{\text{class}} = \frac{8\pi kT}{\lambda^4},$$

so $$\mathcal{E}_{\text{class}} = \int_{400 \times 10^{-9}\,\text{m}}^{700 \times 10^{-9}\,\text{m}} \frac{8\pi kT}{\lambda^4} d\lambda = -\frac{8\pi kT}{3\lambda^3} \Big|_{400 \times 10^{-9}\,\text{m}}^{700 \times 10^{-9}\,\text{m}}.$$

T/K	$\mathcal{E}/\mathrm{J\,m^{-3}}$	$\mathcal{E}_{class}/\mathrm{J\,m^{-3}}$
(a) 1500	2.136×10^{-6}	2.206
(b) 2500	9.884×10^{-4}	3.676
(c) 5800	3.151×10^{-1}	8.528

The classical values are very different from the accurate Planck values! Try integrating the expressions over $400 - 700\ \mu\mathrm{m}$ or mm to see that the expressions agree reasonably well at longer wavelengths.

P13.3   On the assumption that every collision deactivates the molecule we may write

$$\tau = \frac{1}{z} = \boxed{\frac{kT}{4\sigma p}\left(\frac{\pi m}{kT}\right)^{1/2}}.$$

For HCl, with $m \approx 36\ \mathrm{u}$,

$$\tau \approx \left(\frac{(1.381 \times 10^{-23}\ \mathrm{J\,K^{-1}}) \times (298\ \mathrm{K})}{(4) \times (0.30 \times 10^{-18}\ \mathrm{m^2}) \times (1.013 \times 10^5\ \mathrm{Pa})}\right)$$

$$\times \left(\frac{\pi \times (36) \times (1.661 \times 10^{-27}\ \mathrm{kg})}{(1.381 \times 10^{-23}\ \mathrm{J\,K^{-1}}) \times (298\ \mathrm{K})}\right)^{1/2}$$

$$\approx 2.3 \times 10^{-10}\ \mathrm{s}.$$

$$\delta E \approx h\delta v = \frac{\hbar}{\tau}\ [13.18].$$

The width of the collision-broadened line is therefore approximately

$$\delta v \approx \frac{1}{2\pi\tau} = \frac{1}{(2\pi) \times (2.3 \times 10^{-10}\ \mathrm{s})} \approx \boxed{700\ \mathrm{MHz}}.$$

The Doppler width is approximately 1.3 MHz (Problem 13.2). Since the collision width is proportional to $p\,[\delta v \propto 1/\tau$ and $\tau \propto 1/p]$, the pressure must be reduced by a factor of about $\frac{1.3}{700} = 0.002$ before Doppler broadening begins to dominate collision broadening. Hence, the pressure must be reduced to below $(0.002) \times (760\ \mathrm{Torr}) = \boxed{1\ \mathrm{Torr}}$.

P13.5 $$B = \frac{\hbar}{4\pi cI}\ [13.24]; \quad I = m_{\mathrm{eff}}R^2; \quad R^2 = \frac{\hbar}{4\pi cm_{\mathrm{eff}}B}.$$

$$m_{\mathrm{eff}} = \frac{m_C m_O}{m_C + m_O} = \left(\frac{(12.0000\ \mathrm{u}) \times (15.9949\ \mathrm{u})}{(12.0000\ \mathrm{u}) + (15.9949\ \mathrm{u})}\right) \times (1.66054 \times 10^{-27}\ \mathrm{kg\,u^{-1}})$$

$$= 1.13852 \times 10^{-26}\ \mathrm{kg}.$$

$$\frac{\hbar}{4\pi c} = 2.79932 \times 10^{-44}\ \mathrm{kg\,m}.$$

$$R_0^2 = \frac{2.79932 \times 10^{-44}\ \mathrm{kg\,m}}{(1.13852 \times 10^{-26}\ \mathrm{kg}) \times (1.9314 \times 10^2\ \mathrm{m^{-1}})} = 1.2730\overline{3} \times 10^{-20}\ \mathrm{m^2},$$

$$R_0 = 1.1283 \times 10^{-10} \, \text{m} = \boxed{112.83 \, \text{pm}}.$$

$$R_1^2 = \frac{2.79932 \times 10^{-44} \, \text{kg m}}{(1.13852 \times 10^{-26} \, \text{kg}) \times (1.6116 \times 10^2 \, \text{m}^{-1})} = 1.52565 \times 10^{-20} \, \text{m}^2,$$

$$R_1 = 1.2352 \times 10^{-10} \, \text{m} = \boxed{123.52 \, \text{pm}}.$$

COMMENT.   The change in internuclear distance is roughly 10 per cent, indicating that the rotations and vibrations of molecules are strongly coupled and that it is an oversimplification to consider them independently of each other.

P13.7   The separations between neighboring lines are

20.81, 20.60, 20.64, 20.52, 20.34, 20.37, 20.26   mean: 20.51 cm^{-1}.

Hence $B = \left(\frac{1}{2}\right) \times (20.51 \, \text{cm}^{-1}) = 10.26 \, \text{cm}^{-1}$ and

$$I = \frac{\hbar}{4\pi cB} = \frac{1.05457 \times 10^{-34} \, \text{J s}}{(4\pi) \times (2.99793 \times 10^{10} \, \text{cm s}^{-1} \times (10.26 \, \text{cm}^{-1})} = \boxed{2.728 \times 10^{-47} \, \text{kg m}^2}.$$

$$R = \left(\frac{I}{m_{\text{eff}}}\right)^{1/2} \quad \text{[Table 13.1]} \quad \text{with } m_{\text{eff}} = 1.6266 \times 10^{-27} \, \text{kg [Exercise 13.6(a)]}$$

$$= \left(\frac{2.728 \times 10^{-47} \, \text{kg m}^2}{1.6266 \times 10^{-27} \, \text{kg}}\right)^{1/2} = \boxed{129.5 \, \text{pm}}.$$

COMMENT.   Ascribing the variation of the separations to centrifugal distortion, and not just taking a simple average would result in a more accurate value. Alternatively, the effect of centrifugal distortion could be minimized by plotting the observed separations against J, fitting them to a smooth curve, and extrapolating that curve to $J = 0$. Since $B \propto \dfrac{1}{I}$ and $I \propto m_{\text{eff}}$, $B \propto \dfrac{1}{m_{\text{eff}}}$. Hence, the corresponding lines in $^2\text{H}^{35}\text{Cl}$ will lie at a factor

$$\frac{m_{\text{eff}}(^1\text{H}^{35}\text{Cl})}{m_{\text{eff}}(^2\text{H}^{35}\text{Cl})} = \frac{1.6266}{3.1624} = 0.5144$$

to low frequency of $^1\text{H}^{35}\text{Cl}$ lines. Hence, we expect lines at $\boxed{10.56, \, 21.11, \, 31.67, \ldots \, \text{cm}^{-1}}$.

P13.9 $$R = \left(\frac{\hbar}{4\pi\mu cB}\right)^{1/2} \quad \text{and} \quad \nu = 2cB(J+1) \quad \text{[13.37, with } \nu = c\tilde{\nu}].$$

We use $\mu(\text{CuBr}) \approx \dfrac{(63.55) \times (79.91)}{(63.55) + (79.91)} \text{u} = 35.40 \, \text{u}$

and draw up the following table.

J	13	14	15	
ν/MHz	84421.34	90449.25	96476.72	$\left[B = \dfrac{\nu}{2c(J+1)}\right]$
B/cm^{-1}	0.10057	0.10057	0.10057	

Hence, $R = \left(\dfrac{1.05457 \times 10^{-34}\,\text{J s}}{(4\pi) \times (35.40) \times (1.6605 \times 10^{-27}\,\text{kg}) \times (2.9979 \times 10^{10}\,\text{cm s}^{-1}) \times (0.10057\,\text{cm}^{-1})} \right)^{1/2}$

$= \boxed{218\,\text{pm}}$.

P13.11   Plot frequency against J as in Fig. 13.2.

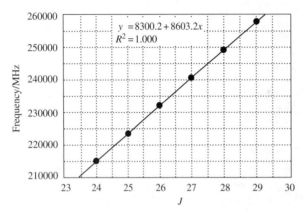

Figure 13.2

The rotational constant is related to the wavenumbers of observed transitions by

$$\tilde{v} = 2B(J + 1) = \frac{v}{c} \quad \text{so} \quad v = 2Bc(J + 1).$$

A plot of v versus J, then, has a slope of $2Bc$. From Fig.13.3, the slope is 8603 MHz, so

$$B = \frac{8603 \times 10^{6}\,\text{s}^{-1}}{2(2.988 \times 10^{8}\,\text{ms}^{-1})} = \boxed{14.35\ \text{m}^{-1}}.$$

The most highly populated energy level is roughly

$$J_{max} \left(\frac{kT}{2hcB} \right)^{1/2} - \frac{1}{2}$$

so $J_{max} = \left(\dfrac{(1.381 \times 10^{-23}\,\text{J K}^{-1}) \times (298\text{K})}{(6.626 \times 10^{-34}\,\text{J s}) \times (8603 \times 10^{6}\,\text{s}^{-1})} \right)^{1/2} - \dfrac{1}{2} = \boxed{26}$ at 298 K

and $J_{max} = \left(\dfrac{(1.381 \times 10^{-23}\,\text{J K}^{-1}) \times (100\,\text{K})}{(6.626 \times 10^{-34}\,\text{J s}) \times (8603 \times 10^{6}\,\text{s}^{-1})} \right)^{1/2} - \dfrac{1}{2} = \boxed{15}$ at 100 K.

P13.13   The Lewis structure is

$$[\ddot{\text{O}}{=}\text{N}{=}\ddot{\text{O}}]^{+}.$$

VSEPR indicates that the ion is $\boxed{\text{linear}}$ and has a center of symmetry. The activity of the modes is consistent with the rule of mutual exclusion; none is both infrared and Raman active. These transitions may be compared to those for CO_2 (Fig. 13.40 of the text) and are consistent with them. The Raman active mode at 1400 cm^{-1} is due to a symmetric stretch ($\tilde{\nu}_1$), that at 2360 cm^{-1} to the antisymmetric stretch ($\tilde{\nu}_3$), and that at 540 cm^{-1} to the two perpendicular bending modes ($\tilde{\nu}_2$). There is a combination band, $\tilde{\nu}_1 + \tilde{\nu}_3 = 3760$ cm$^{-1} \approx 3735$ cm^{-1}, which shows a weak intensity in the infrared.

P13.15 $D_0 = D_e - \tilde{\nu}'$ with $\tilde{\nu}' = \frac{1}{2}\tilde{\nu} - \frac{1}{4}x_e\tilde{\nu}$ [Section 13.11].

(a) ^{1}HCl : $\tilde{\nu}' = \left\{(1494.9) - \left(\frac{1}{4}\right) \times (52.05)\right\}cm^{-1} = 1481.8$ cm^{-1}, or 0.184 eV.

Hence, $D_0 = 5.33 - 0.18 = \boxed{5.15\,\text{eV}}$.

(b) ^{2}HCl: $\dfrac{2m_{\text{eff}}\omega x_e}{\hbar} = a^2$ [13.55], so $\tilde{\nu}x_e \propto \dfrac{1}{m_{\text{eff}}}$ as a is a constant. We also have $D_e = \dfrac{\tilde{\nu}^2}{4x_e\tilde{\nu}}$ [Exercise 13.20(a)], so $\tilde{\nu}^2 \propto \dfrac{1}{m_{\text{eff}}}$, implying $\tilde{\nu} \propto \dfrac{1}{m_{\text{eff}}^{1/2}}$. Reduced masses were calculated in Exercises 13.18(a) and 13.18(b), and we can write

$$\tilde{\nu}(^2\text{HCl}) = \left(\frac{m_{\text{eff}}(^1\text{HCl})}{m_{\text{eff}}(^2\text{HCl})}\right)^{1/2} \times \tilde{\nu}(^1\text{HCl}) = (0.7172) \times (2989.7\,\text{cm}^{-1}) = 2144.2\,\text{cm}^{-1},$$

$$x_e\tilde{\nu}(^2\text{HCl}) = \left(\frac{m_{\text{eff}}(^1\text{HCl})}{m_{\text{eff}}(^2\text{HCl})}\right) \times x_e\tilde{\nu}(^1\text{HCl}) = (0.5144) \times (52.05\,\text{cm}^{-1}) = 26.77\,\text{cm}^{-1},$$

$$\tilde{\nu}'(^2\text{HCl}) = \left(\frac{1}{2}\right) \times (2144.2) - \left(\frac{1}{4}\right) \times (26.77\,\text{cm}^{-1}) = 1065.4\,\text{cm}^{-1},\ 0.132\,\text{eV}.$$

Hence, $D_0(^2\text{HCl}) = (5.33 - 0.132)\,\text{eV} = \boxed{5.20\,\text{eV}}$.

P13.17   **(a)** In the harmonic approximation

$$D_e = D_0 + \frac{1}{2}\tilde{\nu} \text{ so } \tilde{\nu} = 2(D_e - D_0).$$

$$\tilde{\nu} = \frac{2(1.51 \times 10^{-23}\,\text{J} - 2 \times 10^{-26}\,\text{J})}{(6.626 \times 10^{-34}\,\text{J s}) \times (2.998 \times 10^8\,\text{m s}^{-1})} = \boxed{152\,\text{m}^{-1}}.$$

The force constant is related to the vibrational frequency by

$$\omega = \left(\frac{k}{m_{\text{eff}}}\right)^{1/2} = 2\pi\nu = 2\pi c\tilde{\nu} \quad \text{so} \quad k = (2\pi c\tilde{\nu})^2 m_{\text{eff}}.$$

The effective mass is

$$m_{\text{eff}} = \frac{1}{2}m = \frac{1}{2}(4.003\,\text{u}) \times (1.66 \times 10^{-27}\,\text{kg u}^{-1}) = 3.32 \times 10^{-27}\,\text{kg}.$$

$$k = \left[2\pi(2.998 \times 10^8\,\text{m s}^{-1}) \times (152\,\text{m}^{-1})\right]^2 \times (3.32 \times 10^{-27}\,\text{kg})$$

$$= \boxed{2.72 \times 10^{-4}\,\text{kg s}^{-2}}.$$

The moment of inertia is

$$I = m_{\text{eff}}R_e^2 = (3.32 \times 10^{-27}\,\text{kg}) \times (297 \times 10^{-12}\,\text{m})^2 = \boxed{2.93 \times 10^{-46}\,\text{kg m}^2}.$$

The rotational constant is

$$B = \frac{\hbar}{4\pi c I} = \frac{1.0546 \times 10^{-34}\,\text{J s}}{4\pi (2.998 \times 10^8\,\text{m s}^{-1}) \times (2.93 \times 10^{-46}\,\text{kg m}^2)} = \boxed{95.5\,\text{m}^{-1}}.$$

(b) In the Morse potential

$$x_e = \frac{\tilde{\nu}}{4D_e} \quad \text{and} \quad D_e = D_0 + \frac{1}{2}\left(1 - \frac{1}{2}x_e\right)\tilde{\nu} = D_0 + \frac{1}{2}\left(1 - \frac{\tilde{\nu}}{8D_e}\right)\tilde{\nu}.$$

This rearranges to a quadratic equation in $\tilde{\nu}$

$$\frac{\tilde{\nu}^2}{16D_e} - \frac{1}{2}\tilde{\nu} + D_e - D_0 = 0 \quad \text{so} \quad \tilde{\nu} = \frac{\frac{1}{2} - \sqrt{\left(\frac{1}{2}\right)^2 - \frac{4(D_e - D_0)}{16D_e}}}{2(16D_e)^{-1}}.$$

$$\tilde{\nu} = 4D_e\left(1 - \sqrt{\frac{D_0}{D_e}}\right)$$

$$= \frac{4(1.51 \times 10^{-23}\,\text{J})}{(6.626 \times 10^{-34}\,\text{J s}) \times (2.998 \times 10^8\,\text{m s}^{-1})}\left(1 - \sqrt{\frac{2 \times 10^{-26}\,\text{J}}{1.51 \times 10^{-23}\,\text{J}}}\right)$$

$$= \boxed{293\,\text{m}^{-1}}.$$

and $\quad x_e = \dfrac{(293\,\text{m}^{-1}) \times (6.626 \times 10^{-34}\,\text{J s}) \times (2.998 \times 10^8\,\text{m s}^{-1})}{4(1.51 \times 10^{-23}\,\text{J})} = \boxed{0.96}.$

P13.19 **(a)** Follow the flow chart in Fig. 12.7 of the text. CH_3Cl is not linear, it has a C_3 axis (only one), it does not have C_2 axes perpendicular to C_3, it has no σ_h, but does have 3 σ_v planes; so it belongs to $\boxed{C_{3v}}$.

(b) The number of normal modes of a non-linear molecule is $3N - 6$, where N is the number of atoms. So CH_3Cl has $\boxed{\text{nine}}$ normal modes.

(c) To determine the symmetry of the normal modes, consider how the Cartesian axes of each atom are transformed under the symmetry operations of the C_{3v} group; the 15 Cartesian displacements constitute the basis here. All 15 Cartesian axes are left unchanged under the identity, so the character of this operation is 15. Under a C_3 operation, the H atoms are taken into each other, so they do not contribute to the character of C_3. The z axes of the C and Cl atoms, are unchanged, so they contribute 2 to the character of C_3; for these two atoms

$$x \to -\frac{x}{2} + \frac{3^{1/2}\,y}{2} \quad \text{and} \quad y \to -\frac{y}{2} + \frac{3^{1/2}x}{2},$$

so there is a contribution of $-1/2$ to the character from each of these coordinates in each of these atoms. In total, then $\chi = 0$ for C_3. To find the character of σ_v, call one of the σ_v planes the yz plane; it contains C, Cl, and one H atom. The y and z coordinates of these three atoms are unchanged, but the x coordinates are taken into their negatives, contributing $6 - 3 = 3$ to the character for this operation; the other two atoms are interchanged, so they contribute nothing to the character. To find the irreproducible representations that this basis spans, we multiply its characters by the characters of the irreproducible representations, sum those products, and divide the sum by the order h of the group (as in Section 13.5(a)). The table below illustrates the procedure.

	E	$2C_2$	$3\sigma_v$		E	$2C_2$	$3\sigma_v$	sum/h
basis	15	0	3					
A_1	1	1	1	basis $\times A_1$	15	0	3	3
A_2	1	1	-1	basis $\times A_2$	15	0	-3	2
E	2	-1	0	basis $\times$ E	30	0	0	5

Of these 15 modes of motion, 3 are translations (an A_1 and an E) and 3 rotations (an A_2 and an E); we subtract these to leave the vibrations, which span $\boxed{2A_1+A_2+3E}$ (two A_1 modes, one A_2 mode, and 3 doubly degenerate E modes).

(d) Any mode whose symmetry species is the same as that of x, y, or z is infrared active. Thus $\boxed{\text{all but the } A_2 \text{ mode are infrared active}}$.

(e) Only modes whose symmetry species is the same as a quadratic form may be Raman active. Thus $\boxed{\text{all but the } A^2 \text{ mode may be Raman active}}$.

Solutions to theoretical problems

P13.21   The center of mass of a diatomic molecule lies at a distance x from atom A and is such that the masses on either side of it balance

$$m_A x = m_B (R - x)$$

and hence it is at

$$x = \frac{m_B}{m} R \quad m = m_A + m_B.$$

The moment of inertia of the molecule is

$$I = m_A x^2 + m_B (R - x)^2 \text{ [16.27]} = \frac{m_A m_B^2 R^2}{m^2} + \frac{m_B m_A^2 R^2}{m^2} = \frac{m_A m_B}{m} R^2$$

$$= \boxed{m_{\text{eff}} R^2} \text{ since } m_{\text{eff}} = \frac{m_A m_B}{m_A + m_B}.$$

P13.23   Refer to the flow chart in Fig. 12.7 in the text. Yes at the first question (linear?) leads to linear point groups and therefore linear rotors. If the molecule is not linear, then yes at the next question (two or more C_n with $n > 2$?) leads to cubic and icosahedral groups and therefore spherical rotors. If the molecule is not a spherical rotor, yes at the next question leads to symmetric rotors if the highest C_n has $n > 2$; if not, the molecule is an asymmetric rotor.

(a) CH_4: not linear, but more than two C_n ($n > 2$), so $\boxed{\text{spherical rotor}}$.

(b) CH_3CN: not linear, C_3 (only one of them), so $\boxed{\text{symmetric rotor}}$.

(c) CO_2: linear, so $\boxed{\text{linear rotor}}$.

(d) CH_3OH: not linear, no C_n, so asymmetric rotor .

(e) Benzene: not linear, C_6, but only one high-order axis, so symmetric rotor .

(f) Pyridine: not linear, C_2, is highest rotational axis, so asymmetric rotor .

P13.25 $S(v, J) = \left(v + \dfrac{1}{2}\right)\tilde{v} + BJ(J+1)$ [13.61].

$$\Delta S_J^O = \tilde{v} - 2B(2J - 1) \; [\Delta v = 1, \Delta J = -2].$$

$$\Delta S_J^S = \tilde{v} + 2B(2J + 3) \; [\Delta v = 1, \Delta J = +2].$$

The transition of maximum intensity corresponds, approximately, to the transition with the most probable value of J, which was calculated in Problem 13.24,

$$J_{max} = \left(\frac{kT}{2hcB}\right)^{1/2} - \frac{1}{2}.$$

The peak-to-peak separation is then

$$\Delta S = \Delta S_{J_{max}}^S - \Delta S_{J_{max}}^O = 2B(2J_{max} + 3) - \{-2B(2J_{max} - 1)\} = 8B\left(J_{max} + \tfrac{1}{2}\right)$$

$$= 8B\left(\frac{kT}{2hcB}\right)^{1/2} = \left(\frac{32BkT}{hc}\right)^{1/2}.$$

To analyze the data we rearrange the relation to

$$B = \frac{hc(\Delta S)^2}{32kT}$$

and convert to a bond length using $B = \dfrac{\hbar}{4\pi cI}$, with $I = 2m_x R^2$ (Table 13.1) for a linear rotor. This gives

$$R = \left(\frac{\hbar}{8\pi cm_x B}\right)^{1/2} = \left(\frac{1}{\pi c\Delta S}\right) \times \left(\frac{2kT}{m_x}\right)^{1/2}.$$

We can now draw up the following table

	$HgCl_2$	$HgBr_2$	HgI_2
T/K	555	565	565
m_x/u	35.45	79.1	126.90
$\Delta S/\text{cm}^{-1}$	23.8	15.2	11.4
R/pm	227.6	240.7	253.4

Hence, the three bond lengths are approximately 230, 240, and 250 pm .

Solutions to applications

P13.27 (a) Resonance Raman spectroscopy is preferable to vibrational spectroscopy for studying the O—O stretching mode because such a mode would be infrared inactive , or at best only weakly active.

(The mode is sure to be inactive in free O_2, because it would not change the molecule's dipole moment. In a complex in which O_2 is bound, the O—O stretch may change the dipole moment, but it is not certain to do so at all, let alone strongly enough to provide a good signal.)

(b) The vibrational wavenumber is proportional to the frequency, and it depends on the effective mass as follows,

$$\tilde{\nu} \propto \left(\frac{k}{m_{eff}}\right)^{1/2}, \quad \text{so} \quad \frac{\tilde{\nu}(^{18}O_2)}{\tilde{\nu}(^{16}O_2)} = \left(\frac{m_{eff}(^{16}O_2)}{m_{eff}(^{18}O_2)}\right)^{1/2} = \left(\frac{16.0\,u}{18.0\,u}\right)^{1/2} = 0.943,$$

and $\tilde{\nu}(^{18}O_2) = (0.943)(844\,\text{cm}^{-1}) = \boxed{796\,\text{cm}^{-1}}$.

Note the assumption that the effective masses are proportional to the isotopic masses. This assumption is valid in the free molecule, where the effective mass of O_2 is equal to half the mass of the O atom; it is also valid if the O_2 is strongly bound at one end, such that one atom is free and the other is essentially fixed to a very massive unit.

(c) The vibrational wavenumber is proportional to the square root of the force constant. The force constant is itself a measure of the strength of the bond (technically of its stiffness, which correlates with strength), which in turn is characterized by bond order. Simple molecular orbital analysis of O_2, O_2^-, and O_2^{2-} results in bond orders of $\boxed{2, 1.5, \text{and } 1 \text{ respectively}}$. Given decreasing bond order, one would expect decreasing vibrational wavenumbers (and vice versa).

(d) The wavenumber of the O—O stretch is very similar to that of the peroxide anion, suggesting $\boxed{Fe_2^{3+}O_2^{2-}}$.

(e) The detection of two bands due to $^{16}O^{18}O$ implies that the two O atoms occupy non-equivalent positions in the complex. Structures **7** and **8** are consistent with this observation, but structures **5** and **6** are not.

P13.29   According to Problem 10.27(a), the Doppler effect obeys

$$v_{receding} = vf \quad \text{where } f = \left(\frac{1 - s/c}{1 + s/c}\right)^{1/2}.$$

This can be rearranged to yield

$$s = \frac{1 - f^2}{1 + f^2}c.$$

We are given wavelength data, so we use

$$f = \frac{v_{star}}{v} = \frac{\lambda}{\lambda_{star}}.$$

The ratio is:

$$f = \frac{654.2\,\text{nm}}{706.5\,\text{nm}} = 0.9260,$$

so $s = \dfrac{1 - 0.9260^2}{1 + 0.9260^2}c = \boxed{0.0768c} = 2.30 \times 10^7\,\text{m s}^{-1}$.

The broadening of the line is due to local events (collisions) in the distant star. It is temperature dependent and hence yields the surface temperature of the star. Eqn 13.17 relates the observed linewidth to temperature:

$$\delta\lambda_{\text{obs}} = \frac{2\lambda}{c}\left(\frac{2kT\ln 2}{m}\right)^{1/2} \quad \text{so} \quad T = \left(\frac{c\delta\lambda}{2\lambda}\right)^2 \frac{m}{2k\ln 2},$$

$$T = \left(\frac{(2.998 \times 10^8 \text{ m s}^{-1})(61.8 \times 10^{-12} \text{ m})}{2(654.2 \times 10^{-9})}\right)^2 \left[\frac{(47.95 \text{ u})(1.661 \times 10^{-27} \text{ kg u}^{-1})}{2(1.381 \times 10^{-23} \text{ J K}^{-1})\ln 2}\right],$$

$$T = \boxed{8.34 \times 10^5 \text{ K}}.$$

P13.31 $E_J = J(J+1)hcB, \quad g_J = 2J+1.$

$$E_1 - E_0 = 2hcB = hc\left(\frac{1}{\lambda_{\text{shorter}}} - \frac{1}{\lambda_{\text{longer}}}\right).$$

$$B = \frac{1}{2}\left(\frac{1}{\lambda_{\text{shorter}}} - \frac{1}{\lambda_{\text{longer}}}\right) = \frac{1}{2}\left(\frac{1}{\lambda_{\text{shorter}}} - \frac{1}{\lambda_{\text{shorter}} + \Delta\lambda}\right)$$

$$= \frac{1}{2}\left(\frac{1}{\lambda_{\text{shorter}}}\right) \times \left(1 - \frac{1}{1 + (\Delta\lambda/\lambda_{\text{shorter}})}\right)$$

$$= \frac{1}{2}\left(\frac{1}{387.5 \text{ nm}}\right) \times \left(1 - \frac{1}{1 + (0.061/387.5)}\right) \times \left(\frac{10^9 \text{ nm}}{10^2 \text{ cm}}\right),$$

$$B = \boxed{2.031 \text{ cm}^{-1}}.$$

$$\frac{E_1 - E_0}{k} = \frac{2hcB}{k} = \frac{2(6.626 \times 10^{-34} \text{ J s}) \times (3.00 \times 10^{10} \text{ cm s}^{-1}) \times (2.031 \text{ cm}^{-1})}{1.381 \times 10^{-23} \text{ J K}^{-1}}$$

$$= 5.84\bar{7}\text{K}.$$

Intensity of $J' \leftarrow J$ absorption line $I_J \propto gJe^{-E_J/kT}$

$$\frac{I_{\lambda_{\text{longer}}}}{I_{\lambda_{\text{shorter}}}} \simeq \frac{g_1 e^{-E_1/kT}}{g_0 e^{-E_0/kT}} = \frac{g_1}{g_0}e^{-(E_1-E_0)/kT}.$$

Solve for T

$$T = \left(\frac{E_1 - E_0}{k}\right) \times \left(\frac{1}{\ln(g_1 I_{\lambda_{\text{shorter}}}/g_0 I_{\lambda_{\text{longer}}})}\right) = 5.84\bar{7} \text{ K}\left(\frac{1}{\ln(3 \times 4)}\right) = \boxed{2.35 \text{ K}}.$$

P13.33 Temperature effects. At extremely low temperatures (10 K) only the lowest rotational states are populated. No emission spectrum is expected for the cloud and star light microwave absorptions by the cloud are by the lowest rotational states. At higher temperatures additional high-energy lines appear because higher energy rotational states are populated. Circumstellar clouds may exhibit infrared absorptions due to vibrational excitation as well as electronic transitions in the ultraviolet. Ultraviolet absorptions may indicate the photodissocation of carbon monoxide. High temperature clouds exhibit emissions.

Density effects. The density of an interstellar cloud may range from one particle to a billion particles per cm^3. This is still very much a vacuum compared to the laboratory high vacuum of a trillion particles per cm^3. Under such extreme vacuum conditions the half-life of any quantum state is expected to be extremely long and absorption lines should be very narrow. At the higher densities the vast size of nebulae obscures distant stars. High densities and high temperatures may create conditions in which emissions stimulate emissions of the same wavelength by molecules. A cascade of stimulated emissions greatly amplifies normally weak lines—the maser phenomena of Microwave Amplification by Stimulated Emission of Radiation.

Particle velocity effects. Particle velocity can cause Doppler broadening of spectral lines. The effect is extremely small for interstellar clouds at 10 K but is appreciable for clouds near high temperature stars. Outflows of gas from pulsing stars exhibit a red Doppler shift when moving away at high speed and a blue shift when moving toward us.

There will be many more transitions observable in circumstellar gas than in interstellar gas, because many more rotational states will be accessible at the higher temperatures. Higher velocity and density of particles in circumstellar material can be expected to broaden spectral lines compared to those of interstellar material by shortening collisional lifetimes. (Doppler broadening is not likely to be significantly different between circumstellar and interstellar material in the same astronomical neighborhood. The relativistic speeds involved are due to large-scale motions of the expanding universe, compared to which local thermal variations are insignificant.) A temperature of 1000 K is not high enough to significantly populate electronically excited states of CO; such states would have different bond lengths, thereby producing transitions with different rotational constants. Excited vibrational states would be accessible, though, and rotational–vibrational transitions with P and R branches as detailed later in this chapter would be observable in circumstellar but not interstellar material. The rotational constant B for $^{12}C^{16}O$ is $1.691\,cm^{-1}$. The first excited rotational energy level, $J = 1$, with energy $J(J + 1)hcB = 2hcB$, is thermally accessible at about 6 K (based on the rough equation of the rotational energy to thermal energy kT). In interstellar space, only two or three rotational lines would be observable; in circumstellar space (at about 1000 K) the number of transitions would be more like 20.

14 Molecular spectroscopy 2: electronic transitions

Answers to discussion questions

D14.1 The process of the determination of the term symbol for dioxygen, $^3\Sigma_g^-$, is described in Section 14.1(b) and will not be repeated here. The interpretation of the symbol follows: the letter Σ means that the magnitude of the total orbital angular momentum about the internuclear axis is 0; the left superscript 3 means that the component of the total spin angular momentum about the internuclear axis is $1 (2 \times 1 + 1 = 3)$; the subscript g means that the parity of the term is even; and the superscript $-$ means that the molecular wavefunction for O_2 changes sign upon reflection in the plane containing the nuclei.

D14.3 A band head is the convergence of the frequencies of electronic transitions with increasing rotational quantum number, J. They result from the rotational structure superimposed on the vibrational structure of the electronic energy levels of the diatomic molecule. See Figs 14.8 and 14.11. To understand how a band head arises, one must examine the equations describing the transition frequencies (eqns 14.5). As seen from the analysis in Section 14.1(e), convergence can only arise when terms in both $(B' - B)$ and $(B' + B)$ occur in the equation. Since only a term in $(B' - B)$ occurs for the Q branch, no band head can arise for that branch.

D14.5 The overall process associated with fluorescence involves the following steps. The molecule is first promoted from the vibrational ground state of a lower electronic level to a higher vibrational–electronic energy level by absorption of energy from a radiation field. Because of the requirements of the Franck–Condon principle, the transition is to excited vibrational levels of the upper electronic state. See Fig. 14.22. Therefore, the absorption spectrum shows a vibrational structure characteristic of the upper state. The excited state molecule can now lose energy to the surroundings through radiationless transitions and decay to the lowest vibrational level of the upper state. A spontaneous radiative transition now occurs to the lower electronic level and this fluorescence spectrum has a vibrational structure characteristic of the lower state. The fluorescence spectrum is not the mirror image of the absorption spectrum because the vibrational frequencies of the upper and lower states are different due to the difference in their potential energy curves.

D14.7 See Section 14.5 for a detailed description of both the theory and experiment involved in laser action. Here we restrict our discussion to only the most fundamental concepts. The basic requirement for a laser is that it has at least three energy levels. Of these levels, the highest lying state must be capable of being efficiently populated above its thermal equilibrium value by a pulse of radiation. A second state, lower in energy, must be a metastable state with a long enough lifetime for it to accumulate a population greater than its thermal equilibrium value by spontaneous transitions from the higher overpopulated state.

The metastable state must than be capable of undergoing stimulated transitions to a third lower lying state. This last requirement implies not only that the metastable state must have more than its thermal equilibrium population, but also that it must have a higher population than the third lower lying state, namely, that it achieve population inversion. See Figs 14.28 and 14.29 for a description of the three- and four-level lasers. The amplification process occurs when low intensity radiation of frequency equal to the transition frequency between the metastable state and the lower lying state stimulates the transition to the lower lying state and many more photons (higher intensity of the radiation) of that frequency are created. Examples of practical lasers are listed and discussed in *Further information* 14.1.

Solutions to exercises

E14.1(b)   According to Hund's rule, we expect one $1\pi_u$ electron and one $2\pi_g$ electron to be unpaired. Hence $S = 1$ and the multiplicity of the spectroscopic term is $\boxed{3}$. The overall parity is u × g = $\boxed{u}$ since (apart from the complete core), one electron occupies a u orbital another occupies a g orbital.

E14.2(b)   Use the Beer–Lambert law

$$\log \frac{I}{I_0} = -\varepsilon[J]l = (-327\,\mathrm{dm^3\,mol^{-1}\,cm^{-1}}) \times (2.22 \times 10^{-3}\,\mathrm{mol\,dm^{-3}}) \times (0.15\,\mathrm{cm})$$

$$= -0.10\overline{\overline{889}}$$

$$\frac{I}{I_1} = 10^{-0.10\overline{\overline{889}}} = 0.778$$

The reduction in intensity is $\boxed{22.2\ \text{percent}}$

E14.3(b) $$\varepsilon = -\frac{1}{[J]l} \log \frac{I}{I_0} \quad [13.2,\ 13.3]$$

$$= \frac{-1}{(6.67 \times 10^{-4}\,\mathrm{mol\,dm^{-3}}) \times (0.35\,\mathrm{cm})} \log 0.655 = 78\overline{7}\,\mathrm{dm^3\,mol^{-1}\,cm^{-1}}$$

$$= 78\overline{7} \times 10^3\,\mathrm{cm^3\,mol^{-1}\,cm^{-1}} \quad [1\,\mathrm{dm} = 10\,\mathrm{cm}]$$

$$= \boxed{7.9 \times 10^5\ \mathrm{cm^2\,mol^{-1}}}$$

E14.4(b)   The Beer–Lambert law is

$$\log \frac{I}{I_0} = -\varepsilon[J]l \quad \text{so} \quad [J] = \frac{-1}{\varepsilon l} \log \frac{I}{I_0}$$

$$[J] = \frac{-1}{(323\,\mathrm{dm^3\,mol^{-1}\,cm^{-1}} \times (0.750\,\mathrm{cm})} \log(1 - 0.523) = \boxed{1.33 \times 10^{-3}\,\mathrm{mol\,dm^{-3}}}$$

E14.5(b)   Note: a parabolic lineshape is symmetrical, extending an equal distance on either side of its peak. The given data are not consistent with a parabolic lineshape when plotted as a function of either wavelength or wavenumber, for the peak does not fall at the center of either the wavelength or the wavenumber range. The exercise will be solved with the given data assuming a triangular lineshape as a function of wavenumber.

The integrated absorption coefficient is the area under an absorption peak

$$A = \int \varepsilon \, d\tilde{\nu}$$

If the peak is triangular, this area is

$$A = \tfrac{1}{2}(\text{base}) \times (\text{height})$$

$$= \tfrac{1}{2}[(199 \times 10^{-9}\,\text{m})^{-1} - (275 \times 10^{-9}\,\text{m})^{-1}] \times (2.25 \times 10^4\,\text{dm}^3\,\text{mol}^{-1}\,\text{cm}^{-1})$$

$$= 1.5\overline{6} \times 10^{10}\,\text{dm}^3\,\text{m}^{-1}\,\text{mol}^{-1}\,\text{cm}^{-1} = \frac{(1.5\overline{6} \times 10^9\,\text{dm}^3\,\text{m}^{-1}\,\text{mol}^{-1}\,\text{cm}^{-1}) \times (100\,\text{cm}\,\text{m}^{-1})}{10^3\,\text{dm}^3\,\text{m}^{-3}}$$

$$= 1.5\overline{6} \times 10^9\,\text{m}\,\text{mol}^{-1} = \boxed{1.5\overline{6} \times 10^8\,\text{dm}^3\,\text{mol}^{-1}\,\text{cm}^{-2}}$$

E14.6(b) Modeling the π electrons of 1,3,5-hexatriene as free electrons in a linear box yields non-degenerate energy levels of

$$E_n = \frac{n^2 h^2}{8 m_e L^2}$$

The molecule has six π electrons, so the lowest-energy transition is from $n = 3$ to $n = 4$. The length of the box is 5 times the C—C bond distance R. So

$$\Delta E_{\text{linear}} = \frac{(4^2 - 3^3) h^2}{8 m_e (5R)^2}$$

Modelling the π electrons of benzene as free electrons on a ring of radius R yields energy levels of

$$E_{m_l} = \frac{m_l^2 \hbar^2}{2I}$$

where I is the moment of inertia: $I = m_e R^2$. These energy levels are doubly degenerate, except for the non-degenerate $m_l = 0$. The six π electrons fill the $m_l = 0$ and 1 levels, so the lowest-energy transition is from $m_l = 1$ to $m_l = 2$

$$\Delta E_{\text{ring}} = \frac{(2^2 - 1^2) \hbar^2}{2 m_e R^2} = \frac{(2^2 - 1^2) h^2}{8\pi^2 m_e R^2}$$

Comparing the two shows

$$\Delta E_{\text{linear}} = \frac{7}{25} \left(\frac{h^2}{8 m_e R^2} \right) < \Delta E_{\text{ring}} = \frac{3}{\pi^2} \left(\frac{h^2}{8 m_e R^2} \right)$$

Therefore, the lowest-energy absorption will $\boxed{\text{rise}}$ in energy.

E14.7(b) The Beer–Lambert law is

$$\log \frac{I}{I_0} = -\varepsilon[\text{J}]l = \log T$$

so a plot (Figure 14.1) of $\log T$ versus $[\text{J}]$ should give a straight line through the origin with a slope m of $-\varepsilon l$. So $\varepsilon = -m/l$.

The data follow

$[dye]/(mol\,dm^{-3})$	T	$\log T$
0.0010	0.73	−0.1367
0.0050	0.21	−0.6778
0.0100	0.042	−1.3768
0.0500	1.33×10^{-7}	−6.8761

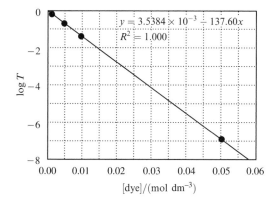

Figure 14.1

The molar absorptivity is

$$\varepsilon = -\frac{-138\,dm^3\,mol^{-1}}{0.250\,cm} = \boxed{522\,dm^3\,mol^{-1}\,cm^{-1}}$$

E14.8(b)   The Beer–Lambert law is

$$\log T = -\varepsilon[J]l \quad so \quad \varepsilon = \frac{-1}{[J]l}\log T$$

$$\varepsilon = \frac{-1}{(0.0155\,mol\,dm^{-3}) \times (0.250\,cm)}\log 0.32 = \boxed{12\overline{8}\,dm^3\,mol^{-1}\,cm^{-1}}$$

Now that we have ε, we can compute T of this solution with any size of cell

$$T = 10^{-\varepsilon[J]l} = 10^{-\{(12\overline{8}\,dm^3\,mol^{-1}\,cm^{-1}) \times (0.0155\,mol\,dm^{-3}) \times (0.450\,cm)\}} = \boxed{0.13}$$

E14.9(b)   The Beer–Lambert law is

$$\log\frac{I}{I_0} = -\varepsilon[J]l \quad so \quad l = -\frac{1}{\varepsilon[J]}\log\frac{I}{I_0}$$

(a) $$l = -\frac{1}{(30\,dm^3\,mol^{-1}\,cm^{-1}) \times (1.0\,mol\,dm^{-3})} \times \log\frac{1}{2} = \boxed{0.010\,cm}$$

(b) $$l = -\frac{1}{(30\,dm^3\,mol^{-1}\,cm^{-1}) \times (1.0\,mol\,dm^{-3})} \times \log 0.10 = \boxed{0.033\,cm}$$

E14.10(b)   The integrated absorption coefficient is the area under an absorption peak

$$A = \int \varepsilon \, d\tilde{v}$$

We are told that ε is a Gaussian function, i.e. a function of the form

$$\varepsilon = \varepsilon_{max} \exp\left(\frac{-x^2}{a^2}\right)$$

where $x = \tilde{v} - \tilde{v}_{max}$ and a is a parameter related to the width of the peak. The integrated absorption coefficient, then, is

$$A = \int_{-\infty}^{\infty} \varepsilon_{max} \exp\left(\frac{-x^2}{a^2}\right) dx = \varepsilon_{max} a \sqrt{\pi}$$

We must relate a to the half-width at half-height, $x_{1/2}$

$$\tfrac{1}{2}\varepsilon_{max} = \varepsilon_{max} \exp\left(\frac{-x_{1/2}^2}{a^2}\right) \quad \text{so} \quad \ln \tfrac{1}{2} = \frac{-x_{1/2}^2}{a^2} \quad \text{and} \quad a = \frac{x_{1/2}}{\sqrt{\ln 2}}$$

So $A = \varepsilon_{max} x_{1/2} \left(\dfrac{\pi}{\ln 2}\right)^{1/2} = (1.54 \times 10^4 \, \text{dm}^3 \, \text{mol}^{-1} \, \text{cm}^{-1}) \times (4233 \, \text{cm}^{-1}) \times \left(\dfrac{\pi}{\ln 2}\right)^{1/2}$

$$= \boxed{1.39 \times 10^8 \;\; \text{dm}^3 \, \text{mol}^{-1} \, \text{cm}^{-2}}$$

In SI base units

$$A = \frac{(1.39 \times 10^8 \, \text{dm}^3 \, \text{mol}^{-1} \, \text{cm}^{-2}) \times (1000 \, \text{cm}^3 \, \text{dm}^{-3})}{100 \, \text{cm} \, \text{m}^{-1}}$$

$$= \boxed{1.39 \times 10^9 \;\; \text{m} \, \text{mol}^{-1}}$$

E14.11(b) F_2^+ is formed when F_2 loses an antibonding electron, so we would expect F_2^+ to have a shorter bond than F_2. The difference in equilibrium bond length between the ground state (F_2) and excited state $(F_2^+ + e^-)$ of the photoionization experiment leads us to expect some vibrational excitation in the upper state. The vertical transition of the photoionization will leave the molecular ion with a stretched bond relative to its equilibrium bond length. A stretched bond means a vibrationally excited molecular ion, hence a $\boxed{\text{stronger}}$ transition to a vibrationally excited state than to the vibrational ground state of the cation.

Solutions to problems

Solutions to numerical problems

P14.1   The potential energy curves for the $X^3\Sigma_g^-$ and $B^3\Sigma_u^-$ electronic states of O_2 are represented schematically in Fig. 14.2 along with the notation used to represent the energy separation of this problem. Curves for the other electronic state of O_2 are not shown. Ignoring rotational structure and anharmonicity we may write

$$\tilde{v}_{00} \approx T_e + \frac{1}{2}(\tilde{v}' - \tilde{v}) = 6.175 \text{ eV} \times \left(\frac{8065.5 \text{ cm}^{-1}}{1 \text{ eV}}\right) + \frac{1}{2}(700 - 1580) \text{ cm}^{-1}$$

$$= \boxed{49\ 36\overline{4} \text{ cm}^{-1}}.$$

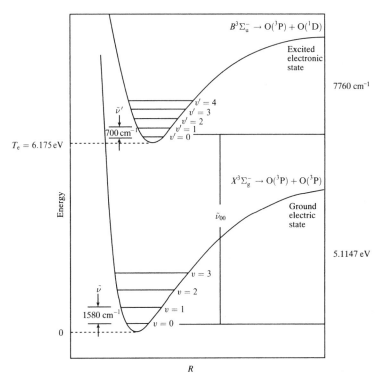

Figure 14.2

COMMENT. Note that the selection rule $\Delta v = \pm 1$ does not apply to vibrational transitions between different electronic states.

Question. What is the percentage change in $\tilde{v}_{00}$ if the anharmonicity constants $x_e\tilde{v}$ (Section 13.11), 12.0730 cm^{-1} and 8.002 cm^{-1} for the ground and excited states, respectively, are included in the analysis?

P14.3   Initially we cannot decide whether the dissociation products are produced in their ground atomic states or excited states. But we note that the two convergence limits are separated by an amount of energy exactly equal to the excitation energy of the bromine atom: 18345 cm^{-1} − 14660 cm^{-1} = 3685 cm^{-1}. Consequently, dissociation at 14660 cm^{-1} must yield bromine atoms in their ground state. Therefore, the possibilities for the dissociation energy are 14660 cm^{-1} or 14660 cm^{-1} − 7598 cm^{-1} = 7062 cm^{-1} depending upon whether the iodine atoms produced are in their ground or excited electronic state.

In order to decide which of these two possibilities is correct we can set up the following Born–Haber cycle.

(1)	$IBr(g)$	$\rightarrow$	$\frac{1}{2}I_2(g) + \frac{1}{2}Br_2(l)$	$\Delta H_1^{\ominus} = -\Delta_f H^{\ominus}(IBr, g)$
(2)	$\frac{1}{2}I_2(s)$	$\rightarrow$	$\frac{1}{2}I_2(g)$	$\Delta H_2^{\ominus} = \frac{1}{2}\Delta_{sub}H^{\ominus}(I_2, s)$
(3)	$\frac{1}{2}Br_2(l)$	$\rightarrow$	$\frac{1}{2}Br_2(g)$	$\Delta H_3^{\ominus} = \frac{1}{2}\Delta_{vap}H^{\ominus}(Br_2, l)$
(4)	$\frac{1}{2}I_2(g)$	$\rightarrow$	$I(g)$	$\Delta H_4^{\ominus} = \frac{1}{2}\Delta H(I-I)$
(5)	$\frac{1}{2}Br_2(s)$	$\rightarrow$	$Br(g)$	$\Delta H_5^{\ominus} = \frac{1}{2}\Delta H(Br-Br)$

$IBr(g)$	$\rightarrow$ $I(g) + Br(g)$	$\Delta H^{\ominus}$

$$\Delta H^{\ominus} = -\Delta_f H^{\ominus}(IBr, g) + \frac{1}{2}\Delta_{sub}H^{\ominus}(I_2, s) + \frac{1}{2}\Delta_{vap}H^{\ominus}(Br_2, l)$$
$$+ \frac{1}{2}\Delta H(I-I) + \frac{1}{2}\Delta H(Br-Br)$$
$$= \left\{-40.79 + \frac{1}{2} \times 62.44 + \frac{1}{2} \times 30.907 + \frac{1}{2} \times 151.24 + \frac{1}{2} \times 192.85\right\} \text{kJ mol}^{-1}$$

[Table 2.7 and data provided]

$$= 177.93 \text{ kJ mol}^{-1} = \boxed{14\,874 \text{ cm}^{-1}}.$$

Comparison to the possibilities $\boxed{14\,660 \text{ cm}^{-1}}$ and 7062 cm^{-1} shows that it is the former that is the correct dissociation energy.

P14.5   We write $\varepsilon = \varepsilon_{max}e^{-x^2} = \varepsilon_{max}e^{-\tilde{\nu}^2/2\Gamma}$ the variable being $\tilde{\nu}$ and Γ being a constant. $\tilde{\nu}$ is measured from the band center, at which $\tilde{\nu} = 0$. $\varepsilon = \frac{1}{2}\varepsilon_{max}$ when $\tilde{\nu}^2 = 2\Gamma \ln 2$. Therefore, the width at half-height is

$$\Delta\tilde{\nu}_{1/2} = 2 \times (2\Gamma \ln 2)^{1/2}, \quad \text{implying that} \quad \Gamma = \frac{\Delta\tilde{\nu}_{1/2}^2}{8 \ln 2}.$$

Now we carry out the integration

$$A = \int \varepsilon d\tilde{\nu} = \varepsilon_{max} \int_{-\infty}^{\infty} e^{-\tilde{\nu}^2/2\Gamma} d\tilde{\nu} = \varepsilon_{max}(2\Gamma\pi)^{1/2} \left[\int_{-\infty}^{\infty} e^{-x^2} dx = \pi^{1/2}\right]$$

$$= \varepsilon_{max} \left(\frac{2\pi\Delta\tilde{\nu}_{1/2}^2}{8 \ln 2}\right)^{1/2} = \left(\frac{\pi}{4 \ln 2}\right)^{1/2} \varepsilon_{max}\Delta\tilde{\nu}_{1/2} = 1.0645\varepsilon_{max}\Delta\tilde{\nu}_{1/2},$$

$A = 1.0645\varepsilon_{max}\Delta\tilde{\nu}_{1/2}$, with $\tilde{\nu}$ centered on $\tilde{\nu}_0$.

Since $\tilde{\nu} = \frac{1}{\lambda}$, $\Delta\tilde{\nu}_{1/2} \approx \frac{\Delta\lambda_{1/2}}{\lambda_0^2} [\lambda \approx \lambda_0]$.

$$A = 1.0645\varepsilon_{max} \left(\frac{\Delta\lambda_{1/2}}{\lambda_0^2}\right).$$

From Fig. 14.6 of the text, we find $\Delta\lambda_{1/2} = 38$ nm with $\lambda_0 = 290$ nm and $\varepsilon_{max} \approx 235$ dm^3 mol^{-1} cm^{-1}; hence

$$A = \frac{1.0645 \times (235 \text{ dm}^3 \text{ mol}^{-1} \text{ cm}^{-1}) \times (38 \times 10^{-7} \text{ cm})}{(290 \times 10^{-7} \text{ cm})^2} = \boxed{1.1 \times 10^6 \text{ dm}^3 \text{ mol}^{-1} \text{ cm}^{-2}}.$$

Since the dipole moment components transform as $A_1(z)$, $B_1(x)$, and $B_2(y)$, excitations from A_1 to A_1, B_1, and B_2 terms are allowed.

P14.7 We use the technique described in Example 13.5, the Birge–Sponer extrapolation method, and plot the difference $\Delta\tilde{v}_v$ against $v + \dfrac{1}{2}$.

We then draw up the following table.

$\Delta\tilde{v}_v$	688.0	665.1	641.5	617.6	591.8	561.2	534.0
$v + \dfrac{1}{2}$	$\dfrac{1}{2}$	$\dfrac{3}{2}$	$\dfrac{5}{2}$	$\dfrac{7}{2}$	$\dfrac{9}{2}$	$\dfrac{11}{2}$	$\dfrac{13}{2}$

$\Delta\tilde{v}_v$	502.1	465.5	428.9	388.2	343.1	300.9	255.0
$v + \dfrac{1}{2}$	$\dfrac{15}{2}$	$\dfrac{17}{2}$	$\dfrac{19}{2}$	$\dfrac{21}{2}$	$\dfrac{23}{2}$	$\dfrac{25}{2}$	$\dfrac{27}{2}$

The data are plotted in Fig. 14.3. Each square corresponds to 25 cm^{-1}. The area under the non-linear extrapolated line is 295 squares; therefore the dissociation energy is 7375 cm^{-1}. The $^3\Sigma_u^- \leftarrow X$ excitation energy (where X denotes the ground state) to $v = 0$ is 49357.6 cm^{-1} which corresponds to 6.12 eV. The $^3\Sigma_u^-$ dissociation energy for

$$O_2(^3\Sigma_u^-) \rightarrow O + O^*$$

is 7375 cm^{-1}, or 0.91 eV. Therefore, the energy of

$$O_2(X) \rightarrow O + O^*$$

is 6.12 eV $+$ 0.91 eV $=$ 7.03 eV. Since $O^* \rightarrow O$ is -190 kJ mol^{-1}, corresponding to -1.97 eV, the energy of

$$O_2(X) \rightarrow 2O$$

is 7.03 eV $-$ 1.97 eV $= \boxed{5.06 \text{ eV}}$.

COMMENT. This value of the dissociation energy is close to the experimental value of 5.08 eV quoted by Herzberg [*Further reading*, Chapters 13 and 14], but differs somewhat from the value obtained in Problem 14.2. The difficulty arises from the Birge–Sponer extrapolation, which works best when the experimental data fit a linear extrapolation curve as in Example 13.5. A glance at Figure 14.3 shows that the plot

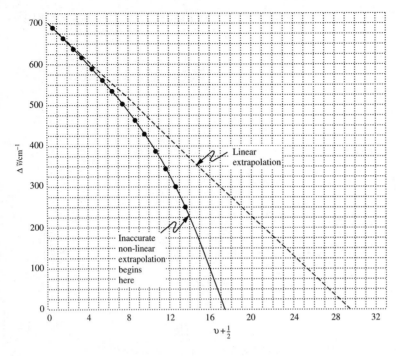

Figure 14.3

of the data is far from linear; hence, it is not surprising that the extrapolation here does not compare well to the extrapolation quoted in Problem 14.2. The extrapolation can be improved by using a quadratic or higher terms in the formula for ΔG [Chapter 13].

P14.9   Draw up a table like the following.

Hydrocarbon	$h\nu_{max}/eV$	E_{HOMO}/eV^*
Benzene	4.184	−9.7506
Biphenyl	3.654	−8.9169
Naphthalene	3.452	−8.8352
Phenanthrene	3.288	−8.7397
Pyrene	2.989	−8.2489
Anthracene	2.890	−8.2477

*Semi-empirical, PM3 level, PC Spartan Pro™.

Figure 14.4 shows a good correlation: $r^2 = 0.972$.

P14.11   (a) The molar concentration corresponding to 1 molecule per cubic µm is:

$$\frac{n}{V} = \frac{1}{6.022 \times 10^{23} \text{ mol}^{-1}} \times \frac{(10^6 \text{ µm m}^{-1})^3}{(1.0 \text{ µm}^3)(10 \text{ dm m}^{-1})^3} = \boxed{1.7 \times 10^{-9} \text{ mol dm}^{-3}},$$

i.e. nanomolar concentrations.

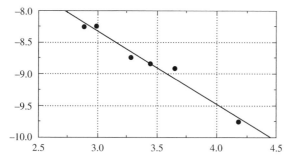

Figure 14.4

(b) An impurity of a compound of molar mass 100 g mol^{-1} present at 1.0×10^{-7} kg per 1.00 kg water can be expected to be present at a level of N molecules per cubic μm where N is

$$N = \frac{1.0 \times 10^{-7} \text{ kg impurity}}{1.00 \text{ kg water}} \times \frac{6.022 \times 10^{23} \text{mol}^{-1}}{100 \times 10^{-3} \text{ kg impurity mol}^{-1}}$$

$$\times (1.0 \times 10^3 \text{ kg water m}^{-3}) \times (10^{-6} \text{m})^3,$$

$$N = \boxed{6.0 \times 10^2}.$$

Pure as it seems, the solvent is much too contaminated for single-molecule spectroscopy.

Solutions to theoretical problems

P14.13   We need to establish whether the transition dipole moments

$$\boldsymbol{\mu}_{\text{fi}} = \int \psi_{\text{f}}^* \boldsymbol{\mu} \psi_{\text{i}} \, d\tau \; [13.13]$$

connecting the states 1 and 2 and the states 1 and 3 are zero or nonzero. The particle in a box wavefunctions are $\psi_n = (2/L)^{1/2} \sin(n\pi x/L)$ [9.5].

Thus $\mu_{2,1} \propto \displaystyle\int \sin\left(\frac{2\pi x}{L}\right) x \sin\left(\frac{\pi x}{L}\right) dx \propto \int x \left[\cos\left(\frac{\pi x}{L}\right) - \cos\left(\frac{3\pi x}{L}\right)\right] dx$

and $\mu_{3,1} \propto \displaystyle\int \sin\left(\frac{3\pi x}{L}\right) x \sin\left(\frac{\pi x}{L}\right) dx \propto \int x \left[\cos\left(\frac{2\pi x}{L}\right) - \cos\left(\frac{4\pi x}{L}\right)\right] dx$

having used $\sin\alpha \sin\beta = \frac{1}{2}\cos(\alpha - \beta) - \frac{1}{2}\cos(\alpha + \beta)$. Both of these integrals can be evaluated using the standard form

$$\int x(\cos ax) \, dx = \frac{1}{a^2}\cos ax + \frac{x}{a}\sin ax.$$

$$\int_0^L x\cos\left(\frac{\pi x}{L}\right) dx = \frac{1}{(\pi/L)^2}\cos\left(\frac{\pi x}{L}\right)\Big|_0^L + \frac{x}{(\pi/L)}\sin\left(\frac{\pi x}{L}\right)\Big|_0^L = -2\left(\frac{L}{\pi}\right)^2 \neq 0,$$

$$\int_L^0 x\cos\left(\frac{3\pi x}{L}\right) dx = \frac{1}{(3\pi/L)^2}\cos\left(\frac{3\pi x}{L}\right)\Big|_0^L + \frac{x}{(3\pi/L)}\sin\left(\frac{3\pi x}{L}\right)\Big|_0^L = -2\left(\frac{L}{3\pi}\right)^2 \neq 0.$$

Thus $\mu_{2,1} \neq 0$.

In a similar manner, $\mu_{3,1} = 0$.

COMMENT. A general formula for μ_{fi} applicable to all possible particle in a box transitions may be derived. The result is $(n = f, m = i)$

$$\mu_{nm} = -\frac{eL}{\pi^2}\left[\frac{\cos(n-m)\pi - 1}{(n-m)^2} - \frac{\cos(n+m)\pi - 1}{(n+m)^2}\right].$$

For m and n both even or both odd numbers, $\mu_{nm} = 0$; if one is even and the other odd, $\mu_{nm} \neq 0$. See also Problem 14.17.

Question. Can you establish the general relation for μ_{nm} above?

P14.15   We need to determine how the oscillator strength (Problem 14.16) depends on the length of the chain. We assume that wavefunctions of the conjugated electrons in the linear polyene can be approximated by the wavefunctions of a particle in a one-dimensional box. Then

$$f = \frac{8\pi^2 m_e \nu}{3he^2}|\mu_{fi}^2| \text{ [Problem 14.16]}.$$

$$\mu_x = -e\int_0^L \psi_{n'}(x)x\psi_n(x)dx, \quad \psi_n = \left(\frac{2}{L}\right)^{1/2}\sin\left(\frac{n\pi x}{L}\right)$$

$$= -\frac{2e}{L}\int_0^L x\sin\left(\frac{n'\pi x}{L}\right)\sin\left(\frac{n\pi x}{L}\right)dx$$

$$= \begin{cases} 0 & \text{if } n' = n+2. \\ +\left(\frac{8eL}{\pi^2}\right)\frac{n(n+1)}{(2n+1)^2} & \text{if } n' = n+1. \end{cases}$$

The integral is standard, but may also be evaluated using $2\sin A \sin B = \cos(A-B) - \cos(A+B)$ as in Problem 14.13.

$$h\nu = E_{n+1} - E_n = (2n+1)\frac{h^2}{8m_e L^2}.$$

Therefore, for the transition $n+1 \leftarrow n$,

$$f = \left(\frac{8\pi^2}{3}\right)\left(\frac{m_e}{he^2}\right)\left(\frac{h}{8m_e L^2}\right)(2n+1)\left(\frac{8eL}{\pi^2}\right)^2\frac{n^2(n+1)^2}{(2n+1)^4} = \left(\frac{64}{3\pi^2}\right)\left[\frac{n^2(n+1)^2}{(2n+1)^3}\right].$$

Therefore, $f \propto \dfrac{n^2(n+1)^2}{(2n+1)^3}$.

The value of n depends on the number of bonds: each π bond supplies two π electrons and so n increases by 1. For large n,

$$f \propto \frac{n^4}{8n^3} \rightarrow \frac{n}{8} \quad \text{and} \quad f \propto n.$$

Therefore, for the longest wavelength transitions f increases as the chain length is increased. The energy of the transition is proportional to $(2n + 1)/L^2$, but, as $n \propto L$, this energy is proportional to $1/L$.

Since $E_n = \dfrac{n^2 h^2}{8m_e L^2}$, $\Delta E = \dfrac{(2n + 1)h^2}{8m_e L^2} [\Delta n = +1]$

but $L = 2nd$ is the length of the chain (Exercise 14.6(a)), with d the carbon–carbon interatomic distance. Hence

$$\Delta E = \frac{((L/2d) + 1)h^2}{8m_e L^2} \approx \frac{h^2}{16 m_e dL} \propto \frac{1}{L}.$$

Therefore, the transition moves toward the red as $\boxed{L \text{ is increased}}$ and the apparent color of the dye $\boxed{\text{shifts towards blue}}$.

P14.17 $\mu = -eSR$ [given].

$$S = \left[1 + \frac{R}{a_0} + \frac{1}{3}\left(\frac{R}{a_0}\right)^2\right] e^{-R/a_0} \text{ [Problem 11.3]}.$$

$$f = \frac{8\pi^2 m_e \nu}{3he^2} \mu^2 = \frac{8\pi^2 m_e \nu}{3h} R^2 S^2 = \frac{8\pi^2 m_e \nu a_0^2}{3h} \left(\frac{RS}{a_0}\right)^2 = \boxed{\left(\frac{RS}{a_0}\right)^2 f_0}.$$

We then draw up the following table.

R/a_0	0	1	2	3	4	5	6	7	8
f/f_0	0	0.737	1.376	1.093	0.573	0.233	0.080	0.024	0.007

These points are plotted in Fig. 14.5.

The maximum in f occurs at the maximum of RS,

$$\frac{d}{dR}(RS) = S + R\frac{dS}{dR} = \left[1 + \frac{R}{a_0} - \frac{1}{3}\left(\frac{R}{a_0}\right)^3\right] e^{-R/a_0} = 0 \quad \text{at } R = R^*.$$

That is, $1 + \dfrac{R^*}{a_0} - \dfrac{1}{3}\left(\dfrac{R^*}{a_0}\right)^3 = 0.$

This equation may be solved either numerically or analytically (see Abramowitz and Stegun, *Handbook of mathematical functions*, Section 3.8.2), and $R^* = 2.10380a_0$.

As $R \to 0$, the transition becomes $s \to s$, which is forbidden. As $R \to \infty$, the electron is confined to a single atom because its wavefunction does not extend to the other.

P14.19   The fluorescence spectrum gives the vibrational splitting of the lower state. The wavelengths stated correspond to the wavenumbers 22 730, 24 390, 25 640, 27 030 cm^{-1}, indicating spacings of 1660, 1250, and 1390 cm^{-1}. The absorption spectrum spacing gives the separation of the vibrational levels of the upper state. The wavenumbers of the absorption peaks are 27 800, 29 000, 30 300, and 32 800 cm^{-1}. The vibrational spacings are therefore 1200, 1300, and 2500 cm^{-1}.

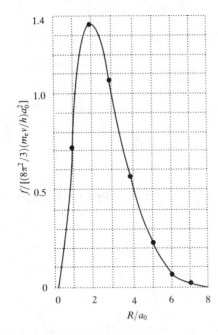

Figure 14.5

P14.21   Use the Clebsch–Gordan series [Chapter 10] to compound the two resultant angular momenta, and impose the conservation of angular momentum on the composite system.

(a) O_2 has $S = 1$ [it is a spin triplet]. The configuration of an O atom is $[\text{He}]2s^2 2p^4$, which is equivalent to a Ne atom with two electron-like 'holes'. The atom may therefore exist as a spin singlet or as a spin triplet. Since $S_1 = 1$ and $S_2 = 0$ or $S_1 = 1$ and $S_2 = 1$ may each combine to give a resultant with $S = 1$, both may be the products of the reaction. Hence multiplicities $\boxed{3 + 1}$ and $\boxed{3 + 3}$ may be expected.

(b) $N_2, S = 0$. The configuration of an N atom is $[\text{He}]\,2s^2 2p^3$. The atoms may have $S = \frac{3}{2}$ or $\frac{1}{2}$. Then we note that $S_1 = \frac{3}{2}$ and $S_1 = \frac{3}{2}$ can combine to give $S = 0$; $S_1 = \frac{1}{2}$ and $S_2 = \frac{1}{2}$ can also combine to give $S = 0$ (but $S_1 = \frac{3}{2}$ and $S_2 = \frac{1}{2}$ cannot). Hence, the multiplicities $\boxed{4 + 4}$ and $\boxed{2 + 2}$ may be expected.

Solutions to applications

P14.23   Fraction transmitted to the retina is

$$(1 - 0.30) \times (1 - 0.25) \times (1 - 0.09) \times 0.57 = 0.272.$$

The number of photons focused on the retina in 0.1 s is

$$0.272 \times 40\,\text{mm}^2 \times 0.1\,\text{s} \times 4 \times 10^3 \text{mm}^{-2}\,\text{s}^{-1} = \boxed{4.4 \times 10^3}.$$ More than what one might have guessed.

P14.25   The integrated absorption coefficient is

$$A = \int \varepsilon(\tilde{\nu}) \, d\tilde{\nu} \quad [13.5].$$

If we can express ε as an analytical function of $\tilde{\nu}$, we can carry out the integration analytically. Following the hint in the problem, we seek to fit ε to an exponential function, which means that a plot of $\ln \varepsilon$ versus $\tilde{\nu}$ ought to be a straight line (Fig. 14.6). So if

$$\ln \varepsilon = m\tilde{\nu} + b, \quad \text{then} \quad \varepsilon = \exp(m\tilde{\nu}) \exp(b)$$

and $A = (e^{b}/m) \exp(m\tilde{\nu})$ (evaluated at the limits of integration). We draw up the following table and find the best-fit line.

λ/nm	$\varepsilon/(dm^3 \, mol^{-1} \, cm^{-1})$	$\tilde{\nu}/cm^{-1}$	$\ln \varepsilon/(\, dm^3 \, mol^{-1} \, cm^{-1})$
292.0	1512	34248	4.69
296.3	865	33748	4.13
300.8	477	33248	3.54
305.4	257	32748	2.92
310.1	135.9	32248	2.28
315.0	69.5	31746	1.61
320.0	34.5	31250	0.912

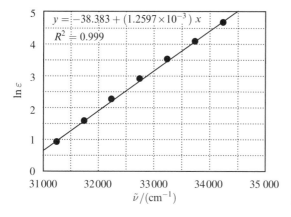

Figure 14.6

So $A = \dfrac{e^{-38.383}}{1.26 \times 10^{-3} \, cm} \left[\exp\left(\dfrac{1.26 \times 10^{-3} \, cm}{290 \times 10^{-7} \, cm} \right) - \exp\left(\dfrac{1.26 \times 10^{-3} \, cm}{320 \times 10^{-7} \, cm} \right) \right] dm^3 \, mol^{-1} \, cm^{-1}$

$= \boxed{1.24 \times 10^5 \, dm^3 \, mol^{-1} \, cm^{-2}}.$

P14.27 **(a)** The integrated absorption coefficient is (specializing to a triangular lineshape)

$$A = \int \varepsilon(\tilde{\nu}) \, d\tilde{\nu} = (1/2)\varepsilon_{max} \Delta\tilde{\nu}$$

$$= (1/2) \times (150 \, dm^3 \, mol^{-1} \, cm^{-1}) \times (34\,483 - 31\,250) \, cm^{-1},$$

$$\boxed{A = 2.42 \times 10^5 \, dm^3 \, mol^{-1} \, cm^{-2}}$$

(b) The concentration of gas under these conditions is

$$c = \frac{n}{V} = \frac{p}{RT} = \frac{2.4 \, \text{Torr}}{(62.364 \, \text{Torr dm}^3 \, mol^{-1} \, K^{-1}) \times (373 \, K)} = 1.03 \times 10^{-4} \, mol \, dm^{-3}.$$

Over 99 per cent of these gas molecules are monomers, so we take this concentration to be that of CH_3I (If 1 of every 100 of the original monomers turned to dimers, each produces 0.5 dimers; remaining monomers represent 99 of 99.5 molecules.) Beer's law states

$$A = \varepsilon c l = (150 \, dm^3 \, mol^{-1} \, cm^{-1}) \times (1.03 \times 10^{-4} \, mol \, dm^{-3}) \times (12.0 \, cm) = \boxed{0.185}.$$

(c) The concentration of gas under these conditions is

$$c = \frac{n}{V} = \frac{p}{RT} = \frac{100 \, \text{Torr}}{(62.364 \, \text{Torr dm}^3 \, mol^{-1} \, K^{-1}) \times (373 \, K)} = 4.30 \times 10^{-3} \, mol \, dm^{-3}.$$

Since 18 per cent of these CH_3I units are in dimers (forming 9 per cent as many molecules as were originally present as monomers), the monomer concentration is only 82/91 of this value or $3.87 \times 10^{-3} \, mol \, dm^{-3}$. Beer's law is

$$A = \varepsilon c l = (150 \, dm^3 \, mol^{-1} \, cm^{-1}) \times (3.87 \times 10^{-3} \, mol \, dm^{-3}) \times (12.0 \, cm) = \boxed{6.97}.$$

If this absorbance were measured, the molar absorption coefficient inferred from it without consideration of the dimerization would be

$$\varepsilon = A/cl = 6.97/((4.30 \times 10^{-1} \, mol \, dm^{-3}) \times (12.0 \, cm))$$

$$= \boxed{135 \, dm^3 \, mol^{-1} \, cm^{-1}},$$

an apparent drop of 10 per cent compared to the low-pressure value.

P14.29 In Fig.14.7

$$\Delta E_{11} = \frac{hc}{\lambda_{11}} = \frac{hc}{386.4 \, nm} = 5.1409 \times 10^{-19} \, J = 3.2087 \, eV$$

and

$$\Delta E_{00} = \frac{hc}{\lambda_{00}} = \frac{hc}{387.6 \, nm} = 5.1250 \times 10^{-19} \, J = 3.1987 \, eV.$$

Energy of excited singlet, $S_1 : E_1(\upsilon, J) = V_1 + (\upsilon + 1/2)\tilde{\nu}_1 hc + J(J+1)\tilde{B}_1 hc.$

Energy of excited singlet, $S_0 : E_0(\upsilon, J) = V_0 + (\upsilon + 1/2)\tilde{\nu}_0 hc + J(J+1)\tilde{B}_0 hc.$

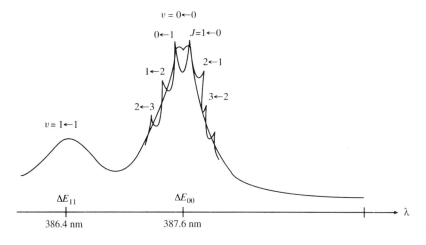

Figure 14.7

The midpoint of the 0–0 band corresponds to the forbidden Q branch ($\Delta J = 0$) with $J = 0$ and $\upsilon = 0 \leftarrow 0$.

$$\Delta E_{00} = E_1(0,0) - E_0(0,0) = (V_1 - V_0) + \tfrac{1}{2}(\tilde{\nu}_1 - \tilde{\nu}_0)hc. \tag{1}$$

The midpoint of the 1–1 band corresponds to the forbidden Q branch ($\Delta J = 0$) with $J = 0$ and $\upsilon = 1 \leftarrow 1$.

$$\Delta E_{11} = E_1(1,0) - E_0(1,0) = (V_1 - V_0) + \tfrac{3}{2}(\tilde{\nu}_1 - \tilde{\nu}_0)hc. \tag{2}$$

Multiplying eqn 1 by three and subtracting eqn 2 gives

$$3\Delta E_{00} - \Delta E_{11} = 2(V_1 - V_0).$$

$$V_1 - V_0 = \tfrac{1}{2}(3\Delta E_{00} - \Delta E_{11})$$

$$= \tfrac{1}{2}\{3(5.1250) - (5.1409)\}10^{-19}\text{J}$$

$$= 5.1171 \times 10^{-19}\,\text{J} = \boxed{3.193\text{eV}}. \tag{3}$$

This is the potential energy difference between S_0 and S_1.

Equations (1) and (3) may be solved for $\tilde{\nu}_1 - \tilde{\nu}_0$.

$$\tilde{\nu}_1 - \tilde{\nu}_0 = 2\{\Delta E_{00} - (V_1 - V_0)\}$$

$$= 2\{5.1250 - 5.1171\}10^{-19}\,\text{J}/hc$$

$$= 1.5800 \times 10^{-21}\,\text{J} = 0.0098615\,\text{eV} = \boxed{79.538\text{ cm}^{-1}}.$$

The $\tilde{\nu}_1$ value can be determined by analyzing the band head data for which $J + 1 \leftarrow J$.

$$\Delta E_{10}(J) = E_1(0,J) - E_0(1,J+1)$$

$$= V_1 - V_0 + \tfrac{1}{2}(\tilde{\nu}_1 - 3\tilde{\nu}_0)hc + J(J+1)\tilde{B}_1 hc - (J+1) \times (J+2)\tilde{B}_0 hc.$$

$$\Delta E_{00}(J) = V_1 - V_0 + \tfrac{1}{2}(\tilde{\nu}_1 - \tilde{\nu}_0)hc + J(J+1)\tilde{B}_1 hc - (J+1) \times (J+2)\tilde{B}_0 hc.$$

Therefore,

$$\Delta E_{00}(J) - \Delta E_{10}(J) = \tilde{\nu}_0 hc.$$

$$\Delta E_{00}(J_{\text{head}}) = \frac{hc}{388.3 \text{ nm}} = 5.1158 \times 10^{-19} \text{ J}.$$

$$\Delta E_{10}(J_{\text{head}}) = \frac{hc}{421.6 \text{ nm}} = 4.7117 \times 10^{-19} \text{ J}.$$

$$\tilde{\nu}_0 = \frac{\Delta E_{00}(J) - \Delta E_{10}(J)}{hc}$$

$$= \frac{(5.1158 - 4.7117) \times 10^{-19} \text{ J}}{hc}$$

$$= \frac{4.0410 \times 10^{-20} \text{ J}}{hc} = 0.25222 \text{ eV} = \boxed{2034.3 \text{ cm}^{-1}}.$$

$$\tilde{\nu}_1 = \tilde{\nu}_0 + 79.538 \text{ cm}^{-1}$$

$$= (2034.3 + 79.538) \text{ cm}^{-1} = \boxed{2113.8 \text{ cm}^{-1} = \frac{4.1990 \times 10^{-20} \text{ J}}{hc}}.$$

$$\frac{I_{1-1}}{I_{0-0}} \approx \frac{e^{-E_1(1,0)/kT_{\text{eff}}}}{e^{-E_1(0,0)/kT_{\text{eff}}}} = e^{-(E_1(1,0)-E_1(0,0))/kT_{\text{eff}}}$$

$$\approx e^{-hc\tilde{\nu}_1/kT_{\text{eff}}}.$$

$$\ln\left(\frac{I_{1-1}}{I_{0-0}}\right) = -\frac{hc\tilde{\nu}_1}{kT_{\text{eff}}}.$$

$$T_{\text{eff}} = \frac{hc\tilde{\nu}_1}{k \ln\left(\dfrac{I_{0-0}}{I_{1-1}}\right)} = \frac{4.1990 \times 10^{-20} \text{ J}}{(1.38066 \times 10^{-23} \text{ J K}^{-1}) \ln(10)} = \boxed{1321 \text{ K}}.$$

The relative population of the $\upsilon = 0$ and $\upsilon = 1$ vibrational states is the inverse of the relative intensities of the transitions from those states; hence $\dfrac{1}{0.1} = \boxed{10}$.

It would seem that with such a high effective temperature more than eight of the rotational levels of the S_1 state should have a significant population. But the spectra of molecules in comets are never as clearly resolved as those obtained in the laboratory and that is most probably the reason why additional rotational structure does not appear in these spectra.

Molecular spectroscopy 3: magnetic resonance

Answers to discussion questions

D15.1 Detailed discussions of the origins of the local, neighboring group, and solvent contributions to the shielding constant can be found in Sections 15.5(d), (e), and (f) as well as the books listed under *Further reading*. Here we will merely summarize the major features.

The local contribution is essentially the contribution of the electrons in the atom that contains the nucleus being observed. It can be expressed as a sum of a diamagnetic and paramagnetic part, that is, $\sigma(\text{local}) = \sigma_d + \sigma_p$. The diamagnetic part arises because the applied field generates a circulation of charge in the ground state of the atom. In turn, the circulating charge generates a magnetic field. The direction of this field can be found through Lenz's law, which states that the induced magnetic field must be opposite in direction to the field producing it. Thus it shields the nucleus. The diamagnetic contribution is roughly proportional to the electron density on the atom and it is the only contribution for closed shell free atoms and for distributions of charge that have spherical or cylindrical symmetry. The local paramagnetic contribution is somewhat harder to visualize since there is no simple and basic principle analogous to Lenz's law that can be used to explain the effect. The applied field adds a term to the hamiltonian of the atom that mixes excited electronic states into the ground state and any theoretical calculation of the effect requires detailed knowledge of the excited state wave functions. It is to be noted that the paramagnetic contribution does not require that the atom or molecule be paramagnetic. It is paramagnetic only in the sense that it results in an induced field in the same direction as the applied field.

The neighboring group contributions arise in a manner similar to the local contributions. Both diamagnetic and paramagnetic currents are induced in the neighboring atoms and these currents result in shielding contributions to the nucleus of the atom being observed. However, there are some differences: The magnitude of the effect is much smaller because the induced currents in neighboring atoms are much farther away. It also depends on the anisotropy of the magnetic susceptibility (see Chapter 20) of the neighboring group as shown in eqn 15.23. Only anisotropic susceptibilities result in a contribution.

Solvents can influence the local field in many different ways. Detailed theoretical calculations of the effect are difficult due to the complex nature of the solute–solvent interaction. Polar solvent–polar solute interactions are an electric field effect that usually causes deshielding of the solute protons.

Solvent magnetic antisotropy can cause shielding or deshielding, for example, for solutes in benzene solution. In addition, there are a variety of specific chemical interactions between solvent and solute that can affect the chemical shift. See the references listed under *Further reading* for more details.

D15.3 Both spin–lattice and spin–spin relaxation are caused by fluctuating magnetic and electric fields at the nucleus in question and these fields result from the random thermal motions present in the solution or other form of matter. These random motions can be a result of a number of processes and it is hard to summarize all that could be important. In theory every known nuclear interaction coupled with every type of motion can contribute to relaxation and detailed treatments can be exceedingly complex. However, they all depend on the magnetogyric ratio of the atom in question and the magnetogyric ratio of the proton is much larger than that of ^{13}C. Hence the interaction of the proton with fluctuating local magnetic fields caused by the presence of neighboring magnetic nuclei will be greater, and the relaxation will be quicker, corresponding to a shorter relaxation time for protons. Another consideration is the structure of compounds containing carbon and hydrogen. Typically the C atoms are in the interior of the molecule bonded to other C atoms, 99% of which are nonmagnetic, so the primary relaxation effects are due to bonded protons. Protons are on the outside of the molecule and are subject to many more interactions and hence faster relaxation.

D15.5 Spin–spin couplings in NMR are due to a polarization mechanism that is transmitted through bonds. The following description applies to the coupling between the protons in a H_X—C—H_Y group as is typically found in organic compounds. See Figs. 15.20–15.22 of the text. On H_X, the Fermi contact interaction causes the spins of its proton and electron to be aligned antiparallel. The spin of the electron from C in the H_X—C bond is then aligned antiparallel to the electron from H_X due to the Pauli exclusion principle. The spin of the C electron in the bond H_Y is then aligned parallel to the electron from H_X because of Hund's rule. Finally, the alignment is transmitted through the second bond in the same manner as the first. This progression of alignments (antiparallel × antiparallel × parallel × antiparallel × antiparallel) yields an overall energetically favorable parallel alignment of the two proton nuclear spins. Therefore, in this case the coupling constant, $^2J_{HH}$ is negative in sign.

The hyperfine structure in the ESR spectrum of an atomic or molecular system is a result of two interactions: an anisotropic dipolar coupling between the net spin of the unpaired electrons and the nuclear spins and also an isotropic coupling due to the Fermi contact interaction. In solution, only the Fermi contact interaction contributes to the splitting as the dipolar contribution averages to zero in a rapidly tumbling system. In the case of π-electron radicals, such as $C_6H_6^-$, no hyperfine interaction between the unpaired electron and the ring protons might have been expected. The protons lie in the nodal plane of the molecular orbital occupied by the unpaired electron, so any hyperfine structure cannot be explained by a simple Fermi contact interaction, which requires an unpaired electron density at the proton. However, an indirect spin polarization mechanism, similar to that used to explain spin–spin couplings in NMR, can account for the existence of proton hyperfine interactions in the ESR spectra of these systems. Refer to Fig. 15.6 of the text. Because of Hund's rule, the unpaired electron and the first electron in the C—H bond (the one from the C atom), will tend to align parallel to each other. The second electron in the C—H bond (the one from H) will then align antiparallel to the first by the Pauli principle, and finally the Fermi contact interaction will align the proton and electron on H antiparallel. The net result (parallel × antiparallel × antiparallel) is that the spins of the unpaired electron and the proton are aligned parallel and effectively they have detected each other.

Solutions to exercises

E15.1(b)   For ^{19}F, $\dfrac{\mu}{\mu_N} = 2.62835$, $g = 5.2567$

$$\nu = \nu_L = \frac{\gamma \mathscr{B}}{2\pi} \text{ with } \gamma = \frac{g_I \mu_N}{\hbar}$$

Hence, $\nu = \dfrac{g_I \mu_N \mathscr{B}}{h} = \dfrac{(5.2567) \times (5.0508 \times 10^{-27} \,\mathrm{J\,T^{-1}}) \times (16.2\,\mathrm{T})}{(6.626 \times 10^{-34}\,\mathrm{J\,s})}$

$$= 6.49 \times 10^8 \,\mathrm{s^{-1}} = \boxed{649\,\mathrm{MHz}}$$

E15.2(b) $E_{m_I} = -\gamma \hbar \mathscr{B} m_I = -g_I \mu_N \mathscr{B} m_I$
$m_I = 1, 0, -1$

$$E_{m_I} = -(0.404) \times (5.0508 \times 10^{-27}\,\mathrm{J\,T^{-1}}) \times (11.50\,\mathrm{T}) m_I$$

$$= -\left(2.34\overline{66} \times 10^{-26}\,\mathrm{J}\right) m_I$$

$$\boxed{-2.35 \times 10^{-26}\,\mathrm{J},\ 0,\ +2.35 \times 10^{-26}\,\mathrm{J}}$$

E15.3(b)   The energy separation between the two levels is

$$\Delta E = h\nu \text{ where } \nu = \frac{\gamma \mathscr{B}}{2\pi} = \frac{(1.93 \times 10^7 \,T^{-1}\,s^{-1}) \times (15.4\,T)}{2\pi}$$

$$= 4.73 \times 10^7 \,\mathrm{s^{-1}} = \boxed{47.3\,\mathrm{MHz}}$$

E15.4(b)   A 600 MHz NMR spectrometer means 600 MHz is the resonance field for protons for which the magnetic field is 14.1 T as shown in Exercise 15.1(a). In high-field NMR it is the field not the frequency that is fixed.

(a) A ^{14}N nucleus has three energy states in a magnetic field corresponding to $m_I = +1, 0, -1$. But
$\Delta E(+1 \to 0) = \Delta E(0 \to -1)$

$$\Delta E = E_{m_I'} - E_{m_I} = -\gamma \hbar \mathscr{B} m_I' - (-\gamma \hbar \mathscr{B} m_I)$$

$$= -\gamma \hbar \mathscr{B}(m_I' - m_I) = -\gamma \hbar \mathscr{B} \Delta m_I$$

The allowed transitions correspond to $\Delta m_I = \pm 1$; hence

$$\Delta E = h\nu = \gamma \hbar \mathscr{B} = g_I \mu_N \mathscr{B} = (0.4036) \times \left(5.051 \times 10^{-27}\mathrm{JT^{-1}}\right) \times (14.1\mathrm{T})$$

$$= \boxed{2.88 \times 10^{-26}\,\mathrm{J}}$$

(b) We assume that the electron g-value in the radical is equal to the free electron g-value, $g_e = 20023$. Then

$$\Delta E = h\nu = g_e \mu_B \mathscr{B} \ [37] = (2.0023) \times (9.274 \times 10^{-24}\,\mathrm{J\,T^{-1}}) \times (0.300\,\mathrm{T})$$

$$= \boxed{5.57 \times 10^{-24}\,\mathrm{J}}$$

COMMENT. The energy level separation for the electron in a free radical in an ESR spectrometer is far greater than that of nuclei in an NMR spectrometer, despite the fact that NMR spectrometers normally operate at much higher magnetic fields.

E15.5(b) $\Delta E = h\nu = \gamma \hbar \mathscr{B} = g_I \mu_N \mathscr{B}$ [Solution to Exercise 15.1(a)]

Hence, $\mathscr{B} = \dfrac{h\nu}{g_I \mu_N} = \dfrac{(6.626 \times 10^{-34}\,\text{J Hz}^{-1}) \times (150.0 \times 10^6\,\text{Hz})}{(5.586) \times (5.051 \times 10^{-27}\,\text{J T}^{-1})} = \boxed{3.523\,\text{T}}$

E15.6(b) In all cases the selection rule $\Delta m_I = \pm 1$ is applied; hence (Exercise 15.4(b)(a))

$$B = \frac{h\nu}{g_I \mu_N} = \frac{6.626 \times 10^{-34}\,\text{J Hz}^{-1}}{5.0508 \times 10^{-27}\,\text{JT}^{-1}} \times \frac{\nu}{g_I}$$

$$= \left(1.3119 \times 10^{-7}\right) \times \frac{\left(\dfrac{\nu}{\text{Hz}}\right)}{g_I}\,\text{T} = (0.13119) \times \frac{\left(\dfrac{\nu}{\text{MHz}}\right)}{g_I}\,\text{T}$$

We can draw up the following table

	$\mathscr{B}/\text{T}$	^{14}N	^{19}F	^{31}P
	g_I	0.40356	5.2567	2.2634
(a)	300 MHz	97.5	7.49	17.4
(b)	750 MHz	244	18.7	43.5

COMMENT. Magnetic fields above 20 T have not yet been obtained for use in NMR spectrometers. As discussed in the solution to Exercise 15.4(b), it is the field, not the frequency, that is fixed in high-field NMR spectrometers. Thus an NMR spectrometer that is called a 300 MHz spectrometer refers to the resonance frequency for protons and has a magnetic field fixed at 7.05 T.

E15.7(b) The relative population difference for spin $-\frac{1}{2}$ nuclei is given by

$$\frac{\delta N}{N} = \frac{N_\alpha - N_\beta}{N_\alpha + N_\beta} \approx \frac{\gamma \hbar \mathscr{B}}{2kT} = \frac{g_I \mu_N \mathscr{B}}{2kT} \quad [\textit{Justification 15.1}]$$

$$= \frac{1.405\,(5.05 \times 10^{-27}\text{J T}^{-1})\,\mathscr{B}}{2\,(1.381 \times 10^{-23}\,\text{J K}^{-1}) \times (298\,\text{K})} = 8.62 \times 10^{-7}\,(\mathscr{B}/\text{T})$$

(a) For 0.50 T $\quad \dfrac{\delta N}{N} = (8.62 \times 10^{-7}) \times (0.50) = \boxed{4.3 \times 10^{-7}}$

(b) For 2.5 T $\quad \dfrac{\delta N}{N} = (8.62 \times 10^{-7}) \times (2.5) = \boxed{2.2 \times 10^{-6}}$

(c) For 15.5 T $\quad \dfrac{\delta N}{N} = (8.62 \times 10^{-7}) \times (15.5) = \boxed{1.34 \times 10^{-5}}$

E15.8(b)   The ground state has

$$m_I = +\frac{1}{2} = \alpha \text{ spin}, \quad m_1 = -\frac{1}{2} = \beta \text{ spin}$$

Hence, with

$$\delta N = N_\alpha - N_\beta$$

$$\frac{\delta N}{N} = \frac{N_\alpha - N_\beta}{N_\alpha + N_\beta} = \frac{N_\alpha - N_\alpha e^{-\Delta E/kT}}{N_\alpha + N_\alpha e^{-\Delta E/kT}} \quad [\textit{Justification 15.1}]$$

$$= \frac{1 - e^{-\Delta E/kT}}{1 + e^{-\Delta E/kT}} \approx \frac{1 - (1 - \Delta E/kT)}{1 + 1} \approx \frac{\Delta E}{2kT} = \frac{g_I \mu_N \mathscr{B}}{2kT} \quad [\text{for } \Delta E \ll kT]$$

$$\delta N = \frac{N g_I \mu_N \mathscr{B}}{2kT} = \frac{Nh\nu}{2kT}$$

Thus, $\delta N \propto \nu$

$$\frac{\delta N(800\,\text{MHz})}{\delta N(60\,\text{MHz})} = \frac{(800\,\text{MHz})}{(60\,\text{MHz})} = \boxed{13}$$

This ratio is not dependent on the nuclide as long as the approximation $\Delta E \ll kT$ holds.

(a) $\delta = \dfrac{\nu - \nu^o}{\nu^o} \times 10^6$ [15.8]

Since both ν and ν^o depend upon the magnetic field in the same manner, namely

$$\nu = \frac{g_I \mu_N \mathscr{B}}{h} \quad \text{and} \quad \nu^o = \frac{g_I \mu_N \mathscr{B}_0}{h} \quad [\text{Exercise 15.1(a)}]$$

δ is $\boxed{\text{independent}}$ of both $\mathscr{B}$ and ν.

(b)   Rearranging [15.18], $\nu - \nu^o = \nu^o \delta \times 10^{-6}$ and we see that the relative chemical shift is

$$\frac{\nu - \nu^o(800\,\text{MHz})}{\nu - \nu^o(60\,\text{MHz})} = \frac{(800\,\text{MHz})}{(60\,\text{MHz})} = \boxed{13}$$

COMMENT.   This direct proportionality between $\nu - \nu^o$ and ν^o is one of the major reasons for operating an NMR spectrometer at the highest frequencies possible.

E15.9(b) $\mathscr{B}_{\text{loc}} = (1 - \sigma)\mathscr{B}$

$$|\Delta\mathscr{B}_{\text{loc}}| = |(\Delta\sigma)|\mathscr{B} \approx |[\delta(\text{CH}_3) - \delta(\text{CH}_2)]|\mathscr{B}$$

$$= |1.16 - 3.36| \times 10^{-6}\mathscr{B} = 2.20 \times 10^{-6}\mathscr{B}$$

(a) $\mathscr{B} = 1.9\,\text{T}, |\Delta\mathscr{B}_{\text{loc}}| = (2.20 \times 10^{-6}) \times (1.9\,\text{T}) = \boxed{4.2 \times 10^{-6}\,\text{T}}$

(b) $\mathscr{B} = 16.5\,\text{T}, |\Delta\mathscr{B}_{\text{loc}}| = (2.20 \times 10^{-6}) \times (16.5\,\text{T}) = \boxed{3.63 \times 10^{-5}\,\text{T}}$

E15.10(b) $\nu - \nu^\circ = \nu^\circ \delta \times 10^{-6}$

$|\Delta\nu| \equiv (\nu - \nu^\circ)(CH_2) - (\nu - \nu^\circ)(CH_3) = \nu(CH_2) - \nu(CH_3)$

$= \nu^\circ[\delta(CH_2) - \delta(CH_3)] \times 10^{-6}$

$= (3.36 - 1.16) \times 10^{-6}\nu^\circ = 2.20 \times 10^{-6}\nu^\circ$

(a) $\nu^\circ = 350\,MHz$ $|\Delta\nu| = (2.20 \times 10^{-6}) \times (350\,MHz) = 770\,Hz$ [Figure 15.1]

(b) $\nu^\circ = 650\,MHz$ $|\Delta\nu| = (2.20 \times 10^{-6}) \times (650\,MHz) = 1.43\,kHz$

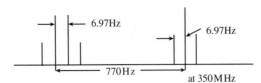

6.97Hz   6.97Hz

770Hz

at 350MHz   **Figure 15.1**

At 650 MHz, the spin–spin splitting remains the same at 6.97 Hz, but as $\Delta\nu$ has increased to 1.43 kHz, the splitting appears narrower on the δ scale.

E15.11(b) The difference in resonance frequencies is

$$\Delta\nu = \left(\nu^\circ \times 10^{-6}\right)\Delta\delta = \left(350\,s^{-1}\right) \times (6.8 - 5.5) = 4.6 \times 10^2\,s^{-1}$$

The signals will be resolvable as long as the conformations have lifetimes greater than

$$\tau = (2\pi\,\Delta\delta)^{-1}$$

The interconversion rate is the reciprocal of the lifetime, so a resolvable signal requires an inter-conversion rate less than

$$\text{rate} = (2\pi\,\Delta\delta) = 2\pi\left(4.6 \times 10^2\,s^{-1}\right) = \boxed{2.9 \times 10^3\,s^{-1}}$$

E15.12(b) $\nu = \dfrac{g_I\mu_N\mathscr{B}}{h}$ [Solution to exercise 15.1(a)]

Hence, $\dfrac{\nu(^{31}P)}{\nu(^1H)} = \dfrac{g(^{31}P)}{g(^1H)}$

or $\nu(^{31}P) = \dfrac{2.2634}{5.5857} \times 500\,MHz = \boxed{203\,MHz}$

The proton resonance consists of 2 lines $\left(2 \times \dfrac{1}{2} + 1\right)$ and the ^{31}P resonance of 5 lines $\left[2 \times \left(4 \times \dfrac{1}{2}\right) + 1\right]$. The intensities are in the ratio 1:4:6:4:1 (Pascal's triangle for four equivalent spin $\frac{1}{2}$ nuclei, Section 15.6). The lines are spaced $\dfrac{5.5857}{2.2634} = 2.47$ times greater in the phosphorus region than the proton region. The spectrum is sketched in Figure 15.2.

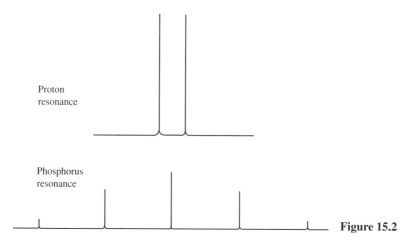

Proton resonance

Phosphorus resonance

Figure 15.2

E15.13(b) Look first at A and M, since they have the largest splitting. The A resonance will be split into a widely spaced triplet (by the two M protons); each peak of that triplet will be split into a less widely spaced sextet (by the five X protons). The M resonance will be split into a widely spaced triplet (by the two A protons); each peak of that triplet will be split into a narrowly spaced sextet (by the five X protons). The X resonance will be split into a less widely spaced triplet (by the two A protons); each peak of that triplet will be split into a narrowly spaced triplet (by the two M protons). (See Figure 15.3.)

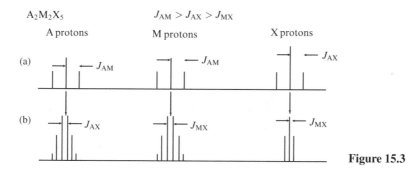

$A_2M_2X_5$ $J_{AM} > J_{AX} > J_{MX}$

A protons   M protons   X protons

(a) J_{AM} J_{AM} J_{AX}

(b) J_{AX} J_{MX} J_{MX}

Figure 15.3

Only the splitting of the central peak of Figure 15.3(a) is shown in Figure 15.3(b).

E15.14(b) **(a)** Since all J_{HF} are equal in this molecule (the CH_2 group is perpendicular to the CF_2 group), the H and F nuclei are both chemically and magnetically equivalent.

(b) Rapid rotation of the PH_3 groups about the Mo–P axes makes the P and H nuclei chemically and magnetically equivalent in both the *cis*- and *trans*-forms.

E15.15(b) Precession in the rotating frame follows

$$\nu_L = \frac{\gamma \mathscr{B}_1}{2\pi} \quad \text{or} \quad \omega_1 = \gamma \mathscr{B}_1$$

Since ω is an angular frequency, the angle through which the magnetization vector rotates is

$$\theta = \gamma \mathscr{B}_1 t = \frac{g_I \mu_N}{\hbar} \mathscr{B}_1 t$$

So $\mathscr{B}_1 = \dfrac{\theta \hbar}{g_I \mu_N t} = \dfrac{(\pi) \times (1.0546 \times 10^{-34}\,\text{J s})}{(5.586) \times (5.0508 \times 10^{-27}\,\text{J T}^{-1}) \times (12.5 \times 10^{-6}\,\text{s})} = \boxed{9.40 \times 10^{-4}\,\text{T}}$

a 90° pulse requires $\frac{1}{2} \times 12.5\,\mu\text{s} = \boxed{6.25\,\mu\text{s}}$

E15.16(b) $\mathscr{B} = \dfrac{h\nu}{g_e \mu_B} = \dfrac{hc}{g_e \mu_B \lambda}$

$$= \frac{(6.626 \times 10^{-34}\,\text{J s}) \times (2.998 \times 10^8\,\text{m s}^{-1})}{(2) \times (9.274 \times 10^{-24}\,\text{J T}^{-1}) \times (8 \times 10^{-3}\,\text{m})} = \boxed{1.\overline{3}\,\text{T}}$$

E15.17(b) The g factor is given by

$$g = \frac{h\nu}{\mu_B \mathscr{B}}; \qquad \frac{h}{\mu_B} = \frac{6.62608 \times 10^{-34}\,\text{J s}}{9.2740 \times 10^{-24}\,\text{J T}^{-1}} = 7.1448 \times 10^{-11}\,\text{T Hz}^{-1} = 71.448\,\text{mT GHz}^{-1}$$

$$g = \frac{71.448\,\text{mT GHz}^{-1} \times 9.2482\,\text{GHz}}{330.02\,\text{mT}} = \boxed{2.0022}$$

E15.18(b) The hyperfine coupling constant for each proton is $\boxed{2.2\,\text{mT}}$, the difference between adjacent lines in the spectrum. The g value is given by

$$g = \frac{h\nu}{\mu_B \mathscr{B}} = \frac{(71.448\,\text{mT GHz}^{-1}) \times (9.332\,\text{GHz})}{334.7\,\text{mT}} = \boxed{1.992}$$

E15.19(b) If the spectrometer has sufficient resolution, it will see a signal split into eight equal parts at $\pm\,1.445 \pm 1.435 \pm 1.055$ mT from the center, namely

$$\boxed{328.865,\ 330.975,\ 331.735,\ 331.755,\ 333.845,\ 333.865,\ 334.625,\ \text{and } 336.735\ \text{mT}}$$

If the spectrometer can only resolve to the nearest 0.1 mT, then the spectrum will appear as a sextet with intensity ratios of 1:1:2:2:1:1. The four central peaks of the more highly resolved spectrum would be the two central peaks of the less resolved spectrum.

E15.20(b) **(a)** If the CH_2 protons have the larger splitting there will be a triplet (1:2:1) of quartets (1:3:3:1). Altogether there will be 12 lines with relative intensities 1(4 lines), 2(2 lines), 3(4 lines), and 6(2 lines). Their positions in the spectrum will be determined by the magnitudes of the two proton splittings which are not given.

(b) If the CD_2 deuterons have the larger splitting there will be a quintet (1:2:3:2:1) of septets (1:3:6:7:6:3:1). Altogether there will be 35 lines with relative intensities 1(4 lines), 2(4 lines), 3(6 lines), 6(8 lines), 7(2 lines), 9(2 lines), 12(4 lines), 14(2 lines), 18(2 lines),and 21(1 line). Their positions in the spectrum will determined by the magnitude of the two deuteron splittings which are not given.

E15.21(b) The hyperfine coupling constant for each proton is $\boxed{2.2\,\text{mT}}$, the difference between adjacent lines in the spectrum. The g value is given by

$$g = \frac{h\nu}{\mu_B \mathscr{B}} \quad \text{so } \mathscr{B} = \frac{h\nu}{\mu_B g}, \quad \frac{h}{\mu_B} = 71.448\,\text{mT GHz}^{-1}$$

(a) $\mathscr{B} = \dfrac{(71.448\,\text{mT GHz}^{-1}) \times (9.312\,\text{GHz})}{2.0024} = \boxed{332.3\,\text{mT}}$

(b) $\mathscr{B} = \dfrac{(71.448\,\text{mT GHz}^{-1}) \times (33.88\,\text{GHz})}{2.0024} = \boxed{1209\,\text{mT}}$

E15.22(b) Two nuclei of spin $\boxed{I = 1}$ give five lines in the intensity ratio 1:2:3:2:1 (Figure 15.4).

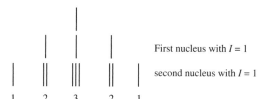

First nucleus with $I = 1$

second nucleus with $I = 1$

1   2   3   2   1   **Figure 15.4**

E15.23(b) The X nucleus produces four lines of equal intensity. Three H nuclei split each into a 1:3:3:1 quartet. The three D nuclei split each line into a septet with relative intensities 1:3:6:7:6:3:1 (see Exercise 15.20(a)). (See Figure 15.5.)

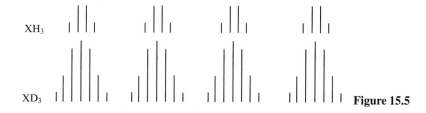

XH_3

XD_3 **Figure 15.5**

Solutions to problems

Solutions to numerical problems

P15.1 $g_I = -3.8260$ (Table 15.2).

$$\mathscr{B}_0 = \frac{h\nu}{g_I \mu_N} = \frac{(6.626 \times 10^{-34}\,\text{J Hz}^{-1}) \times \nu}{(-)(3.8260) \times (5.0508 \times 10^{-27}\,\text{J T}^{-1})} = 3.429 \times 10^{-8}(\nu/\text{Hz})\,\text{T}.$$

Therefore, with $\nu = 300\,\text{MHz}$,

$$\mathscr{B}_0 = (3.429 \times 10^{-8}) \times (300 \times 10^6\,\text{T}) = \boxed{10.3\,\text{T}}.$$

$$\frac{\delta N}{N} \approx \frac{g_I \mu_N B_0}{2kT} \quad \text{[Exercise 15.4(a)]}$$

$$= \frac{(-3.8260) \times (5.0508 \times 10^{-27}\,\text{J}\,\text{T}^{-1}) \times (10.3\,\text{T})}{(2) \times (1.381 \times 10^{-23}\,\text{J}\,\text{K}^{-1}) \times (298\,\text{K})} = \boxed{2.42 \times 10^{-5}}.$$

Since $g_I < 0$ (as for an electron, the magnetic moment is antiparallel to its spin), the $\boxed{\beta}$ state $(m_I = -\tfrac{1}{2})$ lies lower.

P15.3 The envelopes of maxima and minima of the curve are determined by T_2 through eqn 15.30, but the time interval between the maxima of this decaying curve corresponds to the reciprocal of the frequency difference $\Delta\nu$ between the pulse frequency ν_0 and the Larmor frequency ν_L, that is, $\Delta\nu = |\nu_0 - \nu_L|$:

$$\Delta\nu = \frac{1}{0.10\,\text{s}} = 10\,\text{s}^{-1} = 10\,\text{Hz}.$$

Therefore the Larmor frequency is $\boxed{300 \times 10^6\,\text{Hz} \pm 10\,\text{Hz}.}$

According to eqns 15.30 and 15.32 the intensity of the maxima in the FID curve decays exponentially as e^{-t/T_2}. Therefore T_2 corresponds to the time at which the intensity has been reduced to $1/e$ of the original value. In the text figure, this corresponds to a time slightly before the fourth maximum has occurred, or about $\boxed{0.29\,\text{s}}$.

P15.5 It seems reasonable to assume that only staggered conformations can occur. Therefore the equilibria are as shown in Fig. 15.6.

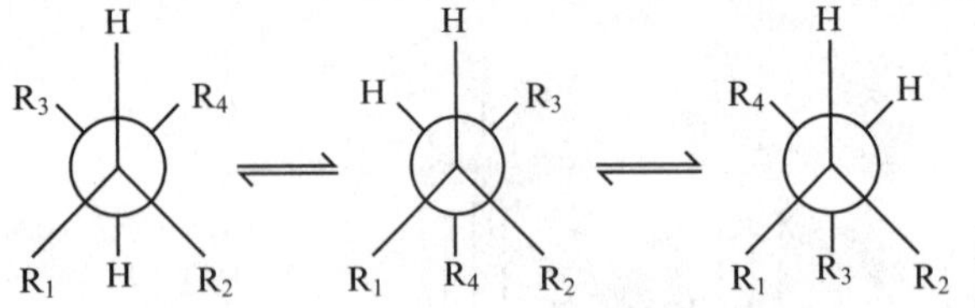

Figure 15.6

When $R_3 = R_4 = H$, all three of the conformations in Fig. 15.6 occur with equal probability; hence

$$^3J_{HH}(\text{methyl}) = \tfrac{1}{3}(^3J_t + 2\,{}^3J_g) \quad \text{[t = trans, g = gauche; CHR}_3\text{R}_4 = \text{methyl]}.$$

Additional methyl groups will avoid being staggered between both R_1 and R_2. Therefore

$$^3J_{HH}(\text{ethyl}) = \tfrac{1}{2}(J_t + J_g) \quad \text{[R}_3 = \text{H, R}_4 = \text{CH}_3\text{]},$$

$$^3J_{HH}(\text{isopropyl}) = J_t \quad \text{[R}_3 = \text{R}_4 = \text{CH}_3\text{]}.$$

We then have three simultaneous equations in two unknowns J_t and J_g.

$$\tfrac{1}{3}(^3J_t + 2\,{}^3J_g) = 7.3\,\text{Hz}, \tag{1}$$

$$\tfrac{1}{2}(^3J_t + {}^3J_g) = 8.0\,\text{Hz}, \tag{2}$$

$$^3J_t = 11.2\,\text{Hz}.$$

The two unknowns are overdetermined. The first two equations yield $^3J_t = 10.1$, $^3J_g = 5.9$. However, if we assume that $^3J_t = 11.2$ as measured directly in the ethyl case then $^3J_g = 5.4$ (eqn 1) or 4.8 (eqn 2), with an average value of 5.1.

Using the original form of the Karplus equation,

$$^3J_t = A\cos^2(180°) + B = 11.2, \qquad ^3J_g = A\cos^2(60°) + B = 5.1$$

or

$$11.2 = A + B, \qquad 5.1 = 0.25A + B.$$

These simultaneous equations yield $A = 6.8$ Hz and $B = 4.8$ Hz. With these values of A and B, the original form of the Karplus equation fits the data exactly (at least to within the error in the values of ^3J_t and ^3J_g and in the measured values reported).

From the form of the Karplus equation in the text [15.27] we see that those values of A, B, and C cannot be determined from the data given, as there are three constants to be determined from only two values of J. However, if we use the values of A, B, and C given in the text, then

$$J_t = 7\,\text{Hz} - 1\,\text{Hz}(\cos 180°) + 5\,\text{Hz}(\cos 360°) = 11\,\text{Hz},$$

$$J_g = 7\,\text{Hz} - 1\,\text{Hz}(\cos 60°) + 5\,\text{Hz}(\cos 120°) = 5\,\text{Hz}.$$

The agreement with the modern form of the Karplus equation is excellent, but not better than the original version. Both fit the data equally well. But the modern version is preferred as it is more generally applicable.

P15.7   The proton COSY spectrum of 1-nitropropane shows that (a) the C_a—H resonance with $\delta = 4.3$ shares a cross-peak with the C_b—H resonance at $\delta = 2.1$ and (b) the C_b—H resonance with $\delta = 2.1$ shares a cross-peak with the C_c—H resonance at $\delta = 1.1$. Off diagonal peaks indicate coupling between H's on various carbons. Thus peaks at (4,2) and (2,4) indicate that the H's on the adjacent CH_2 units are coupled. The peaks at (1,2) and (2,1) indicate that the H's on CH_3 and central CH_2 units are coupled. See Fig. 15.7.

P15.9   Refer to Fig. 15.4 in the solution to Exercise 15.20(a). The width of the CH_3 spectrum is $3a_H = $ 6.9 mT . The width of the CD_3 spectrum is $6a_D$. It seems reasonable to assume, since the hyperfine interaction is an interaction of the magnetic moments of the nuclei with the magnetic moment of the electron, that the strength of the interactions is proportional to the nuclear moments.

$$\mu = g_I \mu_N I \quad \text{or} \quad \mu_z = g_I \mu_N m_I \quad [15.11, 15.10b]$$

and thus nuclear magnetic moments are proportional to the nuclear g-values; hence

$$a_D \approx \frac{0.85745}{5.5857} \times a_H = 0.1535a_H = 0.35\,\text{mT}.$$

Therefore, the overall width is $6a_D = $ 2.1 mT .

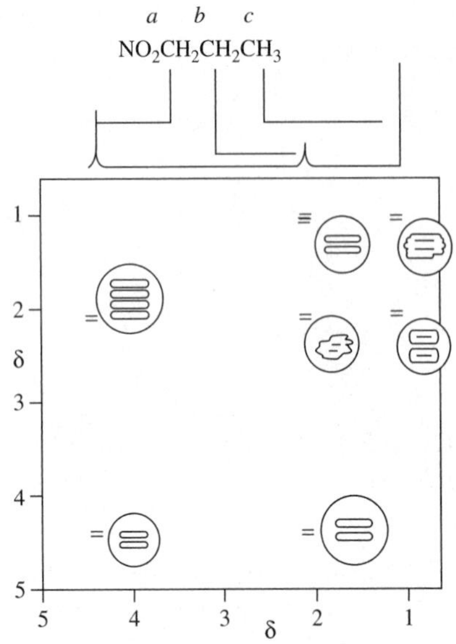

Figure 15.7

P15.11   We write $P(N2s) = \dfrac{5.7\text{mT}}{55.2\text{mT}} = \boxed{0.10}$ (10 per cent of its time);

$$P(N2p_z) = \dfrac{1.3\,\text{mT}}{3.4\,\text{mT}} = \boxed{0.38} \text{ (38 per cent of its time).}$$

The total probability is

(a) $P(N) = 0.10 + 0.38 = \boxed{0.48}$ (48 per cent of its time).

(b) $P(O) = 1 - P(N) = \boxed{0.52}$ (52 per cent of its time).

The hybridization ratio is

$$\dfrac{P(N2p)}{P(B2s)} = \dfrac{0.38}{0.10} = \boxed{3.8}.$$

The unpaired electron therefore occupies an orbital that resembles an sp^3 hybrid on N, in accord with the radical's nonlinear shape.

From the discussion in Section 11.3 we can write

$$a^2 = \dfrac{1 + \cos\phi}{1 - \cos\phi}$$

$$b^2 = 1 - a^2 = \dfrac{-2\cos\phi}{1 - \cos\phi}$$

$$\lambda = \dfrac{b'^2}{a'^2} = \dfrac{-1\cos\phi}{1 + \cos\phi}, \text{ implying that } \cos\phi = \dfrac{\lambda}{2 + \ell}$$

Then, since $\lambda = 3.8$, $\cos\phi = -0.66$, so $\phi = \boxed{131°.}$

Solutions to theoretical problems

P15.13   Use eqn 15.22 and *Illustration* 15.2. For hydrogen itself, we have:

$$\sigma_d = \frac{e^2 \mu_0}{12\pi m_e}\left\langle \frac{1}{r}\right\rangle = \frac{e^2 \mu_0}{12\pi_e a_0}.$$

The only difference in wavefunction (and therefore in the expectation value of $1/r$) between hydrogen and a more general hydrogenic ion is that the latter has a_0/Z where the former has a_0, so:

$$\boxed{\sigma_d = \frac{e^2 \mu_0 Z}{12\pi m_e a_0} = 1.78 \times 10^{-5} Z}.$$

P15.15 $\mathcal{B}_{nuc} = -\dfrac{\gamma \hbar \mu_0 m_I}{4\pi R^3}(1 - 3\cos^2\theta)[15.28] = \dfrac{g_I \mu_N \mu_0}{4\pi R^3} \quad \left[m_I = +\dfrac{1}{2},\ \theta = 0,\ \gamma\hbar = g_I \mu_N\right]$, which rearranges to

$$R = \left(\frac{g_I \mu_N \mu_0}{4\pi \mathcal{B}_{nuc}}\right)^{1/3} = \left(\frac{(5.5857) \times (5.0508 \times 10^{-27}\ \text{J T}^{-1}) \times (4\pi \times 10^{-7}\ \text{T}^2\ \text{J}^{-1}\ \text{m}^3)}{(4\pi) \times (0.715 \times 10^{-3}\ \text{T})}\right)^{1/3}$$

$$= (3.946 \times 10^{-30}\ \text{m}^3)^{1/3} = \boxed{158\ \text{pm}}.$$

P15.17   The shape of spectral line $\mathcal{I}(\omega)$ is related to the free induction decay signal $G(t)$ by

$$\mathcal{I}(\omega) = a\,\text{Re}\int_0^\infty G(t)e^{i\omega t}dt$$

where a is a constant and Re means take the real part of what follows. Calculate the lineshape corresponding to an oscillating, decaying function

$$G(t) = \cos \omega_0 t\, e^{-t/\tau}$$

$$\mathcal{I}(\omega) = a\,\text{Re}\int_0^\infty G(t)e^{i\omega t}\,dt$$

$$= a\,\text{Re}\int_0^\infty \cos \omega_0 t\, e^{-t/r + i\omega t}dt$$

$$= \frac{1}{2}a\,\text{Re}\int_0^\infty (e_i^{-\omega_0 t} + e_i^{\omega_0 t})e^{-t/\tau + i\omega t}dt$$

$$= \frac{1}{2}a\,\text{Re}\int_0^\infty \{e^{i(\omega_0 + \omega + i/\tau)t} + e^{-i(\omega_0 - \omega - i/\tau)t}\}dt$$

$$= -\frac{1}{2}a\text{Re}\left[\frac{1}{i(\omega_0 + \omega + i/\tau)} - \frac{1}{i(\omega_0 - \omega - i/\tau)}\right].$$

When ω and ω_0 are similar to magnetic resonance frequencies (or higher), only the second term in brackets is significant $\left(\text{because} \dfrac{1}{(\omega_0 + \omega)} \ll 1 \text{ but } \dfrac{1}{(\omega_0 - \omega)} \text{ may be large if } \omega \approx \omega_0\right)$. Therefore,

$$\mathcal{I}(\omega) \approx \frac{1}{2} a \operatorname{Re} \frac{1}{\mathrm{i}(\omega_0 - \omega)^2 + 1/\tau}$$

$$= \frac{1}{2} a \operatorname{Re} \frac{1}{(\omega_0 - \omega)^2 + 1/\tau^2}$$

$$\boxed{= \frac{1}{2} \frac{a\tau}{1 + (\omega_0 - \omega)^2 \tau^2}},$$

which is a Lorentzian line centered on ω_0, of amplitude $\frac{1}{2} A\tau$ and width $\dfrac{2}{\tau}$ at half-height.

P15.19   For non-weak fields in which the external magnetic field is comparable to the spin–orbit coupling field of an unpaired electron it is necessary to include a spin–orbit coupling term with coupling constant A [10.41] and apply the second-order perturbation equation [9.65b]. The Hamiltonian is $H = -g_e\gamma_e\mathcal{B}_0 s_z - \gamma_e\mathcal{B}_0 l_z + A\boldsymbol{l} \cdot \boldsymbol{s} = -\gamma_e\boldsymbol{\mathcal{B}} \cdot (g_e\boldsymbol{s} + \boldsymbol{l}) + A\boldsymbol{l} \cdot \boldsymbol{s}$ where vector notation is used for the electron orbital and spin angular momentum. The first-order perturbation equation [9.65a] gives

$$E^{(1)} = -\gamma_e\boldsymbol{\mathcal{B}} \cdot (g_e\langle\boldsymbol{s}\rangle + \langle\boldsymbol{l}\rangle) + A\langle\boldsymbol{l} \cdot \boldsymbol{s}\rangle = -\gamma_e g_e\boldsymbol{\mathcal{B}} \cdot \langle\boldsymbol{s}\rangle - \gamma_e\boldsymbol{\mathcal{B}} \cdot \langle\boldsymbol{l}\rangle + A\langle\boldsymbol{s}\rangle \cdot \langle\boldsymbol{l}\rangle.$$

The expectation value of orbital angular momentum, $\langle\boldsymbol{l}\rangle$, equals zero for real states and $\langle s_z\rangle = m_s$ which gives

$$E^{(1)} = -\gamma_e g_e\boldsymbol{\mathcal{B}} \cdot \langle\boldsymbol{s}\rangle = -\gamma_e g_e\mathcal{B}_0\langle s_z\rangle = -\gamma_e g_e\mathcal{B}_0 m_s\hbar.$$

The second-order perturbation term is written using the ground state '0' and 'n' excited states with the energy difference $\Delta E_{n0} = E_n - E_0$ [9.65b], which is positive.

$$E^{(2)} = -\sum_{n\neq 0} \frac{H_{0n}^{(1)} H_{n0}^{(1)}}{\Delta E_{n0}}$$

$$= -\sum_{n\neq 0} \frac{\langle 0|\{-g_e\gamma_e\mathcal{B}_0 s_z - \gamma_e\mathcal{B}_0 l_z + A\boldsymbol{l} \cdot \boldsymbol{s}\}|n\rangle\langle n|\{-g_e\gamma_e\mathcal{B}_0 s_z - \gamma_e\mathcal{B}_0 l_z + A\boldsymbol{l} \cdot \boldsymbol{s}\}|0\rangle}{\Delta E_{n0}}.$$

The numerator may be expanded and simplified by discarding second-order terms in $\mathcal{B}_0$ and s_z as negligibly small.

$$E^{(2)} = -\sum_{n\neq 0} \frac{\langle 0|\{-\gamma_e\mathcal{B}_0 l_z\}|n\rangle\langle n|A\boldsymbol{l} \cdot \boldsymbol{s}|0\rangle + \langle 0|A\boldsymbol{l} \cdot \boldsymbol{s}|n\rangle\langle n|\{-\gamma_e\mathcal{B}_0 l_z\}|0\rangle}{\Delta E_{n0}}$$

$$= A\gamma_e\mathcal{B}_0 \sum_{n\neq 0} \frac{\langle 0|l_z|n\rangle\langle n|\boldsymbol{l} \cdot \boldsymbol{s}|0\rangle + \langle 0|\boldsymbol{l} \cdot \boldsymbol{s}|n\rangle\langle n|l_z|0\rangle}{\Delta E_{n0}}$$

$$= A\gamma_e B_0 \sum_{n\neq 0} \frac{\langle 0|l_z|n\rangle\langle n|\mathbf{l}\cdot\mathbf{s}|0\rangle + \langle 0|\mathbf{l}\cdot\mathbf{s}|n\rangle\langle n|l_z|0\rangle}{\Delta E_{n0}} = 2A\gamma_e B_0 \sum_{n\neq 0} \frac{\langle 0|l_z|n\rangle\langle n|\mathbf{l}\cdot\mathbf{s}|0\rangle}{\Delta E_{n0}}$$

$$= 2A\gamma_e B_0 \sum_{n\neq 0} \frac{\langle 0|l_z|n\rangle\langle n|l_z s_z|0\rangle}{\Delta E_{n0}} = 2A\gamma_e B_0 \sum_{n\neq 0} \frac{\langle 0|l_z|n\rangle\langle n|l_z m_s \hbar|0\rangle}{\Delta E_{n0}}$$

$$= 2A\gamma_e B_0 m_s \hbar \sum_{n\neq 0} \frac{\langle 0|l_z|n\rangle\langle n|l_z|0\rangle}{\Delta E_{n0}}.$$

The last manipulation uses the assumption that the local field is parallel to the applied field (i.e., $\mathbf{l}\cdot\mathbf{s} = l_z s_z$). Combination of the first- and second-order perturbation estimates gives

$$E^{(\text{spin})} = E^{(1)} + E^{(2)} = -\gamma_e g_e B_0 m_s \hbar + 2A\gamma_e B_0 m_s \hbar \sum_{n\neq 0} \frac{\langle 0|l_z|n\rangle\langle n|l_z|0\rangle}{\Delta E_{n0}}$$

$$= -\left\{g_e - 2A \sum_{n\neq 0} \frac{\langle 0|l_z|n\rangle\langle n|l_z|0\rangle}{\Delta E_{n0}}\right\} \gamma_e B_0 m_s \hbar = -\left\{g_e - 2A \sum_{n\neq 0} \frac{\langle 0|l_z|n\rangle\langle n|l_z|0\rangle}{\Delta E_{n0}}\right\} \gamma_e B_0 s_z.$$

Comparison with the effective spin Hamiltonian, $E^{(\text{spin})} = -g\gamma_e B_0 s_z$, indicates that

$$g = g_e - 2A \sum_{n\neq 0} \frac{\langle 0|l_z|n\rangle\langle n|l_z|0\rangle}{\Delta E_{n0}}.$$

g increases with increasing strength of the spin–orbit coupling (A) and with decreasing excitation energy (ΔE_{n0}). This analysis is presented on p. 434 of P.W. Atkins and R.S. Friedman, *Molecular quantum mechanics*, 3rd edn, Oxford University Press, 1997.

Solutions to applications

P15.21 $$\langle \mathcal{B}_{\text{nucl}}\rangle = \frac{-g_I \mu_N \mu_0 m_I}{4\pi R^3} \frac{\int_0^{\theta_{max}} (1 - 3\cos^2\theta)\sin\theta d\theta}{\int_0^{\theta_{max}} \sin\theta\, d\theta}.$$

The denominator is the normalization constant, and ensures that the total probability of being between 0 and θ_{max} is 1.

$$\langle \mathcal{B}_{\text{nucl}}\rangle = \frac{-g_I \mu_N \mu_0 m_I}{4\pi R^3} \frac{\int_1^{x_{max}} (1 - 3x^2)dx}{\int_1^{x_{max}} dx} \quad [x_{max} = \cos\theta_{max}]$$

$$= \frac{-g_I \mu_N \mu_0 m_I}{4\pi R^3} \times \frac{x_{max}(1 - x_{max}^2)}{x_{max} - 1} = \boxed{\frac{-g_I \mu_N \mu_0 m_I}{4\pi R^3}(\cos^2\theta_{max} + \cos\theta_{max})}.$$

If $\theta_{max} = \pi$ (complete rotation), $\cos\theta_{max} = -1$ and $\langle \mathcal{B}_{\text{nucl}}\rangle = 0$.

If $\theta_{max} = 30°$, $\cos^2 \theta_{max} + \cos \theta_{max} = 1.616$, and

$$\langle \mathcal{B}_{nucl} \rangle = \frac{(5.5857) \times (5.0508 \times 10^{-27} \text{J T}^{-1}) \times (4\pi \times 10^{-7} \text{ T}^2 \text{ J}^{-1} \text{ m}^3) \times (1.616)}{(4\pi) \times (1.58 \times 10^{-10} \text{ m})^3 \times (2)}$$

$$= \boxed{0.58 \text{ mT}}.$$

P15.23   The desired result is the linear equation

$$[I]_0 = \frac{[E]_0 \Delta v}{\delta v} - K,$$

so the first task is to express quantities in terms of $[I]_0$, $[E]_0$, Δv, δv, and K, eliminating terms such as $[I]$, $[EI]$, $[E]$, v_I, v_{EI}, and v. (Note: symbolic mathematical software is helpful here.) Begin with v:

$$v = \frac{[I]}{[I] + [EI]} v_I + \frac{[EI]}{[I] + [EI]} v_{EI} = \frac{[I]_0 - [EI]}{[I]_0} v_I + \frac{[EI]}{[I]_0} v_{EI},$$

where we have used the fact that total I (i.e. free I plus bound I) is the same as initial I. Solve this so it must also be much greater than [EI]:

$$[EI] = \frac{[I]_0 (v - v_I)}{v_{EI} - v_I} = \frac{[I]_0 \delta v}{\Delta v},$$

where in the second equality we notice that the frequency differences that appear are the ones defined in the problem. Now take the equilibrium constant

$$K = \frac{[E][I]}{[EI]} = \frac{([E]_0 - [EI])([I]_0 - [EI])}{[EI]} \approx \frac{([E]_0 - [EI])[I]_0}{[EI]}.$$

We have used the fact that total I is much greater than total E (from the condition that $[I]_0 \gg [E]_0$), so it must also be much greater than [EI], even if all E binds I. Now solve this for $[E]_0$:

$$[E]_0 = \frac{K + [I]_0}{[I]_0} [EI] = \left(\frac{K + [I]_0}{[I]_0} \right) \left(\frac{[I]_0 \delta v}{\Delta v} \right) = \frac{(K + [I]_0) \delta v}{\Delta v}.$$

The expression contains the desired terms and only those terms. Solving for $[I]_0$ yields:

$$\boxed{[I]_0 = \frac{[E]_0 \Delta v}{\delta v} - K},$$

which would result in a straight line with slope $[E]_0 \Delta v$ and y-intercept K if one plots $[I]_0$ against $1/\delta v$.

P15.25   When spin label molecules approach to within 800 pm, orbital overlap of the unpaired electrons and dipolar interactions between magnetic moments cause an exchange coupling interaction between the spins. The electron exchange process occurs at a rate that increases as concentration increases. Thus the process has a lifetime that is too long at low concentrations to affect the 'pure' ESR signal. As the concentration increases, the linewidths increase until the triplet coalesces into a broad singlet. Further increase of the concentration decreases the exchange lifetime and therefore the linewidth of the singlet.

When spin labels within biological membranes are highly mobile, they may approach closely and the exchange interaction may provide the ESR spectra with information that mimics the moderate and high concentration signals below.

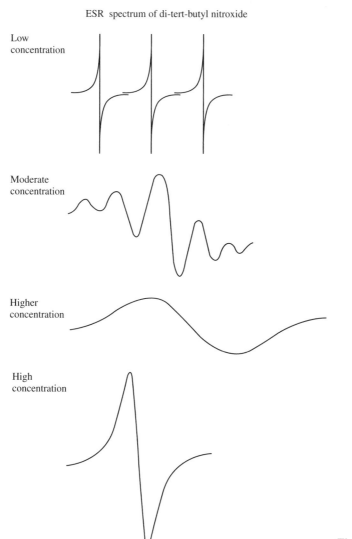

ESR spectrum of di-tert-butyl nitroxide

Low concentration

Moderate concentration

Higher concentration

High concentration

Figure 15.8

P15.27   Assume that the radius of the disk is 1 unit. The volume of each slice is proportional to (length of slice $\times \delta_x$), Fig. 15.9(a).

Length of slice at $x = 2\sin\theta$.

$$x = \cos\theta,$$

$$\theta = \arccos x.$$

x ranges from -1 to $+1$.

Length of slice at $x = 2\sin(\text{arcos}\,x)$.

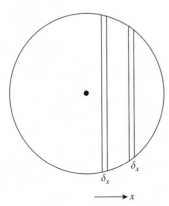

Figure 15.9(a)

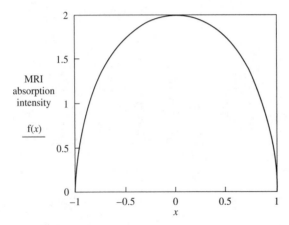

Figure 15.9(b)

Plot $f(x) = 2\sin(\text{arcos}\,x)$ against x between the limits -1 and $+1$. The plot is shown above. The volume at each value of x is proportional to $f(x)$ and the intensity of the MRI signal is proportional to the volume, so Fig. 15.9(b) represents the absorption intensity for the MRI image of the disk.

16 Statistical thermodynamics 1: the concepts

Answers to discussion questions

D16.1 Consider the value of the partition function at the extremes of temperature. The limit of q as T approaches zero, is simply g_0, the degeneracy of the ground state. As T approaches infinity, each term in the sum is simply the degeneracy of the energy level. If the number of levels is infinite, the partition function is infinite as well. In some special cases where we can effectively limit the number of states, the upper limit of the partition function is just the number of states. In general, we see that the molecular partition function gives an indication of the average number of states thermally accessible to a molecule at the temperature of the system.

D16.3 We evaluate β by comparing calculated and experimental values for thermodynamic properties. The calculated values are obtained from the theoretical formulas for these properties, all of which are expressed in terms of the parameter β. So there can be many ways of identifying β, as many as there are thermodynamic properties. One way is through the energy as shown in Section 16.3(b). Another is through the pressure as demonstrated in Example 17.1. Yet another is through the entropy, and this approach to the identification may be the most fundamental. See *Further reading* for elaboration of this method.

D16.5 An ensemble is a set of a large number of imaginary replications of the actual system. These replications are identical in some respects, but not in all respects. For example, in the canonical ensemble, all replications have the same number of particles, the same volume, and the same temperature, but not the same energy. Ensembles are useful in statistical thermodynamics because it is mathematically more tractable to perform an ensemble average to determine the (time averaged) thermodynamic properties than it is to perform an average over time to determine these properties. Recall that macroscopic thermodynamic properties are averages over the time dependent properties of the particles that compose the macroscopic system. In fact, it is taken as a fundamental principle of statistical thermodynamics that the (sufficiently long) time average of every physical observable is equal to its ensemble average. This principle is connected to a famous assumption of Boltzmann's called the ergodic hypothesis. A thorough discussion of these topics would take us far beyond what we need here. See the references under *Further reading*.

Solutions to exercises

E16.1(b) $n_i = \dfrac{N e^{-\beta \varepsilon_i}}{q}$ where $q = \sum_j e^{-\beta \varepsilon_j}$

Thus

$$\frac{n_2}{n_1} = \frac{e^{-\beta\varepsilon_2}}{e^{-\beta\varepsilon_1}} = e^{-\beta(\varepsilon_2 - \varepsilon_1)} = e^{-\beta\Delta\varepsilon} = e^{-\Delta\varepsilon/kT}$$

Given $\dfrac{n_2}{n_1} = \dfrac{1}{2}$, $\Delta\varepsilon = 300\,\text{cm}^{-1}$

$$k = (1.380\,66 \times 10^{-23}\,\text{J K}^{-1}) \times \left(\frac{1\,\text{cm}^{-1}}{1.9864 \times 10^{-23}\,\text{J}}\right) = 0.695\,06\,\text{cm}^{-1}\,\text{K}^{-1}$$

$$\frac{n_2}{n_1} = e^{-\Delta\varepsilon/kT}$$

$$\ln\left(\frac{n_2}{n_1}\right) = -\Delta\varepsilon/kT$$

$$T = \frac{-\Delta\varepsilon}{k\ln(n_2/n_1)} = \frac{\Delta\varepsilon}{k\ln(n_1/n_2)}$$

$$= \frac{300\,\text{cm}^{-1}}{(0.695\,06\,\text{cm}^{-1}\,\text{K}^{-1})\ln(2)} = 622.\overline{7}\,\text{K} \approx \boxed{623\,\text{K}}$$

E16.2(b)   **(a)** $\Lambda = h\left(\dfrac{\beta}{2\pi m}\right)^{1/2}$ [16.19] $= h\left(\dfrac{1}{2\pi mkT}\right)^{1/2}$

$$= (6.626 \times 10^{-34}\,\text{J s})$$

$$\times \left(\frac{1}{(2\pi) \times (39.95) \times (1.6605 \times 10^{-27}\,\text{kg}) \times (1.381 \times 10^{-23}\,\text{J K}^{-1}) \times T}\right)^{1/2}$$

$$= \frac{276\,\text{pm}}{(T/\text{K})^{1/2}}$$

(b) $q = \dfrac{V}{\Lambda^3}$ [16.19] $= \dfrac{(1.00 \times 10^{-6}\,\text{m}^3) \times (T/\text{K})^{3/2}}{(2.76 \times 10^{-10}\,\text{m})^3} = 4.76 \times 10^{22}(T/\text{K})^{3/2}$

(i) $T = 300\,\text{K}$, $\Lambda = 1.59 \times 10^{-11}\,\text{m} = \boxed{15.9\,\text{pm}}$, $q = \boxed{2.47 \times 10^{26}}$,

(ii) $T = 3000\,\text{K}$, $\Lambda = \boxed{5.04\,\text{pm}}$, $q = \boxed{7.82 \times 10^{27}}$

Question. At what temperature does the thermal wavelength of an argon atom become comparable to its diameter?

E16.3(b)   The translational partition function is

$$q_{\text{tr}} = \frac{V}{h^3}(2kT\pi m)^{3/2}$$

so $\dfrac{q_{\text{Xe}}}{q_{\text{He}}} = \left(\dfrac{m_{\text{Xe}}}{m_{\text{He}}}\right)^{3/2} = \left(\dfrac{131.3\,\text{u}}{4.003\,\text{u}}\right)^{3/2} = \boxed{187.9}$

E16.4(b) $q = \displaystyle\sum_{\text{levels}} g_j e^{-\beta\varepsilon_j} = 2 + 3e^{-\beta\varepsilon_1} + 2e^{-\beta\varepsilon_2}$

$$\beta\varepsilon = \frac{hc\tilde{\nu}}{kT} = \frac{1.4388(\tilde{\nu}/\text{cm}^{-1})}{T/\text{K}}$$

Thus $q = 2 + 3e^{-(1.4388 \times 1250/2000)} + 2e^{-(1.4388 \times 1300/2000)}$

$$= 2 + 1.2207 + 0.7850 = \boxed{4.006}$$

E16.5(b) $E = U - U(0) = -\dfrac{N}{q}\dfrac{dq}{d\beta} = -\dfrac{N}{q}\dfrac{d}{d\beta}(2 + 3e^{-\beta\varepsilon_1} + 2e^{-\beta\varepsilon_2})$

$$= -\frac{N}{q}\left(-3\varepsilon_1 e^{-\beta\varepsilon_1} - 2\varepsilon_2 e^{-\beta\varepsilon_2}\right) = \frac{Nhc}{q}\left(3\tilde{v}e^{-\beta hc\tilde{v}_1} + 2\tilde{v}e^{-\beta hc\tilde{v}_2}\right)$$

$$= \left(\frac{N_A hc}{4.006}\right) \times \left\{3(1250\,\text{cm}^{-1}) \times \left(e^{-(1.4388 \times 1250/2000)}\right)\right.$$

$$\left. + 2(1300\,\text{cm}^{-1}) \times \left(e^{-(1.4388 \times 1300/2000)}\right)\right\}$$

$$= \left(\frac{N_A hc}{4.006}\right) \times (2546\,\text{cm}^{-1})$$

$$= (6.022 \times 10^{23}\,\text{mol}^{-1}) \times (6.626 \times 10^{-34}\,\text{J s}) \times (2.9979 \times 10^{10}\,\text{cm s}^{-1}) \times (2546\,\text{cm}^{-1})/4.006$$

$$= \boxed{7.605\,\text{kJ mol}^{-1}}$$

E16.6(b)   In fact there are two upper states, but one upper level. And of course the answer is different if the question asks when 15 percent of the molecules are in the upper level, or if it asks when 15 percent of the molecules are in *each* upper state. The solution below assumes the former.

The relative population of states is given by the Boltzmann distribution

$$\frac{n_2}{n_1} = \exp\left(\frac{-\Delta E}{kT}\right) = \exp\left(\frac{-hc\tilde{v}}{kT}\right) \quad \text{so} \quad \ln\frac{n_2}{n_1} = \frac{-hc\tilde{v}}{kT}$$

Thus $T = \dfrac{-hc\tilde{v}}{k\ln(n_2/n_1)}$

Having 15 percent of the molecules in the upper level means

$$\frac{2n_2}{n_1} = \frac{0.15}{1 - 0.15} \quad \text{so} \quad \frac{n_2}{n_1} = 0.088$$

and $T = \dfrac{-(6.626 \times 10^{-34}\,\text{J s}) \times (2.998 \times 10^{10}\,\text{cm s}^{-1}) \times (360\,\text{cm}^{-1})}{(1.381 \times 10^{-23}\,\text{J K}^{-1}) \times (\ln 0.088)}$

$$= \boxed{21\overline{3}\,\text{K}}$$

E16.7(b)   The energies of the states relative to the energy of the state with $m_I = 0$ are $-\gamma_N\hbar\mathcal{B}, 0, +\gamma_N\hbar\mathcal{B}$, where $\gamma_N\hbar = 2.04 \times 10^{-27}\,\text{J T}^{-1}$. With respect to the lowest level they are $0, \gamma_N\hbar, 2\gamma_N\hbar$.

The partition function is

$$q = \sum_{\text{states}} e^{-E_{\text{state}}/kT}$$

where the energies are measured with respect to the lowest energy. So in this case

$$q = 1 + \exp\left(\frac{-\gamma_N\hbar\mathcal{B}}{kT}\right) + \exp\left(\frac{-2\gamma_N\hbar\mathcal{B}}{kT}\right)$$

As $\mathscr{B}$ is increased at any given T, q decays from $q = 3$ toward $q = 1$ as shown in Figure 16.1(a).

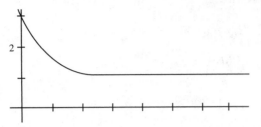

Figure 16.1(a)

The average energy (measured with respect to the lowest state) is

$$\langle E \rangle = \frac{\sum_{\text{states}} E_{\text{state}} e^{-E_{\text{state}}/kT}}{q} = \frac{1 + \gamma_N \hbar \mathscr{B} \exp\left(-\gamma_N \hbar \mathscr{B}/kT\right) + 2\gamma_N \hbar \mathscr{B} \exp\left(-2\gamma_N \hbar \mathscr{B}/kT\right)}{1 + \exp\left(-\gamma_N \hbar \mathscr{B}/kT\right) + \exp\left(-2\gamma_N \hbar \mathscr{B}/kT\right)}$$

The expression for the mean energy measured based on zero spin having zero energy becomes

$$E = \frac{\gamma_N \hbar \mathscr{B} - \gamma_N \hbar \mathscr{B} \exp\left(-2\gamma_N \hbar \mathscr{B}/kT\right)}{1 + \exp\left(-\gamma_N \hbar \mathscr{B}/kT\right) + \exp\left(-2\gamma_N \hbar \mathscr{B}/kT\right)} = \frac{\gamma_N \hbar \mathscr{B} \left(1 - \exp\left(-2\gamma_N \hbar \mathscr{B}/kT\right)\right)}{1 + \exp\left(-\gamma_N \hbar \mathscr{B}/kT\right) + \exp\left(-2\gamma_N \hbar \mathscr{B}/kT\right)}$$

As $\mathscr{B}$ is increased at constant T, the mean energy varies as shown in Figure 16.1(b).

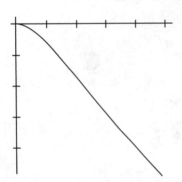

Figure 16.1(b)

The relative populations (with respect to that of the lowest state) are given by the Boltzmann factor

$$\exp\left(\frac{-\Delta E}{kT}\right) = \exp\left(\frac{-\gamma_N \hbar \mathscr{B}}{kT}\right) \quad \text{or} \quad \exp\left(\frac{-2\gamma_N \hbar \mathscr{B}}{kT}\right)$$

Note that $\dfrac{\gamma_N \hbar \mathscr{B}}{k} = \dfrac{(2.04 \times 10^{-27}\,\text{J}\,\text{T}^{-1}) \times (20.0\,\text{T})}{1.381 \times 10^{-23}\,\text{J}\,\text{K}^{-1}} = 2.95 \times 10^{-3}\,\text{K}$

so the populations are

(a) $\quad \exp\left(\dfrac{-2.95 \times 10^{-3}\,\text{K}}{1.0\,\text{K}}\right) = \boxed{0.99\overline{7}}$ $\quad$ and $\quad \exp\left(\dfrac{2(-2.95 \times 10^{-3}\,\text{K})}{1.0\,\text{K}}\right) = \boxed{0.99\overline{4}}$

(b) $\quad \exp\left(\dfrac{-2.95 \times 10^{-3}\,\text{K}}{298}\right) = \boxed{0.99999}$ $\quad$ and $\quad \exp\left(\dfrac{2(-2.95 \times 10^{-3}\,\text{K})}{298}\right) = \boxed{0.99998}$

E16.8(b)   **(a)** The ratio of populations is given by the Boltzmann factor

$$\frac{n_2}{n_1} = \exp\left(\frac{-\Delta E}{kT}\right) = e^{-25.0\,K/T} \quad \text{and} \quad \frac{n_3}{n_1} = e^{-50.0\,K/T}$$

(1) At 1.00 K

$$\frac{n_2}{n_1} = \exp\left(\frac{-25.0\,K}{1.00\,K}\right) = \boxed{1.39 \times 10^{-11}}$$

$$\text{and}\ \frac{n_3}{n_1} = \exp\left(\frac{-50.0\,K}{1.00\,K}\right) = \boxed{1.93 \times 10^{-22}}$$

(2) At 25.0 K

$$\frac{n_2}{n_1} = \exp\left(\frac{-25.0\,K}{25.0\,K}\right) = \boxed{0.368} \quad \text{and} \quad \frac{n_3}{n_1} = \exp\left(\frac{-50.0\,K}{25.0\,K}\right) = \boxed{0.135}$$

(3) At 100 K

$$\frac{n_2}{n_1} = \exp\left(\frac{-25.0\,K}{100\,K}\right) = \boxed{0.779} \quad \text{and} \quad \frac{n_3}{n_1} = \exp\left(\frac{-50.0\,K}{100\,K}\right) = \boxed{0.607}$$

(b) The molecular partition function is

$$q = \sum_{\text{states}} e^{-E_{\text{state}}/kT} = 1 + e^{-25.0\,K/T} + e^{-50.0\,K/T}$$

At 25.0 K, we note that $e^{-25.0\,K/T} = e^{-1}$ and $e^{-50.0\,K/T} = e^{-2}$

$$q = 1 + e^{-1} + e^{-2} = \boxed{1.503}$$

(c) The molar internal energy is

$$U_m = U_m(0) - \frac{N_A}{q}\left(\frac{\partial q}{\partial \beta}\right) \quad \text{where } \beta = (kT)^{-1}$$

So $U_m = U_m(0) - \dfrac{N_A}{q}(-25.0\,K)k\left(e^{-25.0\,K/T} + 2e^{-50.0\,K/T}\right)$

At 25.0 K

$$U_m - U_m(0) = -\frac{(6.022 \times 10^{23}\,\text{mol}^{-1}) \times (-25.0\,K) \times (1.381 \times 10^{-23}\,\text{J K}^{-1})}{1.503}$$

$$\times (e^{-1} + 2e^{-2})$$

$$= \boxed{88.3\,\text{J mol}^{-1}}$$

(d) The molar heat capacity is

$$C_{V,m} = \left(\frac{\partial U_m}{\partial T}\right)_V = N_A(25.0\,K)k\frac{\partial}{\partial T}\frac{1}{q}\left(e^{-25.0\,K/T} + 2e^{-50.0\,K/T}\right)$$

$$= N_A(25.0\,K)k \times \left(\frac{25.0\,K}{qT^2}\left(e^{-25.0\,K/T} + 4e^{-50.0\,K/T}\right)\right.$$

$$\left. -\frac{1}{q^2}\left(e^{-25.0\,K/T} + 2e^{-50.0\,K/T}\right)\frac{\partial q}{\partial T}\right)$$

where $\dfrac{\partial q}{\partial T} = \dfrac{25.0\,\mathrm{K}}{T^2}\left(e^{-25.0\,\mathrm{K}/T} + 2e^{-50.0\,\mathrm{K}/T}\right)$

so $C_{V,\mathrm{m}} = \dfrac{N_A(25.0\,\mathrm{K})^2 k}{T^2 q}\left(e^{-25.0\,\mathrm{K}/T} + 4e^{-50.0\,\mathrm{K}/T} - \dfrac{(e^{-25.0\,\mathrm{K}/T} + 2e^{-50.0\,\mathrm{K}/T})^2}{q}\right)$

At 25.0 K

$$C_{V,\mathrm{m}} = \dfrac{(6.022 \times 10^{23}\,\mathrm{mol}^{-1}) \times (25.0\,\mathrm{K})^2 \times (1.381 \times 10^{-23}\,\mathrm{J\,K}^{-1})}{(25.0\,\mathrm{K})^2 \times (1.503)}$$

$$\times \left(e^{-1} + 4e^{-2} - \dfrac{(e^{-1} + 2e^{-2})^2}{1.503}\right)$$

$$= \boxed{3.53\,\mathrm{J\,K}^{-1}\,\mathrm{mol}^{-1}}$$

(e) The molar entropy is

$$S_{\mathrm{m}} = \dfrac{U_{\mathrm{m}} - U_{\mathrm{m}}(0)}{T} + N_A k \ln q$$

At 25.0 K

$$S_{\mathrm{m}} = \dfrac{88.3\,\mathrm{J\,mol}^{-1}}{25.0\,\mathrm{K}} + (6.022 \times 10^{23}\,\mathrm{mol}^{-1}) \times (1.381 \times 10^{-23}\,\mathrm{J\,K}^{-1})\ln 1.503$$

$$= \boxed{6.92\,\mathrm{J\,K}^{-1}\,\mathrm{mol}^{-1}}$$

E16.9(b) $\quad \dfrac{n_1}{n_0} = \dfrac{g_1 e^{-\varepsilon_1/kT}}{g_0 e^{-\varepsilon_0/kT}} = g_1 e^{-\Delta\varepsilon/kT} = 3e^{-hcB/kT}$

Set $\dfrac{n_1}{n_0} = \dfrac{1}{e}$ and solve for T.

$$\ln\left(\dfrac{1}{e}\right) = \ln 3 + \left(\dfrac{-hcB}{kT}\right)$$

$$T = \dfrac{hcB}{k(1 + \ln 3)}$$

$$= \dfrac{6.626 \times 10^{-34}\,\mathrm{J\,s} \times 2.998 \times 10^{10}\,\mathrm{cm\,s}^{-1} \times 10.593\,\mathrm{cm}^{-1}}{+1.381 \times 10^{-23}\,\mathrm{J\,K}^{-1} \times (1 + 1.0986)}$$

$$= \boxed{7.26\,\mathrm{K}}$$

E16.10(b) The Sackur–Tetrode equation gives the entropy of a monatomic gas as

$$S = nR \ln\left(\dfrac{e^{5/2}kT}{p\Lambda^3}\right) \quad \text{where } \Lambda = \dfrac{h}{\sqrt{2kT\pi m}}$$

(a) At 100 K

$$\Lambda = \dfrac{6.626 \times 10^{-34}\,\mathrm{J\,s}}{\left\{2(1.381 \times 10^{-23}\,\mathrm{J\,K}^{-1}) \times (100\,\mathrm{K}) \times \pi(131.3\,\mathrm{u}) \times (1.66054 \times 10^{-27}\,\mathrm{kg\,u}^{-1})\right\}^{1/2}}$$

$$= 1.52 \times 10^{-11}\,\mathrm{m}$$

and $S_m = (8.3145 \, \text{J K}^{-1} \, \text{mol}^{-1}) \ln \left(\dfrac{e^{5/2}(1.381 \times 10^{-23} \, \text{J K}^{-1}) \times (100 \, \text{K})}{(1.013 \times 10^5 \, \text{Pa}) \times (1.52 \times 10^{-11} \, \text{m})^3} \right)$

$= \boxed{147 \, \text{J K}^{-1} \, \text{mol}^{-1}}$

(b) At 298.15 K

$\Lambda = \dfrac{6.626 \times 10^{-34} \, \text{J s}}{\left\{ 2(1.381 \times 10^{-23} \, \text{J K}^{-1}) \times (298.15 \, \text{K}) \times \pi(131.3 \, \text{u}) \times (1.660\,54 \times 10^{-27} \, \text{kg u}^{-1}) \right\}^{1/2}}$

$= 8.822 \times 10^{-12} \, \text{m}$

and

$S_m = (8.3145 \, \text{J K}^{-1} \, \text{mol}^{-1}) \ln \left(\dfrac{e^{5/2}(1.381 \times 10^{-23} \, \text{J K}^{-1}) \times (298.15 \, \text{K})}{(1.013 \times 10^5 \, \text{Pa}) \times (8.822 \times 10^{-12} \, \text{m})^3} \right)$

$= \boxed{169.6 \, \text{J K}^{-1} \, \text{mol}^{-1}}$

E16.11(b) $q = \dfrac{1}{1 - e^{-\beta \varepsilon}} = \dfrac{1}{1 - e^{-hc\beta \tilde{\nu}}}$

$hc\beta \tilde{\nu} = \dfrac{(1.4388 \, \text{cm K}) \times (321 \, \text{cm}^{-1})}{600 \, \text{K}} = 0.769\overline{76}$

Thus $q = \dfrac{1}{1 - e^{-0.769\overline{76}}} = 1.86\overline{3}$

The internal energy due to vibrational excitation is

$U - U(0) = \dfrac{N\varepsilon e^{-\beta \varepsilon}}{1 - e^{-\beta \varepsilon}}$

$= \dfrac{Nhcve^{-hc\tilde{\nu}\beta}}{1 - e^{-hc\tilde{\nu}\beta}} = \dfrac{Nhc\tilde{\nu}}{e^{hc\tilde{\nu}\beta} - 1} = (0.86\overline{3}) \times (Nhc) \times (321 \, \text{cm}^{-1})$

and hence $\dfrac{S_m}{N_A k} = \dfrac{U - U(0)}{N_A kT} + \ln q = (0.863) \times \left(\dfrac{hc}{kT} \right) \times (321 \, \text{cm}^{-1}) + \ln(1.86\overline{3})$

$= \dfrac{(0.86\overline{3}) \times (1.4388 \, \text{K cm}) \times (321 \, \text{cm}^{-1})}{600 \, \text{K}} + \ln(1.86\overline{3})$

$= 0.66\overline{4} + 0.6219\overline{9} = 1.28\overline{6}$

and $S_m = 1.28\overline{6} R = \boxed{10.7 \, \text{J K}^{-1} \, \text{mol}^{-1}}$

E16.12(b) Inclusion of a factor of $(N!)^{-1}$ is necessary when considering indistinguishable particles. Because of their translational freedom, gases are collections of indistinguishable particles. The factor, then, must be included in calculations on $\boxed{\text{(a) } CO_2 \text{ gas}}$.

Solutions to problems

Solutions to numerical problems

P16.1 Number of configurations of combined system, $W = W_1 W_2$.

$$W = (10^{20}) \times (2 \times 10^{20}) = \boxed{2 \times 10^{40}}.$$

$$S = k \ln W [16.34]; \quad S_1 = k \ln W_1; \quad S_2 = k \ln W_2.$$

$$S = k \ln(2 \times 10^{40}) = k\{\ln 2 + 40 \ln 10\} = 92.8k$$

$$= 92.8 \times (1.381 \times 10^{-23} \text{ J K}^{-1}) = \boxed{1.282 \times 10^{-21} \text{ J K}^{-1}}.$$

$$S_1 = k \ln(10^{20}) = k\{20 \ln 10\} = 46.1k$$

$$= 46.1 \times (1.381 \times 10^{-23} \text{ J K}^{-1}) = \boxed{0.637 \times 10^{-21} \text{ J K}^{-1}}.$$

$$S_2 = k \ln(2 \times 10^{20}) = k\{\ln 2 + 20 \ln 10\} = 46.7k$$

$$= 46.7 \times (1.381 \times 10^{-23} \text{ J K}^{-1}) = \boxed{0.645 \times 10^{-21} \text{ J K}^{-1}}.$$

These results are significant in that they show that the statistical mechanical entropy is an additive property consistent with the thermodynamic result. That is, $S = S_1 + S_2 = (0.637 \times 10^{-21} + 0.645 \times 10^{-21}) \text{ J K}^{-1} = 1.282 \times 10^{-21} \text{ J K}^{-1}$.

P16.3 $S = k \ln W$ [16.34].

Therefore,

$$\left(\frac{\partial S}{\partial U}\right)_V = \frac{k}{W}\left(\frac{\partial W}{\partial U}\right)_V$$

or

$$\left(\frac{\partial W}{\partial U}\right)_V = \frac{W}{k}\left(\frac{\partial S}{\partial U}\right)_V.$$

But from eqn 3.45

$$\left(\frac{\partial U}{\partial S}\right)_V = T.$$

So,

$$\left(\frac{\partial S}{\partial U}\right)_V = \frac{1}{T}.$$

Then

$$\left(\frac{\partial W}{\partial U}\right)_V = \frac{W}{k}\left(\frac{1}{T}\right).$$

Therefore,

$$\frac{\Delta W}{W} \approx \frac{\Delta U}{kT}$$

$$= \frac{100 \times 10^3 \text{ J}}{(1.381 \times 10^{-23} \text{ J K}^{-1}) \times 298 \text{ K}}$$

$$\boxed{= 2.4 \times 10^{25}}$$

P16.5 $q = \dfrac{V}{\Lambda^3}, \qquad \Lambda = \dfrac{h}{(2\pi mkT)^{1/2}} \qquad \left[16.19, \ \beta = \dfrac{1}{kT}\right],$

and hence

$$T = \left(\frac{h^2}{2\pi mk}\right) \times \left(\frac{q}{V}\right)^{2/3}$$

$$= \left(\frac{(6.626 \times 10^{-34} \text{ J s})^2}{(2\pi) \times (39.95) \times (1.6605 \times 10^{27} \text{ kg}) \times (1.381 \times 10^{-23} \text{ J K}^{-1})}\right)$$

$$\times \left(\frac{10}{1.0 \times 10^{-6} \text{ m}^3}\right)^{2/3}$$

$$= \boxed{3.5 \times 10^{-15} \text{ K}} \text{ [a very low temperature]}.$$

The exact partition function in one dimension is

$$q = \sum_{n=1}^{\infty} e^{-(n^2-1)h^2\beta/8mL^2}.$$

For an Ar atom in a cubic box of side 1.0 cm,

$$\frac{h^2\beta}{8mL^2} = \frac{(6.626 \times 10^{-34} \text{ J s})^2}{(8) \times (39.95) \times (1.6605 \times 10^{-27} \text{ kg}) \times (1.381 \times 10^{-23} \text{ J K}^{-1}) \times (3.5 \times 10^{-15} \text{ K}) \times (1.0 \times 10^{-2} \text{ m})^2}$$

$$= 0.17\bar{1}.$$

Then $q = \displaystyle\sum_{n=1}^{\infty} e^{-0.17\bar{1}(n^2-1)} = 1.00 + 0.60 + 0.25 + 0.08 + 0.02 + \cdots = 1.95.$

The partition function for motion in three dimensions is therefore $q = (1.95)^3 = \boxed{7.41}$.

COMMENT. Temperatures as low as 3.5×10^{-15} K have never been achieved. However, a temperature of 2×10^{-8} K has been attained by adiabatic nuclear demagnetization (Chapter 3).

Question. Does the integral approximation apply at 2×10^{-8} K?

P16.7(b)   **(a)** $q = \sum_j g_j e^{-\beta\varepsilon_j} [16.9] = \sum_j g_j e^{-hc\beta\tilde{\nu}_j}.$

We use $hc\beta = \dfrac{1}{207 \text{ cm}^{-1}}$ at 298 K and $\dfrac{1}{3475 \text{ cm}^{-1}}$ at 5000 K. Therefore,

(i) $q = 5 + e^{-4707/207} + 3e^{-4751/207} + 5e^{-10559/207}$

$$= (5) + (1.3 \times 10^{-10}) + (3.2 \times 10^{-10}) + (3.5 \times 10^{-22}) = \boxed{5.00}.$$

(ii) $q = 5 + e^{-4707/3475} + 3e^{-4751/3475} + 5e^{-10559/3475}$

$$= (5) + (0.26) + (0.76) + (0.24) = \boxed{6.26}.$$

(b) $p_j = \dfrac{g_j e^{-\beta \varepsilon_j}}{q} = \dfrac{g_j e^{-hc\beta \tilde{v}_j}}{q}$ [16.7, with degeneracy g_j included]

Therefore, $p_0 = \dfrac{5}{q} = \boxed{1.00}$ at 298 K and $\boxed{0.80}$ at 5000 K.

$$p_2 = \frac{3e^{-4751/207}}{5.00} = \boxed{6.5 \times 10^{-11}} \text{ at 298 K.}$$

$$p_2 = \frac{3e^{-4751/3475}}{6.26} = \boxed{0.12} \text{ at 5000 K.}$$

(c) $S_m = \dfrac{U_m - U_m(0)}{T} + Nk \ln q$ [16.35].

We need $U_m - U_m(0)$, and evaluate it by explicit summation

$$U_m - U_m(0) = E = \frac{N_A}{q} \sum_j g_j \varepsilon_j e^{-\beta \varepsilon_j} \text{ [16.28 with degeneracy } g_j \text{ included].}$$

In terms of wavenumber units

(i) $\dfrac{U_m - U_m(0)}{N_A hc} = \dfrac{1}{5.00}\{0 + 4707 \text{ cm}^{-1} \times e^{-4707/207} + \cdots\} = 4.32 \times 10^{-7} \text{cm}^{-1}$,

(ii) $\dfrac{U_m - U_m(0)}{N_A hc} = \dfrac{1}{6.26}\{0 + 4707 \text{ cm}^{-1} \times e^{-4707/3475} + \cdots\} = 1178 \text{ cm}^{-1}.$

Hence, at 298 K

$$U_m - U_m(0) = 5.17 \times 10^{-6} \text{ J mol}^{-1}$$

and at 5000 K

$$U_m - U_m(0) = 14.10 \text{ kJ mol}^{-1}.$$

It follows that

(i) $S_m = \left(\dfrac{5.17 \times 10^{-6} \text{ J mol}^{-1}}{298 \text{ k}}\right) + (8.314 \text{ J K}^{-1} \text{ mol}^{-1}) \times (\ln 5.00)$

$= \boxed{13.38 \text{ J K}^{-1} \text{ mol}^{-1}}$ [essentially $R \ln 5$].

(ii) $S_m = \left(\dfrac{14.09 \times 10^3 \text{ J mol}^{-1}}{5000\text{K}}\right) + (8.314 \text{ J K}^{-1}\text{mol}^{-1}) \times (\ln 6.26) = \boxed{18.07 \text{ J K}^{-1} \text{ mol}^{-1}}.$

P16.9 $q = \sum_j g_j e^{-\beta \varepsilon_j}$ [16.9] $= \sum_j g_j e^{-hc\beta \tilde{v}_j}.$

$$p_i = \frac{g_i e^{-\beta \varepsilon_i}}{q} \text{ [16.7]} = \frac{g_i e^{-hc\beta \tilde{v}_i}}{q}.$$

We measure energies from the lower states, and write

$$q = 2 + 2e^{-hc\beta\tilde{\nu}} = 2 + 2e^{-(1.4388\times121.1)/(T/K)} = 2 + 2e^{-174.2/(T/K)}.$$

This function is plotted in Fig. 16.2.

(a) At 300 K,

$$p_0 = \frac{2}{q} = \frac{1}{1 + e^{-174.2/300}} = \boxed{0.64},$$

$$p_1 = 1 - p_0 = \boxed{0.36}.$$

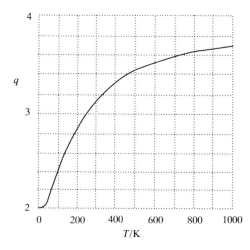

q

T/K

Figure 16.2

(b) The electronic contribution to U_m in wavenumber units is

$$\frac{U_m - U_m(0)}{N_A hc} = -\frac{1}{hcq}\frac{dq}{d\beta} \ [16.31a] = \frac{2\tilde{\nu}e^{-hc\beta\tilde{\nu}}}{q}$$

$$= \frac{(121.1 \text{ cm}^{-1}) \times (e^{-174.2/300})}{1 + e^{-174.2/300}} = 43.45 \text{ cm}^{-1}$$

which corresponds to $\boxed{0.52 \text{ kJ mol}^{-1}}$.

For the electronic contribution to the molar entropy, we need q and $U_m - U_m(0)$ at 500 K as well as at 300 K. These are

	300 K	500 K
$U_m - U_m(0)$	0.518 kJ mol^{-1}	0.599 kJ mol^{-1}
q	3.120	3.412

Then we form

$$S_m = \frac{U_m - U_m(0)}{T} + R \ln q \; [16.35].$$

At 300 K $\quad S_m = \left(\frac{518 \text{ J mol}^{-1}}{300 \text{ K}}\right) + (8.3141 \text{ J K}^{-1}\text{mol}^{-1}) \times (\ln 3.120) = \boxed{11.2 \text{ J K}^{-1} \text{ mol}^{-1}}.$

At 1500 K $\quad S_m = \left(\frac{518 \text{ J mol}^{-1}}{500 \text{ K}}\right) + (8.3141 \text{ J K}^{-1}\text{mol}^{-1}) \times (\ln 3.412) = \boxed{11.4 \text{ J K}^{-1} \text{ mol}^{-1}}.$

P16.11 $\qquad q = \sum_i e^{-\beta \varepsilon_i} = \sum_i e^{-hc\beta \tilde{\nu}_i} \; [16.8]$

At 100 K, $hc\beta = \dfrac{1}{69.50 \text{ cm}^{-1}}$ and, at 298 K, $hc\beta = \dfrac{1}{207.22 \text{ cm}^{-1}}$. Therefore, at 100 K,

(a) $q = 1 + e^{-213.30/69.50} + e^{-435.39/69.50} + e^{-636.27/69.50} + e^{-845.93/69.50} = \boxed{1.049}$

and at 298 K

(b) $q = 1 + e^{-213.30/207.22} + e^{-425.39/207.22} + e^{-636.27/207.22} + e^{-845.93/207.22} = \boxed{1.55}.$

In each case, $p_i = \dfrac{e^{-hc\beta \tilde{\nu}_i}}{q} \; [16.7].$

$$p_0 = \frac{1}{q} = \text{(a)} \boxed{0.953}, \quad \text{(b)} \boxed{0.645}.$$

$$p_1 = \frac{e^{-hc\beta \tilde{\nu}_1}}{q} = \text{(a)} \boxed{0.044}, \quad \text{(b)} \boxed{0.230}.$$

$$p_2 = \frac{e^{-hc\beta \tilde{\nu}_2}}{q} = \text{(a)} \boxed{0.002}, \quad \text{(b)} \boxed{0.083}.$$

For the molar entropy we need to form $U_m - U_m(0)$ by explicit summation:

$$U_m - U_m(0) = \frac{N_A}{q} \sum_i \varepsilon_i e^{-\beta \varepsilon_i} = \frac{N_A}{q} \sum_i hc\tilde{\nu}_i e^{-hc\beta \tilde{\nu}_i} \; [16.29, \; 16.30]$$

$$= \boxed{123 \text{ J mol}^{-1} \text{ (at 100 K)}}, \boxed{1348 \text{ J mol}^{-1} \text{ (at 298 K)}}.$$

$$S_m = \frac{U_m - U_m(0)}{T} + R \ln q \; [16.35].$$

(a) $S_m = \dfrac{123 \text{ J mol}^{-1}}{100 \text{ K}} + R \ln 1.049 = \boxed{1.63 \text{ J K}^{-1} \text{ mol}^{-1}}.$

(b) $S_m = \dfrac{1348 \text{ J mol}^{-1}}{298 \text{ K}} + R \ln 1.55 = \boxed{8.17 \text{ J K}^{-1} \text{ mol}^{-1}}.$

Solutions to theoretical problems

P16.13 **(a)** $W = \dfrac{N!}{n_1! n_2! \cdots} \; [16.1] = \dfrac{5!}{0! 5! 0! 0! 0!} = \boxed{1}.$

(b) We draw up the following table.

0	ε	2ε	3ε	4ε	5ε	$W = \dfrac{N!}{n_1!n_2!\cdots}$
4	0	0	0	0	1	5
3	1	0	0	1	0	20
3	0	1	1	0	0	20
2	2	0	1	0	0	30
2	1	2	0	0	0	30
1	3	1	0	0	0	20
0	5	0	0	0	0	1

The most probable configurations are $\boxed{\{2, 2, 0, 1, 0, 0\}}$ and $\boxed{\{2, 1, 2, 0, 0, 0\}}$ jointly.

P16.15 **(a)** $\dfrac{n_j}{n_0} = e^{-\beta(\varepsilon_j - \varepsilon_0)} = e^{-\beta j\varepsilon}$, which implies that $-j\beta\varepsilon = \ln n_j - \ln n_0$ and therefore that $\boxed{\ln n_j = \ln n_0 - \dfrac{j\varepsilon}{kT}}$.

Therefore, a plot of $\ln n_j$ against j should be a straight line with slope $-\dfrac{\varepsilon}{kT}$. Alternatively, plot $\ln p_j$ against j, since

$$\boxed{\ln p_j = \text{const} - \frac{j\varepsilon}{kT}}.$$

We draw up the following table using the information in Problem 16.8.

j	0	1	2	3	
n_j	4	2	2	1	[most probable configuration]
$\ln n_j$	1.39	0.69	0.69	0	

These are points plotted in Fig. 16.3 (full line). The slope is -0.46 and, since $\dfrac{\varepsilon}{hc} = 50\,\text{cm}^{-1}$, the slope corresponds to a temperature

$$T = \frac{(50\,\text{cm}^{-1}) \times (2.998 \times 10^{10}\,\text{cm s}^{-1}) \times (6.626 \times 10^{-34}\,\text{J s})}{(0.46) \times (1.381 \times 10^{-23}\,\text{J K}^{-1})} = \boxed{160\,\text{K}}.$$

(A better estimate, 104 K represented by the dashed line in Fig. 16.3, is found in Problem 16.17.)

(b) Choose one of the weight 2520 configurations and one of the weight 504 configurations, and draw up the following table.

	j	0	1	2	3	4
$W = 2520$	n_j	4	3	1	0	1
	$\ln n_j$	1.39	1.10	0	$-\infty$	0
$W = 504$	n_j	6	0	1	1	1
	$\ln n_j$	1.79	$-\infty$	0	0	0

Inspection confirms that these data give very crooked lines.

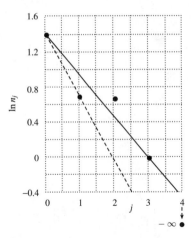

$-\infty$ ● **Figure 16.3**

P16.17 **(a)** $U - U(0) = -N\dfrac{\mathrm{d}\ln q}{\mathrm{d}\beta}$ [16.31b], with $q = \dfrac{1}{1 - \mathrm{e}^{-\beta\varepsilon}}$ [16.12].

$$\frac{\mathrm{d}\ln q}{\mathrm{d}\beta} = \frac{1}{q}\frac{\mathrm{d}q}{\mathrm{d}\beta} = \frac{-\varepsilon\mathrm{e}^{-\beta\varepsilon}}{1 - \mathrm{e}^{-\beta\varepsilon}}.$$

$$a\varepsilon = \frac{U - U(0)}{N} = \frac{\varepsilon\mathrm{e}^{(-\beta\varepsilon)}}{1 - \mathrm{e}^{-\beta\varepsilon}} = \frac{\varepsilon}{\mathrm{e}^{\beta\varepsilon} - 1}.$$

Hence, $\mathrm{e}^{\beta\varepsilon} = \dfrac{1+a}{a}$, implying that, $\beta = \dfrac{1}{\varepsilon}\ln\left(1 + \dfrac{1}{a}\right)$.

For a mean energy of $\varepsilon, a = 1$, $\beta = \dfrac{1}{\varepsilon}\ln 2$, implying that

$$T = \frac{\varepsilon}{k\ln 2}\ln 2 = (50\ \mathrm{cm}^{-1}) \times \left(\frac{hc}{k\ln 2}\right) = \boxed{104\ \mathrm{K}}.$$

(b) $q = \dfrac{1}{1 - \mathrm{e}^{-\beta\varepsilon}} = \dfrac{1}{1 - \left(\dfrac{a}{1+a}\right)} = \boxed{1 + a}$.

(c) $\dfrac{S}{Nk} = \dfrac{U - U(0)}{NkT} + \ln q\ [16.35] = a\beta\varepsilon + \ln q$

$$= a\ln\left(1 + \frac{1}{a}\right) + \ln(1 + a) = a\ln(1 + a) - a\ln a + \ln(1 + a)$$

$$= \boxed{(1 + a)\ln(1 + a) - a\ln a}.$$

When the mean energy is ε, $a = 1$ and then $\boxed{\dfrac{S}{Nk} = 2\ln 2}$.

P16.19 $p = kT\left(\dfrac{\partial \ln Q}{\partial V}\right)_{T,N}$ [17.3]

$$= kT\left(\frac{\partial \ln(q^N/N!)}{\partial V}\right)_{T,N}\ [16.45\mathrm{b}]$$

$$= kT\left(\frac{\partial[N\ln q - \ln N!]}{\partial V}\right)_{T,N} = NkT\left(\frac{\partial \ln q}{\partial V}\right)_{T,N}$$

$$= NkT \left(\frac{\partial \ln(V/\Lambda^3)}{\partial V} \right)_{T,N}$$

$$= NkT \left(\frac{\partial [\ln V - \ln \Lambda^3]}{\partial V} \right)_{T,N} = NkT \left(\frac{\partial \ln V}{\partial V} \right)_{T,N}$$

$$= \frac{NkT}{V} \quad \text{or} \quad \boxed{pV = NkT = nRT}.$$

Solutions to applications

P16.21   At equilibrium $\frac{N(r)/V}{N(r_0)/V} = e^{-\{V(r)-V(r_0)\}/kT}$ [16.6a]

Since $V(r) = -GMm/r$, $V(\infty) = 0$ and [Note: $V(r)$ is potential energy, V is volume]

$$\frac{N(\infty)/V}{N(r_0)/V} = e^{V(r_0)/kT}$$

which says that $N(\infty)/V \propto e^{V(r_0)/kT}$ = constant. This is obviously not the current distribution for planetary atmospheres where $\lim_{r \to \infty} N(r)/V = 0$. Consequently, we may conclude that the earth's atmosphere, or any other planetary atmosphere, cannot be at equilibrium.

P16.23   Each protein binding site can be represented as a distinct box into which a ligand, L, may bind. All possible configurations are shown in the following table and the configuration count of i indistinguishable ligands being placed in n distinguishable sites is seen to be given by the combinatorial: $C(n, i) = \frac{n!}{(n-i)!i!}$

$$C(4, 0) = \frac{4!}{(4-0)!0!} = 1 \text{ conformation}$$

$$C(4, 1) = \frac{4!}{(4-1)!1!} = 4 \text{ conformations}$$

$$C(4, 2) = \frac{4!}{(4-2)!2!} = 6 \text{ conformations}$$

$$C(4, 3) = \frac{4!}{(4-3)!3!} = 4 \text{ conformations}$$

$$C(4, 0) = \frac{4!}{(4-0)!0!} = 1 \text{ conformation}$$

P16.25 $$p_i = \frac{(n-i+1)\sigma s^i}{q} = \frac{(n-i+1)\sigma s^i}{1 + \sum_{i=1}^{n}(n-i+1)\sigma s^i}, \quad \langle i \rangle = \sum_{i=0}^{n} i p_i.$$

(a) The fraction distribution of molecules with i coiled residues depends dramatically upon the value of the stability parameter s. When $s < 1$, low values of i are observed but large i values are observed when $s > 1$. Thus, when $s < 1$, the polypeptide is largely helical and, when $s > 1$, it is more of a random coil. See Fig. 16.4(a).

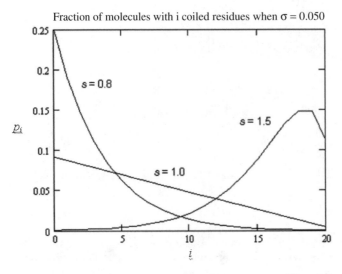

Figure 16.4(a)

(b) The $\langle i \rangle$ plot, Fig. 16.4(b), shows that for $s < 0.5$ the polypeptide model is helical and little changed as s is varied. At the other extreme, for $s > 1.5$ the polypeptide is largely a random coil which changes only slightly with variance of s. The mean number of coiled residues changes rapidly in the middle range of $0.5 < s < 1.5$ giving an overall sigmoidal dependence upon s.

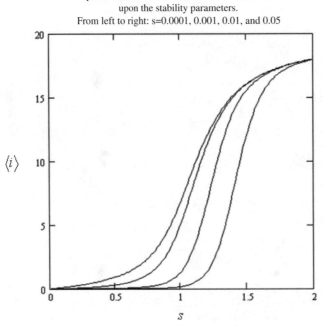

Figure 16.4(b)

17 Statistical thermodynamics 2: applications

Answers to discussion questions

D17.1 An approximation involved in the derivation of all of these expressions is the assumption that the contributions from the different modes of motion are separable. The expression $q^R = kT/hcB$ is the high temperature approximation to the rotational partition function for nonsymmetrical linear rotors. The expression $q^V = kT/hc\tilde{v}$ is the high temperature form of the partition function for one vibrational mode of the molecule in the harmonic approximation. The expression $q^E = g^E$ for the electronic partition function applies at normal temperatures to atoms and molecules with no low lying excited electronic energy levels.

D17.3 Residual entropy is due to the presence of some disorder in the system even at $T = 0$. It is observed in systems where there is very little energy difference—or none—between alternative arrangements of the molecules at very low temperatures. Consequently, the molecules cannot lock into a preferred orderly arrangement and some disorder persists.

D17.5 Equations of state can be thought of as expressions for the pressure of a gas in terms of the state functions, n, V, and T. They are obtained from the expression for the pressure in terms of the canonical partition function given in eqn 17.3.

$$p = kT \left(\frac{\partial \ln Q}{\partial V} \right)_T.$$

Partition functions for perfect and imperfect gases are different. That for the perfect gas is given by $Q = q^N/N!$ with $q = V/\Lambda^3$. There is no one form for imperfect gases. One example is shown in Self-test 17.1. Another which can be shown to lead to the van der Waals equation of state is

$$Q = \frac{1}{N!} \left(\frac{2\pi mkT}{h^2} \right)^{3N/2} (V - Nb)e^{aN^2/kTV}.$$

For the case of the perfect gas there are no molecular features in the partition function, but for imperfect gases there are repulsive and attractive features in the partition function that are related to the structure of the molecules.

D17.7 See *Justification* 17.4 for a derivation of the general expression (eqn 17.54b) for the equilibrium constant in terms of the partition functions and difference in molar energy, $\Delta_r E_0$, of the products and reactants in a chemical reaction. The partition functions are functions of temperature and the ratio of partition functions in eqn 17.54b will therefore vary with temperature. However, the most direct effect of temperature on the equilibrium constant is through the exponential term $e^{-\Delta_r E_0/RT}$. The manner in which both factors affect the magnitudes of the equilibrium constant and its variation with temperature is described in detail for a simple $R \rightleftharpoons P$ gas phase equilibrium in Section 17.8(c) and *Justification* 17.5.

Solutions to exercises

E17.1(b) $C_{V,m} = \frac{1}{2}(3 + v_R^* + 2v_V^*)R$ [17.35]

with a mode active if $T > \theta_M$.

(a) O_3 : $C_{V,m} = \frac{1}{2}(3 + 3 + 0)R = 3R$ [experimental $= 3.7R$]

(b) C_2H_6 : $C_{V,m} = \frac{1}{2}(3 + 3 + 2 \times 1)R = 4R$ [experimental $= 6.3R$]

(c) CO_2 : $C_{V,m} = \frac{1}{2}(3 + 2 + 0)R = \frac{5}{2}R$ [experimental $= 4.5R$]

Consultation of the Herzberg references in *Further reading*, Chapters 13 and 14, turns up only one vibrational mode among these molecules whose frequency is low enough to have a vibrational temperature near room temperature. That mode was in C_2H_6, corresponding to the "internal rotation" of CH_3 groups. The discrepancies between the estimates and the experimental values suggest that there are vibrational modes in each molecule that contribute to the heat capacity—albeit not to the full equipartition value—that our estimates have classified as inactive.

E17.2(b)   The equipartition theorem would predict a contribution to molar heat capacity of $\frac{1}{2}R$ for every translational and rotational degree of freedom and R for each vibrational mode. For an ideal gas, $C_{p,m} = R + C_{V,m}$. So for CO_2

With vibrations

$$C_{V,m}/R = 3\left(\tfrac{1}{2}\right) + 2\left(\tfrac{1}{2}\right) = (3 \times 4 - 6) = 6.5 \quad \text{and} \quad \gamma = \frac{7.5}{6.5} = \boxed{1.15}$$

Without vibrations $C_{V,m}/R = 3\left(\tfrac{1}{2}\right) + 2\left(\tfrac{1}{2}\right) = 2.5 \quad \text{and} \quad \gamma = \frac{3.5}{2.5} = \boxed{1.40}$

Experimental $\gamma = \dfrac{37.11 \text{ J mol}^{-1}\text{K}^{-1}}{(37.11 - 8.3145) \text{ J mol}^{-1}\text{K}^{-1}} = \boxed{1.29}$

The experimental result is closer to that obtained by neglecting vibrations, but not so close that vibrations can be neglected entirely.

E17.3(b)   The rotational partition function of a linear molecule is [Table 17.3]

$$q^R = \frac{0.6950}{\sigma} \times \frac{T/K}{(B/\text{cm}^{-1})} = \frac{(0.6950) \times (T/K)}{2 \times 1.4457} = 0.2404(T/K)$$

(a) At 25 °C: $q^R = (0.2404) \times (298) = \boxed{71.6}$

(b) At 250 °C: $q^R = (0.2404) \times (523) = \boxed{126}$

E17.4(b)   The symmetry number is the order of the rotational subgroup of the group to which a molecule belongs (except for linear molecules, for which $\sigma = 2$ if the molecule has inversion symmetry and 1 otherwise).

(a) CO_2: full group $D_{\infty h}$; subgroup C_2; hence $\sigma = \boxed{2}$

(b) O_3: full group C_{2v}; subgroup C_2; $\sigma = \boxed{2}$

(c) SO_3: full group D_{3h}; subgroup $\{E, C_3, C_3^2, 3C_2\}$; $\sigma = \boxed{6}$

(d) SF_6: full group O_h; subgroup O; $\sigma = \boxed{24}$

(e) Al_2Cl_6: full group D_{2d}; subgroup D_2; $\sigma = \boxed{4}$

E17.5(b)   The rotational partition function of a non-linear molecule is [Table 17.3]

$$q^R = \frac{1.0270}{\sigma} \frac{(T/K)^{3/2}}{(ABC/cm^{-3})^{1/2}} = \frac{1.0270 \times 298^{3/2}}{(2) \times (2.027\,36 \times 0.344\,17 \times 0.293\,535)^{1/2}} [\sigma = 2] = \boxed{5837}$$

The high-temperature approximation is valid if $T > \theta_R$, where

$$\theta_R = \frac{hc(ABC)^{1/3}}{k}$$

$$= \frac{(6.626 \times 10^{-34}\ J\ s) \times (2.998 \times 10^{10}\ cm\ s^{-1}) \times [(2.027\,36) \times (0.344\,17) \times (0.293\,535)\ cm^{-3}]^{1/3}}{1.381 \times 10^{-23}\ J\ K^{-1}}$$

$$= \boxed{0.8479\ K}$$

E17.6(b) $q^R = 5837$ [Exercise 17.5(b)]

All rotational modes of SO_2 are active at 25 °C; therefore

$$U_m^R - U_m^R(0) = E^R = \frac{3}{2}RT$$

$$S_m^R = \frac{E^R}{T} + R\ln q^R$$

$$= \tfrac{3}{2}R + R\ln(5837) = \boxed{84.57\ J\ K^{-1}\ mol^{-1}}$$

E17.7(b)   (a) The partition function is

$$q = \sum_{states} e^{-E_{state}/kT} = \sum_{levels} g e^{-E_{level}/kT}$$

where g is the degeneracy of the level. For rotations of a symmetric rotor such as CH_3CN, the energy levels are $E_J = hc[BJ(J+1) + (A-B)K^2]$ and the degeneracies are $g_{J,K} = 2(2J+1)$ if $K \neq 0$ and $2J+1$ if $K = 0$. The partition function, then, is

$$q = 1 + \sum_{J=1}^{\infty}(2J+1)e^{-\{hcBJ(J+1)/kT\}} \left(1 + 2\sum_{K=1}^{J} e^{-\{hc(A-B)K^2/kT\}}\right)$$

To evaluate this sum explicitly, we set up the following columns in a spreadsheet (values for $A = 5.28 \text{ cm}^{-1}$, $B = 5.2412 \text{ cm}^{-1}$, and $T = 298.15 \text{ K}$)

J	$J(J+1)$	$2J+1$	$e^{-hcBJ(J+1)/kT}$	J term	$e^{-\{hc(A-B)K^2/kT\}}$	K sum	J sum
0	0	1	1	1	1	1	1
1	2	3	0.997	8.832	0.976	2.953	9.832
2	6	5	0.991	23.64	0.908	4.770	33.47
3	12	7	0.982	43.88	0.808	6.381	77.35
⋮	⋮	⋮	⋮	⋮	⋮	⋮	⋮
82	6806	165	4.18×10^{-5}	0.079	8×10^{-71}	11.442	7498.95
83	6972	167	3.27×10^{-5}	0.062	2×10^{-72}	11.442	7499.01

The column labeled K sum is the term in large parentheses, which includes the inner summation. The J sum converges (to 4 significant figures) only at about $J = 80$; the K sum converges much more quickly. But the sum fails to take into account nuclear statistics, so it must be divided by the symmetry number ($\sigma = 3$). At 298 K, $q^R = \boxed{2.50 \times 10^3}$. A similar computation at $T = 500$ K yields $q^R = \boxed{5.43 \times 10^3}$.

(b) The rotational partition function of a nonlinear molecule is [Table 17.3 with $B = C$]

$$q^R = \frac{1.0270}{\sigma} \frac{(T/K)^{3/2}}{(ABC/\text{cm}^{-3})^{1/2}} = \frac{1.0270}{3} \frac{(T/K)^{3/2}}{(5.28 \times 0.307 \times 0.307)^{1/2}} = 0.485 \times (T/K)^{3/2}$$

At 298 K, $q^R = 0.485 \times 298^{3/2} = \boxed{2.50 \times 10^3}$

At 500 K, $q^R = 0.485 \times 500^{3/2} = \boxed{5.43 \times 10^3}$

The high-temperature approximation is certainly valid here.

E17.8(b) The rotational partition function of a nonlinear molecule is [Table 17.3]

$$q^R = \frac{1.0270}{\sigma} \frac{(T/K)^{3/2}}{(ABC/\text{cm}^{-3})^{1/2}} = \frac{1.0270 \times (T/K)^{3/2}}{(3.1752 \times 0.3951 \times 0.3505)^{1/2}} = 1.549 \times (T/K)^{3/2}$$

(a) At 25 °C, $q^R = 1.549 \times (298)^{3/2} = \boxed{7.97 \times 10^3}$

(b) At 100 °C, $q^R = 1.549 \times (373)^{3/2} = \boxed{1.12 \times 10^4}$

E17.9(b) The molar entropy of a collection of oscillators is given by

$$S_m = \frac{U_m - U_m(0)}{T} + k \ln Q \ [17.1] = \frac{N_A \langle \varepsilon \rangle}{T} + R \ln q$$

where $\langle \varepsilon \rangle = \dfrac{hc\tilde{v}}{e^{\beta hc\tilde{v}} - 1} = k\dfrac{\theta_V}{e^{\theta_V/T} - 1}$ [17.28], $q = \dfrac{1}{1 - e^{-\beta hc\tilde{v}}} = \dfrac{1}{1 - e^{-\theta_V/T}}$ [17.19]

and θ_V is the vibrational temperature $hc\tilde{v}/k$. Thus

$$S_m = \frac{R(\theta_V/T)}{e^{\theta_V/T} - 1} - R \ln(1 - e^{-\theta_V/T})$$

A plot of S_m/R versus T/θ_V is shown in Figure 17.1.

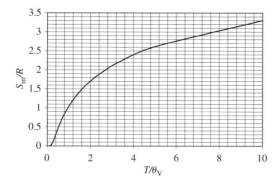

Figure 17.1

The vibrational entropy of ethyne is the sum of contributions of this form from each of its seven normal modes. The table below shows results from a spreadsheet programmed to compute S_m/R at a given temperature for the normal-mode wavenumbers of ethyne.

			$T = 298$ K		$T = 500$ K	
$\tilde{\nu}/cm^{-1}$	θ_V/K	T/θ_V	S_m/R		T/θ_V	S_m/R
612	880	0.336	0.216		0.568	0.554
729	1049	0.284	0.138		0.479	0.425
1974	2839	0.105	0.000 766		0.176	0.0229
3287	4728	0.0630	0.000 002 17		0.106	0.000 818
3374	4853	0.0614	0.000 001 46		0.103	0.000 652

The total vibrational heat capacity is obtained by summing the last column (twice for the first two entries, since they represent doubly degenerate modes).

(a) At 298 K, $S_m = 0.708R = \boxed{5.88 \text{ J mol}^{-1} \text{ K}^{-1}}$

(b) At 500 K, $S_m = 1.982R = \boxed{16.48 \text{ J mol}^{-1} \text{ K}^{-1}}$

E17.10(b) The contributions of rotational and vibrational modes of motion to the molar Gibbs energy depend on the molecular partition functions

$$G_m - G_m(0) = -RT \ln q \quad [17.9; \text{ also see Comment to Exercise 17.6(a)}]$$

The rotational partition function of a nonlinear molecule is given by

$$q^R = \frac{1}{\sigma}\left(\frac{kT}{hc}\right)^{3/2}\left(\frac{\pi}{ABC}\right)^{1/2} = \frac{1.0270}{\sigma}\left(\frac{(T/K)^3}{ABC/cm^{-3}}\right)^{1/2}$$

and the vibrational partition function for each vibrational mode is given by

$$q^V = \frac{1}{1 - e^{-\theta/T}} \quad \text{where } \theta = \frac{hc\tilde{\nu}}{k} = \frac{1.4388\,(\tilde{\nu}/cm^{-1})}{(T/K)}$$

At 298 K $\quad q^R = \dfrac{1.0270}{2} \left(\dfrac{298^3}{(3.553) \times (0.4452) \times (0.3948)} \right)^{1/2} = 3.35 \times 10^3$

and

$$G_m^R - G_m^R(0) = -(8.3145 \text{ J mol}^{-1}\text{K}^{-1}) \times (298 \text{ K}) \ln 3.35 \times 10^3$$

$$= -20.1 \times 10^3 \text{ J mol}^{-1} = \boxed{-20.1 \text{ kJ mol}^{-1}}$$

The vibrational partition functions are so small that we are better off taking

$$\ln q^V = -\ln(1 - e^{-\theta/T}) \approx e^{-\theta/T}$$

$$\ln q_1^V \approx e^{-\{1.4388(1110)/298\}} = 4.70 \times 10^{-3}$$

$$\ln q_2^V \approx e^{-\{1.4388(705)/298\}} = 3.32 \times 10^{-2}$$

$$\ln q_3^V \approx e^{-\{1.4388(1042)/298\}} = 6.53 \times 10^{-3}$$

so $\quad G_m^V - G_m^V(0) = -(8.3145 \text{ J mol}^{-1}\text{K}^{-1}) \times (298 \text{ K})$

$$\times (4.70 \times 10^{-3} + 3.32 \times 10^{-2} + 6.53 \times 10^{-3})$$

$$= -110 \text{ J mol}^{-1} = \boxed{-0.110 \text{ kJ mol}^{-1}}$$

E17.11(b) $\quad q = \sum_j g_j e^{-\beta\varepsilon_j}, \quad$ where $g = (2S + 1) \times \begin{cases} 1 & \text{for } \Sigma \text{ states} \\ 2 & \text{for } \Pi, \Delta, \dots \text{ states} \end{cases}$

The $^3\Sigma$ term is triply degenerate (from spin), and the $^1\Delta$ term is doubly (orbitally) degenerate. Hence

$$q = 3 + 2e^{-\beta\varepsilon}$$

At 400 K

$$\beta\varepsilon = \dfrac{(1.4388 \text{ cm K}) \times (7918.1 \text{ cm}^{-1})}{400 \text{ K}} = 28.48$$

Therefore, the contribution to G_m is

$$G_m - G_m(0) = -RT \ln q \text{ [Table 17.4 for one mole]}$$

$$- RT \ln q = -(8.314 \text{ J K}^{-1} \text{ mol}^{-1}) \times (400 \text{ K}) \times \ln(3 + 2 \times e^{-28.48})$$

$$= -(8.314 \text{ J K}^{-1} \text{ mol}^{-1}) \times (400 \text{ K}) \times (\ln 3) = \boxed{-3.65 \text{ kJ mol}^{-1}}$$

COMMENT. The contribution of the excited state is negligible at this temperature.

E17.12(b) The degeneracy of a species with $S = \frac{5}{2}$ is 6. The electronic contribution to molar entropy is

$$S_m = \frac{U_m - U_m(0)}{T} + R \ln q = R \ln q$$

(The term involving the internal energy is proportional to a temperature-derivative of the partition function, which in turn depends on excited state contributions to the partition function; those contributions are negligible.)

$$S_m = (8.3145 \text{ J mol}^{-1} \text{ K}^{-1}) \ln 6 = \boxed{14.9 \text{ J mol}^{-1} \text{ K}^{-1}}$$

E17.13(b) Use $S_m = R \ln s$ [17.52b]

Draw up the following table

n:	0	1	2			3			4			5	6
			o	m	p	a	b	c	o	m	p		
s	1	6	6	6	3	6	6	2	6	6	3	6	1
S_m/R	0	1.8	1.8	1.8	1.1	1.8	1.8	0.7	1.8	1.8	1.1	1.8	0

where a is the 1,2,3 isomer, b the 1,2,4 isomer, and c the 1,3,5 isomer.

E17.14(b) We need to calculate

$$K = \prod_J \left(\frac{q_{J,m}^{\ominus}}{N_A} \right)^{v_J} \times e^{-\Delta E_0/RT} \text{ [17.54b]} = \frac{q_m^{\ominus}(^{79}\text{Br}_2)q_m^{\ominus}(^{81}\text{Br}_2)}{q_m^{\ominus}(^{79}\text{Br}^{81}\text{Br})^2} e^{-\Delta E_0/RT}$$

Each of these partition functions is a product

$$q_m^{\ominus} = q_m^T q^R q^V q^E$$

with all $q^E = 1$.

The ratio of the translational partition functions is virtually 1 (because the masses nearly cancel; explicit calculation gives 0.999). The same is true of the vibrational partition functions. Although the moments of inertia cancel in the rotational partition functions, the two homonuclear species each have $\sigma = 2$, so

$$\frac{q^R(^{79}\text{Br}_2)q^R(^{81}\text{Br}_2)}{q^R(^{79}\text{Br}^{81}\text{Br})^2} = 0.25$$

The value of ΔE_0 is also very small compared with RT, so

$$K \approx \boxed{0.25}$$

Solutions to problems

Solutions to numerical problems

P17.1

$$q^E = \sum_j g_j e^{-\beta \varepsilon_j} = 2 + 2e^{-\beta \varepsilon}, \quad \varepsilon = \Delta \varepsilon = 121.1 \, \text{cm}^{-1}.$$

$$U_m - U_m(0) = -\frac{N_A}{q^E}\left(\frac{\partial q^E}{\partial \beta}\right)_V = \frac{N_A \varepsilon e^{-\beta \varepsilon}}{q^E} = 1.$$

$$C_{V,m} = -k\beta^2 \left(\frac{\partial U_m}{\partial \beta}\right)_V \quad [17.31a].$$

Let $x = \beta \varepsilon$, then $d\beta = (1/\varepsilon)\,dx$.

$$C_{V,m} = -k\left(\frac{x}{\varepsilon}\right)^2 \varepsilon \frac{\partial}{\partial x}\left(\frac{N_A \varepsilon e^{-x}}{1 + e^{-x}}\right) = -N_A kx^2 \times \frac{\partial}{\partial x}\left(\frac{e^{-x}}{1 + e^{-x}}\right) = R\left(\frac{x^2 e^{-x}}{(1 + e^{-x})^2}\right).$$

Therefore,

$$C_{V,m}/R = \frac{x^2 e^{-x}}{(1 + e^{-x})^2}, \quad x = \beta \varepsilon.$$

We then draw up the following table.

T / K	50	298	500
$(kT/hc)/\text{mol}^{-1}$	34.8	207	348
x	3.48	0.585	0.348
$C_{V,m}/R$	0.351	0.079	0.029
$C_{V,m}/(\text{J K}^{-1}\,\text{mol}^{-1})$	2.91	0.654	0.244

COMMENT. Note that the double degeneracies do not affect the results because the two factors of 2 in q cancel when U is formed. In the range of temperature specified, the electronic contribution to the heat capacity decreases with increasing temperature.

P17.3 The energy expression for a particle on a ring is

$$E = \frac{\hbar^2 m_l^2}{2I} \quad [9.38a].$$

Therefore

$$q = \sum_{m=-\infty}^{\infty} e^{-m_l^2 \hbar^2 / 2IkT} = \sum_{m=-\infty}^{\infty} e^{-\beta m_l^2 \hbar^2 / 2I}.$$

The summation may be approximated by an integration

$$q \approx \frac{1}{\sigma} \int_{-\infty}^{\infty} e^{-m_l^2 \hbar^2 / 2IkT} dm_l = \frac{1}{\sigma} \left(\frac{2IkT}{\hbar^2} \right)^{1/2} \int_{-\infty}^{\infty} e^{-x^2} dx \approx \frac{1}{\sigma} \left(\frac{2\pi IkT}{\hbar^2} \right)^{1/2} = \frac{1}{\sigma} \left(\frac{2\pi I}{\hbar^2 \beta} \right)^{1/2}.$$

$$U - U(0) = -N \frac{\partial \ln q}{\partial \beta} = -N \frac{\partial}{\partial \beta} \ln \frac{1}{\sigma} \left(\frac{2\pi I}{\hbar^2 \beta} \right)^{1/2} = \frac{N}{2\beta} = \frac{1}{2} NkT = \frac{1}{2} RT \quad (N = N_A).$$

$$C_{V,m} = \left(\frac{\partial U_m}{\partial T} \right)_V = \frac{1}{2} R = \boxed{4.2 \text{ J K}^{-1} \text{ mol}^{-1}}.$$

$$S_m = \frac{U_m - U_m(0)}{T} + R \ln q$$

$$= \frac{1}{2} R + R \ln \frac{1}{\sigma} \left(\frac{2\pi IkT}{\hbar^2} \right)^{1/2}$$

$$= \frac{1}{2} R + R \ln \frac{1}{3} \left(\frac{(2\pi) \times (5.341 \times 10^{-47} \text{ kg m}^2) \times (1.381 \times 10^{-23} \text{ J K}^{-1}) \times (298)}{(1.055 \times 10^{-34} \text{ J s})^2} \right)^{1/2}$$

$$= \frac{1}{2} R + 1.31R = 1.81R, \text{ or } \boxed{15 \text{ J K}^{-1} \text{ mol}^{-1}}.$$

P17.5   The absorption lines are the values of differences in adjacent rotational terms. Using eqns 13.25, 13.26, and 13.27, we have

$$F(J+1) - F(J) = \frac{E(J+1) - E(J)}{hc} = 2B(J+1)$$

for $J = 0, 1, \ldots$. Therefore, we can find the rotational constant and reconstruct the energy levels from the data. To make use of all of the data, one would plot the wavenumbers, which represent $F(J+1) - F(J)$, vs. J; from the above equation, the slope of that linear plot is $2B$. Inspection of the data show that the lines in the spectrum are equally spaced with a separation of 21.19 cm^{-1}, so that is the slope:

$$\text{slope} = 21.19 \text{ cm}^{-1} = 2B \quad \text{so} \quad B = 10.59\overline{5} \text{ cm}^{-1}.$$

The partition function is

$$q = \sum_{J=0}^{\infty} (2J+1) e^{-\beta E(J)} \quad \text{where} \quad E(J) = hcBJ(J+1) \text{ [13.25]}$$

and the factor of $2J+1$ is the degeneracy of the energy levels.

At 25°C, $hcB\beta = \dfrac{hcB}{kT} = \dfrac{6.626 \times 10^{-34} \text{ J s} \times 2.998 \times 10^{10} \text{ cm s}^{-1} \times 10.59\overline{5} \text{ cm}^{-1}}{1.381 \times 10^{-23} \text{ J K}^{-1} \times 298.15 \text{ K}} = 0.05112.$

$$q = \sum_{J=0}^{\infty} (2J+1) e^{-0.05112 J(J+1)}$$

$$= 1 + 3e^{-0.05112 \times 1 \times 2} + 5e^{-0.05112 \times 2 \times 3} + 7e^{-0.05112 \times 3 \times 4} + \cdots$$

$$= 1 + 2.708 + 3.679 + 3.791 + 3.238 + \cdots = \boxed{19.90}.$$

P17.7   The molar entropy is given by

$$S_m = \frac{U_m - U_m(0)}{T} + R\left(\ln\frac{q_m}{N_A} - 1\right)$$

where $\dfrac{U_m - U_m(0)}{T} = -N_A\left(\dfrac{\partial \ln q}{\partial \beta}\right)_V$ and $\dfrac{q_m}{N_A} = \dfrac{q_m^T}{N_A}q^R\,q^V q^E$.

The energy term $U_m - U_m(0)$ works out to be

$$U_m - U_m(0) = N_A[\langle\varepsilon^T\rangle + \langle\varepsilon^R\rangle + \langle\varepsilon^V\rangle + \langle\varepsilon^E\rangle].$$

Translation:

$$\frac{q_m^{T\ominus}}{N_A} = 2.561 \times 10^{-2}(T/K)^{5/2} \times (M/\text{g mol}^{-1})^{3/2} \text{ [Table 17.3]}$$

$$= 2.561 \times 10^{-2} \times (298)^{5/2} \times (38.00)^{3/2} = 9.20 \times 10^6$$

and $\langle\varepsilon^T\rangle = \frac{3}{2}kT$.

Rotation of a linear molecule:

$$q^R = \frac{0.6950}{\sigma} \times \frac{T/K}{B/\text{cm}^{-1}} \text{ [Table 17.3]}.$$

The rotational constant is

$$B = \frac{\hbar}{4\pi cI} = \frac{\hbar}{4\pi c\mu R^2}$$

$$= \frac{(1.0546 \times 10^{-34}\,\text{J s}) \times (6.022 \times 10^{23}\,\text{mol}^{-1})}{4\pi(2.998 \times 10^{10}\,\text{cm s}^{-1}) \times (\frac{1}{2} \times 19.00 \times 10^{-3}\,\text{kg mol}^{-1}) \times (190.0 \times 10^{-12}\,\text{m})^2}$$

$$= 0.4915\,\text{cm}^{-1}$$

so $q^R = \dfrac{0.6950}{2} \times \dfrac{298}{0.4915} = 210.\overline{7}$.

Also $\langle\varepsilon^R\rangle = kT$.

Vibration:

$$q^V = \frac{1}{1 - e^{-hc\tilde{\nu}/kT}} = \frac{1}{1 - \exp(-1.4388(\tilde{\nu}/\text{cm}^{-1})/(T/K))} = \frac{1}{1 - \exp(-1.4388(450.0)/298)}$$

$$= 1.129.$$

$$\langle\varepsilon^V\rangle = \frac{hc\tilde{\nu}}{e^{hc\tilde{\nu}/kT} - 1} = \frac{(6.626 \times 10^{-34}\,\text{J s}) \times (2.998 \times 10^{10}\,\text{cm s}^{-1}) \times (450.0\,\text{cm}^{-1})}{\exp(1.4388(450.0)/298) - 1}$$

$$= 1.149 \times 10^{-21}\,\text{J}.$$

The Boltzmann factor for the lowest-lying excited electronic state is

$$\exp\left(\frac{-(1.609\,\text{eV}) \times (1.602 \times 10^{-19}\,\text{J eV}^{-1})}{(1.381 \times 10^{-23}\,\text{J K}^{-1}) \times (298\,\text{K})}\right) = 6 \times 10^{-28}$$

so we may take q^E to equal the degeneracy of the ground state, namely, 2 and $\langle \varepsilon^E \rangle$ to be zero. Putting it all together yields

$$\frac{U_m - U_m(0)}{T} = \frac{N_A}{T}\left(\tfrac{3}{2}kT + kT + 1.149 \times 10^{-21}\,\text{J}\right) = \tfrac{5}{2}R + \frac{N_A(1.149 \times 10^{-21}\,\text{J})}{T}$$

$$= (2.5) \times (8.3145\,\text{J mol}^{-1}\,\text{K}^{-1}) + \frac{(6.022 \times 10^{23}\,\text{mol}^{-1}) \times (1.149 \times 10^{-21}\,\text{J})}{298\,\text{K}}$$

$$= 23.11\,\text{J mol}^{-1}\,\text{K}^{-1}.$$

$$R\left(\ln\frac{q_m}{N_A} - 1\right) = (8.3145\,\text{J mol}^{-1}\,\text{K}^{-1}) \times \{\ln[(9.20 \times 10^6) \times (210.7) \times (1.129) \times (2)] - 1\}$$

$$= 176.3\,\text{J mol}^{-1}\,\text{K}^{-1} \quad \text{and} \quad S_m^\circ = \boxed{199.4\,\text{J mol}^{-1}\,\text{K}^{-1}}.$$

P17.9   **(a)** The probability distribution of rotational energy levels is the Boltzmann factor of each level, weighted by the degeneracy, over the partition function

$$p_J^R(T) = \frac{g(J)e^{-\varepsilon_J/kT}}{q^R} = \frac{(2J+1)e^{-hcBJ(J+1)/kT}}{\displaystyle\sum_{J=0}^{\infty}(2J+1)e^{-hcBJ(J+1)/kT}} \quad [17.13].$$

It is conveniently plotted against J at several temperatures using mathematical software. This distribution at 100 K is shown in Fig. 17.2(a) as both a bar plot and a line plot.

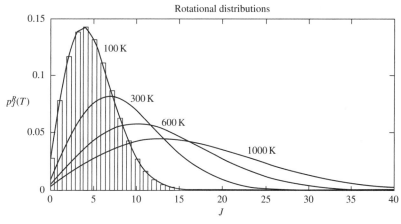

Figure 17.2(a)

The plots show that higher rotational states become more heavily populated at higher temperature. Even at 100 K the most populated state has 4 quanta of rotational energy; it is elevated to 13 quanta at 1000 K.

Values of the vibrational state probability distribution,

$$p_v^V(T) = \frac{e^{-\varepsilon_J/kT}}{q^V} = \frac{e^{-vhc\tilde{\nu}/kT}}{1 - e^{-hc\tilde{\nu}/kT}} \quad [17.19]$$

are conveniently tabulated against v at several temperatures. Computations may be discontinued when values drop below some small number like 10^{-7}.

		$p_v^V(T)$		
v	100 K	300 K	600 K	1000 K
0	1	1	0.095	0.956
1	2.77×10^{-14}	3.02×10^{-5}	5.47×10^{-3}	0.042
2		9.15×10^{-10}	3.01×10^{-5}	1.86×10^{-3}
3			1.65×10^{-7}	8.19×10^{-5}
4				3.61×10^{-6}
5				1.59×10^{-7}

Only the state $v = 0$ is appreciably populated below 1000 K and even at 1000 K only 4% of the molecules have 1 quanta of vibrational energy.

(b) The classical (equipartition) rotational partition function is

$$q_{classical}^R(T) = \frac{kT}{hcB} = \frac{T}{\theta_R} \quad [17.15b].$$ (1)

where θ_R is the rotational temperature. We would expect the partition function to be well approximated by this expression for temperatures much greater than the rotational temperature.

$$\theta_R = \frac{hcB}{k} = \frac{(6.626 \times 10^{-34} \text{ J s}) \times (2.998 \times 10^{10} \text{ cm s}^{-1}) \times (1.931 \text{ cm}^{-1})}{1.381 \times 10^{-23} \text{J K}^{-1}},$$

$$\theta_R = 2.779 \text{ K}.$$

In fact $\theta_R \ll T$ for all temperatures of interest in this problem (100 K or more). Agreement between the classical expression and the explicit sum is indeed good, as Fig. 17.2(b) confirms. The figure displays the percentage deviation $(q_{classical}^R - q^R)100/q^R$. The maximum deviation is about -0.9% at 100 K and the magnitude decreases with increasing temperature.

(c) The translational, rotational, and vibrational contributions to the total energy are specified by eqns 17.25b, 17.26b, and 17.28 respectively. As molar quantities, they are

$$U^T = \tfrac{3}{2}RT, \quad U^R = RT, \quad U^V = \frac{N_A hc\tilde{v}}{e^{hc\tilde{v}/kT} - 1}.$$

The contributions to the difference in energy from its 100 K value are $\Delta U^T(T) = U^T(T) - U^T(100 \text{ K})$, etc. Fig. 17.2(c) shows the individual contributions to $\Delta U(T)$. Translational motion contributes 50% more than the rotational motion because it has 3 quadratic degrees of freedom compared to 2 quadratic degrees of freedom for rotation. Very little change occurs in the vibrational energy because very high temperatures are required to populate $v = 1, 2, \ldots$ states (see Part **a**).

$$C_{V,m}(T) = \left(\frac{\partial U(T)}{\partial T}\right)_V = \left(\frac{\partial}{\partial T}\right)_V (U^T + U^R + U^V)$$

$$= \frac{3}{2}R + R + \frac{dU^V}{dT} = \frac{5}{2}R + \frac{dU^V}{dT}.$$

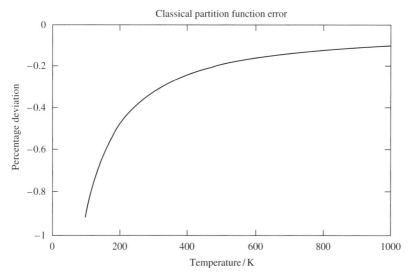

Figure 17.2(b)

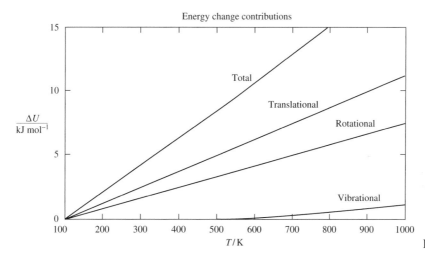

Figure 17.2(c)

The derivative dU^V/dT may be evaluated numerically with numerical software (we advise exploration of the technique) or it may be evaluated analytically using eqn 17.34:

$$C^V_{V,m} = \frac{dU^V}{dT} = R\left\{\frac{\theta_V}{T}\left(\frac{e^{-\theta_V/2T}}{1 - e^{-\theta_V/T}}\right)\right\}^2$$

where $\theta_V = hc\tilde{\nu}/k = 3122$ K. Fig. 17.2(d) shows the ratio of the vibrational contribution to the sum of translational and rotational contributions. Below 300 K, vibrational motion makes a small, perhaps negligible, contribution to the heat capacity. The contribution is about 10% at 600 K and grows with increasing temperature.

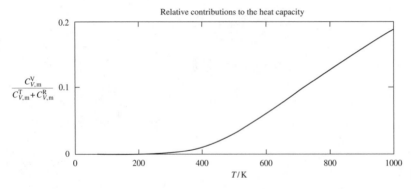

Relative contributions to the heat capacity

Figure 17.2(d)

The change with temperature of molar entropy may be evaluated by numerical integration with mathematical software.

$$\Delta S(T) = S(T) - S(100\,\text{K}) = \int_{100\,\text{K}}^{T} \frac{C_{p,\text{m}}(T)\mathrm{d}T}{T}\ \ [3.18]$$

$$= \int_{100\,\text{K}}^{T} \frac{C_{V,\text{m}}(T) + R}{T}\mathrm{d}T\ \ [2.48]$$

$$= \int_{100\,\text{K}}^{T} \frac{\frac{7}{2}R + C_{V,\text{m}}^{V}(T)}{T}\mathrm{d}T.$$

$$\Delta S(T) = \underbrace{\frac{7}{2}R\ln\left(\frac{T}{100\,\text{K}}\right)}_{\Delta S^{T+R}(T)} + \underbrace{\int_{100\,\text{K}}^{T} \frac{C_{V,\text{m}}^{V}(T)}{T}\mathrm{d}T}_{\Delta S^{V}(T)}$$

Fig. 17.2(e) shows the ratio of the vibrational contribution to the sum of translational and rotational contributions. Even at the highest temperature the vibrational contribution to the entropy change is less than 2.5% of the contributions from translational and rotational motion. The vibrational contribution is negligible at low temperature.

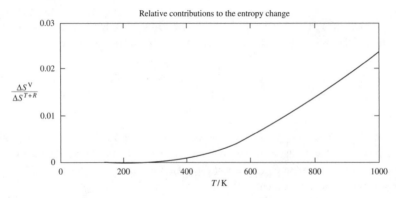

Relative contributions to the entropy change

Figure 17.2(e)

P17.11 $H_2O + DCl \rightleftharpoons HDO + HCl.$

$$K = \frac{q^{\ominus}(HDO)q^{\ominus}(HCl)}{q^{\ominus}(H_2O)q^{\ominus}(DCl)} e^{-\beta \Delta E_0} \text{ [17.54; } N_A \text{ factors cancel].}$$

Use partition function expressions from Table 17.3. The ratio of translational partition functions is

$$\frac{q_m^T(HDO)q_m^T(HCl)}{q_m^T(H_2O)q_m^T(DCl)} = \left(\frac{M(HDO)M(HCl)}{M(H_2O)M(DCl)}\right)^{3/2} = \left(\frac{19.02 \times 36.46}{18.02 \times 37.46}\right)^{3/2} = 1.041.$$

The ratio of rotational partition functions is

$$\frac{q^R(HDO)q^R(HCl)}{q^R(H_2O)q^R(DCl)} = \frac{\sigma(H_2O)}{1} \frac{(A(H_2O)B(H_2O)C(H_2O)/cm^{-3})^{1/2}B(DCl)/cm^{-1}}{(A(HDO)B(HDO)C(HDO)/cm^{-3})^{1/2}B(HCl)/cm^{-1}}$$

$$= 2 \times \frac{(27.88 \times 14.51 \times 9.29)^{1/2} \times 5.449}{(23.38 \times 9.102 \times 6.417)^{1/2} \times 10.59} = 1.707$$

($\sigma = 2$ for H_2O; $\sigma = 1$ for the other molecules).

The ratio of vibrational partition functions (call it Q) is

$$\frac{q^V(HDO)q^V(HCl)}{q^V(H_2O)q^V(DCl)} = \frac{q(2726.7)q(1402.2)q(3707.5)q(2991)}{q(3656.7)q(1594.8)q(3755.8)q(2145)} = Q$$

where $q(x) = \dfrac{1}{1 - e^{-1.4388x/(T/K)}}.$

$$\frac{\Delta E_0}{hc} = \frac{1}{2}\{(2726.7 + 1402.2 + 3707.5 + 2991) - (3656.7 + 1594.8 + 3755.8 + 2145)\} \text{ cm}^{-1}$$

$$= -162 \text{ cm}^{-1}.$$

So the exponent in the energy term is

$$-\beta \Delta E_0 = -\frac{\Delta E_0}{kT} = -\frac{hc}{k} \times \frac{\Delta E_0}{hc} \times \frac{1}{T} = -\frac{1.4388 \times (-162)}{T/K} = +\frac{233}{T/K}.$$

Therefore, $K = 1.041 \times 1.707 \times Q \times e^{233/(T/K)} = 1.777\, Q e^{233/(T/K)}.$

We then draw up the following table (using a computer)

T/K	100	200	300	400	500	600	700	800	900	1000
K	18.3	5.70	3.87	3.19	2.85	2.65	2.51	2.41	2.34	2.29

and specifically $K = \boxed{3.89}$ at **(a)** 298 K and $\boxed{2.41}$ at **(b)** 800 K.

Solutions to theoretical problems

P17.13   **(a)** θ_V and θ_R are the constant factors in the numerators of the negative exponents in the sums that are the partition functions for vibration and rotation. They have the dimensions of temperature, which occurs in the denominator of the exponents. So high temperature means $T \gg \theta_V$ or θ_R and only then does the exponential become substantial. Thus θ_V and θ_R are measures of the temperature at which higher vibrational and rotational states, respectively, become significantly populated:

$$\theta_R = \frac{hc\beta}{k} = \frac{(2.998 \times 10^{10}\,\text{cm s}^{-1}) \times (6.626 \times 10^{-34}\,\text{J s}) \times (60.864\,\text{cm}^{-1})}{(1.381 \times 10^{-23}\,\text{J K}^{-1})} = \boxed{87.55\,\text{K}}$$

and

$$\theta_V = \frac{hc\tilde{\nu}}{k} = \frac{(6.626 \times 10^{-34}\,\text{J s}) \times (4400.39\,\text{cm}^{-1}) \times (2.998 \times 10^{10}\,\text{cm s}^{-1})}{(1.381 \times 10^{-23}\,\text{J K}^{-1})} = \boxed{6330\,\text{K}}.$$

(b) and **(c)** These parts of the solution were performed with Mathcad 7.0 and are reproduced on the following pages.

Objective: To calculate the equilibrium constant $K(T)$ and $C_p(T)$ for dihydrogen at high temperature for a system made with n mol H_2 at 1 bar.

$$H_2(g) \rightleftharpoons 2H(g)$$

At equilibrium the degree of dissociation, α, and the equilibrium amounts of H_2 and atomic hydrogen are related by the expressions

$$n_{H_2} = (1 - \alpha)n \quad \text{and} \quad n_H = 2\alpha n.$$

The equilibrium mole fractions are

$$x_{H_2} = (1 - \alpha)n/\{(1 - \alpha)n + 2\alpha n\} = (1 - \alpha)/(1 + \alpha),$$
$$x_H = 2\alpha n/\{(1 - \alpha)n + 2\alpha n\} = 2\alpha/(1 + \alpha).$$

The partial pressures are

$$p_{H_2} = (1 - \alpha)p/(1 + \alpha) \quad \text{and} \quad p_H = 2\alpha p/(1 + \alpha).$$

The equilibrium constant is

$$K(T) = \frac{(p_H/p^{\ominus})^2}{(p_{H_2}/p^{\ominus})} = 4\alpha^2 \frac{(p/p^{\ominus})}{(1 - \alpha^2)} = \frac{4\alpha^2}{(1 - \alpha^2)} \quad \text{where } p = p^{\ominus} = 1 \text{ bar.}$$

The above equation is easily solved for α:

$$\boxed{\alpha = (K/(K + 4))^{1/2}}.$$

The heat capacity at constant volume for the equilibrium mixture is

$$C_V(\text{mixture}) = n_H C_{V,\text{m}}(H) + n_{H_2} C_{V,\text{m}}(H_2).$$

The heat capacity at constant volume per mole of dihydrogen used to prepare the equilibrium mixture is

$$C_V = C_V(\text{mixture})/n = \{n_H C_{V,\text{m}}(H) + n_{H_2} C_{V,\text{m}}(H_2)\}/n$$
$$= \boxed{2\alpha C_{V,\text{m}}(H) + (1 - \alpha)C_{V,\text{m}}(H_2)}.$$

The formula for the heat capacity at constant pressure per mole of dihydrogen used to prepare the equilibrium mixture (C_p) can be deduced from the molar relationship

$$C_{p,m} = C_{V,m} + R.$$

$$C_p = \left\{ n_H C_{p,m}(H) + n_{H_2} C_{p,m}(H_2) \right\} / n$$

$$= \frac{n_H}{n} \left\{ C_{V,m}(H) + R \right\} + \frac{n_{H_2}}{n} \left\{ C_{V,m}(H_2) + R \right\}$$

$$= \frac{n_H C_{V,m}(H) + n_{H_2} C_{V,m}(H_2)}{n} + R \left(\frac{n_H + n_{H_2}}{n} \right)$$

$$= C_V + R(1 + \alpha).$$

Calculations

J = joule	s = second	kJ = 1000 J
mol = mole	g = gram	bar = 1×10^5 Pa
$h = 6.62608 \times 10^{-34}$ J s	$c = 2.9979 \times 10^8$ m s^{-1}	$k = 1.38066 \times 10^{-23}$ J K^{-1}
$R = 8.31451$ J K^{-1} mol^{-1}	$N_A = 6.02214 \times 10^{23}$ mol^{-1}	$p^\ominus = 1$ bar

Molecular properties of H_2:

$$\nu = 4400.39 \text{ cm}^{-1}, \quad B = 60.864 \text{ cm}^{-1}, \quad D = 432.1 \text{ kJ mol}^{-1}.$$

$$m_H = \frac{1 \text{ g mol}^{-1}}{N_A}, \quad m_{H_2} = 2m_H.$$

$$\theta_V = \frac{hc\tilde{\nu}}{k}, \quad \theta_R = \frac{hcB}{k}.$$

Computation of $K(T)$ and $\alpha(T)$

$$N = 200, \quad i = 0, \ldots, N \quad T_i = 500 \text{ K} + \frac{i \times 5500 \text{ K}}{N}.$$

$$\Lambda_{Hi} = \frac{h}{(2\pi m_H k T_i)^{1/2}}, \quad \Lambda_{H2i} = \frac{h}{(2\pi m_{H_2} k T_i)^{1/2}}.$$

$$q_{V_i} = \frac{1}{1 - e^{-(\theta_V/T_i)}}, \quad q_{R_i} = \frac{T_i}{2\theta_R}.$$

$$\boxed{K_{eq_i} = \frac{kT_i(\Lambda_{H_2i})^3 e^{-(D/RT_i)}}{p^\ominus q_{V_i} q_{R_i}(\Lambda_{H_i})^6}}, \quad \alpha_i = \left(\frac{K_{eq_i}}{K_{eq_i} + 4} \right)^{1/2}.$$

See Fig. 17.3(a) and (b).

Heat capacity at constant volume per mole of dihydrogen used to prepare the equilibrium mixture is (see Fig. 17.4(a))

$$C_V(H) = \boxed{1.5R},$$

$$C_V(H_{2_i}) = \boxed{2.5R + \left[\frac{\theta_V}{T_i} \times \frac{e^{-(\theta_V/2T_i)}}{1 - e^{\theta_V/T_i}} \right]^2 R} \quad C_{V_i} = 2\alpha_i C_V(H) + (1 - \alpha_i) C_V(H_{2_i}).$$

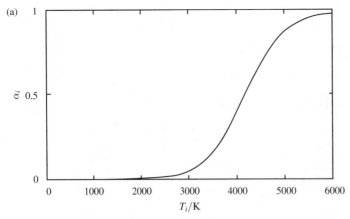

Figure 17.3(a)

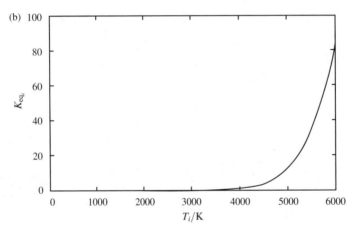

Figure 17.3(b)

The heat capacity at constant pressure per mole of dihydrogen used to prepare the equilibrium mixture is (see Fig. 17.4(b))

$$C_{p_i} = C_{V_i} + R(1 + \alpha_i).$$

P17.15   Eqn. 17.42 relates the second virial coefficient to the pairwise intermolecular potential energy:

$$B = -2\pi N_A \int_0^\infty fr^2 dr \quad \text{where } f = e^{-\beta E_P} - 1.$$

In order to relate the pairwise potential to the van der Waals equation, we must express that equation as a virial series. (See Table 1.7.) The equations are

$$\text{van der Waals } p = \frac{RT}{V_m - b} - \frac{a}{V_m^2}, \quad \text{virial } p = \frac{RT}{V_m}\left(1 + \frac{B}{V_m} + \cdots\right)$$

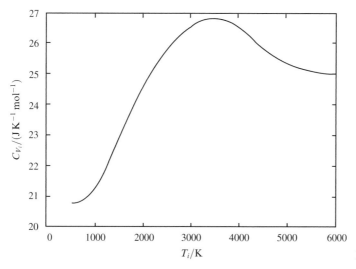

Figure 17.4(a)

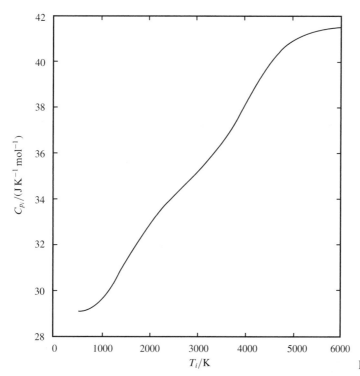

Figure 17.4(b)

Expand the van der Waals equation as a power series in $1/V_m$:

$$p = \frac{RT}{V_m(1 - b/V_m)} - \frac{a}{V_m^2} = \frac{RT}{V_m}\left(1 + \frac{b}{V_m} + \cdots\right) - \frac{a}{V_m^2} \approx \frac{RT}{V_m}\left\{1 + \frac{1}{V_m}\left(b - \frac{a}{RT}\right)\right\}.$$

Thus, the second virial coefficient in terms of van der Waals parameters is

$$B = b - \frac{a}{RT}.$$

The pairwise potential and Mayer f-function are:

$$\text{for } 0 \le r < r_1 \quad E_P \to \infty \quad e^{-\beta E_P} = 0 \quad f = -1$$

$$\text{for } r_1 \le r < r_2 \quad E_P \to -\varepsilon \quad e^{-\beta E_P} = e^{\beta\varepsilon} \quad f = e^{-\beta\varepsilon} - 1 > 0$$

$$\text{for } r_2 \le r \quad E_P \to 0 \quad e^{-\beta E_P} = 1 \quad f = 0$$

So $\dfrac{B}{-2\pi N_A} = \displaystyle\int_0^\infty fr^2 dr = -\int_0^{r_1} r^2 dr + (e^{\beta\varepsilon} - 1)\int_{r_1}^{r_2} r^2 dr$

$$= -\frac{r_1^3}{3} + (e^{\beta\varepsilon} - 1)\left(\frac{r_2^3}{3} - \frac{r_1^3}{3}\right).$$

Expand the exponential, because $\varepsilon \ll kT$, so $\beta\varepsilon \ll 1$.

$$B \approx -2\pi N_A\left\{-\frac{r_1^3}{3} + (1 + \beta\varepsilon - 1)\left(\frac{r_2^3}{3} - \frac{r_1^3}{3}\right)\right\} = \frac{2\pi N_A}{3}\left\{r_1^3 - \frac{\varepsilon(r_2^3 - r_1^3)}{kT}\right\}.$$

Comparing this result to the virial coefficient from the van der Waals equation, we identify

$$b = \frac{2\pi N_A r_1^3}{3} \quad \text{and} \quad a = \frac{2\pi N_A \varepsilon_m(r_2^3 - r_1^3)}{3},$$

where ε_m is ε expressed as a molar quantity. Thus the van der Waals b is proportional to the volume of the hard-sphere (repulsive) part of the potential. The a parameter is more complicated, but it is where the attractive part of the potential appears, including both the depth of the attractive well and the range of distances over which it operates.

Use eqn 17.45 to compute the limiting isothermal Joule–Thomson coefficient

$$\lim_{p \to 0} \mu_T = B - T\frac{dB}{dT}$$

$$= \frac{2\pi N_A}{3}\left\{r_1^3 - \frac{\varepsilon(r_2^3 - r_1^3)}{kT}\right\} - T\frac{2\pi N_A}{3}\left\{\frac{\varepsilon(r_2^3 - r_1^3)}{kT^2}\right\}$$

$$= \frac{2\pi N_A}{3}\left\{r_1^3 - \frac{2\varepsilon(r_2^3 - r_1^3)}{kT}\right\} = b - \frac{2a}{RT}.$$

The Joule–Thomson coefficient itself is [2.55]

$$\mu = -\frac{\mu_T}{C_p} = \frac{2\pi N_A}{3C_p}\left\{\frac{2\varepsilon(r_2^3 - r_1^3)}{kT} - r_1^3\right\} = \frac{b - \frac{2a}{RT}}{C_p}.$$

P17.17 **(a)** Ethene belongs to the D_{2h} point group, whose rotational subgroup includes E and 3 C_2 elements around different axes. So $\sigma = 4$. The rotational partition function of a non-linear molecule is [Table 17.3]

$$q^R = \frac{1.0270}{\sigma} \frac{(T/K)^{3/2}}{(ABC/cm^{-3})^{1/2}} = \frac{1.0270 \times 298.15^{3/2}}{(4) \times (4.828 \times 1.0012 \times 0.8282)^{1/2}} = \boxed{660.6}.$$

(b) Pyridine belongs to the C_{2v} group, the same as water, so $\sigma = 2$.

$$q^R = \frac{1.0270}{\sigma} \frac{(T/K)^{3/2}}{(ABC/cm^{-3})^{1/2}} = \frac{1.0270 \times 298.15^{3/2}}{(2) \times (0.2014 \times 0.1936 \times 0.0987)^{1/2}} = \boxed{4.26 \times 10^4}.$$

P17.19 The partition function of a system with energy levels $\varepsilon(J)$ and degeneracies $g(J)$ is

$$q = \sum_J g(J)e^{-\beta\varepsilon(J)}.$$

The contribution of the heat capacity from this system of states is

$$C_V = -k\beta^2 \left(\frac{\partial U}{\partial \beta}\right)_V \quad [17.31a]$$

where $U - U(0) = -N\left(\frac{\partial \ln q}{\partial \beta}\right)_V = -\frac{N}{q}\left(\frac{\partial q}{\partial \beta}\right)_V.$

Express these quantities in terms of sums over energy levels

$$U - U(0) = -\frac{N}{q}\left(-\sum_J g(J)\varepsilon(J)e^{-\beta\varepsilon(J)}\right) = \frac{N}{q}\sum_J g(J)\varepsilon(J)e^{-\beta\varepsilon(J)}$$

and

$$\frac{C_V}{-k\beta^2} = \left(\frac{\partial U}{\partial \beta}\right)_V = \frac{N}{q}\left(-\sum_J g(J)\varepsilon^2(J)e^{-\beta\varepsilon(J)}\right) - \frac{N}{q^2}\sum_J g(J)\varepsilon(J)e^{-\beta\varepsilon(J)}\left(\frac{\partial q}{\partial \beta}\right)$$

$$= -\frac{N}{q}\sum_J g(J)\varepsilon^2(J)e^{-\beta\varepsilon(J)} + \frac{N}{q^2}\sum_J g(J)\varepsilon(J)e^{-\beta\varepsilon(J)}\sum_{J'} g(J')\varepsilon(J')e^{-\beta\varepsilon(J')}.$$

(1)

Finally a double sum appears, one that has some resemblance to the terms in $\zeta(\beta)$. The fact that $\zeta(\beta)$ is a double sum encourages us to try to express the single sum in C_V as a double sum. We can do so by multiplying it by one in the form $(\sum_{J'} g(J')e^{-\beta\varepsilon(J')})/q$, so

$$\frac{C_V}{-k\beta^2} = -\frac{N}{q^2}\sum_J g(J)\varepsilon^2(J)e^{-\beta\varepsilon(J)}\sum_{J'} g(J')e^{-\beta\varepsilon(J')} + \frac{N}{q^2}\sum_J g(J)\varepsilon(J)e^{-\beta\varepsilon(J)}\sum_{J'} g(J')\varepsilon(J')e^{-\beta\varepsilon(J')}.$$

Now collect terms within each double sum and divide both sides by $-N$:

$$\frac{C_V}{kN\beta^2} = \frac{1}{q^2}\sum_{J,J'} g(J)g(J')\varepsilon^2(J)e^{-\beta[\varepsilon(J)+\varepsilon(J')]} - \frac{1}{q^2}\sum_{J,J'} g(J)g(J')\varepsilon(J)\varepsilon(J')e^{-\beta[\varepsilon(J)+\varepsilon(J')]}.$$

Clearly the two sums could be combined, but it pays to make one observation before doing so. The first sum contains a term $\varepsilon^2(J)$, but all the other factors in that sum are related to J and J' in the same way. Thus, the first sum would not be changed by writing $\varepsilon^2(J')$ instead of $\varepsilon^2(J)$; furthermore, if we add the sum with $\varepsilon^2(J')$ to the sum with $\varepsilon^2(J)$, we would have twice the original sum. Therefore, we can write (finally combining the sums)

$$\frac{C_V}{kN\beta^2} = \frac{1}{2q^2} \sum_{J,J'} g(J)g(J')e^{-\beta[\varepsilon(J)+\varepsilon(J')]}[\varepsilon^2(J) + \varepsilon^2(J') - 2\varepsilon(J)\varepsilon(J')].$$

Recognizing that $\varepsilon^2(J) + \varepsilon^2(J') - 2\varepsilon(J)\varepsilon(J') = [\varepsilon(J) - \varepsilon(J')]^2$, we arrive at

$$C_V = \frac{kN\beta^2}{2}\zeta(\beta).$$

For a linear rotor, the degeneracies are $g(J) = 2J + 1$. The energies are

$$\varepsilon(J) = hcBJ(J + 1) = \theta_R kJ(J + 1)$$

so $\beta\varepsilon(J) = \theta_R J(J + 1)/T$.

The total heat capacity and the contributions of several transitions are plotted in Fig. 17.5. One can evaluate $C_{V,m}/R$ using the following expression, derivable from eqn (1) above. It has the advantage of using single sums rather than double sums.

$$\frac{C_{V,m}}{R} = \frac{1}{q}\sum_J g(J)\beta^2\varepsilon^2(J)e^{-\beta\varepsilon(J)} - \frac{1}{q^2}\left(\sum_J g(J)\beta\varepsilon(J)e^{-\beta\varepsilon(J)}\right)^2.$$

COMMENT. $\zeta(\beta)$ is defined in such a way that J and J' each run independently from 0 to infinity. Thus, identical terms appear twice. (For example, both (0,1) and (1,0) terms appear with identical value in $\zeta(\beta)$. In the plot, though, the (0,1) curve represents both terms.) One could redefine the double sum with an inner sum over J' running from 0 to $J - 1$ and an outer sum over J running from 0 to infinity. In that case, each term appears only once, and the overall factor of 1/2 in C_V would have to be removed.

P17.21   All partition functions other than the electronic partition function of atomic I are unaffected by a magnetic field; hence the relative change in K is due to the relative change in q^E.

$$q^E = \sum_{M_J} e^{-g\mu_B\beta BM_J}, \quad M_J = -\tfrac{3}{2}, -\tfrac{1}{2}, +\tfrac{1}{2}, +\tfrac{3}{2}; \quad g = \tfrac{4}{3}.$$

Since $g\mu_B\beta B \ll 1$ for normally attainable fields, we can expand the exponentials

$$q^E = \sum_{M_J}\left\{1 - g\mu_B\beta BM_J + \frac{1}{2}(g\mu_B\beta BM_J)^2 + \cdots\right\}$$

$$\approx 4 + \frac{1}{2}(g\mu_B\beta B)^2 \sum_{M_J} M_J^2 \left[\sum_{M_J} M_J = 0\right] = 4\left(1 + \frac{10}{9}(\mu_B\beta B)^2\right) \quad \left[g = \frac{4}{3}\right].$$

This partition function appears squared in the numerator of the equilibrium constant expression. (See solution to E17.14(a).) Therefore, if K is the actual equilibrium constant and K^0 is its value when $B = 0$,

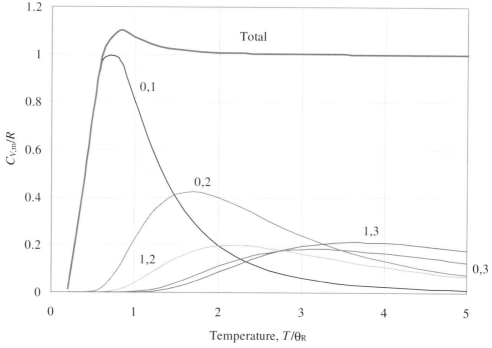

Figure 17.5

we write

$$\frac{K}{K^0} = \left(1 + \frac{10}{9}(\mu_B\beta B)^2\right)^2 \approx 1 + \frac{20}{9}\mu_B^2\beta^2 B^2.$$

For a shift of 1 per cent, we require

$$\tfrac{20}{9}\mu_B^2\beta^2 B^2 \approx 0.01, \quad \text{or} \quad \mu_B\beta B \approx 0.067.$$

Hence

$$B \approx \frac{0.067kT}{\mu_B} = \frac{(0.067) \times (1.381 \times 10^{-23}\,\text{J K}^{-1}) \times (1000\,\text{K})}{9.274 \times 10^{-24}\,\text{J T}^{-1}} \approx \boxed{100\,\text{T}}.$$

Solutions to applications

P17.23 $S = k \ln W$ [16.34].

so $S = k \ln 4^N = Nk \ln 4$

$$= (5 \times 10^8) \times (1.38 \times 10^{-23}\,\text{J K}^{-1}) \times \ln 4 = \boxed{9.57 \times 10^{-15}\,\text{J K}^{-1}}.$$

Question. Is this a large residual entropy? The answer depends on what comparison is made. Multiply the answer by Avogadro's number to obtain the molar residual entropy, $5.76 \times 10^9\,\text{J K}^{-1}\,\text{mol}^{-1}$, surely a large number—but then DNA is a macromolecule. The residual entropy per mole of base pairs may be a more reasonable quantity to compare to molar residual entropies of small molecules. To obtain that answer, divide the molecule's entropy by the number of base pairs before multiplying by N_A. The result is 11.5 J K^{-1} mol^{-1}, a quantity more in line with examples discussed in Section 17.7.

P17.25 The standard molar Gibbs energy is given by

$$G_m^{\ominus} - G_m^{\ominus}(0) = RT \ln \frac{q_m^{\ominus}}{N_A} \quad \text{where} \quad \frac{q_m^{\ominus}}{N_A} = \frac{q_m^{T\ominus}}{N_A} q^R q^V q^E \ [17.53].$$

Translation (see table 17.3 for all partition functions):

$$\frac{q_m^{T\ominus}}{N_A} = 2.561 \times 10^{-2} (T/K)^{5/2} (M/g\ mol^{-1})^{3/2}$$

$$= (2.561 \times 10^{-2}) \times (2000)^{5/2} \times (38.90)^{3/2} = 1.111 \times 10^9.$$

Rotation of a linear molecule:

$$q^R = \frac{kT}{\sigma hcB} = \frac{0.6950}{\sigma} \times \frac{T/K}{B/cm^{-1}}.$$

The rotational constant is

$$B = \frac{\hbar}{4\pi cI} = \frac{\hbar}{4\pi cm_{eff}R^2}$$

where $m_{eff} = \dfrac{m_B m_{Si}}{m_B + m_{Si}} = \dfrac{(10.81) \times (28.09)}{10.81 + 28.09} \times \dfrac{10^{-3} kg\ mol^{-1}}{6.022 \times 10^{23} mol^{-1}} = 1.296 \times 10^{-26} kg.$

$$B = \frac{1.0546 \times 10^{-34} J\ s}{4\pi (2.998 \times 10^{10}\ cm\ s^{-1}) \times (1.296 \times 10^{-26}\ kg) \times (190.5 \times 10^{-12}\ m)^2} = 0.5952\ cm^{-1}$$

so $q^R = \dfrac{0.6950}{1} \times \dfrac{2000}{0.5952} = 2335.$

Vibration:

$$q^V = \frac{1}{1 - e^{-hc\tilde{\nu}/kT}} = \frac{1}{1 - \exp\left(\dfrac{-1.4388(\tilde{\nu}/cm^{-1})}{T/K}\right)} = \frac{1}{1 - \exp\left(\dfrac{-1.4388(772)}{2000}\right)}$$

$$= 2.467.$$

The Boltzmann factor for the lowest-lying electronic excited state is

$$\exp\left(\frac{-(1.4388) \times (8000)}{2000}\right) = 3.2 \times 10^{-3}.$$

The degeneracy of the ground level is 4 (spin degeneracy = 4, orbital degeneracy = 1), and that of the excited level is also 4 (spin degeneracy = 2, orbital degeneracy = 2), so

$$q^E = 4(1 + 3.2 \times 10^{-3}) = 4.013.$$

Putting it all together yields

$$G_m^\ominus - G_m^\ominus(0) = (8.3145 \, \text{J mol}^{-1}\text{K}^{-1}) \times (2000 \, \text{K})$$
$$\times \ln[(1.111 \times 10^9) \times (2335) \times (2.467) \times (4.013)]$$
$$= 5.135 \times 10^5 \, \text{J mol}^{-1} = \boxed{513.5 \, \text{kJ mol}^{-1}}.$$

P17.27   The standard molar Gibbs energy is given by

$$G_m^\ominus - G_m^\ominus(0) = RT \ln \frac{q_m^\ominus}{N_A} \quad \text{where} \quad \frac{q_m^\ominus}{N_A} = \frac{q_m^{T\ominus}}{N_A} q^R q^V q^E \; [17.53]$$

See Table 17.3 for partition function expressions. First, at 10.00 K

$$\text{Translation}: \frac{q_m^{T\ominus}}{N_A} = 2.561 \times 10^{-2}(T/\text{K})^{5/2}(M/\text{g mol}^{-1})^{3/2}$$

$$= (2.561 \times 10^{-2}) \times (10.00)^{5/2} \times (36.033)^{3/2} = 1752.$$

Rotation of a nonlinear molecule:

$$q^R = \frac{1}{\sigma}\left(\frac{kT}{hc}\right)^{3/2}\left(\frac{\pi}{ABC}\right)^{1/2} = \frac{1.0270}{\sigma} \times \frac{(T/\text{K})^{3/2}}{(ABC/\text{cm}^{-3})^{1/2}}.$$

The rotational constants are

$$B = \frac{\hbar}{4\pi cI} \quad \text{so} \quad ABC = \left(\frac{\hbar}{4\pi c}\right)^3 \frac{1}{I_A I_B I_C},$$

$$ABC = \left(\frac{1.0546 \times 10^{-34} \, \text{J s}}{4\pi(2.998 \times 10^{10} \, \text{cm s}^{-1})}\right)^3$$

$$\times \frac{(10^{10}\text{Å m}^{-1})^6}{(39.340) \times (39.032) \times (0.3082) \times (\text{u Å}^2)^3 \times (1.66054 \times 10^{-27} \, \text{kg u}^{-1})^3}$$

$$= 101.2 \, \text{cm}^{-3}$$

so $q^R = \dfrac{1.0270}{2} \times \dfrac{(10.00)^{3/2}}{(101.2)^{1/2}} = 1.614.$

Vibration: for each mode

$$q^V = \frac{1}{1 - e^{-hc\tilde{\nu}/kT}} = \frac{1}{1 - \exp\left(\dfrac{-1.4388(\tilde{\nu}/\text{cm}^{-1})}{T/\text{K}}\right)} = \frac{1}{1 - \exp\left(\dfrac{-1.4388(63.4)}{10.00}\right)}$$

$$= 1.0001$$

Even the lowest-frequency mode has a vibrational partition function of 1; so the stiffer vibrations have q^V even closer to 1. The degeneracy of the electronic ground state is 1, so $q^E = 1$. Putting it all together yields

$$G_m^\ominus - G_m^\ominus(0) = (8.3145 \, \text{J mol}^{-1}\,\text{K}^{-1}) \times (10.00 \, \text{K}) \ln[(1752) \times (1.614) \times (1) \times (1)]$$

$$= \boxed{660.8 \, \text{J mol}^{-1}}.$$

Now at 1000 K

Translation: $\dfrac{q_m^{\ominus}}{N_A} = (2.561 \times 10^{-2}) \times (1000)^{5/2} \times (36.033)^{3/2} = 1.752 \times 10^8$.

Rotation: $q^R = \dfrac{1.0270}{2} \times \dfrac{(1000)^{3/2}}{(101.2)^{1/2}} = 1614$.

Vibration: $q_1^V = \dfrac{1}{1 - \exp\left(-\dfrac{(1.4388) \times (63.4)}{1000}\right)} = 11.47$,

$q_2^V = \dfrac{1}{1 - \exp\left(-\dfrac{(1.4388) \times (1224.5)}{1000}\right)} = 1.207$,

$q_3^V = \dfrac{1}{1 - \exp\left(-\dfrac{(1.4388) \times (2040)}{1000}\right)} = 1.056$,

$q^V = (11.47) \times (1.207) \times (1.056) = 14.62$.

Putting it all together yields

$$G_m^{\ominus} - G_m^{\ominus}(0) = (8.3145 \, \text{J mol}^{-1} \, \text{K}^{-1}) \times (1000 \, \text{K})$$
$$\times \ln[(1.752 \times 10^8) \times (1614) \times (14.62) \times (1)]$$
$$= 2.415 \times 10^5 \, \text{J mol}^{-1} = \boxed{241.5 \, \text{kJ mol}^{-1}}.$$

18 Molecular interactions

Answers to discussion questions

D18.1 Molecules with a permanent separation of electric charge have a permanent dipole moment. In molecules containing atoms of differing electronegativity, the bonding electrons may be displaced in such a way as to produce a net separation of charge in the molecule. Separation of charge may also arise from a difference in atomic radii of the bonded atoms. The separation of charges in the bonds is usually, though not always, in the direction of the more electronegative atom but depends on the precise bonding situation in the molecule as described in Section 18.1(a). A heteronuclear diatomic molecule necessarily has a dipole moment if there is a difference in electronegativity between the atoms, but the situation in polyatomic molecules is more complex. A polyatomic molecule has a permanent dipole moment only if it fulfills certain symmetry requirements as discussed in Section 12.3(a).

An external electric field can distort the electron density in both polar and nonpolar molecules and this results in an induced dipole moment that is proportional to the field. The constant of proportionality is called the polarizability.

D18.3 Dipole moments are not measured directly, but are calculated from a measurement of the relative permittivity, ε_r (dielectric constant) of the medium. Equation 18.15 implies that the dipole moment can be determined from a measurement of ε_r as a function of temperature. This approach is illustrated in Example 18.2. In another method, the relative permittivity of a solution of the polar molecule is measured as a function of concentration. The calculation is again based on the Debye equation, but in a modified form. The values obtained by this method are accurate only to about 10%. See the references listed under *Further reading* for the details of this approach. A third method is based on the relation between relative permittivity and refractive index, eqn 18.17, and thus reduces to a measurement of the refractive index. Accurate values of the dipole moments of gaseous molecules can be obtained from the Stark effect in their microwave spectra.

D18.5 If the A—H bond in the A—H $\cdots$ B arrangement is regarded as formed from the overlap of an orbital on A, ψ_A, and a hydrogen $1s$ orbital ψ_H, and if the lone pair on B occupies an orbital on B, ψ_B, then, when the two molecules are close together, we can build three molecular orbitals from the three basis orbitals:

$$\psi = C_A \psi_A + C_H \psi_H + C_B \psi_B.$$

One of the molecular orbitals is bonding, one almost nonbonding, and the third antibonding. These three orbitals need to accommodate four electrons, two from the A—H bond and two from the lone pair on B.

Two enter the bonding orbital and two the nonbonding orbital, so the net effect is a lowering of the energy, that is, a bond has formed.

D18.7 A molecular beam is a narrow stream of molecules with a narrow spread of velocities and, in some cases, in specific internal states or orientations. Molecular beam studies of non-reactive collisions are used to explore the details of intermolecular interactions with a view to determining the shape of the intermolecular potential.

The primary experimental information from a molecular beam experiment is the fraction of the molecules in the incident beam that is scattered into a particular direction. The fraction is normally expressed in terms of dI, the rate at which molecules are scattered into a cone that represents the area covered by the 'eye' of the detector (Fig. 18.14 of the text). This rate is reported as the differential scattering cross-section, σ, the constant of proportionality between the value of dI and the intensity, I, of the incident beam, the number density of target molecules, N, and the infinitesimal path length dx through the sample:

$$dI = \sigma I N dx.$$

The value of σ (which has the dimensions of area) depends on the impact parameter, b, the initial perpendicular separation of the paths of the colliding molecules (Fig. 18.15), and the details of the intermolecular potential.

The scattering pattern of real molecules, which are not hard spheres, depends on the details of the intermolecular potential, including the anisotropy that is present when the molecules are non-spherical. The scattering also depends on the relative speed of approach of the two particles: a very fast particle might pass through the interaction region without much deflection, whereas a slower one on the same path might be temporarily captured and undergo considerable deflection (Fig. 18.17). The variation of the scattering cross-section with the relative speed of approach therefore gives information about the strength and range of the intermolecular potential.

Another phenomenon that can occur in certain beams is the capturing of one species by another. The vibrational temperature in supersonic beams is so low that van der Waals molecules may be formed, which are complexes of the form AB in which A and B are held together by van der Waals forces or hydrogen bonds. Large numbers of such molecules have been studied spectroscopically, including ArHCl, $(HCl)_2$, $ArCO_2$, and $(H_2O)_2$. More recently, van der Waals clusters of water molecules have been pursued as far as $(H_2O)_6$. The study of their spectroscopic properties gives detailed information about the intermolecular potentials involved.

Solutions to exercises

E18.1(b) A molecule that has a center of symmetry cannot be polar. $SO_3 (D_{3h})$ and $XeF_4 (D_{4h})$ cannot be polar. $\boxed{SF_4}$ (see-saw, C_{2v}) may be polar.

E18.2(b)

$$\mu = (\mu_1^2 + \mu_2^2 + 2\mu_1\mu_2 \cos\theta)^{1/2} \quad [18.2a]$$

$$= [(1.5)^2 + (0.80)^2 + (2) \times (1.5) \times (0.80) \times (\cos 109.5°)]^{1/2}\, D = \boxed{1.4\,D}$$

E18.3(b)   The components of the dipole moment vector are

$$\mu_x = \sum_i q_i x_i = (4e) \times (0) + (-2e) \times (162\,\mathrm{pm})$$

$$+ (-2e) \times (143\,\mathrm{pm}) \times (\cos 30°) = (-572\,\mathrm{pm})e$$

and $\mu_y = \sum_i q_i y_i = (4e) \times (0) + (-2e) \times (0) + (-2e) \times (143\,\mathrm{pm}) \times (\sin 30°) = (-143\,\mathrm{pm})e$

The magnitude is

$$\mu = (\mu_x^2 + \mu_y^2)^{1/2} = ((-570)^2 + (-143)^2)^{1/2}\,\mathrm{pm}\,e = (590\,\mathrm{pm})e$$

$$= (590 \times 10^{-12}\,\mathrm{m}) \times (1.602 \times 10^{-19}\,\mathrm{C}) = \boxed{9.45 \times 10^{-29}\,\mathrm{C\,m}}$$

and the direction is $\theta = \tan^{-1} \dfrac{\mu_y}{\mu_x} = \tan^{-1} \dfrac{-143\,\mathrm{pm}\,e}{-572\,\mathrm{pm}\,e} = \boxed{194.0°}$ from the x-axis (i.e. 14.0° below the negative x-axis).

E18.4(b)   The molar polarization depends on the polarizability through

$$P_{\mathrm{m}} = \frac{N_A}{3\varepsilon_0}\left(\alpha + \frac{\mu^2}{3kT}\right)$$

This is a linear equation in T^{-1} with slope

$$m = \frac{N_A \mu^2}{9\varepsilon_0 k} \quad \text{so} \quad \mu = \left(\frac{9\varepsilon_0 k m}{N_A}\right)^{1/2} = (4.275 \times 10^{-29}\,\mathrm{C\,m}) \times (m/(\mathrm{m^3\,mol^{-1}K}))^{1/2}$$

and with y-intercept

$$b = \frac{N_A \alpha}{3\varepsilon_0} \quad \text{so} \quad \alpha = \frac{3\varepsilon_0 b}{N_A} = (4.411 \times 10^{-35}\,\mathrm{C^2\,m^2\,J^{-1}})b/(\mathrm{m^3\,mol^{-1}})$$

Since the molar polarization is linearly dependent on $\mathrm{T^{-1}}$, we can obtain the slope m and the intercept b

$$m = \frac{P_{\mathrm{m},2} - P_{\mathrm{m},1}}{T_1^{-1} - T_2^{-1}} = \frac{(75.74 - 71.43)\,\mathrm{cm^3\,mol^{-1}}}{(320.0\,\mathrm{K})^{-1} - (421.7\,\mathrm{K})^{-1}} = 5.72 \times 10^3\,\mathrm{cm^3\,mol^{-1}\,K}$$

and $\quad b = P_{\mathrm{m}} - mT^{-1} = 75.74\,\mathrm{cm^3\,mol^{-1}} - (5.72 \times 10^3\,\mathrm{cm^3\,mol^{-1}\,K}) \times (320.0\,\mathrm{K})^{-1}$

$$= 57.9\,\mathrm{cm^3\,mol^{-1}}$$

It follows that

$$\mu = (4.275 \times 10^{-29}\,\mathrm{C\,m}) \times (5.72 \times 10^{-3})^{1/2} = \boxed{3.23 \times 10^{-30}\,\mathrm{C\,m}}$$

and

$$\alpha = (4.411 \times 10^{-35}\,\mathrm{C^2\,m^2\,J^{-1}}) \times (57.9 \times 10^{-6}) = \boxed{2.55 \times 10^{-39}\,\mathrm{C^2\,m^2\,J^{-1}}}$$

E18.5(b) The relative permittivity is related to the molar polarization through

$$\frac{\varepsilon_r - 1}{\varepsilon_r + 2} = \frac{\rho P_m}{M} \equiv C \quad \text{so} \quad \varepsilon_r = \frac{2C + 1}{1 - C},$$

$$C = \frac{(1.92\,\text{g cm}^{-3}) \times (32.16\,\text{cm}^3\,\text{mol}^{-1})}{85.0\,\text{g mol}^{-1}} = 0.726$$

$$\varepsilon_r = \frac{2 \times (0.726) + 1}{1 - 0.726} = \boxed{8.97}$$

E18.6(b) The induced dipole moment is

$$\mu^* = \alpha\varepsilon = 4\pi\varepsilon_0\alpha'\varepsilon$$

$$= 4\pi(8.854 \times 10^{-12}\,\text{J}^{-1}\,\text{C}^2\,\text{m}^{-1}) \times (2.22 \times 10^{-30}\,\text{m}^3) \times (15.0 \times 10^3\,\text{V m}^{-1})$$

$$= \boxed{3.71 \times 10^{-36}\,\text{C m}}$$

E18.7(b) If the permanent dipole moment is negligible, the polarizability can be computed from the molar polarization

$$P_m = \frac{N_A\alpha}{3\varepsilon_0} \quad \text{so} \quad \alpha = \frac{3\varepsilon_0 P_m}{N_A}$$

and the molar polarization from the refractive index

$$\frac{\rho P_m}{M} = \frac{\varepsilon_r - 1}{\varepsilon_r + 2} = \frac{n_r^2 - 1}{n_r^2 + 2} \quad \text{so} \quad \alpha = \frac{3\varepsilon_0 M}{N_A\rho}\left(\frac{n_r^2 - 1}{n_r^2 + 2}\right)$$

$$\alpha = \frac{3 \times (8.854 \times 10^{-12}\,\text{J}^{-1}\,\text{C}^2\,\text{m}^{-1}) \times (65.5\,\text{g mol}^{-1})}{(6.022 \times 10^{23}\,\text{mol}^{-1}) \times (2.99 \times 10^6\,\text{g m}^{-3})} \times \left(\frac{1.622^2 - 1}{1.622^2 + 2}\right)$$

$$= \boxed{3.40 \times 10^{-40}\,\text{C}^2\,\text{m}^2\,\text{J}^{-1}}$$

E18.8(b) The solution to Exercise 18.7(a) showed that

$$\alpha = \left(\frac{3\varepsilon_0 M}{\rho N_A}\right) \times \left(\frac{n_r^2 - 1}{n_r^2 + 2}\right) \quad \text{or} \quad \alpha' = \left(\frac{3M}{4\pi\rho N_A}\right) \times \left(\frac{n_r^2 - 1}{n_r^2 + 2}\right)$$

which may be solved for n_r to yield

$$n_r = \left(\frac{\beta' + 2\alpha'}{\beta' - \alpha'}\right)^{1/2} \quad \text{with} \quad \beta' = \frac{3M}{4\pi\rho N_A}$$

$$\beta' = \frac{(3) \times (72.3\,\text{g mol}^{-1})}{(4\pi) \times (0.865 \times 10^6\,\text{g m}^{-3}) \times (6.022 \times 10^{23}\,\text{mol}^{-1})} = 3.31\bar{4} \times 10^{-29}\,\text{m}^3$$

$$n_r = \left(\frac{33.1\bar{4} + 2 \times 2.2}{33.1\bar{4} - 2.2}\right)^{1/2} = \boxed{1.10}$$

E18.9(b)   The relative permittivity is related to the molar polarization through

$$\frac{\varepsilon_r - 1}{\varepsilon_r + 2} = \frac{\rho P_m}{M} \equiv C \quad \text{so} \quad \varepsilon_r = \frac{2C + 1}{1 - C}$$

The molar polarization depends on the polarizability through

$$P_m = \frac{N_A}{3\varepsilon_0}\left(\alpha + \frac{\mu^2}{3kT}\right) \quad \text{so} \quad C = \frac{\rho N_A}{3\varepsilon_0 M}\left(4\pi\varepsilon_0\alpha' + \frac{\mu^2}{3kT}\right)$$

$$C = \frac{(1491 \text{ kg m}^{-3}) \times (6.022 \times 10^{23} \text{ mol}^{-1})}{3(8.854 \times 10^{-12} \text{ J}^{-1}\text{ C}^2\text{ m}^{-1}) \times (157.01 \times 10^{-3} \text{ kg mol}^{-1})}$$

$$\times \left(4\pi(8.854 \times 10^{-12}\text{ J}^{-1}\text{ C}^2\text{ m}^{-1}) \times (1.5 \times 10^{-29}\text{ m}^3)\right.$$

$$+ \frac{(5.17 \times 10^{-30}\text{ C m})^2}{3(1.381 \times 10^{-23}\text{ J K}^{-1}) \times (298\text{ K})}\bigg)$$

$$C = 0.83 \quad \text{and} \quad \varepsilon_r = \frac{2(0.83) + 1}{1 - 0.83} = \boxed{16}$$

E18.10(b) $$V_m = \frac{M}{\rho} = \frac{18.02 \text{ g mol}^{-1}}{999.4 \times 10^3 \text{ g m}^{-3}} = 1.803 \times 10^{-5}\text{ m}^3\text{ mol}^{-1}$$

$$\frac{2\gamma V_m}{rRT} = \frac{2\left(7.275 \times 10^{-2}\text{ N m}^{-1}\right) \times \left(1.803 \times 10^{-5}\text{ m}^3\text{ mol}^{-1}\right)}{\left(20.0 \times 10^{-9}\text{ m}\right) \times \left(8.314\text{ J K}^{-1}\text{ mol}^{-1}\right) \times (308.2\text{ K})}$$

$$= 5.11\bar{9} \times 10^{-2}$$

$$p = (5.623\text{ kPa})\,e^{0.0511\bar{9}} = \boxed{5.92\text{ kPa}}$$

E18.11(b) $$\gamma = \frac{1}{2}\rho ghr = \frac{1}{2}\left(0.9956\text{ g cm}^{-3}\right) \times \left(9.807\text{ m s}^{-2}\right) \times \left(9.11 \times 10^{-2}\text{ m}\right)$$

$$\times \left(0.16 \times 10^{-3}\text{ m}\right) \times \left(\frac{1000\text{ kg m}^{-3}}{\text{g cm}^{-3}}\right)$$

$$= \boxed{7.12 \times 10^{-2}\text{ N m}^{-1}}$$

E18.12(b) $$p_{in} - p_{out} = \frac{2\gamma}{r}\text{ [18.38]} = \frac{(2) \times (22.39 \times 10^{-3}\text{ N m}^{-1})}{2.20 \times 10^{-7}\text{ m}} = \boxed{2.04 \times 10^5\text{ Pa}}$$

Solutions to problems

Solutions to numerical problems

P18.1   The positive (H) end of the dipole will lie closer to the (negative) anion. The electric field generated by a dipole is

$$\mathcal{E} = \left(\frac{\mu}{4\pi\varepsilon_0}\right) \times \left(\frac{2}{r^3}\right)\text{ [18.21]}$$

$$= \frac{(2) \times (1.85) \times (3.34 \times 10^{-30}\text{ C m})}{(4\pi) \times (8.854 \times 10^{-12}\text{ J}^{-1}\text{ C}^2\text{ m}^{-1}) \times r^3} = \frac{1.11 \times 10^{-19}\text{ V m}^{-1}}{(r/\text{m})^3} = \frac{1.11 \times 10^8\text{ V m}^{-1}}{(r/\text{nm})^3}.$$

(a) $\mathscr{E} = \boxed{1.1 \times 10^8 \text{ V m}^{-1}}$ when $r = 1.0 \text{ nm}$.

(b) $\mathscr{E} = \dfrac{1.11 \times 10^8 \text{ V m}^{-1}}{0.3^3} = \boxed{4 \times 10^9 \text{ V m}^{-1}}$ for $r = 0.3 \text{ nm}$.

(c) $\mathscr{E} = \dfrac{1.11 \times 10^8 \text{ V m}^{-1}}{30^3} = \boxed{4 \text{ kV m}^{-1}}$ for $r = 30 \text{ nm}$.

P18.3 The equations relating dipole moment and polarizability volume to the experimental quantities ε_r and ρ are

$$P_m = \left(\frac{M}{\rho}\right) \times \left(\frac{\varepsilon_r - 1}{\varepsilon_r + 2}\right) \text{ [18.14] and } P_m = \frac{4\pi}{3} N_A \alpha' + \frac{N_A \mu^2}{9\varepsilon_0 kT} \text{ [18.15, with } \alpha = 4\pi\varepsilon_0 \alpha'].$$

Therefore, we draw up the following table (with $M = 119.4 \text{ g mol}^{-1}$).

$\theta/°C$	-80	-70	-60	-40	-20	0	20
T/K	193	203	213	233	253	273	293
$1000/(T/K)$	5.18	4.93	4.69	4.29	3.95	3.66	3.41
ε_r	3.1	3.1	7.0	6.5	6.0	5.5	5.0
$\dfrac{\varepsilon_r - 1}{\varepsilon_r + 2}$	0.41	0.41	0.67	0.65	0.63	0.60	0.57
$\rho/\text{g cm}^{-3}$	1.65	1.64	1.64	1.61	1.57	1.53	1.50
$P_m/(\text{cm}^3 \text{ mol}^{-1})$	29.8	29.9	48.5	48.0	47.5	56.8	45.4

P_m is plotted against $1/T$ in Fig. 18.1.

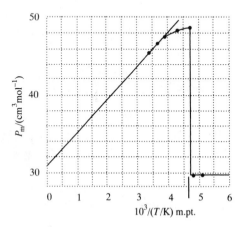

Figure 18.1

The (dangerously unreliable) intercept is ≈ 30 and the slope is $\approx 4.5 \times 10^3$. It follows that

$$\alpha' = \frac{(3) \times (30 \text{ cm}^3 \text{ mol}^{-1})}{(4\pi) \times (6.022 \times 10^{23} \text{ mol}^{-1})} = \boxed{1.2 \times 10^{-23} \text{ cm}^3}.$$

To determine μ we need

$$\mu = \left(\frac{9\varepsilon_0 k}{N_A}\right)^{1/2} \times (\text{slope} \times \text{cm}^3 \text{ mol}^{-1} \text{ K})^{1/2}$$

$$= \left\{\left(\frac{(9) \times (8.854 \times 10^{-12} \text{ J}^{-1} \text{ C}^2 \text{ m}^{-1}) \times (1.381 \times 10^{-23} \text{ J K}^{-1})}{6.022 \times 10^{-23} \text{ mol}^{-1}}\right)^{1/2} \right.$$

$$\left. \times (\text{slope} \times \text{cm}^3 \text{ mol}^{-1} \text{ K})^{1/2}\right\}$$

$$= (4.275 \times 10^{-29} \text{ C}) \times \left(\frac{\text{mol}}{\text{K m}}\right)^{1/2} \times (\text{slope} \times \text{cm}^3 \text{ mol}^{-1} \text{ K})^{1/2}$$

$$= (4.275 \times 10^{-29} \text{ C}) \times (\text{slope} \times \text{cm}^3 \text{ m}^{-1})^{1/2}$$

$$= (4.275 \times 10^{-32} \text{ C m}) \times (\text{slope})^{1/2} = (1.282 \times 10^{-2} \text{ D}) \times (\text{slope})^{1/2}$$

$$= (1.282 \times 10^{-2} \text{ D}) \times (4.5 \times 10^3)^{1/2} = \boxed{0.86 \text{ D}}.$$

The sharp decrease in P_m occurs at the freezing point of chloroform ($-63°C$), indicating that the dipole reorientation term no longer contributes. Note that P_m for the solid corresponds to the extrapolated, dipole-free, value of P_m, so the extrapolation is less hazardous than it looks.

P18.5
$$P_m = \frac{4\pi}{3}N_A\alpha' + \frac{N_A\mu^2}{9\varepsilon_0 kT} \quad [18.15, \text{with } \alpha = 4\pi\varepsilon_0\alpha'].$$

Therefore, draw up the following table.

T/K	292.2	309.0	333.0	387.0	413.0	446.0
$1000/(T/K)$	3.42	3.24	3.00	2.58	2.42	2.24
$P_m/(\text{cm}^3 \text{ mol}^{-1})$	57.57	55.01	51.22	44.99	42.51	39.59

The points are plotted in Fig. 18.2.

The extrapolated (least squares) intercept lies at $5.65 \text{ cm}^3 \text{ mol}^{-1}$ (not shown in the figure), and the least squares slope is $1.52 \times 10^4 \text{ cm}^3 \text{ K}^{-1} \text{ mol}^{-1}$. It follows that

$$\alpha' = \frac{3P_m(\text{at intercept})}{4\pi N_A} = \frac{3 \times 5.65 \text{ cm}^3 \text{ mol}^{-1}}{4\pi \times 6.022 \times 10^{23} \text{ mol}^{-1}}$$

$$= \boxed{2.24 \times 10^{-24} \text{ cm}^3}.$$

$$\mu = 1.282 \times 10^{-2} \text{ D} \times (1.52 \times 10^4)^{1/2} \text{ [from Problem 18.3]} = \boxed{1.58 \text{ D}}.$$

The high-frequency contribution to the molar polarization, P'_m, at 273 K may be calculated from the refractive index:

$$P'_m = \left(\frac{M}{\rho}\right) \times \left(\frac{\varepsilon_r - 1}{\varepsilon_r + 2}\right) \quad [18.14] = \left(\frac{M}{\rho}\right) \times \left(\frac{n_r^2 - 1}{n_r^2 + 2}\right).$$

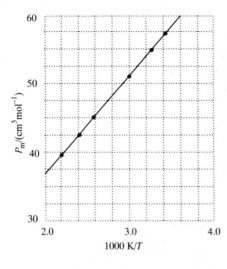

Figure 18.2

Assuming that ammonia under these conditions (1.00 atm pressure assumed) can be considered a perfect gas, we have

$$\rho = \frac{pM}{RT}$$

and $\dfrac{M}{\rho} = \dfrac{RT}{p} = \dfrac{82.06 \text{ cm}^3 \text{ atm K}^{-1} \text{ mol}^{-1} \times 273 \text{ K}}{1.00 \text{ atm}} = 2.24 \times 10^4 \text{ cm}^3 \text{ mol}^{-1}.$

Then $P'_m = 2.24 \times 10^4 \text{ cm}^3 \text{ mol}^{-1} \times \left\{ \dfrac{(1.000379)^2 - 1}{(1.000379)^2 + 2} \right\} = \boxed{5.66 \text{ cm}^3 \text{ mol}^{-1}}.$

If we assume that the high-frequency contribution to P_m remains the same at 292.2 K then we have

$$\frac{N_A \mu^2}{9\varepsilon_0 kT} = P_m - P_m' = (57.57 - 5.66) \text{ cm}^3 \text{ mol}^{-1}$$

$$= 51.91 \text{ cm}^3 \text{ mol}^{-1} = 5.191 \times 10^{-5} \text{ m}^3 \text{ mol}^{-1}.$$

Solving for μ we have

$$\mu = \left(\frac{9\varepsilon_0 k}{N_A} \right)^{1/2} T^{1/2} (P_m - P'_m)^{1/2}.$$

The factor $\left(\dfrac{9\varepsilon_0 k}{N_A} \right)^{1/2}$ has been calculated in Problem 18.3 and is 4.275×10^{-29} C $\times$ (mol/K m)$^{1/2}$.

Therefore $\mu = 4.275 \times 10^{-29}$ C $\times \left(\dfrac{\text{mol}}{\text{K m}} \right)^{1/2} \times (292.2 \text{ K})^{1/2} \times (5.191 \times 10^{-5})^{1/2} (\text{m}^3/\text{mol})^{1/2}$

$$= 5.26 \times 10^{-30} \text{ C m} = \boxed{1.58 \text{ D}}.$$

The agreement is exact!

P18.7   **(a)** The depth of the well in energy units is

$$\varepsilon = hcD_e = \boxed{1.51 \times 10^{-23} \text{ J}}.$$

The distance at which the potential is zero is given by

$$R_e = 2^{1/6} r_0 \quad \text{so} \quad r_0 = R_e 2^{-1/6} = 2^{-1/6}(297 \text{ pm}) = \boxed{265 \text{ pm}}.$$

(b) In Fig. 18.3 both potentials were plotted with respect to the bottom of the well, so the Lennard-Jones potential is the usual L-J potential plus ε.

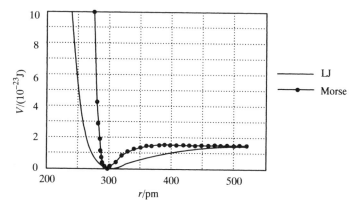

Figure 18.3

Note that the Lennard-Jones potential has a much softer repulsive branch than the Morse.

P18.9   Neglecting the permanent dipole moment contribution,

$$P_m = \frac{N_A \alpha}{3\varepsilon_0} \text{ [18.15]}$$

$$= \frac{(6.022 \times 10^{23} \text{ mol}^{-1}) \times (3.59 \times 10^{-40} \text{ J}^{-1} \text{ C}^2 \text{ m}^2)}{3(8.854 \times 10^{-12} \text{ J}^{-1} \text{ C}^2 \text{ m}^{-1})}$$

$$= 8.14 \times 10^{-6} \text{ m}^3 \text{ mol}^{-1} = \boxed{8.14 \text{ cm}^3 \text{ mol}^{-1}}.$$

$$\frac{\varepsilon_r - 1}{\varepsilon_r + 2} = \frac{\rho P_m}{M} \text{ [18.16]}$$

$$= \frac{(0.7914 \text{ g cm}^{-3}) \times (8.14 \text{ cm}^3 \text{ mol}^{-1})}{32.04 \text{ g mol}^{-1}} = 0.201.$$

$$\varepsilon_r - 1 = 0.201\varepsilon_r + 0.402; \quad \boxed{\varepsilon_r = 1.76}.$$

$$n_r = \varepsilon_r^{1/2} \text{ [18.17]} = (1.76)^{1/2} = \boxed{1.33}.$$

The neglect of the permanent dipole moment contribution means that the results are applicable only to the case for which the applied field has a much larger frequency than the rotational frequency. Since red light has a frequency of 4.3×10^{14} and a typical rotational frequency is about 1×10^{12} Hz, the results apply in the visible.

Solutions to theoretical problems

P18.11 Exercise 18.7 showed

$$\alpha = \left(\frac{3\varepsilon_0 M}{\rho N_A}\right) \times \left(\frac{n_r^2 - 1}{n_r^2 + 2}\right) \quad \text{or} \quad \alpha' = \left(\frac{3M}{4\pi \rho N_A}\right) \times \left(\frac{n_r^2 - 1}{n_r^2 + 2}\right).$$

Therefore, $\dfrac{n_r^2 - 1}{n_r^2 + 2} = \dfrac{4\pi \alpha' N_A \rho}{3M}$.

Solving for n_r, $n_r = \left(\dfrac{1 + \dfrac{8\pi \alpha' \rho N_A}{3M}}{1 - \dfrac{4\pi \alpha' \rho N_A}{3M}}\right)^{1/2} = \left(\dfrac{1 + \dfrac{8\pi \alpha' p}{3kT}}{1 - \dfrac{4\pi \alpha' p}{3kT}}\right)^{1/2} \quad \left[\text{for a gas, } \rho = \dfrac{M}{V_m} = \dfrac{Mp}{RT}\right]$

$$\approx \left[\left(1 + \frac{8\pi \alpha' p}{3kT}\right) \times \left(1 + \frac{4\pi \alpha' p}{3kT}\right)\right]^{1/2} \quad \left[\frac{1}{1-x} \approx 1 + x\right]$$

$$\approx \left(1 + \frac{12\pi \alpha' p}{3kT} + \cdots\right)^{1/2} \approx 1 + \frac{2\pi \alpha' p}{kT} \quad \left[(1+x)^{1/2} \approx 1 + \frac{1}{2}x\right].$$

Hence, $\boxed{n_r = 1 + \text{const.} \times p}$, with constant $= \dfrac{2\pi \alpha'}{kT}$. From the first line above,

$$\alpha' = \left(\frac{3M}{4\pi N_A \rho}\right) \times \left(\frac{n_r^2 - 1}{n_r^2 + 2}\right) = \boxed{\left(\frac{3kT}{4\pi p}\right) \times \left(\frac{n_r^2 - 1}{n_r^2 + 2}\right)}.$$

P18.13 Consider a single molecule surrounded by $N - 1(\approx N)$ others in a container of volume V. The number of molecules in a spherical shell of thickness dr at a distance r is $4\pi r^2 \times (N/V)\, dr$. Therefore, the interaction energy is

$$u = \int_a^R 4\pi r^2 \times \left(\frac{N}{V}\right) \times \left(\frac{-C_6}{r^6}\right) dr = \frac{-4\pi N C_6}{V} \int_a^R \frac{dr}{r^4}$$

where R is the radius of the container and d the molecular diameter (the distance of closest approach). Therefore,

$$u = \left(\frac{4\pi}{3}\right) \times \left(\frac{N}{V}\right)(C_6) \times \left(\frac{1}{R^3} - \frac{1}{d^3}\right) \approx \frac{-4\pi N C_6}{3V d^3}$$

because $d \ll R$. The mutual pairwise interaction energy of all N molecules is $U = \frac{1}{2}Nu$ (the $\frac{1}{2}$ appears because each pair must be counted only once, i.e. A with B but not A with B and B with A). Therefore,

$$U = \boxed{\frac{-2\pi N^2 C_6}{3V d^3}}.$$

For a van der Waals gas, $\dfrac{n^2 a}{V^2} = \left(\dfrac{\partial U}{\partial V}\right)_T = \dfrac{2\pi N^2 C_6}{3V^2 d^3}$ and therefore $a = \boxed{\dfrac{2\pi N_A^2 C_6}{3d^3}}$ $[N = nN_A]$.

P18.15   The number of molecules in a volume element $d\tau$ is $\mathcal{N} \, d\tau / V = \mathcal{N} d\tau$. The energy of interaction of these molecules with one at a distance r is $V \mathcal{N} \, d\tau$. The total interaction energy, taking into account the entire sample volume, is therefore

$$u = \int V \mathcal{N} \, d\tau = \mathcal{N} \int V \, d\tau \quad [V \text{ is the interaction energy, not the volume}].$$

The total interaction energy of a sample of N molecules is $\frac{1}{2} N u$ (the $\frac{1}{2}$ is included to avoid double counting), and so the cohesive energy density is

$$\mathcal{U} = -\frac{U}{V} = \frac{-\frac{1}{2} N u}{V} = -\frac{1}{2} \mathcal{N} u = -\frac{1}{2} \mathcal{N}^2 \int V \, d\tau.$$

For $V = -C_6 / r^6$ and $d\tau = 4\pi r^2 \, dr$,

$$-\frac{U}{V} = 2\pi \mathcal{N}^2 C_6 \int_a^\infty \frac{dr}{r^4} = \frac{2\pi}{3} \times \frac{\mathcal{N}^2 C_6}{d^3}.$$

However, $\mathcal{N} = N_A \rho / M$, where M is the molar mass; therefore

$$\boxed{\mathcal{U} = \left(\frac{2\pi}{3}\right) \times \left(\frac{N_A \rho}{M}\right)^2 \times \left(\frac{C_6}{d^3}\right).}$$

P18.17   Once again (as in Problem 18.16) we can write

$$\theta(v) = \begin{cases} \pi - 2 \arcsin\left(\dfrac{b}{R_1 + R_2(v)}\right) & b \le R_1 + R_2(v) \\ 0 & b > R_1 + R_2(v) \end{cases}$$

but R_2 depends on v

$$R_2(v) = R_2 e^{-v/v^*}.$$

Therefore, with $R_1 = \frac{1}{2} R_2$ and $b = \frac{1}{2} R_2$,

(a) $\theta(v) = \pi - 2 \arcsin\left(\dfrac{1}{1 + 2 e^{-v/v^*}}\right).$

(The restriction $b \le R_1 + R_2(v)$ transforms into $\frac{1}{2} R_2 \le \frac{1}{2} R_2 + R_2 e^{-v/v^*}$, which is valid for all v.)
This function is plotted as curve a in Fig. 18.4.

The kinetic energy of approach is $E = \frac{1}{2} m v^2$, and so

(b) $\theta(E) = \pi - 2 \arcsin\left(\dfrac{1}{1 + 2 e^{-(E/E^*)^{1/2}}}\right)$ with $E^* = \frac{1}{2} m v^{*2}$. This function is plotted as curve b in Fig. 18.4.

Solutions to applications

P18.19   **(a)** The energy of induced-dipole–induced-dipole interactions can be approximated by the London formula (eqn 18.25):

$$V = -\frac{C}{r^6} = -\frac{3\alpha_1'\alpha_2'}{2r^6}\frac{I_1 I_2}{I_1 + I_2} = -\frac{3\alpha'^2 I}{4r^6}$$

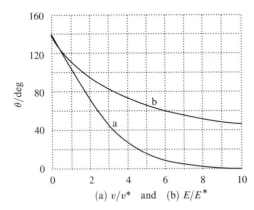

(a) v/v^* and   (b) E/E^* **Figure 18.4**

where the second equality uses the fact that the interaction is between two of the same molecule. For two phenyl groups, we have

$$V = -\frac{3(1.04 \times 10^{-29}\ \text{m}^3)^2 (5.0\ \text{eV})(1.602 \times 10^{-19}\ \text{J\,eV}^{-1})}{4(1.0 \times 10^{-9}\ \text{m})^6} = 6.6 \times 10^{-23}\ \text{J}$$

or $\boxed{-39\ \text{J mol}^{-1}}$.

(b) The potential energy is everywhere negative. We can obtain the distance dependence of the force by taking

$$F = -\frac{dV}{dr} = -\frac{6C}{r^7}.$$

This force is everywhere attractive (i.e. it works against increasing the distance between interacting groups). The force $\boxed{\text{approaches zero as the distance becomes very large}}$; there is no finite distance at which the dispersion force is zero. (Of course, if one takes into account repulsive forces, then the net force is zero at a distance at which the attractive and repulsive forces balance.)

P18.21   **(a)** The dipole moment computed for *trans-N*-methylacetamide is

$$\mu = (3.092\ \text{D}) \times (3.336 \times 10^{-30}\ \text{C m D}^{-1}) = \boxed{1.03 \times 10^{-29}\ \text{C m}}$$

(semi-empirical, PM3 level, PC Spartan Pro™). The dipole is oriented mainly along the carbonyl group. The interaction energy of two parallel dipoles is given by eqn 18.22:

$$V = \frac{\mu_1 \mu_2 f(\theta)}{4\pi\varepsilon_0 r^3} \quad \text{where } f(\theta) = 1 - 3\cos^2\theta$$

and r is the distance between the dipoles and θ the angle between the direction of the dipoles and the line that joins them. The angular dependence is shown in Fig. 18.5. Note that $V(\theta)$ is at a minimum for $\theta = 0°$ and $180°$ while it is at a maximum for $90°$ and $270°$.

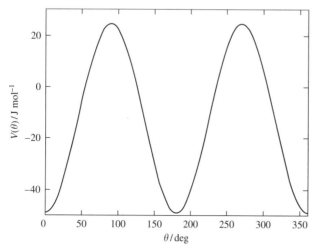

Figure 18.5

(b) If the dipoles are separated by 3.0 nm, then the maximum energy of interaction is:

$$V_{max} = \frac{(1.031 \times 10^{-29} \, \text{C m})^2}{4\pi(8.854 \times 10^{-12} \, \text{J}^{-1} \, \text{C}^2 \, \text{m}^{-1}) \times (3.0 \times 10^{-9} \, \text{m})^3} = \boxed{3.5\overline{5} \times 10^{-23} \, \text{J}}.$$

In molar units

$$V_{max} = (3.5\overline{5} \times 10^{-23} \, \text{J}) \times (6.022 \times 10^{23} \, \text{mol}^{-1}) = 21 \, \text{J mol}^{-1} = 2.1 \times 10^{-2} \, \text{kJ mol}^{-1}.$$

Thus, dipole–dipole interactions at this distance are $\boxed{\text{dwarfed by hydrogen bonding interactions}}$.

However, the typical hydrogen bond length is much shorter, so this may not be a fair comparison.

P18.23   Here is a solution using MathCad.

(a) $\text{Data} := \begin{pmatrix} 7.36 & 8.37 & 8.3 & 7.47 & 7.25 & 6.73 & 8.52 & 7.87 & 7.53 \\ 3.53 & 4.24 & 4.09 & 3.45 & 2.96 & 2.89 & 4.39 & 4.03 & 3.80 \\ 1.00 & 1.80 & 1.70 & 1.35 & 1.60 & 1.60 & 1.95 & 1.60 & 1.60 \end{pmatrix}$

$\text{log}A := (\text{Data}^T)^{\langle 0 \rangle}$ $S := (\text{Data}^T)^{\langle 1 \rangle}$ $W := (\text{Data}^T)^{\langle 2 \rangle}$ $\text{Mxy} := \text{augment}(S, W)$

$\text{info} := \text{regress}(\text{Mxy}, \text{log}A, 1)$ $b := \text{submatrix}(\text{info}, 3, 5, 0, 0)$ $b = \begin{pmatrix} 0.957 \\ 0.362 \\ 3.59 \end{pmatrix} \begin{matrix} b_0 \\ b_1 \\ b_2 \end{matrix}$

(b) $W := 1.5$ Estimate for Given/Find Solve Bank
     $S := 4.84$ $\text{log}A := 7.60$
     Given $\text{log}A = b_0 + b_1 \cdot S + b_2 \cdot W$ $W := \text{Find}(W)$ $W = 1.362$

Answers to discussion questions

D19.1 Number average is the value obtained by weighting each molar mass by the number of molecules with that mass (eqn 19.1)

$$\overline{M_n} = \frac{1}{N} \sum_i N_i M_i.$$

In this expression, N_i is the number of molecules of molar mass M_i and N is the total number of molecules. Measurements of the osmotic pressures of macromolecular solutions yield the number average molar mass.

Weight average is the value obtained by weighting each molar mass by the mass of each one present (eqn 19.2)

$$\overline{M_w} = \frac{1}{m} \sum_i m_i M_i = \frac{\sum_i N_i M_i^2}{\sum_i N_i M_i} \quad [19.3].$$

In this expression, m_i is the total mass of molecules with molar mass M_i and m is the total mass of the sample. Light scattering experiments give the weight average molar mass.

Z-average molar mass is defined through the formula (eqn 19.4)

$$\overline{M_z} = \frac{\sum_i N_i M_i^3}{\sum_i N_i M_i^2}.$$

The Z-average molar mass is obtained from sedimentation equilibria experiments.

D19.3 Contour length: the length of the macromolecule measured along its backbone, the length of all its monomer units placed end to end. This is the stretched-out length of the macromolecule, but with bond angles maintained within the monomer units. It is proportional to the number of monomer units, N, and to the length of each unit (eqn 19.30).

Root mean square separation: one measure of the average separation of the ends of a random coil. It is the square root of the mean value of R^2, where R is the separation of the two ends of the coil. This mean value is calculated by weighting each possible value of R^2 with the probability, f (eqn 19.27), of that value of R occurring. It is proportional to $N^{1/2}$ and the length of each unit (eqn 19.31).

Radius of gyration: the radius of a thin hollow spherical shell of the same mass and moment of inertia as the macromolecule. In general, it is not easy to visualize this distance geometrically. However, for the simple case of a molecule consisting of a chain of identical atoms this quantity can be visualized as the root mean square distance of the atoms from the center of mass. It also depends on $N^{1/2}$, but is smaller than the root mean square separation by a factor of $(1/6)^{1/2}$ (eqn 19.33).

D19.5   For a molecular mechanics calculation, potential energy functions are chosen for all the interactions between the atoms in the molecule; the calculation itself is a mathematical procedure that locates the energy minima (local and global) of the molecule as a function of bond distances and bond angles. Because only the potential energy is included in the calculation, contributions to the total energy from the kinetic energy are excluded in the result. The global minimum of a molecular mechanics calculation is a snapshot of the molecular structure at $T = 0$. No equations of motion are solved in a molecular mechanics calculation. The structure of a macromolecule (or any molecule, for that matter) can, in principle, be determined by solving the time independent Schrödinger equation for the molecule with methods similar to those described in Chapter 11. But, due to the very large size of macromolecules, these methods may be impractical and, due to approximations to make them tractable, inaccurate.

In a molecular dynamics calculation, equations of motion are integrated to determine the trajectories of all atoms in the molecule. The equations of motion can, in principle, be either classical (Newton's laws of motion) or quantum mechanical. But, in practice, due to the very large number of atoms in a macromolecule, Newton's equations of motion are used. Quantum mechanical methods are too time consuming, complicated, and at this stage too inaccurate to be popular in the field of polymer chemistry.

D19.7   A surfactant is a species that is active at the interface of two phases or substances, such as the interface between hydrophilic and hydrophobic phases. A surfactant accumulates at the interface and modifies the properties of the surface, in particular, decreasing its surface tension. A typical surfactant consists of a long hydrocarbon tail and other non-polar materials, and a hydrophilic head group, such as the carboxylate group, $-CO_2^-$, that dissolves in a polar solvent, typically water. In other words, a surfactant is an amphipathic substance, meaning that it has both hydrophobic and hydrophilic regions.

How does the surfactant decrease the surface tension? Surface tension is a result of cohesive forces and the solute molecules must weaken the attractive forces between solvent molecules. Thus molecules with bulky hydrophobic regions such as fatty acids can decrease the surface tension because they attract solvent molecules less strongly than solvent molecules attract each other. See Section 19.15(b) for an analysis of the thermodynamics involved in this process.

D19.9   A Langmuir–Blodgett (LB) film is a monolayer or multilayer film that has been placed upon a substrate by transferring a surface film from a liquid to the substrate. A Langmuir trough, shown in Fig. 19.1(a), is designed to perform the transfer. A surface film of water-insoluble, film-forming molecules is assembled upon the water by mechanical compression. Dipping and withdrawing the substrate affects monolayer transfer. Repeated dipping produces multilayers. Weak van der Waals forces hold the monolayers together.

Self-assembled monolayers (SAMs) do not require assembly by mechanical compression. SAMs form from charged materials that have adsorption–desorption properties that promote self-assembly as shown in Fig. 19.1(b). The substrate is simply immersed in a dispersion of the charged materials, withdrawn, and rinsed. Films are held together with either strong ionic bonds or covalent bonds.

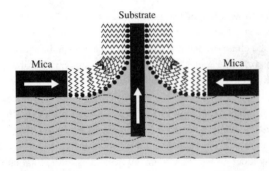

Figure 19.1(a)

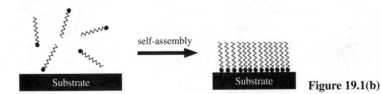

Figure 19.1(b)

Both methods yield well-organized monolayers but LB films upon water provide better organizational control than is possible with spontaneous self-assembled films. However, not requiring mechanical compression, SAMs are much more versatile. The strong bonding of SAMs gives long-lasting, stable films in contrast to the less stable van der Waals LB films.

Solutions to exercises

E19.1(b)   The number-average molar mass is (eqn 19.1)

$$\overline{M}_n = \frac{1}{N} \sum N_i M_i = \frac{[3 \times (62) + 2 \times (78)]\,\text{kg mol}^{-1}}{5} = \boxed{68\,\text{kg mol}^{-1}}$$

The mass-average molar mass is (eqn 19.3)

$$\overline{M}_w = \frac{\sum N_i M_i^2}{\sum N_i M_i} = \frac{3 \times (62)^2 + 2 \times (78)^2}{3 \times (62) + 2 \times (78)}\,\text{kg mol}^{-1} = \boxed{69\,\text{kg mol}^{-1}}$$

E19.2(b)   For a random coil, the radius of gyration is (19.33)

$$R_g = l(N/6)^{1/2} \quad \text{so} \quad N = 6(R_g/l)^2 = 6 \times (18.9\,\text{nm}/0.450\,\text{nm})^2 = \boxed{1.06 \times 10^4}$$

E19.3(b)   (a) Osmometry gives the number-average molar mass, so

$$\overline{M}_n = \frac{N_1 M_1 + N_2 M_2}{N_1 + N_2} = \frac{(m_1/M_1)\,M_1 + (m_2/M_2)\,M_2}{(m_1/M_1) + (m_2/M_2)} = \frac{m_1 + m_2}{(m_1/M_1) + (m_2/M_2)}$$

$$= \frac{100\,\text{g}}{\left(\dfrac{25\,\text{g}}{22\,\text{kg mol}^{-1}}\right) + \left(\dfrac{75\,\text{g}}{22/3\,\text{kg mol}^{-1}}\right)}\,[\text{assume 100 g of solution}] = \boxed{8.8\,\text{kg mol}^{-1}}$$

(b) Light-scattering gives the mass-average molar mass, so

$$\overline{M}_w = \frac{m_1 M_1 + m_2 M_2}{m_1 + m_2} = \frac{(25) \times (22) + (75) \times (22/3)}{25 + 75} \, \text{kg mol}^{-1} = \boxed{11 \, \text{kg mol}^{-1}}$$

E19.4(b) The formula for the rotational correlation time is

$$\tau = \frac{4\pi a^3 \eta}{3kT}$$

$$\eta(H_2O, 20\,°C) = 1.00 \times 10^{-3} \, \text{kg m}^{-1}\,\text{s}^{-1} \, [\textit{CRC Handbook}]$$

$$\tau = \frac{4\pi \times (4.5 \times 10^{-9}\text{m})^3 \times 1.00 \times 10^{-3} \, \text{kg m}^{-1}\,\text{s}^{-1}}{3 \times 1.381 \times 10^{-23} \, \text{J K}^{-1} \times 293 \, \text{K}} = \boxed{9.4 \times 10^{-8} \, \text{s}}$$

E19.5(b) The effective mass of the particles is

$$m_{\text{eff}} = bm = (1 - \rho v_s)m \, [19.14] = m - \rho v_s m = v\rho_p - v\rho = v(\rho_p - \rho)$$

where v is the particle volume and ρ_p is the particle density. Equating the forces

$$m_{\text{eff}} r\omega^2 = fs = 6\pi \eta a s \, [19.15, 19.12]$$

or $v(\rho_p - \rho)r\omega^2 = \frac{4}{3}\pi a^3(\rho_p - \rho)r\omega^2 = 6\pi \eta a s$

Solving for s yields

$$s = \frac{2a^2(\rho_p - \rho)r\omega^2}{9\eta}$$

Thus, the relative rates of sedimentation are $\dfrac{s_2}{s_1} = \dfrac{a_2^2(\rho_p - \rho)_2}{a_1^2(\rho_p - \rho)_1} = \left(\dfrac{a_2}{a_1}\right)^2 \dfrac{(\rho_p - \rho)_2}{(\rho_p - \rho)_1}.$

The value of this ratio depends on the density of the solution. For example, in a dilute aqueous solution with $\rho = 1.01 \, \text{g cm}^{-3}$, the difference in polymer densities matters in that the factor involving densities is significantly different than 1:

$$\frac{s_2}{s_1} = (8.4)^2 \frac{(1.10 - 0.794)}{(1.18 - 0.794)} = \boxed{56}$$

In a less dense organic solution, for example a dilute solution in octane with $\rho = 0.71 \, \text{g cm}^{-3}$, the density difference has a smaller effect, for the factor involving densities is closer to 1:

$$\frac{s_2}{s_1} = (8.4)^2 \frac{(1.10 - 0.71)_2}{(1.18 - 0.71)_1} = \boxed{59}$$

In both cases, the larger particle sediments faster.

E19.6(b) The molar mass is related to the sedimentation constant through eqns 19.19 and 19.14:

$$\overline{M} = \frac{SRT}{bD} = \frac{SRT}{(1 - \rho v_s)D}$$

where we have assumed the data refer to aqueous solution at 298 K.

$$\overline{M}_n = \frac{(7.46 \times 10^{-13}\,\text{s}) \times (8.3145\,\text{J}\,\text{K}^{-1}\,\text{mol}^{-1}) \times (298\,\text{K})}{[1 - (1000\,\text{kg}\,\text{m}^{-3}) \times (8.01 \times 10^{-4}\,\text{m}^3\,\text{kg}^{-1})] \times (7.72 \times 10^{-11}\,\text{m}^2\,\text{s}^{-1})}$$

$$= \boxed{120\,\text{kg}\,\text{mol}^{-1}}$$

E19.7(b)   See the solution to Exercise 19.5(b). In place of the centrifugal force $m_{\text{eff}}r^2$ we have the gravitational force $m_{\text{eff}}g$. The rest of the analysis is similar, leading to

$$s = \frac{2a^2(\rho_p - \rho)g}{9\eta} = \frac{(2) \times (15.5 \times 10^{-6}\,\text{m})^2 \times (1250 - 1000)\,\text{kg}\,\text{m}^{-3} \times (9.81\,\text{m}\,\text{s}^{-2})}{(9) \times (8.9 \times 10^{-4}\,\text{kg}\,\text{m}^{-1}\text{s}^{-1})}$$

$$= \boxed{1.47 \times 10^{-4}\,\text{m}\,\text{s}^{-1}}$$

E19.8(b)   The molar mass is related to the sedimentation constant through eqns 19.19 and 19.14:

$$\overline{M} = \frac{SRT}{bD} = \frac{SRT}{(1 - \rho v_s)D}$$

Assuming that the data refer to an aqueous solution,

$$\overline{M} = \frac{(5.1 \times 10^{-13}\,\text{s}) \times (8.3145\,\text{J}\;\text{K}^{-1}\,\text{mol}^{-1}) \times (293\,\text{K})}{[1 - (0.997\,\text{g}\,\text{cm}^{-3}) \times (0.721\,\text{cm}^3\text{g}^{-1})] \times (7.9 \times 10^{-11}\,\text{m}^2\,\text{s}^{-1})} = \boxed{56\,\text{kg}\,\text{mol}^{-1}}$$

E19.9(b)   In a sedimentation experiment, the weight-average molar mass is given by (eqn 19.20)

$$\overline{M}_w = \frac{2RT}{(r_2^2 - r_1^2)b\omega^2} \ln \frac{c_2}{c_1} \qquad \text{so} \qquad \ln \frac{c_2}{c_1} = \frac{\overline{M}_w(r_2^2 - r_1^2)b\omega^2}{2RT}$$

This implies that

$$\ln c = \frac{\overline{M}_w r^2 b\omega^2}{2RT} + \text{constant}$$

so the plot of $\ln c$ versus r^2 has a slope m equal to

$$m = \frac{\overline{M}_w b\omega^2}{2RT} \quad \text{and} \quad \overline{M}_w = \frac{2RTm}{b\omega^2}$$

$$\overline{M}_w = \frac{2 \times (8.3145\,\text{J}\,\text{K}^{-1}\,\text{mol}^{-1}) \times (293\,\text{K}) \times (821\,\text{cm}^{-2}) \times (100\,\text{cm}\,\text{m}^{-1})^2}{[1 - (1000\,\text{kg}\,\text{m}^{-3}) \times (7.2 \times 10^{-4}\,\text{m}^3\,\text{kg}^{-1})] \times [(1080\,\text{s}^{-1}) \times (2\pi)]^2}$$

$$= \boxed{3.1 \times 10^3\,\text{kg}\,\text{mol}^{-1}}$$

E19.10(b)   The centrifugal acceleration is

$$a = r\omega^2 \quad \text{so} \quad a/g = r\omega^2/g$$

$$a/g = \frac{(5.50\,\text{cm}) \times [2\pi \times (1.32 \times 10^3\,\text{s}^{-1})]^2}{(100\,\text{cm}\;\text{m}^{-1}) \times (9.81\,\text{m}\,\text{s}^{-2})} = \boxed{3.86 \times 10^5}$$

E19.11(b) For a random coil, the rms separation is [19.31]

$$R_{rms} = N^{1/2}l = (1200)^{1/2} \times (1.125\,nm) = \boxed{38.97\ nm}$$

E19.12(b) Polypropylene is $-(CH(CH_3)CH_2)-_N$, where N is given by

$$N = \frac{M_{polymer}}{M_{monomer}} = \frac{174\,kg\,mol^{-1}}{42.1 \times 10^{-3}\,kg\,mol^{-1}} = 4.13 \times 10^3$$

The repeat length l is the length of two C–C bonds. The contour length is [19.30]

$$R_c = Nl = (4.13 \times 10^3) \times (2 \times 1.53 \times 10^{-10}m) = \boxed{1.26 \times 10^{-6}\ m}$$

The rms seperation is [19.31]

$$R_{rms} = lN^{1/2} = (2 \times 1.53 \times 10^{-10}\ m) \times (4.13 \times 10^3)^{1/2} = \boxed{1.97 \times 10^{-8}\ m} = 19.7\,nm$$

Solutions to problems

Solutions to numerical problems

P19.1
$$S = \frac{s}{r\omega^2} \ [19.16].$$

Since $s = \dfrac{dr}{dt}, \dfrac{s}{r} = \dfrac{1}{r}\dfrac{dr}{dt} = \dfrac{d\ln r}{dt}$

and, if we plot $\ln r$ against t, the slope gives S through

$$S = \frac{1}{\omega^2}\frac{d\ln r}{dt}.$$

The data are as follows

$t/$ min	15.5	29.1	36.4	58.2
r/cm	5.05	5.09	5.12	5.19
$\ln(r/cm)$	1.619	1.627	1.633	1.647

The points are plotted in Fig. 19.2.

The least-squares slope is $6.62 \times 10^{-4}\,min^{-1}$, so

$$S = \frac{6.62 \times 10^{-4}\,min^{-1}}{\omega^2} = \frac{(6.62 \times 10^{-4}\,min^{-1}) \times (1\,min/60\,s)}{(2\pi \times 4.5 \times 10^4/60\,s)^2} = 4.9\overline{7} \times 10^{-13}\ s \text{ or } \boxed{5.0\ Sv}.$$

P19.3
$$[\eta] = \lim_{c \to 0}\left(\frac{\eta/\eta_0 - 1}{c}\right) \ [19.23].$$

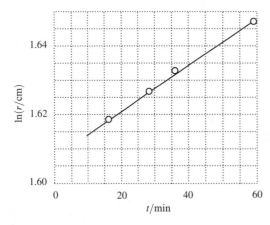

Figure 19.2

We see that the y-intercept of a plot of the right-hand side against c, extrapolated to $c = 0$, gives $[\eta]$. We begin by constructing the following table using $\eta_0 = 0.985$ g m^{-1} s^{-1}.

$c/(\text{g dm}^{-3})$		1.32	2.89	5.73	9.17
$\left(\dfrac{\eta\big/\eta_0 - 1}{c}\right)\bigg/(\text{dm}^3\,\text{g}^{-1})$		0.0731	0.0755	0.0771	0.0825

The points are plotted in Fig. 19.3. The least-squares intercept is at 0.0716, so $[\eta] = \boxed{0.0716\ \text{dm}^3\ \text{g}^{-1}}$.

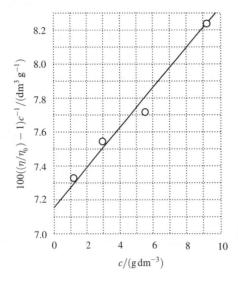

Figure 19.3

P19.5   We follow the procedure of Example 19.5. Also compare to Problems 19.3 and 19.4.

$$[\eta] = \lim_{c \to 0} \left(\frac{\eta/\eta_0 - 1}{c} \right) \ [19.23] \quad \text{and} \quad [\eta] = K\overline{M}_V^a \ [19.25]$$

with K and a from Table 19.4. We draw up the following table using $\eta_0 = 0.647 \times 10^{-3} \ \text{kg m}^{-1} \text{s}^{-1}$.

$c/(\text{g/100 cm}^3)$		0	0.2	0.4	0.6	0.8	1.0
$\eta/(10^{-3} \ \text{kg m}^{-1} \text{s}^{-1})$		0.647	0.690	0.733	0.777	0.821	0.865
$((\eta/\eta_0 - 1)/c)/(100 \ \text{cm}^3 \ \text{g}^{-1})$	—		0.332	0.332	0.335	0.336	0.337

The values are plotted in Fig 19.4, and the y-intercept is 0.330.

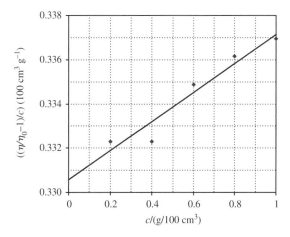

Figure 19.4

Hence $[\eta] = (0.330) \times (100 \ \text{cm}^3 \ \text{g}^{-1}) = 33.0 \ \text{cm}^3 \ \text{g}^{-1}$

and $\dfrac{\overline{M}_V}{\text{g mol}^{-1}} = \left(\dfrac{33.0 \ \text{cm}^3 \text{g}^{-1}}{8.3 \times 10^{-2} \ \text{cm}^3 \ \text{g}^{-1}} \right)^{1/0.50} = 158 \times 10^3.$

That is, $M = \boxed{158 \ \text{kg mol}^{-1}}$.

P19.7   The empirical Mark–Kuhn–Houwink–Sakurada equation [19.25] is

$$[\eta] = K\overline{M}_V^a.$$

As the constant a may be non-integral the molar mass here is to be interpreted as unitless, that is, as $\overline{M}_V/(\text{g mol}^{-1})$. The units of K are then the same as those of $[\eta]$.

We fit the data to the above equation and obtain K and a from the fitting procedure. The plot is shown in Fig 19.5.

$$\boxed{K = 0.0117 \ \text{cm}^3 \ \text{g}^{-1}} \quad \text{and} \quad \boxed{a = 0.717}.$$

$$y = 0.01167x^{0.71661}, \quad R = 0.99983$$

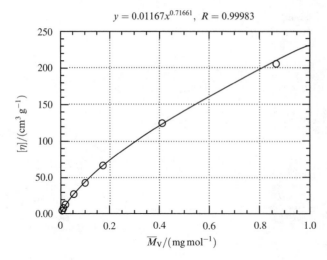

Figure 19.5

(Many plotting programs can fit a power series directly. If not, the equation can be transformed into a linear one

$$\ln[\eta] = \ln K + a \ln M_V$$

so a plot of $\ln[\eta]$ versus $\ln \overline{M}_V$ will have a slope of a and a y-intercept of $\ln K$.)

COMMENT. This value for a is not much different from that for polystyrene in benzene listed in Table 19.4. This is somewhat surprising as one would expect both the K and a values to be solvent-dependent. THF is not chemically similar to benzene. On the other hand, benzene and cyclohexane are very much alike, yet the values of K and a as determined in Example 19.5 are markedly different from those in Table 19.4 for polystyrene in cyclohexane.

P19.9 See Section 5.5(e) and Example 5.4.

$$\frac{h}{c} = \frac{RT}{\rho g \overline{M}_n} + \frac{BRT}{\rho g \overline{M}_n^2} \cdot c \text{ [Example 5.4]}.$$

We plot h/c against c. Draw up the following table.

$c/(\text{g}/100\ \text{cm}^3)$	0.200	0.400	0.600	0.800	1.00
h/cm	0.48	1.12	1.86	2.76	3.88
$\dfrac{h}{c}/(100\ \text{cm}^4\ \text{g}^{-1})$	2.4	2.80	3.10	3.45	3.88

The points are plotted in Fig. 19.6, and give a least-squares intercept at $2.04\bar{3}$ and a slope $1.80\bar{5}$.

Therefore, $RT/\rho g \overline{M}_n = (2.04\bar{3}) \times (100\ \text{cm}^4\ \text{g}^{-1}) = 2.04\bar{3} \times 10^{-3}\ \text{m}^4\ \text{kg}^{-1}$ and hence

$$\overline{M}_n = \frac{(8.314\ \text{J K}^{-1}\ \text{mol}^{-1}) \times (298\ \text{K})}{(0.798 \times 10^3\ \text{kg m}^{-3}) \times (9.81\ \text{m s}^{-2}) \times (2.04\bar{3} \times 10^{-3}\ \text{m}^4\ \text{kg}^{-1})} = \boxed{155\ \text{kg mol}^{-1}}.$$

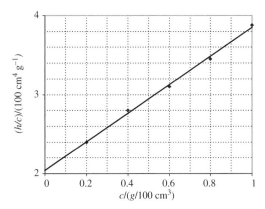

Figure 19.6

From the slope,

$$\frac{BRT}{\rho g \overline{M}_n^2} = (1.80\overline{5}) \times \left(\frac{100 \text{ cm}^4 \text{ g}^{-1}}{\text{g}/(100 \text{ cm}^3)}\right) = 1.80\overline{5} \times 10^4 \text{ cm}^7 \text{ g}^{-2} = 1.80\overline{5} \times 10^{-4} \text{ m}^7 \text{ kg}^{-2}$$

and hence

$$B = \left(\frac{\rho g \overline{M}_n}{RT}\right) \times \overline{M}_n \times (1.80\overline{5} \times 10^{-4} \text{ m}^7 \text{ kg}^{-2})$$

$$= \frac{(155 \text{ kg mol}^{-1}) \times (1.805 \times 10^{-4} \text{m}^7 \text{ kg}^{-2})}{2.04\overline{3} \times 10^{-3} \text{ m}^4 \text{ kg}^{-1}}$$

$$= \boxed{13.7 \text{ m}^3 \text{ mol}^{-1}}.$$

Solutions to theoretical problems

P19.11   See the discussion of radius of gyration in Section 19.8(a). For a random coil $R_g \propto N^{1/2} \propto M^{1/2}$. For a rigid rod, the radius of gyration is proportional to the length of the rod, which is in turn proportional to the number of polymer units, N, and therefore also proportional to M. Therefore, poly(γ-benzyl-L-glutamate) is rod-like whereas polystyrene is a random coil (in butanol).

P19.13 $\quad dN \propto e^{-(M-\overline{M})^2/2\gamma} dM.$

Call the constant of proportionality K, and evaluate it by requiring that $\int dN = N$.

Let $M - \overline{M} = (2\gamma)^{1/2}x \quad$ so $\quad dM = (2\gamma)^{1/2}dx$

and $N = \int_0^\infty K e^{-(M-\overline{M})^2/2\gamma} dM = K(2\gamma)^{1/2} \int_{-a}^\infty e^{-x^2} dx$ where $a = \overline{M}/(2\gamma)^{1/2}$.

Note that the point $x = 0$ represents $M = \overline{M}$, and $x = -a$ represents $M = 0$. In a narrow distribution, the number of molecules with masses much different than the mean falls off rapidly as one moves away from the mean; therefore, $dN \approx 0$ at $M \leq 0$ (that is, at $x \leq -a$). Therefore

$$N \approx K(2\gamma)^{1/2} \int_{-\infty}^{\infty} e^{-x^2} dx = K(2\gamma)^{1/2} \pi^{1/2}.$$

Hence, $K = \dfrac{N}{(2\pi\gamma)^{1/2}}$. It then follows from turning eqn 19.1 into an integral that

$$\overline{M}_n = \frac{1}{N} \int M dN = \frac{1}{(2\pi\gamma)^{1/2}} \int_0^{\infty} M e^{-(M-\overline{M})^2/2\gamma} dM$$

$$= \frac{1}{(2\pi\gamma)^{1/2}} \int_0^{\infty} [(2\gamma)^{1/2}x + \overline{M}] e^{-x^2} (2\gamma)^{1/2} dM$$

$$= \left(\frac{2\gamma}{\pi}\right)^{1/2} \int_{-a}^{\infty} \left(x e^{-x^2} + \frac{\overline{M}}{(2\gamma)^{1/2}} e^{-x^2}\right) dx.$$

Once again extending the lower limit of integration to $-\infty$ adds negligibly to the integral, so

$$\overline{M}_n \approx \left(\frac{2\gamma}{\pi}\right)^{1/2} \times \left[1 + \left(\frac{\pi}{2\gamma}\right)^{1/2} \overline{M}\right] = \boxed{\overline{M} + \left(\frac{2\gamma}{\pi}\right)^{1/2}}.$$

P19.15   (a) Following *Justification* 19.4, we have

$$R_{rms}^2 = \int_0^{\infty} R^2 f dR$$

with $f = 4\pi \left(\dfrac{a}{\pi^{1/2}}\right)^3 R^2 e^{-a^2 R^2}$, $a = \left(\dfrac{3}{2Nl^2}\right)^{1/2}$ [19.27].

Therefore, $R_{rms}^2 = 4\pi \left(\dfrac{a}{\pi^{1/2}}\right)^3 \int_0^{\infty} R^4 e^{-a^2 R^2} dR = 4\pi \left(\dfrac{a}{\pi^{1/2}}\right)^3 \times \left(\dfrac{3}{8}\right) \times \left(\dfrac{\pi}{a^{10}}\right)^{1/2}$

$$= \frac{3}{2a^2} = Nl^2.$$

Hence, $R_{rms} = \boxed{lN^{1/2}}$.

(b) The mean separation is

$$R_{mean} = \int_0^{\infty} R f \, dr = 4\pi \left(\frac{a}{\pi^{1/2}}\right)^3 \int_0^{\infty} R^3 e^{-a^2 R^2} dR$$

$$= 4\pi \left(\frac{a}{\pi^{1/2}}\right)^3 \times \left(\frac{1}{2a^4}\right) = \frac{2}{a\pi^{1/2}} = \boxed{\left(\frac{8N}{3\pi}\right)^{1/2} l}.$$

(c) The most probable separation is the value of R for which f is a maximum, so set $df/dR = 0$ and solve for R.

$$\frac{df}{dR} = 4\pi \left(\frac{a}{\pi^{1/2}}\right)^3 \{2R - 2a^2 R^3\} e^{-a^2 R^2} = 0 \quad \text{when } a^2 R^2 = 1.$$

Therefore, the most probable separation is

$$R^* = \frac{1}{a} = \boxed{l\left(\frac{2}{3}N\right)^{1/2}}.$$

When $N = 4000$ and $l = 154$ pm,

(a) $R_{\text{rms}} = \boxed{9.74 \text{ nm}}$; (b) $R_{\text{mean}} = \boxed{8.97 \text{ nm}}$; (c) $R^* = \boxed{7.95 \text{ nm}}$.

P19.17   We use the definition of the radius of gyration given in Problem 19.19, namely,

$$R_g^2 = \frac{1}{N} \sum_j R_j^2.$$

(a) For a sphere of uniform density, the center of mass is at the center of the sphere. We may visualize the sphere as a collection of a very large number, N, of small particles distributed with equal number density throughout the sphere. Then the summation above may be replaced with an integration.

$$R_g^2 = \frac{1}{N} \frac{N \int_0^a r^2 P(r) dr}{\int_0^a P(r) dr}.$$

$P(r)$ is the probability per unit distance that a small particle will be found at distance r from the center, that is, within a spherical shell of volume $4\pi r^2 dr$. Hence, $P(r) = 4\pi r^2 dr$. If $P(r)$ were normalized, the integral in the numerator would represent the average value of r^2, so N times that integral replaces the sum. The denominator enforces normalization. Hence

$$R_g^2 = \frac{\int_0^a r^2 P(r) dr}{\int_0^a P(r) dr} = \frac{\int_0^a 4\pi r^4 dr}{\int_0^a 4\pi r^2 dr} = \frac{\frac{1}{5}a^5}{\frac{1}{3}a^3} = \frac{3}{5}a^2, \quad \boxed{R_g = \left(\frac{3}{5}\right)^{1/2} a}.$$

(b) For a long straight rod of uniform density the center of mass is at the center of the rod and $P(z)$ is constant for a rod of uniform radius; hence,

$$R_g^2 = \frac{2 \int_0^{l/2} z^2 dz}{2 \int_0^{l/2} dz} = \frac{\frac{1}{3}\left(\frac{1}{2}l\right)^3}{\frac{1}{2}l} = \frac{1}{12}l^2, \quad \boxed{R_g = \frac{l}{2\sqrt{3}}}.$$

COMMENT.   The radius of the rod does not enter into the result. In fact, the distribution function is $P(r,z)$, the probability that a small particle will be found at a distance r from the central axis of the rod and z along that axis from the center, that is, within a squat cylindrical shell of volume $2\pi r dr dz$. Integration radially outward from the axis is the same in numerator and denominator.

For a spherical macromolecule, the specific volume is

$$v_s = \frac{V}{m} = \frac{4\pi a^3}{3} \times \frac{N_A}{M} \quad \text{so} \quad a = \left(\frac{3 v_s M}{4\pi N_A}\right)^{1/3}$$

and

$$R_g = \left(\frac{3}{5}\right)^{1/2} \times \left(\frac{3v_s M}{4\pi N_A}\right)^{1/3}$$

$$= \left(\frac{3}{5}\right)^{1/2} \times \left(\frac{(3v_s/\text{cm}^3\,\text{g}^{-1}) \times \text{cm}^3\,\text{g}^{-1} \times (M/\text{g mol}^{-1}) \times \text{g mol}^{-1}}{(4\pi) \times (6.022 \times 10^{23}\,\text{mol}^{-1})}\right)^{1/3}$$

$$= (5.690 \times 10^{-9}) \times (v_s/\text{cm}^3\,\text{g}^{-1})^{1/3} \times (M/\text{g mol}^{-1})^{1/3}\,\text{cm}$$

$$= (5.690 \times 10^{-11}\text{m}) \times \{(v_s/\text{cm}^3\,\text{g}^{-1}) \times (M/\text{g mol}^{-1})\}^{1/3}.$$

That is $R_g/\text{nm} = \boxed{0.05690 \times \{(v_s/\text{cm}^3\text{g}^{-1}) \times (M/\text{g mol}^{-1})^{1/3}\}}$.

When $M = 100$ kg mol^{-1} and $v_s = 0.750$ cm^3 g^{-1},

$$R_g/\text{nm} = (0.05690) \times \{0.750 \times 1.00 \times 10^5\}^{1/3} = \boxed{2.40}.$$

For a rod, $v_{mol} = \pi a^2 l$, so

$$R_g = \frac{v_{mol}}{2\pi a^2 \sqrt{3}} = \frac{v_s M}{N_A} \times \frac{1}{2\pi a^2 \sqrt{3}}$$

$$= \frac{(0.750\ \text{cm}^3\ \text{g}^{-1}) \times (1.00 \times 10^5\ \text{g mol}^{-1})}{(6.022 \times 10^{23}\ \text{mol}^{-1}) \times (2\pi) \times (0.5 \times 10^{-7}\ \text{cm})^2 \times \sqrt{3}}$$

$$= 4.6 \times 10^{-6}\ \text{cm} = \boxed{46\ \text{nm}}.$$

COMMENT. R_g may also be defined through the relation

$$R_g^2 = \frac{\sum\limits_i m_i r_i^2}{\sum\limits_i m_i}.$$

Question. Does this definition lead to the same formulas for the radii of gyration of the sphere and the rod as those derived above?

P19.19   Refer to Fig. 19.7.

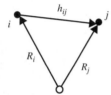

Figure 19.7

The definition in the text (eqn 19.32) is

$$R_g = \frac{1}{N}\left(\frac{1}{2}\sum_{ij} h_{ij}^2\right)^{1/2} \quad \text{so} \quad R_g^2 = \frac{1}{2N^2}\sum_{ij} h_{ij}^2 = \frac{1}{2N^2}\sum_i\sum_j h_{ij}^2.$$

The scalar quantity h_{ij} can be written as the dot product $\boldsymbol{h}_{ij} \cdot \boldsymbol{h}_{ij}$. If we refer all our measurements to a common origin (which we will later specify as the center of mass), the interatomic vectors $\boldsymbol{h}_{ij}$ can be expressed in terms of vectors from the origin: $\boldsymbol{h}_{ij} = \boldsymbol{R}_j - \boldsymbol{R}_i$. (If this is not apparent, note that $\boldsymbol{R}_i + \boldsymbol{h}_{ij} = \boldsymbol{R}_j$.) Therefore

$$R_g^2 = \frac{1}{2N^2} \sum_i \sum_j (\boldsymbol{R}_j - \boldsymbol{R}_i) \cdot (\boldsymbol{R}_j - \boldsymbol{R}_i)$$

$$= \frac{1}{2N^2} \sum_i \sum_j (\boldsymbol{R}_j \cdot \boldsymbol{R}_j + \boldsymbol{R}_i \cdot \boldsymbol{R}_i - 2\boldsymbol{R}_i \cdot \boldsymbol{R}_j) = \frac{1}{2N^2} \sum_i \sum_j (R_j^2 + R_i^2 - 2\boldsymbol{R}_i \cdot \boldsymbol{R}_j).$$

Look at the sums over the squared terms:

$$\sum_i \sum_j R_j^2 = \sum_i \sum_j R_i^2 = N \sum_j R_j^2.$$

Hence $R_g^2 = \dfrac{1}{N} \sum_j R_j^2 - \dfrac{1}{N^2} \sum_i \sum_j \boldsymbol{R}_i \cdot \boldsymbol{R}_j = \dfrac{1}{N} \sum_j R_j^2 - \dfrac{1}{N^2} \sum_i \boldsymbol{R}_i \cdot \sum_j \boldsymbol{R}_j.$

If we choose the origin of our coordinate system to be the center of mass, then

$$\sum_i \boldsymbol{R}_i = \sum_j \boldsymbol{R}_j = 0 \quad \text{and} \quad R_g^2 = \frac{1}{N} \sum_j R_j^2,$$

for the center of mass is the point in the center of the distribution such that all vectors from that point to identical individual masses sum to zero.

P19.21   Write $t = aT$, then

$$\left(\frac{\partial t}{\partial T} \right)_l = a \quad \text{and, using P19.20,} \quad \left(\frac{\partial U}{\partial l} \right)_T = t - aT = 0.$$

Thus the internal energy is independent of the extension. Therefore

$$t = aT = T \left(\frac{\partial t}{\partial T} \right)_l = \boxed{-T \left(\frac{\partial S}{\partial l} \right)_T} \quad \text{[P19.20]}$$

and the tension is proportional to the variation of entropy with extension. Extension reduces the disorder of the chains, and they tend to revert to their disorderly (non-extended) state.

Solutions to applications

P19.23   The center of the sphere cannot approach more closely than $2a$; hence the excluded volume is

$$v_p = \frac{4}{3} \pi (2a)^3 = 8 \left(\frac{4}{3} \pi a^3 \right) = \boxed{8 v_{\text{mol}}}$$

where v_{mol} is a molecular volume.

The osmotic virial coefficient, B (see eqn 5.41), arises largely from the effect of excluded volume. If we imagine a solution of a macromolecule being built by the successive addition of macromolecules to the solvent, each one being excluded by the ones that preceded it, then the value of B turns out to be (P19.18)

$$B = \frac{1}{2} N_A v_p$$

where v_p is the excluded volume due to a single molecule.

$$B(\text{BSV}) = \frac{1}{2} N_A \times \frac{32}{3} \pi a^3 = \frac{16}{3} \pi a^3 N_A$$

$$= \left(\frac{16\pi}{3}\right) \times (6.022 \times 10^{23}\ \text{mol}^{-1}) \times (14.0 \times 10^{-9}\ \text{m})^3 = \boxed{28\ \text{m}^3\ \text{mol}^{-1}}.$$

$$B(\text{Hb}) = \left(\frac{16\pi}{3}\right) \times (6.022 \times 10^{23}\ \text{mol}^{-1}) \times (3.2 \times 10^{-9}\text{m})^3 = \boxed{0.33\ \text{m}^3\ \text{mol}^{-1}}.$$

Since $\Pi = RT\,[\text{J}] + BRT[\text{J}]^2 + \cdots$ [5.41], if we write $\Pi^\circ = RT[\text{J}]$, then $\dfrac{\Pi - \Pi^\circ}{\Pi^\circ} \approx \dfrac{BRT[\text{J}]^2}{RT[\text{J}]} = B[\text{J}].$

For BSV,

$$[\text{J}] = \left(\frac{1.0\,\text{g}}{M}\right) \times (10\,\text{dm}^{-3}) = \frac{10\,\text{g\,dm}^{-3}}{1.07 \times 10^7\,\text{g mol}^{-1}} = 9.35 \times 10^{-7} \text{mol dm}^{-3}$$

$$= 9.35 \times 10^{-4}\ \text{mol m}^{-3}$$

and

$$\frac{\Pi - \Pi^\circ}{\Pi^\circ} = (28\,\text{m}^3\,\text{mol}^{-1}) \times (9.35 \times 10^{-4}\,\text{mol m}^{-3}) = 2.6 \times 10^{-2}\ \text{corresponding to}\ \boxed{2.6\ \text{per cent}}.$$

For Hb, $[\text{J}] = \dfrac{10\,\text{g dm}^{-3}}{66.5 \times 10^3\,\text{g mol}^{-1}} = 0.15\ \text{mol m}^{-3}$

and $\dfrac{\Pi - \Pi^\circ}{\Pi^\circ} = (0.15\,\text{mol m}^{-3}) \times (0.33\,\text{m}^3\,\text{mol}^{-1}) = 5.0 \times 10^{-2}$ which corresponds to $\boxed{5\ \text{per cent}}$.

P19.25 **(a)** We seek an expression for a ratio of scattering intensities of a macromolecule in two different conformations, a rigid rod or a closed circle. The dependence on scattering angle θ is contained in the Rayleigh ratio R_θ. The definition of this quantity, in eqn 19.7, may be inverted to give an expression for the scattering intensity at scattering angle θ

$$I_\theta = R_\theta I_0 \frac{\sin^2 \phi}{r^2},$$

where ϕ is an angle related to the polarization of the incident light and r is the distance between sample and detector. Thus, for any given scattering angle, the ratio of scattered intensity of two conformations is the same as the ratio of their Rayleigh ratios:

$$\frac{I_{\text{rod}}}{I_{\text{cc}}} = \frac{R_{\text{rod}}}{R_{\text{cc}}} = \frac{P_{\text{rod}}}{P_{\text{cc}}}.$$

The last equality stems from eqn 19.8, which related the Rayleigh ratios to a number of angle-independent factors that would be the same for both conformations, and the structure factor (P_θ)

that depends on both conformation and scattering angle. Finally, eqn 19.9 gives an approximate value of the structure factor as a function of the macromolecule's radius of gyration R_g, the wavelength of light, and the scattering angle:

$$P_\theta \approx 1 - \frac{16\pi^2 R_g^2 \sin^2\left(\frac{1}{2}\theta\right)}{3\lambda^2} = \frac{3\lambda^2 - 16\pi^2 R_g^2 \sin^2\left(\frac{1}{2}\theta\right)}{3\lambda^2}.$$

The radius of gyration of a rod of length l is

$$R_{rod} = l/(12)^{1/2} \text{ [Section 19.8(a)]}.$$

For a closed circle, the radius of gyration, which is the rms distance from the center of mass [P19.19], is simply the radius of a circle whose circumference is l:

$$l = 2\pi R_{cc} \quad \text{so} \quad R_{cc} = \frac{l}{2\pi}.$$

The intensity ratio is:

$$\frac{I_{rod}}{I_{cc}} = \frac{3\lambda^2 - \frac{4}{3}\pi^2 l^2 \sin^2\left(\frac{1}{2}\theta\right)}{3\lambda^2 - 4l^2 \sin^2\left(\frac{1}{2}\theta\right)}.$$

Putting the numbers in yields:

$\theta/°$	20	45	90
I_{rod}/I_{cc}	0.976	0.876	0.514

(b) I would work at a detection angle at which the ratio is smallest, i.e. most different from unity, provided I had sufficient intensity to make accurate measurements. Of the angles considered in part (a), $90°$ is the best choice. With the help of a spreadsheet or symbolic mathematical program, the ratio can be computed for a large range of scattering angles and plotted (Fig. 19.8).

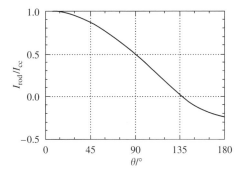

Figure 19.8

A look at the results of such a calculation shows that both the intensity ratio and the intensities themselves decrease with increasing scattering angle from 0° through 180°, that of the closed circle conformation changing much more slowly than that of the rod. Note: the approximation used above yields negative numbers for P_{rod} at large scattering angles; this is because the approximation, which depends on the molecule being much smaller than the wavelength, is shaky at best, particularly at large angles.

P19.27   The molar mass is given by eqn 19.19

$$\overline{M}_n = \frac{SRT}{bD} = \frac{SRT}{(1 - \rho v_s)D} \quad [19.14, \text{ for } b]$$

$$= \frac{(4.5 \times 10^{-13}\text{s}) \times (8.314 \text{ J K}^{-1} \text{ mol}^{-1}) \times (293 \text{ K})}{(1 - 0.75 \times 0.998) \times (6.3 \times 10^{-11} \text{ m}^2 \text{ s}^{-1})} = \boxed{69 \text{ kg mol}^{-1}}.$$

Now combine $f = 6\pi a \eta$ [19.12] with $f = kT/D$ [19.11]:

$$a = \frac{kT}{6\pi \eta D} = \frac{(1.381 \times 10^{-23} \text{ J K}^{-1}) \times (293 \text{ K})}{(6\pi) \times (1.00 \times 10^{-3} \text{ kg m}^{-1}\text{s}^{-1}) \times (6.3 \times 10^{-11} \text{ m}^2 \text{ s}^{-1})} = \boxed{3.4 \text{ nm}}.$$

P19.29   The isoelectric point is the pH at which the protein has no charge. At that point, then, its drift speed under electrophoresis, s, vanishes. Plot the drift speed against pH and extrapolate the line to $s = 0$. The plot is shown in Fig. 19.9.

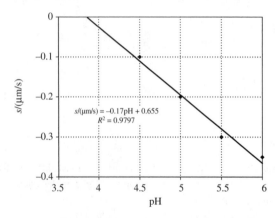

$s/(\mu\text{m/s}) = -0.17\text{pH} + 0.655$
$R^2 = 0.9797$

Figure 19.9

Isoelectric pH is the x-intercept on the graph, that is, the value of x at which $y = 0$. One can find this by solving the fit equation:

$$s/(\mu\text{m/s}) = -0.17\text{pH} + 0.655 = 0$$

so pH = $\boxed{3.85\overline{5}}$.

COMMENT.   One could obtain the result to about ± 0.05 pH by reading the value directly from the graph.

P19.31   (a) The data are plotted in Fig. 19.10. Both samples give rise to tolerably linear curves, so we estimate the melting point by interpolation using the best-fit straight line.
   The best-fit equation has the form $T_m/K = mf + b$, and we want T_m when $f = 0.40$:

$$c_{\text{salt}} = 1.0 \times 10^{-2} \text{ mol dm}^{-3} : \quad T_m = (39.7 \times 0.40 + 324) \text{ K} = \boxed{340 \text{ K}}.$$

$$c_{\text{salt}} = 0.15 \text{ mol dm}^{-3} : \quad T_m = (39.7 \times 0.40 + 344) \text{ K} = \boxed{360 \text{ K}}.$$

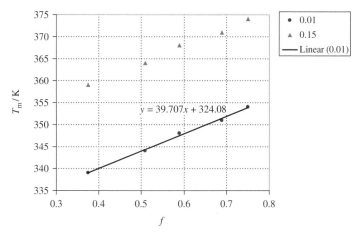

Figure 19.10

(b) The slopes are the same for both samples. The different concentrations of dissolved salt simply offset the melting temperatures by a constant amount. The greater the concentration, the higher the melting point. This behavior is not what is typically observed with small molecules, where the presence of dissolved impurities disrupts freezing and *depresses* the freezing point. The dissolved ions can interact with charged regions of the macromolecule that might otherwise experience unfavorable intramolecular interactions. For example, if two regions bearing negative charge would have to approach each other in the absence of dissolved salts, the incorporation of a cation very close to each region and an anion in between them would turn an unfavorable interaction into a favorable one. (See Fig. 19.11).

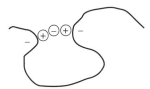

Figure 19.11

The melting points are greater at both larger fractions of G—C base pairs and at larger salt concentrations. T_m increases with the number of G—C base pairs because this pair is held togethar with the three hydrogen bonds in the double helical structure, whereas the A—T pair is held with two hydrogen bonds (see Section 19.11). The ΔH_m contribution is greater for the G—C pair. Low salt concentrations destabilize the double helix by inadequately contributing to the attractive forces between the solution and the sugar-phosphate backbone of the double helix. This makes it easier for a base to rotate out from the center of the double helix.

P19.33   The peaks are separated by 104 g mol^{-1}, so this is the molar mass of the repeating unit of the polymer. This peak separation is consistent with the identification of the polymer as polystyrene, for the repeating group of $CH_2CH(C_6H_5)$ (8 C atoms and 8 H atoms) has a molar mass of $8 \times (12 + 1)$ g mol$^{-1} = 104$ g mol^{-1}. A consistent difference between peaks suggests a pure system and points away from different numbers of subunits of different molecular weight (such as the *t*-butyl initiators) being incorporated into the polymer molecules. The most intense peak has a molar mass equal to that of *n* repeating groups plus that of a silver cation plus that of terminal groups:

$$M(\text{peak}) = nM(\text{repeat}) + M(\text{Ag}^+) + M(\text{terminal}).$$

If both ends of the polymer have terminal t-butyl groups, then

$$M(\text{terminal}) = 2M(t\text{-butyl}) = 2(4 \times 12 + 9) \text{ g mol}^{-1} = 114 \text{ g mol}^{-1}$$

and $n = \dfrac{M(\text{peak}) - M(\text{Ag}^+) - M(\text{terminal})}{M(\text{repeat})} = \dfrac{25598 - 108 - 114}{104} = \boxed{244}$.

P19.35   The procedure is that described in Problem 19.7. The data are fitted to the Mark–Kuhn–Houwink–Sakurada equation.

$$[\eta] = K\overline{M}_{\text{v}}^{a} \text{ [19.25].}$$

The values obtained for the parameters are

$$K = \boxed{2.38 \times 10^{-3} \text{ cm}^3 \text{ g}^{-1}} \quad \text{and} \quad a = \boxed{0.955}.$$

This K value is smaller than any in Table 19.4 or that in Problem 19.7. The value for a is quite close to 1. When $a = 1$ exactly, the molar mass, $\overline{M}_{\text{v}}$ corresponds to the weight average molar mass, $\overline{M}_{\text{w}}$.

COMMENT.   The magnitude of the constant a reflects the stiffness of the polymer chain as a result of π-orbital interactions between heterocyclic rings.

20 Materials 2: the solid state

Answers to discussion questions

D20.1 Lattice planes are labeled by their Miller indices h, k, and l, where h, k, and l refer respectively to the reciprocals of the smallest intersection distances (in units of the lengths of the unit cell, a, b, and c) of the plane along the x, y, and z axes.

D20.3 If the overall amplitude of a wave diffracted by planes (hkl) is zero, that plane is said to be absent in the diffraction pattern.

When the phase difference between adjacent planes in the set of planes (hkl) is π, destructive interference between the waves diffracted from the planes can occur and this will diminish the intensity of the diffracted wave. This is illustrated in Fig. 20.21 in the text. The overall intensity of a diffracted wave from a plane (hkl) is determined from a calculation of the structure factor, F_{hkl}, which is a function of the positions (hence, of the Miller indices) and of the scattering factors of the atoms in the crystal (see eqn 20.7). If F_{hkl} is zero for the plane (hkl), that plane is absent. See Example 20.3.

D20.5 The majority of metals crystallize in structures that can be interpreted as the closest packing arrangements of hard spheres. These are the cubic close-packed (ccp) and hexagonal close-packed (hcp) structures. In these models, 74% of the volume of the unit cell is occupied by the atoms (packing fraction = 0.74). Most of the remaining metallic elements crystallize in the body-centered cubic (bcc) arrangement, which is not too much different from the close-packed structures in terms of the efficiency of the use of space (packing fraction 0.68 in the hard sphere model). Polonium is an exception; it crystallizes in the simple cubic structure, which has a packing fraction of 0.52. See the solution to Problem 20.24 for a derivation of all the packing fractions in cubic systems. If atoms were truly hard spheres, we would expect that all metals would crystallize in either the ccp or hcp close-packed structures. The fact that a significant number crystallize in other structures is proof that a simple hard sphere model is an inaccurate representation of the interactions between the atoms. Covalent bonding between the atoms may influence the structure.

D20.7 Because enantiomers give almost identical diffraction patterns it is difficult to distinguish between them. But absolute configurations can be obtained from an analysis of small differences in diffraction intensities by a method developed by J.M. Bijvoet. The method makes use of extra phase shifts that occur when the frequency of the X-rays approaches an absorption frequency of atoms in the compound. The phase shifts are called anomalous scattering and result in different intensities in the diffraction patterns of different enantiomers. See Section 23.7(b) of the 7th edition of this text for an explanation of the origin of this anomalous phase shift. The incorporation of heavy atoms into the compound makes the observation of the extra phase shift easier to observe, but with very sensitive modern diffractometers this is no longer strictly necessary.

D20.9   The Fermi–Dirac distribution is a version of the Boltzmann distribution that takes into account the effect of the Pauli exclusion principle. It can therefore be used to calculate the population, P, of a state of given energy in a many-electron system at a temperature T:

$$P = \frac{1}{e^{(E-\mu)/kT} + 1}.$$

In this expression, μ is the Fermi energy, or chemical potential, the energy of the level for which $P = 1/2$. The Fermi energy should be distinguished from the Fermi level, which is the energy of the highest occupied state at $T = 0$. See Fig. 20.54 of the text.

From thermodynamics (Chapter 3) we know that $dU = -p\,dV + T\,dS + \mu\,dn$ for a one-component system. This may also be written $dU = -p\,dV + T\,dS + \mu\,dN$, and this μ is the chemical potential per particle that appears in the F–D distribution law. The term in dU containing μ is the chemical work and gives the change in internal energy with change in the number of particles. Thus, μ has a wider significance than its interpretation as a partial molar Gibbs energy and it is not surprising that it occurs in the F–D expression in comparison to the energy of the particle. The Helmholtz energy, A, and μ are related through $dA = -p\,dV - S\,dT + \mu\,dN$, and so μ also gives the change in the Helmholtz energy with change in number of particles. To fully understand how the chemical potential μ enters into the F–D expression for P, we must examine its derivation (see *Further reading*) which makes use of the relation between μ and A and of that between A and the partition function for F–D particles.

Solutions to exercises

E20.1(b) $\left(\frac{1}{2}, 0, \frac{1}{2}\right)$ is the midpoint of a face. All face midpoints are alike, including $\boxed{\left(\frac{1}{2}, \frac{1}{2}, 0\right) \text{ and } \left(0, \frac{1}{2}, \frac{1}{2}\right)}$. There are six faces to each cube, but each face is shared by two cubes. So other face midpoints can be described by one of these three sets of coordinates on an adjacent unit cell.

E20.2(b)   Taking reciprocals of the coordinates yields $\left(1, \frac{1}{3}, -1\right)$ and $\left(\frac{1}{2}, \frac{1}{3}, \frac{1}{4}\right)$ respectively. Clearing the fractions yields the Miller indices $\boxed{(31\bar{3}) \text{ and } (643)}$

E20.3(b)   The distance between planes in a cubic lattice is

$$d_{hkl} = \frac{a}{(h^2 + k^2 + l^2)^{1/2}}$$

This is the distance between the origin and the plane which intersects coordinate axes at $(h/a, k/a, l/a)$.

$$d_{121} = \frac{523\,\text{pm}}{(1 + 2^2 + 1)^{1/2}} = \boxed{214\,\text{pm}}$$

$$d_{221} = \frac{523\,\text{pm}}{(2^2 + 2^2 + 1)^{1/2}} = \boxed{174\,\text{pm}}$$

$$d_{244} = \frac{523\,\text{pm}}{(2^2 + 4^2 + 4^2)^{1/2}} = \boxed{87.2\,\text{pm}}$$

E20.4(b)   The Bragg law is

$$n\lambda = 2d \sin \theta$$

Assuming the angle given is for a first-order reflection, the wavelength must be

$$\lambda = 2(128.2 \, \text{pm}) \sin 19.76° = \boxed{86.7 \, \text{pm}}$$

E20.5(b)   Combining the Bragg law with Miller indices yields, for a cubic cell

$$\sin \theta_{hkl} = \frac{\lambda}{2a} (h^2 + k^2 + l^2)^{1/2}$$

In a face-centered cubic lattice, h, k, and l must be all odd or all even. So the first three reflections would be from the (1 1 1), (2 0 0), and (2 2 0) planes. In an fcc cell, the face diagonal of the cube is $4R$, where R is the atomic radius. The relationship of the side of the unit cell to R is therefore

$$(4R)^2 = a^2 + a^2 = 2a^2 \quad \text{so} \quad a = \frac{4R}{\sqrt{2}}$$

Now we evaluate

$$\frac{\lambda}{2a} = \frac{\lambda}{4\sqrt{2}R} = \frac{154 \, \text{pm}}{4\sqrt{2}(144 \, \text{pm})} = 0.189$$

We set up the following table

hkl	$\sin \theta$	$\theta/°$	$2\theta/°$
111	0.327	19.1	38.2
200	0.378	22.2	44.4
220	0.535	32.3	64.6

E20.6(b)   In a circular camera, the distance between adjacent lines is $D = R\Delta(2\theta)$, where R is the radius of the camera (distance from sample to film) and θ is the diffraction angle. Combining these quantities with the Bragg law ($\lambda = 2d \sin \theta$, relating the glancing angle to the wavelength and separation of planes), we get

$$D = 2R\Delta\theta = 2R\Delta\left(\sin^{-1}\frac{\lambda}{2d}\right)$$

$$= 2(5.74 \, \text{cm}) \times \left(\sin^{-1}\frac{96.035}{2(82.3 \, \text{pm})} - \sin^{-1}\frac{95.401 \, \text{pm}}{2(82.3 \, \text{pm})}\right) = \boxed{0.054 \, \text{cm}}$$

E20.7(b)   The volume of a hexagonal unit cell is the area of the base times the height c. The base is equivalent to two equilateral triangles of side a. The altitude of such a triangle is $a \sin 60°$. So the volume is

$$V = 2 \left(\tfrac{1}{2}a \times a \sin 60°\right) c = a^2 c \sin 60° = (1692.9 \, \text{pm})^2 \times (506.96 \, \text{pm}) \times \sin 60°$$

$$= 1.2582 \times 10^9 \, \text{pm}^3 = \boxed{1.2582 \, \text{nm}^3}$$

E20.8(b)   The volume of an orthorhombic unit cell is

$$V = abc = (589\,\text{pm}) \times (822\,\text{pm}) \times (798\,\text{pm}) = \frac{3.86 \times 10^8\,\text{pm}^3}{(10^{10}\,\text{pm cm}^{-1})^3} = 3.86 \times 10^{-22}\text{cm}^3$$

The mass per formula unit is

$$m = \frac{135.01\,\text{g mol}^{-1}}{6.022 \times 10^{23}\,\text{mol}^{-1}} = 2.24 \times 10^{-22}\,\text{g}$$

The density is related to the mass m per formula unit, the volume V of the unit cell, and the number N of formula units per unit cell as follows

$$\rho = \frac{Nm}{V} \quad \text{so} \quad N = \frac{\rho V}{m} = \frac{(2.9\,\text{g cm}^{-3}) \times (3.86 \times 10^{-22}\,\text{cm}^3)}{2.24 \times 10^{-22}\,\text{g}} = \boxed{5}$$

A more accurate density, then, is

$$\rho = \frac{5(2.24 \times 10^{-22}\,\text{g})}{3.86 \times 10^{-22}\,\text{cm}^3} = \boxed{2.90\,\text{g cm}^{-3}}$$

E20.9(b)   The distance between the origin and the plane which intersects coordinate axes at $(h/a, k/b, l/c)$ is given by

$$d_{hkl} = \left(\frac{h^2}{a^2} + \frac{k^2}{b^2} + \frac{l^2}{c^2} \right)^{-1/2} = \left(\frac{3^2}{(679\,\text{pm})^2} + \frac{2^2}{(879\,\text{pm})^2} + \frac{2^2}{(860\,\text{pm})^2} \right)^{-1/2}$$

$$d_{322} = \boxed{182\,\text{pm}}$$

E20.10(b)   The fact that the 111 reflection is the third one implies that the cubic lattice is simple, where all indices give reflections. The 111 reflection would be the first reflection in a face-centered cubic cell and would be absent from a body-centered cubic.

The Bragg law

$$\sin\theta_{hkl} = \frac{\lambda}{2a}(h^2 + k^2 + l^2)^{1/2}$$

can be used to compute the cell length

$$a = \frac{\lambda}{2\sin\theta_{hkl}}(h^2 + k^2 + l^2)^{1/2} = \frac{137\,\text{pm}}{2\sin 17.7°}(1^2 + 1^2 + 1^2)^{1/2} = 390\,\text{pm}$$

With the cell length, we can predict the glancing angles for the other reflections expected from a simple cubic

$$\theta_{hkl} = \sin^{-1}\left(\frac{\lambda}{2a}(h^2 + k^2 + l^2)^{1/2} \right) = \sin^{-1}(0.176(h^2 + k^2 + l^2)^{1/2})$$

$$\theta_{100} = \sin^{-1}(0.176(1^2 + 0 + 0)^{1/2}) = 10.1° \text{ (checks)}$$

$$\theta_{110} = \sin^{-1}(0.176(1^2 + 1^2 + 0)^{1/2}) = 14.4° \text{ (checks)}$$

$$\theta_{200} = \sin^{-1}(0.176(2^2 + 0 + 0)^{1/2}) = 20.6° \text{ (checks)}$$

These angles predicted for a simple cubic fit those observed, confirming the hypothesis of a simple lattice; the reflections are due to the $\boxed{(100), (110), (111), \text{ and } (200)}$ planes.

E20.11(b) The Bragg law relates the glancing angle to the separation of planes and the wavelength of radiation

$$\lambda = 2d \sin \theta \quad \text{so} \quad \theta = \sin^{-1} \frac{\lambda}{2d}$$

The distance between the origin and plane which intersects coordinate axes at $(h/a, k/b, l/c)$ is given by

$$d_{hkl} = \left(\frac{h^2}{a^2} + \frac{k^2}{b^2} + \frac{l^2}{c^2} \right)^{-1/2}$$

So we can draw up the following table

hkl	d_{hkl}/pm	$\theta_{hkl}/^\circ$
100	574.1	4.166
010	796.8	3.000
111	339.5	7.057

E20.12(b) All of the reflections present have $h + k + l$ even, and all of the even $h + k + l$ are present. The unit cell, then, is $\boxed{\text{body-centered cubic}}$

E20.13(b) The structure factor is given by

$$F_{hkl} = \sum_i f_i e^{i\phi_i} \quad \text{where} \quad \phi_i = 2\pi(hx_i + ky_i + lz_i)$$

All eight of the vertices of the cube are shared by eight cubes, so each vertex has a scattering factor of $f/8$.

The coordinates of all vertices are integers, so the phase ϕ is a multiple of 2π and $e^{i\phi} = 1$. The body-center point belongs exclusively to one unit cell, so its scattering factor is f. The phase is

$$\phi = 2\pi \left(\tfrac{1}{2}h + \tfrac{1}{2}k + \tfrac{1}{2}l \right) = \pi(h + k + l)$$

When $h + k + l$ is even, ϕ is a multiple of 2π and $e^{i\phi} = 1$; when $h + k + l$ is odd, ϕ is π + a multiple of 2π and $e^{i\phi} = -1$. So $e^{i\phi} = (-1)^{h+k+l}$ and

$$F_{hkl} = 8(f/8)(1) + f(-1)^{h+k+l}$$

$$= \boxed{2f \text{ for } h + k + l \text{ even} \quad \text{and} \quad 0 \text{ for } h + k + l \text{ odd}}$$

E20.14(b) There are two smaller (white) triangles to each larger (gray) triangle. Let the area of the larger triangle be A and the area of the smaller triangle be a. Since $b = \tfrac{1}{2}B$(base) and $h = \tfrac{1}{2}H$(height), $a = \tfrac{1}{4}A$. The white space is then $2NA/4$, for N of the larger triangles. The total space is then $(NA + (NA/2)) = 3NA/2$. Therefore the fraction filled is $NA/(3NA/2) = \boxed{2/3}$

E20.15(b)   See Figure 20.1.

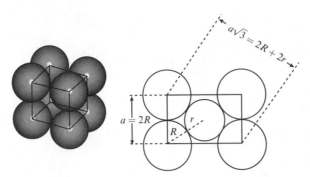

Figure 20.1

The body diagonal of a cube is $a\sqrt{3}$. Hence

$$a\sqrt{3} = 2R + 2r \quad \text{or} \quad \sqrt{3}R = R + r \quad [a = 2R]$$

$$\frac{r}{R} = \boxed{0.732}$$

E20.16(b)   The ionic radius of K^+ is 138 pm when it is 6-fold coordinated, 151 pm when it is 8-fold coordinated.

  (a) The smallest ion that can have 6-fold coordination with it has a radius of $\left(\sqrt{2} - 1\right) \times (138\,\text{pm}) =$ $\boxed{57\,\text{pm}}$.

  (b) The smallest ion that can have 8-fold coordination with it has a radius of $\left(\sqrt{3} - 1\right) \times (151\,\text{pm}) =$ $\boxed{111\,\text{pm}}$.

E20.17(b)   The diagonal of the face that has a lattice point in its center is equal to $4r$, where r is the radius of the atom. The relationship between this diagonal and the edge length a is

$$4r = a\sqrt{2} \quad \text{so} \quad a = 2\sqrt{2}r$$

The volume of the unit cell is a^3, and each cell contains 2 atoms. (Each of the 8 vertices is shared among 8 cells; each of the 2 face points is shared by 2 cells.) So the packing fraction is

$$\frac{2V_{\text{atom}}}{V_{\text{cell}}} = \frac{2(4/3)\pi r^3}{(2\sqrt{2}r)^3} = \frac{\pi}{3(2)^{3/2}} = \boxed{0.370}$$

E20.18(b)   The volume of an atomic crystal is proportional to the cube of the atomic radius divided by the packing fraction. The packing fraction for hcp, a close-packed structure, is 0.740; for bcc, it is 0.680. So for titanium

$$\frac{V_{\text{bcc}}}{V_{\text{hcp}}} = \frac{0.740}{0.680}\left(\frac{122\,\text{pm}}{126\,\text{pm}}\right)^3 = 0.99$$

The bcc structure has a smaller volume, so the transition involves a $\boxed{\text{contraction}}$. (Actually, the data are not precise enough to be sure of this. 122 could mean 122.49 and 126 could mean 125.51, in which case an expansion would occur.)

E20.19(b) Draw points corresponding to the vectors joining each pair of atoms. Heavier atoms give more intense contributions than light atoms. Remember that there are two vectors joining any pair of atoms ($\overrightarrow{AB}$ and $\overleftarrow{AB}$); don't forget the AA zero vectors for the center point of the diagram. See Figure 20.2 for C_6H_6.

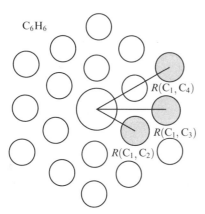

C_6H_6

$R(C_1, C_4)$

$R(C_1, C_3)$

$R(C_1, C_2)$

Figure 20.2

E20.20(b) Combine $E = \frac{1}{2}kT$ and $E = \frac{1}{2}mv^2 = \frac{h^2}{2m\lambda^2}$, to obtain

$$\lambda = \frac{h}{(mkT)^{1/2}} = \frac{6.626 \times 10^{-34}\,\text{J s}}{[(1.675 \times 10^{-27}\,\text{kg}) \times (1.381 \times 10^{-23}\,\text{J K}^{-1}) \times (300\,\text{K})]^{1/2}} = \boxed{252\,\text{pm}}$$

E20.21(b) The lattice enthalpy is the difference in enthalpy between an ionic solid and the corresponding isolated ions. In this exercise, it is the enthalpy corresponding to the process

$$\text{MgBr}_2(\text{s}) \rightarrow \text{Mg}^{2+}(\text{g}) + 2\text{Br}^-(\text{g})$$

The standard lattice enthalpy can be computed from the standard enthalpies given in the exercise by considering the formation of $\text{MgBr}_2(\text{s})$ from its elements as occuring through the following steps: sublimation of $\text{Mg}(\text{s})$, removing two electrons from $\text{Mg}(\text{g})$, vaporization of $\text{Br}_2(\text{l})$, atomization of $\text{Br}_2(\text{g})$, electron attachment to $\text{Br}(\text{g})$, and formation of the solid MgBr_2 lattice from gaseous ions

$$\Delta_f H^\ominus(\text{MgBr}_2, \text{s}) = \Delta_{\text{sub}} H^\ominus(\text{Mg}, \text{s}) + \Delta_{\text{ion}} H^\ominus(\text{Mg}, \text{g}) + \Delta_{\text{vap}} H^\ominus(\text{Br}_2, \text{l})$$
$$+ \Delta_{\text{at}} H^\ominus(\text{Br}_2, \text{g}) + 2\Delta_{\text{eg}} H^\ominus(\text{Br}, \text{g}) - \Delta_L H^\ominus(\text{MgBr}_2, \text{s})$$

So the lattice enthalpy is

$$\Delta_L H^\ominus(\text{MgBr}_2, \text{s}) = \Delta_{\text{sub}} H^\ominus(\text{Mg}, \text{s}) + \Delta_{\text{ion}} H^\ominus(\text{Mg}, \text{g}) + \Delta_{\text{vap}} H^\ominus(\text{Br}_2, \text{l})$$
$$+ \Delta_{\text{at}} H^\ominus(\text{Br}_2, \text{g}) + 2\Delta_{\text{eg}} H^\ominus(\text{Br}, \text{g}) - \Delta_f H^\ominus(\text{MgBr}_2, \text{s})$$

$$\Delta_L H^\ominus(\text{MgBr}_2, \text{s}) = [148 + 2187 + 31 + 193 - 2(331) + 524]\,\text{kJ mol}^{-1} = \boxed{2421\,\text{kJ mol}^{-1}}$$

E20.22(b)   Tension reduces the disorder in the rubber chains; hence, if the rubber is sufficiently stretched, crystallization may occur at temperatures above the normal crystallization temperature. In unstretched rubber the random thermal motion of the chain segments prevents crystallization. In stretched rubber these random thermal motions are drastically reduced. At higher temperatures the random motions may still have been sufficient to prevent crystallization even in the stretched rubber, but lowering the temperature to $0\,°C$ may have resulted in a transition to the crystalline form. Since it is random motion of the chains that resists the stretching force and allows the rubber to respond to forced dimensional changes, this ability ceases when the motion ceases. Hence, the seals failed.

> **COMMENT.** The solution to the problem of the cause of the *Challenger* disaster was the final achievement, just before his death, of Richard Feynman, a Nobel prize winner in physics and a person who loved to solve problems. He was an outspoken person who abhorred sham, especially in science and technology. Feynman concluded his personal report on the disaster by saying, 'For a successful technology, reality must take precedence over public relations, for nature cannot be fooled' (James Gleick, *Genius: The Life and Science of Richard Feynman*. Pantheon Books, New York (1992).)

E20.23(b)   Young's modulus is defined as:

$$E = \frac{\text{normal stress}}{\text{normal strain}}$$

where stress is deforming force per unit area and strain is a fractional deformation. Here the deforming force is gravitational, mg, acting across the cross-sectional area of the wire, πr^2. So the strain induced in the exercise is

$$\text{strain} = \frac{\text{stress}}{E} = \frac{mg}{\pi(d/2)^2 E} = \frac{4mg}{\pi d^2 E} = \frac{4(10.0\,\text{kg})(9.8\,\text{m s}^{-2})}{\pi(0.10 \times 10^{-3}\,\text{m})^2(215 \times 10^9\,\text{Pa})} = \boxed{5.8 \times 10^{-2}}$$

The wire would stretch by 5.8%.

E20.24(b)   Poisson's ratio is defined as:

$$\nu_P = \frac{\text{transverse strain}}{\text{normal strain}}$$

where normal strain is the fractional deformation along the direction of the deforming force and transverse strain is the fractional deformation in the directions transverse to the deforming force. Here the length of a cube of lead is stretched by 2.0 percent, resulting in a contraction by 0.41×2.0 percent, or 0.82 percent, in the width and height of the cube. The relative change in volume is:

$$\frac{V + \Delta V}{V} = (1.020)(0.9918)(0.9918) = 1.003$$

and the absolute change is:

$$\Delta V = (1.003 - 1)(1.0\,\text{dm}^3) = \boxed{0.003\,\text{dm}^3}$$

E20.25(b)   p-type; the dopant, gallium, belongs to Group 13, whereas germanium belongs to Group 14.

E20.26(b) $E_g = h\nu_{\min}$ and $\nu_{\min} = E_g/h = \dfrac{1.12\,\text{eV}}{6.626 \times 10^{-34}\,\text{J s}} \left(\dfrac{1.602 \times 10^{-19}\,\text{J}}{1\,\text{eV}} \right) = \boxed{2.71 \times 10^{14}\,\text{Hz}}$

E20.27(b) $m = g_e\{S(S+1)\}^{1/2}\mu_B$ [20.34, with S in place of s]

Therefore, since $m = 4.00\mu_B$

$$S(S+1) = \left(\tfrac{1}{4}\right) \times (4.00)^2 = 4.00, \quad \text{implying that} \quad S = 1.56$$

Thus $S \approx \tfrac{3}{2}$, implying three unpaired spins.

In actuality most Mn^{2+} compounds have $\boxed{5}$ unpaired spins.

E20.28(b) $\chi_m = \chi V_m = \dfrac{\chi M}{\rho} = \dfrac{(-7.9 \times 10^{-6}) \times (84.15\,\text{g mol}^{-1})}{0.811\,\text{g cm}^{-3}}$

$$= \boxed{-8.2 \times 10^{-4}\,\text{cm}^3\,\text{mol}^{-1}} = \boxed{-8.2 \times 10^{-10}\,\text{m}^3\,\text{mol}^{-1}}$$

E20.29(b) The molar susceptibility is given by

$$\chi_m = \frac{N_A g_e^2 \mu_0 \mu_B^2 S(S+1)}{3kT}$$

NO_2 is an odd-electron species, so it must contain at least one unpaired spin; in its ground state it has one unpaired spin, so $S = \tfrac{1}{2}$. Therefore,

$$\chi_m = (6.022 \times 10^{23}\,\text{mol}^{-1}) \times (2.0023)^2 \times (4\pi \times 10^{-7}\,\text{T}^2\,\text{J}^{-1}\text{m}^3)$$

$$\times \frac{(9.274 \times 10^{-24}\,\text{J T}^{-1})^2 \times \left(\tfrac{1}{2}\right) \times \left(\tfrac{1}{2}+1\right)}{3(1.381 \times 10^{-23}\,\text{J K}^{-1}) \times (298\,\text{K})}$$

$$= \boxed{1.58 \times 10^{-8}\,\text{m}^3\,\text{mol}^{-1}}$$

The expression above does not indicate any pressure-dependence in the molar susceptibility. However, the observed decrease in susceptibility with increased pressure is consistent with the fact that NO_2 has a tendency to dimerize, and that dimerization is favored by higher pressure. The dimer has no unpaired electrons, so the dimerization reaction effectively reduced the number of paramagnetic species.

E20.30(b) The molar susceptibility is given by

$$\chi_m = \frac{N_A g_e^2 \mu_0 \mu_B^2 S(S+1)}{3kT} \quad \text{so} \quad S(S+1) = \frac{3kT\chi_m}{N_A g_e^2 \mu_0 \mu_B^2}$$

$$S(S+1) = \frac{3(1.381 \times 10^{-23}\,\text{J K}^{-1}) \times (298\,\text{K})}{(6.022 \times 10^{23}\,\text{mol}^{-1}) \times (2.0023)^2}$$

$$\times \frac{(6.00 \times 10^{-8}\,\text{m}^3\,\text{mol}^{-1})}{(4\pi \times 10^{-7}\,\text{T}^2\,\text{J}^{-1}\,\text{m}^3) \times (9.274 \times 10^{-24}\,\text{J T}^{-1})^2}$$

$$= 2.84 \quad \text{so} \quad S = \frac{-1 + \sqrt{1 + 4(2.84)}}{2} = 1.26$$

corresponding to $\boxed{2.52}$ effective unpaired spins. The theoretical number is $\boxed{2}$. The magnetic moments in a crystal are close together, and they interact rather strongly. The discrepancy is most likely due to an interaction among the magnetic moments.

E20.31(b) The molar susceptibility is given by

$$\chi_m = \frac{N_A g_e^2 \mu_0 \mu_B^2 S(S+1)}{3kT}$$

Mn^{2+} has five unpaired spins, so $S = 2.5$ and

$$\chi_m = \frac{(6.022 \times 10^{23} \text{ mol}^{-1}) \times (2.0023)^2 \times (4\pi \times 10^{-7} \text{ T}^2 \text{ J}^{-1} \text{ m}^3)}{3(1.381 \times 10^{-23} \text{ J K}^{-1})}$$

$$\times \frac{(9.274 \times 10^{-24} \text{ J T}^{-1})^2 \times (2.5) \times (2.5+1)}{(298 \text{ K})}$$

$$= \boxed{1.85 \times 10^{-7} \text{ m}^3 \text{ mol}^{-1}}$$

E20.32(b) The orientational energy of an electron spin system in a magnetic field is

$$E = g_e \mu_B M_S \mathcal{B}$$

The Boltzmann distribution says that the population ratio r of the various states is proportional to

$$r = \exp\left(\frac{-\Delta E}{kT}\right)$$

where ΔE is the difference between them. For a system with $S = 1$, the M_S states are 0 and ± 1. So between adjacent states

$$r = \exp\left(\frac{-g_e \mu_B M_S \mathcal{B}}{kT}\right) = \exp\left(\frac{-(2.0023) \times (9.274 \times 10^{-24} \text{ J T}^{-1}) \times (1) \times (15.0 \text{ T})}{(1.381 \times 10^{-23} \text{ J K}^{-1}) \times (298 \text{ K})}\right)$$

$$= \boxed{0.935}$$

The population of the highest-energy state is r^2 times that of the lowest; $r^2 = \boxed{0.873}$.

Solutions to problems

Solutions to numerical problems

P20.1
$$\lambda = 2d_{hkl} \sin\theta_{hkl} = \frac{2a \sin\theta_{hkl}}{(h^2 + k^2 + l^2)^{1/2}} \text{ [eqn 20.5, inserting eqn 20.2]}$$

$$= 2a \sin 6.0° = 0.209a.$$

In an NaCl unit cell (Fig. 20.3) the number of formula units is 4 (each corner ion is shared by 8 cells, each edge ion by 4, and each face ion by 2).

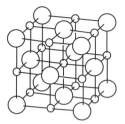

Figure 20.3

Therefore,

$$\rho = \frac{NM}{VN_A} = \frac{4M}{a^3N_A}, \quad \text{implying that} \quad a = \left(\frac{4M}{\rho N_A}\right)^{1/3} \quad \text{[Exercise 20.8(a)]}.$$

$$a = \left(\frac{(4) \times (58.44\,\text{g mol}^{-1})}{(2.17 \times 10^{16}\,\text{g m}^{-3}) \times (6.022 \times 10^{23}\,\text{mol}^{-1})}\right)^{1/3} = 563.\overline{5}\,\text{pm}$$

and hence $\lambda = (0.209) \times (563.\overline{5}\,\text{pm}) = \boxed{118\,\text{pm}}$.

P20.3 See Fig. 20.23 of the text or Fig. 20.1 of this manual. The length of an edge in the fcc lattice of these compounds is

$$a = 2(r_+ + r_-).$$

Then

(1) $a(NaCl) = 2(r_{Na^+} + r_{Cl^-}) = 562.8\,\text{pm}$; (2) $a(KCl) = 2(r_{K^+} + r_{Cl^-}) = 627.7\,\text{pm}$;

(3) $a(NaBr) = 2(r_{Na^+} + r_{Br^-}) = 596.2\,\text{pm}$; (4) $a(KBr) = 2(r_{K^+} + r_{Br^-}) = 658.6\,\text{pm}$.

If the ionic radii of all the ions are constant then

    (1) + (4) = (2) + (3).

    (1) + (4) = (562.8 + 658.6) pm = 1221.4 pm.

    (2) + (3) = (627.7 + 596.2) pm = 1223.9 pm.

The difference is slight; $\boxed{\text{hence the data support}}$ the constancy of the radii of the ions.

P20.5 For the three given reflections

$$\sin 19.076° = 0.32682, \quad \sin 22.171° = 0.37737, \quad \sin 32.256° = 0.53370.$$

For cubic lattices $\sin\theta_{hkl} = \dfrac{\lambda(h^2 + k^2 + l^2)^{1/2}}{2a}$ [20.5 with 20.2].

First consider the possibility of simple cubic; the first three reflections are (100), (110), and (111). (See Fig. 20.22 of the text.)

$$\frac{\sin\theta(100)}{\sin\theta(110)} = \frac{1}{\sqrt{2}} \neq \frac{0.32682}{0.37737} \quad \text{[not simple cubic]}.$$

Consider next the possibility of body-centered cubic; the first three reflections are (110), (200), and (211).

$$\frac{\sin\theta(110)}{\sin\theta(200)} = \frac{\sqrt{2}}{\sqrt{4}} = \frac{1}{\sqrt{2}} \neq \frac{0.32682}{0.37737} \text{ (not bcc)}.$$

Consider finally face-centered cubic; the first three reflections are (111), (200), and (220).

$$\frac{\sin\theta(111)}{\sin\theta(200)} = \frac{\sqrt{3}}{\sqrt{4}} = 0.86603$$

which compares very favorably to $0.32682/0.37737 = 0.86605$. Therefore, the lattice is $\boxed{\text{face-centered cubic}}$.

This conclusion may easily be confirmed in the same manner using the second and third reflection.

$$a = \frac{\lambda}{2\sin\theta}(h^2 + k^2 + l^2)^{1/2} = \left(\frac{154.18\,\text{pm}}{(2)\times(0.32682)}\right) \times \sqrt{3} = \boxed{408.55\,\text{pm}}.$$

$$\rho = \frac{NM}{N_A V} \text{ [Exercise 20.8(a)]} = \frac{(4)\times(107.87\,\text{g mol}^{-1})}{(6.0221\times10^{23}\,\text{mol}^{-1})\times(4.0855\times10^{-8}\,\text{cm})^3}$$

$$= \boxed{10.507\,\text{g cm}^{-1}}.$$

This compares favorably to the value listed in the *Data section*.

P20.7 $\lambda = 2a\sin\theta_{100}$ as $d_{100} = a$.

Therefore, $a = \dfrac{\lambda}{2\sin\theta_{100}}$ and

$$\frac{a(\text{KCl})}{a(\text{NaCl})} = \frac{\sin\theta_{100}(\text{NaCl})}{\sin\theta_{100}(\text{KCl})} = \frac{\sin 6°0'}{\sin 5°23'} = 1.114.$$

Therefore, $a(\text{KCl}) = (1.114)\times(564\,\text{pm}) = \boxed{628\,\text{pm}}$.

The relative densities calculated from these unit cell dimensions are

$$\frac{\rho(\text{KCl})}{\rho(\text{NaCl})} = \left(\frac{M(\text{KCl})}{M(\text{NaCl})}\right) \times \left(\frac{a(\text{NaCl})}{a(\text{KCl})}\right)^3 = \left(\frac{74.55}{58.44}\right) \times \left(\frac{564\,\text{pm}}{628\,\text{pm}}\right)^3 = 0.924.$$

Experimentally

$$\frac{\rho(\text{KCl})}{\rho(\text{NaCl})} = \frac{1.99\,\text{g cm}^{-3}}{2.17\,\text{g cm}^{-3}} = 0.917.$$

and the measurements $\boxed{\text{are broadly consistent}}$.

P20.9   As demonstrated in *Justification* 20.3 of the text, close-packed spheres fill 0.7404 of the total volume of the crystal. Therefore 1 cm³ of close-packed carbon atoms would contain

$$\frac{0.74040 \text{ cm}^3}{\left(\frac{4}{3}\pi r^3\right)} = 3.838 \times 10^{23} \text{ atoms}$$

$$\left(r = \left(\frac{154.45}{2}\right) \text{ pm} = 77.225 \text{ pm} = 77.225 \times 10^{-10} \text{ cm}\right).$$

Hence the close-packed density would be

$$\rho = \frac{\text{mass in 1 cm}^3}{1 \text{ cm}^3} = \frac{(3.838 \times 10^{23} \text{ atom}) \times (12.01 \text{ u/atom}) \times (1.6605 \times 10^{-24} \text{ g u}^{-1})}{1 \text{ cm}^3}$$

$$= \boxed{7.654 \text{ g cm}^{-3}}.$$

The diamond structure (solution to Exercise 20.17(a)) is a very open structure, which is dictated by the tetrahedral bonding of the carbon atoms. As a result many atoms that would be touching each other in a normal fcc structure do not in diamond; for example, the C atom in the center of a face does not touch the C atoms at the corners of the face.

P20.11 $$\rho = \frac{m \text{ (unit cell)}}{V \text{ (unit cell)}} = \frac{(2) \times (M(\text{CH}_2\text{CH}_2))/N_A}{abc}$$

$$= \frac{(2) \times (28.05 \text{ g mol}^{-1})}{(6.022 \times 10^{23} \text{ mol}^{-1}) \times [(740 \times 493 \times 253) \times 10^{-39}] \text{ m}^3}$$

$$= 1.01 \times 10^6 \text{ g m}^{-3} = \boxed{1.01 \text{ g cm}^{-3}}.$$

P20.13   **(a)** When there is only one pair of identical atoms, the Wierl equation reduces to

$$I(\theta) = f^2 \frac{\sin sR}{sR} \quad \text{where} \quad s = \frac{4\pi}{\lambda} \sin \frac{1}{2}\theta.$$

Extrema occur at $sR = \sin sR/\cos sR = \tan sR$ and this equation may be solved either graphically or numerically to give the extrema values shown in Fig. 20.4(a).
The angles of extrema are calculated using the Br_2 bond length of 228.3 pm (Table 13.2), the equation $\theta = 2\sin^{-1}\left(\frac{sR\lambda}{4\pi R}\right)$, and sR extrema values shown in Fig. 20.4(a).

Neutron diffraction: $\theta_{\text{1st max}} = 0$,

$$\theta_{\text{1st min}} = 2\sin^{-1}\left(\frac{(0.9534 \times 3\pi/2)(78 \text{ pm})}{4\pi(229.0 \text{ pm})}\right) = \boxed{14.0°},$$

$$\theta_{\text{2nd max}} = 2\sin^{-1}\left(\frac{(0.9836 \times 5\pi/2)(78 \text{ pm})}{4\pi(229.0 \text{ pm})}\right) = \boxed{24.2°}.$$

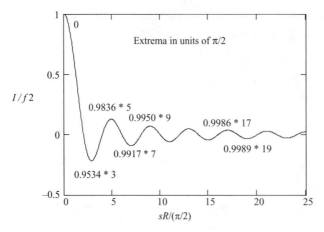

Figure 20.4(a)

Electron diffraction: $\theta_{1st\ max} = 0$,

$$\theta_{1st\ min} = 2 \sin^{-1} \left(\frac{(0.9534 \times 3\pi/2)(4.0\ pm)}{4\pi(229.0\ pm)} \right) = \boxed{0.72°},$$

$$\theta_{2nd\ max} = 2 \sin^{-1} \left(\frac{(0.9836 \times 5\pi/2)(4\ pm)}{4\pi(229.0\ pm)} \right) = \boxed{1.23°}.$$

(b)
$$I = \sum_{i,j} f_i f_j \frac{\sin sR_{i,j}}{sR_{i,j}}, \quad s = \frac{4\pi}{\lambda} \sin \frac{1}{2}\theta$$

$$= 4f_C f_{Cl} \frac{\sin sR_{CCl}}{sR_{CCl}} + 6f_{Cl}^2 \frac{\sin sR_{ClCl}}{sR_{ClCl}} \text{[4C—Cl pairs, 6Cl—Cl pairs]}$$

$$= (4) \times (6) \times (17) \times (f^2) \times \left(\frac{\sin x}{x} \right) + (6) \times (17) \times (f^2) \frac{\sin(\frac{8}{3})^{1/2}x}{(\frac{8}{3})^{1/2}x} [x = sR_{CCl}].$$

$$\frac{I}{f^2} = (408) \times \frac{\sin x}{x} + (1062) \frac{\sin \left(\frac{8}{3} \right)^{1/2} x}{x}.$$

This function is plotted in Fig. 20.4(b).
We find x_{max} and x_{min} from the graph, and s_{max} and s_{min} from the data. Then, since $x = sR_{CCl}$, we can take the ratio x/s to find the bond length R_{CCl}. The calculation of s requires the wavelength of the electron beam.

$$\frac{p^2}{2m_e} = eV \quad \text{or} \quad p = (2m_e eV)^{1/2}.$$

From the de Broglie relation [8.12],

$$\lambda = \frac{h}{p} = \frac{h}{(2m_e eV)^{1/2}}$$

$$= \frac{6.626 \times 10^{-34}\ J\ s}{\{2 \times (9.109 \times 10^{-31}\ kg)(1.609 \times 10^{-19}\ C)(1.00 \times 10^4\ V)\}}$$

$$= 12.2\ pm.$$

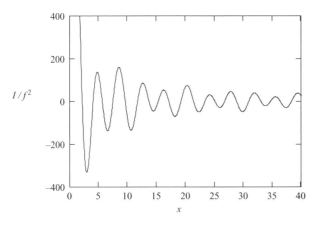

Figure 20.4(b)

We draw up the following table.

	Maxima			Minima			
θ(expt.)	$3°0'$	$5°22'$	$7°54'$	$1°46'$	$4°6'$	$6°40'$	$9°10'$
s/pm^{-1}	0.0270	0.0482	0.0710	0.0159	0.0368	0.0599	0.0819
x(calc.)	4.77	8.52	12.6	2.89	6.52	10.6	14.5
$(x/s)/\text{pm}$	177	177	177	177	177	177	177

Hence, $\boxed{R_{\text{CCl}} = 177 \text{ pm}}$ and the experimental diffraction pattern is consistent with tetrahedral geometry.

P20.15   The volume per unit cell is

$$V = abc = (3.6881 \text{ nm}) \times (0.9402 \text{ nm}) \times (1.7652 \text{ nm}) = 6.121 \text{ nm}^3 = 6.121 \times 10^{-21} \text{ cm}^3.$$

The mass per unit cell is 8 times the mass of the formula unit, $RuN_2C_{28}H_{44}S_4$, for which the molar mass is

$$M = \{101.07 + 2(14.007) + 28(12.011) + 44(1.008) + 4(32.066)\} \text{ g mol}^{-1} = 638.01 \text{ g mol}^{-1}.$$

The density is

$$\rho = \frac{m}{V} = \frac{8M}{N_A V} = \frac{8(638.01 \text{ g mol}^{-1})}{(6.022 \times 10^{23} \text{ mol}^{-1}) \times (6.121 \times 10^{-21} \text{ cm}^3)} = \boxed{1.385 \text{ g cm}^{-3}}.$$

The osmium analog has a molar mass of 727.1 g mol^{-1}. If the volume of the crystal changes negligibly with the substitution, then the densities of the complexes are in proportion to their molar masses,

$$\rho_{os} = \frac{727.1}{638.01} (1.385 \text{ g cm}^{-3}) = \boxed{1.578 \text{ g cm}^{-3}}.$$

P20.17 $G = G_0 e^{-E_g/2kT}$

$$\ln(G/S) = \ln(G_0/S) - \left(E_g/2k\right) \times 1/T.$$

Thus, the slope of a $\ln(G)$ against $1/T$ plot equals $-E_g/2k$. The data has minimal uncertainty so the slope can be calculated by the two-point difference method. Alternatively, a linear regression fit of $(1/T, \ln(G/S))$ data points gives the slope.

$$\text{slope} = \frac{\Delta \ln(G/S)}{\Delta (1/T)} = \frac{\ln(.0847) - \ln(2.86)}{(1/312 \text{ K}) - (1/420 \text{ K})} = -4270 \text{ K},$$

$$E_g = -2k \times (\text{slope}) = -2 \times (1.381 \times 10^{-23} \text{ J K}) \times (-4270) = 1.18 \times 10^{-19} \text{ J}.$$

This is equivalent to 71.0 kJ/mol or $\boxed{0.736 \text{ eV}}$.

P20.19   The molar magnetic susceptibility is given by

$$\chi_m = \frac{N_A g_e^2 \mu_0 \mu_B^2 S(S+1)}{3kT} = (6.3001 \times 10^{-6}) \times \frac{S(S+1)}{T/K} \text{ m}^3 \text{ mol}^{-1} \text{ [Illustration 20.3]}.$$

$$\text{For } S = 2, \ \chi_m = \frac{(6.3001 \times 10^{-6}) \times (2) \times (2+1)}{298} \text{ m}^3 \text{ mol}^{-1} = \boxed{0.127 \times 10^{-6} \text{ m}^3 \text{ mol}^{-1}}.$$

$$\text{For } S = 3, \ \chi_m = \frac{(6.3001 \times 10^{-6}) \times (3) \times (3+1)}{298} \text{ m}^3 \text{ mol}^{-1} = \boxed{0.254 \times 10^{-6} \text{ m}^3 \text{ mol}^{-1}}.$$

$$\text{For } S = 4, \ \chi_m = \frac{(6.3001 \times 10^{-6}) \times (4) \times (4+1)}{298} \text{ m}^3 \text{ mol}^{-1} = \boxed{0.423 \times 10^{-6} \text{ m}^3 \text{ mol}^{-1}}.$$

Instead of a single value of S we use an average weighted by the Boltzmann factor

$$\exp \left(\frac{-50 \times 10^3 \text{ J mol}^{-1}}{(8.3145 \text{ J mol}^{-1}\text{K}^{-1}) \times (298 \text{ K})} \right) = 1.7 \times 10^{-9}.$$

Thus the $S = 2$ and $S = 4$ forms are present in negligible quantities compared to the $S = 3$ form. The compound's susceptibility, then, is that of the $S = 3$ form, namely $\boxed{0.254 \times 10^{-6} \text{ m}^3 \text{ mol}^{-1}}$.

P20.21   If the unit cell volume does not change upon substitution of Ca for Y, then, the density of the superconductor and that of the Y-only compound will be proportional to their molar masses.

$$M_{\text{super}} = [2(200.59) + 2(137.327) + (1-x) \times (88.906) + x(40.078)$$
$$+ 2(63.546) + 7.55(15.999)] \text{ g mol}^{-1},$$

$$M_{\text{super}}/(\text{g mol}^{-1}) = 1012.6 - 48.828x.$$

The molar mass of the Y-only compound is 1012.6 g mol^{-1}, and the ratio of their densities is

$$\frac{\rho_{\text{super}}}{\rho_{\text{Y-only}}} = \frac{1012.6 - 48.828x}{1012.6} = 1 - 0.04822x \quad \text{so } x = \frac{1}{0.04822} \left(1 - \frac{\rho_{\text{super}}}{\rho_{\text{Y-only}}} \right).$$

The density of the Y-only compound is its mass over its volume. The volume is

$$V_{\text{Y-only}} = a^2 c = (0.38606 \text{ nm})^2 \times (2.8915 \text{ nm}) = 0.43096 \text{ nm}^3 = 0.43096 \times 10^{-21} \text{ cm}^3,$$

so the density

$$\rho_{\text{Y-only}} = \frac{2M}{N_A V} = \frac{2(1012.6 \text{ g mol}^{-1})}{(6.022 \times 10^{23} \text{ mol}^{-1}) \times (0.43096 \times 10^{-21} \text{ cm}^3)} = 7.804 \text{ g cm}^{-3}.$$

The extent of Ca substitution is

$$x = \frac{1}{0.04822}\left(1 - \frac{7.651}{7.804}\right) = \boxed{0.41}.$$

COMMENT.   The precision of this method depends strongly on just how constant the lattice volume really is.

Solutions to theoretical problems

P20.23   If the sides of the unit cell define the vectors $\boldsymbol{a}$, $\boldsymbol{b}$, and $\boldsymbol{c}$, then its volume is $V = \boldsymbol{a} \cdot \boldsymbol{b} \times \boldsymbol{c}$ [given]. Introduce the orthogonal set of unit vectors $\hat{\boldsymbol{i}}, \hat{\boldsymbol{j}}, \hat{\boldsymbol{k}}$ so that

$$\boldsymbol{a} = a_x\hat{\boldsymbol{i}} + a_y\hat{\boldsymbol{j}} + a_z\hat{\boldsymbol{k}},$$

$$\boldsymbol{b} = b_x\hat{\boldsymbol{i}} + b_y\hat{\boldsymbol{j}} + b_z\hat{\boldsymbol{k}},$$

$$\boldsymbol{c} = c_x\hat{\boldsymbol{i}} + c_y\hat{\boldsymbol{j}} + c_z\hat{\boldsymbol{k}}.$$

Then $V = \boldsymbol{a} \cdot \boldsymbol{b} \times \boldsymbol{c} = \begin{vmatrix} a_x & a_y & a_z \\ b_x & b_y & b_z \\ c_x & c_y & c_z \end{vmatrix}.$

Therefore

$$V^2 = \begin{vmatrix} a_x & a_y & a_z \\ b_x & b_y & b_z \\ c_x & c_y & c_z \end{vmatrix}\begin{vmatrix} a_x & a_y & a_z \\ b_x & b_y & b_z \\ c_x & c_y & c_z \end{vmatrix}$$

$$= \begin{vmatrix} a_x & a_y & a_z \\ b_x & b_y & b_z \\ c_x & c_y & c_z \end{vmatrix}\begin{vmatrix} a_x & a_y & a_z \\ b_x & b_y & b_z \\ c_x & c_y & c_z \end{vmatrix}$$

[interchange rows and columns, no change in value]

$$= \begin{vmatrix} a_xa_x + a_ya_y + a_za_z & a_xb_x + a_yb_y + a_zb_z & a_xc_x + a_yc_y + a_zc_z \\ b_xa_x + b_ya_y + b_za_z & b_xb_x + b_yb_y + b_zb_z & b_xc_x + b_yc_y + b_zc_z \\ c_xa_x + c_ya_y + c_za_z & c_xb_x + c_yb_y + c_zb_z & c_xc_x + c_yc_y + c_zc_z \end{vmatrix}$$

$$= \begin{vmatrix} a^2 & \boldsymbol{a} \cdot \boldsymbol{b} & \boldsymbol{a} \cdot \boldsymbol{c} \\ \boldsymbol{b} \cdot \boldsymbol{a} & b^2 & \boldsymbol{b} \cdot \boldsymbol{c} \\ \boldsymbol{c} \cdot \boldsymbol{a} & \boldsymbol{c} \cdot \boldsymbol{b} & c^2 \end{vmatrix} = \begin{vmatrix} a^2 & ab\cos\gamma & ac\cos\beta \\ ab\cos\gamma & b^2 & bc\cos\alpha \\ ac\cos\beta & bc\cos\alpha & c^2 \end{vmatrix}$$

$$= a^2b^2c^2(1 - \cos^2\alpha - \cos^2\beta - \cos^2\gamma + 2\cos\alpha\cos\beta\cos\gamma)^{1/2}.$$

Hence $\boxed{V = abc(1 - \cos^2\alpha - \cos^2\beta - \cos^2\gamma + 2\cos\alpha\cos\beta\cos\gamma)^{1/2}}.$

For a monoclinic cell, $\alpha = \gamma = 90°$

$$V = abc(1 - \cos^2\beta)^{1/2} = \boxed{abc\sin\beta}.$$

For an orthorhombic cell, $\alpha = \beta = \gamma = 90°$, and

$$V = \boxed{abc}.$$

P20.25 The four values of $hx + ky + lz$ that occur in the exponential functions of F have the values 0, 5/2, 3, and 7/2, and so

$$F_{hkl} \propto 1 + e^{5i\pi} + e^{6i\pi} + e^{7i\pi} = 1 - 1 + 1 - 1 = \boxed{0}.$$

P20.27 According to eqn 20.18,

$$G = \frac{E}{2(1 + \nu_P)} \quad \text{and} \quad K = \frac{E}{3(1 - 2\nu_P)}.$$

Substituting the Lamé-constant expressions for E and ν_P into the right-hand side of these relationships yields

$$G = \frac{(\mu(3\lambda + 2\mu))/(\lambda + \mu)}{2\left(1 + \dfrac{\lambda}{2(\lambda + \mu)}\right)} \quad \text{and} \quad K = \frac{\dfrac{\mu(3\lambda + 2\mu)}{\lambda + \mu}}{3\left(1 - \dfrac{\lambda}{\lambda + \mu}\right)}.$$

Expanding leads to:

$$G = \frac{(\mu(3\lambda + 2\mu)/\lambda + \mu)}{2\left((2\lambda + 2\mu + \lambda)/(2(\lambda + \mu))\right)} = \frac{\mu(3\lambda + 2\mu)}{3\lambda + 2\mu} = \boxed{\mu}$$

$$\text{and} \quad K = \frac{\mu(3\lambda + 2\mu)/(\lambda + \mu)}{3(\lambda + \mu - \lambda)/(\lambda + \mu)} = \frac{\mu(3\lambda + 2\mu)}{3\mu} = \boxed{\frac{3\lambda + 2\mu}{3}},$$

as the problem asks us to prove.

P20.29 Permitted states at the low energy edge of the band must have a relatively long characteristic wavelength while the permitted states at the high energy edge of the band must have a relatively short characteristic wavelength. There are few wavefunctions that have these characteristics so the density of states is lowest at the edges. This is analogous to the MO picture that shows a few bonding MOs that lack nodes and few antibonding MOs that have the maximum number of nodes.

Another insightful view is provided by consideration of the spatially periodic potential that the electron experiences within a crystal. The periodicity demands that the electron wavefunction be a periodic function of the position vector $\vec{r}$. We can approximate it with a Bloch wave: $\psi \propto e^{i\vec{k}\cdot\vec{r}}$ where $\vec{k} = k_x\hat{i} + k_y\hat{j} + k_z\hat{k}$ is called the wavenumber vector. This is a bold, 'free' electron approximation and in the spirit of searching for a conceptual explanation, not an accurate solution, suppose that the wavefunction satisfies a Hamiltonian in which the potential can be neglected: $\hat{H} = -(\hbar^2/2m)\nabla^2$. The eigenvalues of the Bloch wave are: $E = \hbar^2|\vec{k}|^2/2m$. The Bloch wave is periodic when the components of the wave number vector are multiples of a basic repeating unit. Writing the repeating unit as $2\pi/L$ where L is a length that depends upon the structure of the unit cell, we find: $k_x = 2n_x\pi/L$ where $n_x = 0, \pm 1, \pm 2, \ldots$. Similar equations can be written for k_y and k_z and with substitution the eigenvalues become: $E = 1/2m\,(2\pi\hbar/L)^2\left(n_x^2 + n_y^2 + n_z^2\right)$. This equation suggests that the density of states for energy level E can be visually evaluated by looking at a plot of permitted n_x, n_y, n_z values as shown in Fig. 20.5. The number of n_x, n_y, n_z values within a thin, spherical shell around the origin equals the density of states that have energy E. Three shells, labeled 1, 2, and 3, are shown in the graph. All have the same width but their energies increase with their distance from the origin. It is obvious that the low energy shell 1 has a much lower density of states than the intermediate energy shell 2. The sphere of shell 3 has been cut into the shape determined by the periodic potential pattern of the crystal and, because of this phenomenon, it also has a lower density of states than the intermediate energy shell 2. The general concept is that the low energy and high energy edges of a band have lower density of states than that of the band center.

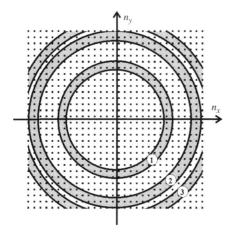

Figure 20.5

P20.31
$$\xi = \frac{-e^2}{6m_e}\langle r^2 \rangle.$$

$$\langle r^2 \rangle = \int_0^\infty r^2 \psi^2 d\tau \quad \text{with } \psi = \left(\frac{1}{\pi a_0^3}\right)^{1/2} e^{-r/a_0}$$

$$= 4\pi \int_0^\infty r^4 \psi^2 \, dr \quad [d\tau = 4\pi r^2 \, dr]$$

$$= \frac{4}{a_0^3} \int_0^\infty r^4 e^{-2r/a_0} dr = 3a_0^2 \quad \left[\int_0^\infty x^n e^{-ax} \, dx = \frac{n!}{a^{n+1}}\right].$$

Therefore, $\boxed{\xi = \dfrac{-e^2 a_0^2}{2M_e}}$.

Then, since $\chi_m = N_A \mu_0 \xi$ [20.32, $m = 0$],

$$\chi_m = \boxed{\frac{-N_A \mu_0 e^2 a_0^2}{2m_e}}.$$

P20.33   If the proportion of molecules in the upper level is P, where they have a magnetic moment of $2\mu_B$ (which replaces $\{S(S+1)\}^{1/2}\mu_B$ in eqn 20.35), the molar susceptibility

$$\chi_m = \frac{(6.3001 \times 10^{-6}) \times [S(S+1)]}{T/K} \text{ m}^3 \text{ mol}^{-1} \quad [\textit{Illustration 20.3}]$$

is changed to

$$\chi_m = \frac{(6.3001 \times 10^{-6}) \times (4) \times P}{T/K} \, m^3 mol^{-1} \, [2^2 \text{ replaces } S(S+1)] = \frac{25.2P}{T/K} \times 10^{-6} \, m^3 \, mol^{-1}.$$

The proportion of molecules in the upper state is

$$P = \frac{e^{-hc\tilde{\nu}/kT}}{1 + e^{-hc\tilde{\nu}/kT}} \, [\text{Boltzmann distribution}] = \frac{1}{1 + e^{hc\tilde{\nu}/kT}}$$

and $\frac{hc\tilde{\nu}}{kT} = \frac{(1.4388 \, cm \, K) \times (121 \, cm^{-1})}{T} = \frac{174}{T/K}.$

Therefore, $\chi_m = \frac{25.2 \times 10^{-6} \, m^3 \, mol^{-1}}{(T/K) \times (1 + e^{174/(T/K)})}.$

This function is plotted in Fig. 20.6.

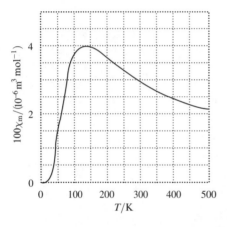

Figure 20.6

COMMENT. The explanation of the magnetic properties of NO is more complicated and subtle than indicated by the solution here. In fact the full solution for this case was one of the important triumphs of the quantum theory of magnetism which was developed about 1930. See J.H. van Vleck, *The theory of electric and magnetic susceptibilities*. Oxford University Press (1932).

Solutions to applications

P20.35 The X-ray diffraction pattern of fibrous B-DNA (Figure 20.26) is discussed in *Impact* I20.1. Figures 20.27 and 20.28 provide definitions of the helical tilt angle α and the base-layer spacing h. The helical pitch p is the vertical rise per turn of the helix. The characteristic X-shape of the diffraction pattern is that of a helix with incident radiation (Cu $K\alpha$ 0.1542 nm) perpendicular to the cylindrical axis. An angle $\theta = 2.6°$ between the line of the incident radiation and the line from sample to the first spot on the X gives $p = \lambda/\sin\theta = 0.1542 \, nm/\sin(2.6°) = 3.4 \, nm$. 10 spots (counting two 'missing fourth' spots) along the X diagonal indicate that there 10 base-planes per turn of the helix with each accounting for a turn of 36°. The very large spot is at a distance $(1/h)$ which is 10 times the distance $1/p$ shown in

Fig. 20.7(a). Consequently, $h = 0.34$ nm. The missing fourth spots on the X diagonals indicate two coaxial sugar-phosphate backbones that are separated by $3p/8$ along the axis. The periodic h spacing of the large, very electron-dense phosphorus atoms causes the $1/h$ spots to be very intense. The fact that the fibrous X-ray sample was saturated with water suggests that the phosphates are to the outside.

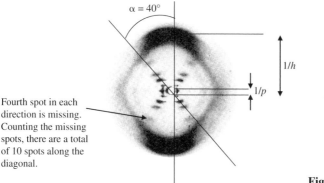

Fourth spot in each direction is missing. Counting the missing spots, there are a total of 10 spots along the diagonal.

Figure 20.7(a)

Figure 20.7(b) shows the two-dimensional zig-zag projection of the helical sugar–phosphate backbone. It serves to define the projection length l, perpendicular distance d between backbone planes, and the helix radius r. Examination of the right triangle that shows the definition of α yields

$$\tan(\alpha) = \frac{p}{4r} \quad \text{or} \quad r = \frac{p}{4 \tan(\alpha)} = \frac{3.4 \text{ nm}}{4 \tan(40°)} = 1.0 \text{ nm.}$$

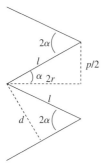

Figure 20.7(b)

Examination of the right triangle containing the angle α also shows that $l \sin(\alpha) = p/2$ while the right triangle containing the angle 2α shows that $l \sin(2\alpha) = d$. Dividing these two equations yields

$$\frac{\sin(2\alpha)}{\sin(\alpha)} = \frac{2d}{p} \quad \text{or} \quad \frac{2 \sin(\alpha) \cos(\alpha)}{\sin(\alpha)} = \frac{2d}{p} \quad \text{or} \quad \cos(\alpha) = \frac{d}{p}.$$

$$d = p \cos(\alpha) = (3.5 \text{ nm}) \cos(40°) = 2.6 \text{ nm.}$$

Finishing,

$$l = \frac{p}{2 \sin(\alpha)} = \frac{3.4 \text{ nm}}{2 \sin(40°)} = 2.6 \text{ nm.}$$

P20.37 Refer to Fig. 20.8.

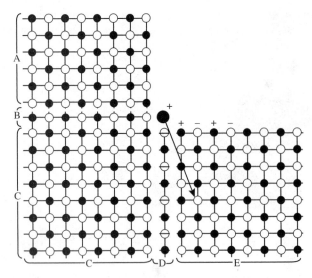

Figure 20.8

Evaluate the sum of $\pm(1/r_i)$, where r_i is the distance from the ion i to the ion of interest, taking $+(1/r)$ for ions of like charge and $-(1/r)$ for ions of opposite charge. The array has been divided into five zones. Zones B and D can be summed analytically to give $-\ln 2 = -0.69$. The summation over the other zones, each of which gives the same result, is tedious because of the very slow convergence of the sum. Unless you make a very clever choice of the sequence of ions (grouping them so that their contributions almost cancel), you will find the following values for arrays of different sizes

10×10	20×20	50×50	100×100	200×200
0.259	0.273	0.283	0.286	0.289

The final figure is in good agreement with the analytical value, 0.289 259 7...

For a cation above a flat surface, the energy (relative to the energy at infinity, and in multiples of $e^2/(4\pi\varepsilon r_0)$ where r_0 is the lattice spacing (200 pm)), is

$$\text{Zone C} + \text{D} + \text{E} = 0.29 - 0.69 + 0.29 = \boxed{-0.11}$$

which implies an attractive state.

PART 3 Change

21 Molecules in motion

Answers to discussion questions

D21.1 (a) See Section 24.1 which discusses the collision theory of gas phase reactions. Their rate depends on the number of collisions having a relative kinetic energy above a certain critical value ε_a. The relative kinetic energy, in turn, depends on the relative velocities of the colliding molecules. The rate of reaction also depends on the number of collisions per unit volume per unit time, or collision density, Z_{AB}, the formula for which is derived in *Justification* 24.1. Z_{AB} also depends on the average relative velocity of the colliding molecules.

(b) A complete analysis of the composition of planetary atmospheres is a complicated process. See the solution to Problem 21.35 for detailed calculations on the depletion of the Earth's atmosphere and on the atmosphere of planets in general. Also see the solution to Problem 16.21 which deals with the inherent instability of planetary atmospheres. The simple answer to this question, though, is that light molecules are more likely to have velocities in excess of the escape velocity than are heavy molecules. Therefore, heavy molecules will remain in the atmosphere much longer than light molecules, though all will eventually escape, unless there is a source of replenishment.

D21.3 Gases are very dilute systems and on average the molecules are very far apart from each other except when they collide. So what little resistance there is to flow in a gaseous fluid is almost entirely due to the collisions between molecules. The frequency of collisions increases with increasing temperature (see eqns 21.11b and 24.8); hence the viscosity of gases increases with temperature. In liquids, on the other hand, the molecules are very close to each other, which results in there being strong forces of attraction between them that resist their movement relative to each other. However, as the temperature increases, more and more molecules are likely to have sufficient kinetic energy to overcome the forces of attraction, resulting in decreased viscosity.

D21.5 (a) This is Fick's first law of diffusion in one dimension written in terms of concentrations rather than activities; hence, it applies strictly only to ideal solutions.

(b) In addition to the restriction to ideal solutions as in (a), the derivation of this expression uses the additional approximation that the frictional retarding force on a moving particle is proportional to the first power of the speed of the particle (as opposed to a more general functional relation).

(c) The restrictions of parts (a) and (b) still apply, as well as a third, which is the assumption that the particle is spherical.

D21.7 Because the drift speed governs the rate at which charge is transported, we might expect the conductivity to decrease with increasing solution viscosity and ion size. Experiments confirm these predictions for bulky ions, but not for small ions. For example, the molar conductivities of the alkali metal ions increase from Li^+ to Cs^+ (Table 21.6) even though the ionic radii increase. The paradox is resolved when we realize that the radius a in the Stokes formula is the hydrodynamic radius (or 'Stokes radius') of the ion, its effective radius in the solution taking into account all the H_2O molecules it carries in its hydration sphere. Small ions give rise to stronger electric fields than large ones, so small ions are more extensively solvated than big ions. Thus, an ion of small ionic radius may have a large hydrodynamic radius because it drags many solvent molecules through the solution as it migrates. The hydrating H_2O molecules are often very labile, however, and NMR and isotope studies have shown that the exchange between the coordination sphere of the ion and the bulk solvent is very rapid.

The proton, although it is very small, has a very high molar conductivity (Table 21.6)! Proton and $^{17}O-NMR$ show that the times characteristic of protons hopping from one molecule to the next are about 1.5 ps, which is comparable to the time that inelastic neutron scattering shows it takes a water molecule to reorientate through about 1 rad $(1 - 2\ \text{ps})$.

Solutions to exercises

E21.1(b) **(a)** The mean speed of a gas molecule is

$$\bar{c} = \left(\frac{8RT}{\pi M}\right)^{1/2}$$

so $\dfrac{\bar{c}(\text{He})}{\bar{c}(\text{Hg})} = \left(\dfrac{M(\text{Hg})}{M(\text{He})}\right)^{1/2} = \left(\dfrac{200.59}{4.003}\right)^{1/2} = \boxed{7.079}$

(b) The mean kinetic energy of a gas molecule is $\frac{1}{2} mc^2$, where c is the root mean square speed

$$c = \left(\frac{3RT}{M}\right)^{1/2}$$

So $\frac{1}{2} mc^2$ is independent of mass, and the ratio of mean kinetic energies of He and Hg is $\boxed{1}$

E21.2(b) **(a)** The mean speed can be calculated from the formula derived in Example 21.1.

$$\bar{c} = \left(\frac{8RT}{\pi M}\right)^{1/2} = \left(\frac{8 \times (8.314\,\text{J K}^{-1}\text{mol}^{-1}) \times (298\,\text{K})}{\pi \times (28.02 \times 10^{-3}\,\text{kg mol}^{-1})}\right)^{1/2} = \boxed{4.75 \times 10^2\,\text{m s}^{-1}}$$

(b) The mean free path is calculated from $\lambda = kT/(2^{1/2}\sigma p)$ [21.13] with $\sigma = \pi d^2 = \pi \times (3.95 \times 10^{-10}\,\text{m})^2 = 4.90 \times 10^{-19}\,\text{m}^2$

Then, $\lambda = \dfrac{(1.381 \times 10^{-23}\,\text{J K}^{-1}) \times (298\,\text{K})}{2^{1/2} \times (4.90 \times 10^{-19}\,\text{m}^2) \times (1 \times 10^{-9}\,\text{Torr}) \times \left(\frac{1\,\text{atm}}{760\,\text{Torr}}\right) \times \left(\frac{1.013 \times 10^5\,\text{Pa}}{1\,\text{atm}}\right)}$

$= \boxed{4 \times 10^4\,\text{m}}$

(c) The collision frequency could be calculated from eqn 21.11, but is most easily obtained from eqn 21.12, since λ and $\bar{c}$ have already been calculated

$$z = \frac{\bar{c}}{\lambda} = \frac{4.75 \times 10^2\,\text{m s}^{-1}}{4.\overline{46} \times 10^4\,\text{m}} = \boxed{1 \times 10^{-2}\,\text{s}^{-1}}$$

Thus there are 100 s between collisions, which is a very long time compared to the usual timescale of molecular events. The mean free path is much larger than the dimensions of the pumping apparatus used to generate the very low pressure.

E21.3(b) $p = \dfrac{kT}{2^{1/2}\sigma\lambda}$ [21.13]

$$\sigma = \pi\, d^2, \quad d = \left(\frac{\sigma}{\pi}\right)^{1/2} = \left(\frac{0.36\,\text{nm}^2}{\pi}\right)^{1/2} = 0.34\,\text{nm}$$

$$p = \frac{\left(1.381 \times 10^{-23}\,\text{J K}^{-1}\right) \times (298\,\text{K})}{\left(2^{1/2}\right) \times \left(0.36 \times 10^{-18}\,\text{m}^2\right) \times \left(0.34 \times 10^{-9}\,\text{m}\right)} = \boxed{2.4 \times 10^7\,\text{Pa}}$$

This pressure corresponds to about 240 atm, which is comparable to the pressure in a compressed gas cylinder in which argon gas is normally stored.

E21.4(b)   The mean free path is

$$\lambda = \frac{kT}{2^{1/2}\sigma p} = \frac{\left(1.381 \times 10^{-23}\,\text{J K}^{-1}\right) \times (217\,\text{K})}{2^{1/2}\left[0.43 \times \left(10^{-9}\,\text{m}\right)^2\right] \times \left(12.1 \times 10^3\,\text{Pa atm}^{-1}\right)} = \boxed{4.1 \times 10^{-7}\,\text{m}}$$

E21.5(b)   Obtain data from Exercise 21.4(b)

The expression for z obtained in Exercise 21.5(a) is $z = [16/(\pi mkT)]^{1/2}\,\sigma p$

Substituting $\sigma = 0.43\,\text{nm}^2$, $p = 12.1 \times 10^3\,\text{Pa}$, $m = (28.02\,\text{u})$, and $T = 217\text{K}$ we obtain

$$z = \frac{4 \times \left(0.43 \times 10^{-18}\,\text{m}^2\right) \times \left(12.1 \times 10^3\,\text{Pa}\right)}{\left[\pi \times (28.02) \times \left(1.6605 \times 10^{-27}\,\text{kg}\right) \times \left(1.381 \times 10^{-23}\,\text{J K}^{-1}\right) \times (217\,\text{K})\right]^{1/2}}$$

$$= \boxed{9.9 \times 10^8\,\text{s}^{-1}}$$

E21.6(b)   The mean free path is

$$\lambda = \frac{kT}{2^{1/2}\sigma p} = \frac{\left(1.381 \times 10^{-23}\,\text{J K}^{-1}\right) \times (25 + 273)\,\text{K}}{2^{1/2}\left[0.52 \times \left(10^{-9}\,\text{m}\right)^2\right]p} = \frac{5.5\bar{0} \times 10^{-3}\,\text{m Pa}}{p}$$

(a) $\lambda = \dfrac{5.5\bar{0} \times 10^{-3}\,\text{m Pa}}{(15\,\text{atm}) \times \left(1.013 \times 10^5\,\text{Pa atm}^{-1}\right)} = \boxed{3.7 \times 10^{-9}\,\text{m}}$

(b) $\lambda = \dfrac{5.5\bar{0} \times 10^{-3}\,\text{m Pa}}{(1.0\,\text{bar}) \times \left(10^5\,\text{Pa bar}^{-1}\right)} = \boxed{5.5 \times 10^{-8}\,\text{m}}$

(c) $\lambda = \dfrac{5.5\bar{0} \times 10^{-3}\,\text{m Pa}}{(1.0\,\text{Torr}) \times \left(1.013 \times 10^5\,\text{Pa atm}^{-1}/760\,\text{Torr atm}^{-1}\right)} = \boxed{4.1 \times 10^{-5}\,\text{m}}$

E21.7(b)   The fraction F of molecules in the speed range from 200 to $250\,\text{m s}^{-1}$ is

$$F = \int_{200\,\text{m s}^{-1}}^{250\,\text{m s}^{-1}} f(v)\,dv$$

where $f(v)$ is the Maxwell distribution. This can be approximated by

$$F \approx f(v) \Delta v = 4\pi \left(\frac{M}{2\pi RT}\right)^{3/2} v^2 \exp\left(\frac{-Mv^2}{2RT}\right) \Delta v,$$

with $f(v)$ evaluated in the middle of the range

$$F \approx 4\pi \left(\frac{44.0 \times 10^{-3}\,\mathrm{kg\,mol^{-1}}}{2\pi\,(8.3145\,\mathrm{J\,K^{-1}\,mol^{-1}}) \times (300\,\mathrm{K})}\right)^{3/2} \times \left(225\,\mathrm{m\,s^{-1}}\right)^2$$

$$\times \exp\left(\frac{-\left(44.0 \times 10^{-3}\,\mathrm{kg\,mol^{-1}}\right) \times \left(225\,\mathrm{m\,s^{-1}}\right)^2}{2\,(8.3145\,\mathrm{J\,K^{-1}\,mol^{-1}}) \times (300\,\mathrm{K})}\right) \times \left(50\,\mathrm{m\,s^{-1}}\right),$$

$$\boxed{F \approx 9.6 \times 10^{-2}}$$

COMMENT. The approximation we have employed, taking $f(v)$ to be nearly constant over a narrow range of speeds, might not be accurate enough, for that range of speeds includes about 10 percent of the molecules. You may wish to do the integration without this approximation (a considerably more complicated process) to see how much difference there is.

E21.8(b)   The number of collisions is

$$N = Z_W At = \frac{pAt}{(2\pi mkT)^{1/2}}$$

$$= \frac{(111\,\mathrm{Pa}) \times \left(3.5 \times 10^{-3}\,\mathrm{m}\right) \times \left(4.0 \times 10^{-2}\,\mathrm{m}\right) \times (10\,\mathrm{s})}{\left\{2\pi \times (4.00\,\mathrm{u}) \times \left(1.66 \times 10^{-27}\,\mathrm{kg\,u^{-1}}\right) \times \left(1.381 \times 10^{-23}\,\mathrm{J\,K^{-1}}\right) \times (1500\,\mathrm{K})\right\}^{1/2}}$$

$$= \boxed{5.3 \times 10^{21}}$$

E21.9(b)   The mass of the sample in the effusion cell decreases by the mass of the gas which effuses out of it. That mass is the molecular mass times the number of molecules that effuse out

$$\Delta m = mN = mZ_W At = \frac{mpAt}{(2\pi mkT)^{1/2}} = pAt\left(\frac{m}{2\pi kT}\right)^{1/2} = pAt\left(\frac{M}{2\pi RT}\right)^{1/2}$$

$$= (0.224\,\mathrm{Pa}) \times \pi \times \left(\tfrac{1}{2} \times 3.00 \times 10^{-3}\,\mathrm{m}\right)^2 \times (24.00\,\mathrm{h}) \times \left(3600\,\mathrm{s\,h^{-1}}\right)$$

$$\times \left\{\frac{300 \times 10^{-3}\,\mathrm{kg\,mol^{-1}}}{2\pi \times (8.3145\,\mathrm{J\,K^{-1}\,mol^{-1}}) \times (450\,\mathrm{K})}\right\}^{1/2}$$

$$= \boxed{4.98 \times 10^{-4}\,\mathrm{kg}}$$

E21.10(b) The time dependence of the pressure of a gas effusing without replenishment is

$$p = p_0 e^{-t/\tau} \text{ where } \tau \propto \sqrt{m}$$

The time t it takes for the pressure to go from any initial pressure p_0 to a prescribed fraction of that pressure fp_0 is

$$t = \tau \ln \frac{fp_0}{p_0} = \tau \ln f$$

so the time is proportional to τ and therefore also to $\sqrt{m}$. Therefore, the ratio of times it takes two different gases to go from the same initial pressure to the same final pressure is related to their molar masses as follows

$$\frac{t_1}{t_2} = \left(\frac{M_1}{M_2}\right)^{1/2} \qquad \text{and} \qquad M_2 = M_1 \left(\frac{t_2}{t_1}\right)^2$$

So $\quad M_{\text{fluorocarbon}} = \left(28.01 \text{ g mol}^{-1}\right) \times \left(\dfrac{82.3\,\text{s}}{18.5\,\text{s}}\right)^2 = \boxed{554 \text{ g mol}^{-1}}$

E21.11(b) The time dependence of the pressure of a gas effusion without replenishment is

$$p = p_0 \, e^{-t/\tau} \qquad \text{so} \qquad t = \tau \ln p_0/p$$

where $\tau = \dfrac{V}{A_0}\left(\dfrac{2\pi m}{kT}\right)^{1/2} = \dfrac{V}{A_0}\left(\dfrac{2\pi M}{RT}\right)^{1/2}$

$$= \left(\frac{22.0\,\text{m}^3}{\pi \times (0.50 \times 10^{-3}\,\text{m})^2}\right) \times \left(\frac{2\pi \times (28.0 \times 10^{-3}\,\text{kg mol}^{-1})}{(8.3145\,\text{J K}^{-1}\,\text{mol}^{-1}) \times (293\,\text{K})}\right)^{1/2} = 2.4 \times 10^5\,\text{s}$$

so $\quad t = (8.6 \times 10^5\,\text{s}) \ln \dfrac{122\,\text{kPa}}{105\,\text{kPa}} = \boxed{1.5 \times 10^4\,\text{s}}$

E21.12(b) The flux is

$$J = -\kappa \frac{dT}{dz} = -\frac{1}{3}\lambda C_{V,\text{m}} \bar{c}\,[X]\,\frac{dT}{dz}$$

where the minus sign indicates flow toward lower temperature and

$$\lambda = \frac{1}{\sqrt{2}N\sigma}, \quad \bar{c} = \left(\frac{8kT}{\pi m}\right)^{1/2} = \left(\frac{8RT}{\pi M}\right)^{1/2}, \quad \text{and } [M] = n/V = N/N_A$$

So $\quad J = -\dfrac{2C_{V,\text{m}}}{3\sigma N_A}\left(\dfrac{RT}{\pi M}\right)^{1/2}\dfrac{dT}{dz}$

$$= \left(\frac{2 \times (28.832 - 8.3145)\,\text{J K}^{-1}\,\text{mol}^{-1}}{3 \times \left[0.27 \times (10^{-9}\,\text{m})^2\right] \times (6.022 \times 10^{23}\,\text{mol}^{-1})}\right)$$

$$\times \left(\frac{(8.3145\,\text{J K}^{-1}\,\text{mol}^{-1}) \times (260\,\text{K})}{\pi \times (2.016 \times 10^{-3}\,\text{kg mol}^{-1})}\right)^{1/2} \times (3.5\,\text{K m}^{-1})$$

$$= \boxed{0.17\,\text{J m}^{-2}\,\text{s}^{-1}}$$

E21.13(b) The thermal conductivity is

$$\kappa = \frac{1}{3}\lambda C_{V,m}\bar{c}\,[X] = \frac{2C_{V,m}}{3\sigma N_A}\left(\frac{RT}{\pi M}\right)^{1/2} \quad \text{so } \sigma = \frac{2C_{V,m}}{3\kappa N_A}\left(\frac{RT}{\pi M}\right)^{1/2}$$

$$\kappa = \left(0.240\,\text{mJ cm}^{-2}\,\text{s}^{-1}\right) \times \left(\text{K cm}^{-1}\right)^{-1} = 0.240 \times 10^{-1}\,\text{J m}^{-1}\,\text{s}^{-1}\,\text{K}^{-1}$$

$$\text{so } \sigma = \left(\frac{2 \times (29.125 - 8.3145)\,\text{J K}^{-1}\,\text{mol}^{-1}}{3 \times \left(0.240 \times 10^{-1}\,\text{J m}^{-1}\,\text{s}^{-1}\,\text{K}^{-1}\right) \times \left(6.022 \times 10^{23}\,\text{mol}^{-1}\right)}\right)$$

$$\times \left(\frac{\left(8.3145\,\text{J K}^{-1}\,\text{mol}^{-1}\right) \times (298\,\text{K})}{\pi \times \left(28.013 \times 10^{-3}\,\text{kg mol}^{-1}\right)}\right)^{1/2}$$

$$= \boxed{1.61 \times 10^{-19}\,\text{m}^2}$$

E21.14(b) Assuming the space between sheets is filled with air, the flux is

$$J = -k\frac{dT}{dz} = \left[\left(0.241 \times 10^{-3}\,\text{J cm}^{-2}\,\text{s}^{-1}\right) \times \left(\text{K cm}^{-1}\right)^{-1}\right] \times \left(\frac{[50 - (-10)\,\text{K}]}{10.0\,\text{cm}}\right)$$

$$= 1.4\overline{5} \times 10^{-3}\,\text{J cm}^{-2}\,\text{s}^{-1}.$$

So the rate of energy transfer and energy loss is

$$JA = (1.4\overline{5} \times 10^{-3}\,\text{J cm}^{-2}\,\text{s}^{-1}) \times (1.50\,\text{m}^2) \times (100\,\text{cm m}^{-1})^2 = \boxed{22\,\text{J s}^{-1}}$$

E21.15(b) The coefficient of viscosity is

$$\eta = \frac{1}{3}\lambda m N\bar{c} = \frac{2}{3\sigma}\left(\frac{mkT}{\pi}\right)^{1/2} \quad \text{so} \quad \sigma = \frac{2}{3\eta}\left(\frac{mkT}{\pi}\right)^{1/2}$$

$$\eta = 1.66\,\mu\text{P} = 166 \times 10^{-7}\,\text{kg m}^{-1}\,\text{s}^{-1}$$

$$\text{so } \sigma = \left(\frac{2}{3 \times \left(166 \times 10^{-7}\,\text{kg m}^{-1}\,\text{s}^{-1}\right)}\right)$$

$$\times \left(\frac{\left(28.01 \times 10^{-3}\,\text{kg mol}^{-1}\right) \times \left(1.381 \times 10^{-23}\,\text{J K}^{-1}\right) \times (273\,\text{K})}{\pi \times \left(6.022 \times 10^{23}\,\text{mol}^{-1}\right)}\right)^{1/2}$$

$$= \boxed{3.00 \times 10^{-19}\,\text{m}^2}$$

E21.16(b) The rate of fluid flow through a tube is described by

$$\frac{dV}{dt} = \frac{(p_{in}^2 - p_{out}^2)\,\pi r^4}{16l\eta p_0} \quad \text{so} \quad p_{in} = \left(\frac{16l\eta p_0}{\pi r^4}\frac{dV}{dt} + p_{out}^2\right)^{1/2}$$

Several of the parameters need to be converted to SI units

$$r = \tfrac{1}{2}(15 \times 10^{-3}\,\text{m}) = 7.5 \times 10^{-3}\,\text{m}$$

and $\dfrac{dV}{dt} = 8.70\,\text{cm}^3 \times \left(10^{-2}\,\text{m}\,\text{cm}^{-1}\right)^3\,\text{s}^{-1} = 8.70 \times 10^{-6}\,\text{m}^3\,\text{s}^{-1}$.

Also, we have the viscosity at 293 K from the table. According to the $T^{1/2}$ temperature dependence, the viscosity at 300 K ought to be

$$\eta\,(300\,\text{K}) = \eta\,(293\,\text{K}) \times \left(\dfrac{300\,\text{K}}{293\,\text{K}}\right)^{1/2} = (176 \times 10^{-7}\,\text{kg}\,\text{m}^{-1}\,\text{s}^{-1}) \times \left(\dfrac{300}{293}\right)^{1/2}$$

$$= 1.78 \times 10^{-7}\,\text{kg}\,\text{m}^{-1}\,\text{s}^{-1}$$

$$p_{\text{in}} = \left\{\left(\dfrac{16\,(10.5\,\text{m}) \times \left(178 \times 10^{-7}\,\text{kg}\,\text{m}^{-1}\text{s}^{-1}\right) \times \left(1.00 \times 10^5\,\text{Pa}\right)}{\pi \times \left(7.5 \times 10^{-3}\,\text{m}\right)^4}\right)\right.$$

$$\left. \times \,(8.70 \times 10^{-6}\,\text{m}^3\,\text{s}^{-1}) + (1.00 \times 10^5\,\text{Pa})^2\right\}^{1/2}$$

$$= \boxed{1.00 \times 10^5\,\text{Pa}}$$

COMMENT. For the exercise as stated the answer is not sensitive to the viscosity. The flow rate is so low that the inlet pressure would equal the outlet pressure (to the precision of the data) whether the viscosity were that of N_2 at 300 K or 293 K, or even liquid water at 293 K!

E21.17(b) The coefficient of viscosity is

$$\eta = \dfrac{1}{3}\lambda m N \bar{c} = \dfrac{2}{3\sigma}\left(\dfrac{mkT}{\pi}\right)^{1/2}$$

$$= \left(\dfrac{2}{3\left[0.88 \times \left(10^{-9}\,\text{m}\right)^2\right]}\right) \times \left(\dfrac{\left(78.12 \times 10^{-3}\,\text{kg}\,\text{mol}^{-1}\right) \times \left(1.381 \times 10^{-23}\,\text{J}\,\text{K}^{-1}\right) T}{\pi \times \left(6.022 \times 10^{23}\,\text{mol}^{-1}\right)}\right)^{1/2}$$

$$= 5.7\overline{2} \times 10^{-7} \times (T/\text{K})^{1/2}\,\text{kg}\,\text{m}^{-1}\,\text{s}^{-1}$$

(a) At 273 K $\eta = \left(5.7\overline{2} \times 10^{-7}\right) \times (273)^{1/2}\,\text{kg}\,\text{m}^{-1}\,\text{s}^{-1} = \boxed{0.95 \times 10^{-5}\,\text{kg}\,\text{m}^{-1}\,\text{s}^{-1}}$

(b) At 298 K $\eta = \left(5.7\overline{2} \times 10^{-7}\right) \times (298)^{1/2}\,\text{kg}\,\text{m}^{-1}\,\text{s}^{-1} = \boxed{0.99 \times 10^{-5}\,\text{kg}\,\text{m}^{-1}\,\text{s}^{-1}}$

(c) At 1000 K $\eta = \left(5.7\overline{2} \times 10^{-7}\right) \times (1000)^{1/2}\,\text{kg}\,\text{m}^{-1}\,\text{s}^{-1} = \boxed{1.81 \times 10^{-5}\,\text{kg}\,\text{m}^{-1}\,\text{s}^{-1}}$

E21.18(b) The thermal conductivity is

$$k = \tfrac{1}{3}\lambda C_{V,\text{m}}\bar{c}\,[\text{X}] = \dfrac{2C_{V,\text{m}}}{3\sigma N_{\text{A}}}\left(\dfrac{RT}{\pi M}\right)^{1/2}$$

(a) $$\kappa = \left(\dfrac{2 \times \left[(20.786 - 8.3145)\,\text{J}\,\text{K}^{-1}\,\text{mol}^{-1}\right]}{3\left[0.24 \times \left(10^{-9}\,\text{m}\right)^2\right] \times \left(6.022 \times 10^{23}\,\text{mol}^{-1}\right)}\right)$$

$$\times \left(\dfrac{\left(8.3145\,\text{J}\,\text{K}^{-1}\,\text{mol}^{-1}\right) \times (300\,\text{K})}{\pi\,\left(20.18 \times 10^{-3}\,\text{kg}\,\text{mol}^{-1}\right)}\right)^{1/2}$$

$$= \boxed{0.011\,\overline{4}\,\text{J}\,\text{m}^{-1}\,\text{s}^{-1}\,\text{K}^{-1}}$$

The flux is

$$J = -\kappa \frac{dT}{dz} = \left(0.011\overline{4}\,J\,m^{-1}\,s^{-1}\,K^{-1}\right) \times \left(\frac{(305 - 295)\,K}{0.15\,m}\right) = 0.76\,J\,m^{-2}\,s^{-1}$$

so the rate of energy loss is

$$JA = \left(0.76\,J\,m^{-2}\,s^{-1}\right) \times (0.15\,m)^2 = \boxed{0.017\,J\,s^{-1}}$$

(b)

$$\kappa = \left(\frac{2 \times \left[(29.125 - 8.3145)\,J\,K^{-1}\,mol^{-1}\right]}{3\left[0.43 \times \left(10^{-9}\,m\right)^2\right] \times \left(6.022 \times 10^{23}\,mol^{-1}\right)}\right)$$

$$\times \left(\frac{8.3145\,J\,K^{-1}\,mol^{-1}) \times (300\,K)}{\pi\left(28.013 \times 10^{-3}\,kg\,mol^{-1}\right)}\right)^{1/2}$$

$$= \boxed{9.0 \times 10^{-3}\,J\,m^{-1}\,s^{-1}\,K^{-1}}$$

The flux is

$$J = -\kappa \frac{dT}{dz} = \left(9.0 \times 10^{-3}\,J\,m^{-1}\,s^{-1}\,K^{-1}\right) \times \left(\frac{(305 - 295)\,K}{0.15\,m}\right) = 0.60\,J\,m^{-2}\,s^{-1}$$

so the rate of energy loss is

$$JA = \left(0.60\,J\,m^{-2}\,s^{-1}\right) \times (0.15\,m)^2 = \boxed{0.014\,J\,s^{-1}}$$

E21.19(b) The rate of fluid flow through a tube is described by

$$\frac{dV}{dt} = \frac{\left(p_{in}^2 - p_{out}^2\right)\pi r^4}{16l\eta p_0}$$

so the rate is inversely proportional to the viscosity, and the time required for a given volume of gas to flow through the same tube under identical pressure conditions is directly proportional to the viscosity

$$\frac{t_1}{t_2} = \frac{\eta_1}{\eta_2} \quad \text{so } \eta_2 = \frac{\eta_1 t_2}{t_1}$$

$$\eta_{CFC} = \frac{(208\,\mu P) \times (18.0\,s)}{72.0\,s} = \boxed{52.0\,\mu P} = 52.0 \times 10^{-7}\,kg\,m^{-1}\,s^{-1}$$

The coefficient of viscosity is

$$\eta = \frac{1}{3}\lambda m N \bar{c} = \left(\frac{2}{3\sigma}\right) \times \left(\frac{mkT}{\pi}\right)^{1/2} = \left(\frac{2}{3\pi d^2}\right) \times \left(\frac{mkT}{\pi}\right)^{1/2}$$

so the molecular diameter is

$$d = \left(\frac{2}{3\pi\eta}\right)^{1/2} \times \left(\frac{mkT}{\pi}\right)^{1/4}$$

$$= \left(\frac{2}{3\pi\left(52.0 \times 10^{-7}\,\text{kg}\,\text{m}^{-1}\,\text{s}^{-1}\right)}\right)^{1/2}$$

$$\times \left(\frac{\left(200 \times 10^{-3}\,\text{kg}\,\text{mol}^{-1}\right) \times \left(1.381 \times 10^{-23}\,\text{J}\,\text{K}^{-1}\right) \times (298\,\text{K})}{\pi \times \left(6.022 \times 10^{23}\,\text{mol}^{-1}\right)}\right)^{1/4}$$

$$= 9.23 \times 10^{-10}\,\text{m} = \boxed{923\,\text{pm}}$$

E21.20(b)
$$\kappa = \frac{1}{3}\lambda C_{V,\text{m}}\bar{c}\,[X] = \frac{2C_{V,\text{m}}}{3\sigma N_A}\left(\frac{RT}{\pi M}\right)^{1/2}$$

$$= \left(\frac{2 \times (29.125 - 8.3145)\,\text{J}\,\text{K}^{-1}\,\text{mol}^{-1}}{3\left[0.43 \times \left(10^{-9}\,\text{m}\right)^2\right] \times \left(6.022 \times 10^{23}\,\text{mol}^{-1}\right)}\right) \times \left(\frac{\left(8.3145\,\text{J}\,\text{K}^{-1}\,\text{mol}^{-1}\right) \times (300\,\text{K})}{\pi \times \left(28.013 \times 10^{-3}\,\text{kg}\,\text{mol}^{-1}\right)}\right)^{1/2}$$

$$= \boxed{9.0 \times 10^{-3}\,\text{J}\,\text{m}^{-1}\,\text{s}^{-1}\,\text{K}^{-1}}$$

E21.21(b) The diffusion constant is

$$D = \frac{1}{3}\lambda\bar{c} = \frac{2(RT)^{3/2}}{3\sigma p N_A (\pi M)^{1/2}}$$

$$= \frac{2\left[\left(8.3145\,\text{J}\,\text{K}^{-1}\,\text{mol}^{-1}\right) \times (298\,\text{K})\right]^{3/2}}{3\left[0.43 \times \left(10^{-9}\,\text{m}\right)^2\right]p\left(6.022 \times 10^{23}\,\text{mol}^{-1}\right) \times \left\{\pi\left(28.013 \times 10^{-3}\,\text{kg}\,\text{mol}^{-1}\right)\right\}^{1/2}}$$

$$= \frac{1.07\,\text{m}^2\,\text{s}^{-1}}{p/\text{Pa}}$$

The flux due to diffusion is

$$J = -D\frac{d[X]}{dx} = -D\frac{d}{dx}\left(\frac{n}{V}\right) = -\left(\frac{D}{RT}\right)\frac{dp}{dx}$$

where the minus sign indicates flow from high pressure to low. So for a pressure gradient of $0.10\,\text{atm}\,\text{cm}^{-1}$

$$J = \left(\frac{D/(\text{m}^2\,\text{s}^{-1})}{\left(8.3145\,\text{J}\,\text{K}^{-1}\,\text{mol}^{-1}\right) \times (298\,\text{K})}\right) \times \left(0.20 \times 10^5\,\text{Pa}\,\text{m}^{-1}\right)$$

$$= \left(8.1\,\text{mol}\,\text{m}^{-2}\,\text{s}^{-1}\right) \times \left(D/(\text{m}^2\,\text{s}^{-1})\right)$$

(a) $$D = \frac{1.07\,\text{m}^2\,\text{s}^{-1}}{10.0} = \boxed{0.107\,\text{m}^2\,\text{s}^{-1}}$$

and $$J = \left(8.1\,\text{mol}\,\text{m}^{-2}\,\text{s}^{-1}\right) \times (0.107) = \boxed{0.87\,\text{mol}\,\text{m}^{-2}\,\text{s}^{-1}}$$

(b) $$D = \frac{1.07\,\text{m}^2\,\text{s}^{-1}}{100 \times 10^3} = \boxed{1.07 \times 10^{-5}\,\text{m}^2\,\text{s}^{-1}}$$

and $J = (8.1\,\text{mol}\,\text{m}^{-2}\,\text{s}^{-1}) \times (1.07 \times 10^{-5}) = \boxed{8.7 \times 10^{-5}\,\text{mol}\,\text{m}^{-2}\,\text{s}^{-1}}$

(c) $$D = \frac{1.07\,\text{m}^2\,\text{s}^{-1}}{15.0 \times 10^6} = \boxed{7.13 \times 10^{-8}\,\text{m}^2\,\text{s}^{-1}}$$

and $J = (8.1\,\text{mol}\,\text{m}^{-2}\,\text{s}^{-1}) \times (7.13 \times 10^{-8}) = \boxed{5.8 \times 10^{-7}\,\text{mol}\,\text{m}^{-2}\,\text{s}^{-1}}$

E21.22(b) Molar ionic conductivity is related to mobility by

$$\lambda = zuF = (1) \times \left(4.24 \times 10^{-8}\,\text{m}^2\,\text{s}^{-1}\,\text{V}^{-1}\right) \times \left(96485\,\text{C}\,\text{mol}^{-1}\right)$$

$$= \boxed{4.09 \times 10^{-3}\,\text{S}\,\text{m}^2\,\text{mol}^{-1}}$$

E21.23(b) The drift speed is given by

$$s = u\varepsilon = \frac{u\Delta\phi}{l} = \frac{\left(4.01 \times 10^{-8}\,\text{m}^2\,\text{s}^{-1}\,\text{V}^{-1}\right) \times (12.0\,\text{V})}{1.00 \times 10^{-2}\,\text{m}} = \boxed{4.81 \times 10^{-5}\,\text{m}\,\text{s}^{-1}}$$

E21.24(b) The limiting transport number for Cl^- in aqueous NaCl at 25°C is

$$t_-^\circ = \frac{u_-}{u_+ + u_-} = \frac{7.91}{5.19 + 7.91} = \boxed{0.604}$$

(The mobilities are in $10^{-8}\,\text{m}^2\,\text{s}^{-1}\,\text{V}^{-1}$.)

E21.25(b) The limiting molar conductivity of a dissolved salt is the sum of that of its ions, so

$$\Lambda_{\text{m}}^\circ\left(\text{MgI}_2\right) = \lambda\left(\text{Mg}^{2+}\right) + 2\lambda\left(\text{I}^-\right) = \Lambda_{\text{m}}^\circ\left(\text{Mg}(\text{C}_2\text{H}_3\text{O}_2)_2\right) + 2\Lambda_{\text{m}}^\circ\left(\text{NaI}\right) - 2\Lambda_{\text{m}}^\circ\left(\text{NaC}_2\text{H}_3\text{O}_2\right)$$

$$= (18.78 + 2\,(12.69) - 2\,(9.10))\,\text{mS}\,\text{m}^2\,\text{mol}^{-1} = \boxed{25.96\,\text{mS}\,\text{m}^2\,\text{mol}^{-1}}$$

E21.26(b) Molar ionic conductivity is related to mobility by

$$\lambda = z\mu F \quad \text{so} \quad u = \frac{\lambda}{zF}$$

$$\text{F}^- : \quad u = \frac{5.54 \times 10^{-3}\,\text{S}\,\text{m}^2\,\text{mol}^{-1}}{(1) \times \left(96\,485\,\text{C}\,\text{mol}^{-1}\right)} = \boxed{5.74 \times 10^{-8}\,\text{m}^2\,\text{V}^{-1}\,\text{s}^{-1}}$$

$$\text{Cl}^- : \quad u = \frac{7.635 \times 10^{-3}\,\text{S}\,\text{m}^2\,\text{mol}^{-1}}{(1) \times \left(96\,485\,\text{C}\,\text{mol}^{-1}\right)} = \boxed{7.913 \times 10^{-8}\,\text{m}^2\,\text{V}^{-1}\,\text{s}^{-1}}$$

$$\text{Br}^- : \quad u = \frac{7.81 \times 10^{-3}\,\text{S}\,\text{m}^2\,\text{mol}^{-1}}{(1) \times \left(96\,485\,\text{C}\,\text{mol}^{-1}\right)} = \boxed{8.09 \times 10^{-8}\,\text{m}^2\,\text{V}^{-1}\,\text{s}^{-1}}$$

E21.27(b) The diffusion constant is related to the mobility by

$$D = \frac{uRT}{zF} = \frac{\left(4.24 \times 10^{-8}\,\mathrm{m^2\,s^{-1}\,V^{-1}}\right) \times \left(8.3145\,\mathrm{J\,K^{-1}\,mol^{-1}}\right) \times (298\,\mathrm{K})}{(1) \times \left(96\,485\,\mathrm{C\,mol^{-1}}\right)}$$

$$= \boxed{1.09 \times 10^{-9}\,\mathrm{m^2\,s^{-1}}}$$

E21.28(b) The mean square displacement for diffusion in one dimension is

$$\left\langle x^2 \right\rangle = 2Dt$$

In fact, this is also the mean square displacement in any direction in two- or three-dimensional diffusion from a concentrated source. In three dimensions

$$r^2 = x^2 + y^2 + z^2 \quad \text{so} \quad \left\langle r^2 \right\rangle = \left\langle x^2 \right\rangle + \left\langle y^2 \right\rangle + \left\langle z^2 \right\rangle = 3\left\langle x^2 \right\rangle = 6Dt$$

So the time it takes to travel a distance $\sqrt{\left\langle r^2 \right\rangle}$ is

$$t = \frac{\left\langle r^2 \right\rangle}{6D} = \frac{\left(1.0 \times 10^{-2}\,\mathrm{m}\right)^2}{6\left(4.05 \times 10^{-9}\,\mathrm{m^2\,s^{-1}}\right)} = \boxed{4.1 \times 10^3\,\mathrm{s}}$$

E21.29(b) The diffusion constant is related to the viscosity of the medium and the size of the diffusing molecule as follows

$$D = \frac{kT}{6\pi\eta a} \quad \text{so} \quad a = \frac{kT}{6\pi\eta D} = \frac{\left(1.381 \times 10^{-23}\,\mathrm{J\,K^{-1}}\right) \times (298\,\mathrm{K})}{6\pi\left(1.00 \times 10^{-3}\,\mathrm{kg\,m^{-1}\,s^{-1}}\right) \times \left(1.055 \times 10^{-9}\,\mathrm{m^2\,s^{-1}}\right)}$$

$$a = 2.07 \times 10^{-10}\,\mathrm{m} = \boxed{207\,\mathrm{pm}}$$

E21.30(b) The Einstein–Smoluchowski equation related the diffusion constant to the unit jump distance and time

$$D = \frac{\lambda^2}{2\tau} \quad \text{so} \quad \tau = \frac{\lambda^2}{2D}$$

If the jump distance is about one molecular diameter, or two effective molecular radii, then the jump distance can be obtained by use of the Stokes–Einstein equation

$$D = \frac{kT}{6\pi\eta a} = \frac{kT}{3\pi\eta\lambda} \quad \text{so} \quad \lambda = \frac{kT}{3\pi\eta D}$$

$$\text{and } \tau = \frac{(kT)^2}{18\,(\pi\eta)^2\,D^3} = \frac{\left[\left(1.381 \times 10^{-23}\,\mathrm{J\,K^{-1}}\right) \times (298\,\mathrm{K})\right]^2}{18\left[\pi\left(0.387 \times 10^{-3}\,\mathrm{kg\,m^{-1}\,s^{-1}}\right)\right]^2 \times \left(3.17 \times 10^{-9}\,\mathrm{m^2\,s^{-1}}\right)^3}$$

$$= \boxed{200 \times 10^{-11}\,\mathrm{s}} = 20\,\mathrm{ps}$$

E21.31(b) The mean square displacement is (from Exercise 21.28(b))

$$\left\langle r^2 \right\rangle = 6Dt \quad \text{so} \quad t = \frac{\left\langle r^2 \right\rangle}{6D} = \frac{\left(1.0 \times 10^{-6}\,\mathrm{m}\right)^2}{6\left(1.0 \times 10^{-11}\,\mathrm{m^2\,s^{-1}}\right)} = \boxed{1.7 \times 10^{-2}\,\mathrm{s}}$$

Solutions to problems

Solutions to numerical problems

P21.1 The time in seconds for a disk to rotate $360°$ is the inverse of the frequency. The time for it to advance $2°$ is $(2°/360°)/v$. This is the time required for slots in neighboring disks to coincide. For an atom to pass through all neighbouring slots it must have the speed $v_x = \dfrac{1.0\,\text{cm}}{(2/360)/v} = 180\,v\,\text{cm} = 180(v/\text{Hz})\,\text{cm s}^{-1}$.

Hence, the distributions of the x-component of velocity are

v/Hz	20	40	80	100	120
$v_x/(\text{cm s}^{-1})$	3600	7200	14400	18000	21600
I (40 K)	0.846	0.513	0.069	0.015	0.002
I (100 K)	0.592	0.485	0.217	0.119	0.057

Theoretically, the velocity distribution in the x-direction is

$$f(v_x) = \left(\frac{m}{2\pi kT}\right)^{1/2} e^{-mv_x^2/2kT} \quad [21.6, \text{ with } M/R = m/k].$$

Therefore, as $I \propto f$, $I \propto \left(\dfrac{1}{T}\right)^{1/2} e^{-mv_x^2/2kT}$.

Since $\dfrac{mv_x^2}{2kT} = \dfrac{83.8 \times (1.6605 \times 10^{-27}\,\text{kg}) \times \{1.80(v/\text{Hz})\,\text{m s}^{-1}\}^2}{(2) \times (1.381 \times 10^{-23}\,\text{J K}^{-1}) \times (T)} = \dfrac{1.63 \times 10^{-2}(v/\text{Hz})^2}{T/\text{K}}$, we can

write $I \propto \left(\dfrac{1}{T/\text{K}}\right)^{1/2} e^{-1.63 \times 10^{-2}(v/\text{Hz})^2/(T/\text{K})}$ and draw up the following table, obtaining the constant of proportionality by fitting I to the value at $T = 40\,\text{K}$, $v = 80\,\text{Hz}$

v/Hz	20	40	80	100	120
$I/(40\,\text{K})$	0.80	0.49	(0.069)	0.016	0.003
$I/(100\,\text{K})$	0.56	0.46	0.209	0.116	0.057

in fair agreement with the experimental data.

P21.3 $\langle X \rangle = \dfrac{1}{N} \sum_i N_i X_i$ [see Problem 21.2].

(a) $\langle h \rangle = \dfrac{1}{53}\{1.80\,\text{m} + 2 \times (1.82\,\text{m}) + \cdots + 1.98\,\text{m}\} = \boxed{1.89\,\text{m}}$.

(b)
$$\langle h^2 \rangle = \frac{1}{53}\left\{(1.80\,\text{m})^2 + 2 \times (1.82\,\text{m})^2 + \cdots + (1.98\,\text{m})^2\right\} = 3.57\,\text{m}^2$$

$$\sqrt{\langle h^2 \rangle} = \boxed{1.89\,\text{m}}.$$

P21.5 The number of molecules that escape in unit time is the number per unit time that would have collided with a wall section of area A equal to the area of the small hole. That is,

$$\frac{dN}{dt} = -Z_W A = \frac{-Ap}{(2\pi mkT)^{1/2}} \quad [21.14]$$

where p is the (constant) vapor pressure of the solid. The change in the number of molecules inside the cell in an interval Δt is therefore $\Delta N = -Z_W A \Delta t$, and so the mass loss is

$$\Delta w = \Delta N m = -Ap \left(\frac{m}{2\pi kT}\right)^{1/2} \Delta t = -Ap \left(\frac{m}{2\pi RT}\right)^{1/2} \Delta t.$$

Therefore, the vapor pressure of the substance in the cell is

$$p = \left(\frac{-\Delta w}{A \Delta t}\right) \times \left(\frac{2\pi RT}{M}\right)^{1/2}.$$

For the vapor pressure of germanium

$$p = \left(\frac{4.3 \times 10^{-8} \text{ kg}}{\pi \times (5.0 \times 10^{-4}\text{m})^2 \times (7200 \text{ s})}\right) \times \left(\frac{(2\pi) \times (8.314 \text{ J K}^{-1} \text{ mol}^{-1}) \times (1273 \text{ K})}{72.6 \times 10^{-3} \text{ kg mol}^{-1}}\right)^{1/2}$$

$$= 7.3 \times 10^{-3} \text{ Pa, or } \boxed{7.3 \text{ mPa}}.$$

P21.7 The atomic current is the number of atoms emerging from the slit per second, which is $Z_W A$ with $A = 1 \times 10^{-7} \text{ m}^2$. We use

$$Z_W = \frac{p}{(2\pi mkT)^{1/2}} \quad [21.14]$$

$$= \frac{p/\text{Pa}}{\left[(2\pi) \times (M/\text{g mol}^{-1}) \times (1.6605 \times 10^{-27} \text{ kg}) \times (1.381 \times 10^{-23} \text{ J K}^{-1}) \times (380 \text{ K})\right]^{1/2}}$$

$$= \left(1.35 \times 10^{23} \text{ m}^{-2} \text{ s}^{-1}\right) \times \left(\frac{p/\text{Pa}}{(M/\text{g mol}^{-1})^{1/2}}\right).$$

(a) Cadmium:

$$Z_W A = \left(1.35 \times 10^{23} \text{ m}^{-2} \text{ s}^{-1}\right) \times \left(1 \times 10^{-7} \text{ m}^2\right) \times \left(\frac{0.13}{(112.4)^{1/2}}\right) = \boxed{2 \times 10^{14} \text{ s}^{-1}}.$$

(b) Mercury:

$$Z_W A = \left(1.35 \times 10^{23} \text{ m}^{-2} \text{ s}^{-1}\right) \times \left(1 \times 10^{-7} \text{ m}^2\right) \times \left(\frac{152}{(200.6)^{1/2}}\right) = \boxed{1 \times 10^{17} \text{ s}^{-1}}.$$

P21.9
$$\Lambda_m = \Lambda_m^\circ - \mathscr{K}c^{1/2} \text{ [21.29]}, \quad \Lambda_m = \frac{C}{cR} \text{ [21.28]}$$

where $C = 20.63 \text{ m}^{-1}$ ($C = \kappa^* R^*$, where κ^* and R^* are the conductivity and resistance of a standard solution, respectively).

Therefore, we draw up the following table.

c/M		0.0005	0.001	0.005	0.010	0.020	0.050
$(c/M)^{1/2}$		0.224	0.032	0.071	0.100	0.141	0.224
R/Ω		3314	1669	342.1	174.1	89.08	37.14
$\Lambda_m/(\text{mS m}^2\text{ mol}^{-1})$	12.45	12.36	12.06	11.85	11.58	11.11

The values of Λ_m are plotted against $c^{1/2}$ in Fig. 21.1.

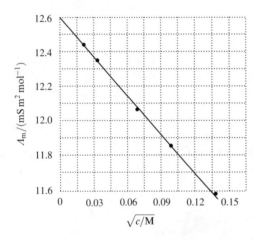

Figure 21.1

The limiting value is $\Lambda_m^\circ = \boxed{12.6 \text{ mS m}^2\text{ mol}^{-1}}$. The slope is -7.30; hence

$$\mathscr{K} = \boxed{7.30 \text{ mS m}^2\text{ mol}^{-1}\text{ M}^{-1/2}}.$$

(a) $\Lambda_m = (5.01 + 7.68) \text{ mS m}^2\text{mol}^{-1} - (+7.30 \text{ mS m}^2\text{mol}^{-1}) \times (0.010)^{1/2}$

$\qquad = \boxed{11.96 \text{ mS m}^2\text{ mol}^{-1}}.$

(b) $\kappa = c\Lambda_m = (10 \text{ mol m}^{-3}) \times (11.96 \text{ mS m}^2\text{ mol}^{-1}) = 119.6 \text{ mS m}^2\text{ m}^{-3} = \boxed{119.6 \text{ mS m}^{-1}}.$

(c) $R = \dfrac{C}{\kappa} = \dfrac{20.63 \text{ m}^{-1}}{119.6 \text{ mS m}^{-1}} = \boxed{172.5 \Omega}.$

P21.11 $s = u\mathscr{E}$ [21.42] with $\mathscr{E} = \dfrac{10\,\text{V}}{1.00\,\text{cm}} = 10\,\text{V cm}^{-1}$.

$s(\text{Li}^+) = (4.01 \times 10^{-4}\,\text{cm}^2\,\text{s}^{-1}\,\text{V}^{-1}) \times (10\,\text{V cm}^{-1}) = \boxed{4.0 \times 10^{-3}\,\text{cm s}^{-1}}$.

$s(\text{Na}^+) = (5.19 \times 10^{-4}\,\text{cm}^2\,\text{s}^{-1}\,\text{V}^{-1}) \times (10\,\text{V cm}^{-1}) = \boxed{5.2 \times 10^{-3}\,\text{cm s}^{-1}}$.

$s(\text{K}^+) = (7.62 \times 10^{-4}\,\text{cm}^2\,\text{s}^{-1}\,\text{V}^{-1}) \times (10\,\text{V cm}^{-1}) = \boxed{7.6 \times 10^{-3}\,\text{cm s}^{-1}}$.

$t = d/s$ with $d = 1.0\,\text{cm}$:

$t(\text{Li}^+) = \dfrac{1.0\,\text{cm}}{4.0 \times 10^{-3}\,\text{cm s}^{-1}} = \boxed{250\,\text{s}}, \quad t(\text{Na}^+) = \boxed{190\,\text{s}}, \quad t(\text{K}^+) = \boxed{130\,\text{s}}$.

(a) For the distance moved during a half-cycle, write

$$d = \int_0^{1/2\nu} s\,dt = \int_0^{1/2\nu} u\mathscr{E}\,dt = u\varepsilon_0 \int_0^{1/2\nu} \sin(2\pi\nu t)\,dt \quad [\mathscr{E} = \mathscr{E}_0 \sin(2\pi\nu t)]$$

$$= \dfrac{u\mathscr{E}_0}{\pi\nu} = \dfrac{u \times (10\,\text{V cm}^{-1})}{\pi \times (1.0 \times 10^3\,\text{s}^{-1})} \quad [\text{assume } \mathscr{E}_0 = 10\,\text{V}] = 3.18 \times 10^{-3} u\,\text{V s cm}^{-1}.$$

That is, $d/\text{cm} = (3.18 \times 10^{-3}) \times (u/\text{cm}^2\,\text{V}^{-1}\,\text{s}^{-1})$. Hence,

$$d(\text{Li}^+) = (3.18 \times 10^{-3}) \times (4.0 \times 10^{-4}\,\text{cm}) = \boxed{1.3 \times 10^{-6}\,\text{cm}},$$

$$d(\text{Na}^+) = \boxed{1.7 \times 10^{-6}\,\text{cm}}, \quad d(\text{K}^+) = \boxed{2.4 \times 10^{-6}\,\text{cm}}.$$

(b) These correspond to about $\boxed{43}$, $\boxed{55}$, and $\boxed{81}$ solvent molecule diameters respectively.

P21.13 $t = \dfrac{zcVF}{I\Delta t} = \dfrac{zcAFl}{I\Delta t}$ [21.52]

$$= \left(\dfrac{(21\,\text{mol m}^{-3}) \times (\pi) \times (2.073 \times 10^{-3}\,\text{m})^2 \times (9.6485 \times 10^4\,\text{C mol}^{-1})}{18.2 \times 10^{-3}\,\text{A}} \right) \times \left(\dfrac{l}{\Delta t} \right)$$

$$= (1.50 \times 10^3\,\text{m}^{-1}\,\text{s}) \times \left(\dfrac{l}{\Delta t} \right) = (1.50) \times \left(\dfrac{l/\text{mm}}{\Delta t/\text{s}} \right).$$

Then we draw up the following table.

$\Delta t/\text{s}$	200	400	600	800	1000
l/mm	64	128	192	254	318
t_+	0.48	0.48	0.48	0.48	0.48
$t_- = 1 - t_+$	0.52	0.52	0.52	0.52	0.52

Hence, we conclude that $t_+ = \boxed{0.48}$ and $t_- = \boxed{0.52}$. For the mobility of K^+ we use

$$t_+ = \frac{\lambda_+}{\Lambda_m^\circ} \text{ [21.50]} = \frac{u_+ F}{\Lambda_m^\circ} \text{ [21.44]}$$

to obtain

$$u_+ = \frac{t_+^\circ \Lambda_m^\circ}{F} = \frac{(0.48) \times (149.9 \, \text{S cm}^2 \, \text{mol}^{-1})}{9.6485 \times 10^4 \, \text{C mol}^{-1}} = \boxed{7.5 \times 10^{-4} \, \text{cm}^2 \, \text{s}^{-1} \, \text{V}^{-1}},$$

$$\lambda_+ = t_+ \Lambda_m^\circ \text{ [21.50]} = (0.48) \times (149.9 \, \text{S cm}^2 \, \text{mol}^{-1}) = \boxed{72 \, \text{S cm}^2 \, \text{mol}^{-1}}.$$

P21.15
$$\mathscr{F} = -\frac{RT}{c} \times \frac{dc}{dx} \text{ [21.58]}.$$

$$\frac{dc}{dx} = \frac{(0.05 - 0.10) \, \text{M}}{0.10 \, \text{m}} = -0.50 \, \text{M m}^{-1} \text{ [linear gradation]}.$$

$$RT = 2.48 \times 10^3 \, \text{J mol}^{-1} = 2.48 \times 10^3 \, \text{N m mol}^{-1}.$$

(a) $\mathscr{F} = \left(\dfrac{-2.48 \, \text{kN m mol}^{-1}}{0.10 \, \text{M}} \right) \times (-0.50 \, \text{M m}^{-1}) = \boxed{12 \, \text{kN mol}^{-1}}, \boxed{2.1 \times 10^{-20} \, \text{N molecule}^{-1}}.$

(b) $\mathscr{F} = \left(\dfrac{-2.48 \, \text{kN m mol}^{-1}}{0.075 \, \text{M}} \right) \times (-0.50 \, \text{M m}^{-1}) = \boxed{17 \, \text{kN mol}^{-1}}, \boxed{2.8 \times 10^{-20} \, \text{N molecule}^{-1}}.$

(c) $\mathscr{F} = \left(\dfrac{-2.48 \, \text{kN m mol}^{-1}}{0.05 \, \text{M}} \right) \times (-0.50 \, \text{M m}^{-1}) = \boxed{25 \, \text{kN mol}^{-1}}, \boxed{4.1 \times 10^{-20} \, \text{N molecule}^{-1}}.$

P21.17   If diffusion is analogous to viscosity [Section 21.5, eqn 21.26] in that it is also an activation energy controlled process, then we expect

$$D \propto e^{-E_a/RT}.$$

Therefore, if the diffusion constant is D at T and D' at T',

$$E_a = -\frac{R \ln \left(\dfrac{D'}{D} \right)}{\left(\dfrac{1}{T'} - \dfrac{1}{T} \right)} = -\frac{(8.314 \, \text{J K}^{-1} \, \text{mol}^{-1}) \times \ln \left(\dfrac{2.89}{2.05} \right)}{\dfrac{1}{298 \, \text{K}} - \dfrac{1}{273 \, \text{K}}} = 9.3 \, \text{kJ mol}^{-1}.$$

That is, the activation energy for diffusion is $\boxed{9.3 \, \text{kJ mol}^{-1}}$.

P21.19 $\langle x^2 \rangle = 2Dt$ [21.83], $\quad D = \dfrac{kT}{6\pi a \eta}$ [21.67].

Hence,

$$\eta = \frac{kT}{6\pi D a} = \frac{kTt}{3\pi a \langle x^2 \rangle} = \frac{1.381 \times 10^{-23} \, \text{J K}^{-1} \times (298.15 \, \text{K}) \times t}{(3\pi) \times (2.12 \times 10^{-7} \, \text{m}) \times \langle x^2 \rangle}$$

$$= \left(2.06 \times 10^{-15} \, \text{J m}^{-1} \right) \times \left(\frac{t}{\langle x^2 \rangle} \right)$$

and therefore $\eta/\left(\mathrm{kg\,m^{-1}\,s^{-1}}\right) = \dfrac{2.06 \times 10^{-11}\,(t/\mathrm{s})}{\left(\langle x^2 \rangle/\mathrm{cm}^2\right)}.$

We draw up the following table.

t/s	30	60	90	120
$10^8\langle x^2\rangle/\mathrm{cm}^2$		88.2	113.4	128	144
$10^3\eta/\left(\mathrm{kg\,m^{-1}s^{-1}}\right)$		0.701	1.09	1.45	1.72

Hence, the mean value is $\boxed{1.2 \times 10^{-3}\,\mathrm{kg\,m^{-1}\,s^{-1}}}$.

P21.21   The viscosity of a perfect gas is

$$\eta = \tfrac{1}{3}\mathcal{N}\,m\lambda\bar{c} = \frac{m\bar{c}}{3\sigma\sqrt{2}} = \frac{2}{3\sigma}\left(\frac{mkT}{\pi}\right)^{1/2} \quad \text{so} \quad \sigma = \frac{2}{3\eta}\left(\frac{mkT}{\pi}\right)^{1/2}.$$

The mass is

$$m = \frac{17.03 \times 10^{-3}\,\mathrm{kg\,mol^{-1}}}{6.022 \times 10^{23}\,\mathrm{mol^{-1}}} = 2.828 \times 10^{-26}\,\mathrm{kg}.$$

(a) $\sigma = \dfrac{2}{3\left(9.08 \times 10^{-6}\,\mathrm{kg\,m^{-1}\,s^{-1}}\right)}$

$$\times \left(\frac{\left(2.828 \times 10^{-26}\,\mathrm{kg}\right) \times \left(1.381 \times 10^{-23}\,\mathrm{J\,K^{-1}}\right) \times (270\,\mathrm{K})}{\pi}\right)^{1/2}$$

$$= 4.25 \times 10^{-19}\,\mathrm{m}^2 = \pi d^2 \quad \text{so} \quad d = \left(\frac{4.25 \times 10^{-19}\,\mathrm{m}^2}{\pi}\right)^{1/2} = \boxed{3.68 \times 10^{-10}\,\mathrm{m}}.$$

(b) $\sigma = \dfrac{2}{3\left(17.49 \times 10^{-6}\,\mathrm{kg\,m^{-1}\,s^{-1}}\right)}$

$$\times \left(\frac{\left(2.828 \times 10^{-26}\,\mathrm{kg}\right) \times \left(1.381 \times 10^{-23}\,\mathrm{J\,K^{-1}}\right) \times (490\,\mathrm{K})}{\pi}\right)^{1/2}$$

$$= 2.97 \times 10^{-19}\,\mathrm{m}^2 = \pi d^2 \quad \text{so} \quad d = \left(\frac{2.97 \times 10^{-19}\,\mathrm{m}^2}{\pi}\right)^{1/2} = \boxed{3.07 \times 10^{-10}\,\mathrm{m}}.$$

COMMENT.   The change in diameter with temperature can be interpreted in two ways. First, it shows the approximate nature of the concept of molecular diameter, with different values resulting from measurements of different quantities. Second, it is consistent with the idea that, at higher temperatures, more forceful collisions contract a molecule's perimeter.

Solutions to theoretical problems

P21.23 The most probable speed of a gas molecule corresponds to the condition that the Maxwell distribution be a maximum (it has no minimum); hence we find it by setting the first derivative of the function to zero and solve for the value of v for which this condition holds.

$$f(v) = 4\pi \left(\frac{m}{2\pi kT}\right)^{3/2} v^2 e^{-mv^2/2kT} = \text{const} \times v^2 e^{-mv^2/2kT} \quad \left[\frac{M}{R} = \frac{m}{k}\right].$$

$$\frac{df(v)}{ds} = 0 \quad \text{when} \quad \left(2 - \frac{mv^2}{kT}\right) = 0.$$

So, $\boxed{v \text{ (most probable)} = c^* = \left(\frac{2kT}{m}\right)^{1/2} = \left(\frac{2RT}{M}\right)^{1/2}}.$

The average kinetic energy corresponds to the average of $\frac{1}{2}mv^2$. The average is obtained by determining
$\langle v^2 \rangle = \int_0^\infty v^2 f(v) dv = 4\pi (m/2\pi)^{3/2} \times (1/kT)^{3/2} \int_0^\infty v^4 e^{-mv^2/2kT} \, dv.$

The integral evaluates to $(3/8)\pi^{1/2}(m/2kT)^{-5/2}$. Then

$$\langle v^2 \rangle = 4\pi \left(\frac{m}{2\pi}\right)^{3/2} \times \left(\frac{1}{kT}\right)^{3/2} \times \left(\frac{3}{8}\pi^{1/2}\right) \times \left(\frac{2kT}{m}\right)^{5/2} = \frac{3kT}{m};$$

thus $\langle \varepsilon \rangle = \frac{1}{2}m\langle v^2 \rangle = \frac{3}{2}kT.$

P21.25 Write the mean velocity initially as a; then in the emerging beam $\langle v_x \rangle = K \int_0^a v_x f(v_x) dv_x$ where K is a constant that ensures that the distribution in the emergent beam is also normalized. That is,
$1 = K \int_0^a f(v_x) dv_x = K (m/2\pi kT)^{1/2} \int_0^a e^{-mv_x^2/2kT} \, dv_x.$ This integral cannot be evaluated analytically but it can be related to the error function by defining

$$x^2 = \frac{mv_x^2}{2kT}$$

which gives $dv_x = (2kT/m)^{1/2} \, dx.$ Then

$$1 = K \left(\frac{m}{2\pi kT}\right)^{1/2} \left(\frac{2kT}{m}\right)^{1/2} \int_0^b e^{-x^2} \, dx \quad \left[b = (m/2kT)^{1/2} \times a\right]$$

$$= \frac{K}{\pi^{1/2}} \int_0^b e^{-x^2} \, dx = \frac{1}{2}K\text{erf}(b)$$

where erf (z) is the error function [Table 9.2]: erf $(z) = (2/\pi^{1/2}) \int_0^z e^{-x^2} \, dx.$

Therefore, $K = \dfrac{2}{\text{erf}(b)}.$

The mean velocity of the emerging beam is

$$\langle v_x \rangle = K \left(\frac{m}{2\pi kT}\right)^{1/2} \int_0^a v_x e^{-mv_x^2/2kT} dv_x = K \left(\frac{m}{2\pi kT}\right)^{1/2} \left(\frac{-kT}{m}\right) \int_0^a \frac{d}{dv_x} \left(e^{-mv_x^2/2kT} dv_x\right)$$

$$= -K \left(\frac{kT}{2m\pi}\right)^{1/2} \left(e^{-ma^2/2kT} - 1\right).$$

Now use $a = \langle v_x \rangle_{\text{initial}} = (2kT/m\pi)^{1/2}$.

This expression for the average magnitude of the one-dimensional velocity in the x direction may be obtained from

$$\langle v_x \rangle = 2 \int_0^\infty v_x f(v_x) dv_x = 2 \int_0^\infty v_x \left(\frac{m}{2\pi kT} \right)^{1/2} e^{-mv_x^2/2kT} dv_x$$

$$= \left(\frac{m}{2\pi kT} \right)^{1/2} \left(\frac{2kT}{m} \right) = \left(\frac{2kT}{m\pi} \right)^{1/2}.$$

It may also be obtained very quickly by setting $a = \infty$ in the expression for $\langle v_x \rangle$ in the emergent beam with $\text{erf}(b) = \text{erf}(\infty) = 1$.

Substituting $a = (2kT/m\pi)^{1/2}$ into $\langle v_x \rangle$ in the emergent beam, $e^{-ma^2/2kT} = e^{-1/\pi}$ and $\text{erf}(b) = \text{erf}(1/\pi^{1/2})$.

Therefore, $\langle v_x \rangle = \left(\dfrac{2kT}{m\pi} \right)^{1/2} \times \dfrac{1 - e^{-1/\pi}}{\text{erf}\left(\dfrac{1}{\pi^{1/2}} \right)}$.

From tables of the error function (expanded version of Table 9.2), or from readily available software, or by interpolating Table 9.2,

$$\text{erf}\left(\frac{1}{\pi^{1/2}} \right) = \text{erf}(0.56) = 0.57 \text{ and } e^{-1/\pi} = 0.73.$$

Therefore, $\langle v_x \rangle = \boxed{0.47 \langle v_x \rangle_{\text{initial}}}$.

P21.27   The most probable speed, c^*, was evaluated in Problem 21.23 and is

$$c^* = v(\text{most probable}) = \left(\frac{2kT}{m} \right)^{1/2}.$$

Consider a range of speeds Δv around c^* and nc^*; then, with $v = c^*$,

$$\frac{f(nc^*)}{f(c^*)} = \frac{(nc^*)^2 e^{-mn^2 c^{*2}/2kT}}{c^{*2} e^{-mc^{*2}/2kT}} \text{ [21.4]} = n^2 e^{-(n^2-1)mc^{*2}/2kT} = \boxed{n^2 e^{(1-n^2)}}.$$

Therefore, $\dfrac{f(3c^*)}{f(c^*)} = 9 \times e^{-8} = \boxed{3.02 \times 10^{-3}}$, $\dfrac{f(4c^*)}{f(c^*)} = 16 \times e^{-15} = \boxed{4.9 \times 10^{-6}}$.

P21.29   The current I_j carried by an ion j is proportional to its concentration c_j, mobility u_j, and charge number $|z_j|$ (Section 21.7). Therefore

$$I_j = Ac_j u_j z_j$$

where A is a constant. The total current passing through a solution is

$$I = \sum_j I_j = A \sum_j c_j u_j z_j.$$

The transport number of the ion j is therefore

$$t_j = \frac{I_j}{I} = \frac{Ac_j u_j z_j}{A \sum_j c_j u_j z_j} \frac{c_j u_j z_j}{\sum_j c_j u_j z_j}.$$

If there are two cations in the mixture

$$\frac{t'}{t''} = \boxed{\frac{c'u'z'}{c''u''z''}} = \frac{c'u'}{c''u''} \quad \text{if } z' = z''.$$

P21.31
$$p(x) = \frac{N!}{\left\{ \frac{1}{2}(N+s) \right\}! \left\{ \frac{1}{2}(N-s) \right\}! 2^N} \quad \text{[Justification 21.7]}, \quad s = \frac{x}{\lambda}.$$

$$p(6d) = \frac{N!}{\left\{ \frac{1}{2}(N+6) \right\}! \left\{ \frac{1}{2}(N-6) \right\}! 2^N}.$$

(a) $N = 4$, $p(6\lambda) = \boxed{0}$ $(m! = \infty$ for $m < 0)$.

(b) $N = 6$, $p(6\lambda) = \dfrac{6!}{6!0!2^6} = \dfrac{1}{2^6} = \dfrac{1}{64} = \boxed{0.0616}$.

(c) $N = 12$, $p(6\lambda) = \dfrac{12!}{9!3!2^{12}} = \dfrac{12 \times 11 \times 10}{3 \times 2 \times 2^{12}} = \boxed{0.054}$.

(NB $0! = 1$).

P21.33
$$P = \frac{N!}{\{\frac{1}{2}(N+n)\}!\{\frac{1}{2}(N-n)\}!2^N}.$$

The intermediate mathematical manipulations of Justification 21.7 begin with the above expression. Simplification of the expression proceeds by taking the natural logarithm of the expression, applying Stirling's approximation to each term that has the $\ln(x!)$ form, checking for term cancellations, and simplification using basic logarithm properties.

Stirling's approximation: $\ln x! = \ln(2\pi)^{1/2} + (x + \frac{1}{2}) \ln x - x$.

Basic logarithm properties: $\ln(x \times y) = \ln x + \ln y$

$$\ln(x/y) = \ln x - \ln y$$

$$\ln(x^y) = y \ln x$$

Taking the natural logarithm and applying Stirling's formula gives

$$\ln P = \ln \left\{ \frac{N!}{\{\frac{1}{2}(N+n)\}!\{\frac{1}{2}(N-n)\}!2^N} \right\}$$

$$= \ln N! - \ln(\{\frac{1}{2}(N+n)\}!) - \ln(\{\frac{1}{2}(N-n)\}!) - \ln 2^N$$

$$= \ln(2\pi)^{1/2} + (N + \frac{1}{2}) \ln N - N$$

$$- [\ln(2\pi)^{1/2} + \{\frac{1}{2}(N+n) + \frac{1}{2}\} \ln\{\frac{1}{2}(N+n)\} - \frac{1}{2}(N+n)]$$

$$- [\ln(2\pi)^{1/2} + \{\frac{1}{2}(N-n) + \frac{1}{2}\} \ln\{\frac{1}{2}(N-n)\} - \frac{1}{2}(N-n)] - \ln 2^N.$$

$$\ln P = (N + \tfrac{1}{2}) \ln N - \ln(2\pi)^{1/2} - \ln 2^N$$

$$- \{\tfrac{1}{2}(N + n) + \tfrac{1}{2}\} \ln\{\tfrac{1}{2}(N + n)\} - \{\tfrac{1}{2}(N - n) + \tfrac{1}{2}\} \ln\{\tfrac{1}{2}(N - n)\}$$

$$= \ln \left\{ \frac{(N/2)^{N+\frac{1}{2}}}{\pi^{1/2}} \right\} - \tfrac{1}{2}\{N + n + 1\} \ln \left\{ \frac{N}{2}\left(1 + \frac{n}{N}\right) \right\} - \tfrac{1}{2}\{N - n + 1\} \ln \left\{ \frac{N}{2}\left(1 - \frac{n}{N}\right) \right\}$$

$$= \ln \left\{ \frac{(N/2)^{N+\frac{1}{2}}}{\pi^{1/2}} \right\} - \tfrac{1}{2}\{N + n + 1\} \left\{ \ln \left(\frac{N}{2}\right) + \ln \left(1 + \frac{n}{N}\right) \right\}$$

$$- \tfrac{1}{2}\{N - n + 1\} \left\{ \ln \left(\frac{N}{2}\right) + \ln \left(1 - \frac{n}{N}\right) \right\}$$

$$= \ln \left\{ \frac{(N/2)^{N+\frac{1}{2}}}{\pi^{1/2}} \right\} - \tfrac{1}{2}\{N + \cancel{n} + 1\} \ln \left(\frac{N}{2}\right) - \tfrac{1}{2}\{N + n + 1\} \ln \left(1 + \frac{n}{N}\right)$$

$$- \tfrac{1}{2}\{N - \cancel{n} + 1\} \ln \left(\frac{N}{2}\right) - \tfrac{1}{2}\{N - n + 1\} \ln \left(1 - \frac{n}{N}\right)$$

$$= \ln \left\{ \frac{(N/2)^{N+\frac{1}{2}}}{\pi^{1/2}} \right\} - \{N + 1\} \ln \left(\frac{N}{2}\right) - \tfrac{1}{2}\{N + n + 1\} \ln \left(1 + \frac{n}{N}\right)$$

$$- \tfrac{1}{2}\{N - n + 1\} \ln \left(1 - \frac{n}{N}\right)$$

$$= \ln \left\{ \frac{(N/2)^{N+\frac{1}{2}}}{(N/2)^{N+1}\pi^{1/2}} \right\} - \tfrac{1}{2}\{N + n + 1\} \ln \left(1 + \frac{n}{N}\right) - \tfrac{1}{2}\{N - n + 1\} \ln \left(1 - \frac{n}{N}\right).$$

$$\ln P = \ln \left(\frac{2}{\pi N}\right)^{1/2} - \tfrac{1}{2}\{N + n + 1\} \ln \left(1 + \frac{n}{N}\right) - \tfrac{1}{2}\{N - n + 1\} \ln \left(1 - \frac{n}{N}\right).$$

Solutions to applications

P21.35 The work required for a mass, m, to go from a distance r from the center of a planet of mass m' to infinity is

$$w = \int_r^\infty F \, dr$$

where F is the force of gravity and is given by Newton's law of universal gravitation, which is

$$F = \frac{Gmm'}{r^2}.$$

G is the gravitational constant (not to be confused with g). Then

$$w' = \int_r^\infty \frac{Gmm'}{r^2} \, dr = \frac{Gmm'}{r}.$$

Since, according to Newton's second law of motion, $F = mg$, we may make the identification

$$g = \frac{Gm'}{r^2}.$$

Thus, $w = grm$. This is the kinetic energy that the particle must have in order to escape the planet's gravitational attraction at a distance r from the planet's center; hence $w = \frac{1}{2}mv^2 = mgr$.

$$v_e = (2g R_p)^{1/2} \quad [R_p = \text{radius of planet}]$$

which is the escape velocity.

(a) $v_e = [(2) \times (9.81 \,\mathrm{m\,s^{-2}}) \times (6.37 \times 10^6 \,\mathrm{m})]^{1/2} = \boxed{11.2 \,\mathrm{km\,s^{-1}}}$

(b) $g(\text{Mars}) = \dfrac{m(\text{Mars})}{m(\text{Earth})} \times \dfrac{R(\text{Earth})^2}{R(\text{Mars})^2} \times g(\text{Earth}) = (0.108) \times \left(\dfrac{6.37}{3.38}\right)^2 \times (9.81 \,\mathrm{m\,s^{-2}})$

$$= 3.76 \,\mathrm{m\,s^{-2}}.$$

Hence, $v_e = [(2) \times (3.76 \,\mathrm{m\,s^{-2}}) \times (3.38 \times 10^6 \,\mathrm{m})]^{1/2} = \boxed{5.0 \,\mathrm{km\,s^{-1}}}$.

Since $\bar{c} = (8RT/\pi M)^{1/2}$, $T = \pi M \bar{c}^2 / 8R$

and we can draw up the following table.

$10^{-3}T/K$	H_2	He	O_2	
Earth	11.9	23.7	190	$[\bar{c} = 11.2 \,\mathrm{km\,s^{-1}}]$
Mars	2.4	4.8	38	$[\bar{c} = 5.0 \,\mathrm{km\,s^{-1}}]$

In order to calculate the proportion of molecules that have speeds exceeding the escape velocity, v_e, we must integrate the Maxwell distribution [21.4] from v_e to infinity.

$$P = \int_{v_e}^{\infty} f(v)dv = \int_{v_e}^{\infty} 4\pi \left(\frac{m}{2\pi kT}\right)^{3/2} v^2 e^{-mv^2/2kT} dv \quad \left[\frac{M}{R} = \frac{m}{k}\right]$$

This integral cannot be evaluated analytically and must be expressed in terms of the error function. We proceed as follows.

Defining $\beta = m/2kT$ and $y^2 = \beta v^2$ gives $v = \beta^{-1/2}y$, $v^2 = \beta^{-1}y^2$, $v_e = \beta^{-1/2}y_e$,

$y_e = \beta^{1/2}v_e$, and $dv = \beta^{-1/2}dy$.

$$P = 4\pi \left(\frac{\beta}{\pi}\right)^{3/2} \beta^{-1}\beta^{-1/2} \int_{\beta^{1/2}v_e}^{\infty} y^2 e^{-y^2} dy = \frac{4}{\pi^{1/2}} \int_{\beta^{1/2}v_e}^{\infty} y^2 e^{-y^2} dy$$

$$= \frac{4}{\pi^{1/2}} \left[\int_{0}^{\infty} y^2 e^{-y^2} dy - \int_{0}^{\beta^{1/2}v_e} y^2 e^{-y^2} dy\right].$$

The first integral can be evaluated analytically; the second cannot.

$$\int_{0}^{\infty} y^2 e^{-y^2} dy = \frac{\pi^{1/2}}{4}; \text{ hence}$$

$$P = 1 - \frac{2}{\pi^{1/2}} \int_{0}^{\beta^{1/2}v_e} y e^{-y^2} (2y dy) = 1 - \frac{2}{\pi^{1/2}} \int_{0}^{\beta^{1/2}v_e} y d(-e^{-y^2}).$$

This integral may be evaluated by parts

$$P = 1 - \frac{2}{\pi^{1/2}} \left[y(-e^{-y^2}) \Big|_0^{\beta^{1/2}v_e} - \int_0^{\beta^{1/2}v_e} (-e^{-y^2}) dy \right].$$

$$P = 1 + 2 \left(\frac{\beta}{\pi} \right)^{1/2} v_e e^{-\beta v_e^2} - \frac{2}{\pi^{1/2}} \int_0^{\beta^{1/2}v_e} e^{-y^2} dy = 1 + 2 \left(\frac{\beta}{\pi} \right)^{1/2} v_e e^{-\beta v_e^2} - \text{erf}(\beta^{1/2}v_e)$$

$$= \text{erfc}(\beta^{1/2}v_e) + 2 \left(\frac{\beta}{\pi} \right)^{1/2} v_e e^{-\beta v_e^2} \quad [\text{erfc}(z) = 1 - \text{erf}(z)].$$

From $\beta = \dfrac{m}{2kT} = \dfrac{M}{2RT}$ and $v_e = (2gR_p)^{1/2}$,

$$\beta^{1/2}v_e = \left(\frac{MgR_p}{RT} \right)^{1/2}.$$

For H_2 on Earth at 240 K

$$\beta^{1/2}v_e = \left(\frac{(0.002016 \,\text{kg mol}^{-1}) \times (9.807 \,\text{m s}^{-2}) \times (6.37 \times 10^6 \,\text{m})}{(8.314 \,\text{J K}^{-1} \,\text{mol}^{-1}) \times (240 \,\text{K})} \right)^{1/2} = 7.94,$$

$$P = \text{erfc}(7.94) + 2 \left(\frac{7.94}{\pi^{1/2}} \right) e^{-(7.94)^2} = (2.9 \times 10^{-29}) + (3.7 \times 10^{-27}) = \boxed{3.7 \times 10^{-27}}.$$

At 1500 K

$$\beta^{1/2}v_e = \left(\frac{(0.002016 \,\text{kg mol}^{-1}) \times (9.807 \,\text{m s}^{-2}) \times (6.37 \times 10^6 \,\text{m})}{(8.314 \,\text{J K}^{-1} \,\text{mol}^{-1}) \times (1500 \,\text{K})} \right)^{1/2} = 3.18,$$

$$P = \text{erfc}(3.18) + 2 \left(\frac{3.18}{\pi^{1/2}} \right) e^{-(3.18)^2} = (6.9 \times 10^{-6}) + (1.4\bar{6} \times 10^{-4}) = \boxed{1.5 \times 10^{-4}}.$$

For H_2 on Mars at 240 K

$$\beta^{1/2}v_e = \left(\frac{(0.002016 \,\text{kg mol}^{-1}) \times (3.76 \,\text{m s}^{-2}) \times (3.38 \times 10^6 \,\text{m})}{(8.314 \,\text{J K}^{-1} \,\text{mol}^{-1}) \times (240 \,\text{K})} \right)^{1/2} = 3.58,$$

$$P = \text{erfc}(3.58) + 2 \left(\frac{3.58}{\pi^{1/2}} \right) e^{-(3.58)^2} = (4.13 \times 10^{-7}) + (1.1\bar{0} \times 10^{-5}) = \boxed{1.1 \times 10^{-5}}.$$

At 1500 K, $\beta^{1/2}v_e = 1.43$,

$$P = \text{erfc}(1.43) + (1.128) \times (1.43) \times e^{-(1.43)^2} = 0.0431 + 0.20\bar{9} = \boxed{0.25}.$$

For He on Earth at 240 K

$$\beta^{1/2}v_e = \left(\frac{(0.004003 \,\text{kg mol}^{-1}) \times (9.807 \,\text{m s}^{-2}) \times (6.37 \times 10^6 \,\text{m})}{(8.314 \,\text{J K}^{-1} \,\text{mol}^{-1}) \times (240 \,\text{K})} \right)^{1/2} = 11.1\bar{9},$$

$$P = \text{erfc}(11.2) + (1.128) \times (11.2) \times e^{-(11.2)^2} = 0 + (4 \times 10^{-54}) = \boxed{4 \times 10^{-54}}.$$

At 1500 K, $\beta^{1/2}v_e = 4.48$,

$$P = \text{erfc}(4.48) + (1.128) \times (4.48) \times e^{-(4.48)^2} = (2.36 \times 10^{-10}) + (9.7\bar{1} \times 10^{-9})$$

$$= \boxed{1.0 \times 10^{-8}}.$$

For He on Mars at 240 K

$$\beta^{1/2}v_e = \left(\frac{(0.004003\,\text{kg mol}^{-1}) \times (3.76\,\text{m s}^{-2}) \times (3.38 \times 10^6\,\text{m})}{(8.314\,\text{J K}^{-1}\,\text{mol}^{-1}) \times (240\,\text{K})}\right)^{1/2} = 5.05,$$

$$P = \text{erfc}(5.05) + (1.128) \times (5.05) \times e^{-(5.05)^2} = (9.21 \times 10^{-13}) + (4.7\bar{9} \times 10^{-11})$$

$$= \boxed{4.9 \times 10^{-11}}.$$

At 1500 K, $\beta^{1/2}v_e = 2.02$,

$$P = \text{erfc}(2.02) + (1.128) \times (2.02) \times e^{-(2.02)^2} = (4.28 \times 10^{-3}) + (0.040\bar{1}) = \boxed{0.444}.$$

For O_2 on Earth it is clear that $P \approx 0$ at both temperatures.

For O_2 on Mars at 240 K, $\beta^{1/2}v_e = 14.3$,

$$P = \text{erfc}(14.3) + (1.128) \times (14.3) \times e^{-(14.3)^2} = 0 + (2.5 \times 10^{-88}) = \boxed{2.5 \times 10^{-88}} \approx 0.$$

At 1500 K, $\beta^{1/2}v_e = 5.71$,

$$P = \text{erfc}(5.71) + (1.128) \times (5.71) \times e^{-(5.71)^2} = (6.7 \times 10^{-6}) + (4.46 \times 10^{-14})$$

$$= \boxed{4.5 \times 10^{-14}}.$$

Based on these numbers alone, it would appear that H_2 and He would be depleted from the atmosphere of both Earth and Mars only after many (millions?) years; that the rate on Mars, though still slow, would be many orders of magnitude larger than on Earth; that O_2 would be retained on Earth indefinitely; and that the rate of O_2 depletion on Mars would be very slow (billions of years?), though not totally negligible. The temperatures of both planets may have been higher in past times than they are now.

In the analysis of the data, we must remember that the proportions, P, are not rates of depletion, though the rates should be roughly proportional to P.

The results of the calculations are summarized in the following table.

	240 K			1500 K		
	H_2	He	O_2	H_2	He	O_2
P(Earth)	3.7×10^{-27}	4×10^{-54}	0	1.5×10^{-4}	1.0×10^{-8}	0
P(Mars)	1.1×10^{-5}	4.9×10^{-11}	0	0.25	0.044	4.5×10^{14}

P21.37 Dry atmospheric air is 78.08% N_2, 20.95% O_2, 0.93% Ar, 0.03% CO_2, plus traces of other gases. Nitrogen, oxygen, and carbon dioxide contribute 99.06% of the molecules in a volume with each molecule contributing an average rotational energy equal to kT. The rotational energy density is given by

$$\rho_R = \frac{E_R}{V} = \frac{0.9906N(\varepsilon^R)}{V} = \frac{0.9906(\varepsilon^R)pN_A}{RT}$$

$$= \frac{0.9906kT \, pN_A}{RT} = 0.9906p$$

$$= 0.9906(1.013 \times 10^5 \, \text{Pa}) = 0.1004 \, \text{J cm}^{-1}.$$

The total energy density (translational plus rotational) is

$$\rho_T = \rho_K + \rho_R = 0.15 \, \text{J cm}^{-3} + 0.10 \, \text{J cm}^{-3}$$

$$\rho_T = 0.25 \, \text{J cm}^{-3}.$$

P21.39 For order of magnitude calculations we restrict our assumed values to powers of 10 of the base units. Thus

$$\rho = 1 \, \text{g cm}^{-3} = 1 \times 10^3 \, \text{kg m}^{-3},$$

$$\eta(\text{air}) = 1 \times 10^{-5} \, \text{kg m}^{-1} \, \text{s}^{-1} \text{ [see comment and question below]}.$$

We need the diffusion constant

$$D = \frac{kT}{6\pi\eta a}.$$

a is calculated from the volume of the virus which is assumed to be spherical

$$V = \frac{m}{\rho} \approx \frac{(1 \times 10^5 \, \text{u}) \times (1 \times 10^{-27} \, \text{kg u}^{-1})}{1 \times 10^3 \, \text{kg m}^3} \approx 1 \times 10^{-25} \, \text{m}^3.$$

$$V = \tfrac{4}{3}\pi a^3.$$

$$a \approx \left(\frac{V}{4}\right)^{1/3} \approx \left(\frac{1 \times 10^{-25} \, \text{m}^3}{4}\right)^{1/3} \approx 1 \times 10^{-8} \text{m}.$$

$$D \approx \left(\frac{(1 \times 10^{-23} \, \text{J K}^{-1}) \times (300 \, \text{K})}{(6\pi) \times (1 \times 10^{-5} \, \text{kg m}^{-1} \, \text{s}^{-1})(1 \times 10^{-5}\text{m})}\right) \approx 1 \times 10^{-9} \, \text{m}^2 \, \text{s}^{-1}.$$

For three-dimensional diffusion,

$$t = \frac{\langle r^2 \rangle}{6D} \approx \frac{1 \, \text{m}^2}{1 \times 10^{-8} \, \text{m}^2 \, \text{s}^{-1}} \approx \boxed{10^8 \, \text{s}}.$$

Therefore it does not seem likely that a cold could be caught by the process of diffusion.

COMMENT. In a Fermi calculation only those values of physical quantities that can be determined by scientific common sense should be used. Perhaps the value for $\eta(\text{air})$ used above does not fit that description.

Question. Can you obtain the value of $\eta(\text{air})$ by a Fermi calculation based on the relation in Table 21.3?

P21.41 $c(x,t) = c_0 + (c_s - c_0)\{1 - \text{erf}(\xi)\}$ where $\xi(x,t) = \dfrac{x}{(4Dt)^{1/2}}$.

In order for $c(x,t)$ to be the correct solution of this diffusion problem it must satisfy the boundary condition, the initial condition, and the diffusion equation (eqn 21.68). According to *Justification 9.4*,

$$\text{erf}(\xi) = 1 - \frac{2}{\pi^{1/2}} \int_{\xi}^{\infty} e^{y^2}\,dy.$$

At the boundary $x = 0$, $\xi = 0$, and $\text{erf}(0) = 1 - (2/\pi^{1/2}) \int_0^{\infty} e^{-y^2}\,dy = 1 - \left(2/\pi^{1/2}\right) \times \left(\pi^{1/2}/2\right) = 0$.

Thus, $c(0,t) = c_0 + (c_s - c_0)\{1 - 0\} = c_s$. The boundary condition is satisfied. At the initial time $(t = 0)$, $\xi(x,0) = \infty$ and $\text{erf}(\infty) = 1$. Thus, $c(x,0) = c_0 + (c_s - c_0)\{1 - 1\} = c_0$. The initial condition is satisfied. We must find the analytical forms for $\partial c/\partial t$ and $\partial^2 c/\partial x^2$. If they are proportional with a constant of proportionality equal to D, $c(x,t)$ satisfies the diffusion equation.

$$\frac{\partial c(x,t)}{\partial x} = D\left[\frac{1}{2}\frac{(c_s - c_0)x}{\sqrt{\pi}(Dt)^{3/2}}e^{-x^2/4Dt}\right].$$

$$\frac{\partial^2 c(x,t)}{\partial x^2} = \left[\frac{1}{2}\frac{(c_s - c_0)x}{\sqrt{\pi}(Dt)^{3/2}}e^{-x^2/4Dt}\right].$$

The constant of proportionality between the partials equals D and we conclude that the suggested solution satisfies the diffusion equation.

Diffusion through alveoli sites (about 1 cell thick) of oxygen and carbon dioxide between lungs and blood capillaries (also about 1 cell thick) occurs through about 0.075 mm (the diameter of a red blood cell). So we will examine diffusion profiles for $0 \le x \le 0.1$ mm. The largest distance suggests that the

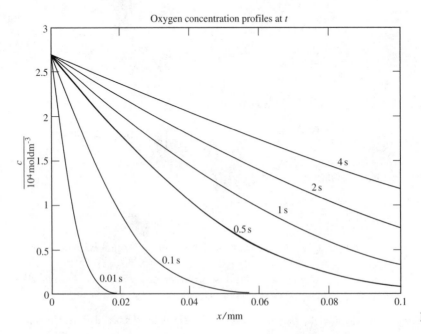

Oxygen concentration profiles at t

x/mm

Figure 21.2

longest time that must be examined is estimated with eqn 21.82.

$$t_{max} \simeq \frac{\pi x_{max}^2}{4D} = \frac{\pi (1 \times 10^{-4}\,\text{m})^2}{4(2.10 \times 10^{-9}\,\text{m}^2\,\text{s}^{-1})} = 3.74\,\text{s}.$$

Figure 21.2 shows oxygen concentration distributions for times between 0.01 s and 4.0 s.

Illustration 5.2 uses Henry's law to show that the equilibrium concentration of oxygen in water equals 2.9×10^{-4} mol dm^{-3}. We use this as an estimate for c_s and take c_0 to equal zero.

22 The rates of chemical reactions

Answers to discussion questions

D22.1 The timescales of atomic processes are rapid indeed: according to the following table, a nanosecond is an eternity. Note that the times given here are in some way typical values for times that may vary over two or three orders of magnitude. For example, vibrational wavenumbers can range from about 4400 cm^{-1} (for H$_2$) to 100 cm^{-1} (for I$_2$) and even lower, with a corresponding range of associated times. Radiative decay rates of electronic states can vary even more widely: Times associated with phosphorescence can be in the millisecond and even second range. A large number of timescales for physical, chemical, and biological processes on the atomic and molecular scale are reported in Figure 2 of A.H. Zewail, *Femtochemistry: atomic-scale dynamics of the chemical bond. J. Phys. Chem. A* **104**, 5660 (2000).

Process	t/ns	Reference
Radiative decay of electronic excited state	1×10^{1}	Section 13.3b
Rotational motion	3×10^{-2}	$B \approx 1$ cm^{-1}
Vibrational motion	3×10^{-5}	$\tilde{\nu} \approx 1000$ cm^{-1}
Proton transfer (in water)	2×10^{-5}	Section 21.7a
Initial chemical reaction of vision*	1×10^{-4}	Impact I14.1
Energy transfer in photosynthesis†	1×10^{-3}	Impact I23.2
Electron transfer in photosynthesis	3×10^{-3}	Impact I23.2
Polypeptide helix–coil transition	2×10^{2}	Impact I22.1
Collision frequency in liquids	4×10^{-4}	Section 21.1b‡

*Photoisomerization of retinal from 11-*cis* to all-*trans*.
†Time from absorption until electron transfer to adjacent pigment.
‡Use formula for gas collision frequency at 300 K, parameters for benzene from *Data section*, and density of liquid benzene.

Radiative decay of excited electronic states can range from about 10^{-9} s to 10^{-4} s—even longer for phosphorescence involving 'forbidden' decay paths. Molecular rotational motion takes place on a scale of 10^{-12} to 10^{-9} s. Molecular vibrations are faster still, about 10^{-14} to 10^{-12} s. The mean time between collisions in liquids is similarly short, 10^{-14} to 10^{-13} s. Proton transfer reactions occur on a timescale of about 10^{-10} to 10^{-9} s. *Impact* I14.1 describes several events in vision, including the 200-fs photoisomerization that gets the process started. *Impact* I23.2 lists timescales of several energy-transfer and electron-transfer steps in photosynthesis. Initial energy transfer (to a nearby pigment) has a timescale of around 10^{-13} to 10^{-11} s, with longer-range transfer (to the reaction center) taking about 10^{-10} s. Immediate electron transfer is also very fast (about 3 ps), with ultimate transfer (leading to oxidation of water and reduction of plastoquinone) taking from 10^{-10} to 10^{-3} s. *Impact* I22.1 discusses helix–coil transitions, including experimental measurements of timescales of tens or hundreds of microseconds (10^{-5} to 10^{-4} s) for formation of tightly packed cores. The rate-determining step for the helix–coil transition of small poly-peptides has a relaxation time of about 160 ns in contrast to the faster 50 ns relaxation time of large protein.

D22.3   The determination of a rate law is simplified by the isolation method in which the concentrations of all the reactants except one are in large excess. If B is in large excess, for example, then to a good approximation its concentration is constant throughout the reaction. Although the true rate law might be $v = k[A][B]$, we can approximate [B] by $[B]_0$ and write

$$v = k'[A], \quad \text{where } k' = k[B]_0 \text{ [22.10]}$$

which has the form of a first-order rate law. Because the true rate law has been forced into first-order form by assuming that the concentration of B is constant, it is called a pseudo first-order rate law. The dependence of the rate on the concentration of each of the reactants may be found by isolating them in turn (by having all the other substances present in large excess) and so constructing a picture of the overall rate law.

In the method of initial rates, which is often used in conjunction with the isolation method, the rate is measured at the beginning of the reaction for several different initial concentrations of reactants. We shall suppose that the rate law for a reaction with A isolated is $v = k[A]^a$; then its initial rate, v_0 is given by the initial values of the concentration of A, and we write $v_0 = k[A]_0^a$. Taking logarithms gives

$$\log v_0 = \log k + a \log[A]_0. \text{ [22.11]}$$

For a series of initial concentrations, a plot of the logarithms of the initial rates against the logarithms of the initial concentrations of A should be a straight line with slope a.

The method of initial rates might not reveal the full rate law, for the products may participate in the reaction and affect the rate. For example, products participate in the synthesis of HBr, where the full rate law depends on the concentration of HBr. To avoid this difficulty, the rate law should be fitted to the data throughout the reaction. The fitting may be done, in simple cases at least, by using a proposed rate law to predict the concentration of any component at any time, and comparing it with the data.

Because rate laws are differential equations, we must integrate them if we want to find the concentrations as a function of time. Even the most complex rate laws may be integrated numerically. However, in a number of simple cases analytical solutions are easily obtained, and prove to be very useful. These are summarized in Table 22.3. In order to determine the rate law, one plots the right-hand side of the integrated rate laws shown in the table against t in order to see which of them results in a straight line through the origin. The one that does is the correct rate law.

D22.5   The rate-determining step is not just the slowest step: it must be slow *and* be a crucial gateway for the formation of products. If a faster reaction can also lead to products, then the slowest step is irrelevant because the slow reaction can then be side-stepped. The rate-determining step is like a slow ferry crossing between two fast highways: the overall rate at which traffic can reach its destination is determined by the rate at which it can make the ferry crossing.

If the first step in a mechanism is the slowest step with the highest activation energy, then it is rate-determining, and the overall reaction rate is equal to the rate of the first step because all subsequent steps are so fast that once the first intermediate is formed it results immediately in the formation of products. Once over the initial barrier, the intermediates cascade into products. However, a rate-determining step may also stem from the low concentration of a crucial reactant or catalyst and need not correspond to the step with highest activation barrier. A rate-determining step arising from the low activity of a crucial enzyme can sometimes be identified by determining whether or not the reactants and products for that step are in equilibrium: if the reaction is not at equilibrium it suggests that the step may be slow enough to be rate-determining.

D22.7

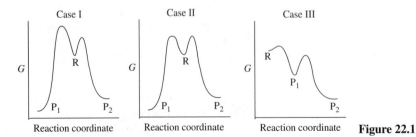

Figure 22.1

Simple diagrams of Gibbs energy against reaction coordinate are useful for distinguishing between kinetic and thermodynamic control of a reaction. For the simple parallel reactions $R \rightarrow P_1$ and $R \rightarrow P_2$, shown in Fig. 22.1 cases I and II, the product P_1 is thermodynamically favored because the Gibbs energy decreases to a greater extent for its formation. However, the rate at which each product appears does not depend upon thermodynamic favorability. Rate constants depend upon activation energy. In case I the activation energy for the formation of P_1 is much larger than that for formation of P_2. At low and moderate temperature the large activation energy may not be readily available and P_1 either cannot form or forms at a slow rate. The much smaller activation energy for P_2 formation is available and, consequently, P_2 is produced even though it is not the thermodynamically favored product. This is kinetic control. In this case, $[P_2]/[P_1] = k_2/k_1 > 1$ [22.46].

The activation energies for the parallel reactions are equal in case II and, consequently, the two products appear at identical rates. If the reactions are irreversible, $[P_2]/[P_1] = k_2/k_1 = 1$ at all times. The results are very different for reversible reactions. The activation energy for $P_1 \rightarrow R$ is much larger than that for $P_2 \rightarrow R$ and P_1 accumulates as the more rapid $P_2 \rightarrow R \rightarrow P_1$ occurs. Eventually the ratio $[P_2]/[P_1]$ approaches the equilibrium value for which

$$\left(\frac{[P_2]}{[P_1]} \right)_{eq} = e^{-(\Delta G_2 - \Delta G_1)/RT} < 1.$$

This is thermodynamic control.

Case III represents an interesting consecutive reaction series $R \rightarrow P_1 \rightarrow P_2$. The first step has relatively low activation energy and P_1 rapidly appears. However, the relatively large activation energy for the second step is not available at low and moderate temperatures. By using low or moderate temperatures and short reaction times it is possible to produce more of the thermodynamically less favorable P_1. This is kinetic control. High temperatures and long reaction times will yield the thermodynamically favored P_2.

The ratio of reaction products is determined by relative reaction rates in kinetic controlled reactions. Favorable conditions include short reaction times, lower temperatures, and irreversible reactions. Thermodynamic control is favored by long reaction times, higher temperatures, and reversible reactions. The ratio of products depends on the relative stability of products for thermodynamically controlled reactions.

D22.9   The primary isotope effect is the change in rate constant of a reaction in which the breaking of a bond involving the isotope occurs. The reaction coordinate in a C—H bond breaking process corresponds to the stretching of that bond. The vibrational energy of the stretching depends upon the effective mass of the C and H atoms. See eqn 13.50. Upon deuteration, the zero point energy of the bond is lowered due to the greater mass of the deuterium atom. However, the height of the energy barrier is not much changed because the relevant vibration in the activated complex has a very low force constant (bonding in the complex is very weak), so there is little zero point energy associated with the complex and little change in its zero point energy upon deuteration. The net effect is an increase in the activation energy of the reaction. We then expect that the rate constant for the reaction will be lowered in the deuterated molecule and that is what is observed. See the derivation leading to eqns 22.51–22.53 for a quantitative description of the effect.

A secondary kinetic isotope effect is the reduction in the rate of a reaction involving the bonded isotope even though the bond is not broken in the reaction. The cause is again related to the change in zero point energy that occurs upon replacement of an atom with its isotope, but in this case it arises from the differences in zero point energies between reactants and an activated complex with significantly different structure. See *Illustration* 22.3 for an example of the estimation of the magnitude of the effect in a heterolytic dissociation reaction.

If the rate of a reaction is altered by isotopic substitution it implies that the substituted site plays an important role in the mechanism of the reaction. For example, an observed effect on the rate can identify bond breaking events in the rate-determining step of the mechanism. On the other hand, if no isotope effect is observed, the site of the isotopic substitution may play no critical role in the mechanism of the reaction.

Solutions to exercises

E22.1(b) $v = -\dfrac{d[A]}{dt} = -\dfrac{1}{3}\dfrac{d[B]}{dt} = \dfrac{d[C]}{dt} = \dfrac{1}{2}\dfrac{d[D]}{dt} = 1.00\,\text{mol dm}^{-3}\,\text{s}^{-1}$, so

Rate of consumption of A $= \boxed{1.0\,\text{mol dm}^{-3}\,\text{s}^{-1}}$

Rate of consumption of B $= \boxed{3.0\,\text{mol dm}^{-3}\,\text{s}^{-1}}$

Rate of formation of C $= \boxed{1.0\,\text{mol dm}^{-3}\,\text{s}^{-1}}$

Rate of formation of D $= \boxed{2.0\,\text{mol dm}^{-3}\,\text{s}^{-1}}$

E22.2(b)   Rate of consumption of B $= -\dfrac{d[B]}{dt} = \boxed{1.00\,\text{mol dm}^{-3}\,\text{s}^{-1}}$

Rate of reaction $= -\dfrac{1}{3}\dfrac{d[B]}{dt} = \boxed{0.33\,\text{mol dm}^{-3}\,\text{s}^{-1}} = \dfrac{d[C]}{dt} = \dfrac{1}{2}\dfrac{d[D]}{dt} = -\dfrac{d[A]}{dt}$

Rate of formation of C $= \boxed{0.33\,\text{mol dm}^{-3}\,\text{s}^{-1}}$

Rate of formation of D $= \boxed{0.66\,\text{mol dm}^{-3}\,\text{s}^{-1}}$

Rate of consumption of A $= \boxed{0.33\,\text{mol dm}^{-3}\,\text{s}^{-1}}$

E22.3(b)   The dimensions of k are

$$\frac{\text{dim of } v}{(\text{dim of } [A]) \times (\text{dim of } [B])^2} = \frac{\text{amount} \times \text{length}^{-3} \times \text{time}^{-1}}{(\text{amount} \times \text{length}^{-3})^3}$$

$$= \text{length}^6 \times \text{amount}^{-2} \times \text{time}^{-1}$$

In mol, dm, s units, the units of k are $\boxed{\text{dm}^6 \text{ mol}^{-2} \text{ s}^{-1}}$

(a) $v = -\dfrac{d[A]}{dt} = k[A][B]^2$ so $\boxed{\dfrac{d[A]}{dt} = -k[A][B]^2}$

(b) $v = \dfrac{d[C]}{dt}$ so $\boxed{\dfrac{d[C]}{dt} = k[A][B]^2}$

E22.4(b)   The dimensions of k are

$$\frac{\text{dim of } v}{\text{dim of } [A] \times \text{dim of } [B] \times (\text{dim of } [C])^{-1}} = \frac{\text{amount} \times \text{length}^{-3} \times \text{time}^{-1}}{\text{amount} \times \text{length}^{-3}} = \text{time}^{-1}$$

The units of k are $\boxed{\text{s}^{-1}}$

$$v = \frac{d[C]}{dt} = \boxed{k[A][B][C]^{-1}}$$

E22.5(b)   The rate law is

$$v = k[A]^a \propto p^a = \{p_0(1-f)\}^a$$

where a is the reaction order, and f the fraction reacted (so that $1-f$ is the fraction remaining). Thus

$$\frac{v_1}{v_2} = \frac{\{p_0(1-f_1)\}^a}{\{p_0(1-f_2)\}^a} = \left(\frac{1-f_1}{1-f_2}\right)^a \quad \text{and} \quad a = \frac{\ln(v_1/v_2)}{\ln\left(\dfrac{1-f_1}{1-f_2}\right)} = \frac{\ln(9.71/7.67)}{\ln\left(\dfrac{1-0.100}{1-0.200}\right)} = \boxed{2.00}$$

E22.6(b)   The half-life changes with concentration, so we know the reaction order is not 1. That the half-life increases with decreasing concentration indicates a reaction order <1. Inspection of the data shows the half-life roughly proportional to concentration, which would indicate a reaction order of 0 according to Table 22.3. More quantitatively, if the reaction order is 0, then

$$t_{1/2} \propto p \quad \text{and} \quad \frac{t_{1/2}^{(1)}}{t_{1/2}^{(2)}} = \frac{p_1}{p_2}$$

We check to see if this relationship holds

$$\frac{t_{1/2}^{(1)}}{t_{1/2}^{(2)}} = \frac{340 \text{ s}}{178 \text{ s}} = 1.91 \quad \text{and} \quad \frac{p_1}{p_2} = \frac{55.5 \text{ kPa}}{28.9 \text{ kPa}} = 1.92$$

so the reaction order is $\boxed{0}$.

E22.7(b)   The rate law is

$$v = -\frac{1}{2}\frac{d[A]}{dt} = k[A]$$

The half-life formula in eqn 22.13 is based on the assumption that

$$-\frac{d[A]}{dt} = k[A].$$

That is, it would be accurate to take the half-life from the table and say

$$t_{1/2} = \frac{\ln 2}{k'}$$

where $k' = 2k$. Thus

$$t_{1/2} = \frac{\ln 2}{2(2.78 \times 10^{-7}\,\text{s}^{-1})} = \boxed{1.80 \times 10^6\,\text{s}}$$

Likewise, we modify the integrated rate law (eqn 22.12b), noting that pressure is proportional to concentration:

$$p = p_0 e^{-2kt}$$

(a) Therefore, after 10 h, we have

$$p = (32.1\,\text{kPa})\exp[-2 \times (2.78 \times 10^{-7}\,\text{s}^{-1}) \times (3.6 \times 10^4\,\text{s})] = \boxed{31.5\,\text{kPa}}$$

(b) After 50 h,

$$p = (32.1\,\text{kPa})\exp[-2 \times (2.78 \times 10^{-7}\,\text{s}^{-1}) \times (1.8 \times 10^5\,\text{s})] = \boxed{29.0\,\text{kPa}}$$

E22.8(b) From Table 22.3, we see that for $A + 2B \rightarrow P$ the integrated rate law is

$$kt = \frac{1}{[B]_0 - 2[A]_0}\ln\left[\frac{[A]_0([B]_0 - 2[P])}{([A]_0 - [P])[B]_0}\right]$$

(a) Substituting the data after solving for k

$$k = \frac{1}{(3.6 \times 10^3\,\text{s}) \times (0.080 - 2 \times 0.075) \times (\text{mol dm}^{-3})} \times \ln\left[\frac{(0.075 \times (0.080 - 0.060)}{(0.075 - 0.030) \times 0.080}\right]$$

$$= \boxed{3.4\overline{7} \times 10^{-3}\,\text{dm}^3\,\text{mol}^{-1}\,\text{s}^{-1}}$$

(b) The half-life in terms of A is the time when $[A] = [A]_0/2 = [P]$, so

$$t_{1/2}(A) = \frac{1}{k([B]_0 - 2[A]_0)}\ln\left[\frac{[A]_0([B]_0 - (2[A]_0/2))}{([A]_0[B]_0/2)}\right]$$

which reduces to

$$t_{1/2}(A) = \frac{1}{k([B]_0 - 2[A]_0)}\ln\left(2 - \frac{2[A]_0}{[B]_0}\right)$$

$$= \frac{1}{(3.4\overline{7} \times 10^{-3}\,\text{dm}^3\,\text{mol}^{-1}\,\text{s}^{-1}) \times (-0.070\,\text{mol dm}^{-3})} \times \ln\left(2 - \frac{0.150}{0.080}\right)$$

$$= 856\overline{1}\,\text{s} = \boxed{2.4\,\text{h}}$$

The half-life in terms of B is the time when $[B] = [B]_0/2$ and $[P] = [B]_0/4$:

$$t_{1/2}(B) = \frac{1}{k([B]_0 - 2[A]_0)} \ln \left[\frac{[A]_0 \left([B]_0 - \dfrac{[B]_0}{2} \right)}{\left([A]_0 - \dfrac{[B]_0}{4} \right) [B]_0} \right]$$

which reduces to

$$t_{1/2}(B) = \frac{1}{k([B]_0 - 2[A]_0)} \ln \left(\frac{[A]_0/2}{[A]_0 - [B]_0/4} \right)$$

$$= \frac{1}{(3.4\overline{7} \times 10^{-3}\, \mathrm{dm^3\, mol^{-1}\, s^{-1}}) \times (-0.070\, \mathrm{mol\, dm^{-3}})} \times \ln \left(\frac{0.075/2}{0.075 - (0.080/4)} \right)$$

$$= 157\overline{6}\, \mathrm{s} = \boxed{0.44\, \mathrm{h}}$$

E22.9(b)   **(a)** The dimensions of a second-order constant are

$$\frac{\mathrm{dim\ of\ } v}{(\mathrm{dim\ of\ } [A])^2} = \frac{\mathrm{amount} \times \mathrm{length}^{-3} \times \mathrm{time}^{-1}}{(\mathrm{amount} \times \mathrm{length}^{-3})^2} = \mathrm{length}^3 \times \mathrm{amount}^{-1} \times \mathrm{time}^{-1}$$

In molecule, m, s units, the units of k are $\boxed{\mathrm{m^3\ molecule^{-1}\ s^{-1}}}$

The dimensions of a third-order rate constant are

$$\frac{\mathrm{dim\ of\ } v}{(\mathrm{dim\ of\ } [A])^3} = \frac{\mathrm{amount} \times \mathrm{length}^{-3} \times \mathrm{time}^{-1}}{(\mathrm{amount} \times \mathrm{length}^{-3})^3} = \mathrm{length}^6 \times \mathrm{amount}^{-2} \times \mathrm{time}^{-1}$$

In molecule, m, s units, the units of k are $\boxed{\mathrm{m^6\ molecule^{-2}\ s^{-1}}}$

COMMENT. Technically, "molecule" is not a unit, so a number of molecules is simply a number of individual objects, that is, a pure number. In the chemical kinetics literature, it is common to see rate constants given in molecular units reported in units of $\mathrm{m^3\ s^{-1}}$, $\mathrm{m^6\ s^{-1}}$, $\mathrm{cm^3\ s^{-1}}$, etc.

(b) The dimensions of a second-order rate constant in pressure units are

$$\frac{\mathrm{dim\ of\ } v}{(\mathrm{dim\ of\ } p)^2} = \frac{\mathrm{pressure} \times \mathrm{time}^{-1}}{(\mathrm{pressure})^2} = \mathrm{pressure}^{-1} \times \mathrm{time}^{-1}$$

In SI units, the pressure unit is $\mathrm{N\, m^{-2}} = \mathrm{Pa}$, so the units of k are $\boxed{\mathrm{Pa^{-1}\ s^{-1}}}$

The dimensions of a third-order rate constant in pressure units are

$$\frac{\mathrm{dim\ of\ } v}{(\mathrm{dim\ of\ } p)^3} = \frac{\mathrm{pressure} \times \mathrm{time}^{-1}}{(\mathrm{pressure})^3} = \mathrm{pressure}^{-2} \times \mathrm{time}^{-1}$$

In SI pressure units, the units of k are $\boxed{\mathrm{Pa^{-2}\ s^{-1}}}$.

E22.10(b) The integrated rate law is

$$kt = \frac{1}{[B]_0 - 2[A]_0} \ln \frac{[A]_0([B]_0 - 2[C])}{([A]_0 - [C])[B]_0} \quad \text{[Table 22.3]}$$

Solving for [C] yields, after some rearranging

$$[C] = \frac{[A]_0[B]_0\{\exp[kt([B]_0 - 2[A]_0)] - 1\}}{[B]_0 \exp[kt([B]_0 - 2[A]_0)] - 2[A]_0}$$

so $\frac{[C]}{\text{mol dm}^{-3}} = \frac{(0.025) \times (0.150) \times (e^{0.21 \times (0.100) \times t/s} - 1)}{(0.150) \times e^{0.21 \times (0.100) \times t/s} - 2 \times (0.025)} = \frac{(3.7\overline{5} \times 10^{-3}) \times (e^{0.021 \times t/s} - 1)}{(0.150) \times e^{0.021 \times t/s} - (0.050)}$

(a) $[C] = \frac{(3.7\overline{5} \times 10^{-3}) \times (e^{0.21} - 1)}{(0.150) \times e^{0.21} - (0.050)}$ mol dm^{-3} = $\boxed{6.5 \times 10^{-3} \text{ mol dm}^{-3}}$

(b) $[C] = \frac{(3.7\overline{5} \times 10^{-3}) \times (e^{12.6} - 1)}{(0.150) \times e^{12.6} - (0.050)}$ mol dm^{-3} = $\boxed{0.025 \text{ mol dm}^{-3}}$

E22.11(b) The rate law is

$$v = -\frac{1}{2}\frac{d[A]}{dt} = k[A]^3$$

which integrates to

$$2kt = \frac{1}{2}\left(\frac{1}{[A]^2} - \frac{1}{[A]_0^2}\right) \quad \text{so} \quad t = \frac{1}{4k}\left(\frac{1}{[A]^2} - \frac{1}{[A]_0^2}\right),$$

$$t = \left(\frac{1}{4(3.50 \times 10^{-4} \text{ dm}^6 \text{ mol}^{-2} \text{ s}^{-1})}\right) \times \left(\frac{1}{(0.021 \text{ mol dm}^{-3})^2} - \frac{1}{(0.077 \text{ mol dm}^{-3})^2}\right)$$

$$= \boxed{1.5 \times 10^6 \text{ s}}$$

E22.12(b) A reaction nth-order in A has the following rate law

$$-\frac{d[A]}{dt} = k[A]^n \quad \text{so} \quad \frac{d[A]}{[A]^n} = -k\,dt = [A]^{-n}\,d[A]$$

Integration yields

$$\frac{[A]^{1-n} - [A]_0^{1-n}}{1 - n} = -kt$$

Let $t_{1/3}$ be the time at which $[A] = [A]_0/3$,

so $-kt_{1/3} = \dfrac{(\frac{1}{3}[A]_0)^{1-n} - [A]_0^{1-n}}{1 - n} = \dfrac{[A]_0^{1-n}[(\frac{1}{3})^{1-n} - 1]}{1 - n}$

and $t_{1/3} = \boxed{\dfrac{3^{n-1} - 1}{k(n - 1)}[A]_0^{1-n}}$

E22.13(b) The equilibrium constant of the reaction is the ratio of rate constants of the forward and reverse reactions:

$$K = \frac{k_f}{k_r} \quad \text{so} \quad k_f = Kk_r.$$

The relaxation time for the temperature jump is (Example 22.4):

$$\tau = \{k_f + k_r([B] + [C])\}^{-1} \quad \text{so} \quad k_f = \tau^{-1} - k_r([B] + [C])$$

Setting these two expressions for k_f equal yields

$$Kk_r = \tau^{-1} - k_r([B] + [C]) \quad \text{so} \quad k_r = \frac{1}{\tau(K + [B] + [C])}$$

Hence

$$k_r = \frac{1}{(3.0 \times 10^{-6}\,\text{s}) \times (2.0 \times 10^{-16} + 2.0 \times 10^{-4} + 2.0 \times 10^{-4})\,\text{mol dm}^{-3}}$$

$$= \boxed{8.3 \times 10^8\,\text{dm}^3\,\text{mol}^{-1}\,\text{s}^{-1}}$$

and $k_f = (2.0 \times 10^{-16}\,\text{mol dm}^{-3}) \times (8.3 \times 10^8\,\text{dm}^3\,\text{mol}^{-1}\,\text{s}^{-1}) = \boxed{1.7 \times 10^{-7}\,\text{s}^{-1}}$

E22.14(b) The rate constant is given by

$$k = A \exp\left(\frac{-E_a}{RT}\right) \quad [22.31]$$

so at 24 °C it is

$$1.70 \times 10^{-2}\,\text{dm}^3\,\text{mol}^{-1}\,\text{s}^{-1} = A \exp\left(\frac{-E_a}{(8.3145\,\text{J K}^{-1}\,\text{mol}^{-1}) \times [(24 + 273)\,\text{K}]}\right)$$

and at 37 °C it is

$$2.01 \times 10^{-2}\,\text{dm}^3\,\text{mol}^{-1}\,\text{s}^{-1} = A \exp\left(\frac{-E_a}{(8.3145\,\text{J K}^{-1}\,\text{mol}^{-1}) \times [(37 + 273)\,\text{K}]}\right)$$

Dividing the two rate constants yields

$$\frac{1.70 \times 10^{-2}}{2.01 \times 10^{-2}} = \exp\left[\left(\frac{-E_a}{8.3145\,\text{J K}^{-1}\,\text{mol}^{-1}}\right) \times \left(\frac{1}{297\,\text{K}} - \frac{1}{310\,\text{K}}\right)\right]$$

so $\ln\left(\dfrac{1.70 \times 10^{-2}}{2.01 \times 10^{-2}}\right) = \left(\dfrac{-E_a}{8.3145\,\text{J K}^{-1}\,\text{mol}^{-1}}\right) \times \left(\dfrac{1}{297\,\text{K}} - \dfrac{1}{310\,\text{K}}\right)$

and $E_a = -\left(\dfrac{1}{297\,\text{K}} - \dfrac{1}{310\,\text{K}}\right)^{-1} \ln\left(\dfrac{1.70 \times 10^{-2}}{2.01 \times 10^{-2}}\right) \times (8.3145\,\text{J K}^{-1}\,\text{mol}^{-1})$

$$= 9.9 \times 10^3\,\text{J mol}^{-1} = \boxed{9.9\,\text{kJ mol}^{-1}}$$

With the activation energy in hand, the prefactor can be computed from either rate constant value

$$A = k \exp\left(\frac{E_a}{RT}\right) = (1.70 \times 10^{-2}\,\text{dm}^3\,\text{mol}^{-1}\,\text{s}^{-1}) \times \exp\left(\frac{9.9 \times 10^3\,\text{J mol}^{-1}}{(8.3145\,\text{J K}^{-1}\,\text{mol}^{-1}) \times (297\,\text{K})}\right)$$

$$= \boxed{0.94\,\text{dm}^3\,\text{mol}^{-1}\,\text{s}^{-1}}$$

E22.15(b) **(a)** Assuming that the rate-determining step is the scission of a C—H bond, the ratio of rate constants for the tritiated versus protonated reactant should be

$$\frac{k_T}{k_H} = e^{-\lambda}, \quad \text{where} \quad \lambda = \left(\frac{\hbar k_f^{1/2}}{2k_B T}\right) \times \left(\frac{1}{\mu_{CH}^{1/2}} - \frac{1}{\mu_{CD}^{1/2}}\right) \quad [22.53 \text{ with } hc\tilde{\nu} = \hbar\omega = \hbar(k/\mu)^{1/2}]$$

The reduced masses will be roughly 1 u and 3 u respectively, for the protons and ^{3}H nuclei are far lighter than the rest of the molecule to which they are attached. So

$$\lambda \approx \frac{(1.0546 \times 10^{-34}\,\mathrm{J\,s}) \times (450\,\mathrm{N\,m^{-1}})^{1/2}}{2 \times (1.381 \times 10^{-23}\,\mathrm{J\,K^{-1}}) \times (298\,\mathrm{K})}$$

$$\times \left(\frac{1}{(1\,\mathrm{u})^{1/2}} - \frac{1}{(3\,\mathrm{u})^{1/2}}\right) \times (1.66 \times 10^{-27}\,\mathrm{kg\,u^{-1}})^{-1/2}$$

$$\approx 2.8$$

so $\dfrac{k_\mathrm{T}}{k_\mathrm{H}} \approx e^{-2.8} = \boxed{0.06 \approx 1/16}$

(b) The analogous expression for ^{16}O and ^{18}O requires reduced masses for C—^{16}O and C—^{18}O bonds. These reduced masses could vary rather widely depending on the size of the whole molecule, but in no case will they be terribly different for the two isotopes. Take ^{12}CO, for example:

$$\mu_{16} = \frac{(16.0\,\mathrm{u}) \times (12.0\,\mathrm{u})}{(16.0 + 12.0)\,\mathrm{u}} = 6.86\,\mathrm{u} \quad \text{and} \quad \mu_{18} = \frac{(18.0\,\mathrm{u}) \times (12.0\,\mathrm{u})}{(18.0 + 12.0)\,\mathrm{u}} = 7.20\,\mathrm{u}$$

$$\lambda = \frac{(1.0546 \times 10^{-34}\,\mathrm{J\,s}) \times (1750\,\mathrm{N\,m^{-1}})^{1/2}}{2 \times (1.381 \times 10^{-23}\,\mathrm{J\,K^{-1}}) \times (298\,\mathrm{K})}$$

$$\times \left(\frac{1}{(6.86\,\mathrm{u})^{1/2}} - \frac{1}{(7.20\,\mathrm{u})^{1/2}}\right) \times (1.66 \times 10^{-27}\,\mathrm{kg\,u^{-1}})^{-1/2}$$

$$= 0.12$$

so $\dfrac{k_{18}}{k_{16}} = e^{-0.12} = \boxed{0.89}$

At the other extreme, the O atoms could be attached to heavy fragments such that the effective mass of the relevant vibration approximates the mass of the oxygen isotope. That is, $\mu_{16} \approx 16\,\mathrm{u}$ and $\mu_{18} \approx 18\,\mathrm{u}$

so $\lambda \approx 0.19$ so $\dfrac{k_{18}}{k_{16}} = e^{-0.19} = \boxed{0.83}$

E22.16(b) $\dfrac{1}{k} = \dfrac{k'_\mathrm{a}}{k_\mathrm{a}k_\mathrm{b}} + \dfrac{1}{k_\mathrm{a}p_\mathrm{A}}$ [analogous to 22.67]

Therefore, for two different pressures we have

$$\frac{1}{k} - \frac{1}{k'} = \frac{1}{k_\mathrm{a}}\left(\frac{1}{p} - \frac{1}{p'}\right),$$

so $k_\mathrm{a} = \left(\dfrac{1}{p} - \dfrac{1}{p'}\right)\left(\dfrac{1}{k} - \dfrac{1}{k'}\right)^{-1}$

$$= \left(\frac{1}{1.09 \times 10^3\,\mathrm{Pa}} - \frac{1}{25\,\mathrm{Pa}}\right) \times \left(\frac{1}{1.7 \times 10^{-3}\,\mathrm{s^{-1}}} - \frac{1}{2.2 \times 10^{-4}\,\mathrm{s^{-1}}}\right)^{-1}$$

$$= \boxed{9.9 \times 10^{-6}\,\mathrm{s^{-1}\,Pa^{-1}}} = \boxed{9.9\,\mathrm{s^{-1}\,MPa^{-1}}}$$

Solutions to problems

Solutions to numerical problems

P22.1   A simple but practical approach is to make an initial guess at the order by observing whether the half-life of the reaction appears to depend on concentration. If it does not, the reaction is first-order; if it does, it may be second-order. Examination of the data shows that the first half-life is roughly 45 minutes, but that the second is about double the first. (Compare the $0 \rightarrow 50.0$ minute data to the $50.0 \rightarrow 150$ minute data.) Therefore, assume second-order and confirm by plotting $1/[A]$ against time. If the reaction is second-order, it will obey

$$\frac{1}{[A]} = kt + \frac{1}{[A]_0} \quad [22.15b].$$

We draw up the following table ($A = NH_4CNO$).

t/min	0	20.0	50.0	65.0	150
$m(\text{urea})/\text{g}$	0	7.0	12.1	13.8	17.7
$m(A)/\text{g}$	22.9	15.9	10.8	9.1	5.2
$[A]/(\text{mol dm}^{-3})$	0.381	0.265	0.180	0.152	0.0866
$[A]^{-1}/(\text{dm}^3\,\text{mol}^{-1})$	2.62	3.78	5.56	6.60	11.5

The data are plotted in Fig. 22.2 and fit closely to a straight line. Hence, the reaction is $\boxed{\text{second-order}}$. The rate constant is the slope: $\boxed{k = 0.059\overline{4}\ \text{dm}^3\,\text{mol}^{-1}\,\text{min}^{-1}}$.

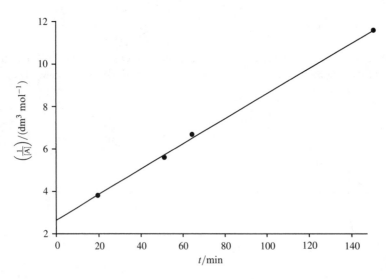

Figure 22.2

To find $[A]$ at 300 min, use eqn 22.15c:

$$[A] = \frac{[A]_0}{1 + kt[A]_0} = \frac{0.382\ \text{mol dm}^{-3}}{1 + (0.059\overline{4}) \times (300) \times (0.382)} = 0.0489\ \text{mol dm}^{-3}.$$

The mass of NH_4CNO left after 300 minutes is

$$m = (0.048\overline{9}\,\text{mol dm}^{-3}) \times (1.00\,\text{dm}^3) \times (60.06\,\text{g mol}^{-1}) = \boxed{2.94\,\text{g}}.$$

P22.3 The procedure adopted in the solutions to Problems 22.1 and 22.2 is employed here. Examination of the data indicates the half-life is independent of concentration and that the reaction is therefore first-order. That is confirmed by a plot of $\ln\left(\frac{[A]}{[A]_0}\right)$ against time (eqn 22.12b). We draw up the following table (A = nitrile).

$t/(10^3\,\text{s})$	0	2.00	4.00	6.00	8.00	10.00	12.00
$[A]/(\text{mol dm}^{-3})$	1.10	0.86	0.67	0.52	0.41	0.32	0.25
$\frac{[A]}{[A]_0}$	1	0.78	0.61	0.47	0.37	0.29	0.23
$\ln\left(\frac{[A]}{[A]_0}\right)$	0	-0.246	-0.496	-0.749	-0.987	-1.235	-1.482

A least-squares fit to a linear equation gives $k = -\text{slope} = \boxed{1.2\overline{3} \times 10^{-4}\,\text{s}^{-1}}$ with a correlation coefficient of 1.000.

P22.5 As described in Example 22.5, if the rate constant obeys the Arrhenius equation [22.29], a plot of $\ln k$ against $1/T$ should yield a straight line with slope $-E_a/R$. However, since data are available only at three temperatures, we use the two-point method, that is,

$$\ln\frac{k_2}{k_1} = -\frac{E_a}{R}\left(\frac{1}{T_2} - \frac{1}{T_1}\right)$$

which yields $E_a = \dfrac{-R\ln(k_2/k_1)}{((1/T_2)-(1/T_1))}$.

For the pair $\theta = 0°C$ and $40°C$,

$$E_a = \frac{-R\ln(576/2.46)}{((1/313\,\text{K})-(1/273\,\text{K}))} = 9.69 \times 10^4\,\text{J mol}^{-1}.$$

For the pair $\theta = 20°C$ and $40°C$,

$$E_a = \frac{-R\ln(576/45.1)}{((1/313\,\text{K})-(1/293\,\text{K}))} = 9.71 \times 10^4\,\text{J mol}^{-1}.$$

The agreement of these values of E_a indicates that the rate constant data fits the Arrhenius equation and that the activation energy is $\boxed{9.70 \times 10^4\,\text{J mol}^{-1}}$.

P22.7 The data for this experiment do not extend much beyond one half-life. Therefore the half-life method of predicting the order of the reaction as described in the solutions to Problems 22.1 and 22.2 cannot be used here. However, a similar method based on three-quarters lives will work. For a first-order reaction, we may write (analogous to the derivation of eqn 22.13)

$$kt_{3/4} = -\ln\frac{\frac{3}{4}[A]_0}{[A]_0} = -\ln\frac{3}{4} = \ln\frac{4}{3} = 0.288 \quad\text{or}\quad t_{3/4} = \frac{0.288}{k}.$$

Thus the three-quarters life (or any given fractional life) is also independent of concentration for a first-order reaction. Examination of the data shows that the first three-quarters life (time to $[A] = 0.237$ mol dm^{-3}) is about 80 min and by interpolation the second (time to $[A] = 0.178$ mol dm^{-3}) is also about 80 min. Therefore the reaction is first-order and the rate constant is approximately

$$k = \frac{0.288}{t_{3/4}} \approx \frac{0.288}{80 \text{ min}} = 3.6 \times 10^{-3} \text{ min}^{-1}.$$

A least-squares fit of the data to the first-order integrated rate law [22.12b] gives the slightly more accurate result, $k = \boxed{3.65 \times 10^{-3} \text{ min}^{-1}}$. The half-life is

$$t_{1/2} = \frac{\ln 2}{k} = \frac{\ln 2}{3.65 \times 10^{-3} \text{ min}^{-1}} = \boxed{190 \text{ min}}.$$

The average lifetime is calculated form

$$\frac{[A]}{[A]_0} = e^{-kt} \text{ [22.12b]}.$$

which has the form of a distribution function. The ratio $\frac{[A]}{[A]_0}$ is the fraction of sucrose molecules that have lived to time t. The average lifetime is then

$$\langle t \rangle = \frac{\int_0^\infty te^{-kt}dt}{\int_0^\infty e^{-kt}dt} = \frac{1}{k} = \boxed{274 \text{ min}}.$$

The denominator ensures normalization of the distribution function.

COMMENT. The average lifetime is also called the relaxation time. Compare to eqn 22.28. Note that the average lifetime is not the half-life. The latter is 190 minutes. Also note that $2 \times t_{3/4} \neq t_{1/2}$.

P22.9 The data do not extend much beyond one half-life; therefore, we cannot see whether the **half**-life is constant over the course of the reaction as a preliminary step in guessing a reaction order. In a first-order reaction, however, not only the half-life but any other similarly defined fractional lifetime remains constant. (That is a property of the exponential function.) In this problem, we can see that the $\frac{2}{3}$-life is **not** constant. (It takes less than 1.6 ms for [ClO] to drop from the first recorded value (8.49 μmol dm^{-3}) by more than $\frac{1}{3}$ of that value (to 5.79 μmol dm^{-3}); it takes more than 4.0 more ms for the concentration to drop by not even $\frac{1}{3}$ of **that** value (to 3.95 μmol dm^{-3}). So our working assumption is that the reaction is not first-order but second-order. Draw up the following table.

t/ms	$[ClO]/(\mu$mol $dm^{-3})$	$(1/[ClO])/(dm^3 \mu mol^{-1})$
0.12	8.49	0.118
0.62	8.09	0.124
0.96	7.10	0.141
1.60	5.79	0.173
3.20	5.20	0.192
4.00	4.77	0.210
5.75	3.95	0.253

The plot of [ClO] vs. t in Fig. 22.3 yields a reasonable straight line; the linear least squares fit is:

$$(1/[\text{ClO}])/(\text{dm}^3\mu\text{mol}^{-1}) = 0.118 + 0.0237(t/\text{ms}) \quad R^2 = 0.974.$$

The rate constant is equal to the slope

$$k = 0.0237\,\text{dm}^3\mu\text{mol}^{-1}\,\text{ms}^{-1} = \boxed{2.37 \times 10^7\,\text{dm}^3\,\text{mol}^{-1}\,\text{s}^{-1}}$$

The half-life depends on the initial concentration (eqn 22.16):

$$t_{1/2} = \frac{1}{k[\text{ClO}]_0} = \frac{1}{(2.37 \times 10^{-7}\,\text{dm}^3\,\text{mol}^{-1}\,\text{s}^{-1})(8.47 \times 10^{-6}\,\text{mol dm}^{-3})} = \boxed{4.98 \times 10^{-3}\,\text{s}}.$$

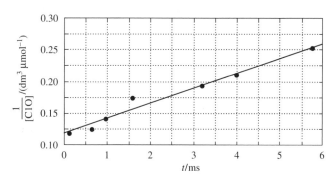

Figure 22.3

P22.11 $\quad$ $A + B \rightarrow P, \quad \dfrac{d[P]}{dt} = k[A]^m[B]^n$

and, for a short interval δt,

$$\delta[P] \approx k[A]^m[B]^n\delta t$$

Therefore, since $\delta[P] = [P]_t - [P]_0 = [P]_t$,

$$\frac{[P]}{[A]} = k[A]^{m-1}[B]^n\delta t.$$

$\dfrac{[\text{Chloropropane}]}{[\text{Propene}]}$ is independent of [Propene], implying that $m = 1$.

$$\frac{[\text{Chloropropane}]}{[\text{HCl}]} = \begin{cases} p(\text{HCl}) & 10 & 7.5 & 5.0 \\ & 0.05 & 0.03 & 0.01 \end{cases}$$

These results suggest that the ratio is roughly proportional to $p(\text{HCl})^2$, and therefore that $m = 3$ when A is identified with HCl. The rate law is therefore

$$\frac{d[\text{Chloropropane}]}{dt} = k[\text{Propane}][\text{HCl}]^3$$

and the reaction is $\boxed{\text{first-order}}$ in propene and $\boxed{\text{third-order}}$ in HCl.

P22.13
$$2HCl \rightleftharpoons (HCl)_2, \quad K_1 \quad [(HCl)_2] = K_1[HCl]^2$$

$$HCl + CH_3CH=CH_2 \rightleftharpoons complex \quad K_2 \quad [complex] = K_2[HCl][CH_3CH=CH_2]$$

$$(HCl)_2 + complex \rightarrow CH_3CHClCH_3 + 2HCl \ k$$

$$rate = \frac{d[CH_3CHClCH_3]}{dt} = k[(HCl)_2][complex].$$

Both $(HCl)_2$ and the complex are intermediates, so substitute for them using equilibrium expressions:

$$rate = k[(HCl)_2][complex] = k(K_1[HCl]^2)(K_2[HCl][CH_3CH=CH_2])$$

$$= \boxed{kK_1K_2[HCl]^3[CH_3CH=CH_2]}$$

which is third-order in HCl and first-order in propene. One approach to experimental verification is to look for evidence of proposed intermediates, using infrared spectroscopy to search for $(HCl)_2$, for example.

P22.15 We can estimate the activation energy of the overall reaction by proceeding as in P22.5:

$$E_{a,eff} = \frac{-R \ln \left(k_{eff}/k'_{eff} \right)}{((1/T) - (1/T'))} = \frac{-R \ln 3}{(1/292 \text{ K}) - (1/343 \text{ K})} = \boxed{-18 \text{ kJ mol}^{-1}}.$$

To relate this quantity to the rate constants and equilibrium constants of the mechanism (P22.13), we identify the effective rate constant as $k_{eff} = kK_1K_2$ and apply the general definition of activation energy (eqn 22.30):

$$E_{a,eff} = RT^2 \frac{d \ln k_{eff}}{dT} = RT^2 \frac{d \ln k_{eff}}{d(1/T)} \frac{d(1/T)}{dT} = -R \frac{d \ln k_{eff}}{d(1/T)}.$$

This form is useful because rate constants and equilibrium constants are often more readily differentiated when considered as functions of $1/T$ rather than functions of T, as in this case:

$$\ln k_{eff} = \ln k + \ln K_1 + \ln K_2$$

so $E_{a,eff} = -R \dfrac{d \ln k_{eff}}{d(1/T)} = -R \dfrac{d \ln k}{d(1/T)} - R \dfrac{d \ln K_1}{d(1/T)} - R \dfrac{d \ln K_2}{d(1/T)} = E_a + \Delta_r H_1 + \Delta_r H_2$

since $\dfrac{d \ln K}{d(1/T)} = \dfrac{-\Delta_r H}{R}$ [van't Hoff equation, 7.23b].

Hence $E_a = E_{a,eff} - \Delta_r H_1 - \Delta_r H_2 = (-18 + 14 + 14) \text{ kJ mol}^{-1} = \boxed{+10 \text{ kJ mol}^{-1}}$.

P22.17
$$\frac{1}{k} = \frac{k'_a}{k_a k_b} + \frac{1}{k_a p} \text{ [analogous to 22.67].}$$

We expect a straight line when $\dfrac{1}{k}$ is plotted against $\dfrac{1}{p}$. We draw up the following table.

p/Torr	84.1	11.0	2.89	0.569	0.120	0.067
$1/(p/\text{Torr})$	0.012	0.091	0.346	1.76	8.33	14.9
$10^{-4}/(k/\text{s}^{-1})$	0.336	0.448	0.629	1.17	2.55	3.30

These points are plotted in Fig. 22.4. There are marked deviations at low pressures, indicating that the Lindemann theory is deficient in that region.

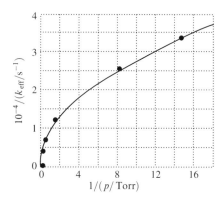

Figure 22.4

P22.19   The reasoning that led to eqn 22.46 holds as long as the rate laws for the two products have the same reaction orders:

$$\frac{[P]_1}{[P]_2} = \frac{k_1}{k_2} = \frac{A_1 e^{-E_{a,1}/RT}}{A_2 e^{-E_{a,2}/RT}} = \frac{A_1}{A_2} e^{-(E_{a,1}-E_{a,2})/RT}.$$

Then, since $E_{a,1} > E_{a,2}$, the exponent in the exponential function is negative, and it gets less negative as the temperature increases. Therefore, the exponential function itself increases and the $\boxed{\text{product concentration ratio also increases}}$.

COMMENT. A qualitative argument can be made that leads to the same conclusion, provided one understands that the activation energy is a measurement of the strength of a reaction's temperature dependence. (See eqn. 22.30.) Since $E_{a,1} > E_{a,2}$, the rate of reaction 1 increases faster with increasing temperature than does the rate of reaction 2.

Solutions to theoretical problems

P22.21 $A \rightleftharpoons B$

$$\frac{d[A]}{dt} = -k[A] + k'[B] \quad \text{and} \quad \frac{d[B]}{dt} = -k'[B] + k[A].$$

At all times, $[A] + [B] = [A]_0 + [B]_0$.

Therefore, $[B] = [A]_0 + [B]_0 - [A]$.

$$\frac{d[A]}{dt} = -k[A] + k'\{[A]_0 + [B]_0 - [A]\} = -(k + k')[A] + k'([A]_0 + [B]_0).$$

To solve, one must integrate

$$\int \frac{d[A]}{(k + k')[A] - k'([A]_0 + [B]_0)} = -\int dt.$$

The solution is $[A] = \dfrac{k'([A]_0 + [B]_0) + (k[A]_0 - k'[B]_0)e^{-(k+k')t}}{k + k'}$.

The final composition is found by setting $t = \infty$:

$$[A]_\infty = \left(\frac{k'}{k+k'}\right) \times ([A]_0 + [B]_0).$$

and $[B]_\infty = [A]_0 + [B]_0 - [A]_\infty = \left(\dfrac{k}{k+k'}\right) \times ([A]_0 + [B]_0).$

Note that $\boxed{\dfrac{[B]_\infty}{[A]_\infty} = \dfrac{k}{k'}}$.

P22.23 $\dfrac{d[A]}{dt} = -2k[A]^2[B], \quad 2A + B \rightarrow P.$

(a) Let $[P] = x$ at t, then $[A] = A_0 - 2x$ and $[B] = B_0 - x = \frac{A_0}{2} - x$. Therefore,

$$\frac{d[A]}{dt} = -2\frac{dx}{dt} = -2k(A_0 - 2x)^2 \times (B_0 - x),$$

$$\frac{dx}{dt} = k(A_0 - 2x)^2 \times \left(\frac{1}{2}A_0 - x\right) = \frac{1}{2}k(A_0 - 2x)^3,$$

$$\frac{1}{2}kt = \int_0^x \frac{dx}{(A_0 - 2x)^3} = \frac{1}{4} \times \left[\left(\frac{1}{A_0 - 2x}\right)^2 - \left(\frac{1}{A_0}\right)^2\right].$$

Therefore, $\boxed{kt = \dfrac{2x(A_0 - x)}{A_0^2(A_0 - 2x)^2}}$.

(b) Now $B_0 = A_0$, so

$$\frac{dx}{dt} = k(A_0 - 2x)^2 \times (B_0 - x) = k(A_0 - 2x)^2 \times (A_0 - x),$$

$$kt = \int_0^x \frac{dx}{(A_0 - 2x)^2 \times (A_0 - x)}.$$

We proceed by the method of partial fractions (which is employed in the general case too), and look for the values of α, β, and γ such that

$$\frac{1}{(A_0 - 2x)^2 \times (A_0 - x)} = \frac{\alpha}{(A_0 - 2x)^2} + \frac{\beta}{A_0 - 2x} + \frac{\gamma}{A_0 - x}.$$

This requires that

$$\alpha(A_0 - x) + \beta(A_0 - 2x) \times (A_0 - x) + \gamma(A_0 - 2x)^2 = 1.$$

Expand and gather terms by powers of x:

$$(A_0\alpha + A_0^2\beta + A_0^2\gamma) - (\alpha + 3\beta A_0 + 4\gamma A_0)x + (2\beta + 4\gamma)x^2 = 1.$$

This must be true for all x; therefore

$$A_0\alpha + A_0^2\beta + A_0^2\gamma = 1,$$

$$\alpha + 3A_0\beta + 3A_0\gamma = 0,$$

$$2\beta + 4\gamma = 0.$$

These solve to give $\alpha = \dfrac{2}{A_0}$, $\beta = \dfrac{-2}{A_0^2}$, and $\gamma = \dfrac{1}{A_0^2}$.

Therefore,

$$kt = \int_0^x \left(\frac{(2/A_0)}{(A_0 - 2x)^2} - \frac{(2/A_0^2)}{A_0 - 2x} + \frac{(1/A_0^2)}{A_0 - x} \right) dx$$

$$= \left. \left(\frac{(1/A_0)}{A_0 - 2x} + \frac{1}{A_0^2}\ln(A_0 - 2x) - \frac{1}{A_0^2}\ln(A_0 - x) \right) \right|_0^x$$

$$= \boxed{\left(\frac{2x}{(A_0^2(A_0 - 2x))} \right) + \left(\frac{1}{A_0^2} \right)\ln\left(\frac{A_0 - 2x}{A_0 - x} \right)}.$$

P22.25   The rate law $\dfrac{d[A]}{dt} = -k[A]^n$ for $n \neq 1$ integrates to

$$kt = \left(\frac{1}{n-1} \right) \times \left(\frac{1}{[A]^{n-1}} - \frac{1}{[A]_0^{n-1}} \right) \quad [E22.12(a)].$$

At $t = t_{1/2}$, $\quad kt_{1/2} = \left(\dfrac{1}{n-1} \right)\left[\left(\dfrac{2}{[A]_0} \right)^{n-1} - \left(\dfrac{1}{[A]_0} \right)^{n-1} \right].$

At $t = t_{3/4}$, $\quad kt_{3/4} = \left(\dfrac{1}{n-1} \right)\left[\left(\dfrac{4}{3[A]_0} \right)^{n-1} - \left(\dfrac{1}{[A]_0} \right)^{n-1} \right].$

Hence, $\dfrac{t_{1/2}}{t_{3/4}} = \boxed{\dfrac{2^{n-1} - 1}{\left(\frac{4}{3} \right)^{n-1} - 1}}.$

P22.27 $v = k([A]_0 - x)([B]_0 + x).$

$$\frac{dv}{dx} = k([A]_0 - x) - k([B]_0 + x).$$

The extrema correspond to $\dfrac{dv}{dx} = 0$, or

$$[A]_0 - x = [B]_0 + x \quad \text{or} \quad 2x = [A]_0 - [B]_0 \quad \text{or} \quad x = \frac{[A]_0 - [B]_0}{2}.$$

Substitute into v to obtain

$$v_{max} = k\left(\frac{[A]_0}{2} + \frac{[B]_0}{2}\right) \times \left(\frac{[B]_0}{2} + \frac{[A]_0}{2}\right) = \boxed{k\left(\frac{[A]_0 + [B]}{2}\right)^2}.$$

Since v and x cannot be negative in the reaction,

$$\boxed{[B]_0 \le [A]_0}.$$

To see the variation of v with x, let $[B]_0 = [A]_0$. The rate equation becomes

$$v = k([A]_0 - x)([A]_0 + x) = k([A]_0^2 - x^2) = k[A]_0^2 - kx^2$$

or $\dfrac{v}{k[A]_0^2} = \left(1 - \dfrac{x^2}{[A]_0^2}\right) = \left(1 + \dfrac{x}{[A]_0}\right)\left(1 - \dfrac{x}{[A]_0}\right).$

Thus we plot $\dfrac{v}{k[A_0]} = \left(1 - \dfrac{x^2}{[A_0]^2}\right) = (1 - X^2)$ against $\dfrac{x}{[A_0]} = X$ from $X = 0$.

The plot is shown in Fig. 22.5 in which $X = \dfrac{x}{[A]_0}$. $\boxed{\dfrac{x}{[A]_0} \le 1 \text{ corresponds to reality}}.$

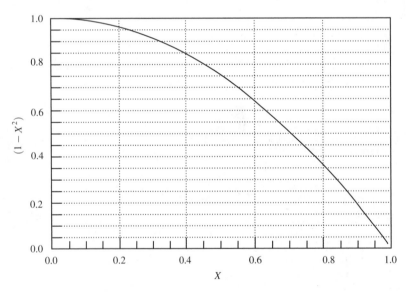

Figure 22.5

Solutions to applications

P22.29   The integrated rate law is

$$[^{14}C] = [^{14}C]_0 e^{-kt} \quad [22.12b] \quad \text{with } k = \frac{\ln 2}{t_{1/2}} \quad [22.13].$$

Solve for t.

$$t = \frac{1}{k} \ln \frac{[^{14}C]_0}{[^{14}C]} = \frac{t_{1/2}}{\ln 2} \ln \frac{[^{14}C]_0}{[^{14}C]} = \left(\frac{5730 \text{ y}}{\ln 2} \right) \times \ln \left(\frac{1.00}{0.72} \right) = \boxed{2720 \text{ y}}.$$

P22.31   A simple but practical approach is to make an initial guess at the order by observing whether the half-life of the reaction appears to depend on concentration. If it does not, the reaction is first-order; if it does, it may be second-order. Examination of the data shows that the half-life is roughly 90 minutes, but it is not exactly constant. (Compare the 60 → 150 minute data to the 150 → 240 minute data; in both intervals the concentration drops by roughly half. Then examine the 30 → 120 minute interval, where the concentration drops by less than half.) If the reaction is first-order, it will obey

$$\ln \left(\frac{c}{c_0} \right) = -kt \quad [22.12b].$$

If it is second-order, it will obey

$$\frac{1}{c} = kt + \frac{1}{c_0} \quad [22.15b].$$

See whether a first-order plot of $\ln c$ vs. time or a second-order plot of $1/c$ vs. time has a substantially better fit. We draw up the following table.

t/ min	30	60	120	150	240	360	480
c/(ng cm^{-3})	699	622	413	292	152	60	24
(ng cm^{-3})/c	0.00143	0.00161	0.00242	0.00342	0.00658	0.0167	0.0412
$\ln \{c/(\text{ng cm}^{-3})\}$	6.550	6.433	6.023	5.677	5.024	4.094	3.178

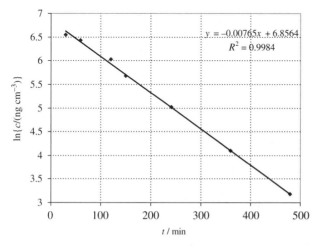

Figure 22.6(a)

The data are plotted in Figs. 22.6(a) and (b). The first-order plot fits closely to a straight line with just a hint of curvature near the outset. The second-order plot, conversely, is strongly curved throughout. Hence,

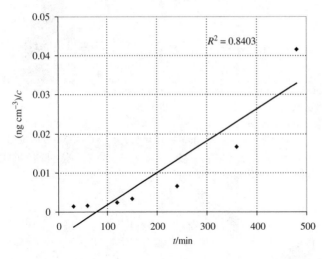

Figure 22.6(b)

the reaction is $\boxed{\text{first-order}}$. The rate constant is the slope of the first-order plot: $k = \boxed{0.00765 \text{ min}^{-1}} = \boxed{0.459 \text{ h}^{-1}}$.

The half-life is (eqn 22.13)

$$t_{1/2} = \frac{\ln 2}{k} = \frac{\ln 2}{0.459 \text{ h}^{-1}} = \boxed{1.51 \text{ h}} = \boxed{91 \text{ min}}.$$

COMMENT. As noted in the problem, the drug concentration is a result of absorption and elimination of the drug, two processes with distinct rates. Elimination is characteristically slower, so the later data points reflect elimination only, for absorption is effectively complete by then. The earlier data points, by contrast, reflect both absorption and elimination. It is, therefore, not surprising that the early points do not adhere so closely to the line so well defined by the later data.

P22.33 (a) For the mechanism

$$hhhh\ldots \underset{k_a'}{\overset{k_a}{\rightleftharpoons}} hchh\ldots$$

$$hchh\ldots \underset{k_b'}{\overset{k_b}{\rightleftharpoons}} cccc\ldots$$

the rate equations are

$$\frac{d[hhhh\ldots]}{dt} = -k_a[hhhh\ldots] + k_a'[hchh\ldots],$$

$$\frac{d[hchh\ldots]}{dt} = k_a[hhhh\ldots] - k_a'[hchh\ldots] - k_b[hchh\ldots] + k_b'[cccc\ldots],$$

$$\frac{d[cccc\ldots]}{dt} = k_b[hchh\ldots] - k_b'[cccc\ldots].$$

(b) Apply the steady-state approximation to the intermediate:

$$\frac{d[hchh\ldots]}{dt} = k_a[hhhh\ldots] - k_a'[hchh\ldots] - k_b[hchh\ldots] + k_b'[cccc\ldots] = 0$$

so $[hchh\ldots] = \dfrac{k_a[hhhh\ldots] + k_b'[cccc\ldots]}{k_a' + k_b}$.

Therefore, $\dfrac{d[hhhh\ldots]}{dt} = -\dfrac{k_a k_b}{k_a' + k_b}[hhhh\ldots] + \dfrac{k_a' k_b'}{k_a' + k_b}[cccc\ldots]$.

This rate expression may be compared to that given in the text [Section 22.4] for the mechanism $A \underset{k'}{\overset{k}{\rightleftharpoons}} B$.

Here $hhhh\ldots \underset{k_{\text{eff}}'}{\overset{k_{\text{eff}}}{\rightleftharpoons}} cccc\ldots$ with $k_{\text{eff}} = \dfrac{k_a k_b}{k_a' + k_b}$, $k_{\text{eff}}' = \dfrac{k_a' k_b'}{k_a' + k_b}$.

(c) It is difficult to make conclusive inferences about intermediates from kinetic data alone. For example, if rate measurements show formation of coils from helices with a single rate constant, they tell us nearly nothing about the mechanism. The rate law

$$\frac{d[cccc\ldots]}{dt} = k[hhhh\ldots]$$

is consistent with a single-step mechanism, with a two-step mechanism with a rate-determining second step, and with a two-step mechanism with a steady-state intermediate. Even if kinetic monitoring of the product shows production with two rate constants, the rate constants could belong to competing paths or to steps of a single reaction path. The best evidence for an intermediate's participation in a reaction is detection of the intermediate, or at least detection of structural features that can belong to a proposed intermediate but not reactant or product.

P22.35   We assume a pre-equilibrium (as the initial step is fast), and write

$$K = \frac{[\text{unstable helix}]}{[A][B]}, \quad \text{implying that } [\text{unstable helix}] = K[A][B].$$

The rate-determining step then gives

$$v = \frac{d[\text{double helix}]}{dt} = k_2[\text{unstable helix}] = k_2 K[A][B] = \boxed{k[A]\,[B]}\,[k = k_2 K].$$

The equilibrium constant is the outcome of the two processes

$$A + B \underset{k_1'}{\overset{k_1}{\rightleftharpoons}} \text{unstable helix}, \quad K = \frac{k_1}{k_1'}$$

Therefore, with $v = k[A][B]$, $\boxed{k = \dfrac{k_1 k_2}{k_1'}}$.

P22.37   The Arrhenius expression for the rate constant is

$$k = Ae^{-E_a/RT} \text{ [22.31] so } \ln k = \ln A - E_a/RT \text{ [22.29]}.$$

A plot of $\ln k$ versus $1/T$ will have slope $-E_a/R$ and y-intercept $\ln A$. The transformed data and plot (Fig. 22.6) follow.

T/K	295	223	218	213	206	200	195
$10^{-6}k/(\text{dm}^3\,\text{mol}^{-1}\,\text{s}^{-1})$	3.55	0.494	0.452	0.379	0.295	0.241	0.217
$\ln k/(\text{dm}^3\,\text{mol}^{-1}\,\text{s}^{-1})$	15.08	13.11	13.02	12.85	12.59	12.39	12.29
$10^{-3}\,\text{K}/T$	3.39	4.48	4.59	4.69	4.85	5.00	5.13

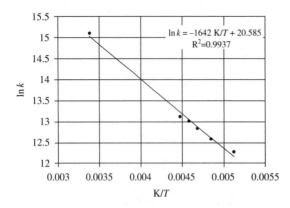

Figure 22.7

So $E_a = -(8.3145\,\text{J K}^{-1}\,\text{mol}^{-1}) \times (-1642\,\text{K}) = 1.37 \times 10^4\,\text{J mol}^{-1} = \boxed{13.7\,\text{kJ mol}^{-1}}$

and $A = e^{20.585}\,\text{dm}^3\,\text{mol}^{-1}\,\text{s}^{-1} = \boxed{8.7 \times 10^8\,\text{dm}^3\,\text{mol}^{-1}\,\text{s}^{-1}}$

P22.39   The rate constants are:

$$k = A \exp\left(\frac{-E_a}{RT}\right) \quad [22.31].$$

$$k_1 = (1.13 \times 10^9\,\text{dm}^3\,\text{s}^{-1}\,\text{mol}^{-1}) \exp\left(\frac{-14.1 \times 10^3\,\text{J mol}^{-1}}{(8.3145\,\text{J K}^{-1}\,\text{mol}^{-1}) \times (298\,\text{K})}\right)$$

$$= \boxed{3.82 \times 10^6\,\text{dm}^3\,\text{mol}^{-1}\,\text{s}^{-1}},$$

$$k_2 = (6.0 \times 10^8\,\text{dm}^3\,\text{s}^{-1}\,\text{mol}^{-1}) \exp\left(\frac{-17.5 \times 10^3\,\text{J mol}^{-1}}{(8.3145\,\text{J K}^{-1}\,\text{mol}^{-1}) \times (298\,\text{K})}\right)$$

$$= \boxed{5.1 \times 10^5\,\text{dm}^3\,\text{mol}^{-1}\,\text{s}^{-1}}$$

$$k_3 = (1.01 \times 10^9\,\text{dm}^3\,\text{s}^{-1}\,\text{mol}^{-1}) \exp\left(\frac{-13.6 \times 10^3\,\text{J mol}^{-1}}{(8.3145\,\text{J K}^{-1}\,\text{mol}^{-1}) \times (298\,\text{K})}\right)$$

$$= \boxed{4.17 \times 10^6\,\text{dm}^3\,\text{mol}^{-1}\,\text{s}^{-1}}.$$

Compared to reaction 1, reaction 2 shows a significant kinetic isotope effect whereas reaction 3 shows practically none. This difference should not be surprising: in reaction 2 a C—D bond is broken, whereas in reaction 3 the D atom is simply along for the ride already attached to the O atom. Compare the measured isotope effect of 0.13 to that expected in reaction 2.

$$\frac{k_2}{k_1} = \exp\left(-\frac{\hbar k_f^{1/2}}{2k_BT}\left(\frac{1}{\mu_{CH}^{1/2}} - \frac{1}{\mu_{CD}^{1/2}}\right)\right) \quad [E22.15(a)].$$

We take $\mu_{CH} \approx m_H$ and $\mu_{CD} \approx m_D \approx 2m_H$, so

$$\frac{k_2}{k_1} = \exp\left(\left(\frac{-(1.0546 \times 10^{-34}\,\text{J s}) \times (500\,\text{kg s}^{-2})^{1/2}}{2(1.381 \times 10^{-23}\,\text{J K}^{-1}) \times (298\,\text{K})}\right) \times \left(1 - \frac{1}{2^{1/2}}\right)\right.$$
$$\left. \times \left(\frac{6.022 \times 10^{23}\,\text{mol}^{-1}}{1 \times 10^{-3}\,\text{kg mol}^{-1}}\right)^{1/2}\right)$$

$$= \boxed{0.13}$$

in agreement with the experimental value.

Answers to discussion questions

D23.1 (a) (1) $AH \rightarrow A\cdot + H\cdot$ Initiation [radicals formed]

(2) $A\cdot \rightarrow B\cdot + C$ Propagation [new radicals formed]

(3) $AH + B\cdot \rightarrow A\cdot + D$ Propagation [new radicals formed]

(4) $A\cdot + B\cdot \rightarrow P$ Termination [non-radical product formed]

(b) (1) $A_2 \rightarrow A\cdot + A\cdot$ Initiation [radicals formed]

(2) $A\cdot \rightarrow B\cdot + C$ Propagation [radicals formed]

(3) $A\cdot + P \rightarrow B\cdot$ Retardation [product destroyed, but chain not terminated]

(4) $A\cdot + B\cdot \rightarrow P$ Termination [non-radical product formed]

D23.3 The Michaelis–Menten mechanism of enzyme activity models the enzyme with one active site that, weakly and reversibly, binds a substrate in homogeneous solution. It is a three-step mechanism. The first and second steps are the reversible formation of the enzyme–substrate complex (ES). The third step is the decay of the complex into the product. The steady-state approximation is applied to the concentration of the intermediate (ES) and its use simplifies the derivation of the final rate expression. However, the justification for the use of the approximation with this mechanism is suspect, in that both rate constants for the reversible steps may not be as large, in comparison to the rate constant for the decay to products, as they need to be for the approximation to be valid. The simplest form of the mechanism applies only when $k_b \gg k_a'$. Nevertheless, the form of the rate equation obtained does seem to match the principal experimental features of enzyme-catalyzed reactions; it explains why there is a maximum in the reaction rate and provides a mechanistic understanding of the turnover number. The model may be expanded to include multisubstrate reaction rate and provides a mechanistic understanding of the turnover number. The model may be expanded to include multisubstrate reactions and inhibition.

D23.5 The primary quantum yield is associated with the primary photochemical event in the overall photochemical process which may involve secondary events as well. An example that illustrates both kinds of events is the photolysis of HI described in Section 23.8(a). The primary quantum yield is defined as the ratio of the number of primary events to the number of photons absorbed (eqn 23.28) and its value can never exceed one. However, in reactions described by complex mechanisms, the overall quantum yield, which is the number of reactant molecules consumed in both primary and secondary processes per photon absorbed, can easily exceed one. Experimental procedures for the determination of the overall quantum yield involve measurements of the intensity of the radiation used, defined here as the number of photons generated and directed at the reacting sample, and of the amount of product formed. This ratio is the overall quantum yield. See Example 23.5. In addition to chemical reactions, the concept of the quantum yield enters into the description of other kinds of photochemical processes, such as fluorescence and phosphorescence, and in each case there are techniques specific to the process for the determination of the quantum yield.

D23.7 The Förster theory of resonance energy transfer examines the interaction between an induced oscillating dipole moment in chromophore S, the energy donor, with a second chromophore Q, the energy acceptor. The oscillating dipole moment of S is induced by incident electromagnetic radiation and the chromophores are separated by distance R. S transfers the excitation energy of the radiation to Q via a mechanism in which its oscillating dipole moment induces an oscillating dipole moment in Q. Resonance energy transfer can be efficient when R is short (typically less than about 9 nm) and when the absorption spectrum of the acceptor overlaps with the emission spectrum of the donor.

Fluorescence resonance energy transfer (FRET) experiments commonly use the fluorescent spectrum and relaxation times of the Förster donor and acceptor chromophores to find the distances between fluorescent dyes at labeled sites in protein, DNA, RNA, etc. FRET is a type of spectroscopic 'ruler'. The computation uses either experimental quantum yields or relaxation lifetimes to calculate the efficiency of resonance energy transfer E_T.

$$E_T = 1 - \frac{\phi_f}{\phi_{f,0}} = 1 - \frac{\tau}{\tau_0} \quad [23.37].$$

E_T is used to calculate R.

$$E_T = \frac{R_0^6}{R_0^6 + R^6} \quad \text{or} \quad R = R_0 \left(\frac{1 - E_T}{E_T} \right)^{1/6} \quad [23.38].$$

Solutions to exercises

In the following exercises and problems, it is recommended that rate constants are labeled with the number of the step in the proposed reaction mechanism and that any reverse steps are labeled similarly but with a prime.

E23.1(b) The intermediates are NO and NO_3 and we apply the steady-state approximation to each of their concentrations

$$k_2 [NO_2] [NO_3] - k_3 [NO] [N_2O_5] = 0$$

$$k_1 [N_2O_5] - k_1' [NO_2] [NO_3] - k_2 [NO_2] [NO_3] = 0$$

$$\text{Rate} = -\frac{1}{2} \frac{d [N_2O_5]}{dt}$$

$$\frac{d [N_2O_5]}{dt} = -k_1 [N_2O_5] + k_1' [NO_2][NO_3] - k_3 [NO][N_2O_5]$$

From the steady-state equations

$$k_3 [NO] [N_2O_5] = k_2 [NO_2] [NO_3]$$

$$[NO_2] [NO_3] = \frac{k_1 [N_2O_5]}{k_1' + k_2}$$

Substituting,

$$\frac{d [N_2O_5]}{dt} = -k_1 [N_2O_5] + \frac{k_1' k_1}{k_1' + k_2} [N_2O_5] - \frac{k_2 k_1}{k_1' + k_2} [N_2O_5] = -\frac{2k_1 k_2}{k_1' + k_2} [N_2O_5]$$

$$\text{Rate} = \frac{k_1 k_2}{k_1' + k_2} [N_2O_5] = k [N_2O_5]$$

E23.2(b) $$\frac{d [R]}{dt} = 2k_1 [R_2] - k_2 [R] [R_2] + k_3 [R'] - 2k_4 [R]^2$$

$$\frac{d [R']}{dt} = k_2 [R] [R_2] - k_3 [R']$$

Apply the steady-state approximation to both equations

$$2k_1 [R_2] - k_2 [R] [R_2] + k_3 [R'] - 2k_4 [R]^2 = 0$$
$$k_2 [R] [R_2] - k_3 [R'] = 0$$

The second solves to $[R'] = \dfrac{k_2}{k_3}[R][R_2]$

and then the first solves to $[R] = \left(\dfrac{k_1}{k_4} [R_2] \right)^{1/2}$

Therefore, $\dfrac{d [R_2]}{dt} = -k_1 [R_2] - k_2 [R_2] [R] = \boxed{ -k_1 [R_2] - k_2 \left(\dfrac{k_1}{k_4} \right)^{1/2} [R_2]^{3/2} }$

E23.3(b)   **(a)** The figure suggests that a chain-branching explosion $\boxed{\text{does not occur}}$ at temperatures as low as 700 K. There may, however, be a thermal explosion regime at pressures in excess of 10^6 Pa.
(b) The lower limit seems to occur when

$$\log (p/\text{Pa}) = 2.1 \quad \text{so} \quad p = 10^{2.1} \text{ Pa} = \boxed{1.3 \times 10^2 \text{ Pa}}$$

There does not seem to be a pressure above which a steady reaction occurs. Rather the chain-branching explosion range seems to run into the thermal explosion range around

$$\log (p/\text{Pa}) = 4.5 \quad \text{so} \quad p = 10^{4.5} \text{ Pa} = \boxed{3 \times 10^4 \text{ pa}}$$

E23.4(b)   The rate of production of the product is

$$\frac{d [BH^+]}{dt} = k_2 [HAH^+] [B]$$

HAH^+ is an intermediate involved in a rapid pre-equilibrium

$$\frac{\left[HAH^+\right]}{[HA]\left[H^+\right]} = \frac{k_1}{k_1'} \text{ so } \left[HAH^+\right] = \frac{k_1\,[HA]\left[H^+\right]}{k_1'}$$

and $\dfrac{d\left[BH^+\right]}{dt} = \boxed{\dfrac{k_1 k_2}{k_1'}\,[HA]\left[H^+\right][B]}$

This rate law can be made independent of $[H^+]$ if the source of H^+ is the acid HA, for then H^+ is given by another equilibrium

$$\frac{[H^+][A^-]}{[HA]} = K_a = \frac{[H^+]^2}{[HA]} \text{ so } [H^+] = (K_a[HA])^{1/2}$$

and $\dfrac{d[BH^+]}{dt} = \boxed{\dfrac{k_1 k_2 K_a^{1/2}}{k_1'}[HA]^{3/2}[B]}$

E23.5(b) A_2 appears in the initiation step only.

$$\frac{d[A_2]}{dt} = -k_1\,[A_2]$$

Consequently, the rate of consumption of $[A_2]$ is first order in A_2 and the rate is independent of intermediate concentrations.

E23.6(b)   The maximum velocity is $k_b\,[E]_0$ and the velocity in general is

$$v = k\,[E]_0 = \frac{k_b\,[S]\,[E]_0}{K_M + [S]} \text{ so } v_{max} = k_b\,[E]_0 = \frac{K_M + [S]}{[S]}v$$

$$v_{max} = \frac{(0.042 + 0.890)\,\text{mol dm}^{-3}}{0.890\,\text{mol dm}^{-3}}(2.45 \times 10^{-4}\,\text{mol dm}^{-3}\text{s}^{-1}) = \boxed{2.57 \times 10^{-4}\,\text{mol dm}^{-3}\,\text{s}^{-1}}$$

E23.7(b)   The quantum yield tells us that each mole of photons absorbed causes 1.2×10^2 moles of A to react; the stoichiometry tells us that 1 mole of B is formed for every mole of A which reacts. From the yield of 1.77 mmol B, we infer that 1.77 mmol A reacted, caused by the absorption of $1.77 \times 10^{-3}\,\text{mol}/(1.2 \times 10^2\,\text{mol Einstein}^{-1}) = \boxed{1.5 \times 10^{-5}\ \text{moles of photons}}$

E23.8(b)   The quantum efficiency is defined as the amount of reacting molecules n_A divided by the amount of photons absorbed n_{abs}. The fraction of photons absorbed f_{abs} is one minus the fraction transmitted f_{trans}; and the amount of photons emitted n_{photon} can be inferred from the energy of the light source (power P times time t) and the energy of the photons (hc/λ).

$$\Phi = \frac{n_A h c N_A}{(1 - f_{trans})\lambda P t}$$

$$= \frac{(0.324\,\text{mol}) \times (6.626 \times 10^{-34}\,\text{J s}) \times (2.998 \times 10^8\,\text{m s}^{-1}) \times (6.022 \times 10^{23}\,\text{mol}^{-1})}{(1 - 0.257) \times (320 \times 10^{-9}\,\text{m}) \times (87.5\,\text{W}) \times (28.0\,\text{min}) \times (60\,\text{s min}^{-1})}$$

$$= \boxed{1.11}$$

Solutions to problems

Solutions to numerical problems

P23.1 $H + NO_2 \rightarrow OH + NO, \quad k_1 = 2.9 \times 10^{10} \, \text{dm}^3 \, \text{mol}^{-1} \, \text{s}^{-1}.$

$OH + OH \rightarrow H_2O + O, \quad k_2 = 1.55 \times 10^9 \, \text{dm}^3 \, \text{mol}^{-1} \, \text{s}^{-1}.$

$O + OH \rightarrow O_2 + H, \quad k_3 = 1.1 \times 10^{10} \, \text{dm}^3 \, \text{mol}^{-1} \, \text{s}^{-1}.$

$[H]_0 = 4.5 \times 10^{-10} \, \text{mol cm}^{-3}, \quad [NO_2]_0 = 5.6 \times 10^{-10} \, \text{mol cm}^{-3}.$

$$\frac{d[O]}{dt} = k_2[OH]^2 + k_3[O][OH], \quad \frac{d[O_2]}{dt} = k_3[O][OH],$$

$$\frac{d[OH]}{dt} = k_1[H][NO_2] - 2k_2[OH]^2 - k_3[O][OH], \quad \frac{d[NO_2]}{dt} = -k_1[H][NO_2],$$

$$\frac{d[H]}{dt} = k_3[O][OH] - k_1[H][NO_2].$$

These equations serve to show how even a simple sequence of reactions leads to a complicated set of nonlinear differential equations. Since we are interested in the time behavior of the composition we may not invoke the steady-state assumption. The only thing left is to use a computer and to integrate the equations numerically. The outcome of this is the set of curves shown in Fig. 23.1 (they have been sketched from the original reference). The similarity to an A → B → C scheme should be noticed (and expected), and the general features can be analysed quite simply in terms of the underlying reactions.

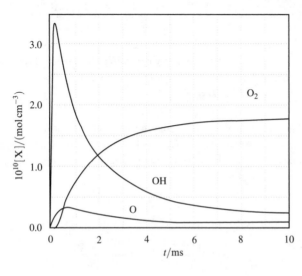

Figure 23.1

P23.3   The roles are

(1) $N_2O \rightarrow N_2 + O$ initiation,
(2) $O + SiH_4 \rightarrow SiH_3 + OH$ propagation [or transfer],
(3) $OH + SiH_4 \rightarrow SiH_3 + H_2O$ propagation [or transfer],
(4) $SiH_3 + N_2O \rightarrow SiH_3O + N_2$ propagation,
(5) $SiH_3O + SiH_4 \rightarrow SiH_3OH + SiH_3$ propagation,
(6) $SiH_3 + SiH_3O \rightarrow (H_3Si)_2O$ termination.

The rate of silane consumption is

$$\frac{d[SiH_4]}{dt} = -k_2[SiH_4][O] - k_3[SiH_4][OH] - k_5[SiH_3O][SiH_4].$$

Steady-state approximation (SSA) for O

$$\frac{d[O]}{dt} = k_1[N_2O] - k_2[SiH_4][O] \approx 0 \quad \text{so} \quad [O] = \frac{k_1[N_2O]}{k_2[SiH_4]}.$$

SSA for OH

$$\frac{d[OH]}{dt} = k_2[SiH_4][O] - k_3[OH][SiH_4] \approx 0 = k_1[N_2O] - k_3[OH][SiH_4]$$

so $[OH] = \dfrac{k_1[N_2O]}{k_3[SiH_4]}.$

SSA for SiH_3O and SiH_3

$$\frac{d[SiH_3O]}{dt} = k_4[SiH_3][N_2O] - k_5[SiH_3O][SiH_4] - k_6[SiH_3O][SiH_3] = 0,$$

$$\frac{d[SiH_3]}{dt} = k_2[SiH_4][O] + k_3[SiH_4][OH] - k_4[SiH_3][N_2O]$$

$$+ k_5[SiH_3O][SiH_4] - k_6[SiH_3O][SiH_3]$$

$$= 2k_1[N_2O] - k_4[SiH_3][N_2O] + k_5[SiH_3O][SiH_4] - k_6[SiH_3O][SiH_3] \approx 0.$$

Adding these expressions together yields

$$0 = 2k_1[N_2O] - 2k_6[SiH_3O][SiH_3] \quad \text{so} \quad [SiH_3] = \frac{k_1[N_2O]}{k_6[SiH_3O]}$$

and subtracting them gives

$$0 = 2k_1[N_2O] - k_4[SiH_3][N_2O] + k_5[SiH_3O][SiH_4].$$

Solve for $[SiH_3O]$:

$$0 = 2k_1[N_2O] - \frac{k_1 k_4[N_2O]^2}{k_6[SiH_3O]} + k_5[SiH_3O][SiH_4]$$

$$= 2k_1 k_6[SiH_3O][N_2O] - k_1 k_4[N_2O]^2 + k_5 k_6[SiH_3O]^2[SiH_4].$$

$$[SiH_3O] = \frac{-2k_1k_6[N_2O] \pm (4k_1^2k_6^2[N_2O]^2 + 4k_1k_4k_5k_6[N_2O]^2[SiH_4])^{1/2}}{2k_5k_6[SiH_4]}$$

$$= \frac{k_1[N_2O]}{k_5[SiH_4]}\left[-1 + \left(1 + \frac{k_4k_5[SiH_4]}{k_1k_6}\right)^{1/2}\right].$$

If k_1 is small, then

$$[SiH_3O] \approx \frac{k_1[N_2O]}{k_5[SiH_4]}\left(\frac{k_4k_5[SiH_4]}{k_1k_6}\right)^{1/2} = [N_2O]\left(\frac{k_1k_4}{k_5k_6[SiH_4]}\right)^{1/2}.$$

Putting it all together yields

$$\frac{d[SiH_4]}{dt} = -2k_1[N_2O] - k_5[SiH_4][N_2O]\left(\frac{k_1k_4}{k_5k_6[SiH_4]}\right)^{1/2} \approx \boxed{\left(\frac{k_1k_4k_5}{k_6}\right)^{1/2}[N_2O][SiH_4]^{1/2}}.$$

P23.5

$$\frac{d[HI]}{dt} = 2k_b[I\cdot]^2[H_2]. \quad (1)$$

$$\frac{d[I\cdot]}{dt} = 2k_a[I_2] - 2k_a'[I\cdot]^2 - 2k_b[I\cdot]^2[H_2].$$

In the steady-state approximation for $[I\cdot]$,

$$\frac{d[I\cdot]}{dt} = 0 = 2k_a[I_2] - 2k_a'[I\cdot]_{SS}^2 - 2k_b[I\cdot]_{SS}^2[H_2]$$

$$[I\cdot]_{SS}^2 = \frac{k_a}{k_a' + k_b[H_2]}[I_2]. \quad (2)$$

Substitution of (2) into (1) gives

$$\frac{d[HI]}{dt} = \frac{2k_bk_a[I_2][H_2]}{k_a' + k_b[H_2]}.$$

This simple rate law is observed when step (b) is rate-determining so that step (a) is a rapid equilibrium and $[I\cdot]$ is in an approximate steady state. This is equivalent to $k_b[H_2] \ll k_a'$ and hence,

$$\frac{d[HI]}{dt} = 2k_bK[I_2][H_2]$$

P23.7 (a) $\frac{I_f}{I_0} = e^{-t/\tau_0}$ [23.31]   or $\ln\left(\frac{I_f}{I_0}\right) = -\frac{t}{\tau_0}$.

A plot of $\ln(I_f/I_0)$ against t should be linear with a slope equal to $-1/\tau_0$ (i.e. $\tau_0 = -1/\text{slope}$) and an intercept equal to zero. Consequently, we make the plot to determine whether it is linear. If it is linear (it is), we do a linear regression fit with a zero intercept and use the regression slope to calculate τ_0. See Fig. 23.2. Alternatively, average the experimental values of $(1/t)\ln(I_f/I_0)$ and check that the standard deviation is a small fraction of the average (it is). The average equals $-1/\tau_0$ (i.e. $\tau_0 = -1/\text{average}$).

Slope $= -0.150$ ns^{-1},

$\tau_0 = -(-0.150 \text{ ns}^{-1})^{-1}$,

$\tau_0 = \boxed{6.67 \text{ ns}}$.

(b) $k_f = \phi_f/\tau_0$ [23.34] $= 0.70/(6.67 \text{ ns})$

$$k_f = \boxed{0.105 \text{ ns}^{-1}}.$$

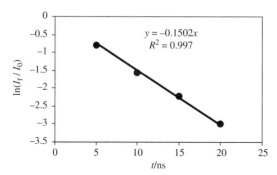

Figure 23.2

P23.9   Since $I_f = k_f[S^*]_t = k_f[S^*]_0\, e^{t/\tau_0}$, we surmise that a graph of $\ln(I_f/I_0)$ against t should be linear with a slope equal to $-1/\tau_0$ in the absence of a quencher. The plot is in fact linear with a regression slope equal to $-1.004 \times 10^5 \text{ s}^{-1}$,

$$\tau_0 = \frac{1}{1.004 \times 10^5 \text{s}^{-1}} = 9.96\,\mu\text{s}.$$

In the presence of a quencher, a graph of $\ln(I_f/I_0)$ against t is still linear but with a slope equal to $-1/\tau$. This plot is found to be linear with a regression slope equal to $-1.788 \times 10^5 \text{s}^{-1}$.

$$\tau = \frac{1}{1.788 \times 10^5 \text{ s}^{-1}} = 5.59\,\mu\text{s}.$$

$$\frac{1}{\tau} = \frac{1}{\tau_0} + k_Q[Q]\ [23.36].$$

$$k_q = \frac{\tau^{-1} - \tau_0^{-1}}{[N_2]} = \frac{RT(\tau^{-1} - \tau_0^{-1})}{p_{N_2}}$$

$$= \frac{(0.08206 \text{ dm}^3 \text{ atm K}^{-1} \text{ mol}^{-1})(300 \text{ K})(1.788 - 1.004)10^5 \text{ s}^{-1}}{9.74 \times 10^{-4} \text{ atm}}.$$

$$k_q = \boxed{1.98 \times 10^9 \text{ dm}^3 \text{ mol}^{-1} \text{ s}^{-1}}.$$

Solutions to theoretical problems

P23.11 $$\frac{d[CH_3CH_3]}{dt} = -k_a[CH_3CH_3] - k_b[CH_3][CH_3CH_3] - k_d[CH_3CH_3][H] + k_e[CH_3CH_2][H].$$

We apply the steady-state approximation to the three intermediates CH_3, CH_3CH_2, and H.

$$\frac{d[CH_3]}{dt} = 2k_a[CH_3CH_3] - k_b[CH_3CH_3][CH_3] = 0$$

which implies that $[CH_3] = \dfrac{2k_a}{k_b}$.

$$\frac{d[CH_3CH_2]}{dt} = k_b[CH_3][CH_3CH_3] - k_c[CH_3CH_2]$$

$$+ k_d[CH_3CH_3][H] - k_e[CH_3CH_2][H] = 0.$$

$$\frac{d[H]}{dt} = k_c[CH_3CH_2] - k_d[CH_3CH_3][H] - k_e[CH_3CH_2][H] = 0.$$

These three equations give

$$[H] = \frac{k_c}{k_e + k_d \frac{[CH_3CH_3]}{[CH_3CH_2]}},$$

$$[CH_3CH_2]^2 - \left(\frac{k_a}{k_c}\right)[CH_3CH_3][CH_3CH_2] - \left(\frac{k_a k_d}{k_c k_e}\right)[CH_3CH_3]^2 = 0,$$

$$\text{or } [CH_3CH_2] - \left\{\left(\frac{k_a}{2k_c}\right) + \left[\left(\frac{k_a}{2k_c}\right)^2 + \left(\frac{k_a k_d}{k_c k_e}\right)\right]^{1/2}\right\}[CH_3CH_3]$$

which implies that

$$[H] = \frac{k_c}{k_e + (k_d/\kappa)}, \quad \kappa = \left(\frac{k_a}{2k_c}\right) + \left[\left(\frac{k_a}{2k_c}\right)^2 + \left(\frac{k_a k_d}{k_c k_e}\right)\right]^{1/2}.$$

If k_a is small in the sense that only the lowest order need be retained,

$$[CH_3CH_2] \approx \left(\frac{k_a k_d}{k_c k_e}\right)^{1/2}[CH_3CH_3],$$

$$[H] \approx \frac{k_c}{k_e + k_d (k_c k_e/k_a k_d)^{1/2}} \approx \left(\frac{k_a k_c}{k_d k_e}\right)^{1/2}.$$

The rate of production of ethene is therefore

$$\frac{d[CH_2CH_2]}{dt} = k_c[CH_3CH_2] = \left(\frac{k_a k_c k_d}{k_e}\right)^{1/2}[CH_3CH_3].$$

The rate of production of ethene is equal to the rate of consumption of ethane (the intermediates all have low concentrations), so

$$\frac{d[CH_3CH_3]}{dt} = -k[CH_3CH_3], \quad k = \left(\frac{k_a k_c k_d}{k_e}\right)^{1/2}.$$

Different orders may arise if the reaction is sensitized so that k_a is increased.

P23.13
$$\langle \overline{M} \rangle_N = \frac{M}{1-p} \quad \text{[eqn 23.8a with } \langle \overline{M} \rangle_N = \langle n \rangle M].$$

The probability P_n that a polymer consists of n monomers is equal to the probability that it has $n-1$ reacted end groups and one unreacted end group. The former probability is p^{n-1}; the latter $1-p$.

Therefore, the total probability of finding an n-mer is

$$P_n = p^{n-1}(1-p).$$

$$\langle M^2 \rangle N = M^2 \langle n^2 \rangle = M^2 \sum_n n^2 P_n = M^2(1-p) \sum_n n^2 p^{n-1} = M^2(1-p)\frac{d}{dp}p\frac{d}{dp}\sum_n p^n$$

$$= M^2(1-p)\frac{d}{dp}p\frac{d}{dp}(1-p)^{-1} = \frac{M^2(1+p)}{(1-p)^2}.$$

We see that $\langle n^2 \rangle = \frac{1+p}{(1-p)^2}$

and that $\langle M^2 \rangle_N - \langle \bar{M} \rangle_{N^2} = M^2\left(\frac{1+p}{(1-p)^2} - \frac{1}{(1-p)^2}\right) = \frac{pM^2}{(1-p)^2}.$

Hence, $\boxed{\delta M = \dfrac{p^{1/2}M}{1-p}}.$

The time dependence is obtained from

$$p = \frac{kt[A]_0}{1 + kt[A]_0} \quad [23.7]$$

and $\dfrac{1}{1-p} = 1 + kt[A]_0$ [23.8b].

Hence $\dfrac{p^{1/2}}{1-p} = p^{1/2}(1 + kt[A]_0) = \{kt[A]_0(1 + kt[A]_0)\}^{1/2}$

and $\delta M = \boxed{M\{kt[A]_0(1 + kt[A]_0)\}^{1/2}}.$

P23.15   In termination by disproportionation, the radicals do not combine. The average number of monomers in a polymer molecule equals the number in the radical, the kinetic chain length, v.

$$\langle n \rangle = v = \boxed{k[\cdot M][I]^{-1/2}} \quad [23.14].$$

P23.17   **(a)** $A + P \rightarrow P + P$ autocatalytic step, $v = k[A][P]$.

Let $[A] = [A]_0 - x$ and $[P] = [P]_0 + x$.

We substitute these definitions into the rate expression, simplify, and integrate.

$$v = -\frac{d[A]}{dt} = k[A][P]$$

$$-\frac{d([A]_0 - x)}{dt} = k([A]_0 - x)([P]_0 + x).$$

$$\frac{dx}{([A]_0 - x)([P]_0 + x)} = k\,dt.$$

$$\frac{1}{[A]_0 + [P]_0}\left(\frac{1}{[A]_0 - x} + \frac{1}{[P]_0 + x}\right)dx = k\,dt.$$

$$\frac{1}{[A]_0 + [P]_0} \int_0^x \left(\frac{1}{[A]_0 - x} + \frac{1}{[P]_0 + x} \right) dx = k \int_0^t dt.$$

$$\frac{1}{[A]_0 + [P]_0} \left\{ \ln \left(\frac{[A]_0}{[A]_0 - x} \right) + \ln \left(\frac{[P]_0 + x}{[P]_0} \right) \right\} = kt.$$

$$\ln \left\{ \left(\frac{[A]_0}{[P]_0} \right) \left(\frac{[P]_0 + x}{[A]_0 - x} \right) \right\} = k([A]_0 + [P]_0)t.$$

$$\ln \left\{ \left(\frac{[A]_0}{[P]_0} \right) \left(\frac{[P]}{[A]_0 + [P]_0 - [P]} \right) \right\} = k([A]_0 + [P]_0)t$$

$$\ln \left\{ \left(\frac{1}{b} \right) \left(\frac{[P]}{[A]_0 + [P]_0 - [P]} \right) \right\} = at \quad \text{where} \quad a = k([A]_0 + [P]_0) \quad \text{and} \quad b = \frac{[P]_0}{[A]_0}.$$

$$\frac{[P]}{[A]_0 + [P]_0 - [P]} = be^{at}.$$

$$[P] = ([A]_0 + [P]_0)be^{at} - be^{at}[P].$$

$$(1 + be^{at})[P] = [P]_0 \left(1 + \frac{[A]_0}{[P]_0} \right) be^{at} = [P]_0 \left(1 + \frac{1}{b} \right) be^{at} = [P]_0(b + 1)e^{at}.$$

$$\boxed{\frac{[P]}{[P]_0} = (b + 1) \frac{e^{at}}{1 + be^{at}}}$$

(b) See Figure 23.3(a).

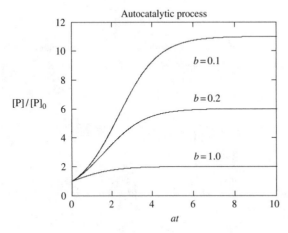

Figure 23.3a

The growth to [P] reaches a maximum at very long times. As $t \to \infty$, the exponential term in the denominator of $[P]/[P]_0 = (b+1)(e^{at}/(1 + be^{at}))$ becomes so large that the denominator becomes be^{at}. Thus, $([P]/[P]_0)_{max} = (b+1)(e^{at}/be^{at}) = (b + 1)/b$ where $b = [P]_0/[A]_0$ and this maximum occurs as $t \to \infty$.

The autocatalytic curve $[P]/[P]_0 = (b+1)(e^{at}/(1 + be^{at}))$ has a shape that is very similar to that of the first-order process $([P]/[A]_0) = 1 - e^{-kt}$. However, $[P]_{max} = [A]_0$ at $t \to \infty$ for the first-order

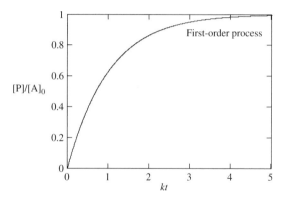

Figure 23.3b

process whereas $[P]_{max} = (1 + 1/b)[P]_0$ for the autocatalytic mechanism. In a series of experiments at fixed $[A]_0$ and assorted $[P]_0$, only the autocatalytic mechanism will show variation in $[P]_{max}$. Another difference is that the autocatalytic curve is initially concave up, which gives an overall sigmoidal curve, whereas the first-order curve is concave down. See Fig. 23.3(b).

(c) Let $[P]_{v_{max}}$ be the concentration of P at which the reaction rate is a maximum and let t_{max} be the corresponding time.

$$v = k[A][P] = k([A]_0 - x)([P]_0 + x)$$

$$= k\{[A]_0[P]_0 + ([A]_0 - [P]_0)x - x^2\}.$$

$$\frac{dv}{dt} = k([A]_0 - [P]_0 - 2x).$$

The reaction rate is a maximum when $dv/dt = 0$. This occurs when

$$x = [P]_{v_{max}} - [P]_0 = \frac{[A]_0 - [P]_0}{2} \quad \text{or} \quad \frac{[P]_{v_{max}}}{[P]_0} = \frac{b+1}{2b}.$$

Substitution into the final equation of part **(a)** gives

$$\frac{[P]_{v_{max}}}{[P]_0} = \frac{b+1}{2b} = (b+1)\frac{e^{at_{max}}}{1 + be^{at_{max}}}.$$

Solving for t_{max},

$$1 + be^{at_{max}} = 2be^{at_{max}}.$$

$$e^{at_{max}} = b^{-1}.$$

$$at_{max} = \ln(b^{-1}) = -\ln(b).$$

$$t_{max} = \boxed{-\frac{1}{a}\ln(b)}.$$

(d) $\quad \dfrac{d[P]}{dt} = k[A]^2[P].$

$$[A] = A_0 - x, \quad [P] = P_0 + x, \quad \dfrac{d[P]}{dt} = \dfrac{dx}{dt} = k(A_0 - x)^2(P_0 + x)$$

$$\int_0^x \dfrac{dx}{(A_0 - x)^2(P_0 + x)} = kt$$

Solve the integral by partial fractions

$$\dfrac{1}{(A_0 - x)^2(P_0 + x)} = \dfrac{\alpha}{(A_0 - x)^2} + \dfrac{\beta}{A_0 - x} + \dfrac{\gamma}{P_0 + x}$$

$$= \dfrac{\alpha(P_0 + x) + \beta(A_0 - x)(P_0 + x) + \gamma(A_0 - x)^2}{(A_0 - x)^2(P_0 + x)}.$$

$$\left. \begin{array}{c} P_0\alpha + A_0 P_0 \beta + A_0^2 \gamma = 1 \\ \alpha + (A_0 - P_0)\beta - 2A_0\gamma = 0 \\ -\beta + \gamma = 0 \end{array} \right\}.$$

This set of simultaneous equations solves to

$$\alpha = \dfrac{1}{A_0 + P_0}, \quad \beta = \gamma = \dfrac{\alpha}{A_0 + P_0}.$$

Therefore,

$$kt = \left(\dfrac{1}{A_0 + P_0}\right) \int_0^x \left[\left(\dfrac{1}{A_0 - x}\right)^2 + \left(\dfrac{1}{A_0 + P_0}\right)\left(\dfrac{1}{A_0 - x} + \dfrac{1}{P_0 - x}\right)\right] dx$$

$$= \left(\dfrac{1}{A_0 + P_0}\right) \left\{ \left(\dfrac{1}{A_0 - x}\right) - \left(\dfrac{1}{A_0}\right) + \left(\dfrac{1}{A_0 + P_0}\right) \left[\ln\left(\dfrac{A_0}{A_0 - x}\right) + \ln\left(\dfrac{P_0 + x}{P_0}\right)\right]\right\}$$

$$= \left(\dfrac{1}{A_0 + P_0}\right) \left[\left(\dfrac{x}{A_0(A_0 - x)}\right) + \left(\dfrac{1}{A_0 + P_0}\right) \ln\left(\dfrac{A_0(P_0 + x)}{(A_0 - x)P_0}\right)\right].$$

Therefore with $y = \dfrac{x}{A_0}$ and $p = \dfrac{P_0}{A_0}$,

$$\boxed{A_0(A_0 + P_0)kt = \left(\dfrac{y}{1 - y}\right) + \left(\dfrac{1}{1 - p}\right) \ln\left(\dfrac{p + y}{p(1 - y)}\right).}$$

The maximum rate occurs at

$$\dfrac{dv_P}{dt} = 0, \quad v_P = k[A]^2[P]$$

and hence at the solution of

$$2k\left(\dfrac{d[A]}{dt}\right)[A][P] + k[A]^2\dfrac{d[P]}{dt} = 0.$$

$$-2k[A][P]v_P + k[A]^2 v_P = 0 \quad [\text{as } v_A = -v_P].$$

$$k[A]([A] - 2[p])v_P = 0.$$

That is, the rate is a maximum when $[A] = 2[P]$, which occurs at

$$A_0 - x = 2P_0 + 2x, \quad \text{or} \quad x = \tfrac{1}{3}(A_0 - 2P_0); \quad y = \tfrac{1}{3}(1 - 2p).$$

Substituting this condition into the integrated rate law gives

$$A_0(A_0 + P_0)kt_{max} = \left(\frac{1}{1+p}\right)\left(\frac{1}{2}(1 - 2p) + \ln\frac{1}{2p}\right)$$

or $\boxed{(A_0 + P_0)^2 kt_{max} = \tfrac{1}{2} - p - \ln 2p}$.

(e) $\dfrac{d[P]}{dt} = k[A][P]^2.$

$$\frac{dx}{dt} = k(A_0 - x)(P_0 + x)^2 \quad [x = P - P_0].$$

$$kt = \int_0^x \frac{dx}{(A_0 - x)(P_0 + x)^2}.$$

Integrate by partial fractions (as in part **(d)**)

$$kt = \left(\frac{1}{A_0 + P_0}\right)\int_0^x \left\{\left(\frac{1}{P_0 + x}\right)^2 + \left(\frac{1}{A_0 + P_0}\right)\left[\frac{1}{P_0 + x} + \frac{1}{A_0 - x}\right]\right\} dx$$

$$= \left(\frac{1}{A_0 + P_0}\right)\left\{\left(\frac{1}{P_0} - \frac{1}{P_0 + x}\right) + \left(\frac{1}{A_0 + P_0}\right)\left[\ln\left(\frac{P_0 + x}{P_0}\right) + \ln\left(\frac{A_0}{A_0 - x}\right)\right]\right\}$$

$$= \left(\frac{1}{A_0 + P_0}\right)\left[\left(\frac{x}{P_0(P_0 + x)}\right) + \left(\frac{1}{A_0 + P_0}\right)\ln\left(\frac{(P_0 + x)A_0}{P_0(A_0 - x)}\right)\right].$$

Therefore, with $y = \dfrac{x}{[A]_0}$ and $p = \dfrac{P_0}{A_0}$,

$$A_0(A_0 + P_0)kt = \boxed{\left(\frac{y}{p(p + y)}\right) + \left(\frac{1}{1+p}\right)\ln\left(\frac{p + y}{p(1 - y)}\right)}.$$

The rate is maximum when

$$\frac{dv_P}{dt} = 2k[A][P]\left(\frac{d[P]}{dt}\right) + k\left(\frac{d[A]}{dt}\right)[P]^2$$

$$= 2k[A][P]v_P - k[P]^2 v_P = k[P](2[A] - [P])v_P = 0.$$

That is, at $[A] = \tfrac{1}{2}[P]$.

On substitution of this condition into the integrated rate law, we find

$$A_0(A_0 + P_0)kt_{max} = \left(\frac{2 - p}{2p(1 + p)}\right) + \left(\frac{1}{1+p}\right)\ln\frac{2}{p}$$

or $\boxed{(A_0 + P_0)^2 kt_{max} = \dfrac{2 - p}{2p} + \ln\dfrac{2}{p}}$.

P23.19

$A \to 2R$ I.

$A + R \to R + B$ k_2.

$R + R \to R_2$ k_3.

$$\frac{d[A]}{dt} = \boxed{-I - k_2[A][R]}, \quad \frac{d[R]}{dt} = 2I - 2k_3[R]^2 = 0.$$

The latter implies that $[R] = \left(\dfrac{I}{k_3}\right)^{1/2}$, and so

$$\frac{d[A]}{dt} = \boxed{-I - k_2\left(\frac{I}{k_3}\right)^{1/2}}[A],$$

$$\frac{d[B]}{dt} = k_2[A][R] = k_2\left(\frac{I}{k_3}\right)^{1/2}[A].$$

Therefore, only the combination $\dfrac{k_2}{k_3^{1/2}}$ may be determined if the reaction attains a steady state.

COMMENT. If the reaction can be monitored at short enough times so that termination is negligible compared to initiation, then $[R] \approx 2It$ and $\frac{d[B]}{dt} \approx k_2It\,[A]$. So monitoring B sheds light on just k_2.

P23.21

$$\frac{d[Cr(CO)_5]}{dt} = I - k_2[Cr(CO)_5][CO] - k_3[Cr(CO)_5][M] + k_4[Cr(CO)_5M] = 0 \text{ [steady state]}.$$

Hence, $[Cr(CO)_5] = \dfrac{I + k_4[Cr(CO)_5M]}{k_2[CO] + k_3[M]}$.

$$\frac{d[Cr(CO)_5M]}{dt} = k_3[Cr(CO)_5][M] - k_4[Cr(CO)_5M].$$

Substituting for $[Cr(CO)_5]$ from above,

$$\frac{d[Cr(CO)_5M]}{dt} = \frac{k_3I[M] - k_2k_4[Cr(CO)_5M[CO]}{k_2[CO] + k_3[M]} = -f[Cr(CO)_5M]$$

if $f = \boxed{\dfrac{k_2k_4[CO]}{k_2[CO] + k_3[M]}}$

and we have taken $k_3I[M] \ll k_2k_4[Cr(CO)_5M][CO]$. Therefore,

$$\frac{1}{f} = \frac{1}{k_4} + \frac{k_3[M]}{k_2k_4[CO]}$$

and a graph of $1/f$ against $[M]$ should be a straight line.

Solutions to applications

P23.23   **(a)** The mechanism considered is

$$E + S \underset{k_a'}{\overset{k_a}{\rightleftharpoons}} (ES) \underset{k_a'}{\overset{k_a}{\rightleftharpoons}} P + E.$$

We apply the steady-state approximation to [(ES)].

$$\frac{d[ES]}{dt} = k_a[E][S] - k_a'[(ES)] - k_b[(ES)] + k_b'[E][P] = 0.$$

Substituting $[E] = [E]_0 - [(ES)]$ we obtain

$$k_a([E]_0 - [(ES)])[S] - k_a'[(ES)] - k_b[(ES)] + k_b'([E]_0 - [(ES)])[P] = 0.$$

$$(-k_a[S] - k_a' - k_b - k_b'[P])[(ES)] + k_a[E]_0[S] - k_b'[E]_0[P] = 0.$$

$$[(ES)] = \frac{k_a[E]_0[S] + k_b'[E]_0[P]}{k_a[S] + k_a' + k_b + k_b'[P]} = \frac{[E]_0[S] + (k_b'/k_a)\,[E]_0[P]}{K_M + [S] + (k_b'/k_a)\,[P]} \quad \left[K_M = \frac{k_a' + k_b}{k_a}\right].$$

Then, $\dfrac{d[P]}{dt} = k_b[(ES)] - k_b'[P][E] = k_b \dfrac{[E]_0[S] + (k_b'/k_a)\,[E]_0[P]}{K_M + [S] + (k_b'/k_a)\,[P]} - k_b'[P]$

$$\times \left([E]_0 - \frac{[E]_0[S] + (k_b'/k_a)\,[E]_0[P]}{K_M + [S] + (k_b'/k_a)\,[P]}\right)$$

$$= \frac{k_b\left[[E]_0[S] + (k_b'/k_a)\,[E]_0[P]\right] - k_b'[E]_0[P]K_M}{K_M + [S] + (k_b'/k_a)\,[P]}.$$

Substituting for K_M in the numerator and rearranging

$$\boxed{\frac{d[P]}{dt} = \frac{k_b[E]_0[S] + (k_a'k_b'/k_a)\,[E]_0[P]}{K_M + [S] + (k_b'/k_a)\,[P]}} \quad \left[v = \frac{d[P]}{dt}\right].$$

(b) For large concentrations of substrate, such that $[S] \gg K_M$ and $[S] \gg [P]$,

$$\boxed{\frac{d[P]}{dt} = k_b[E]_0}$$

which is the same as for the unmodified mechanism. For $[S] \gg K_M$, but $[S] \approx [P]$

$$\boxed{\frac{d[P]}{dt} = k_b[E]_0 \left\{\frac{[S] - (k/k_b)[P]}{[S] + (k/k_a')[P]}\right\}} \quad k = \frac{k_a'k_b'}{k_a}.$$

For $[S] \to 0$, $\dfrac{d[P]}{dt} = \dfrac{-k_a'k_b'[E]_0[P]}{k_a' + k_b + k_b'[P]} = \boxed{\dfrac{-k_a'[E]_0[P]}{K_P + [P]}}$

where $k_P = \dfrac{k_a' + k_b}{k_b'}$.

COMMENT. The negative sign in the expression for d[P]/dt for the case [S] → 0 is to be interpreted to mean that the mechanism in this case is the reverse of the mechanism for the case [P] → 0. The roles of P and S are interchanged.

Question. Can you demonstrate the last statement in the comment above?

P23.25   **(a)**

[ATP]/(μmol dm^{-3})	0.60	0.80	1.4	2.0	3.0
v/(μmol dm^{-3} s^{-1})	0.81	0.97	1.30	1.47	1.69
v/[ATP]/ s^{-1}	1.35	1.21	0.929	0.735	0.563

$$v = \frac{v_{max}}{1 + K_M[S]_0} \quad [23.21].$$

Taking the inverse and multiplying by $v_{max}v$, we find that

$$v_{max} = v + K_M \frac{v}{[S]_0}.$$

Thus,

$$v = v_{max} - K_M \frac{v}{[S]_0} \quad \text{(Eadie–Hofstee plot)} \quad \text{or} \quad \boxed{\frac{v}{[S]_0} = \frac{v_{max}}{K_M} - \frac{v}{K_M}}.$$

(b) The regression slope and intercept of the Eadie–Hofstee data plot of v against $v/[S]_0$ gives $-K_M$ and v_{max}, respectively. Alternatively, the regression slope and intercept of an alternative form of the Eadie–Hofstee data plot of $v/[S]_0$ against v gives $-1/K_M$, and v_{max}/K_M, respectively. The slope and intercept of the latter plot can be used in the calculation of K_M and v_{max}.

(c) We draw up the following table, which includes data rows required for a Eadie–Hofstee plot (v against $v/[S]_0$). The linear regression fit is found for the plot as seen in Fig. 23.4.

$$v_{max} = \boxed{2.30 \ \mu\text{mol dm}^{-3} \ \text{s}^{-1}} \quad \text{and} \quad K_M = \boxed{1.10 \ \mu\text{mol dm}^{-3}}.$$

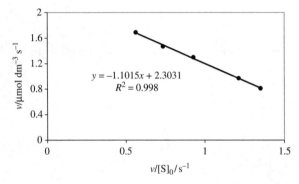

$y = -1.1015x + 2.3031$
$R^2 = 0.998$

Figure 23.4

P23.27   When using reaction rates v, the Lineweaver–Burk plot without inhibition [23.22] has the form:

$$\frac{1}{v} = \frac{1}{v_{max}} + \left(\frac{K_M}{v_{max}}\right)\frac{1}{[S]_0}$$

where the intercept and slope are simple functions of v_{max} and K_M. When using reaction rates relative to a specific, non-inhibited rate ($v_{rel} = v/v_{reference}$), the Lineweaver–Burk plot without inhibition has the same basic form:

$$\frac{1}{v_{rel}} = \frac{1}{v_{max,rel}} + \left(\frac{K_M}{v_{max,rel}}\right)\frac{1}{[S]_0}$$

The linear regression fit of the non-inhibited Lineweaver–Burk data plot is

$$\frac{1}{v_{rel}} = 0.797 + (2.17)\frac{1}{[CBGP]_0/10^{-2} \text{ mol dm}^{-3}}, \quad R^2 = 0.980.$$

Consequently, $v_{max,rel} = 1/\text{intercept} = 1/0.797 = 1.25$ and

$K_M = \text{slope} \times v_{max,rel} = (2.17 \times 10^{-2} \text{ mol dm}^{-3}) \times (1.25) = 2.71 \times 10^{-2} \text{ mol dm}^{-3}$.

The Lineweaver–Burk plot with inhibition has the basic form

$$\frac{1}{v_{rel}} = \frac{\alpha'}{v_{max,rel}} + \left(\frac{\alpha K_M}{v_{max,rel}}\right)\frac{1}{[S]_0}.$$

The linear regression fit of the Lineweaver–Burk data plot for phenylbutyrate ion inhibition is

$$\frac{1}{v_{rel}} = 1.02 + (6.01)\frac{1}{[CBGP]_0/10^{-2} \text{ mol dm}^{-3}}, \quad R^2 = 0.972.$$

Therefore, $\alpha' = \text{intercept} \times v_{max,rel} = 1.02 \times 1.25 = 1.28$ and $\alpha = \text{slope} \times v_{max,rel}/K_M = (6.01 \times 10^{-2} \text{ mol dm}^{-3}) \times (1.25)/(2.71 \times 10^{-2} \text{ mol dm}^{-3}) = 2.77$. Since both $\alpha > 1$ and $\alpha' \sim 1$ (see Section 23.6(c)), we conclude that phenylbutyrate ion is a competitive inhibitor of carboxypeptidase. The linear regression fit of the Lineweaver–Burk data plot for benzoate ion inhibition is

$$\frac{1}{v_{rel}} = 3.75 + (3.01)\frac{1}{[CBGP]_0/10^{-2} \text{ mol dm}^{-3}}, \quad R^2 = 0.999.$$

Therefore, $\alpha' = \text{intercept} \times v_{max,rel} = 3.75 \times 1.25 = 4.69$ and $\alpha = \text{slope} \times v_{max,rel}/K_M = (3.01 \times 10^{-2} \text{ mol dm}^{-3}) \times (1.25)/(2.71 \times 10^{-2} \text{ mol dm}^{-3}) = 1.39$. Since both $\alpha \sim 1$ and $\alpha' > 1$, we conclude that benzoate ion is an $\boxed{\text{uncompetitive}}$ inhibitor of carboxypeptidase.

P23.29
$$E_T = 1 - \frac{\phi_f}{\phi_{f,0}} = 1 - \frac{\tau}{\tau_0} \quad [23.37].$$

$$E_T = \frac{R_0^6}{R_0^6 + R^6} \quad [23.38].$$

Equating these two expressions for E_T and solving for R gives

$$\frac{R_0^6}{R_0^6 + R^6} = 1 - \frac{\tau}{\tau_0}.$$

$$\frac{R_0^6 + R^6}{R_0^6} = \frac{1}{1 - (\tau/\tau_0)}.$$

$$\left(\frac{R}{R_0}\right)^6 = \frac{1}{1 - (\tau/\tau_0)} - 1 = \frac{\tau/\tau_0}{1 - (\tau/\tau_0)} \quad \text{or} \quad R = R_0 \left(\frac{\tau/\tau_0}{1 - (\tau/\tau_0)}\right)^{1/6}.$$

$\tau/\tau_0 = 10 \text{ ps}/10^3 \text{ ps} = 0.010$ and $R = 5.6 \text{ nm} \left(\frac{0.010}{1-0.010}\right)^{1/6} = \boxed{2.6 \text{ nm}}$.

P23.31 *Hypothesis*: The 1270 nm emission band is the emission of the first excited state of $O_2(a^1\Delta_g^+)$ as it returns to the O_2 ground state $(^3\Sigma_g^-)$. The singlet oxygen is produced by porphyrin photosensitization.

$$\underset{\text{porphyrin}}{P} \xrightarrow{h\nu} P^* + {}^3O_2 \rightarrow P + {}^1O_2 \xrightarrow{-h\nu} {}^3O_2$$

Test of hypothesis: It is well known that the dioxygen state $a^1\Delta_g^+$ is 0.977 eV (1270 nm) above the ground state. If the hypothesis is correct the emission intensity should be proportional to both the concentration of dissolved oxygen and to the intensity of the porphyrin absorption.

P23.33 **(a)** $k_2 = 6.2 \times 10^{-34} \text{cm}^6 \text{ molecule}^{-2} \text{ s}^{-1}.$

$k_4 = 8.0 \times 10^{-15} \text{ cm}^3 \text{ molecule}^{-1} \text{ s}^{-1}.$

The concentration of atomic oxygen will be very, very small making a binary collision between atomic oxygen extremely unlikely. In fact, the reaction

$$O + O + M \rightarrow O_2 + M \quad v = k_5[O]^2[M]$$

is ternary which makes it even less likely. Rate terms in k_5 may be safely omitted from consideration.

(b) For all practical purposes $d[O_2]/dt = 0$ because very little dioxygen reacts to form either atomic oxygen or ozone. Using 'molecules per cm^3', or simply cm^{-3}, as the concentration unit, we find that

$$[O_2] = \frac{N_A p}{RT} = \frac{N_A(10 \text{ Torr})}{R(298 \text{ K})} = 3.239 \times 10^{17} \text{ molecules cm}^{-3}.$$

$$\frac{d[O]}{dt} = 2k_1[O_2] - k_2[O][O_2]^3 + k_3[O_3] - k_4[O][O_3],$$

$$= a_2 - a_2[O] + a_3[O_3] - a_4[O][O_3],$$

$$\frac{d[O_3]}{dt} = k_2[O][O_2]^2 - k_3[O_3] - k_4[O][O_3]$$

$$= a_2[O] - a_3[O_3] - a_4[O][O_3],$$

where $a_1 = 2k_1[O_2] = 6.478 \times 10^9\,\text{s}^{-1}\,\text{cm}^{-3}$,

$$a_2 = k_2[O_2]^2 = 65.036\,\text{s}^{-1},$$

$$a_3 = k_3 = 0.016\,\text{s}^{-1},$$

$$a_4 = k_4 = 8.0 \times 10^{-15}\,\text{cm}^3\,\text{s}^{-1}.$$

(c) We break the time period in two. The early period encompasses the first 0.05 s with the initial conditions $[O]_0 = [O_3]_0 = 0$. The second period covers the remainder of the 4 hours with the initial conditions provided by the $[O]_{0.05\,\text{s}} = [O_3]_{0.05\,\text{s}}$ values of the early period. Numerical integration of the coupled differential equations yields the concentrations of Figs 23.5(a), (b), and (c).

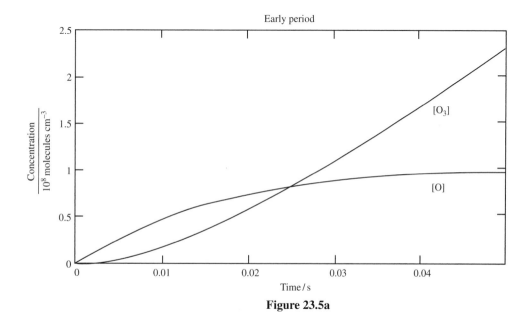

Figure 23.5a

During the early period ($t < 0.05$ s) UV radiation causes the formation of a small amount of atomic oxygen (less than 10^8 molecules cm^{-3}). A steady-state is approached in which a near balance is established between production of atomic oxygen by dioxygen dissociation and usage of atomic oxygen to produce ozone. There is, however, a 100-fold growth in the atomic oxygen concentration over the next 4 h so it is not a perfectly steady state.

After 4 h of photochemistry the percentage ozone is 0.0123%.

$$\text{Percentage ozone} \sim \left(\frac{3.97 \times 10^{13}\,\text{cm}^{-3}}{3.24 \times 10^{17}\,\text{cm}^{-3}} \right) 100 \boxed{\sim 0.123\%}.$$

Ozone production does not require low pressure in the Chapman model. Changing the oxygen pressure to 100 Torr gives 0.025% ozone after 4 h but the increased collision rate reduces atomic oxygen to 1/5'th the value at 10 Torr. However, a pressure increase may require the inclusion in the mechanism of the step $O_3 + M \rightarrow O + O_2 + M$. This would reduce ozone production.

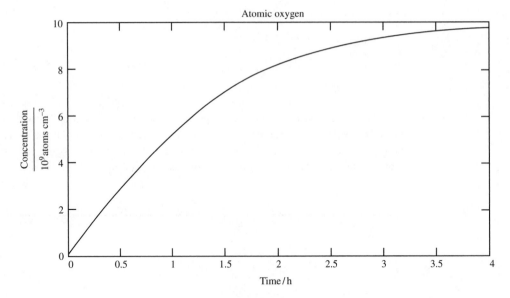

Figure 23.5b

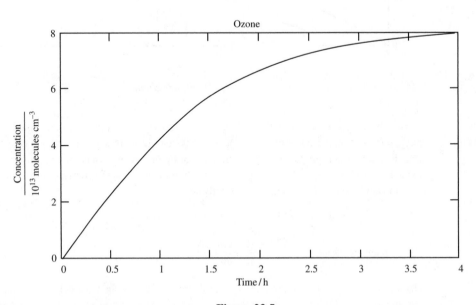

Figure 23.5c

P23.35   In this solution the notation differs slightly from that in the problem. k_1 to k_4 are replaced by k_a to k_d, respectively, with k_4' replaced by k_{-d}. k_7 is replaced by k_e (k_6 does not figure in the solution).

(a)

$$2NO \rightarrow N_2O + O \qquad k_a \qquad \text{initiation}$$
$$O + NO \rightarrow O_2 + N \qquad k_b \qquad \text{propagation}$$
$$N + NO \rightarrow N_2 + O \qquad k_c \qquad \text{propagation}$$
$$2O + M \rightarrow O_2 + M \qquad k_d \qquad \text{termination}$$
$$O_2 + M \rightarrow 2O + M \qquad k_{-d} \qquad \text{initiation}$$

(b)

$$\frac{d[NO]}{dt} = -2k_a[NO]^2 - k_b[O][NO] - k_c[N][NO].$$

To determine the steady-state concentration of N, $[N]_{SS}$, write the rate expression for $d[N]/dt$ and set it equal to zero.

$$\frac{d[N]}{dt} = k_b[O][NO] - k_c[N]_{SS}[NO] = 0.$$

$$[N]_{SS} = \frac{k_b}{k_c}[O].$$

Substitution of $[N]_{SS}$ into the expression for $d[NO]/dt$ indicates that, under steady-state conditions for [N], $v_b = v_c$ and

$$\boxed{\frac{d[NO]}{dt} = -2k_a[NO]^2 - 2k_b[O][NO]}$$

[N] steady-state conditions:
If the propagation step is much more rapid than initiation, the last term predominates

$$\boxed{\frac{d[NO]}{dt} = -2k_b[O][NO]}.$$

[N] is in steady-state and initiation is very slow:
If oxygen atoms and molecules are in equilibrium

$$2O + M \underset{k_{-d}}{\overset{k_d}{\rightleftharpoons}} O_2 + M.$$

$$K_{O/O_2} = \frac{k_d}{k_{-d}} = \frac{[O_2]}{[O]^2}.$$

$$[O] = \left(\frac{k_{-d}[O_2]}{k_d}\right)^{1/2}.$$

Substitution into the previous rate expression yields

$$\boxed{\frac{d[NO]}{dt} = -2k_b\left(\frac{k_{-d}}{k_d}\right)^{1/2}[O_2]^{1/2}[NO]}$$

[N] is in steady-state; initiation is very slow; atomic and molecular oxygen are in equilibrium.

(c) Since $k \propto e^{-E_a/RT}$ where E_a is the activation energy, we may write the individual rate constants in the form $k_i \propto e^{-E_i/RT}$ where the subscript 'a' has been dropped and 'i' represents the ith elementary step with activation energy E_i. Substitution of such expressions in the last equation of part **(b)** yields

$$\frac{d[NO]}{dt} \propto e^{-E_b/RT} \left(\frac{e^{-E_{-d}/RT}}{e^{-E_d/RT}} \right)^{1/2} [O_2]^{1/2}[NO]$$

$$\propto e^{-\left(\frac{E_b + (1/2)E_{-d} - (1/2)E_d}{RT} \right)} [O_2]^{1/2}[NO].$$

We conclude that the effective activation energy, $E_{a,eff}$, is given by

$$\boxed{E_{a,eff} = E_b + \tfrac{1}{2}E_{-d} + \tfrac{1}{2}E_d}.$$

(d) Using the estimate that activation energy is approximately equal to the bond energies that must be broken

$$E_{a,eff} \approx B(NO) + \tfrac{1}{2}B(O_2) - \tfrac{1}{2}B(O)$$

$$\approx 630.57 \text{ kJ mol}^{-1} + \tfrac{1}{2}(498.36 \text{ kJ mol}^{-1}) - \tfrac{1}{2}(0)$$

$$\approx 879.75 \text{ kJ mol}^{-1}$$

where this is the unimolecular bond-breakage estimate of activation energies.

The previous estimate of $E_{a,eff}$ may be much too high because the activation energy of step **(b)** is probably being greatly overestimated. A more realistic estimate of E_b would be that difference between the energies of the NO bond that must be broken and the O_2 bond that is formed. Then,

$$E_{a,eff} \approx \{B(NO) - B(O_2)\} + \tfrac{1}{2}B(O)$$

$$\approx B(NO) - \tfrac{1}{2}B(O_2)$$

$$\approx 630.57 \text{ kJ mol}^{-1} - \tfrac{1}{2}(498.36 \text{ kJ mol}^{-1})$$

$$\approx \boxed{381.39 \text{ kJ mol}^{-1}}.$$

The energy of the activated complex of step **(b)** is the difference between bond-breakage and bond-formation energies.

It is interesting to compare these estimates with the value based upon E_i values that have been determined by experiment

$$E_{a,eff} = (161 \text{ kJ mol}^{-1}) + \tfrac{1}{2}(493 \text{ kJ mol}^{-1}) - \tfrac{1}{2}(14 \text{ kJ mol}^{-1})$$

$$= 401 \text{ kJ mol}^{-1}.$$

This value is based upon experimental activation energies for the elementary steps.

(e) We now eliminate the assumption of O/O_2 equilibrium and assume that both $[N]$ and $[O]$ are at steady-state value. From part **(b)**, $[N_{SS}] = k_b[O]_{SS}/k_c$.

$$\frac{d[O]}{dt} = k_a[NO]^2 - k_b[O]_{SS} + k_c[N] + k_c[N]_{SS}[NO] - 2k_d[O]_{SS}^2[M] + 2k_{-d}[O_2][M] = 0.$$

$$k_a[NO]^2 - k_b[O]_{SS}[NO] + k_b[O]_{SS}[NO] - 2k_d[O]_{SS}^2[M] + 2k_{-d}[O_2][M] = 0.$$

$$k_a[NO]^2 - 2k_d[O]_{SS}^2[M] + 2k_{-d}[O_2][M] = 0.$$

At very low values of $[O_2]$ the last term is negligible so that

$$[O]_{SS} \approx \left(\frac{k_a}{2k_d[M]}\right)^{1/2} [NO].$$

Substitution of the expression for $[N]_{SS}$ and $[O]_{SS}$ into the expression for $d[NO]/dt$ (top of part **(b)**)) gives

$$\frac{d[NO]}{dt} = -2k_a[NO]^2 - k_b[O]_{SS}[NO] - k_c\left(\frac{k_b[O]_{SS}}{k_c}\right)[NO]$$

$$= -2k_a[NO]^2 - 2k_b[O]_{SS}[NO]$$

$$= -2k_a[NO]^2 - 2k_b\left(\frac{k_a}{2k_d[M]}\right)^{1/2} [NO]^2.$$

If propagation is much more rapid than initiation so that $k_b\left(\dfrac{k_a}{2k_d[M]}\right)^{1/2} \gg k_a$, this expression becomes

$$\boxed{\frac{d[NO]}{dt} = -2k_b\left(\frac{k_a}{2k_d[M]}\right)^{1/2} [NO]^2.}$$

(f) $NO + O_2 \rightarrow O + NO_2$ k_e initiation.

$$\frac{d[NO]}{dt} = -2k_a[NO]^2 - k_b[O][NO] - k_c[N][NO] - k_e[NO][O_2]$$

if the conversion has proceeded to the extent that $[O_2]$ has become significant and $k_a[NO][O_2] \gg 2k_a[NO]^2$

$$\frac{d[NO]}{dt} = -k_b[O][NO] - k_c[N][NO] - k_e[NO][O_2].$$

Applying the steady-state approximation to both $[N]$ and $[O]$ gives $[N]_{SS} = k_b[O]_{SS}/k_c$ and

$$\frac{d[O]}{dt} = k_a[NO]^2 - k_b[O]_{SS}[NO] + k_c[N]_{SS}[NO] - 2k_d[O]_{SS}^2[M] + 2k_{-d}[O^2][M]$$

$$+ k_e[O_2][NO] = 0 \quad \text{and} \quad - k_b[O]_{SS}[NO] + k_c[N]_{SS}[NO] = 0$$

Thus $2k_d[O]_{SS}^2[M] - 2k_{-d}[O_2][M] - k_e[O_2][NO] = 0$.

At high concentrations of O_2 species 'M' is likely to be O_2 and $k_d[O]_{SS}^2[NO] \gg k_b[O]_{SS}[NO]$. The value of k_{-d} is so small that it can be neglected.

$$2k_d[O]_{SS}^2[M] \approx k_e[O_2][NO].$$

$$[O]_{SS} = \left(\frac{k_e}{2k_d[M]}\right)^{1/2} [O_2]^{1/2}[NO]^{1/2}.$$

Substitution of $[N]_{SS}$ and $[O]_{SS}$ into the expression for $d[NO]/dt$ gives

$$\frac{d[NO]}{dt} = -k_b[O]_{SS} - k_b[O]_{SS}[NO] - k_c[N]_{SS}[NO] - k_e[NO][O_2]$$

$$= -2k_b[O]_{SS}[NO] - k_e[NO][O_2]$$

$$= -2k_b\left(\frac{k_e}{2k_d[M]}\right)^{1/2} [O_2]^{1/2}[NO]^{3/2} - k_e[NO][O_2].$$

If propagation is much more rapid than initiation, the expression becomes

$$\boxed{\frac{d[NO]}{dt} = -2k_b\left(\frac{k_e}{2k_d[M]}\right)^{1/2} [O_2]^{1/2}[NO]^{3/2}}.$$

$$E_{a,\text{eff}} = E_b + \tfrac{1}{2}E_e - \tfrac{1}{2}E_d.$$

Using the experimental values of E_i, $E_{a,\text{eff}}$ is estimated to be given by

$$E_{a,\text{eff}} = 161 \text{ kJ mol}^{-1} + \tfrac{1}{2}(198 \text{ kJ mol}^{-1}) - \tfrac{1}{2}(14 \text{ kJ mol}^{-1})$$

$$= \boxed{253 \text{ kJ mol}^{-1}}.$$

This value is consistent with the low range of the experimental values of $E_{a,\text{eff}}$, whereas the value found in part (**b**) is consistent with the high experimental values.

24 Molecular reaction dynamics

Answers to discussion questions

D24.1 The harpoon mechanism accounts for the large steric factor of reactions of the kind $K + Br_2 \rightarrow KBr + Br$ in beams. It is supposed that an electron hops across from K to Br_2 when they are within a certain distance, and then the two resulting ions are drawn together by their mutual Coulombic attraction.

D24.3 The Eyring equation (eqn 24.53) results from activated complex theory, which is an attempt to account for the rate constants of bimolecular reactions of the form $A + B \rightleftharpoons C^{\ddagger} \rightarrow P$ in terms of the formation of an activated complex. In the formulation of the theory, it is assumed that the activated complex and the reactants are in equilibrium, and the concentration of activated complex is calculated in terms of an equilibrium constant, which in turn is calculated from the partition functions of the reactants and a postulated form of the activated complex. It is further supposed that one normal mode of the activated complex, the one corresponding to displacement along the reaction coordinate, has a very low force constant and displacement along this normal mode leads to products provided that the complex enters a certain configuration of its atoms, which is known as the transition state. The derivation of the equilibrium constant from the partition functions leads to eqn 24.51 and in turn to eqn 24.53, the Eyring equation. See Section 24.4 for a more complete discussion of a complicated subject.

D24.5 *Infrared chemiluminescence*. Chemical reactions may yield products in excited states. The emission of radiation as the molecules decay to lower energy states is called chemiluminescence. If the emission is from vibrationally excited states, then it is infrared chemiluminescence. The vibrationally excited product molecule in the example of Figure 24.13 in the text is CO. By studying the intensities of the infrared emission spectrum, the populations of the vibrational states in the product CO may be determined and this information allows us to determine the relative rates of formation of CO in these excited states.

Multi-photon ionization (MPI). Multi-photon absorption is the absorption of two or more photons by the molecule in its transition to a higher electronic state. The frequencies of the photons satisfy the condition

$$\Delta E = h\nu_1 + h\nu_2 + \cdots$$

which is similar to the frequency condition for one-photon absorption. However, multi-photon selection rules are different from one-photon selection rules. Therefore, multi-photon processes allow examination of energy states that otherwise could not be reached. In multi-photon ionization, the second or third photon takes the molecule into the energy continuum above its highest lying energy state. This technique is especially useful for the study of weakly fluorescing molecules.

Resonant multi-photon ionization (REMPI). This is a variant of MPI described above, in which one or more photons promote a molecule to an electronically excited state and then additional photons generate ions from the excited state. The power of this method in the study of chemical reactions is its selectivity. In a chemically reacting system, individual reactants and products can be chosen by tuning the frequency of the laser generating the radiation to the electronic absorption band of specific molecules.

Reaction product imaging. In this technique, product ions are accelerated by an electric field toward a phosphorescent screen and the light emitted from the screen is imaged by a charge-coupled device. The significance of this experiment to the study of chemical reactions is that it allows for a detailed analysis of the angular distribution of products.

Femtosecond spectroscopy. See Section 24.9 for a more detailed discussion. Until recently, because of their exceedingly short lifetimes, there have been no direct observations of the activated complexes postulated to exist in the transition state of chemical reactions. But, after the development of femtosecond pulsed lasers, species resembling activated complexes can now be studied spectroscopically. Transitions to and from the activated complex have been observed and such experiments have greatly extended our knowledge of the dynamics of chemical reactions.

D24.7 The Rb atom must hit the I side of CH_3I in order to produce $RbI + CH_3$. The orientation of CH_3I can be controlled by exciting rotations about the CI axis with linearly polarized light; the optimal orientation aims the I side of CH_3I at the direction of approach of the beam of Rb atoms. Two possible alignments of the reactant beams are shown in Fig. 24.1. In the top depiction, the beams are antiparallel, thereby maximizing the likelihood of collision and the volume within which collision can occur (but also putting each beam source in the path of the other beam). In the lower depiction, the beam paths are at right angles, thereby minimizing the region in which the beams collide, but facilitating the study of that well-defined collision volume by a 'probe' laser at right angles to both beams.

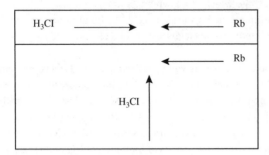

Figure 24.1

Solutions to exercises

E24.1(b) The collision frequency is

$$z = \frac{2^{1/2}\sigma \langle \bar{c}\rangle p}{kT} \quad \text{where } \sigma = \pi d^2 = 4\pi r^2 \text{ and } \langle \bar{c}\rangle = \left(\frac{8RT}{\pi M}\right)^{1/2}$$

so $z = \dfrac{2^{1/2}p}{kT}(4\pi r^2)\left(\dfrac{8RT}{\pi M}\right)^{1/2} = \dfrac{16pN_A r^2 \pi^{1/2}}{(RTM)^{1/2}}$

$= \dfrac{16 \times (100 \times 10^3\,\text{Pa}) \times (6.022 \times 10^{23}\,\text{mol}^{-1}) \times (180 \times 10^{-12}\,\text{m})^2 \times (\pi)^{1/2}}{[(8.3145\,\text{J K}^{-1}\text{mol}^{-1}) \times (298\,\text{K}) \times (28.01 \times 10^{-3}\,\text{kg mol}^{-1})]^{1/2}}$

$= \boxed{6.64 \times 10^9\,\text{s}^{-1}}$

The collision density is

$$Z_{AA} = \frac{1}{2}zN/V = \frac{zp}{2kT} = \frac{(6.64 \times 10^9\,\text{s}^{-1}) \times (100 \times 10^3\,\text{Pa})}{2(1.381 \times 10^{-23}\,\text{J K}^{-1}) \times (298\,\text{K})} = \boxed{8.07 \times 10^{34}\,\text{m}^{-3}\text{s}^{-1}}$$

Raising the temperature at constant volume means raising the pressure in proportion to the temperature

$$Z_{AA} \propto \sqrt{T}$$

so the percent increase in z and Z_{AA} due to a 10 K increase in temperature is $\boxed{1.6\ \text{percent}}$, same as Exercise 24.1(a).

E24.2(b)   The appropriate fraction is given by

$$f = \exp\left(\frac{-E_a}{RT}\right)$$

The values in question are

(a)   (i) $f = \exp\left(\dfrac{-15 \times 10^3\,\text{J mol}^{-1}}{(8.3145\,\text{J K}^{-1}\,\text{mol}^{-1}) \times (300\,\text{K})}\right) = \boxed{2.4 \times 10^{-3}}$

 (ii) $f = \exp\left(\dfrac{-15 \times 10^3\,\text{J mol}^{-1}}{(8.3145\,\text{J K}^{-1}\,\text{mol}^{-1}) \times (800\,\text{K})}\right) = \boxed{0.10}$

(b)   (i) $f = \exp\left(\dfrac{-150 \times 10^3\,\text{J mol}^{-1}}{(8.3145\,\text{J K}^{-1}\,\text{mol}^{-1}) \times (300\,\text{K})}\right) = \boxed{7.7 \times 10^{-27}}$

 (ii) $f = \exp\left(\dfrac{-150 \times 10^3\,\text{J mol}^{-1}}{(8.3145\,\text{J K}^{-1}\,\text{mol}^{-1}) \times (800\,\text{K})}\right) = \boxed{1.6 \times 10^{-10}}$

E24.3(b)   A straightforward approach would be to compute $f = \exp(-E_a/RT)$ at the new temperature and compare it to that at the old temperature. An approximate approach would be to note that f changes from $f_0 = \exp(-E_a/RT)$ to $f = \exp(-E_a/RT(1+x))$, where x is the fractional increase in the temperature. If x is small, the exponent changes from $-E_a/RT$ to approximately $(-E_a/RT)(1-x)$ and f changes from $\exp(-E_a/RT)$ to $\exp(-E_a(1-x)/RT) = \exp(-E_a/RT)\left[\exp(-E_a/RT)\right]^{-x} = f_0 f_0^{-x}$. Thus the new Boltzmann factor is the old one times a factor of f_0^{-x}. The factor of increase is

(a)   (i) $f_0^{-x} = (2.4 \times 10^{-3})^{-10/300} = \boxed{1.2}$

 (ii) $f_0^{-x} = (0.10)^{-10/800} = \boxed{1.03}$

(b) (i) $f_0^{-x} = (7.7 \times 10^{-27})^{-10/300} = \boxed{7.4}$

(ii) $f_0^{-x} = (1.6 \times 10^{-10})^{-10/800} = \boxed{1.3}$

E24.4(b) The reaction rate is given by

$$v = P\sigma \left(\frac{8k_BT}{\pi\mu}\right)^{1/2} N_A \exp(-E_a/RT)[D_2][Br_2]$$

so, in the absence of any estimate of the reaction probability P, the rate constant is

$$k = \sigma \left(\frac{8k_BT}{\pi\mu}\right)^{1/2} N_A \exp(-E_a/RT)$$

$$= [0.30 \times (10^{-9}\,\text{m})^2] \times \left(\frac{8(1.381 \times 10^{-23}\,\text{J K}^{-1}) \times (450\,\text{K})}{\pi(3.930\,\text{u}) \times (1.66 \times 10^{-27}\,\text{kg u}^{-1})}\right)^{1/2}$$

$$\times (6.022 \times 10^{23}\,\text{mol}^{-1}) \exp\left(\frac{-200 \times 10^3\,\text{J mol}^{-1}}{(8.3145\,\text{J K}^{-1}\,\text{mol}^{-1}) \times (450\,\text{K})}\right)$$

$$= 1.71 \times 10^{-15}\,\text{m}^3\,\text{mol}^{-1}\text{s}^{-1} = \boxed{1.7 \times 10^{12}\,\text{dm}^3\,\text{mol}^{-1}\,\text{s}^{-1}}$$

E24.5(b) The rate constant is

$$k_d = 4\pi R^* D N_A$$

where D is the sum of two diffusion constants. So

$$k_d = 4\pi (0.50 \times 10^{-9}\,\text{m}) \times (2 \times 4.2 \times 10^{-9}\,\text{m}^2\,\text{s}^{-1}) \times (6.022 \times 10^{23}\,\text{mol}^{-1})$$

$$= \boxed{3.2 \times 10^7\,\text{m}^3\,\text{mol}^{-1}\,\text{s}^{-1}}$$

In more common units, this is

$$k_d = \boxed{3.2 \times 10^{10}\,\text{dm}^3\,\text{mol}^{-1}\,\text{s}^{-1}}$$

E24.6(b) **(a)** A diffusion-controlled rate constant in decylbenzene is

$$k_d = \frac{8RT}{3\eta} = \frac{8 \times (8.3145\,\text{J K}^{-1}\,\text{mol}^{-1}) \times (298\,\text{K})}{3 \times (3.36 \times 10^{-3}\,\text{kg m}^{-1}\,\text{s}^{-1})} = \boxed{1.97 \times 10^6\,\text{m}^3\,\text{mol}^{-1}\,\text{s}^{-1}}$$

(b) In concentrated sulfuric acid

$$k_d = \frac{8RT}{3\eta} = \frac{8 \times (8.3145\,\text{J K}^{-1}\,\text{mol}^{-1}) \times (298\,\text{K})}{3 \times (27 \times 10^{-3}\,\text{kg m}^{-1}\,\text{s}^{-1})} = \boxed{2.4 \times 10^5\,\text{m}^3\,\text{mol}^{-1}\,\text{s}^{-1}}$$

E24.7(b) The diffusion-controlled rate constant is

$$k_d = \frac{8RT}{3\eta} = \frac{8 \times (8.3145\,\text{J K}^{-1}\,\text{mol}^{-1}) \times (298\,\text{K})}{3 \times (0.601 \times 10^{-3}\,\text{kg m}^{-1}\,\text{s}^{-1})} = \boxed{1.10 \times 10^7\,\text{m}^3\,\text{mol}^{-1}\,\text{s}^{-1}}$$

In more common units, $k_d = \boxed{1.10 \times 10^{10} \text{ dm}^3 \text{ mol}^{-1} \text{ s}^{-1}}$

The recombination reaction has a rate of

$$v = k_d[A][B] \qquad \text{with } [A] = [B]$$

so the half-life is given by

$$t_{1/2} = \frac{1}{k[A]_0} = \frac{1}{(1.10 \times 10^{10} \text{ dm}^3 \text{ mol}^{-1} \text{ s}^{-1}) \times (1.8 \times 10^{-3} \text{ mol dm}^{-3})} = \boxed{5.05 \times 10^{-8} \text{ s}}$$

E24.8(b)   The reactive cross-section σ^* is related to the collision cross-section σ by

$$\sigma^* = P\sigma \quad \text{so} \quad P = \sigma^*/\sigma.$$

The collision cross-section σ is related to effective molecular diameters by

$$\sigma = \pi d^2 \quad \text{so} \quad d = (\sigma/\pi)^{1/2}$$

Now $\sigma_{AB} = \pi d_{AB}^2 = \pi \left[\tfrac{1}{2}(d_A + d_B) \right]^2 = \tfrac{1}{4} \left(\sigma_{AA}^{1/2} + \sigma_{BB}^{1/2} \right)^2$

so $\quad P = \dfrac{\sigma^*}{\tfrac{1}{4} \left(\sigma_{AA}^{1/2} + \sigma_{BB}^{1/2} \right)^2}$

$$= \frac{8.7 \times 10^{-22} \text{ m}}{\tfrac{1}{4}[((0.88)^{1/2} + (0.40)^{1/2}) \times 10^{-9} \text{ m}]^2} = \boxed{1.41 \times 10^{-3}}$$

E24.9(b)   The diffusion-controlled rate constant is

$$k_d = \frac{8RT}{3\eta} = \frac{8 \times (8.3145 \text{ J K}^{-1} \text{ mol}^{-1}) \times (293 \text{ K})}{3 \times (1.27 \times 10^{-3} \text{ kg m}^{-1} \text{ s}^{-1})} = 5.12 \times 10^6 \text{ m}^3 \text{ mol}^{-1} \text{ s}^{-1}$$

In more common units, $k_d = 5.12 \times 10^9 \text{ dm}^3 \text{ mol}^{-1} \text{ s}^{-1}$.

The recombination reaction has a rate of

$$v = k_d[A][B] = (5.12 \times 10^9 \text{ dm}^3 \text{ mol}^{-1} \text{ s}^{-1}) \times (0.200 \text{ mol dm}^{-3}) \times (0.150 \text{ mol dm}^{-3})$$

$$= \boxed{1.54 \times 10^8 \text{ mol dm}^{-3} \text{ s}^{-1}}$$

E24.10(b) The enthalpy of activation for a reaction in solution is

$$\Delta^{\ddagger}H = E_a - RT = (8.3145 \text{ J K}^{-1} \text{ mol}^{-1}) \times (6134 \text{ K}) - (8.3145 \text{ J K}^{-1} \text{ mol}^{-1}) \times (298 \text{ K})$$

$$= 4.852 \times 10^4 \text{ J mol}^{-1} = \boxed{48.52 \text{ kJ mol}^{-1}}$$

The entropy of activation is

$$\Delta^{\ddagger}S = R\left(\ln\frac{A}{B} - 1\right) \quad \text{where } B = \frac{kRT^2}{hp^{\ominus}}$$

$$B = \frac{(1.381 \times 10^{-23}\,\text{J K}^{-1}) \times (8.3145\,\text{J K}^{-1}\,\text{mol}^{-1}) \times (298\,\text{K})^2}{(6.626 \times 10^{-34}\,\text{J s}) \times (1.00 \times 10^5\,\text{Pa})}$$

$$= 1.54 \times 10^{11}\,\text{m}^3\,\text{mol}^{-1}\,\text{s}^{-1}$$

so $\Delta^{\ddagger}S = (8.3145\,\text{J K}^{-1}\,\text{mol}^{-1}) \times \left(\ln\dfrac{8.72 \times 10^{12}\,\text{dm}^3\,\text{mol}^{-1}\,\text{s}^{-1}}{(1000\,\text{dm}^3\,\text{m}^{-3}) \times (1.54 \times 10^{11}\,\text{m}^3\,\text{mol}^{-1}\,\text{s}^{-1})} - 1\right)$

$$= \boxed{-32.2\,\text{J K}^{-1}\,\text{mol}^{-1}}$$

COMMENT. In this connection, the enthalpy of activation is often referred to as "energy" of activation.

E24.11(b) The Gibbs energy of activation is related to the rate constant by

$$k_2 = B\exp\left(\frac{-\Delta^{\ddagger}G}{RT}\right) \quad \text{where } B = \frac{kRT^2}{hp^{\ominus}} \quad \text{so} \quad \Delta^{\ddagger}G = -RT\ln\frac{k_2}{B}$$

$$k_2 = (6.45 \times 10^{13}\,\text{dm}^3\,\text{mol}^{-1}\,\text{s}^{-1})e^{-\{(5375\,\text{K})/(298\,\text{K})\}} = 9.47 \times 10^5\,\text{dm}^3\,\text{mol}^{-1}\,\text{s}^{-1}$$

$$= 947\,\text{m}^3\,\text{mol}^{-1}\,\text{s}^{-1}$$

Using the value of B computed in Exercise 27.13(b), we obtain

$$\Delta^{\ddagger}G = -(8.3145 \times 10^{-3}\,\text{kJ K}^{-1}\,\text{mol}^{-1}) \times (298\,\text{K}) \times \ln\left(\frac{947\,\text{m}^3\,\text{mol}^{-1}\,\text{s}^{-1}}{1.54 \times 10^{11}\,\text{m}^3\,\text{mol}^{-1}\,\text{s}^{-1}}\right)$$

$$= \boxed{46.8\,\text{kJ mol}^{-1}}$$

E24.12(b) The entropy of activation for a bimolecular reaction in the gas phase is

$$\Delta^{\ddagger}S = R\left(\ln\frac{A}{B} - 2\right) \quad \text{where } B = \frac{kRT^2}{hp^{\ominus}}$$

$$B = \frac{(1.381 \times 10^{-23}\,\text{J K}^{-1}) \times (8.3145\,\text{J K}^{-1}\,\text{mol}^{-1}) \times [(55 + 273)\,\text{K}]^2}{(6.626 \times 10^{-34}\,\text{J s}) \times (1.00 \times 10^5\,\text{Pa})}$$

$$= 1.86 \times 10^{11}\,\text{m}^3\,\text{mol}^{-1}\,\text{s}^{-1}$$

The rate constant is

$$k_2 = A\exp\left(\frac{-E_a}{RT}\right) \quad \text{so} \quad A = k_2\exp\left(\frac{E_a}{RT}\right)$$

$$A = (0.23\,\text{m}^3\,\text{mol}^{-1}\,\text{s}^{-1}) \times \exp\left(\frac{49.6 \times 10^3\,\text{J mol}^{-1}}{(8.3145\,\text{J K}^{-1}\,\text{mol}^{-1}) \times (328\,\text{K})}\right)$$

$$= 1.8 \times 10^7\,\text{m}^3\,\text{mol}^{-1}\,\text{s}^{-1}$$

and $\Delta^{\ddagger}S = (8.3145 \, \text{J K}^{-1} \, \text{mol}^{-1}) \times \left(\ln \left(\dfrac{1.8 \times 10^7 \, \text{m}^3 \, \text{mol}^{-1} \, \text{s}^{-1}}{1.86 \times 10^{11} \, \text{m}^3 \, \text{mol}^{-1} \, \text{s}^{-1}} \right) - 2 \right)$

$= \boxed{-93 \, \text{J K}^{-1} \, \text{mol}^{-1}}$

E24.13(b) The entropy of activation for a bimolecular reaction in the gas phase is

$$\Delta^{\ddagger}S = R \left(\ln \frac{A}{B} - 2 \right) \quad \text{where } B = \frac{kRT^2}{hp^{\ominus}}$$

For the collision of structureless particles, the rate constant is

$$k_2 = N_A \left(\frac{8kT}{\pi \mu} \right)^{1/2} \sigma \exp \left(\frac{-\Delta E_0}{RT} \right)$$

so the prefactor is

$$A = N_A \left(\frac{8kT}{\pi \mu} \right)^{1/2} \sigma = 4N_A \left(\frac{RT}{\pi M} \right)^{1/2} \sigma$$

where we have used the fact that $\mu = \frac{1}{2}m$ for identical particles and $k/m = R/M$. So

$$A = 4 \times (6.022 \times 10^{23} \, \text{mol}^{-1}) \times \left(\frac{(8.3145 \, \text{J K}^{-1} \, \text{mol}^{-1}) \times (500 \, \text{K})}{\pi \times (78 \times 10^{-3} \, \text{kg mol}^{-1})} \right)^{1/2} \times (0.68 \times 10^{-18} \, \text{m}^2)$$

$$= 2.13 \times 10^8 \, \text{m}^3 \, \text{mol}^{-1} \, \text{s}^{-1}$$

$$B = \frac{(1.381 \times 10^{-23} \, \text{J K}^{-1}) \times (8.3145 \, \text{J K}^{-1} \, \text{mol}^{-1}) \times (500 \, \text{K})^2}{(6.626 \times 10^{-34} \, \text{J s}) \times (1.00 \times 10^5 \, \text{Pa})}$$

$$= 4.33 \times 10^{11} \, \text{m}^3 \, \text{mol}^{-1} \, \text{s}^{-1}$$

and $\Delta^{\ddagger}S = (8.3145 \, \text{J K}^{-1} \, \text{mol}^{-1}) \times \left(\ln \left(\dfrac{2.13 \times 10^8 \, \text{m}^3 \, \text{mol}^{-1} \, \text{s}^{-1}}{4.33 \times 10^{11} \, \text{m}^3 \, \text{mol}^{-1} \, \text{s}^{-1}} \right) - 2 \right)$

$= \boxed{-80.0 \, \text{J K}^{-1} \, \text{mol}^{-1}}$

E24.14(b) **(a)** The entropy of activation for a unimolecular gas-phase reaction is

$$\Delta^{\ddagger}S = R \left(\ln \frac{A}{B} - 1 \right) \quad \text{where } B = 1.54 \times 10^{11} \, \text{m}^3 \, \text{mol}^{-1} \, \text{s}^{-1} \, \text{[See Exercise 24.14(a)]}$$

so $\Delta^{\ddagger}S = (8.3145 \, \text{J K}^{-1} \, \text{mol}^{-1})$

$$\times \left(\ln \left(\frac{2.3 \times 10^{13} \, \text{dm}^3 \, \text{mol}^{-1} \, \text{s}^{-1}}{(1000 \, \text{dm}^3 \, \text{m}^{-3}) \times (1.54 \times 10^{11} \, \text{m}^3 \, \text{mol}^{-1} \, \text{s}^{-1})} \right) - 1 \right)$$

$$= \boxed{-24.1 \, \text{J K}^{-1} \, \text{mol}^{-1}}$$

(b) The enthalpy of activation is

$$\Delta^{\ddagger}H = E_a - RT = 30.0 \times 10^3 \, \text{J mol}^{-1} - (8.3145 \, \text{J K}^{-1} \, \text{mol}^{-1}) \times (298 \, \text{K})$$

$$= 27.5 \times 10^3 \, \text{J mol}^{-1} = \boxed{27.5 \, \text{kJ mol}^{-1}}$$

(c) The Gibbs energy of activation is

$$\Delta^{\ddagger}G = \Delta^{\ddagger}H - T\Delta^{\ddagger}S = 27.5\,\text{kJ mol}^{-1} - (298\,\text{K}) \times (-24.1 \times 10^{-3}\,\text{kJ K}^{-1}\,\text{mol}^{-1})$$

$$= \boxed{34.7\,\text{kJ mol}^{-1}}$$

E24.15(b) The dependence of a rate constant on ionic strength is given by

$$\log k_2 = \log k_2^{\circ} + 2Az_{\text{A}}z_{\text{B}}I^{1/2}$$

At infinite dilution, $I = 0$ and $k_2 = k_2^{\circ}$, so we must find

$$\log k_2^{\circ} = \log k_2 - 2Az_{\text{A}}z_{\text{B}}I^{1/2} = \log(1.55) - 2 \times (0.509) \times (+1) \times (+1) \times (0.0241)^{1/2}$$

$$= 0.0323 \quad \text{and} \quad \boxed{k_2^{\circ} = 1.08\,\text{dm}^6\,\text{mol}^{-2}\text{min}^{-1}}$$

E24.16(b) Equation 24.84 holds for a donor–acceptor pair separated by a constant distance, assuming that the reorganization energy is constant:

$$\ln k_{\text{et}} = -\frac{(\Delta_r G^{\ominus})^2}{4\lambda RT} - \frac{\Delta_r G^{\ominus}}{2RT} + \text{constant},$$

or equivalently

$$\ln k_{\text{et}} = -\frac{(\Delta_r G^{\ominus})^2}{4\lambda kT} - \frac{\Delta_r G^{\ominus}}{2kT} + \text{constant},$$

if energies are expressed as molecular rather than molar quantities. Two sets of rate constants and reaction Gibbs energies can be used to generate two equations (eqn 24.84 applied to the two sets) in two unknowns: λ and the constant.

$$\ln k_{\text{et},1} + \frac{(\Delta_r G_1^{\ominus})^2}{4\lambda kT} + \frac{\Delta_r G_1^{\ominus}}{2kT} = \text{constant} = \ln k_{\text{et},2} + \frac{(\Delta_r G_2^{\ominus})^2}{4\lambda kT} + \frac{\Delta_r G_2^{\ominus}}{2kT},$$

so

$$\frac{(\Delta_r G_1^{\ominus})^2 - (\Delta_r G_2^{\ominus})^2}{\Delta\lambda kT} = \ln \frac{k_{\text{et},2}}{k_{\text{et},1}} + \frac{\Delta_r G_2^{\ominus} - \Delta_r G_1^{\ominus}}{2kT}$$

and $\lambda = $

$$\lambda = \frac{(\Delta_r G_1^{\ominus})^2 - (\Delta_r G_2^{\ominus})^2}{4\left(kT \ln(k_{\text{et},2}/k_{\text{et},1}) + (\Delta_r G_2^{\ominus} - \Delta_r G_1^{\ominus}/2)\right)}$$

$$\lambda = \frac{(-0.665\,\text{eV})^2 - (-0.975\,\text{eV})^2}{\dfrac{4(1.381 \times 10^{-23}\,\text{J K}^{-1})(298\,\text{K})}{1.602 \times 10^{-19}\,\text{J eV}^{-1}} \ln \dfrac{3.33 \times 10^6}{2.02 \times 10^5} - 2(0.975 - 0.665)\,\text{eV}} = \boxed{1.53\overline{1}\,\text{eV}}$$

If we knew the activation Gibbs energy, we could use eqn 24.81 to compute $\langle H_{\text{DA}} \rangle$ from either rate constant, and we *can* compute the activation Gibbs energy from eqn 24.82:

$$\Delta^{\ddagger}G = \frac{(\Delta_r G^{\ominus} + \lambda)^2}{4\lambda} = \frac{[(-0.665 + 1.53\overline{1})\,\text{eV}]^2}{4(1.53\overline{1}\,\text{eV})} = 0.122\,\text{eV}.$$

Now $k_{\text{et}} = \dfrac{2\langle H_{\text{DA}} \rangle^2}{h} \left(\dfrac{\pi^3}{4\lambda kT}\right)^{1/2} \exp\left(\dfrac{-\Delta^{\ddagger}G}{kT}\right)$

so $\langle H_{DA} \rangle = \left(\dfrac{hk_{et}}{2} \right)^{1/2} \left(\dfrac{4\lambda kT}{\pi^3} \right)^{1/4} \exp \left(\dfrac{\Delta^{\ddagger}G}{2kT} \right),$

$\langle H_{DA} \rangle = \left(\dfrac{(6.626 \times 10^{-34} \, \text{J s})(2.02 \times 10^5 \, \text{s}^{-1})}{2} \right)^{1/2}$

$\times \left(\dfrac{4(1.53\overline{1} \, \text{eV})(1.602 \times 10^{-19} \, \text{J eV}^{-1})(1.381 \times 10^{-23} \, \text{J K}^{-1})(298 \, \text{K})}{\pi^3} \right)^{1/4}$

$\times \exp \left(\dfrac{(0.122 \, \text{eV})(1.602 \times 10^{-19} \, \text{J eV}^{-1})}{2(1.381 \times 10^{-23} \, \text{J K}^{-1})(298 \, \text{K})} \right) = \boxed{9.39 \times 10^{-24} \, \text{J}}$

E24.17(b) Equation 24.83 applies. In Exercise 24.17(a), we found the parameter β to equal $12 \, \text{nm}^{-1}$, so:

$$\ln k_{et}/\text{s}^{-1} = -\beta r + \text{constant} \quad \text{so} \quad \text{constant} = \ln k_{et}/\text{s}^{-1} + \beta r,$$

and constant $= \text{In} \, 2.02 \times 10^5 + (12 \, \text{nm}^{-1})(1.11 \, \text{nm}) = 25.$

Taking the exponential of eqn 24.83 yields:

$$k_{et} = e^{-\beta r + \text{constant}} \, \text{s}^{-1} = e^{-(12/\text{nm})(1.48 \, \text{nm})+25} \, \text{s}^{-1} = \boxed{1.4 \times 10^3 \, \text{s}^{-1}}.$$

Solutions to problems

Solutions to numerical problems

P24.1 $A = N_A \sigma^* \left(\dfrac{8kT}{\pi\mu} \right)$ [Section 24.1 and Exercise 24.13(a); $\mu = \frac{1}{2}m(CH_3)$]

$= (\sigma^*) \times (6.022 \times 10^{23} \, \text{mol}^{-1}) \times \left(\dfrac{(8) \times (1.381 \times 10^{-23} \, \text{J K}^{-1}) \times (298 \, \text{K})}{(\pi) \times (1/2) \times (15.03 \, \text{u}) \times (1.6605 \times 10^{-27} \, \text{kg/u})} \right)^{1/2}$

$= (5.52 \times 10^{26}) \times (\sigma^* \, \text{mol}^{-1} \, \text{m s}^{-1}).$

(a) $\sigma^* = \dfrac{2.4 \times 10^{10} \, \text{mol}^{-1} \, \text{dm}^3 \, \text{s}^{-1}}{5.52 \times 10^{26} \, \text{mol}^{-1} \, \text{m s}^{-1}} = \dfrac{2.4 \times 10^7 \, \text{mol}^{-1} \, \text{m}^3 \, \text{s}^{-1}}{5.52 \times 10^{26} \, \text{mol}^{-1} \, \text{m s}^{-1}} = \boxed{4.4 \times 10^{-20} \, \text{m}^2}.$

(b) Take $\sigma \approx \pi d^2$ and estimate d as $2 \times$ bond length; therefore

$$\sigma = (\pi) \times (154 \times 2 \times 10^{-12} \, \text{m})^2 = 3.0 \times 10^{-19} \, \text{m}^2.$$

Hence $P = \dfrac{\sigma^*}{\sigma} = \dfrac{4.3\overline{5} \times 10^{-20}}{3.0 \times 10^{-19}} = \boxed{0.15}.$

P24.3   For radical recombination it has been found experimentally that $E_a \approx 0$. The maximum rate of recombination is obtained when $P = 1$ (or more), and then

$$k_2 = A = \sigma^* N_A \left(\dfrac{8kT}{\pi\mu} \right)^{1/2} = 4\sigma^* N_A \left(\dfrac{kT}{\pi m} \right)^{1/2} \quad [\mu = \tfrac{1}{2}m].$$

$$\sigma^* \approx \pi d^2 = \pi \times (308 \times 10^{-12} \, \text{m})^2 = 3.0 \times 10^{-19} \, \text{m}^2.$$

Hence

$$k_2 = (4) \times (3.0 \times 10^{-19}\,m^2) \times (6.022 \times 10^{23}\,mol^{-1})$$

$$\times \left(\frac{(1.381 \times 10^{-23}\,J\,K^{-1}) \times (298\,K)}{(\pi) \times (15.03\,u) \times (1.6605 \times 10^{-27}\,kg/u)} \right)^{1/2}$$

$$= 1.7 \times 10^8\,m^3\,mol^{-1}\,s^{-1} = \boxed{1.7 \times 10^{11}\,M^{-1}\,s^{-1}}.$$

The rate constant is for the rate law

$$v = k_2[CH_3]^2.$$

Therefore $d[CH_3]/dt = 2k_2[CH_3]^2$

and its solution is $\dfrac{1}{[CH_3]} - \dfrac{1}{[CH_3]_0} = 2k_2t.$

For 90 per cent recombination, $[CH_3] = 0.10 \times [CH_3]_0$, which occurs when

$$2k_2t = \frac{9}{[CH_3]_0} \quad \text{or} \quad t = \frac{9}{2k_2[CH_3]_0}.$$

The mole fractions of CH_3 radicals in which $10\,mol\%$ of ethane is dissociated is

$$\frac{(2) \times (0.10)}{1 + 0.10} = 0.18.$$

The initial partial pressure of CH_3 radicals is thus

$$p_0 = 0.18p = 1.8 \times 10^4\,Pa$$

and $[CH_3]_0 = \dfrac{1.8 \times 10^4\,Pa}{RT}.$

Therefore $t = \dfrac{9RT}{(2k_2) \times (1.8 \times 10^4\,Pa)} = \dfrac{(9) \times (8.314\,J\,K^{-1}\,mol^{-1}) \times (298\,K)}{(1.7 \times 10^8\,m^3\,mol^{-1}\,s^{-1}) \times (3.6 \times 10^4\,Pa)}$

$$= \boxed{3.6\,ns}.$$

P24.5 $\log k_2 = \log k_2^\circ + 2Az_Az_BI^{1/2}$ with $A = 0.509\,(mol\,dm^{-3})^{-1/2}$ [24.69].

This expression suggests that we should plot $\log k$ against $I^{1/2}$ determine z_B from the slope, since we know that $|z_A| = 1$. We draw up the following table.

$I/(mol\,dm^{-3})$	0.0025	0.0037	0.0045	0.0065	0.0085
$(I/(mol\,dm^{-3}))^{1/2}$	0.050	0.061	0.067	0.081	0.092
$\log(k_2/(dm^3\,mol^{-1}\,s^{-1}))$	0.021	0.049	0.064	0.072	0.100

These points are plotted in Fig. 24.2.

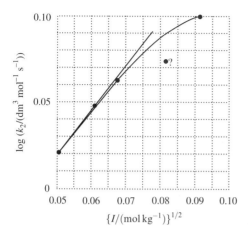

Figure 24.2

The slope of the limiting line in Fig. 24.2 is ≈ 2.5. Since this slope is equal to $2\,A z_A z_B \times (\text{mol dm}^{-3})^{1/2} = 1.018\, z_A z_B$, we have $z_A z_B \approx 2.5$. But $|z_A| = 1$, and so $|z_B| = 2$. Furthermore, z_A and z_B have the same sign because $z_A z_B > 0$. (The data refer to I^- and $S_2O_8^{2-}$.)

P24.7
$$\frac{\sigma^*}{\sigma} \approx \left(\frac{e^2}{4\pi\varepsilon_0 d(I - E_{ea})}\right)^2 \quad \text{[Example 24.2]}.$$

Taking $\sigma = \pi d^2$ gives

$$\sigma^* \approx \pi \left(\frac{e^2}{4\pi\varepsilon_0[I(M) - E_{ea}(X_2)]}\right)^2 = \frac{6.5\,\text{nm}^2}{(I - E_{ea})/\text{eV}}.$$

Thus, σ^* is predicted to increase as $I - E_{ea}$ decreases. The data let us construct the following table.

σ^*/nm^2	Cl_2	Br_2	I_2
Na	0.45	0.42	0.56
K	0.72	0.68	0.97
Rb	0.77	0.72	1.05
Cs	0.97	0.90	1.34

All values of σ^* in the table are smaller than the experimental ones, but they do show the correct trends down the columns. The variation with E_{ea} across the table is not so good, possibly because the electron affinities used here are poor estimates.

Question. Can you find better values of electron affinities and do they improve the horizontal trends in the table?

P24.9 (a)

$$\frac{d[F_2O]}{dt} = -k_1[F_2O]^2 - k_2[F][F_2O], \tag{1}$$

$$\frac{d[F]}{dt} = k_1[F_2O]^2 - k_2[F][F_2O] + 2k_3[OF]^2 - 2k_4[F]^2[F_2O], \tag{2}$$

$$\frac{d[OF]}{dt} = k_1[F_2O]^2 + k_2[F][F_2O] - 2k_3[OF]^2. \tag{3}$$

Applying the steady-state approximation to both [F] and [OF] and adding the resulting equations gives

$$
\begin{array}{ll}
k_1[F_2O]^2 - k_2[F]_{SS}[F_2O] + 2k_3[OF]_{SS}^2 \quad -2k_4[F]_{SS}^2[F_2O] & = 0 \\
k_1[F_2O]^2 + k_2[F]_{SS}[F_2O] - 2k_3[OF]_{SS}^2 & = 0 \\
\hline
\end{array}
$$

$$2k_1\,[F_2O]^2 \qquad\qquad\qquad -2k_4\,[F]_{SS}^2\,[F_2O] = 0$$

Solving for $[F]_{SS}$ gives

$$[F]_{SS} = \left(\frac{k_1}{k_4}[F_2O]\right)^{1/2}.$$

Substituting (4) into (1)

$$\frac{d\,[F_2O]}{dt} = k_1\,[F_2O]^2 - k_2\left(\frac{k_1}{k_4}\right)^{1/2}[F_2O]^{3/2}$$

or

$$\boxed{-\frac{d[F_2O]}{dt} = k_1[F_2O]^2 + k_2\left(\frac{k_1}{k_4}\right)^{1/2}[F_2O]^{3/2}.}$$

Comparison with the experimental rate law reveals that they are consistent when we make the following identifications.

$$k = k_1 = 7.8 \times 10^{13} e^{-E_1/RT}\ \text{dm}^3\ \text{mol}^{-1}\ \text{s}^{-1},$$

$$E_1 = (19350\ \text{K})R = 160.9\ \text{kJ mol}^{-1},$$

$$k' = k_2\left(\frac{k_1}{k_4}\right)^{1/2} = 2.3 \times 10^{10} e^{-E'/RT}\ \text{dm}^3\ \text{mol}^{-1}\ \text{s}^{-1},$$

$$E' = (16910\ \text{K})R = 140.6\ \text{kJ mol}^{-1}.$$

(b)

$$
\begin{array}{lll}
\tfrac{1}{2}O_2 + F_2 \to F_2O & \Delta_f H(F_2O) = 24.41\ \text{kJ mol}^{-1} \\
2F \to F_2 & \Delta H = -D(F-F) = -160.6\ \text{kJ mol}^{-1} \\
O \to \tfrac{1}{2}O_2 & \Delta H = -\tfrac{1}{2}D(O-O) = -249.1\ \text{kJ mol}^{-1} \\
\hline
2F + O \to F_2O &
\end{array}
$$

$$\Delta H(\text{FO–F}) + \Delta H(\text{O–F}) = -\left[\Delta_f H(\text{F}_2\text{O}) - D(\text{F–F}) - \tfrac{1}{2}D(\text{O–O})\right]$$

$$= -(24.41 - 160.6 - 249.1) \text{ kJ mol}^{-1}$$

$$= 385.3 \text{ kJ mol}^{-1}.$$

We estimate that $\boxed{\Delta H(\text{FO–F}) \approx E_1 = 160.9\,\text{kJ mol}^{-1}}$.
Then

$$\Delta H(\text{O–F}) = 385.3 \text{ kJ mol}^{-1} - \Delta H(\text{FO–F})$$

$$\approx (385.3 - 160.9) \text{ kJ mol}^{-1}.$$

$$\boxed{\Delta H(\text{O–F}) \approx 224.4\,\text{kJ mol}^{-1}.}$$

In order to determine the activation energy of reaction (2) we assume that each rate is expressed in Arrhenius form, then

$$\ln k' = \ln k_2 + \tfrac{1}{2}\ln k_1 - \tfrac{1}{2}\ln k_4$$

or

$$\ln A' - \frac{E'}{RT} = \ln A_2 - \frac{E_2}{RT} + \tfrac{1}{2}\ln A_1 - \tfrac{1}{2}\frac{E_1}{RT} - \tfrac{1}{2}\ln A_4 + \tfrac{1}{2}\frac{E_4}{RT}.$$

Differentiating with respect to T, we obtain

$$E' = E_2 + \tfrac{1}{2}E_1 - \tfrac{1}{2}E_4 = 140.6 \text{ kJ mol}^{-1}$$

or

$$E_2 - \tfrac{1}{2}E_4 = E' - \tfrac{1}{2}E_1 = (140.6 - 80.4) \text{ kJ mol}^{-1}$$
$$= 60.2 \text{ kJ mol}^{-1}.$$

E_4 is expected to be small since reaction (4) is termolecular, so we set $E_4 \approx 0$; then

$$E_2 \approx \boxed{60\,\text{kJ mol}^{-1}}.$$

P24.11 Linear regression analysis of ln(rate constant) against $1/T$ yields the following results:

$$\ln(k/22.4\,\text{dm}^3\,\text{mol}^{-1}\text{min}^{-1}) = C + B/T$$

where $C = 34.36$,   standard deviation $= 0.36$,

$\quad B = -23227$ K,   standard deviation $= 252$ K,

$\quad R = \boxed{0.99976}$ [good fit].

$$\ln(k'/22.4\,\text{dm}^3\,\text{mol}^{-1}\text{min}^{-1}) = C_2 + B_2/T$$

where $C' = 28.30$,   standard deviation $= 0.84$,
  $B' = -21065$ K,   standard deviation $= 582$ K,
  $R = \boxed{0.99848}$ [good fit].

The regression parameters can be used in the calculation of the pre-exponential factor (A) and the activation energy (E_a) using $\ln k = \ln A - E_a/RT$.

$$\ln A = C + \ln(22.4) = 37.47,$$

$$A = 1.87 \times 10^{16}\,\text{dm}^3\,\text{mol}^{-1}\,\text{min}^{-1} = \boxed{3.12 \times 10^{14}\,\text{dm}^3\,\text{mol}^{-1}\text{s}^{-1}}.$$

$$E_a = -RB = -(8.3145\,\text{J K}^{-1}\text{mol}^{-1}) \times (-23227\,\text{K}) \times \left(\frac{10^{-3}\,\text{kJ}}{\text{J}}\right)$$

$$= \boxed{193\,\text{kJ mol}^{-1}}.$$

$$\ln A' = C' + \ln(22.4) = 31.41,$$

$$A' = 4.37 \times 10^{13}\,\text{dm}^3\,\text{mol}^{-1}\,\text{min}^{-1} = \boxed{7.29 \times 10^{11}\,\text{dm}^3\,\text{mol}^{-1}\,\text{s}^{-1}}.$$

$$E_a' = -RB' = -(8.3145\,\text{J K}^{-1}\text{mol}^{-1}) \times (-21065\,\text{K}) \times \left(\frac{10^{-3}\,\text{kJ}}{\text{J}}\right)$$

$$= \boxed{175\,\text{kJ mol}^{-1}}.$$

To summarize

	$A/(\text{dm}^3\,\text{mol}^{-1}\text{s}^{-1})$	$E_a/(\text{kJ mol}^{-1})$
k	$3.12 \times 10^{14}(= A)$	193
k'	$7.29 \times 10^{11}(= A')$	175

Both sets of data, k and k', fit the Arrhenius equation very well and hence are consistent with the collision theory of bimolecular gas-phase reactions which provides an equation 24.19 compatible with the Arrhenius equation. The numerical values for k' and A may be compared to the results of Exercise 24.7(a) and are in rough agreement at 647 K, as is the value of E_a.

Solutions to theoretical problems

P24.13 $$[J]^* = k \int_0^t [J]e^{-kt}dt + [J]e^{-kt}\,[24.40],$$

$$\frac{\partial [J]^*}{\partial t} = k[J]e^{-kt} + \frac{\partial [J]}{\partial t}e^{-kt} - k[J]e^{-kt} = \left(\frac{\partial [J]}{\partial t}\right)e^{-kt},$$

$$\frac{\partial^2 [J]^*}{\partial x^2} = k \int_0^t \left(\frac{\partial^2 [J]}{\partial x^2}\right)e^{-kt}dt + \left(\frac{\partial^2 [J]}{\partial x^2}\right)e^{-kt}.$$

Then, since

$$D \frac{\partial^2 [J]}{\partial x^2} = \frac{\partial [J]}{\partial t} \quad [24.39, k = 0],$$

we find that

$$D \frac{\partial^2 [J]^*}{\partial x^2} = k \int_0^t \left(\frac{\partial [J]^*}{\partial t} \right) e^{-kt} dt + \left(\frac{\partial [J]}{\partial t} \right) e^{-kt}$$

$$= k \int_0^t \left(\frac{\partial [J]^*}{\partial t} \right) dt + \frac{\partial [J]^*}{\partial t} = k[J]^* + \frac{\partial [J]^*}{\partial t}$$

which rearranges to eqn 24.39. When $t = 0$, $[J]^* = [J]$, and so the same initial conditions are satisfied. (The same boundary conditions are also satisfied.)

P24.15

$$\frac{q_m^{\ominus T}}{N_A} = 2.561 \times 10^{-2} (T/K)^{5/2} (M/\text{g mol}^{-1})^{3/2} \quad \text{[Table 17.3]}.$$

For $T \approx 300\,\text{K}$, $M \approx 50\,\text{g mol}^{-1}$, $\dfrac{q_m^{\ominus T}}{N_A} \approx \boxed{1.4 \times 10^7}$,

$$q^R(\text{nonlinear}) = \frac{1.0270}{\sigma} \times \frac{(T/K)^{3/2}}{(ABC/\text{cm}^{-3})^{1/2}} \quad \text{[Table 17.3]}.$$

For $T \approx 300\,\text{K}$, $A \approx B \approx C = 2\,\text{cm}^{-1}$, $\sigma \approx 2$ [Section 13.5], $q^R(\text{NL}) \approx \boxed{900}$,

$$q^R(\text{linear}) = \frac{0.6950}{\sigma} \times \frac{(T/K)}{(B/\text{cm}^{-1})} \quad \text{[Table 17.3]}.$$

For $T \approx 300\,\text{K}$, $B \approx 1\,\text{cm}^{-1}$, $\sigma \approx 1$ [Section 13.5], $q^R(L) \approx \boxed{200}$.

$$q^V \approx \boxed{1} \quad \text{and} \quad q^E \approx \boxed{1} \quad \text{[Table 17.3]},$$

$$k_2 = \frac{\kappa kT}{h} \bar{K}^{\ddagger} \quad [24.53]$$

$$= \left(\frac{\kappa kT}{h} \right) \times \left(\frac{RT}{p} \right) \times \left(\frac{N_A \bar{q}_C^{\ominus}}{q_A^{\ominus} q_B^{\ominus}} \right) e^{-\Delta E_0/RT} \quad [24.51] \approx A e^{-E_a/RT}.$$

We then use

$$\frac{q_A^{\ominus}}{N_A} = \frac{q_A^{\ominus T}}{N_A} \approx 1.4 \times 10^7 \text{[above]},$$

$$\frac{q_B^{\ominus}}{N_A} = \frac{q_B^{\ominus T}}{N_A} \approx 1.4 \times 10^7 \text{[above]},$$

$$\frac{\bar{q}_C^{\ominus}}{N_A} = \frac{q_C^{\ominus T} q^R(L)}{N_A} \approx (2^{3/2}) \times (1.4 \times 10^7) \times (200 \text{ [above]}) = 7.9 \times 10^9,$$

[The factor of $2^{3/2}$ comes from $m_C = m_A + m_B \approx 2m_A$ and $q^T \propto m^{3/2}$.]

$$\frac{RT}{p^\ominus} \approx = \frac{(8.314\,\text{J K}^{-1}\,\text{mol}^{-1}) \times (300\,\text{K})}{10^5\,\text{Pa}} = 2.5 \times 10^{-2}\,\text{m}^3\,\text{mol}^{-1},$$

$$\frac{\kappa kT}{h} \approx \frac{kT}{h} = \frac{(1.381 \times 10^{-23}\,\text{J K}^{-1}) \times (300\text{K})}{6.626 \times 10^{-34}\,\text{J s}} = 6.25 \times 10^{12}\,\text{s}^{-1}.$$

Therefore, the pre-exponential factor

$$A \approx \frac{(6.25 \times 10^{12}\,\text{s}^{-1}) \times (2.5 \times 10^{-12}\,\text{m}^3\,\text{mol}^{-1}) \times (7.9 \times 10^9)}{(1.4 \times 10^7)^2}$$

$$\approx 6.3 \times 10^6\,\text{m}^3\,\text{mol}^{-1}\,\text{s}^{-1} \quad \text{or} \quad \boxed{6.3 \times 10^9\,\text{dm}^3\,\text{mol}^{-1}\,\text{s}^{-1}}.$$

If all three species are nonlinear,

$$\frac{q_A^\ominus}{N_A} \approx (1.4 \times 10^7) \times (900) = 1.3 \times 10^{10} \approx \frac{q_B^\ominus}{N_A}.$$

$$\frac{q_A^\ominus}{N_A} \approx (2^{3/2}) \times (1.4 \times 10^7) \times (900) = 3.6 \times 10^{10}$$

$$A \approx \frac{(6.25 \times 10^{12}\,\text{s}^{-1}) \times (2.5 \times 10^{-2}\,\text{m}^3\,\text{mol}^{-1}) \times (3.6 \times 10^{10})}{(1.3 \times 10^{10})^2}$$

$$\approx 33\,\text{m}^3\,\text{mol}^{-1}\,\text{s}^{-1} \quad \text{or} \quad \boxed{3.3 \times 10^4\,\text{dm}^3\,\text{mol}^{-1}\,\text{s}^{-1}}.$$

Therefore, $P = \dfrac{A(\text{NL})}{A(\text{L})} = \dfrac{3.3 \times 10^4}{6.3 \times 10^9} = \boxed{5.2 \times 10^{-6}}$.

These numerical values may be compared to those given in Table 24.1 and in Example 24.1. They lie within the range found experimentally.

P24.17   We consider the y-direction to be the direction of diffusion. Hence, for the activated atom the vibrational mode in this direction is lost. Therefore,

$$q^\ddagger = q_Z^{\ddagger V} q_x^{\ddagger V} \text{ for the activated atom, and}$$

$$q = q_x^V q_y^V q_z^V \text{ for an atom at the bottom of a well.}$$

For classical vibration, $q^V \approx kT/h\nu$ [Section 24.4].

The diffusion process described is unimolecular, hence first-order, and therefore analogously to the second-order case of Section 24.4 [also see Problem 24.4] we may write

$$-\frac{d[x]}{dt} = k^\ddagger [x]^\ddagger = \nu K^\ddagger [x] = k_1 [x] \quad \left[K^\ddagger = \frac{[x]^\ddagger}{[x]} \right].$$

Thus

$$k_1 = \nu K^\ddagger = \nu \left(\frac{kT}{h\nu} \right) \times \left(\frac{q^\ddagger}{q} \right) e^{-\beta \Delta E_0} \quad \left[\beta = \frac{1}{RT} \text{ here} \right]$$

where $q^{\ddagger}$ and q are the (vibrational) partition functions at the top and foot of the well respectively. Therefore

$$k_1 = \frac{kT}{h} \left(\frac{(kT/h\nu^{\ddagger})^2}{(kT/h\nu)^3} \right) e^{-\beta \Delta E_0} = \boxed{\frac{\nu^3}{\nu'^2} e^{-\beta \Delta E_0}}.$$

(a) $\nu^{\ddagger} = \nu$; $k_1 = \nu e^{-\beta \Delta E_0}$. Assume $\Delta E_0 \approx E_a$; hence

$$k_1 \approx 10^{11} \, \text{Hz} \, e^{-60 \times 10^3/(8.314 \times 500)} = 5.4 \times 10^4 \, \text{s}^{-1}.$$

But $D = \dfrac{\lambda^2}{2\tau} \approx \dfrac{1}{2}\lambda^2 k_1 \left[21.85; \tau = \dfrac{1}{k_1}, \text{Problem 22.10} \right]$

$$= \frac{1}{2} \times (316 \, \text{pm})^2 \times 5.4 \times 10^4 \, \text{s}^{-1} = \boxed{2.7 \times 10^{-15} \, \text{m}^2 \, \text{s}^{-1}}.$$

(b) $\nu^{\ddagger} = \dfrac{1}{2}\nu$; $k_1 = 4\nu e^{-\beta \Delta E_0} = 2.2 \times 10^5 \, \text{s}^{-1}$;

$$D = (4) \times (2.7 \times 10^{-15} \, \text{m}^2 \, \text{s}^{-1}) = \boxed{1.1 \times 10^{-14} \, \text{m}^2 \, \text{s}^{-1}}.$$

P24.19   The change in intensity of the beam, dI, is proportional to the number of scatterers per unit volume, $\mathcal{N}_S$, the intensity of the beam, I, and the path length dl. The constant of proportionality is defined as the collision cross-section σ. Therefore,

$$dI = -\sigma \mathcal{N}_s I dl \quad \text{or} \quad d \ln I = -\sigma \mathcal{N}_s dl.$$

If the incident intensity (at $l = 0$) is I_0 and the emergent intensity is I, we can write

$$\ln \frac{I}{I_0} = -\sigma \mathcal{N}_s l \quad \text{or} \quad I = \boxed{I_0 e^{-\sigma \mathcal{N}_s l}}.$$

P24.21 $A + B \rightarrow C^{\ddagger} \rightarrow P.$

$$k_2 = \left(\kappa \frac{kT}{h} \right) \times \left(\frac{N_A RT}{p^{\ominus}} \right) \frac{q_{C^{\ddagger}}^{\ominus}}{q_A^{\ominus} q_B^{\ominus}} e^{-\Delta E_0/RT} \quad [24.52].$$

We assume that the only factor that changes between the atomic and molecular case is the ratio of the partition functions.

(1) For collisions between atoms

$$q_A^{\ominus} = q_A^{\text{T}} \approx 10^{26},$$

$$q_B^{\ominus} = q_B^{\text{T}} \approx 10^{26},$$

$$q_C^{\ominus} = (q_C^{\text{R}})^2 q_C^{\text{V}} q_C^{\text{T}} \approx (10^{1.5})^2 \times (1) \times (10^{26}) \approx 10^{29},$$

$$k_2(\text{atoms}) \propto \frac{10^{29}}{10^{26} \times 10^{26}} = 10^{-23}.$$

(2) For collisions between nonlinear molecules

$$q_A^{\ominus} = (q_A^R)^3 (q_A^V)^{3N-6} (q_A^T) \approx (10^{1.5})^3 \times (1) \times (10^{26}) \approx 3 \times 10^{30},$$

$$q_B^{\ominus} = (q_B^R)^3 (q_B^V)^{3N'-6} (q_B^T) \approx 3 \times 10^{30},$$

$$q_C^{\ominus} = (q_C^R)^3 (q_C^V)^{3(N+N')-6} (q_C^T) \approx 3 \times 10^{30}.$$

$$k_2(\text{molecules}) \propto \frac{3 \times 10^{30}}{1 \times 10^{61}} = 3 \times 10^{-31}.$$

Therefore $k_2(\text{atoms})/k_2(\text{molecules}) \approx \dfrac{10^{-23}}{3 \times 10^{-31}} \approx \boxed{3 \times 10^7}$.

Solutions to applications

P24.23   Collision theory gives for a rate constant with no energy barrier

$$k = P\sigma \left(\frac{8kT}{\pi\mu} \right)^{1/2} N_A \quad \text{so} \quad P = \frac{k}{\sigma N_A} \left(\frac{\pi\mu}{8kT} \right)^{1/2}.$$

$$P = \frac{k/(\text{dm}^3 \, \text{mol}^{-1} \, \text{s}^{-1}) \times (10^{-3} \, \text{m}^3 \, \text{dm}^{-3})}{(\sigma/\text{nm}^2) \times (10^{-9} \, \text{m})^2 \times (6.022 \times 10^{23} \, \text{mol}^{-1})}$$

$$\times \left(\frac{\pi \times (\mu/\text{u}) \times (1.66 \times 10^{-27} \, \text{kg})}{8 \times (1.381 \times 10^{-23} \, \text{J K}^{-1}) \times (298 \, \text{K})} \right)^{1/2}$$

$$= \frac{(6.61 \times 10^{-13}) k/(\text{dm}^3 \, \text{mol}^{-1} \, \text{s}^{-1})}{(\sigma/\text{nm}^2) \times (\mu/\text{u})^{1/2}}.$$

The collision cross-section is

$$\sigma_{AB} = \pi d_{AB}^2 \quad \text{where} \quad d_{AB} = \tfrac{1}{2}(d_A + d_B) = \frac{\sigma_A^{1/2} + \sigma_B^{1/2}}{2\pi^{1/2}} \quad \text{so} \quad \sigma_{AB} = \frac{(\sigma_A^{1/2} + \sigma_B^{1/2})^2}{4}.$$

The collision cross-section for O_2 is listed in the *Data section*. We would not be far wrong if we took that of the ethyl radical to equal that of ethene; similarly, we will take that of cyclohexyl to equal that of benzene. For O_2 with ethyl

$$\sigma = \frac{(0.40^{1/2} + 0.64^{1/2})^2}{4} \, \text{nm}^2 = 0.51 \, \text{nm}^2,$$

$$\mu = \frac{m_O m_e}{m_O + m_e} = \frac{(32.0 \, \text{u}) \times (29.1 \, \text{u})}{(32.0 + 29.1) \, \text{u}} = 15.2 \, \text{u},$$

so $P = \dfrac{(6.61 \times 10^{-13}) \times (4.7 \times 10^9)}{(0.51) \times (15.2)^{1/2}} = \boxed{1.6 \times 10^{-3}}$

For O_2 with cyclohexyl

$$\sigma = \frac{(0.40^{1/2} + 0.88^{1/2})^2}{4} \, \text{nm}^2 = 0.62 \, \text{nm}^2,$$

$$\mu = \frac{m_O m_C}{m_O + m_C} = \frac{(32.0 \, \text{u}) \times (77.1 \, \text{u})}{(32.0 + 77.1) \, \text{u}} = 22.6 \, \text{u},$$

so $P = \dfrac{(6.61 \times 10^{-13}) \times (8.4 \times 10^9)}{(0.62) \times (22.6)^{1/2}} = \boxed{1.8 \times 10^{-3}}$.

P24.25   Equation 24.69 may be written in the form:

$$z_A^2 = \frac{1}{2A} \frac{\log(k_2/k_2^\circ)}{I^{1/2}}$$ where we have used $z_A = z_B$ for the cationic protein. This equation suggests that z_A can be determined through analysis that uses the mean value of $(\log(k_2/k_2^\circ))/I^{1/2}$ for several experiments over a range of various ionic strengths.

$$z_A = \sqrt{\frac{1}{2A} \, \text{mean} \left\{ \frac{\log\left(k_2/k_2^\circ\right)}{I^{1/2}} \right\}}.$$

We draw up a table that contains data rows needed for the computation.

I	0.0100	0.0150	0.0200	0.0250	0.0300	0.0350
k/k°	8.10	13.30	20.50	27.80	38.10	52.00
$\log(k/k^\circ)/I^{0.5}$	9.08	9.18	9.28	9.13	9.13	9.17

mean $\{ \log(k/k^\circ)/I^{0.5} \} = 9.17$,

$$z_A = \sqrt{\frac{1}{2A} \, \text{mean} \left\{ \frac{\log(k_2/k_2^\circ)}{I^{1/2}} \right\}} = \sqrt{\frac{9.16}{2(0.509)}} = \boxed{+3.0}$$

where we have used the positive root because the protein is cationic.

P24.27   Does eqn 24.83,

$$\ln k_{et} = -\beta r + \text{constant},$$

apply to these data? Draw the following table.

r/nm	k_{et}/s^{-1}	$\ln k_{et}/\text{s}^{-1}$
0.48	1.58×10^{12}	28.1
0.95	3.98×10^9	22.1
0.96	1.00×10^9	20.7
1.23	1.58×10^8	18.9
1.35	3.98×10^7	17.5
2.24	6.31×10^1	4.14

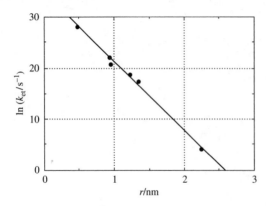

Figure 24.3

and plot $\ln k_{et}$ vs. r (Fig. 24.3).

The data fall on a good straight line, so the equation $\boxed{\text{appears to apply}}$. The least squares linear fit equation is

$$\ln k_{et}/s = 34.7 - 13.4 r/\text{nm}, \quad R^2 (\text{correlation coefficient}) = 0.991$$

so we identify $\boxed{\beta = 13.4\,\text{nm}^{-1}}$.

P24.29   azurin(red) + cytochrome c(ox) $\rightarrow$ azurin(ox) + cytochrome c(red).

$$E^\circ = E_R^\circ - E_L^\circ = 0.260\,\text{V} - 0.304\,\text{V} = -0.044\,\text{V}.$$

$$K = e^{vFE^\circ/RT}\,[7.30] = e^{1(96485.3\,\text{C mol}^{-1})(-0.044\,\text{V})/(8.31451\,\text{J K}^{-1}\,\text{mol}^{-1})(298.15\,\text{K})} = e^{-1.713} = 0.180.$$

$$k_{obs} = (k_{DD}k_{AA}K)^{1/2}\,[24.86].$$

$$k_{DD} = \frac{k_{obs}^2}{k_{AA}K} = \frac{\left(1.6 \times 10^3\,\text{dm}^3\,\text{mol}^{-1}\,\text{s}^{-1}\right)^2}{\left(1.5 \times 10^2\,\text{dm}^3\,\text{mol}^{-1}\,\text{s}^{-1}\right)(0.180)} = \boxed{9.5 \times 10^4\,\text{dm}^3\,\text{mol}^{-1}\,\text{s}^{-1}}.$$

25 Processes at solid surfaces

Answers to discussion questions

D25.1 **(a)** A terrace is a flat layer of atoms on a surface. There can be more than one terrace on a surface, each at a different height. Steps are the joints between the terraces; the height of the step can be constant or variable.

(b) The motion of one section of a crystal past another (a dislocation) results in steps and terraces. See Figures 25.2 and 25.3 of the text. A special kind of dislocation is the screw dislocation shown in Fig. 25.3. Imagine a cut in the crystal, with the atoms to the left of the cut pushed up through a distance of one unit cell. The surface defect formed by a screw dislocation is a step, possibly with kinks, where growth can occur. The incoming particles lie in ranks on the ramp, and successive ranks reform the step at an angle to its initial position. As deposition continues the step rotates around the screw axis, and is not eliminated. Growth may therefore continue indefinitely. Several layers of deposition may occur, and the edges of the spirals might be cliffs several atoms high (Fig. 25.4).

Propagating spiral edges can also give rise to flat terraces. Terraces are formed if growth occurs simultaneously at neighboring left- and right-handed screw dislocations (Fig. 25.5). Successive tables of atoms may form as counter-rotating defects collide on successive circuits, and the terraces formed may then fill up by further deposition at their edges to give flat crystal planes.

D25.3 *Langmuir isotherm*. This isotherm applies under the following conditions:

1. Adsorption cannot proceed beyond monolayer coverage.
2. All sites are equivalent and the surface is uniform.
3. The ability of a molecule to adsorb at a given site is independent of the occupation of neighboring sites.

BET isotherm. Condition number 1 above is removed. This isotherm applies to multi-layer coverage.

Temkin isotherm. Condition number 2 is removed and it is assumed that the energetically most favorable sites are occupied first. The Temkin isotherm corresponds to supposing that the adsorption enthalpy changes linearly with pressure.

Freundlich isotherm. Condition 2 is again removed, but this isotherm corresponds to a logarithmic change in the adsorption enthalpy with pressure.

D25.5 In the Langmuir–Hinshelwood mechanism of surface-catalyzed reactions, the reaction takes place by encounters between molecular fragments and atoms already adsorbed on the surface. We therefore expect the rate law to be second-order in the extent of surface coverage:

$$A + B \to P \quad v = k\theta_A\theta_B$$

Insertion of the appropriate isotherms for A and B then gives the reaction rate in terms of the partial pressures of the reactants. For example, if A and B follow Langmuir isotherms (eqn 25.4), and adsorb without dissociation, then it follows that the rate law is

$$v = \frac{kK_AK_Bp_Ap_B}{(1 + K_Ap_A + K_Bp_B)^2}.$$

The parameters K in the isotherms and the rate constant k are all temperature dependent, so the overall temperature dependence of the rate may be strongly non-Arrhenius (in the sense that the reaction rate is unlikely to be proportional to $\exp(-E_a/RT)$).

In the Eley–Rideal mechanism (ER mechanism) of a surface-catalyzed reaction, a gas-phase molecule collides with another molecule already adsorbed on the surface. The rate of formation of product is expected to be proportional to the partial pressure, p_B of the non-adsorbed gas B and the extent of surface coverage, θ_A, of the adsorbed gas A. It follows that the rate law should be

$$A + B \to P \quad v = kp_A\theta_B.$$

The rate constant, k, might be much larger than for the uncatalyzed gas-phase reaction because the reaction on the surface has a low activation energy and the adsorption itself is often not activated.

If we know the adsorption isotherm for A, we can express the rate law in terms of its partial pressure, p_A. For example, if the adsorption of A follows a Langmuir isotherm in the pressure range of interest, then the rate law would be

$$v = \frac{kKp_Ap_B}{1 + Kp_A}.$$

If A were a diatomic molecule that adsorbed as atoms, we would substitute the isotherm given in eqn 25.6 instead.

According to eqn 25.27, when the partial pressure of A is high (in the sense $Kp_A \gg 1$, there is almost complete surface coverage, and the rate is equal to kp_B. Now the rate-determining step is the collision of B with the adsorbed fragments. When the pressure of A is low ($Kp_A \ll 1$), perhaps because of its reaction, the rate is equal to kKp_Ap_B; and now the extent of surface coverage is important in the determination of the rate.

In the Mars van Krevelen mechanism of catalytic oxidation, for example, in the partial oxidation of propene to propenal, the first stage is the adsorption of the propene molecule with loss of a hydrogen to form the allyl radical, $CH_2=CHCH_2$. An O atom in the surface can now transfer to this radical, leading to the formation of acrolein (propenal, $CH_2=CHCHO$) and its desorption from the surface. The H atom also escapes with a surface O atom, and goes on to form H_2O, which leaves the surface. The surface is left with vacancies and metal ions in lower oxidation states. These vacancies are attacked by O_2 molecules in the overlying gas, which then chemisorb as O_2^- ions, so reforming the catalyst. This sequence of events involves great upheavals of the surface, and some materials break up under the stress.

D25.7   Zeolites are microporous aluminosilicates, in which the surface effectively extends deep inside the solid. M^{n+} cations and H_2O molecules can bind inside the cavities, or pores, of the Al—O—Si framework (see

Fig. 25.29 of the text). Small neutral molecules, such as CO_2, NH_3, and hydrocarbons (including aromatic compounds), can also adsorb to the internal surfaces and this partially accounts for the utility of zeolites as catalysts.

Like enzymes, a zeolite catalyst with a specific composition and structure is very selective toward certain reactants and products because only molecules of certain sizes can enter and exit the pores in which catalysis occurs. It is also possible that zeolites derive their selectivity from the ability to bind and to stabilize only transition states that fit properly in the pores.

D25.9   The net current density at an electrode is j; j_0 is the exchange current density; α is the transfer coefficient; f is the ratio F/RT; and η is the overpotential.

(a) $j = j_0 f \eta$ is the current density in the low overpotential limit.
(b) $j = j_0 e^{(1-\alpha)f\eta}$ applies when the overpotential is large and positive.
(c) $j = -j_0 e^{-\alpha f\eta}$ applies when the overpotential is large and negative.

D25.11   The principles of operation of a fuel cell are very much the same as those of a conventional galvanic cell. Both employ a spontaneous electrochemical reaction to produce an electric current that can be used as a power source for external devices. The main difference between fuel cells and ordinary cells is that the reacting substance is a material that we normally classify as a fuel and it is continuously supplied to the cell from an external source. We wish to obtain large currents from fuel cells and in order to accomplish that goal a number of obstacles limiting the rate of reaction have to be overcome. One way to increase the rate of the reaction in the cell is to use a catalytic surface with a large effective surface area to increase the current density. Operating the cells at high temperatures can increase reaction rates and in some cases molten electrolytes and electrodes are employed.

Solutions to exercises

E25.1(b)   The number of collisions of gas molecules per unit surface area is

$$Z_W = \frac{N_A p}{(2\pi M R T)^{1/2}}$$

(a) For N_2

(i) $Z_W = \dfrac{(6.022 \times 10^{23}\,\text{mol}^{-1}) \times (10.0\,\text{Pa})}{(2\pi \times (28.013 \times 10^{-3}\,\text{kg mol}^{-1}) \times (8.3145\,\text{J K}^{-1}\,\text{mol}^{-1}) \times (298\,\text{K}))^{1/2}}$

$= 2.88 \times 10^{23}\,\text{m}^{-2}\,\text{s}^{-1}$

$= \boxed{2.88 \times 10^{19}\,\text{cm}^{-2}\,\text{s}^{-1}}$

(ii) $Z_W = \dfrac{(6.022 \times 10^{23}\,\text{mol}^{-1}) \times (0.150 \times 10^{-6}\,\text{Torr}) \times (1.01 \times 10^5\,\text{Pa}/760\,\text{Torr})}{(2\pi \times (28.013 \times 10^{-3}\,\text{kg mol}^{-1}) \times (8.3145\,\text{J K}^{-1}\,\text{mol}^{-1}) \times (298\,\text{K}))^{1/2}}$

$= 5.75 \times 10^{17}\,\text{m}^{-2}\,\text{s}^{-1}$

$= \boxed{5.75 \times 10^{13}\,\text{cm}^{-2}\,\text{s}^{-1}}$

(b) For methane

(i) $Z_W = \dfrac{(6.022 \times 10^{23}\,\text{mol}^{-1}) \times (10.0\,\text{Pa})}{(2\pi \times (16.04 \times 10^{-3}\,\text{kg mol}^{-1}) \times (8.3145\,\text{J K}^{-1}\,\text{mol}^{-1}) \times (298\,\text{K}))^{1/2}}$

$= 3.81 \times 10^{23}\,\text{m}^{-2}\,\text{s}^{-1}$

$= \boxed{3.81 \times 10^{19}\,\text{cm}^{-2}\,\text{s}^{-1}}$

(ii) $Z_W = \dfrac{(6.022 \times 10^{23}\,\text{mol}^{-1}) \times (0.150 \times 10^{-6}\,\text{Torr}) \times (1.01 \times 10^5\,\text{Pa}/760\,\text{Torr})}{(2\pi \times (16.04 \times 10^{-3}\,\text{kg mol}^{-1}) \times (8.3145\,\text{J K}^{-1}\,\text{mol}^{-1}) \times (298\,\text{K}))^{1/2}}$

$= 7.60 \times 10^{17}\,\text{m}^{-2}\,\text{s}^{-1}$

$= \boxed{7.60 \times 10^{13}\,\text{cm}^{-2}\,\text{s}^{-1}}$

E25.2(b) The number of collisions of gas molecules per unit surface area is

$$Z_W = \dfrac{N_A p}{(2\pi MRT)^{1/2}} \quad \text{so} \quad p = \dfrac{Z_W A (2\pi MRT)^{1/2}}{N_A A}$$

$$p = \dfrac{(5.00 \times 10^{19}\,\text{s}^{-1})}{(6.022 \times 10^{23}\,\text{mol}^{-1}) \times \pi \times (1/2 \times 2.0 \times 10^{-3}\,\text{m})^2}$$
$$\times (2\pi \times (28.013 \times 10^{-3}\,\text{kg mol}^{-1}) \times (8.3145\,\text{J mol}^{-1}\,\text{K}^{-1}) \times (525\,\text{K}))^{1/2}$$

$$= \boxed{7.3 \times 10^2\,\text{Pa}}$$

E25.3(b) The number of collisions of gas molecules per unit surface area is

$$Z_W = \dfrac{N_A p}{(2\pi M\,RT)^{1/2}}$$

so the rate of collision per Fe atom will be $Z_W A$ where A is the area per Fe atom. The exposed surface consists of faces of the bcc unit cell, with one atom per face. So the area per Fe is

$$A = c^2 \quad \text{and} \quad \text{rate} = Z_W A = \dfrac{N_A p c^2}{(2\pi M\,RT)^{1/2}}$$

where c is the length of the unit cell. So

$$\text{rate} = \dfrac{(6.022 \times 10^{23}\,\text{mol}^{-1}) \times (24\,\text{Pa}) \times (145 \times 10^{-12}\,\text{m})^2}{(2\pi \times (4.003 \times 10^{-3}\,\text{kg mol}^{-1}) \times (8.3145\,\text{J K}^{-1}\,\text{mol}^{-1}) \times (100\,\text{K}))^{1/2}}$$

$$= \boxed{6.6 \times 10^4\,\text{s}^{-1}}$$

E25.4(b) The number of CO molecules adsorbed on the catalyst is

$$N = nN_A = \dfrac{pVN_A}{RT} = \dfrac{(1.00\,\text{atm}) \times (4.25 \times 10^{-3}\,\text{dm}^3) \times (6.022 \times 10^{23}\,\text{mol}^{-1})}{(0.08206\,\text{dm}^3\,\text{atm K}^{-1}\,\text{mol}^{-1}) \times (273\,\text{K})}$$

$$= 1.14 \times 10^{20}$$

The area of the surface must be the same as that of the molecules spread into a monolayer, namely, the number of molecules times each one's effective area

$$A = Na = (1.14 \times 10^{20}) \times (0.165 \times 10^{-18} \, \text{m}^2) = \boxed{18.8 \, \text{m}^2}$$

E25.5(b) If the adsorption follows the Langmuir isotherm, then

$$\theta = \frac{Kp}{1 + Kp} \quad \text{so} \quad K = \frac{\theta}{p(1 - \theta)} = \frac{V/V_{\text{mon}}}{p(1 - V/V_{\text{mon}})}$$

Setting this expression at one pressure equal to that at another pressure allows solution for V_{mon}

$$\frac{V_1/V_{\text{mon}}}{p_1(1 - V_1/V_{\text{mon}})} = \frac{V_2/V_{\text{mon}}}{p_2(1 - V_2/V_{\text{mon}})} \quad \text{so} \quad \frac{p_1(V_{\text{mon}} - V_1)}{V_1} = \frac{p_2(V_{\text{mon}} - V_2)}{V_2}$$

$$V_{\text{mon}} = \frac{p_1 - p_2}{p_1/V_1 - p_2/V_2} = \frac{(52.4 - 104) \, \text{kPa}}{(52.4/1.60 - 104/2.73) \, \text{kPa cm}^{-3}} = \boxed{9.7 \, \text{cm}^3}$$

E25.6(b) The mean lifetime of a chemisorbed molecule is comparable to its half-life:

$$t_{1/2} = \tau_0 \exp\left(\frac{E_d}{RT}\right) \approx (10^{-14} \, \text{s}) \exp\left(\frac{155 \times 10^3 \, \text{J mol}^{-1}}{(8.3145 \, \text{J K}^{-1} \, \text{mol}^{-1}) \times (500 \, \text{K})}\right) = \boxed{200 \, \text{s}}$$

E25.7(b) The desorption rate constant is related to the half-life by

$$t = (\ln 2)/k_d \quad \text{so} \quad k_d = (\ln 2)/t$$

The desorption rate constant is related to its Arrhenius parameters by

$$k_d = A \exp\left(\frac{-E_d}{RT}\right) \quad \text{so} \quad \ln k_d = \ln A - \frac{E_d}{RT}$$

and $$E_d = \frac{(\ln k_1 - \ln k_2)R}{T_2^{-1} - T_1^{-1}} = \frac{(\ln 1.35 - \ln 1) \times (8.3145 \, \text{J K}^{-1} \, \text{mol}^{-1})}{(600 \, \text{K})^{-1} - (1000 \, \text{K})^{-1}}$$

$$E_d = \boxed{3.7 \times 10^3 \, \text{J mol}^{-1}}$$

E25.8(b) The Langmuir isotherm is

$$\theta = \frac{Kp}{1 + Kp} \quad \text{so} \quad p = \frac{\theta}{K(1 - \theta)}$$

(a) $$p = \frac{0.20}{(0.777 \, \text{kPa}^{-1}) \times (1 - 0.20)} = \boxed{0.32 \, \text{kPa}}$$

(b) $$p = \frac{0.75}{(0.777 \, \text{kPa}^{-1}) \times (1 - 0.75)} = \boxed{3.9 \, \text{kPa}}$$

E25.9(b) The Langmuir isotherm is

$$\theta = \frac{Kp}{1 + Kp}$$

We are looking for θ, so we must first find K or m_{mon}

$$K = \frac{\theta}{p(1 - \theta)} = \frac{m/m_{mon}}{p(1 - m/m_{mon})}$$

Setting this expression at one pressure equal to that at another pressure allows solution for m_{mon}

$$\frac{m_1/m_{mon}}{p_1(1 - m_1/m_{mon})} = \frac{m_2/m_{mon}}{p_2(1 - m_2/m_{mon})} \quad \text{so} \quad \frac{p_1(m_{mon} - m_1)}{m_1} = \frac{p_2(m_{mon} - m_2)}{m_2}$$

$$m_{mon} = \frac{p_1 - p_2}{p_1/m_1 - p_2/m_2} = \frac{(36.0 - 4.0)\,\text{kPa}}{(36.0/0.63 - 4.0/0.21)\,\text{kPa mg}^{-1}} = 0.84\,\text{mg}$$

So $\theta_1 = 0.63/0.84 = \boxed{0.75}$ and $\theta_2 = 0.21/0.84 = \boxed{0.25}$

E25.10(b) The mean lifetime of a chemisorbed molecule is comparable to its half-life

$$t_{1/2} = \tau_0 \exp\left(\frac{E_d}{RT}\right)$$

(a) At 400 K : $t_{1/2} = (0.12 \times 10^{-12}\,\text{s}) \exp\left(\frac{20 \times 10^3\,\text{J mol}^{-1}}{(8.3145\,\text{J K}^{-1}\text{mol}^{-1}) \times (400\,\text{K})}\right)$

$$= \boxed{4.9 \times 10^{-11}\,\text{s}}$$

At 800 K : $t_{1/2} = (0.12 \times 10^{-12}\,\text{s}) \exp\left(\frac{20 \times 10^3\,\text{J mol}^{-1}}{(8.3145\,\text{J K}^{-1}\text{mol}^{-1}) \times (800\,\text{K})}\right)$

$$= \boxed{2.4 \times 10^{-12}\,\text{s}}$$

(b) At 400 K : $t_{1/2} = (0.12 \times 10^{-12}\,\text{s}) \exp\left(\frac{200 \times 10^3\,\text{J mol}^{-1}}{(8.3145\,\text{J K}^{-1}\text{mol}^{-1}) \times (400\,\text{K})}\right)$

$$= \boxed{1.6 \times 10^{13}\,\text{s}}$$

At 800 K : $t_{1/2} = (0.12 \times 10^{-12}\,\text{s}) \exp\left(\frac{200 \times 10^3\,\text{J mol}^{-1}}{(8.3145\,\text{J K}^{-1}\text{mol}^{-1}) \times (800\,\text{K})}\right)$

$$= \boxed{1.4\,\text{s}}$$

E25.11(b) The Langmuir isotherm is

$$\theta = \frac{Kp}{1 + Kp} \quad \text{so} \quad p = \frac{\theta}{K(1 - \theta)}$$

For constant fractional adsorption

$$pK = \text{constant} \quad \text{so} \quad p_1 K_1 = p_2 K_2 \quad \text{and} \quad p_2 = p_1\frac{K_1}{K_2}$$

But $K \propto \exp\left(\dfrac{-\Delta_{ad}H^{\ominus}}{RT}\right)$ so $\dfrac{K_1}{K_2} = \exp\left(\dfrac{-\Delta_{ad}H^{\ominus}}{R}\left(\dfrac{1}{T_1} - \dfrac{1}{T_2}\right)\right)$

$$p_2 = p_1 \exp\left(\dfrac{-\Delta_{ad}H^{\ominus}}{R}\left(\dfrac{1}{T_1} - \dfrac{1}{T_2}\right)\right)$$

$$= (8.86\,\text{kPa}) \times \exp\left(\left(\dfrac{-12.2 \times 10^3\,\text{J mol}^{-1}}{8.3145\,\text{J K}^{-1}\,\text{mol}^{-1}}\right) \times \left(\dfrac{1}{298\,\text{K}} - \dfrac{1}{318\,\text{K}}\right)\right) = \boxed{6.50\,\text{kPa}}$$

E25.12(b) The Langmuir isotherm would be

(a) $\theta = \dfrac{Kp}{1 + Kp}$

(b) $\theta = \dfrac{(Kp)^{1/2}}{1 + (Kp)^{1/2}}$

(c) $\theta = \dfrac{(Kp)^{1/3}}{1 + (Kp)^{1/3}}$

A plot of θ versus p at low pressures (where the denominator is approximately 1) would show progressively weaker dependence on p for dissociation into two or three fragments.

E25.13(b) The Langmuir isotherm is

$$\theta = \dfrac{Kp}{1 + Kp} \quad\text{so}\quad p = \dfrac{\theta}{K(1 - \theta)}$$

For constant fractional adsorption

$$pK = \text{constant} \quad\text{so}\quad p_1 K_1 = p_2 K_2 \quad\text{and}\quad \dfrac{p_2}{p_1} = \dfrac{K_1}{K_2}$$

But $K \propto \exp\left(\dfrac{-\Delta_{ad}H^{\ominus}}{RT}\right)$ so $\dfrac{p_2}{p_1} = \exp\left(\dfrac{-\Delta_{ad}H^{\ominus}}{R}\left(\dfrac{1}{T_1} - \dfrac{1}{T_2}\right)\right)$

and $\Delta_{ad}H^{\ominus} = R\left(\dfrac{1}{T_1} - \dfrac{1}{T_2}\right)^{-1} \ln\dfrac{p_1}{p_2}$,

$$\Delta_{ad}H^{\ominus} = (8.3145\,\text{J K}^{-1}\,\text{mol}^{-1}) \times \left(\dfrac{1}{180\,\text{K}} - \dfrac{1}{240\,\text{K}}\right)^{-1} \times \left(\ln\dfrac{350\,\text{kPa}}{1.02 \times 10^3\,\text{kPa}}\right)$$

$$= -6.40 \times 10^4\,\text{J mol}^{-1} = \boxed{-6.40\,\text{kJ mol}^{-1}}$$

E25.14(b) The time required for a given quantity of gas to desorb is related to the activation energy for desorption by

$$t \propto \exp\left(\dfrac{E_d}{RT}\right) \quad\text{so}\quad \dfrac{t_1}{t_2} = \exp\left(\dfrac{E_d}{R}\left(\dfrac{1}{T_1} - \dfrac{1}{T_2}\right)\right)$$

and $E_d = R\left(\dfrac{1}{T_1} - \dfrac{1}{T_2}\right)^{-1} \ln\dfrac{t_1}{t_2}$

$$E_d = (8.3145\,\text{J K}^{-1}\,\text{mol}^{-1}) \times \left(\dfrac{1}{873\,\text{K}} - \dfrac{1}{1012\,\text{K}}\right)^{-1} \times \left(\ln\dfrac{1856\,\text{s}}{8.44\,\text{s}}\right)$$

$$= \boxed{2.85 \times 10^5\,\text{J mol}^{-1}}$$

(a) The same desorption at 298 K would take

$$t = (1856 \text{ s}) \times \exp\left(\left(\frac{2.85 \times 10^5 \text{ J mol}^{-1}}{8.3145 \text{ J K}^{-1}\text{mol}^{-1}}\right) \times \left(\frac{1}{298 \text{ K}} - \frac{1}{873 \text{ K}}\right)\right) = \boxed{1.48 \times 10^{36} \text{ s}}$$

(b) The same desorption at 1500 K would take

$$t = (8.44 \text{ s}) \times \exp\left(\left(\frac{2.85 \times 10^5 \text{ J mol}^{-1}}{8.3145 \text{ J K}^{-1} \text{ mol}^{-1}}\right) \times \left(\frac{1}{1500 \text{ K}} - \frac{1}{1012 \text{ K}}\right)\right)$$

$$= \boxed{1.38 \times 10^{-4} \text{ s}}$$

E25.15(b) Disregarding signs, the electric field is the gradient of the electrical potential

$$\varepsilon = \frac{d\Delta\varphi}{dx} \approx \frac{\Delta\phi}{d} = \frac{\sigma}{\varepsilon} = \frac{\sigma}{\varepsilon_r \varepsilon_0} = \frac{0.12 \text{ C m}^{-2}}{(48) \times (8.854 \times 10^{-12} \text{ J}^{-1} \text{ C}^2 \text{ m}^{-1})} = \boxed{2.8 \times 10^8 \text{ V m}^{-1}}$$

E25.16(b) In the high overpotential limit

$$j = j_0 e^{(1-\alpha)f\eta} \quad \text{so} \quad \frac{j_1}{j_2} = e^{(1-\alpha)f(\eta_1 - \eta_2)} \quad \text{where} \quad f = \frac{F}{RT} = \frac{1}{25.69 \text{ mV}}$$

The overpotential η_2 is

$$\eta_2 = \eta_1 + \frac{1}{f(1-\alpha)} \ln\frac{j_2}{j_1} = 105 \text{ mV} + \left(\frac{25.69 \text{ mV}}{1 - 0.42}\right) \times \ln\left(\frac{72 \text{ mA cm}^{-2}}{17.0 \text{ mA cm}^{-2}}\right)$$

$$= \boxed{16\overline{7} \text{ mV}}$$

E25.17(b) In the high overpotential limit

$$j = j_0 e^{(1-\alpha)f\eta} \quad \text{so} \quad j_0 = j e^{(\alpha-1)f\eta}$$

$$j_0 = (17.0 \text{ mA cm}^{-2}) \times e^{\{(0.42-1) \times (105 \text{ mV})/(25.69 \text{ mV})\}} = \boxed{1.6 \text{ mA cm}^{-2}}$$

E25.18(b) In the high overpotential limit

$$j = j_0 e^{(1-\alpha)f\eta} \quad \text{so} \quad \frac{j_1}{j_2} = e^{(1-\alpha)f(\eta_1 - \eta_2)} \quad \text{and} \quad j_2 = j_1 e^{(1-\alpha)f(\eta_2 - \eta_1)}.$$

So the current density at 0.60 V

$$j_2 = (1.22 \text{ mA cm}^{-2}) \times e^{\{(1-0.50) \times (0.60 \text{ V} - 0.50 \text{ V})/(0.025 69 \text{ V})\}} = \boxed{8.5 \text{ mA cm}^{-2}}$$

Note: the exercise says the data refer to the same material and at the same temperature as the previous Exercise (25.18(a)), yet the results for the current density at the same overpotential differ by a factor of over 5!

E25.19(b) **(a)** The Butler–Volmer equation gives

$$j = j_0(e^{(1-\alpha)f\eta} - e^{-\alpha f\eta})$$

$$= (2.5 \times 10^{-3}\,\text{A cm}^{-2}) \times \left(e^{\{1-0.58)\times(0.30\,\text{V})/(0.025\,69\,\text{V})\}} - e^{-\{(0.58)\times(0.30\,\text{V})/0.025\,69\,\text{V})\}}\right)$$

$$= \boxed{0.34\,\text{A cm}^{-2}}$$

(b) According to the Tafel equation

$$j = j_0 e^{(1-\alpha)f\eta}$$

$$= (2.5 \times 10^{-3}\,\text{A cm}^{-2})e^{\{(1-0.58)\times(0.30\,\text{V})/(0.025\,69\,\text{V})\}} = \boxed{0.34\,\text{A cm}^{-2}}$$

The validity of the Tafel equation improves as the overpotential increases.

E25.20(b) The limiting current density is

$$j_{\lim} = \frac{zFDc}{\delta}$$

but the diffusivity is related to the ionic conductivity (Chapter 21)

$$D = \frac{\lambda RT}{z^2 F^2} \quad \text{so} \quad j_{\lim} = \frac{c\lambda}{\delta z f}$$

$$j_{\lim} = \frac{(1.5\,\text{mol m}^{-3}) \times (10.60 \times 10^{-3}\,\text{S m}^2\,\text{mol}^{-1}) \times (0.025\,69\,\text{V})}{(0.32 \times 10^{-3}\,\text{m}) \times (+1)}$$

$$= \boxed{1.3\,\text{A m}^{-2}}$$

E25.21(b) For the iron electrode $E^{\ominus} = -0.44\text{V}$ (Table 7.2) and the Nernst equation for this electrode (section 7.7a) is

$$E = E^{\ominus} - \frac{RT}{\nu F}\ln\left(\frac{1}{[\text{Fe}^{2+}]}\right) \quad \nu = 2$$

Since the hydrogen overpotential is 0.60 V evolution of H_2 will begin when the potential of the Fe electrode reaches -0.60 V. Thus

$$-0.60\text{V} = -0.44\text{V} + \frac{0.025\,69\text{V}}{2}\ln[\text{Fe}^{2+}]$$

$$\ln[\text{Fe}^{2+}] = \frac{-0.16\,\text{V}}{0.0128\,\text{V}} = -12.\bar{5}$$

$$[\text{Fe}^{2+}] = \boxed{4 \times 10^{-6}\,\text{mol dm}^{-3}}$$

COMMENT. Essentially all Fe^{2+} has been removed by deposition before evolution of H_2 begins

E25.22(b) The zero-current potential of the electrode is given by the Nernst equation

$$E = E^\ominus - \frac{RT}{\nu F} \ln Q = E^\ominus - \frac{1}{f} \ln \frac{a\left(\text{Fe}^{2+}\right)}{a\left(\text{Fe}^{3+}\right)} = 0.77\,\text{V} - \frac{1}{f} \ln \frac{a\left(\text{Fe}^{2+}\right)}{a\left(\text{Fe}^{3+}\right)}$$

The Butler–Volmer equation gives

$$j = j_0(e^{(1-\alpha)f\eta} - e^{-\alpha f\eta}) = j_0(e^{(0.42)f\eta} - e^{-0.58f\eta})$$

where η is the overpotential, defined as the working potential E' minus the zero-current potential E.

$$\eta = E' - 0.77\,\text{V} + \frac{1}{f} \ln \frac{a\left(\text{Fe}^{2+}\right)}{a\left(\text{Fe}^{3+}\right)} = E' - 0.77\,\text{V} + \frac{1}{f} \ln r,$$

where r is the ratio of activities; so

$$j = j_0(e^{(0.42)E'/f}e^{\{(0.42)\times(-0.77\,\text{V})/(0.025\,69\,\text{V})\}}r^{0.42}$$

$$- e^{(-0.58)E'/f}e^{\{(-0.58)\times(-0.77\,\text{V})/(0.025\,69\,\text{V})\}}r^{-0.58})$$

Specializing to the condition that the ions have equal activities yields

$$\boxed{j = (2.5\,\text{mA cm}^{-2}) \times [(e^{(0.42)E'/f} \times (3.4\overline{1} \times 10^{-6}) - e^{(-0.58)E'/f} \times (3.5\overline{5} \times 10^{7})]}$$

E25.23(b) Note. The exercise did not supply values for j_0 or α. Assuming $\alpha = 0.5$, only j/j_0 is calculated. From Exercise 25.22(b)

$$j = j_0(e^{(0.50)E'/f}e^{-(0.50)E^\ominus/f}r^{0.50} - e^{(-0.50)E'/f}e^{(0.50)E^\ominus/f}r^{-0.50})$$

$$= 2j_0 \sinh\left[\tfrac{1}{2}f\,E' - \tfrac{1}{2}f\,E^\ominus + \tfrac{1}{2}\ln r\right],$$

so, if the working potential is set at 0.50 V, then

$$j = 2j_0 \sinh\left[\tfrac{1}{2}(0.91\,\text{V})/(0.02569\,\text{V}) + \tfrac{1}{2}\ln r\right]$$

$$j/j_0 = 2\,\sinh\left(8.4\overline{8} + \tfrac{1}{2}\ln r\right)$$

At $r = 0.1$: $j/j_0 = 2\sinh\left(8.4\overline{8} + \tfrac{1}{2}\ln 0.10\right) = 1.5 \times 10^3\,\text{mA cm}^{-2} = \boxed{1.5\,\text{A cm}^{-2}}$

At $r = 1$: $j/j_0 = 2\sinh(8.4\overline{8} + 0.0) = 4.8 \times 10^3\,\text{mA cm}^{-2} = \boxed{4.8\,\text{A cm}^{-2}}$

At $r = 10$: $j/j_0 = 2\sinh\left(8.4\overline{8} + \tfrac{1}{2}\ln 10\right) = 1.5 \times 10^4\,\text{mA cm}^{-2} = \boxed{15\,\text{A cm}^{-2}}$

E25.24(b) The potential needed to sustain a given current depends on the activities of the reactants, but the *over* potential does not. The Butler–Volmer equation says

$$j = j_0(e^{(1-\alpha)f\eta} - e^{-\alpha f\eta})$$

This cannot be solved analytically for η, but in the high-overpotential limit it reduces to the Tafel equation

$$j = j_0 e^{(1-\alpha)f\eta} \quad \text{so} \quad \eta = \frac{1}{(1-\alpha)f} \ln \frac{j}{j_0} = \frac{0.025\,69\,\text{V}}{1-0.75} \ln \frac{15\,\text{mA cm}^{-2}}{4.0 \times 10^{-2}\,\text{mA cm}^{-2}}$$

$$\eta = \boxed{0.61\,\text{V}}$$

This is a sufficiently large overpotential to justify use of the Tafel equation.

E25.25(b) The number of singly charged particles transported per unit time per unit area at equilibrium is the exchange current density divided by the charge

$$N = \frac{j_0}{e}$$

The frequency f of participation per atom on an electrode is

$$f = Na$$

where a is the effective area of an atom on the electrode surface.

For the Cu, $H_2|H^+$ electrode

$$N = \frac{j_0}{e} = \frac{1.0 \times 10^{-6}\,\text{A cm}^{-2}}{1.602 \times 10^{-19}\,\text{C}} = \boxed{6.2 \times 10^{12}\,\text{s}^{-1}\text{cm}^{-2}}$$

$$f = Na = (6.2 \times 10^{12}\,\text{s}^{-1}\,\text{cm}^{-2}) \times (260 \times 10^{-10}\,\text{cm})^2$$

$$= \boxed{4.2 \times 10^{-3}\,\text{s}^{-1}}$$

For the $Pt|Ce^{4+}, Ce^{3+}$ electrode

$$N = \frac{j_0}{e} = \frac{4.0 \times 10^{-5}\,\text{A cm}^{-2}}{1.602 \times 10^{-19}\,\text{C}} = \boxed{2.5 \times 10^{14}\,\text{s}^{-1}\,\text{cm}^{-2}}$$

The frequency f of participation per atom on an electrode is

$$f = Na = (2.5 \times 10^{14}\,\text{s}^{-1}\,\text{cm}^{-2}) \times (260 \times 10^{-10}\,\text{cm})^2 = \boxed{0.17\,\text{s}^{-1}}$$

E25.26(b) The resistance R of an ohmic resistor is

$$R = \frac{\text{potential}}{\text{current}} = \frac{\eta}{jA}$$

where A is the surface area of the electrode. The overpotential in the low overpotential limit is

$$\eta = \frac{j}{f\,j_0} \quad \text{so} \quad R = \frac{1}{f\,j_0 A}$$

(a) $R = \dfrac{0.025\,69\text{ V}}{(5.0 \times 10^{-12}\text{ A cm}^{-2}) \times (1.0\,\text{cm}^2)} = 5.1 \times 10^9\,\Omega = \boxed{5.1\text{ G}\Omega}$

(b) $R = \dfrac{0.025\,69\text{ V}}{(2.5 \times 10^{-3}\text{ A cm}^{-2}) \times (1.0\,\text{cm}^2)} = \boxed{10\,\Omega}$

E25.27(b) No reduction of cations to metal will occur until the cathode potential is dropped below the zero-current potential for the reduction of Ni^{2+} (-0.23 V at unit activity). Deposition of Ni will occur at an appreciable rate after the potential drops significantly below this value; however, the deposition of Fe will begin (albeit slowly) after the potential is brought below -0.44 V. If the goal is to deposit pure Ni, then the Ni will be deposited rather slowly at just above -0.44 V; then the Fe can be deposited rapidly by dropping the potential well below -0.44 V.

E25.28(b) As was noted in Exercise 25.18(a), an overpotential of 0.6 V or so is necessary to obtain significant deposition or evolution, so H_2 is evolved from acid solution at a potential of about -0.6 V. The reduction potential of Cd^{2+} is more positive than this (-0.40 V), so Cd will deposit (albeit slowly) from Cd^{2+} before H_2 evolution.

E25.29(b) Zn can be deposited if the H^+ discharge current is less than about 1 mA cm^{-2}. The exchange current, according to the high negative overpotential limit, is

$$j = j_0 e^{-\alpha f \eta}$$

At the standard potential for reduction of Zn^{2+} (-0.76 V)

$$j = (0.79\text{ mA cm}^{-2}) \times e^{-\{(0.5) \times (-0.76\text{ V})/(0.025\,69\text{ V})\}} = 2.1 \times 10^9\text{ mA cm}^{-2}$$

$\boxed{\text{much too large to allow deposition}}$. (That is, H_2 would begin being evolved, and fast, long before Zn began to deposit.)

E25.30(b) Fe can be deposited if the H^+ discharge current is less than about 1 mA cm^{-2}. The exchange current, according to the high negative overpotential limit, is

$$j = j_0 e^{-\alpha f \eta}$$

At the standard potential for reduction of Fe^{2+} (-0.44 V)

$$j = (1 \times 10^{-6}\text{ A cm}^{-2}) \times e^{-\{(0.5) \times (-0.44\text{ V})/(0.025\,69\text{ V})\}} = 5.2 \times 10^{-3}\text{ A cm}^{-2}$$

$\boxed{\text{a bit too large to allow deposition}}$. (That is, H_2 would begin being evolved at a moderate rate before Fe began to deposit.)

E25.31(b) The lead acid battery half-cells are

$$Pb^{4+} + 2e^- \rightarrow Pb^{2+} \qquad\qquad\qquad 1.67\text{ V}$$

$$\text{and } PbSO_4 + 2e^- \rightarrow Pb + SO_4^{2-} \qquad\qquad -0.36\text{ V,}$$

for a total of $E^{\ominus} = \boxed{2.03\text{ V}}$. Power is

$$P = IV = (100 \times 10^{-3}\,\text{A}) \times (2.03\,\text{V}) = \boxed{0.203\text{ W}}$$

if the cell were operating at its zero-current potential yet producing 100mA.

E25.32(b) Two electrons are lost in the corrosion of each zinc atom, so the number of zinc atoms lost is half the number of electrons which flow per unit time, i.e. half the current divided by the electron charge. The volume taken up by those zinc atoms is their number divided by number density; their number density is their mass density divided by molar mass times Avogadro's number. Dividing the volume of the corroded zinc over the surface from which they are corroded gives the linear corrosion rate; this affects the calculation by changing the current to the current density. So the rate of corrosion is

$$\text{rate} = \frac{jM}{2e\rho N_A} = \frac{(2.0\,\text{A m}^{-2}) \times (65.39 \times 10^{-3}\,\text{kg mol}^{-1})}{2(1.602 \times 10^{-19}\,\text{C}) \times (7133\,\text{kg m}^{-3}) \times (6.022 \times 10^{23}\,\text{mol}^{-1})}$$

$$= 9.5 \times 10^{-11}\,\text{m s}^{-1}$$

$$= (9.5 \times 10^{-11}\,\text{m s}^{-1}) \times (10^{3}\,\text{mm m}^{-1}) \times (3600 \times 24 \times 365\,\text{s y}^{-1})$$

$$= \boxed{3.0\ \text{mm y}^{-1}}$$

Solutions to problems

Solutions to numerical problems

P25.1   Refer to Fig. 25.1.

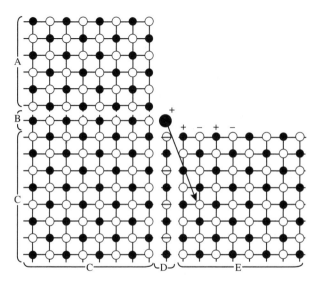

Figure 25.1

Evaluate the sum of $\pm\dfrac{1}{r_i}$, where r_i is the distance from the ion i to the ion of interest, taking $+\dfrac{1}{r}$ for ions of like charge and $-\dfrac{1}{r}$ for ions of opposite charge. The array has been divided into five zones. Zones B and D can be summed analytically to give $-\ln 2 = -0.69$. The summation over the other zones, each of which gives the same result, is tedious because of the very slow convergence of the sum. Unless you make a very clever choice of the sequence of ions (grouping them so that their contributions almost cancel), you will find the following values for arrays of different sizes.

10×10	20×20	50×50	100×100	200×200
0.259	0.273	0.283	0.286	0.289

The final figure is in good agreement with the analytical value, 0.289 259 7...

(a) For a cation above a flat surface, the energy (relative to the energy at infinity, and in multiples of $e^2/4\pi\varepsilon r_0$ where r_0 is the lattice spacing (200 pm)), is

$$\text{Zone } C + D + E = 0.29 - 0.69 + 0.29 = \boxed{-0.11}$$

which implies an attractive state.

(b) For a cation at the foot of a high cliff, the energy is

$$\text{Zone } A + B + C + D + E = 3 \times 0.29 + 2 \times (-0.69) = \boxed{-0.51}$$

which is significantly more attractive. Hence, the latter is the more likely settling point (if potential energy considerations such as these are dominant).

P25.3   Refer to Fig. 25.2.

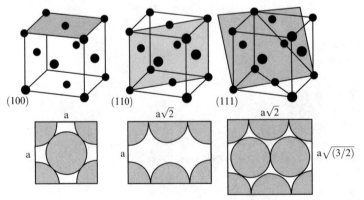

Figure 25.2

The (100) and (110) faces each expose two atoms, and the (111) face exposes four. The areas of the faces of each cell are **(a)** $(352 \text{ pm})^2 = 1.24 \times 10^{-15} \text{ cm}^2$, **(b)** $\sqrt{2} \times (352 \text{ pm})^2 = 1.75 \times 10^{-15} \text{ cm}^2$, and **(c)** $\sqrt{3} \times (352 \text{ pm})^2 = 2.15 \times 10^{-15} \text{ cm}^2$. The numbers of atoms exposed per square centimetre are therefore

(a) $\dfrac{2}{1.24 \times 10^{-15} \text{ cm}^2} = \boxed{1.61 \times 10^{15} \text{ cm}^{-2}}$.

(b) $\dfrac{2}{1.75 \times 10^{-15} \text{ cm}^2} = \boxed{1.14 \times 10^{15} \text{ cm}^{-2}}$.

(c) $\dfrac{4}{2.15 \times 10^{-15} \text{ cm}^2} = \boxed{1.86 \times 10^{15} \text{ cm}^{-2}}$.

For the collision frequencies calculated in Exercise 25.1(a), the frequency of collision per atom is calculated by dividing the values given there by the number densities just calculated. We can therefore draw up the following table.

$Z/(\text{atom}^{-1}\,\text{s}^{-1})$	Hydrogen		Propane	
	100 Pa	10^{-7} Torr	100 Pa	10^{-7} Torr
(100)	6.8×10^5	8.7×10^{-2}	1.4×10^5	1.9×10^{-2}
(110)	9.6×10^5	1.2×10^{-1}	2.0×10^5	2.7×10^{-2}
(111)	5.9×10^5	7.5×10^{-2}	1.2×10^5	1.7×10^{-2}

P25.5

$$\dfrac{V}{V_{\text{mon}}} = \dfrac{cz}{(1-z)\{1-(1-c)z\}} \quad \left[25.8, \text{ BET isotherm, } z = \dfrac{p}{p^*} \right].$$

This rearranges to

$$\dfrac{z}{(1-z)V} = \dfrac{1}{cV_{\text{mon}}} + \dfrac{(c-1)z}{cV_{\text{mon}}}.$$

Therefore a plot of the left-hand side against z should result in a straight line if the data obeys the BET isotherm. We draw up the following tables.

(a) 0°C, $p^* = 3222$ Torr.

$p/$Torr	105	282	492	594	620	755	798
$10^3 z$	32.6	87.5	152.7	184.4	192.4	234.3	247.7
$10^3 z/(1-z)(V/\text{cm}^3)$	3.04	7.10	12.1	14.1	15.4	17.7	20.0

(b) 18°C, $p^* = 6148$ Torr.

$p/$Torr	39.5	62.7	108	219	466	555	601	765
$10^3 z$	6.4	10.2	17.6	35.6	75.8	90.3	97.8	124.4
$10^3 z/(1-z)(V/\text{cm}^3)$	0.70	1.05	1.74	3.27	6.36	7.58	8.09	10.8

The points are plotted in Fig. 25.3, but we analyse the data by a least-squares procedure.

The intercepts are at **(a)** 0.466 and **(b)** 0.303. Hence

$$\dfrac{1}{cV_{\text{mon}}} = \text{(a) } 0.466 \times 10^{-3} \text{ cm}^{-3}, \quad \text{(b) } 0.303 \times 10^{-3} \text{ cm}^{-3}.$$

The slopes of the lines are **(a)** 76.10 and **(b)** 79.54. Hence

$$\frac{c-1}{cV_{mon}} = \text{(a) } 76.10 \times 10^{-3} \text{ cm}^{-3}, \quad \text{(b) } 79.54 \times 10^{-3} \text{ cm}^{-3}.$$

Solving the equations gives

$$c - 1 = \text{(a)}163.\bar{3}, \text{ (b)}262.\bar{5}$$

and hence

$$c = \text{(a) } \boxed{164}, \text{ (b) } \boxed{264}; \quad V_{mon} = \text{(a) } \boxed{13.1 \text{ cm}^3}, \text{ (b) } \boxed{12.5 \text{ cm}^3}.$$

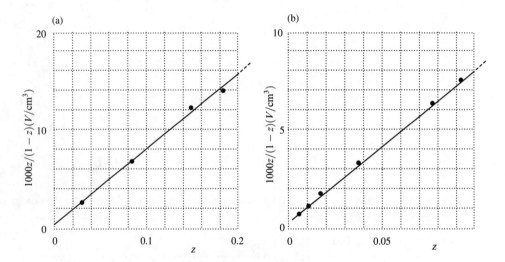

Figure 25.3

P25.7 $\theta = c_1 p^{1/c_2}.$

We adapt this isotherm to a liquid by noting that $w_a \propto \theta$ and replacing p by [A], the concentration of the acid. Then $w_a = c_1[A]^{1/c_2}$ (with c_1, c_2 modified constants), and hence

$$\log w_a = \log c_1 + \frac{1}{c_2} \times \log[A].$$

We draw up the following table.

[A]/(mol dm^{-3})	0.05	0.10	0.15	0.20	0.25
log([A]/mol dm^{-3})	−1.30	−1.00	−0.30	−0.00	0.18
log(w_a/g)	−1.40	−1.22	−0.92	−0.80	−0.72

These points are plotted in Fig 25.4.

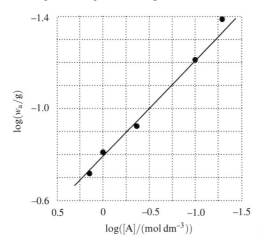

Figure 25.4

They fall on a reasonably straight line with slope 0.42 and intercept -0.80. Therefore, $c_2 = 1/0.42 = \boxed{2.4}$ and $c_1 = \boxed{0.16}$. (The units of c_1 are bizarre: $c_1 = 0.16\,\text{g mol}^{-0.42}\,\text{dm}^{1.26}$.)

P25.9   Taking the log of the isotherm gives

$$\ln c_{\text{ads}} = \ln K + (\ln c_{\text{sol}})/n$$

so a plot of $\ln c_{\text{ads}}$ versus $\ln c_{\text{sol}}$ would have a slope of $1/n_\infty$ and a y-intercept of $\ln K$. The transformed data and plot are shown in Fig. 25.5.

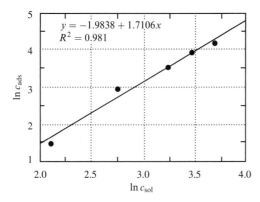

Figure 25.5

$c_{\text{sol}}/(\text{mg g}^{-1})$	8.26	15.65	25.43	31.74	40.00
$c_{\text{ads}}/(\text{mg g}^{-1})$	4.4	19.2	35.2	52.0	67.2
$\ln c_{\text{sol}}$	2.11	2.75	3.24	3.46	3.69
$\ln c_{\text{ads}}$	1.48	2.95	3.56	3.95	4.21

$$K = e^{-1.9838}\,\text{mg g}^{-1} = \boxed{0.138\,\text{mg g}^{-1}} \quad \text{and} \quad n = 1/1.71 = \boxed{0.58}.$$

In order to express this information in terms of fractional coverage, the amount of adsorbate correspond-ing to monolayer coverage must be known. This saturation point, however, has no special significance in the Freundlich isotherm (i.e. it does not correspond to any limiting case).

P25.11   The Langmuir isotherm is

$$\theta = \frac{Kp}{1 + Kp} = \frac{n}{n_\infty} \text{ so } n(1 + Kp) = n_\infty Kp \text{ and } \frac{p}{n} = \frac{p}{n_\infty} + \frac{1}{Kn_\infty}.$$

So a plot of p/n against p should be a straight line with slope $1/n_\infty$ and y-intercept $1/Kn_\infty$. The transformed data and plot (Fig. 25.6) follow

$p/$kPa	31.00	38.22	53.03	76.38	101.97	130.47	165.06	182.41	205.75	219.91
$n/(\text{mol kg}^{-1})$	1.00	1.17	1.54	2.04	2.49	2.90	3.22	3.30	3.35	3.36
$(p/n)/(\text{kPa mol}^{-1}\text{ kg})$	31.00	32.67	34.44	37.44	40.95	44.99	51.26	55.28	61.42	65.45

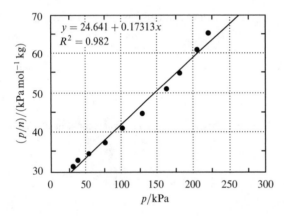

Figure 25.6

$$n_\infty = \frac{1}{0.17313 \text{ mol}^{-1}\text{ kg}} = \boxed{5.78 \text{ mol kg}^{-1}}.$$

The y-intercept is

$$b = \frac{1}{Kn_\infty} \text{ so } K = \frac{1}{bn_\infty} = \frac{1}{(24.641 \text{ kPa mol}^{-1}\text{ kg}) \times (5.78 \text{ mol kg}^{-1})},$$

$$K = 7.02 \times 10^{-3} \text{ kPa}^{-1} = \boxed{7.02 \text{ Pa}^{-1}}.$$

P25.13 $\ln j = \ln j_0 + (1 - \alpha)f\eta$ [25.45].

Draw up the following table.

$\eta/$mV	50	100	150	200	250
$\ln(j/\text{mA cm}^{-2})$	0.98	2.19	3.40	4.61	5.81

The points are plotted in Fig. 25.7.

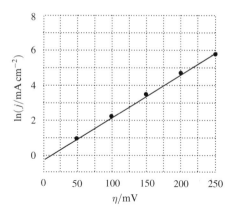

Figure 25.7

The intercept is at -0.25, and so $j_0/(\text{mA cm}^{-2}) = e^{-0.25} = \boxed{0.78}$. The slope is 0.0243, and so $(1 - \alpha)F/RT = 0.0243 \text{ mV}^{-1}$. It follows that $1 - \alpha = 0.62$, and so $\boxed{\alpha = 0.38}$. If η were large but negative,

$$|j| \approx j_0 e^{-\alpha f \eta} [25.46] = (0.78 \text{ mA cm}^{-2}) \times \left(e^{-0.38\eta/25.7 \text{ mV}}\right)$$

$$= (0.78 \text{ mA cm}^{-2}) \times \left(e^{-0.015(\eta/\text{mV})}\right)$$

and we draw up the following table.

η/mV	-50	-100	-150	-200	-250
$j/(\text{mA cm}^{-2})$	1.65	3.50	7.40	15.7	33.2

P25.15 $j_{\text{lim}} = \dfrac{zFDc}{\delta}$ [25.57a], and so $\delta = \dfrac{FDc}{j_{\text{lim}}}$ $[z = 1]$

Therefore,

$$\delta = \frac{(9.65 \times 10^4 \text{ C mol}^{-1}) \times (1.14 \times 10^{-9} \text{ m}^2 \text{ s}^{-1}) \times (0.66 \text{ mol m}^{-3})}{28.9 \times 10^{-2} \text{ A m}^{-2}}$$

$$= 2.5 \times 10^{-4} \text{ m, or } \boxed{0.25 \text{ mm}}.$$

P25.17 $E' = E - \left(\dfrac{4RT}{F}\right) \ln\left(\dfrac{I}{A\bar{j}}\right) - IR_{\text{s}}$ [25.62].

$P = I E' = IE - aI \ln\left(\dfrac{I}{I_0}\right) - I^2 R_{\text{s}}$ where $a = \dfrac{4RT}{F}$ and $I_0 = A\bar{j}$. For maximum power,

$$\frac{\mathrm{d}p}{\mathrm{d}I} = E = -a \ln\left(\frac{I}{I_0}\right) - a - 2IR_{\text{s}} = 0$$

which requires

$$\ln\left(\frac{I}{I_0}\right) = \left(\frac{E}{a} - 1\right) - \frac{2I\,R_s}{a}.$$

This expression may be written

$$\ln\left(\frac{I}{I_0}\right) = c_1 - c_2 I; \quad c_1 = \frac{E}{a} - 1, \quad c_2 = \frac{2R_s}{a} = \frac{F\,R_s}{2RT}.$$

For the present calculation, use the data in Problem 25.16. Then

$$I_0 = A\bar{j} = (5\text{ cm}^2) \times (1\text{ mA cm}^{-2}) = 5\text{ mA},$$

$$c_1 = \frac{(1.10\text{ V})}{(4) \times (0.0257\text{ V})} - 1 = 10.7,$$

$$c_2 = \frac{(3.7\bar{5}\ \Omega)}{(2) \times (0.0257\text{ V})} = 73\ \Omega\text{ V}^{-1} = 73\text{ A}^{-1}.$$

That is, $\ln(0.20I/\text{mA}) = 10.7 - 0.073(I/\text{mA})$.

We then draw up the following table.

I/mA	103	104	105	106	107
$\ln(0.20I/\text{mA})$	3.025	3.034	3.044	3.054	3.063
$10.7 - 0.073(I/\text{mA})$	3.181	3.108	3.035	2.962	2.889

The two sets of points are plotted in Fig 25.8.

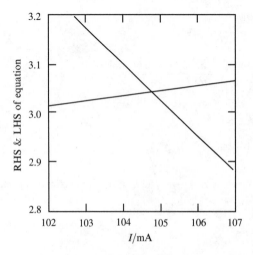

Figure 25.8

The lines intersect at $I = 105$ mA, which therefore corresponds to the current at which maximum power is delivered. The power at this current is

$$P = (105 \text{ mA}) \times (1.10 \text{ V}) - (0.103 \text{ V}) \times (105 \text{ mA}) \times \ln\left(\frac{105}{5}\right) - (105 \text{ mA})^2 \times (3.7\overline{5} \ \Omega)$$

$$= 41 \text{ mW}.$$

P25.19 $r_D = \left(\varepsilon RT / 2\rho F^2 I b^{\ominus}\right)^{1/2}$ [19.46]

where $I = \frac{1}{2} \sum_i z_i^2 (b_i / b^{\ominus})$, $b^{\ominus} = 1 \text{ mol kg}^{-1}$.

For NaCl: $I b^{\ominus} = b_{\text{NaCl}} \approx [\text{NaCl}]$ assuming 100 per cent dissociation.

For Na_2SO_4 : $I b^{\ominus} = \frac{1}{2} \left((1)^2 (2 b_{\text{Na}_2\text{SO}_4}) + (2)^2 b_{\text{Na}_2\text{SO}_4}\right)$

$$= 3 b_{\text{Na}_2\text{SO}_4} \approx 3[\text{Na}_2\text{SO}_4], \text{ assuming 100 per cent dissociation.}$$

$$r_D \approx \left(\frac{78.54 \times (8.854 \times 10^{-12} \text{ J}^{-1} \text{ C}^2 \text{ m}^{-1}) \times (8.315 \text{ J K}^{-1} \text{ mol}^{-1}) \times (298.15 \text{ K})}{2 \times (1.00 \text{ g cm}^{-3}) \times (10^{-3} \text{ kg/g}) \times (10^6 \text{ cm}^3/\text{m}^3) \times (96485 \text{ C mol}^{-1})^2}\right)^{1/2} \times \left(\frac{1}{Ib}\right)^{1/2}$$

$$\approx \frac{3.043 \times 10^{-10} \text{ m mol}^{1/2} \text{ kg}^{-1/2}}{(I b^{\ominus})^{1/2}}$$

$$\approx \frac{304.3 \text{ pm mol}^{1/2} \text{ kg}^{-1/2}}{(I b^{\ominus})^{1/2}}.$$

These equations can be used to produce the graph of r_D against b_{salt} shown in Fig. 25.9. Note the contraction of the double layer with increasing ionic strength.

P25.21   **(a)** The accompanying Tafel plot (Fig. 25.10) of $\ln j$ against E shows no region of linearity so the Tafel equation cannot be used to determine j_0 and α.

(b) $M_{\text{sol}} \overset{K_1}{\rightleftharpoons} M_{\text{ads}}$,

$M_{\text{ads}} + H^+ + e^- \overset{K_2}{\rightleftharpoons} MH_{\text{ads}}$,

$2 MH_{\text{ads}} \xrightarrow[\text{rate-determining}]{k_3} HMMH$.

Assuming that the dimerization is rate-determining, two electrons are transferred per molecule of HMMH and $z = 2$. It is also reasonable to suppose that the first two reactions are at quasi-equilibrium. According to reaction 3, the current density is proportional to the square of the functional surface coverage by MH_{ads}, θ_{MH},

$$j = z F k_3 \theta_{\text{MH}}^2,$$

$$\ln j = \ln(z F k_3) + 2 \ln \theta_{\text{MH}}.$$

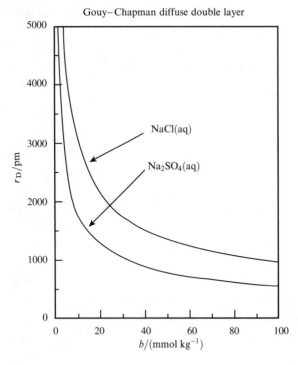

Figure 25.9

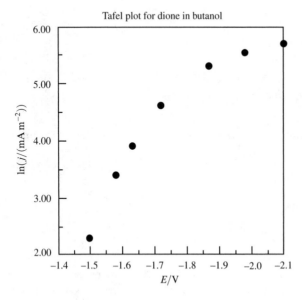

Figure 25.10

The characteristics of this equation differ from those of the Tafel equatioin at high negative overpotentials

$$\ln j = \ln j_0 - \alpha f \eta \quad [25.47].$$

At low concentration of M the value of θ_{MH} changes with the overpotential in a non-exponential manner. This makes $\ln j$ non-linear throughout the potential range.

Solutions to theoretical problems

P25.23   Refer to Fig. 25.11.

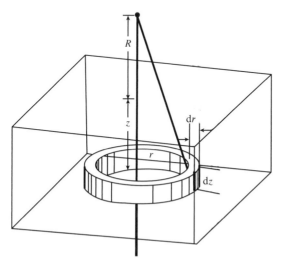

Figure 25.11

Let the number density of atoms in the solid be $\mathcal{N}$. Then the number in the annulus between r and $r + dr$ and thickness dz at a depth z below the surface is $2\pi \mathcal{N} r \, dr \, dz$. The interaction energy of these atoms and the single adsorbate atom at a height R above the surface is

$$dU = \frac{-2\pi \mathcal{N} r \, dr \, dz \, C_6}{\{(R+z)^2 + r^2\}^3}$$

if the individual atoms interact as $-C_6/d^6$, with $d^2 = (R+z)^2 + r^2$. The total interaction energy of the atom with the semi-infinite slab of uniform density is therefore

$$U = -2\pi \mathcal{N} C_6 \int_0^\infty dr \int_0^\infty dz \frac{r}{\{(R+z)^2 + r^2\}^3}.$$

We then use

$$\int_0^\infty \frac{r \, dr}{(a^2 + r^2)^3} = \frac{1}{2} \int_0^\infty \frac{d(r^2)}{(a^2 + r^2)^3} = \frac{1}{2} \int_0^\infty \frac{dx}{(a^2 + x)^3} = \frac{1}{4a^4}$$

and obtain

$$U = -\frac{1}{2} \pi \mathcal{N} C_6 \int_0^\infty \frac{dz}{(R+z)^4} = \boxed{\frac{\pi \mathcal{N} C_6}{6R^3}}.$$

This result confirms that $U \propto 1/R^3$. (A shorter procedure is to use a dimensional argument, but we need the explicit expression in the following.) When

$$V = 4\varepsilon \left[\left(\frac{\sigma}{R} \right)^{12} - \left(\frac{\sigma}{R} \right)^{6} \right] = \frac{C_{12}}{R^{12}} - \frac{C_6}{R^6},$$

we also need the contribution from C_{12}

$$U' = 2\pi\mathcal{N}C_{12} \int_0^\infty dr \int_0^\infty dz \frac{r}{\{(R+z)^2 + r^2\}^6} = 2\pi\mathcal{N} C_{12} \times \frac{1}{10} \int_0^\infty \frac{dz}{(R+z)^{10}} = \frac{2\pi\mathcal{N}C_{12}}{90R^9}$$

and therefore the total interaction energy is

$$U = \frac{2\pi\mathcal{N}C_{12}}{90R^9} - \frac{\pi\mathcal{N}C_6}{6R^3}.$$

We can express this result in terms of ε and σ by noting that $C_{12} = 4\varepsilon\sigma^{12}$ and $C_6 = 4\varepsilon\sigma^6$, for then

$$U = 8\pi\varepsilon\sigma^3\mathcal{N} \left[\frac{1}{90} \left(\frac{\sigma}{R} \right)^9 - \frac{1}{12} \left(\frac{\sigma}{R} \right)^3 \right].$$

For the position of equilibrium, we look for the value of R for which $dU/dR = 0$,

$$\frac{dU}{dR} = 8\pi\varepsilon\sigma^3\mathcal{N} \left[-\frac{1}{10} \left(\frac{\sigma^9}{R^{10}} \right) + \frac{1}{4} \left(\frac{\sigma^3}{R^4} \right) \right] = 0.$$

Therefore, $\sigma^9/10R^{10} = \sigma^3/4R^4$ which implies that $R = \left(\frac{2}{5} \right)^{1/6} \sigma = \boxed{0.858\sigma}$. For $\sigma = 342\,\text{pm}$, $R \approx \boxed{294\,\text{pm}}$.

P25.25
$$d\mu' = -c_2 \left(\frac{RT}{\sigma} \right) dV_\text{a}$$

which implies that

$$\frac{d\mu'}{d\ln p} = \left(\frac{-c_2 RT}{\sigma} \right) \times \left(\frac{dV_\text{a}}{d\ln p} \right).$$

However, we established in Problem 25.24 that

$$\frac{d\mu'}{d\ln p} = \frac{-RT\, V_\text{a}}{\sigma}.$$

Therefore,

$$-c_2 \left(\frac{RT}{\sigma} \right) \times \left(\frac{dV_\text{a}}{d\ln p} \right) = \frac{-RTV_\text{a}}{\sigma}, \quad \text{or} \quad c_2\, d\ln V_\text{a} = d\ln p.$$

Hence, $d\ln V_\text{a}^{c_2} = d\ln p$, and therefore $\boxed{V_\text{a} = c_1 p^{1/c_2}}$.

P25.27 For association:

$$\frac{dR}{dt} = k_{on}a_0(R_{eq} - R) \text{ where } a = a_0 \text{ is constant,}$$

$$\frac{dR}{R_{eq} - R} = k_{on}a_0 dt,$$

$$\int_0^R \frac{dR}{R_{eq} - R} = \int_0^t k_{on}a_0 dt = k_{on}a_0 t$$

$$-\ln(R_{eq} - R)|_0^R = k_{on}a_0 t$$

$$-\ln\left(\frac{R_{eq} - R}{R_{eq}}\right) = k_{on}a_0 t,$$

$$\frac{R_{eq} - R}{R_{eq}} = e^{-k_{on}a_0 t},$$

$$R = R_{eq}\left\{1 - e^{-k_{on}a_0 t}\right\},$$

$$\boxed{R(t) = R_{eq}\{1 - e^{-k_{obs}t}\} \text{ where } k_{obs} = k_{on}a_0}.$$

For dissociation:

$$\frac{dR}{dt} = -k_{on}a_0 R,$$

$$\frac{dR}{R} = -k_{on}a_0 dt,$$

$$\int_{R_{eq}}^R \frac{dR}{R} = -\int_0^t k_{on}a_0 dt,$$

$$\ln\left(\frac{R}{R_{eq}}\right) = -k_{on}a_0 dt,$$

$$\boxed{R = R_{eq}e^{-k_{obs}t} \text{ where } k_{obs} = k_{on}a_0}.$$

P25.29 Let η oscillate between η_+ and η_- around a mean value η_0. Then η_- is large and positive (and $\eta_+ > \eta_-$),

$$j \approx j_0 e^{(1-\alpha)\eta f} = j_0 e^{(1/2)\eta f} \quad [\alpha = 0.5]$$

and η varies as depicted in Fig. 25.12(a).

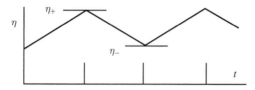

Figure 25.12a

Therefore, j is a chain of increasing and decreasing exponential functions,

$$j = j_0 e^{(\eta_- + \gamma t)f/2} \propto e^{-t/\tau}$$

during the increasing phase of η, where $\tau = 2RT/\gamma F$, γ is a constant, and

$$j = j_0 e^{(\eta_+ - \gamma t)f/2} \propto e^{-t/\tau}$$

during the decreasing phase. This is depicted in Fig.25.12(b).

Figure 25.12b

Solutions to applications

P25.31   For the Langmuir and BET isotherm tests we draw up the following table (using $p^* = 200\,\text{kPa} = 1500\,\text{Torr}$) [Example 25.1, *Illustration* 25.3, and eqn 25.57b].

$p/Torr$		100	200	300	400	500	600
$\dfrac{p}{V}/(\text{Torr cm}^{-3})$		5.59	6.06	6.38	6.58	6.64	6.57
$10^3 z$		67	133	200	267	333	400
$10^3 z/((1-z)(V/\text{cm}^3))$	4.01	4.66	5.32	5.98	6.64	7.30

p/V is plotted against p in Fig. 25.13(a), and $10^3 z/((1-z)V)$ is plotted against z in Fig. 25.13(b).

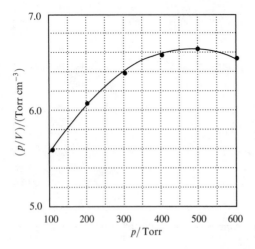

Figure 25.13a

We see that the ⌞ BET isotherm is a much better representation ⌟ of the data than the Langmuir isotherm. The intercept in Fig. 25.13(b) is at 3.33×10^{-3}, and so $1/cV_{\text{mon}} = 3.33 \times 10^{-3}\text{cm}^{-3}$. The slope of the graph is 9.93, and so

$$\frac{c-1}{cV_{\text{mon}}} = 9.93 \times 10^{-3}\,\text{cm}^{-3}.$$

Therefore, $c - 1 = 2.98$, and hence $\boxed{c = 3.98}$, $\boxed{V_{\text{mon}} = 75.4\,\text{cm}^3}$.

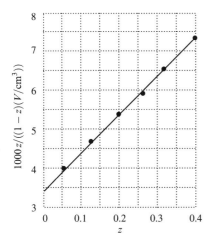

Figure 25.13b

P25.33   **(a)** $\dfrac{1}{q_{\text{VOC, RH}=0}} = \dfrac{1 + bp_{\text{VOC}}}{abc_{\text{VOC}}} = \dfrac{1}{abc_{\text{VOC}}} + \dfrac{1}{a}.$

Parameters of regression fit:

$\theta/°C$	$1/a$	$1/ab$	R	a	b/ppm^{-1}
33.6	9.07	709.8	0.9836	0.110	0.0128
41.5	10.14	890.4	0.9746	0.0986	0.0114
57.4	11.14	1599	0.9943	0.0898	0.00697
76.4	13.58	2063	0.9981	0.0736	0.00658
99	16.82	4012	0.9916	0.0595	0.00419

The linear regression fit is generally good at all temperatures with
$\boxed{R \text{ values in the range } 0.975 \text{ to } 0.991}$.

(b) $\ln a = \ln k_a - \dfrac{\Delta_{\text{ad}}H}{R}\dfrac{1}{T}$

and $\ln b = \ln k_b - \dfrac{\Delta_{\text{b}}H}{R}\dfrac{1}{T}.$

Linear regression analysis of $\ln a$ versus $1/T$ gives the intercept $\ln k_a$ and slope $-\Delta_{\text{ad}}H/R$ while a similar statement can be made for a $\ln b$ versus $1/T$ plot. The temperature must be in Kelvin.
For $\ln a$ versus $1/T$

$\ln k_a = -5.605$, standard deviation $= 0.197$,
$-\Delta_{\text{ad}}H/R = 1043.2$ K, standard deviation $= 65.4$ K,
$R = 0.9942$ [good fit],

$$k_a = e^{-5.605} = \boxed{3.68 \times 10^{-3}},$$

$$\Delta_{ad}H = -(8.31451 \text{ J K}^{-1} \text{ mol}^{-1}) \times (1043.2 \text{ K})$$

$$= \boxed{-8.67 \text{ kJ mol}^{-1}}.$$

For $\ln b$ versus $1/T$

$$\ln(k_b/(\text{ppm}^{-1})) = -10.550, \text{ standard deviation} = 0.713,$$

$$-\Delta_b H/R = 1895.4 \text{ K}, \text{ standard deviation} = 236.8,$$

$$R = 0.9774 \text{ [good fit]},$$

$$k_b = e^{-10.550} \text{ ppm}^{-1} = \boxed{2.62 \times 10^5 \text{ ppm}^{-1}},$$

$$\Delta_b H = -(8.31451 \text{ J K}^{-1} \text{ mol}^{-1}) \times (1895.4 \text{ K}),$$

$$\boxed{\Delta_b H = -15.7 \text{ kJ mol}^{-1}}.$$

(c) k_a may be interpreted to be the maximum adsorption capacity at an adsorption enthalpy of zero, while k_b is the maximum affinity in the case for which the adsorbant–surface bonding enthalpy is zero.

P25.35 (a) K unit : $(g_R \text{ dm}^{-3})^{-1}$ [g_R = mass (grams) of rubber].

K_F unit : $(\text{mg})^{(1-1/n)} g_R^{-1} \text{dm}^{-3/n}$.

K_L unit: $(\text{mg dm}^{-3})^{-1}$.

M unit : $(\text{mg } g_R^{-1})$.

(b) Linear sorption isotherm

$$q = K c_{eq}.$$

$K = q/c_{eq}$ so K is best determined as an average of all q/c_{eq} data pairs.

$$\boxed{K_{av} = 0.126(g_R \text{ dm}^{-3})^{-1}}, \text{ standard deviation} = 0.041(g_R \text{ dm}^{-3})^{-1}.$$

95 per cent confidence limit: $(0.083 - 0.169)(g_R \text{dm}^{-3})^{-1}$.

If this is done as a linear regression, the result is significantly different.

$$K \text{ (linear)} = 0.0813(g_R \text{ dm}^{-3})^{-1}, \text{ standard deviation} = 0.0092(g_R \text{dm}^{-3})^{-1}.$$

$$\boxed{R \text{ (linear)} = 0.9612}$$

Freundlich sorption isotherm: $q = K_F c_{eq}^{1/n}$; using a power regression analysis, we find that

$$\boxed{K_F = 0.164}, \text{ standard deviation} = 0.317.$$

$$\frac{1}{n} = 0.877, \text{ standard deviation} = 0.113; \boxed{n = 1.14}.$$

$$\boxed{R \text{ (Freundlich)} = 0.9682}.$$

Langmuir sorption isotherm

$$q = \frac{K_L M c_{eq}}{1 + K_L c_{eq}}.$$

$$\frac{1}{q} = \left(\frac{1}{K_L M}\right)\left(\frac{1}{c_{eq}}\right) + \frac{1}{M}.$$

$$\frac{1}{K_L M} = 8.089 g_R \text{ dm}^{-3}, \quad \text{standard deviation} = 1.031; \; K_L = -0.00053(g_R \text{ dm}^{-3})^{-1}.$$

$$\frac{1}{M} = -0.0043 g_R \text{ mg}^{-1}, \quad \text{standard deviation} = 0.1985; M = \boxed{-233 \text{ mg } g_R^{-1}}.$$

$$\boxed{R \text{ (Langmuir)} = 0.9690}.$$

All regression fits have nearly the same correlation coefficient so that cannot be used to determine which is the best fit. However, the Langmuir isotherm gives a negative value for K_L. If K_L is to represent an equilibrium constant, which must be positive, the Langmuir description must be rejected. The standard deviation of the slope of the Freundlich isotherm is twice as large as the slope itself. This would seem to be unfavorable. Thus, the $\boxed{\text{linear description seems to be the best}}$, but not excellent choice. However, the Freundlich isotherm is usually preferred for this kind of system, even though that choice is not supported by the data in this case.

(c) $\dfrac{q_{rubber}}{q_{charcoal}} = \dfrac{0.164 c_{eq}^{1.14}}{c_{eq}^{1.6}} = \boxed{0.164 c_{eq}^{-0.46}}.$

The sorption efficiency of ground rubber is much less than that of activated charcoal and drops significantly with increasing concentration. The only advantage of the ground rubber is its exceedingly low cost relative to activated charcoal, which might convert to a lower cost per gram of contaminant adsorbed.

P25.37   **(a)** The electrode potentials of half-reactions (a), (b), and (c) are (Section 25.13):

(a) $E(H_2, H^+) = -0.059 \text{ V pH} = (-7) \times (0.059 \text{ V}) = -0.14 \text{ V}$,

(b) $E(O_2, H^+) = (1.23 \text{ V}) - (0.059 \text{ V})\text{pH} = +0.82 \text{ V}$,

(c) $E(O_2, OH^-) = (0.40 \text{ V}) + (0.059 \text{ V})\text{pOH} = 0.81 \text{ V}$.

$$E(M, M^+) = E^{\ominus}(M, M^+) + \left(\frac{0.059 \text{ V}}{z_+}\right) \log 10^{-6} = E^{\ominus}(M, M^+) - \frac{0.35 \text{ V}}{z_+}.$$

Corrosion will occur if $E(a), E(b), E(c) > E(M, M^+)$.

(i) $E^{\ominus}(Fe, Fe^{2+}) = -0.44 \text{ V}, \quad z_+ = 2$,

$E(Fe, Fe^{2+}) = (-0.44 - 0.18) \text{ V} = -0.62 \text{ V} < E(a, \text{ b, and c})$.

(ii) $E(Cu, Cu^+) = (0.52 - 0.35) \text{ V} = 0.17 \text{ V} \begin{cases} > E(a) \\ < E(\text{b and c}) \end{cases}$,

$E(Cu, Cu^{2+}) = (0.34 - 0.18) \text{ V} = 0.16 \text{ V} \begin{cases} > E(a) \\ < E(\text{b and c}) \end{cases}$.

(iii) $E(Pb, Pb^{2+}) = (-0.13 - 0.18) \text{ V} = -0.31 \text{ V} \begin{cases} > E(a) \\ < E(\text{b and c}) \end{cases}$.

(iv) $E(Al, Al^{3+}) = (-1.66 - 0.12) V = -1.78 V < E(a, b, \text{ and } c)$.

(v) $E(Ag, Ag^+) = (0.80 - 0.35) V = 0.45 V \begin{cases} > E(a) \\ < E(b \text{ and } c) \end{cases}$.

(vi) $E(Cr, Cr^{3+}) = (-0.74 - 0.12) V = -0.86 V < E(a, b, \text{ and } c)$.

(vii) $E(Co, Co^{2+}) = (-0.28 - 0.15) V = -0.43 V < E(a, b, \text{ and } c)$.

Therefore, the metals with a thermodynamic tendency to corrode in moist conditions at pH $= 7$ are $\boxed{Fe, Al, Co, Cr}$ if oxygen is absent, but, if oxygen is present, all seven elements have a tendency to corrode.

(b) A metal has a thermodynamic tendency to corrosion in moist air if the zero-current potential for the reduction of the metal ion is more negative than the reduction potential of the half-reaction $4H^+ + O_2 + 4e^- \rightarrow 2H_2O$, $E^\ominus = 1.23$ V.

The zero-current cell potential is given by the Nernst equation

$$E = E^\ominus - \frac{RT}{\nu F} \ln Q = E^\ominus - \frac{RT}{\nu F} \ln \frac{[M^{z+}]^{\nu/z}}{[H^+]^\nu p(O_2)^{\nu/4}}.$$

We are asked if a tendency to corrode exists at pH 7 ($[H^+] = 10^{-7}$) in moist air ($p(O_2) \approx 0.2$ bar), and are to answer yes if $E \geq 0$ for a metal ion concentration of 10^{-6}, so for $\nu = 4$ and 2+ cations

$$E = 1.23 V - E^\ominus_M - \frac{0.02569 V}{\nu} \ln \frac{(10^{-6})^2}{(1 \times 10^{-7})^4 \times (0.2)} = 0.983 V - E^\ominus_M.$$

In the following, $z = 2$.

For Ni: $E^\ominus = 0.983 V - (-0.23 V) > 0$ $\boxed{corrodes}$.

For Cd: $E^\ominus = 0.983 V - (-0.40 V) > 0$ $\boxed{corrodes}$.

For Mg: $E^\ominus = 0.983 V - (-2.36 V) > 0$ $\boxed{corrodes}$.

For Ti: $E^\ominus = 0.983 V - (-1.63 V) > 0$ $\boxed{corrodes}$.

For Mn: $E^\ominus = 0.983 V - (-1.18 V) > 0$ $\boxed{corrodes}$.

P25.39   Corrosion occurs by way of the reaction

$$Fe + 2H^+ \rightarrow Fe^{2+} + H_2.$$

The half-reactions at the anode and cathode are:

Anode: $Fe \rightarrow Fe^{2+} + 2e^-$,

Cathode: $2H^+ + 2e^- \rightarrow H_2$.

$\Delta\phi_{corr} = (-0.720 \text{ V}) + (0.2802 \text{ V}) = -0.440$ V,

$\Delta\phi_{corr} = \eta(H) + \Delta\phi_e(H)$ [Justification 25.1],

$\Delta\phi_e(H) = (-0.0592 \text{ V}) \times pH = (-0.0592V) \times 3 = -0.17\overline{7}6V$,

$\eta(H) = -\frac{1}{\alpha f} \ln \frac{j_{corr}}{j_0(H)}.$

Then $\Delta\phi_{corr} = -0.440$ V $= -\dfrac{1}{\alpha f}\ln\dfrac{j_{corr}}{j_0(\mathrm{H})} - 0.177\overline{6}$ V

and $\ln\dfrac{j_{corr}}{j_0(\mathrm{H})} = (0.262$ V$) \times \alpha f = (0.262$ V$) \times (18$ V$^{-1}) = 4.7\overline{16}$.

$$j_{corr} = j_0(\mathrm{H}) \times e^{4.71\overline{6}} = (1.0 \times 10^{-7}\text{A cm}^{-2}) \times (112) = 1.1\overline{2} \times 10^{-5} \text{ A cm}^{-2}.$$

Faraday's laws give the amount of iron corroded

$$n = \frac{I_{corr}t}{ZF} = \frac{(1.1\overline{2} \times 10^{-5} \text{ A cm}^{-2}) \times (8.64 \times 10^4 \text{ s d}^{-1})}{(2) \times (9.65 \times 10^4 \text{ C mol}^{-1})} \ 5.0 \times 10^{-6} \text{ mol cm}^{-2}\text{d}^{-1}.$$

$$m = n \times (55.85 \text{ g mol}^{-1}) = (5.0 \times 10^{-6} \text{ mol cm}^{-2}\text{d}^{-1}) \times (55.85 \times 10^3 \text{ mg mol}^{-1})$$

$$= \boxed{0.28 \text{ mg cm}^{-2} \text{ d}^{-1}}.$$